Behaviour of Teleost Fishes

Behaviour of Teleost Fishes

Second edition

Edited by

TONY J. PITCHER

Renewable Resources Assessment Group
Imperial College of Science and Technology
London, UK

CHAPMAN & HALL

London · Glasgow · New York · Tokyo · Melbourne · Madras

Published by Chapman & Hall, 2–6 Boundary Row, London SE1 8HN

Chapman & Hall, 2–6 Boundary Row, London SE1 8HN, UK

Blackie Academic & Professional, Wester Cleddens Road, Bishopbriggs, Glasgow G64, 2NZ, UK

Chapman & Hall, 29 West 35th Street, New York, NY10001, USA

Chapman & Hall Japan, Thomson Publishing Japan, Hirakawacho Nemoto Building, 6F 1-7-11 Hirakawa-cho, Chiyoda-ku, Tokyo 102, Japan

Chapman & Hall Australia, Thomas Nelson Australia, 102 Dodds Street, South Melbourne, Victoria 3205, Australia

Chapman & Hall India, R. Seshadri, 32 Second Main Road, CIT East, Madras 600 035, India

First edition 1986

Second edition 1993

© 1986 Tony J. Pitcher, 1993 Chapman & Hall

Typeset in 10/12 Photina by Thomson Press (India) Ltd, New Delhi

Printed in England by Clays Ltd, St Ives plc

ISBN 0 412 42930 6 (HB) 0 412 42940 3 (PB)

A catalogue record for this book is available from the British Library

Library of Congress Cataloging-in-Publication data available

∞ Printed on permanent acid-free text paper, manufactured in accordance with the proposed ANSI/NISO Z 39.48-199X and ANSI Z 39.48-1984

Contents

Contributors

Horst Bleckmann
Fakultät für Biologie II, Universität Bielefeld, Postfach 8640, 4800 Bielefeld 1, Germany

Patrick W. Colgan
Canadian Museum of Nature, P.O. Box 3443, Station "D", Ottawa, Ontario, Canada K1P 6P4

Roy G. Danzmann
Department of Zoology, University of Guelph, Guelph, Ontario, Canada N1G 2W1

Moira M. Ferguson
Department of Zoology, University of Guelph, Guelph, Ontario, Canada N1G 2W1

Gerard J. FitzGerald
Département de Biologie, Université Laval, Cité Universitaire, Québec, P.Q., Canada G1K 7P4

R. N. Gibson
Scottish Marine Biological Association, Dunstaffnage Marine Research Laboratory, P.O. Box 3, Oban, Argyll, Scotland PA34 4AD

Mart R. Gross
Department of Zoology, University of Toronto, 25 Harbord Street, Toronto, Ontario, Canada M5S 1A1

D. M. Guthrie
90 Ack Lane West, Cheadle Hume, Cheshire SK8 7ES, United Kingdom

Toshiaki J. Hara
Freshwater Institute, 501 University Crescent, Winnipeg, Manitoba, Canada R3T 2N6

Paul J. B. Hart
Department of Zoology, University of Leicester, Leicester LE1 7RH, United Kingdom

A. D. Hawkins
SOAFD Marine Laboratory, P.O. Box 101, Victoria Road, Aberdeen, Scotland
AB9 8DB

Gene S. Helfman
Zoology Department and Institute of Ecology, University of Georgia, Athens,
GA 30602, USA

Felicity A. Huntingford
Department of Zoology, University of Glasgow, Glasgow, Scotland G12 8QQ

Anne E. Magurran
Animal Behaviour Research Group, Department of Zoology, University of
Oxford, South Parks Road, Oxford OX1 3PS, United Kingdom

Manfred Milinski
Universitat Bern, Zoologisches Institut, Abteilung Verhaltensökologie,
Wohlenstrasse 50a, CH-3032 Hinterkappelen, Switzerland

W. R. A. Muntz
Department of Ecology and Evolutionary Biology, Monash University, Clayton,
Victoria 3168, Australia

David L. G. Noakes
Department of Zoology, University of Guelph, Guelph, Ontario, Canada N1G
2W1

Ken O'Hara
Environmental Consultancy Group, School of Life Sciences, Department of
Environmental and Evolutionary Biology, University of Liverpool, P.O. Box
147, Liverpool L69 3BX, United Kingdom.

Julia K. Parrish
Institute for Environmental Studies, University of Washington, FM-12, Seattle,
WA 98195, USA

Jakob Parzefall
Zoologisches Institut und Zoologisches Museum, University of Hamburg,
Martin-Luther-King Platz 3, D-2000 Hamburg 13, Germany

Tony J. Pitcher
Renewable Resources Assessment Group, Imperial College, 8 Prince's
Gardens, London SW7 1NA, United Kingdom

Current address,
Fisheries Centre, 2204 Main Mall, University of British Columbia, Vancouver,
B.C., Canada V6T 124

Robert Craig Sargent
T. H. Morgan School of Biological Sciences, University of Kentucky, Lexington, KY 40506-0225, USA

George F. Turner
Monkey Bay Fisheries Research Station, P.O. Box 27, Monkey Bay, Malawi
Current address,
Department of Zoology, University of Aberdeen, Aberdeen, Scotland AB9 2TN

C. S. Wardle
SOAFD Marine Laboratory, P.O. Box 101, Victoria Road, Aberdeen, Scotland AB9 8DB

Robert J. Wootton
Department of Biological Sciences, The University College of Wales, Aberystwyth, Dyfed SY23 3DA, Wales, United Kingdom.

Preface

This book is about the behaviour of teleosts, a well-defined, highly successful taxonomic group of vertebrate animals sharing a common body plan and forming the vast majority of living bony fishes. There are over 22 000 living species of teleosts, including nearly all the fish of importance in commercial fisheries and aquaculture. Teleosts are represented in just about every conceivable aquatic environment from temporary desert pools to the deep ocean, from soda lakes to sub-zero Antarctic waters. Behaviour forms the primary interface between these effective survival machines and their environment; behavioural plasticity is the key to the success of the teleost fishes.

In the decade before the publication of the first edition of this book (1986) the study of animal behaviour underwent revolutionary changes under the dual impact of the new fields of behavioural ecology and sociobiology. Quantitative, experimentally-verifiable hypotheses about why individual animals behave were formulated for the first time and met with considerable success. Much of the early work in these new fields concentrated on birds and mammals, but material presented in the first edition of this book helped to demonstrate that fish behaviour is not just a simplified version of that seen in birds and mammals, but obeys the same ecological and evolutionary rules. In the five years since the first edition, much of the early theory has matured: optimal solutions to the problems of feeding and mating require subtle trade-offs of energy balance, information about food and predators, and the demands of reproduction. In parallel, there has been a large expansion in the amount of fundamental research work with a primary focus on fish.

The book aims to bring together accounts of the major functional topics in fish behaviour, reviewed in the light of current theory. Reviews at this level are intended to be of use to a broad spectrum of students and research workers interested in fish biology and fisheries. Each chapter commences by setting out fundamental principles often taken for granted in research monographs, moves on to outline the current evidence for the conceptual basis of the field, maps the contemporary research frontiers and finishes with suggestions about where the next advances may be expected. Each chapter closes with a summary of its contents. The book is divided into four sections, encompassing the bases of behaviour, sensory modalities, behavioural ecology and applied topics. Each section has a brief introductory editorial.

Authors have carried out extensive revisions for the second edition, although the scope and structure of each chapter remains similar. Like the first edition, some important behavioural topics such as the central nervous system, learning and migration could not be included in this book. They were omitted from the first edition partly because at that time recent reviews had appeared in those areas. Recent reviews and books covering these topics are listed in the introductory sections where possible.

I am most grateful to Tony Hawkins for providing the splendid drawing for the cover, to Nigel Balmforth and Martin Tribe at Chapman and Hall for their patient encouragement and to John Beddington for making facilities available. Our invaluable copy editor, Chuck Hollingworth, kept all our word processors humming and our copies of Current English Usage well thumbed as we waited to be rewarded with a relieved gasp of dawning comprehension. The author's acknowledgements are given with each chapter.

Tony J. Pitcher, Imperial College, London

Series foreword

Among the fishes, a remarkably wide range of biological adaptations to diverse habitats has evolved. As well as living in the conventional habitats of lakes, ponds, rivers, rock pools and the open sea, fish have solved the problems of life in deserts, in the deep sea, in the cold antarctic, and in warm waters of high alkalinity or of low oxygen. Along with these adaptations, we find the most impressive specializations of morphology, physiology and behaviour. For example we can marvel at the high-speed swimming of the marlins, sailfish and warm-blooded tunas, air-breathing in catfish and lungfish, parental care in the mouth-brooding cichlids and viviparity in many sharks and toothcarps.

Moreover, fish are of considerable importance to the survival of the human species in the form of nutritious and delicious food of numerous kinds. Rational exploitation and management of our global stocks of fishes must rely upon a detailed and precise insight of their biology.

The *Chapman and Hall Fish and Fisheries Series* aims to present timely volumes reviewing important aspects of fish biology. Most volumes will be of interest to research workers in biology, zoology, ecology and physiology but an additional aim is for the books to be accessible to a wide spectrum of non-specialist readers ranging from undergraduates and postgraduates to those with an interest in industrial and commercial aspects of fish and fisheries.

Fundamental work on fish behaviour has been a rapidly moving field and this volume, constituting the seventh in the *Fish and Fisheries Series*, is the second edition of a book reviewing fish behaviour first published in 1986 and comprising 19 chapters and 25 authors.

The behaviour of fishes intimately reflects unique and efficient solutions to the problems raised by their three-dimensional environment. Behavioural plasticity is the key to understanding the success of fishes because behaviour is the interface between fish genes and the hazards of the environment, including food, predators, and the fishing operations of human fisheries. This book focusses on behavioural solutions to gaining information about the environment via the sense organs, the genetic, developmental and motivational bases of fish behaviour, behavioural trade-offs in mating, feeding and evading predation, the effect of environmental constraints such as light and salinity, and concludes with a section on applied fish behaviour in fishing gear and in freshwater fishery management.

Dr Tony J. Pitcher
Editor, Chapman and Hall Fish and Fisheries Series
Special Research Fellow, Imperial College, London

Part one

Bases of Behaviour

INTRODUCTION

In three chapters covering genes, motivation and development, this opening section of the book aims to review the major biological determinants of the behaviours that are observed in adult teleost fish.

Chapter 1 covers the most fundamental area of the functional study of animal behaviour: its genetic basis. Behaviour seems far removed from the direct action of genes in the organism's coded DNA but Roy Danzmann, Moira Ferguson and David Noakes explore the nature of the links between genes and expressed behaviours commencing with a helpful analogy with computer programming. The genetic basis of behaviour is an issue of profound importance in the light of the distinction between causal and functional explanations. The authors argue that, provided we can ascribe differences in behaviour to genetic differences between individuals, it may not be always necessary to unravel the details of causation. Danzmann and his colleagues describe how selection shapes fish behaviour, reviewing work on the inheritance of behaviour in fish including some interesting work on single gene differences, and discuss how we may make sense of the 'chocolate soup and jigsaw puzzle' which nature has put before us. The new techniques of genetic parentage testing using allozymes, mtDNA and nuclear DNA, which are beginning to produce results of immense importance to our evolutionary insight of fish behaviour, are described along with remarkable advances that have been made in this field since the first edition of the book.

Patrick Colgan in Chapter 2 surveys ideas about how to deal with changes in responsiveness: the difficult field of the study of motivation, which he points out began with Aristotle. To some the answers are the pure detailed descriptions of physiology, the 'neurophysiologists nirvana', whereas to others physiological causation is no answer at all, the whole animal's behaviour emerging from its physiology in a holistic sense to meet the needs of survival. According to this view, the functional goal is met by using whatever hardware happens to be available. As in Chapter 1, the split further reflects the distinction between causal and functional explanations. Colgan maps out the internal and external factors which influence the big four motivational systems of hunger, fear, aggression and sex, discussing the development, interaction and extensions of these systems using examples from research on teleost fish. He takes the view that motivation is not 'physiology writ large', and reviews the range of models of motivation used by students of animal behaviour, including the space-state approach, time-sharing, information theory, stochastic and specific models. The author gives us strictures for the design of experiments on fish motivation, taking care to simulate its natural environment and avoiding experiments where 'under precisely controlled conditions, an animal does as he damn pleases'. Patrick Colgan is critical of the current orthodoxy and leaves us with the thought that current analysis has 'scarcely gained a toehold on the terrain to be investigated'.

The development of behaviour is reviewed by Felicity Huntingford in Chapter 3. She opens with a clear description of the development of behaviours in salmonids, one of the basal groups of teleosts, and cichlids, one of the most advanced. The chapter outlines the processes which shape the fishes' growing behaviour repertoire, from the first muscle twitches in the egg, to the range of feeding, antipredator, habitat selection, migration, agonistic, sexual and parental behaviours in the mature adult fish. Huntingford then concentrates on an analysis of the contemporary heuristic classification of factors which influence behavioural development, with detailed examples from experiments. She rejects the distinction between learned and instinctive behaviours as unhelpful. The chapter examines sensitive periods and condition-dependent switches during development, crucial for the downstream migration and reproductive homing of salmonids, reviews the effects of early experience on feeding and social behaviour, discusses the concept of stability and plasticity during development in the face of environmental hazard, and concludes with a consideration of the term 'innate behaviour'. Huntingford rejects the term, but explains why some ethologists continue to use it.

This section does not cover short term causation in the nervous system (Guthrie, 1983), learning (Kieffer and Colgan, 1992), or the interaction of hormones and behaviour (Stacey, 1987).

REFERENCES

Guthrie, D. M. (1983) Integration and control by the central nervous system, in *Control Processes in Fish Physiology* (eds. J. C. Rankin, T. J. Pitcher and R. T. Duggan) Croom Helm, London, pp. 130–154.

Keiffer, J. D. and Colgan, P. (1992) The role of learning in fish behaviour. *Rev. Fish Biol. Fish.*, **2**, 125–143.

Stacey, N. E. (1987) Roles of hormones and pheromones in fish reproductive behaviour, in *Psychobiology of Reproductive Behavior: an evolutionary perspective.* (ed. D. Crews) pp. 28–60.

Chapter one

The genetic basis of fish behaviour

Roy G. Danzmann, Moira M. Ferguson and
David L. G. Noakes

1.1 INTRODUCTION

The basic principles of inheritance, and the corresponding science of genetics, have been well known and widely accepted for at least the past century. There is little question about the relationship between biochemical or even anatomical features and genes, as exemplified by the inheritance of hereditary conditions such as haemophilia, the selective breeding of domesticated species, and recent advances in genetic engineering. It is usually more difficult to accept that something as complex and loosely defined as behaviour can have a similar genetic basis. In this chapter we review the basic principles of genetics as they apply to behaviour and discuss some examples in detail. We also discuss the applications and limitations of recent techniques in this field, particularly as concerns the identification of individual animals and their genetic relatedness.

Most recent behavioural textbooks provide at least a general overview of behavioural genetics, but with few examples of fish species (Huntingford, 1984, is a notable exception). We assume that readers either will be familiar with such an introductory treatment of this field, or can refer to those sources (see also Noakes, 1986). We consider examples from teleost fishes in detail to emphasize their significance to the basic questions of ethology: causation, function, ontogeny, and phylogeny (Tinbergen, 1963). We define behaviour as the patterned output of muscles, glands or nerve cells of intact organisms. Genetics of behaviour includes both heritable differences in behaviour and the mechanisms leading from genes to behaviour within individual animals (Gould, 1974).

Behaviour of Teleost Fishes 2nd edn. Edited by Tony J. Pitcher. Published in 1993 by Chapman & Hall. ISBN   0 412 42930 6 (HB) and 0 412 42940 3 (PB).

Methodology

As a hybrid area of science, behavioural genetics has been characterized by vigour in activity, and a considerable degree of independent assortment and segregation of output. Major reviews and books on the subject (e.g. Manning, 1975, 1976; Ehrman and Parsons, 1976; Burghardt and Bekoff, 1978; Barlow, 1981; Partridge, 1983) have highlighted both empirical research and theoretical advances. There is considerable sophistication in our understanding of molecular genetics (e.g. Allendorf and Ferguson, 1990), and a growing body of hypotheses concerning evolution and function (e.g. Dawkins, 1982, 1986; Futuyama, 1986; Alcock, 1989), with a good deal of activity to link the gap between. To appreciate the field of behavioural genetics it is necessary to understand both the kinds of questions asked, and the techniques available to address those questions.

On the tracks of turtles

How we specify our questions, and how we interpret the evidence to answer those questions, can differ considerably depending upon our viewpoint. The example of specifying the instructions to a computer so as to create a circle on a video screen (Hayes, 1984) is quite instructive. If we take a global perspective, we could instruct the computer to have one point move so as to maintain a constant distance from a second, fixed point. The track of the moving point would be a circle.

If we take a local perspective, the instructions to the computer would be quite different. We can think of this in terms of the computer language LOGO, which gives instructions to a point (the 'turtle'). To such a turtle a circle would consist of a series of repeated instructions to take one step forward and then one turn to the right, until it had turned through a total of 360°.

Ethologists have tended to adopt global views of behaviour; molecular geneticists clearly tend to have local perspectives. Ethologists view animals as defending territories, attracting mates, migrating to feeding areas, or searching for prey. Molecular geneticists are more often concerned with cause-and-effect mechanisms linking genes, enzymes and proteins. Obviously the causal links between genes and behaviour must be in local terms, but the functional aspects of behaviour are global. It is important to bear these differences in mind, and to realize the importance of bridging the gap between these two perspectives.

Sometimes this distinction is drawn between questions of immediate causation or physiological mechanisms (proximate or 'how' questions), and those of evolutionary consequence or function (ultimate or 'why' questions) (Tinbergen, 1963; Alcock, 1989). Particular behavioural genetics studies are not always explicitly directed towards either (or both) of these categories of question, but it is important to keep the distinctions in mind. Genetic questions require genetic evidence, evolutionary questions require evolutionary evidence to provide answers.

1.2   TECHNIQUES FOR BEHAVIOURAL GENETICS RESEARCH

The main techniques of behavioural genetics research remain those established by geneticists for some years: single gene differences and mutations, selection experiments, hybrids and backcrosses, twin comparisons, and quantitative traits. With the exception of twin studies, which are most often applied to humans, the other techniques have often been applied to teleost species. We will briefly review each of these techniques, with some examples from fish studies, and then consider the advantages and limitations of techniques from molecular genetics for identification of individuals.

Single gene differences

What can genes be 'for'?

In principle, this approach is the right stuff of classical genetics, as the simplest and most satisfying relationship between genes and behaviour. Phenotypic differences resulting from single gene differences are easy to trace experimentally, as in Mendel's original studies, and were the basis for the classic 'one gene, one enzyme' causal mechanism model. A comparison of individuals differing only in a single gene should allow us to map cause and effect directly and continuously, from nucleotide sequences to behavioural actions, but this has rarely been the case. Most behavioural traits appear to be associated with differences in a number of gene loci, and many genes appear to have pleiotropic effects on behaviour (Manning, 1976).

Mendel's great insight was to study individual phenotypic characters – such as colour of blossoms, height of plants, colour and texture of seed coat – rather than the overall phenotype of the organism. As Barlow (1981) has suggested, it would be most appropriate to select modal (fixed) action patterns of behaviour for genetic analyses. Undoubtedly much of the complexity and uncertainty in studies of behavoural genetics have resulted from the failure to select independent units of behaviour. Of course, the question of what constitutes basic units of behaviour is neither new nor trivial. Independent units of behaviour are seldom as conspicuous or unambiguous as size or colour. We rarely have a sufficiently detailed knowledge of the genotype of any fish species we study to make other than broad inferences as to genetic similarities or differences between individuals. Other possibilities do exist, however, along different lines, to allow us to ascribe behavioural differences to single gene (or at least a few genes) differences.

The common guppy, *Poecilia reticulata*, and some related species known in the ornamental fish trade as live-bearers (family Poeciliidae), and some species in other families, including most notably goldfish, *Carassius auratus*, and carp, *Cyprinus carpio*, in the family Cyprinidae, have been selectively bred in captivity to produce desired pigmentation patterns on the fins or body, or to elaborate or perpetuate monstrosities (e.g. so-called telescope eyes, lionheads, and veil-tails in goldfish). These are often produced in highly inbred lines, so we can reasonably assume that individuals differing in only one of these pigmentation or other features are probably very similar in genotype. Since these fish are typically raised under carefully regulated conditions in captivity, environmental differences among individual fish are also likely to be small. The social behaviour of guppies, especially male courtship, has been frequently studied (Liley, 1966; Keenleyside, 1979; Endler, 1983). There may also be substantial differences in behaviour, apparently as a result of relatively small genetic differences among males (or females). However, we do not have much information as to how these genetic differences produce the behavioural differences, and since these are artificially selected strains of an inbred species, the differences might not have any significant ecological or evolutionary consequences.

Another possibility is a species kept in large numbers under intensive culture conditions so that individual gene mutants are more likely to be noticed. The fish species kept in sufficiently large numbers to allow screening for behavioural mutants are often those kept for food production, such as rainbow trout, *Oncorhynchus mykiss*. Interestingly, some recent studies of such cases not only provide evidence of single gene differences associated with behavioural differences but also give some clue as to the causal mechanisms. Genetic differences, as measured by isozyme frequencies, have been widely surveyed and established for many different fishes (review, Allendorf and Ferguson, 1990). Geographically or genotypically different stocks of some fish species may also have significant behavioural differences. For example, stocks of rainbow trout differ in swimming performance and responses to water current (Fig. 1.1), brook charr, *Salvelinus fontinalis*, differ in susceptibility to angling by humans, survival and growth, Pacific salmon, *Oncorhynchus* spp., differ in homing ability, and Atlantic salmon, *Salmo salar*, may differ in growth, age at maturity, and other characteristics (Raleigh, 1971; Bams, 1976; Thomas and Donahoo, 1977; Kelso et al., 1981; Thorpe, 1986; Foote *et al.*, 1989; McIsaac and Quinn, 1988), to name but a few examples. These differences are not likely to be one gene, one behaviour, cause–effect relationships, however. The behavioural measures are relatively complex, and the precise genetic constitution of the animals in question is usually not known.

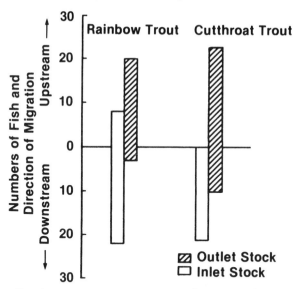

Fig. 1.1 Directional responses to water current of young rainbow trout, *Oncorhynchus mykiss*, and cutthroat trout *O. clarki*, hatched, raised and tested in controlled laboratory conditions. Genetically determined differences in movement with respect to water current correspond to appropriate responses in natural habitat of each stock. In their native habitats, trout from inlet streams have to migrate downstream to reach the lake where they feed; trout from outlet streams have to move upstream to reach the lake. Reproduced with permission from Raleigh (1971).

Nonetheless, the adaptive significance of particular, single gene differences is known for some cases. Rainbow trout differing in lactate dehydrogenase (LDH) phenotype (several LDH isozyme variants are known) have significantly different swimming performances at varying ambient dissolved oxygen conditions (Klar *et al.*, 1979). Since swimming performance is critical to fish, especially salmonids, such differences clearly could be of major ecological and evolutionary significance. The killifish, *Fundulus heteroclitus*, shows comparable differences in swimming endurance at some temperatures, and in time of hatching from the egg membrane, relative to LDH isozyme differences between individuals (DiMichele and Powers, 1982). Far-reaching effects of single gene differences are known for rainbow trout (Allendorf *et al.*, 1983; Ferguson and Danzmann, 1985; Danzmann *et al.*, 1986; Ferguson *et al.*, 1988a). Trout differing in phosphoglucomutase (PGM) liver isozyme differ significantly in time of hatching, developmental rate, and early behaviour. Consequently they differ in size and age of sexual maturity. All these phenotypic differences can have adaptive consequences, and are often cited as central to theoretical arguments concerning life-history tactics (e.g. Noakes *et al.*, 1989; Zimmerer and Kallman, 1989).

A note of caution should be mentioned here. Since single gene differences and single gene effects have such intrinsic appeal, it is tempting to look for examples either to illustrate the principle or as material for study of mechanisms. Schemmel's (1980) study of feeding behaviour in blind and sighted populations of characids, *Astyanax mexicanus*, illustrates a potential limitation. Analysis of feeding behaviour in the initial crosses between these two forms suggested a monofactorial inheritance, but studies of F_2 and backcross generations revealed that these differences in feeding behaviour were controlled by polygenes (Fig. 1.2) (see also Chapter 17 by Parzefall, this volume). Unfortunately, because of limitations of crossings possible in many cases, we may be limited to F_1 generation data, and so we should interpret results in these cases with some caution.

We also know that the action of genes may be affected by a variety of genetic and non-genetic factors. It is often considered that heritability, an estimate of the amount of the phenotypic variance among individuals that may be ascribed to additive genetic variance, can be taken as an absolute indicator of whether some difference is determined by genetic or environmental factors. This is not necessarily the case, for two quite different reasons. Estimates of heritability depend upon the environmental conditions under which the measurements were made. Heritability could be high in one set of conditions and low in another. Furthermore, the effects of genes (and environments) can depend upon the presence or absence of other genes (or environmental factors), and it may not always be obvious how these effects may be related. For example, there is a class of alleles in *Drosophila*, known as temperature-sensitive mutants, whose activity depends upon environmental temperature, as the name implies. There is also evidence that even within a given genotype, alleles may be selectively activated or deactivated at different times during ontogeny (Barlow, 1981). In hybrid charrs (*Salvelinus namaycush* × *S. fontinalis*) for example, there is evidence that maternal and paternal alleles may be activated at different times during ontogeny (Berst *et al.*, 1980). The phenomenon of gene imprinting, in which the activation of genes depends on which parent they came from, is currently receiving a good deal of attention (Sapienza, 1990).

The second major limitation to interpreting the genetic basis of behavioural differences from heritability estimates arises if a particular behavioural action pattern is totally determined by a genetic difference (i.e. one gene, one behaviour), and that particular allele is present in all individuals. For example, a particular trait has been selected for and has been driven to fixation. Since there would be no variability among individuals that could be ascribed to genetic differences, the heritability estimate would be zero. Thus heritability estimates may have some value, and they may have a certain intuitive appeal, but their meaning and limitations must be appreciated.

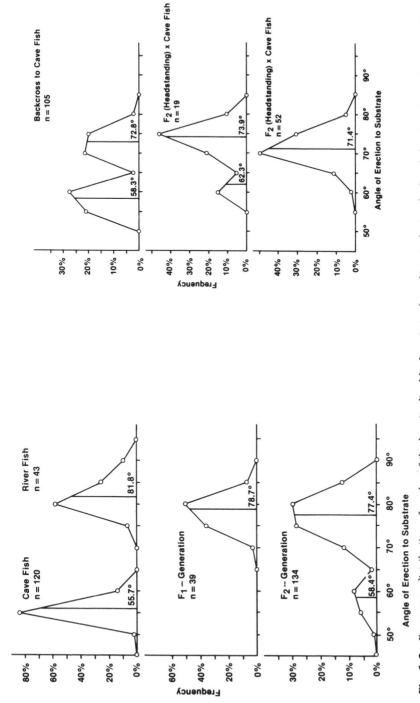

Fig. 1.2 Frequency distribution of angles of the longitudinal body axis to the substrate during feeding behaviour in total darkness by characids, *Astyanax mexicanus*. (left) Results for genetically blind cave fish, visually normal river fish, and their F_1 and F_2 hybrids, showing apparent monofactorial inheritance of this behavioural difference. (right) Results from backcross and F_2 backcross to blind cave fish generations, showing polygenic basis to behavioural difference. Reproduced with permission from Schemmel (1980).

The study of mutations has also been used as a technique, ideally as it might involve single gene differences. The incidence of mutations can be increased by exposing parents or gametes to mutagenic agents such as irradiation or chemicals. Most such mutations tend to be deleterious, or so grossly aberrant as to be clearly maladaptive, and so are of interest primarily for the study of causal mechanisms (Partridge, 1983). Studies of cichlids, *Cichlasoma nigrofasciatum*, and guppies using mutagenic X-irradiation have shown quantitative behavioural differences (Schroder, 1973; Werner and Schroder, 1980). These results confirm the pattern known from other animals (Bentley, 1976), including the polygenic model proposed to account for the observed effects.

That single gene differences should be associated with behavioural differences in fish is hardly suprising. In fact, were it not the case, it would be remarkable indeed, since the correlational and causal links between single genes and behaviour have been widely established in virtually every other kind of animal studied (Ehrman and Parsons, 1976). The remarkable feature of the examples from fishes is the basic and extensive effects that single genes appear to have not only on behaviour in the narrow sense, but also on functional aspects in a more general sense. This latter point is of special interest as it relates to theoretical arguments suggested in 'genes for strategies or tactics' in an evolutionary context (Dawkins, 1982, 1989).

The assumptions in these arguments are necessary and reasonable, but are often justly criticized as not only unrealistic but largely unfounded (Dawkins, 1982). Although it is interesting to ask what we might expect if there were a gene for territoriality or some such behaviour, critics have rightly pointed out that most such behavioural strategies or tactics can hardly be considered as single elements or modal action patterns (Barlow, 1981; Dawkins, 1982, 1986, 1989). The connection between a single gene and behaviour must be an indirect and complex one at best, and some would suggest that it is unreasonable to expect anything other than polygenic influences on behaviour, not to mention the confounding effects of ontogeny, epigenesis, experience, and environmental influences on behaviour. Can we, then, justifiably postulate genes for, say 'territoriality' versus 'female mimicry' in the reproductive behaviour of bluegill sunfish, *Lepomis macrochirus* (Gross, 1984)? Of course, such questions must be answered empirically (see also Chapter 13 by Magurran, this volume). Current evidence in fact suggests that the question is not as unreasonable as it might seem. Obviously, the difference between a male sunfish acting as a territorial resident and one entering a territory as a female mimic is enormous, in both qualitative and quantitative terms. But as the rainbow trout example shows, an entire suite of life-history features can differ as a result of a single gene difference. This could be the result of either a cascade of pleiotropic effects, or a switch-gene effect in development with one locus affecting several other loci.

Selection

What comes naturally . . . or artificially

Evidence from selection experiments relating genes to behaviour is frequently encountered in domestic species, or those whose propagation is controlled by humans. Darwin was well aware of this (1859, 1868), and Mendel used artificially selected strains for his breeding experiments with peas (Bateson, 1913). The bewildering array of domesticated strains of plants and animals, including some fish species, is the most obvious everyday demonstration of the genetic basis for phenotypic variations. Of course no species can escape natural selection, and so similar evidence is potentially present in wild populations (Endler, 1986).

In selection experiments virtually any desired response can be produced, so long as the character subjected to selection has the necessary genetic variance. Detailed analyses of the actual mechanisms producing such responses to artificial selection almost always involve unpredictable twists, especially if behavioural characters are among those being selected. A classic example is Manning's (1976) study of mating speed in *Drosophila melanogaster*. Artificial selection was carried out to produce flies which mated faster than average (= unselected) for one strain, and slower than average for the other. Highly significant differences in mating speed were produced, but the basis for this effect was as much a difference in general locomotor activity and responsiveness as an intrinsic difference in actual sexual behaviour. This effect has been encountered so frequently that it can be viewed as fundamental. A particular behavioural phenotype can be achieved by more than one particular causal mechanism, and it may not always be obvious what that mechanism might be (Manning, 1976).

Evidence of natural selection acting on behavioural phenotypes is typically more subtle, and requires either careful analysis of field data or insightful experiments (Endler, 1986). It could be argued that the adaptive nature of most characteristics of animals, including behaviour, is evidence that natural selection has moulded those characteristics to the particular features of the animal's niche. At worst, such arguments can easily become circular, *post hoc* rationalizations or tautologies, sometimes referred to as 'just-so-stories' (e.g. Gould and Lewontin, 1979). Combined with critical experiments to test specific predictions they can become rigorous and satisfactory analyses.

For example, evolutionary theory can lead to specific predictions as to the patterns of genetic relatedness and behaviour we might expect in particular circumstances. Female guppies carry their developing embryos inside their bodies, and give birth to them as juveniles (Noakes, 1978). Under some conditions, females will cannibalize young, in both laboratory and field conditions. Evolutionary theory would predict that females that avoided

Table 1.1 Schooling behaviour in populations of guppies, *Poecilia reticulata*, from various streams in Trinidad*

Population (stream)	Main fish predators[†]	Schooling behaviour
Guayammare	Characids and cichlids	Well developed
Lower Aripo	Characids and cichlids	Well developed
Petite Curucayo	*Rivulus hartii* (high density)	Intermediate
Upper Aripo	*Rivulus hartii* (intermed. density)	Poorly developed
Paria	*Rivulus hartii* (low density)	Absent

*Source: Seghers (1974a).
[†]*Characids (mainly* Hoplias malabaricus*) and cichlids (mainly* Crenicichla alta*) exert heavy predation pressure on all life stages of the guppy, and so represent a serious, continuing threat. The major impact of* Rivulus hartii *is on immature guppies, and so this predator is a less serious threat. High density of* R. hartii *was more than 50 individuals caught per hour; low density was less than 10 individuals caught per hour.*

cannibalizing their own (genetic) offspring would be favoured. Laboratory tests show that when presented with their own and unrelated young, females preferentially cannibalize young from other females (Loekle *et al.*, 1982). We do not know the mechanism involved, but this difference in behaviour must be related to genetic differences among the fish, as an example of kin recognition (Colgan, 1983).

The guppy has been a particularly rich and rewarding source of correlational and experimental data to test evolutionary hypotheses. The species lives naturally in freshwater streams in Trinidad, with different, isolated populations exposed to varying regimes of a variety of predators. Life-history characteristics (e.g. age and size at maturity, size and number of offspring) vary predictably with the predation regime to which each guppy population is subjected, and these differences are genetically determined (Reznick, 1982a, b). Genetically determined behavioural differences between guppy populations are just as striking. They respond differently to aerial predators, and show varying degrees of development of schooling behaviour, depending upon the predation regime to which they are exposed in nature (Table 1.1) (Seghers, 1974a,b). However, the detailed experimental evidence from both the laboratory and field on the effects of natural and sexual selection on colour patterns of males is one of the most convincing arguments for the genetic basis of adaptive differences in this species (Endler, 1983, 1987). Not only can differences in the details of male colour patterns be correlated with local predation and courtship regimes in the field, but comparable differences can be produced in a few generations in the laboratory in artificial selection experiments simulating the range of conditions known from the field (Fig. 1.3).

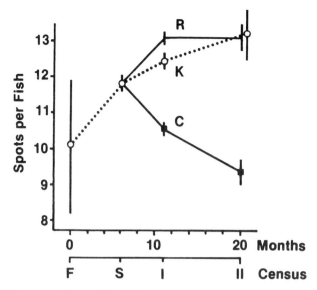

Fig. 1.3 Changes in the number of colour spots on male guppies, *Poecilia reticulata*, in laboratory pond experiments. K, no predators present; R, *Rivulus hartii* present (a weak predator); C, *Crenicichla alta* present (a dangerous predator); F, foundation populations; S, start of experiments, with predators added to R and C ponds only; I and II are dates when censuses were taken of guppies in each pond. Increasing danger from predators produces a decrease in number of colour spots (i.e. a less conspicuous pattern) on males. Points and vertical lines are means and two standard errors, respectively. Reproduced with permission from Endler (1980).

Hybrids and backcrosses

The study of hybrids can be productive both as a first step to demonstrate a genetic basis for a behavioural difference, and as a more specific technique for investigating the details of genetic mechanisms. The intermediacy of hybrids and patterns of simple Mendelian inheritance are typical examples of the evidence from these studies. Hybrids may be produced between inbred strains within the same species, or between species. Crosses between inbred lines often depend on relatively minor, quantitative differences that may be difficult to trace with certainty beyond the F_1 generation, even though the animals are fully interfertile. On the other hand, crosses between distinct species are often limited by hybrid sterility that prevents F_2 or backcross generations, even though in these cases the behavioural differences may be much greater and more clearly distinguished.

Schemmel (1980) has produced a detailed analysis for a behavioural difference between forms of a tropical characid, as mentioned earlier (and in

Chapter 17 by Parzefall, this volume). Results of his F_2 and backcross generations demonstrated that the apparent monofactorial pattern seen in the F_1 generation was produced by a polygenic system. The caution to be drawn from these results is particularly important, since as noted above, hybrid studies are seldom carried beyond the F_1 generation.

Our studies of social behaviour in juvenile charrs, *Salvelinus* spp., have the advantages of a striking contrast in behaviour between distinct species and yet F_2 and backcross generations are easily available (Ferguson and Noakes, 1982, 1983a, b; Ferguson *et al.*, 1983). The social behaviour of one of the parent species, the brook charr, *S. Fontinalis*, consists of aggressive defence of fixed territories. The other parent species, the lake charr, *S. namaycush*, is very much less aggressive, shows no indication of territoriality and tends to move about frequently. Our studies of hybrids and backcrosses showed a general tendency towards intermediacy of social behaviour, but with a

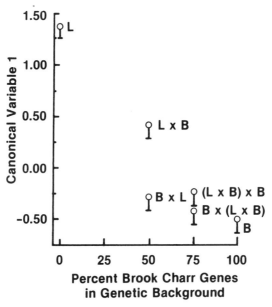

Fig. 1.4 Differences in social behaviour of parental brook charr, *Salvelinus fontinalis*, lake charr, *S. namaycush*, and various hybrid and backcross charrs. The female parent of each cross is named first in each case (B, brook charr, L, lake charr): B × L means the F_1 hybrid from a brook charr female and a lake charr male, for example. The canonical variable 1 (from a discriminant analysis) comprised measures of lateral display, charge and forage by the fish. Brook charr are more aggressive (high levels of display and charge) and forage less than lake charr. Hybrids are intermediate, but closer to the maternal parent, in all cases. Points and vertical lines are means and two standard errors, respectively. Reproduced with permission from Ferguson and Noakes (1983a).

significant maternal influence in all cases (Fig. 1.4). The evidence supported a polygenic basis for these differences, which is not surprising given the complex measure of social behaviour. More interestingly, there was also a general pattern for measures of mobility, aggression and feeding behaviour to covary in these fish, suggesting that perhaps natural selection has favoured the inheritance of integrated suites of behavioural characteristics, perhaps through linkage of the genes involved.

The classic study of courtship in swordtails, *Xiphophorus helleri*, and platyfish, *X. maculatus*, by Clark *et al.*, (1954) dealt largely with the study of hybrids (and is the only behavioural genetics study with fish usually cited in textbooks). They found both quantitative and qualitative differences between male courtship behaviour in these species, with a general tendency towards intermediacy of behaviour in hybrids. The authors suggested a polygenic basis for the differences in male courtship behaviour between these species. This courtship behaviour functions as an effective reproductive isolating mechanism between the species in nature, and under all but the most extreme artificial laboratory conditions (no choice of mates offered, males of one species and females of the other confined together). Even in these extreme cases, fewer that 20% of the infrequent copulation attempts resulted in insemination. Although few other species have been studied in such detail (e.g. Liley, 1966), there is often an assumption that these results typify the general case, i.e. species-typical differences in courtship behaviour, with a polygenic basis of inheritance, are an important reproductive isolating mechanism (Keenleyside, 1979).

1.3 CAUSAL MECHANISMS

Chocolate soup and jigsaw puzzles

Selection (whether natural or artificial) can readily produce shifts in genetically determined behavioural differences, but there is no necessary or even predictable relationship between causal mechanisms and behavioural consequences. One strain of fish may grow efficiently because it has a particular set of enzymes regulating intermediary metabolism, whereas another strain achieves the same result through a decrease in general locomotory activity, sparing energy for metabolic processes. Even if we can establish the causal thread from gene to behaviour, it will obviously include many intermediate steps. Other genes and gene products are quite possibly involved, and for many kinds of behaviour, non-genetic environmental inputs are also likely to have an effect. Given this complexity, it is perhaps surprising that we can make any sense of the system at all!

The difference of a single enzyme in cases such as phenylketonuria in humans has widespread and cumulative phenotypic effects, including some behavioural differences. These cases are understood as a consequence of their nature, as single (typically recessive) mutant alleles, inherited in simple Mendelian fashion. Behaviour is far more removed from the direct action of genes, so we can seldom expect to have such clear-cut examples, yet we believe that the relationship between genes and behaviour can and does exist. It is neither easy nor obvious to parse the continuous, ongoing stream of behavioural output into basic constituent units. The confounding complexity of behavioural measures often degenerates into an indecipherable and unpalatable blend of intermediate states and actions (a chocolate soup).

The causal mechanism may be of no interest or particular importance to someone interested in functional questions. The causal mechanisms in some cases may not even be worth the effort required to decipher them. It may be sufficient to know that differences in courtship behaviour can be attributed to genetic differences between individuals. If the principle can be assumed to be common in different cases, the one more amenable to the study at hand will surely be chosen for study. We have cited several examples under the heading of single genes that may be associated with a causal relationship between genes and behaviour. In a few cases we even know the single gene difference and at least the intermediate gene product involved.

Rainbow trout differing only in LDH or PGM phenotype can vary significantly in swimming performance and age of sexual maturation, for example. These effects could result from any one or several steps in intermediary metabolism regulating oxygen uptake, growth, and ageing in these fish. It is even unclear whether we can ascribe any of the observed effects directly to these genes since they may only serve as markers for other tightly linked genes or regulatory elements that are ultimately the causal agents of the phenotypic differences. To decipher the causal nexus between the expression of a particular LDH or PGM variant and the corresponding differences in physiological and life-history variations is hardly a simple undertaking, and one that may only prove to be of interest for the understanding of reductionist mechanics. Even when the physiological 'blueprint' is fully revealed, there is no guarantee that the connecting behavioural patterns will be more clearly understood. However, such patterns might fit into arguments relating evolutionary shifts in life histories to small, genetically determined changes in early ontogeny (e.g. Noakes, 1981, 1989; Noakes *et al.*, 1989) that also define fairly well-prescribed modal action patterns.

To borrow an analogy from Manning (1976), we are trying to assemble a jigsaw puzzle composed of a large, but unknown, number of pieces. We do not have the completed picture to guide us, and we are presented with random pieces from a variety of experimental and analytical approaches. An extreme strategy might be to try each available piece in every possible combination with every other piece. The futility of this approach can be appreciated only through experience (unless one has the attention span, memory capacity and imagination of a digital computer!). It soon becomes obvious that a much more fruitful strategy must incorporate some effort to formulate an image of what the final picture might be, and an attempt to recognize patterns among the pieces. Dwelling unduly upon the precise size or shape of a given individual piece might by chance provide just the right fit to one piece and reveal a little more of the puzzle. More likely it will produce an expert knowledge of one isolated piece, and a fruitless search for a complementary combination. It is much more likely that we can achieve steady progress if we try to judge what the basic pattern should be, and what kinds of pieces we should group together at different sides of the picture in anticipation of sketching the outlines at some broader scale in a greater perspective.

Some remarkable advances have been made in behavioural genetics and the field continues to prosper. Perhaps one of the most exciting areas that could yield future results would be that of combining the sophisticated techniques of genetic engineering and recombination with behavioural genetic analysis. We would no longer have to content ourselves with deciphering patterns given to us by nature, or searching among random mutations produced by experimental intervention. Single, specified genetic alterations could be employed in an experimental approach to test specific predictions for both causal and functional relationships. The other area of continuing interest and productivity is that of evolutionary aspects. To some this is unfortunately synonymous with the term sociobiology, so perhaps it is more productive to consider this as a way of formulating the overall picture of the jigsaw puzzle. To biologists the question of function must always be a central one, whatever aspects of behaviour we are considering.

As a group, teleost fishes perhaps do not lend themselves particularly to studies of causal mechanisms; nematodes, flies and a few other species of choice appear to have that field solidly to themselves (but see Powers, 1990). The ultimate questions of function seem less likely to emerge so readily from those subjects, so perhaps fish offer more promise in this regard. They are (mostly) not too big to be unmanageable nor too small (microscopic) to be awkward, not too long-lived for our patience nor too short-lived for ontogenetic studies, and with an ecological and evolutionary diversity sufficient to provide material for almost any researcher's interests. Considering the opportunistic nature of most behavioural genetics studies on fish so far, the results have been quite rewarding. Systematic programmes of study would surely be very productive.

1.4 TECHNIQUES AND TECHNOLOGY USED IN PARENTAGE TESTING

The testing of many hypotheses in behavioural ecology requires that the identity of parents be known. However, the behaviour and ecology of many fishes often precludes direct knowledge of parentage. For instance, many species spawn in large assemblages, and in the absence of post-spawning parental care it is difficult to determine the identity of the individuals whose eggs and sperm led to the production of a particular young fish. Furthermore, even if parental care is provided, the care giver is not necessarily the biological parent. Thus, parentage can only be established unambiguously with the transmission of unique genetic elements or markers to offspring. Useful genetic markers must meet two conditions. First, potential parents must differ in the identity of specific genetic elements (i.e. the population under study must be genetically variable). Second, the genetic markers must follow simple genetic rules in that they must be inherited in a predictable way.

Genetic markers differ in their subcellular origin, transmission to progeny, and ease of experimental manipulation. Mitochondrial DNA (mtDNA) markers are maternally and clonally inherited in that an offspring has its mother's mtDNA type. The mtDNA molecule is much smaller than the nuclear genome and therefore is technically easier to manipulate. Nuclear markers can be DNA regions (nucDNA) and the protein products of loci (allozymes); these follow the typical rules of inheritance where both male and female parents transmit alleles to offspring. NucDNA markers can be categorized depending upon how many time they occur in the genome. Some sequences of DNA, termed variable number of tandem repeats or VNTRs, are repeated many times. VNTRs are more complex than unique sequences but have the potential to provide greater numbers of genetic markers because of their inherent variability, and therefore are often referred to as 'fingerprints'.

The ability to detect genetic variation in either the nuclear or mitochondrial genomes, and the use of this variation as a marker of parentage, depends on the molecular approach. DNA variation is detected either by DNA sequencing or by using bacterial enzymes called restriction endonucleases that cut DNA at specific base sequences (generally 4, 5, 6 or 6+ bases). For most applications in maternity or paternity testing, DNA sequencing is too labour intensive, since large regions of DNA would need to be sequenced to detect enough genetic variation among individuals. This requires cloning or polymerase chain reaction (PCR) amplification of several specific regions of nuclear DNA for sequencing (Sambrook *et al.*, 1989; Innis *et al.*, 1990). The technical difficulties and associated costs make such analyses impractical for parentage testing of fish.

Restriction fragment length polymorphisms (RFLPs) detected within a specific region of DNA are less technically demanding, but require the use of specific homologous DNA probes that will only detect a specified region of DNA. NucDNA fragments that are simply generated by digesting the DNA with a restriction enzyme appear as a smear on an agarose or acrylamide gel which is used to separate these fragments. Transferring these fragments to a membrane matrix that binds the DNA (Southern blotting) and then applying a DNA probe (chemiluminescent, chromagen, or radioisotope) that will only bind to a small homologous region of the DNA will greatly reduce the number of fragments detected (Sambrook *et al.*, 1989; Martin *et al.*, 1990). This will allow the specific identification of a smaller number of DNA fragments, whose inheritance can be determined through family crosses. If the probe is sufficiently large and is only homologous to a unique region of DNA, there is a high probability of detecting many more alleles than is possible with a conventional allozyme analysis. Probes to highly repetitive core sequences of DNA may also be used to detect regions of DNA that differ in the number of tandem repeats of the core sequence (VNTRs) (Jeffreys, 1987). By using these probes, many fragments can be detected among individuals, even to the point where one probe may detect several hundred fragments. The Mendelian nature of such fragments is difficult to determine, however (i.e. is the fragment an allele at a polymorphic locus or a single monomorphic locus?), since these tandem repeats may be scattered on different chromosomes throughout the genome.

1.5 APPLICATIONS TO PARENTAGE TESTING

The clonal inheritance of mtDNA makes it an excellent marker of female family lines. It is particularly useful as a marker of hybridization events among fish species when used in combination with nuclear gene markers such as allozymes that will unequivocally establish the hybrid nature of a fish. MtDNA will provide information on the direction of hybridization between species (i.e. which species is the female parent). If hybridization is suspected between closely related species due to the presence of satellite or 'sneaky' males, mtDNA in combination with allozymes will provide information on the frequency of such occurrences in the wild, and will also provide a genetic tag to monitor the survivorship and competitive ability of hybrids (Forbes and Allendorf, 1991; Gyllensten *et al.*, 1985).

MtDNA variation may be used to investigate schooling behaviour, social and territorial interactions, and competition within and among confamilial or closely related families. Since mtDNA is a marker of family lineages, it may be used as a tag to determine whether fish schools are composed of more closely related individuals compared with the entire population. Studies of mtDNA variation can also indicate whether adjacent individuals within spawning aggregations are more closely related than more distantly positioned spawners. It must be remembered, however, that the utility of mtDNA in such studies is directly proportional to the number of distinct mtDNA clones that can be identified in a population. At a bare minimum, therefore, at least two mtDNA clones must be identified, with both clones present in high frequency.

Either allozyme or nucDNA variation is equally applicable to ascertaining the parentage of fish in the wild. Allozymes are currently being used by geneticists to mark hatchery or experimental populations of fish to monitor such factors as the reproductive success of different hatchery strains, and the competitive and agonistic interactions among different family lines within a hatchery strain (Ferguson *et al.*, 1988a; Lane *et al.*, 1990; Seeb *et al.*, 1990). Allozymes can also be used to mark F_1 hybrids between species so that differences in growth rate and developmental rate may be compared between the parental and hybrid stocks when they are reared in a common environment (Ferguson *et al.*, 1988b). Nuclear DNA fingerprints obtained from VNTR probes, although not currently used in fish parentage analyses, have been used in population genetic studies (Turner *et al.*, 1990). However, VNTR variation in almost all teleosts is largely unexplored.

There are ultimately two factors which determine how successful a researcher will be in discriminating the parentage of fish using these markers. These are (1) the number of variable alleles used, and (2) the frequency of the alleles. The probability of unambiguously identifying the parentage of a fish in a population increases with (1) the number of alleles screened, (2) the number of polymorphic loci screened, and (3) allele frequencies at a locus that are approximately equal. For example, if there are two alleles at a locus, a frequency of 0.5 for each allele has the greatest discriminatory utility.

To ascertain the parentage of fish in the wild, the behavioural ecologist may be aided in part by differences in the reproductive modes of different fish species. As a very generalized dichotomy, fish may be categorized as either guarders and bearers (e.g. clutch tenders, nest guarders, external brooders and internal live bearers) or non-guarders (open substratum scatterers and brood hiders) (Balon, 1985). The ability to unambiguously distinguish the parentage of fish varies among these reproductive modes. Generally, for guarders and bearers, the genotypes of several full-sib progeny as well as one of the parental genotypes (that of the guarding or bearing parent) can be ascertained if collections of fish are made when the offspring are still under parental care. This greatly increases the power of genetic markers to distinguish the genotype of the unknown parent. As an example, consider a diploid locus with two alleles A_1 and A_2 that occur at equal frequency in the population. If the genotype of the guarding parent is A_1A_1, then the genotype of the unknown parent can be accurately ascertained by screening the progeny genotypes.

Known parent genotype $= A_1A_1$

Progeny genotypes	Unknown parent genotypes	Frequency of genotypes in excluded parents
A_1A_1	A_1A_1	$A_1A_2 = 0.50 + A_2A_2 = 0.25 = 0.75$
A_1A_1 $\quad$ A_1A_2	A_1A_2	$A_1A_1 = 0.25 + A_2A_2 = 0.25 = 0.50$
A_1A_2	A_2A_2	$A_1A_1 = 0.25 + A_1A_2 = 0.50 = 0.75$

Thus either 75% or 50% of all the fish in the population may be excluded as possible parents. An average estimate of the frequency of fish that may be excluded as parents in a random sampling of progeny and uniparental genotypes for a given gene locus or probe may be called the mean parental exclusion frequency (mPE) and may be calculated as:

$$mPE = \sum_{i=1}^{g} (1 - g_p) \cdot g_p \qquad (1.1)$$

where g_p is the genotypic frequency of the unknown parent. The total possible parentage frequency (tPP) would then be calculated as the product of all possible parentage frequencies for each locus:

$$tPP = \prod_{i=1}^{l} (1 - mPE_l) \qquad (1.2)$$

where mPE_l = mean parental exclusion frequency for the lth polymorphic locus.

With the current example, $mPE = 0.625$. Another way of stating this is to say that on average, approximately 63% of the fish in a population may be excluded as parents using a single allozyme locus or gene probe that detects two alleles in approximately equal frequency if the genotypes of the full-sibs and one parent are known. The frequency of excluded parental genotypes is then simply calculated as $1 - tPP$ (Westneat *et al.*, 1987; Weir 1990).

Figure 1.5 shows the average number of polymorphic loci which need to be screened to reasonably exclude the probability of parentage with different numbers of loci (or probes) screened, assuming the presence of either two alleles (graph A) or of three alleles (graph B) for each locus. The ordinate shows the probability that a fish taken at random from a population may be the parent of the screened fish when the genotype of one of the parents is known. One can see that the number of loci that need to be screened substantially increases at allele frequencies greater than 0.7. For example, if the most common allele occurs at a frequency of 0.9 in the population, 15 diallelic and 14 triallelic loci would need to be screened to have a reasonable probability of excluding greater than 99.5% of the population as possible parents. In contrast, when allele frequencies are closer to equality, only five diallelic or four triallelic loci need to be screened. This threefold difference is substantial and highlights the resolving power that only a few loci may have in discriminating the parentage of guarding fish. However, for a population size of 10 000, one is still left with the sobering conclusion that 30–40 fish could be the unknown parent. Also, the above scenario does not address the problems that may be encountered in live-bearing fish owing to multiple inseminations or multiple paternal contributions to a single brood. For a discussion of the calculation of parental exclusion frequencies involving extra

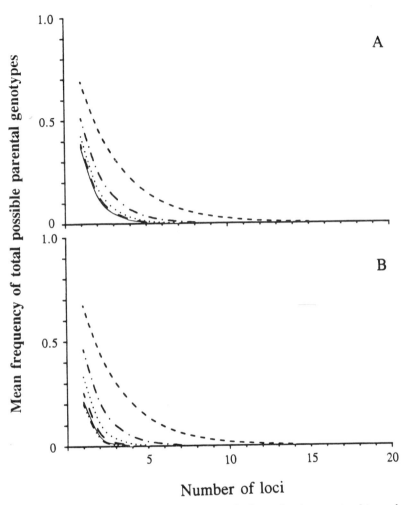

Fig. 1.5 The average number of genetic loci which need to be examined to exclude an unknown fish as a possible parent when the genotype of one parent is known using either (A) diallelic or (B) triallelic loci. Plots are shown assuming all loci have allele frequencies of (0.33–··–··–), (0.5————), (0.6———), (0.7·········), (0.8—·—·—), or (0.9–––––) for the most common allele.

pair copulations and egg-dumping, the reader is directed to Westneat *et al.,* (1987).

The model just outlined does not consider the contribution that mtDNA analysis in combination with allozymes can contribute to discerning the parentage of fish in the wild. In the absence of any type of disequilibrium between allozymes and mtDNA clonal types in the population (Asmussen *et al.,* 1987), the proportion of the female population that can be excluded as a parent is $1 - m_x$, where m_x is the frequency of the xth mtDNA clone (the clone of the sampled fish) in the population. Therefore, as an example, consider a species such as the brown bullhead, *Ictalurus nebulosus*, which shows biparental guarding behaviour (Blumer, 1985a, b). If one were estimating the proportion of parents excluded using two diallelic loci with both alleles occurring at equal frequencies, 86% of the fish in the population could be excluded as potential parents (Fig. 1.5 (A)). If however, two mtDNA clones were present in the population at equal frequency, then 93% (i.e. tPP = tPP$\cdot m_x$) of the fish in the population could be excluded as potential mothers using these two loci. Brown bullheads are likely to be very amenable to such studies as over 20 mtDNA clones have been described in this species (Weider *et al.,* 1989).

In studies where the genotypes of neither parent nor full-sib progeny are known, the number of loci which need to be examined to obtain an estimate of the unknown parental genotypes substantially increases. Such studies may attempt to examine the nature of fish distributions in the wild to see if more closely related individuals are found in closer proximity to one another compared with other fish in the population, or to ascertain whether behavioural interactions among individuals differ in a manner dependent upon their degree of relationship. To reasonably identify a confamilial versus heterofamilial fish in a population may require the average sampling of anywhere from 40 to 325 diallelic loci (Fig. 1.6(A)), or 23 to 297 triallelic loci (Fig. 1.6(B)). These models assume that allele frequencies may range from equality to 0.90 for the most common allele, and are calculated as:

$$\text{mPE} = \sum_{i=1}^{g} \sum_{i=1, i \neq g}^{x} (f_x^2 + 2f_x f_{y \neq x}) g_p \qquad (1.3)$$

where g_p is the population genotypic frequency of the sampled fish at the gth locus, f_x^2 and $2f_x f_{y \neq x}$ are the allele frequencies of genotypes at the gth locus that do not possess any of the alleles present in the gth genotype. For example, if a sampled fish had an A_1A_1 genotype at a diallelic locus, the only genotype that could be excluded as a possible parental genotype would be A_2A_2, since both A_1A_1 and A_1A_2 fish could be parents. For A_1A_2 fish the mPE = 0 since fish with all three genotypes at a diallelic locus could be parents.

Clearly, the number of loci that need to be examined in order to establish

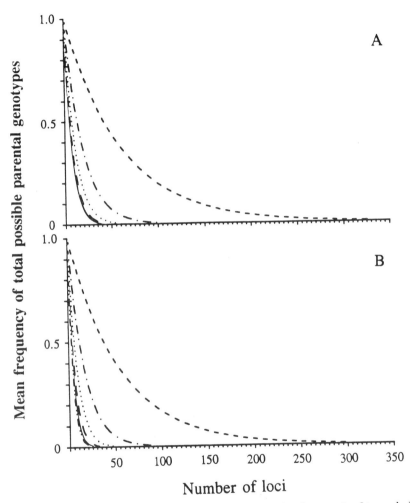

Fig. 1.6 The average number of genetic loci which need to be examined to exclude an unknown fish as a possible parent when the genotype of neither parent is known using either (A) diallelic or (B) triallelic loci. Plots are shown assuming all loci have allele frequencies of 0.33, 0.50, 0.60, 0.70, 0.80, or 0.90. (Curves are coded as in Fig. 1.5).

familial relationships among fish randomly sampled in the wild is beyond the database structures for all fish species at the present time. Most species inventories on genetic polymorphisms are based upon allozyme surveys that typically may detect anywhere from one or two, to one or two dozen polymorphic loci within a population. Generally, two or three alleles are detected at each locus and the frequency of the most common allele at any locus tends to be higher than 0.8 or 0.9. Therefore, allozyme electrophoresis, although possibly useful in inferring parentage in populations where the genotype of one parent and that of the full-sibs is known, will not be useful in examining behaviour–genetic relationships among individuals sampled at random.

The use of molecular DNA technology will greatly increase the amount of variation that may be detected within the nuclear genome of all fish species. Furthermore, many more alleles should be detected within the region of homology of DNA probes as simple restriction fragment length polymorphisms. These probes will greatly reduce the number of loci which need to be examined. VNTR probes are currently widely used in parentage studies, and in many cases rely only on the use of a single probe. Band sharing among fish is used as the basis of inferring parentage. There are several statistical methods which may be used to estimate the probability of parentage (Jeffreys *et al.*, 1985; Jeffreys, 1987; Georges *et al.*, 1988; Kirby, 1990), and include procedures such as 'binning' which classify several fragments of the same mobility class as the same 'allelic' type.

VNTR probes that detect large numbers of RFLPs should prove to be the most economical means of investigating parental exclusion frequencies once the Mendelian nature of the restriction fragments within any species is investigated and verified. For example, consider estimating mPE when the genotype of neither parent is known and it is possible to use a VNTR probe that reliably detects 25 alleles at a locus. If each of these alleles occurs at a frequency of 0.04 in the population, then on average, approximately 85% of the fish in the population could be excluded as possible parents of any fish sampled at random. This value would range from 92.16% of the fish excluded if the sample fish was homozygous at the locus, to 84.64% of the fish excluded if the sample fish was heterozygous. As an example, assume a heterozygote is sampled from the population. Since the frequency of homozygotes in the population is 0.0016 and there are 25 such genotypic classes, the mPE for homozygotes is 0.0369 (homozygote frequency − 2 [the 2 alleles sampled] × 0.0016). Similarly, as each heterozygous class occurs at a frequency of 0.0032 and there are 300 such genotypes, the mPE = 0.8096 (heterozygote freq. − 24 × 0.0032 − 23 × 0.0032). The use of VNTR probes will undoubtedly provide a great deal of genetic information on the relatedness of individuals in a population, but because of the complex fragment patterns observed with such probes, it will also be the type of genetic analysis that is most likely to result in interpretative errors.

The main pitfall in using VNTRs is that careful breeding studies are generally not conducted to examine the Mendelian nature of inheritance of electrophoretic fragments that may be detected with any single probe. Unless such studies are done, it is not possible to discern whether a band is a polymorphic allele, or a monomorphic locus that may have an extra repeat unit. Many bands are also present, which increases the probability that multiple alleles and some monomorphic loci may have the same mobility, leading to erroneous interpretations (Lynch, 1988). Furthermore, since VNTR regions are highly labile with respect to the generation of tandem repeats, the Mendelian genotype of an individual may change from one generation to the next by the random generation of variable tandem repeat numbers in different gametes during meiosis. Incomplete digestion of total DNA may also lead to 'spurious' fragments in different individuals that would be readily misinterpreted as another VNTR fragment. At the present time, therefore, a combination of an allozyme and mtDNA analysis will provide the most parsimonious database for the behavioural ecologist to interpret.

ACKNOWLEDGEMENTS

We thank Tony Pitcher for the opportunity to write this chapter, and for his constructive comments on an earlier draft. We thank all our colleagues, graduate students, research associates and assistants who have contributed to our research and thinking over the years. Our research has been supported by the Department of Fisheries and Oceans of Canada, the Natural Sciences and Engineering Research Council of Canada, the Ontario Ministry of Colleges and Universities, the Ontario Ministry of Natural Resources, Ontario Hydro, and the University of Guelph.

REFERENCES

Alcock, J. (1989) *Animal Behavior, An Evolutionary Approach*, fourth edn., Sinauer, Sunderland, MA, 596 pp.

Allendorf, F. W. and Ferguson, M. M. (1990) Genetics, in *Methods for Fish Biology* (eds C. B. Schreck and P. B. Moyle), American Fisheries Society, Bethesda, MD, pp. 35–63.

Allendorf, F. W., Knudsen, K.L. and Leary, R. F. (1983) Adaptive significance of differences in the tissue–specific expression of a phosphoglucomutase gene in rainbow trout. *Proc. nat. Acad. Sci. U.S.A.*, **80**, 1397–1400.

Asmussen, M. A., Arnoid, J. and Avise, J. C. (1987) Definition and properties of disequilibrium statistics for associations between nuclear and cytoplasmic genotypes. *Genetics*, **115**, 755–68.

Balon, E. K. (1985) Additions and amendments to the classification of reproductive styles in fishes, in *Early Life Histories of Fishes* (ed. E. K. Balon), Dr. W. Junk, Dordrecht, Netherlands, pp. 59–72.

Bams, R. A. (1976) Survival and propensity for homing as affected by presence or absence of locally adapted paternal genes in two transplanted populations of pink salmon (*Oncorhynchus gorbuscha*). *J. Fish. Res. Board Can.*, **33**, 2716–25.

Barlow, G. W. (1981) Genetics and development of behaviour, with special reference to patterned motor output, in *Behavioural Development in Animals and Man* (eds K. Immelmann, G. W. Barlow, L. Petrinovitch and M. Main), Cambridge University Press, Cambridge, pp. 191–251.

Bateson, W. (1913) *Mendel's Principles of Heredity*, Cambridge University Press, Cambridge, 413 pp.

Bentley, D. (1976) Genetic analysis of the nervous system, in *Simpler Networks and Behavior*, (ed. J. C. Fentress), Sinauer, Sunderland, MA, pp. 126–39.

Berst, A. H., Ihssen, P. E., Spangler, G. R., Ayles, G. B. and Martin, G. W. (1980) The splake, a hybrid charr *Salvelinus namaycush* × *S. fontinalis*, in *Charrs, Salmonid Fishes of the Genus Salvelinus* (ed. E. K. Balon), Dr. W. Junk, The Hague, pp. 841–87.

Blumer, L. S. (1985a) Reproductive natural history of the brown bullhead *Ictalurus nebulosus* in Michigan. *Am. Midl. Nat.*, **114**, 318–30.

Blumer, L. S. (1985b). The significance of biparental care in the brown bullhead, *Ictalurus nebulosus*. *Env. Biol. Fishes*, **12**, 231–6.

Burghardt, G. and Bekoff, M. (eds) (1978) *The Development of Behavior: Comparative and Evolutionary Aspects*, Garland, New York, 429 pp.

Clark, E., Aronson, L. R. and Gordon, M. (1954) Mating behaviour patterns in two sympatric species of Xiphophorin fishes: their inheritance and significance in sexual isolation. *Bull. Am. Mus. nat. Hist.*, **103**, 141–225.

Colgan, P. W. (1983) *Comparative Social Recognition*, Wiley-Interscience, New York, 281 pp.

Danzmann, R. G., Ferguson, M. M., Allendorf, F. W., and Knudsen, K. L. (1986) Enzyme heterozygosity and developmental rate in a strain of rainbow trout (*Salmo gairdneri*). *Evolution*, **40**, 86–93.

Darwin, C. R. (1859) *The Origin of Species*, John Murray, London, 460 pp.

Darwin, C. R. (1868) *The Variation of Animals and Plants under Domestication*, John Murray, London, 372 pp.

Dawkins, R. (1982) *The Extended Phenotype. The Gene as the Unit of Selection*, W. H. Freeman, San Francisco, 307 pp.

Dawkins, R. (1986) *The Blind Watchmaker*, Longman, Harlow, 332 pp.

Dawkins, R. (1989) *The Selfish Gene* (new edition), Oxford University Press, Oxford, 352 pp.

DiMichele, L. and Powers, D. A. (1982) Physiological basis for swimming endurance differences between LDH-B genotypes of *Fundulus heteroclitus*. *Science*, **216**, 1014–16.

Ehrman, L. and Parsons, P. A. (1976) *The Genetics of Behavior*, Sinauer, Sunderland, MA, 390 pp.

Endler, J. A. (1980) Natural selection on color patterns in *Poecilia reticulata*. *Evolution*, **34**, 76–91.

Endler, J. A. (1983) Natural and sexual selection on color patterns in Poeciliid fishes. *Env. Biol. Fishes*, **9**, 173–90.

Endler, J. A. (1986) *Natural Selection in the Wild*, Princeton University Press, Princeton, NJ, 335 pp.

Endler, J. A. (1987) Predation, light intensity, and courtship behaviour in *Poecilia reticulata* (Pisces: Poeciliidae). *Anim. Behav.*, **35**, 1376–85.

Ferguson, M. M. and Danzmann, R. G. (1985) Pleiotropic effects of a regulatory gene (Pgml-t) on the social behaviour of juvenile rainbow trout (*Salmo gairdneri*). *Can. J. Zool.*, **63**, 2847–51.

Ferguson, M. M. and Noakes, D. L. G. (1982) Genetics of social behaviour in charrs (*Salvelinus* species). *Anim. Behav.*, **30**, 128–34.

Ferguson, M. M. and Noakes, D. L. G. (1983a) Behaviour-genetics of lake charr (*Salvelinus namaycush*) and brook charr (*Salvelinus fontinalis*): observation of backcross and F$_2$ generations. *Z. Tierpsychol.*, **62**, 72–86.

Ferguson, M. M. and Noakes, D. L. G. (1983b) Movers and stayers: genetic analysis of mobility and positioning in hybrids of lake charr *Salvelinus namaycush* and brook charr *Salvelinus fontinalis*. *Behav. Genet.*, **13**, 213–22.

Ferguson, M. M., Danzmann, R. G., and Allendorf, F. W. (1988a) Adaptive significance of developmental rate in rainbow trout: an experimental test. *Biol. J. Linn. Soc.*, **33**, 205–16.

Ferguson, M. M., Danzmann, R. G. and Allendorf, F. W. (1988b) Developmental success of hybrids between two taxa of salmonid fishes with moderate structural gene divergence. *Can. J. Zool.* **66**, 1389–95.

Ferguson, M. M., Noakes, D. L. G. and Romani, D. (1983) Restricted behavioural plasticity of juvenile lake charr, *Salvelinus namaycush*. *Env. Biol. Fishes*, **8**, 151–6.

Foote, C., Wood, C. C. and Withler, R. E. (1989) Biochemical genetic comparison of sockeye salmon and kokanee, the anadromous and non-anadromous forms of *Oncorhynchus nerka*. *Can. J. Fish. Aquat. Sci.*, **46**, 149–58.

Forbes, S. H., and Allendorf, F. W. (1991) Associations between mitochondrial and nuclear genotypes in cutthroat trout hybrid swarms. *Evolution*, **45**, 1332–49.

Futuyama, D. J. (1986) *Evolutionary Biology*, second ed. Sinauer, Sunderland, MA, 600 pp.

Georges, M., Lequarre, A. S., Castelli, M., Hanset, R. and Vassart, G. (1988) DNA fingerprinting in domestic animals using four different minisatellite probes? *Cytogenet. Cell Genet.*, **7**, 127–131.

Gould, J. L. (1974) Genetics and molecular ethology. *Z. Tierpsychol.*, **36**, 267–192.

Gould, S. J. and Lewontin, R. C. (1979) The spandrels of San Marco and the Panglossian paradigm: a critique of the adaptationist programme. *Proc. R. Soc.*, **205B**, 581–98.

Gross, M. R. (1984) Sunfish, salmon, and the evolution of alternative reproductive strategies and tactics in fishes, in *Fish Reproduction, Strategies and Tactics* (eds G. W. Potts and R. J. Wootton), Academic Press, London, pp. 55–75.

Gyllensten, U., Leary, R. F., Allendorf, F. W. and Wilson, A. C. (1985) Introgression between two cutthroat trout subspecies with substantial karyotypic, nuclear, and mitochondrial genomic divergence. *Genetics*, **111**, 905–15.

Hayes, B. (1984) Computer recreations. *Sci. Am.*, **250**, 14–20.

Huntingford, F. (1984) *The Study of Animal Behaviour*, Chapman and Hall, London, 411 pp.

Innis, M. A., Gelfand, D. H., Sninsky, J. J. and White, T. J. (1990) *PCR Protocols: A Guide to Methods and Applications*, Academic Press, San Diego, CA, 481 pp.

Jeffreys, A. J. (1987) Highly variable minisatellites and DNA fingerprints. *Biochem. Soc. Trans.*, **15**, 309–17.

Jeffreys, A. J., Wilson, V. and Thein, S. L. (1985) Individual-specific fingerprints of human DNA. *Nature, Lond.*, **316**, 76–9.

Keenleyside, M. H. A. (1979) *Diversity and Adaptation in Fish Behaviour*, Springer-Verlag, Berlin, 208 pp.

Kelso, B. W., Northcote, T. G. and Wehrhahn, C. F. (1981) Genetic and environmental aspects of the response to water current by rainbow trout *Salmo gairdneri* originating from inlet and outlet streams of two lakes. *Can. J. Zool.*, **59**, 2177–85.

Kirby, L. T. (1990) *DNA Fingerprinting*, Stockton Press, New York, 365 pp.

Klar, G. T., Stalnaker, C. B. and Farley, T. M. (1979) Comparative physical and physiological performance of rainbow trout, *Salmo gairdneri*, of distinct lactate dehydrogenase B2 phenotypes. *Comp. Biochem. Physiol.*, **63A**, 229–35.

Lane, S., McGregor, A. J., Taylor, S. G. and Gharrett, A. J. (1990) Genetic marking of an Alaskan pink salmon population, with an evaluation of the mark and marking process. *Am. Fish. Soc. Symp.*, **7**, 395–406.

Liley, N. R. (1966) Ethological isolating mechanisms in four sympatric species of Poeciliid fishes. *Behaviour* (Supp.), **13**, 1–197.

Loekle, D. M., Madison, D. M. and Christian, J. J. (1982) Time dependency and kin recognition of cannibalistic behavior among Poeciliid fishes. *Behav. neural Biol.*, **35**, 315–18.

Lynch, M. (1988) Estimation of relatedness by DNA fingerprinting. *Molec. biol. Evol.*, **5**, 584–99.

McIsaac, D. O. and Quinn, T. P. (1988) Evidence for a hereditary component in homing behaviour of chinook salmon *(Oncorhynchus tshawytscha)*, *Can. J. Fish. Aquat. Sci.*, **45**, 2201–5.

Manning, A. W. G. (1975) The place of genetics in the study of behaviour, in *Growing Points in Ethology* (eds P. P. G. Bateson and R. A. Hinde), Cambridge University Press, Cambridge, pp. 327–43.

Manning, A. W. G. (1976) Behaviour genetics and the study of behavioural evolution, in *Function and Evolution in Behaviour* (eds G. P. Baerends, C. M. Beer and A. W. G. Manning), Oxford University Press, Oxford, pp. 71–91.

Martin, R., Hoover, C., Grimme, S., Grogan, C., Holtke, J., and Kessler, C. (1990) A highly sensitive, nonradioactive DNA labelling and detection system. *Biotechniques*, **9**, 762–8.

Noakes, D. L. G. (1978) Ontogeny of behavior in fishes: a survey and suggestions, in *The Development of Behavior: Comparative and Evolutionary Aspects* (eds G. M. Burghardt and M. Bekoff), Garland, New York, pp. 104–25.

Noakes, D. L. G. (1981) Comparative aspects of behavioural development: a philosophy from fishes, in *Behavioural Development. The Bielefeld Interdiscplinary Project* (eds K. Immelmann, G. W. Barlow, L. Petrinovitch and M. Main), Cambridge University Press, Cambridge, pp. 491–508.

Noakes, D. L. G. (1986) The genetic basis of fish behaviour, in *The Behaviour of Teleost Fishes* (ed. T. J. Pitcher), Croom Helm, London, pp. 3–22.

Noakes, D. L. G. (1989) Early life history and behaviour of charrs, in *Biology of Charrs and Masu Salmon* (eds H. Kawanabe, F. Yamazaki and D. L. G. Noakes), *Physiol. Ecol. Japan*, Spec. Vol. **1**, 173–86.

Noakes, D. L. G., Skulason, S. and Snorrason, S. S. (1989) Alternative life-history styles in salmonine fishes with special reference to Arctic charr, *Salvelinus alpinus*, in *Alternative Life-History Styles of Animals* (ed. M. N. Bruton), Kluwer, Dordrecht, Netherlands, pp. 329–46.

Partridge, L. (1983) Genetics and behaviour, in *Animal Behaviour, Vol. 3, Genes, Development and Learning* (eds T. R. Halliday and P. J. B. Slater), W. H. Freeman, San Francisco, pp. 11–51.

Powers, D. (1990) Fish as model systems. *Science*, **246**, 352–8.

Raleigh, R. F. (1971) Innate control of migration of salmon and trout fry from natal gravels to rearing areas. *Ecology*, **52**, 291–7.

Reznick, D. (1982a) Genetic determination of offspring size in the guppy (*Poecilia reticulata*). *Am. Nat.*, **120**, 181–8.

Reznick, D. (1982b) The impact of predation on life history evolution in Trinidadian guppies: genetic basis of observed life history patterns. *Evolution*, **36**, 1236–50.

Sambrook, J., Fritsch, E. F. and Maniatis, T. (1989) *Molecular Cloning, A Laboratory Manual*, 2nd edn, Cold Spring Harbor Lab. Press, New York,

Sapienza, C. (1990) Parental imprinting of genes. *Sci. Am.*, **263**, 52–60.

Schemmel, C. (1980) Studies on the genetics of feeding behavior in the cave fish *Astyanax mexicanus f. anoptichthys*: an example of apparent monofactorial inheritance by polygenes. *Z. Tierpsychol.*, **53**, 9–22.

Schroder, J. J. (ed.) (1973) *Genetics and Mutagenesis of Fish*, Springer-Verlag, New York, 356 pp.

Seeb, L. W., Seeb, J. E. and Gharrett, A. J. (1990) Genetic marking of fish populations, in *Electrophoretic and Isoelectric Focusing Techniques in Fisheries Management* (ed. D. H. Whitmore), CRC Press, Boca Raton, FL, pp. 223–39.

Seghers, B. H. (1974a) Schooling behavior in the guppy (*Poecilia reticulata*): an evolutionary response to predation. *Evolution*, **28**, 486–9.

Seghers, B. H. (1974b) Geographic variation in the responses of guppies (*Poecilia reticulata*) to aerial predators. *Oecologia*, **14**, 93–8.

Thomas, A. E. and Donahoo, M. J. (1977) Differences in swimming performance among strains of rainbow trout *Salmo gairdneri. J. Fish. Res. Bd Can.*, **34**, 304–7.

Thorpe, J. E. (1986) Age at first maturity in Atlantic salmon, *Salmo salar*: freshwater period influences and conflicts with smolting, in *Salmonid age at maturity* (ed. D. J. Meerburg), *Can. Spec. Publs Fish. Aquat. Sci.*, **89**, 7–14.

Tinbergen, N. (1963) On aims and methods of ethology. *Z. Tierpsychol.*, **20**, 410–29.

Turner, B. J., Elder, J. F., jun., Laughlin, T. F. and Davis, W. P. (1990) Genetic variation in clonal vertebrates detected by simple-sequence DNA fingerprinting. *Proc. nat. Acad. Sci. U.S.A.*, **87**, 5653–7.

Weider, L. J., Danzmann, R. G., Murdoch, M. H. and Hebert, P. D. N. (1989) Mitochondrial DNA variability in brown bullheads (*Ictalurus nebulosus*) resident in the Huron–Erie Corridor. Tech. Rep. to Ontario Ministry of the Environment, The Great Lakes Institute, University of Windsor, 28 pp.

Weir, B. S. (1990) *Genetic Data Analysis*, Sinauer, Sunderland, MA, 377 pp.

Werner, M. and Schroder, J. H. (1980) Mutational changes in the courtship activity of male guppies *Poecilia reticulata* after X-irradiation. *Behav. Genet.*, **10**, 427–30.

Westneat, D. F., Frederick, P. C. and Wiley, R. H. (1987) The use of genetic markers to estimate the frequency of successful alternative reproductive tactics. *Behav. Ecol. Sociobiol.*, **21**, 35–45.

Zimmerer, E. J. and Kallman, K. D. (1989) Genetic basis for alternative reproductive tactics in the pygmy swordtail *Xiphophorus nigrensis. Evolution*, **43**, 1298–1302.

Chapter two

The motivational basis of fish behaviour

Patrick Colgan

2.1 INTRODUCTION

For Aristotle, in *Historia Animalium*, the motivation of fish ranged from enjoyment of tasting and eating to madness from pain in pregnancy, and for Francis Day, summarizing for the 1878 Proceedings of the London Zoological Society in the wake of Darwin's epochal *The Expressions of the Emotions in Man and Animals*, fish were variously moved by disgrace, terror, affection, anger, and grief. The study of motivation thus has long been central in the analysis of fish behaviour. In its daily activities of finding food, avoiding predators, fighting, and reproducing, a fish is a motivationally diverse animal. This study focuses on the internal proximate causes of behaviour, variously labelled as drives, instincts, or causal systems (see Colgan, 1989, for a general discussion). Researchers such as Skinnerians, wary of the ontological status of such notions, have emphasized the dynamics of performance and the concomitant controlling variables in the environment. Though the existence and nature of causal systems are indeed empirical matters, conventional wisdom about the economy of nature suggests that they are likely solutions to the common problems encountered by fish, and they are heuristically very convenient. Motivational systems are more than physiology writ large, and require investigation in their own right.

Of Tinbergen's (1963) four aims of ethology (causation, survival value, ontogeny, and evolution), the study of motivation occupies much of the first. Its centrality is seen in the number of allied behavioural topics: physiological substrates (including sensorimotor – see Chapters 4 to 7, this volume – neural and hormonal mechanisms), ontogeny (see Chapter 3, this volume), rhythms, communication, and behavioural ecology (see Chapters 8 to 17, this volume). A common experimental paradigm in motivational research is to measure responses to a fixed external stimulus on several occasions, and then attribute changes in responsiveness to changes in motivation. Thus overt behaviour is

Behaviour of Teleost Fishes 2nd edn. Edited by Tony J. Pitcher. Published in 1993 by Chapman & Hall. ISBN   0 412 42930 6 (HB) and 0 412 42940 3 (PB).

viewed as the outcome of internal and external cues. (The composite effects of external cues such as temperature and substrate, referred to as heterogeneous summation (Tinbergen, 1951), have been shown, for instance, in salmonid spawning behaviour (Fabricius, 1950).) However, on any particular occasion, the situation may become complicated because the presentation of a stimulus or the performance of a response may itself change motivation (e.g. Toates and Birke, in Bateson and Klopfer, 1982). In classical terminology, stimuli can both release and prime responses. Priming has been shown in such cases as the aggressive responses of Siamese fighting fish, *Betta splendens* (Hogan and Bols, 1980). In such cases one may often scrutinize the correlations among various features of responses, such as frequencies, latencies, duration or intensities, in the search for underlying causal mechanisms.

Thus motivational analysis must overcome several hurdles. Although behaviour may be fundamentally deterministic, stochastic models often seem to be the most appropriate analytical tools owing to uncontrollable variation in responsiveness (e.g. Heiligenberg, 1976). Substantiating such models usually requires data too extensive to be easily obtainable since motivational processes are usually non-stationary (i.e. the values of the underlying parameters change with time). Significant differences may exist between individuals (see Chapter 13 by Magurran, this volume). Additionally, many different measures of behaviour are possible, and these often change independently of each other (e.g. in feeding by threespine sticklebacks, *Gasterosteus aculeatus*: Tugendhat, 1960a).

In the shadow of these difficulties, this chapter considers research on the motivation of teleost behaviour, with emphasis on reports appearing in the past two decades. Although the diversity of motivational research reflects the efforts of workers with often quite disparate interests, three key questions are relevant to the investigation of any behaviour: first, how many internal causal factors underlie the behaviour, and how do they interact? Secondly, how are these factors influenced by the appearance of external cues and by deprivation from such cues? Thirdly, how do they change over time, both during the performance of the behaviour and over longer periods? Attempts to answer each question must, of course, involve experimental treatments appropriate to the natural history of the species under study. For convenience of presentation, feeding, fear, reproduction, and aggression, the four major causal systems of behaviour which have been widely discussed, will be reviewed separately, and then their interactions examined. But first the relation of motivation and ontogeny is considered.

2.2   MOTIVATION AND ONTOGENY

A fish may respond differently to the same stimulus on different occasions because of motivational (non-structural) changes or structural changes affecting its capacity to act. The distinction is thus between changes in mechanisms and changes in the activation of mechanisms in performance. Ontogenetic changes in motivation and structure result from both maturation, which involves intrinsic processes, and experience with the environment. Little attention has been paid to motivational ontogeny in fish. In one of the few studies made, the aggressiveness of normally raised *Haplochromis burtoni* (an African cichlid) is increased by a patterned stimulus dummy but not a neutral dummy (Wapler-Leong, 1974), while that of individuals reared as isolates is augmented by the two stimuli. In Atlantic salmon, *Salmo salar*, seasonal changes in the motivation to feed are associated with different patterns of growth and maturation (Metcalfe *et al.*, 1986).

Two major structural changes germane to a consideration of motivation are learning and imprinting. The meaning of **learning** unfortunately varies with the user. It is generally taken to refer to a long-term change in the likelihood of a particular response following a particular stimulus, over successive associations of the stimulus and response ('when followed by appropriate reinforcement' according to reinforcement theorists). Such changes are variously interpreted as the outcome of conditioning of one or more types (Domjan, 1983). **Imprinting** refers to the development of social and habitat attachments as the result of experience during a brief sensitive period in early life. Although imprinting was originally contrasted with traditional learning as investigated in the laboratory, it shares many features with such learning (Bateson, 1987). **Habituation**, the waning of a response after repeated presentation of the same stimulus, lies intermediate between motivational dynamics and learning in its features and time scale. A valid taxonomy of behavioural processes will be possible only when the physiological bases (at present mostly unknown) underlying these phenomena are understood.

Learning research has largely been concerned with testing the validity of general theories about this central process. One key question in such testing revolves around the motivational basis for mechanisms of reinforcement. The pattern of responses in Siamese fighting fish when operantly reinforced to swim through a tunnel, depends on whether the reinforcer is food or the opportunity to display to a conspecific (Hogan *et al.*, 1970; Fig. 2.1). As the number of responses required for a food reinforcement increased, the total number of responses over 12 h increased, whereas the number of reinforcements did not vary significantly. For display, on the other hand, tunnel responses were constant whereas reinforcements fell. It is therefore likely that different mechanisms underlie the two behavioural systems. More recent work with this species (e.g. Bols and Hogan, 1979) emphasizes this conclusion with the finding that a display reinforcer induces both approach and avoidance tendencies.

Social imprinting can involve offspring and sexual attachments (review, Colgan, 1983). Although species recognition by parents of different stages of life is well developed in several cichlid species examined, there is no indication

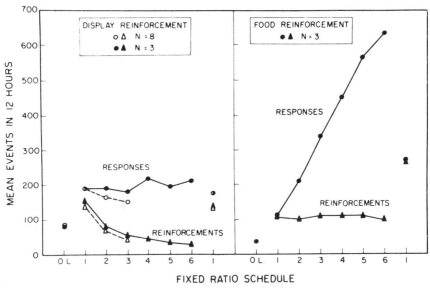

Fig. 2.1 Responses and reinforcements for Siamese fighting fish under various fixed-ratio schedules for display and food. Open symbols represent all eight fish; closed symbols represent three fish completing the series. Reproduced with permission from Hogan *et al.* (1970).

of imprinting by parents on offspring, unlike what occurs in some species of birds. Sexual imprinting (the establishment of sexual preferences as a result of early social experience) has been demonstrated in a variety of tropical-aquarium species, with polymorphic forms being convenient subjects, and experimental designs including cross-fostering and isolation procedures.

Habituation has been studied for many responses. In goldfish, *Carassius auratus*, habituation to an aversive buzzer is slowed by the injection of an endorphin analogue (Olson *et al.*, 1978). (Endorphins constitute an important class of analgesic compounds in vertebrate brains.) In guppies, *Poecilia reticulata*, changes in behaviour associated with repeated exposure to an open field result from the differential habituation of fear and exploratory responses (Warren and Callaghan, 1976). Similarly, components of arousal and fright habituate independently in slippery dicks, *Halichoeres bivittatus* (Laming and Ebbesson, 1984). The bulk of habituation research, however, has dealt with aggression (e.g. paradise fish, *Macropodus opercularis*, Brown and Noakes, 1974; goldfish and roach, *Rutilus rutilus*, Laming and Ennis, 1982; convict cichlids, *Cichlasoma nigrofasciatum*, Gallagher *et al.*, 1972). Siamese fighting fish (Klein *et al.*, 1976; Chantrey, 1978) and sticklebacks (Peeke, 1983) have been favourite species. Figure 2.2 presents some typical habituation curves

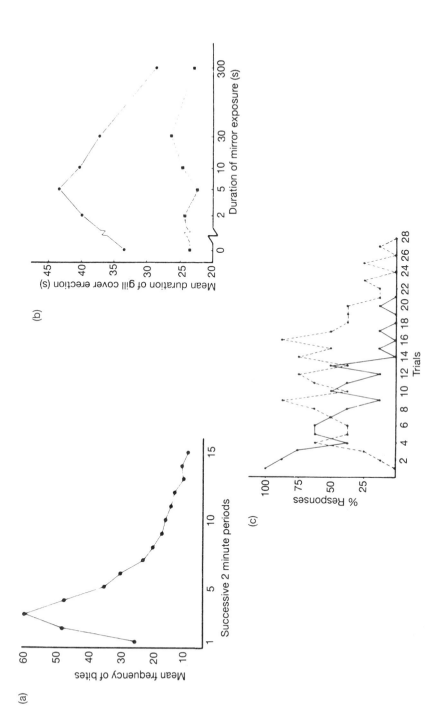

Fig. 2.2 Habituation (a) Biting in sticklebacks (Peeke, 1983). (b) Gill-cover erection in response to two stimulus dummies (●, blue model; ■, green model) following various durations of mirror exposure in Siamese fighting fish (Chantrey, 1978). (c) Fright (solid curve) and arousal (broken curve) responses in goldfish (Laming and Ennis, 1982).

and illustrates the common finding that habituation is often preceded by an initial increment in responding, a 'warm-up effect'. In colonies of bluegill sunfish, *Lepomis macrochirus*, the aggression between males on neighbouring nests habituates and can be reinstated or dishabituated by altering the appearance of the body covering (Colgan *et al.*, 1979).

Thus a variety of ontogenetic processes influence the mechanisms producing behaviour.

2.3   FEEDING

Non-motivational aspects of feeding include the physical aspects of the environment; the densities, distributions, and availabilities of prey types; feeding competitors; and learning. The last involves prey features, preferences, and switching (i.e. the disproportionate consumption of the common food type when several are available: Stephens and Krebs, 1986). Many of these aspects are dealt with by Hart in Chapter 8, this volume. Compared with homeothermic animals, the energy requirements of fish are much lower and hunger rises more slowly. It is widely accepted that feeding motivation includes both a gastric factor based on gut fullness and a systemic factor reflecting metabolic balance (e.g. Holmgren *et al.*, in Rankin *et al.*, 1983). How these factors operate to produce observed feeding behaviour is the question at issue.

Much research has attempted to infer mechanisms underlying feeding performance from aggregate measures such as the amounts of various available food types consumed under different conditions by the end of a meal. Legions of field workers pass the time tallying stomach contents. In operant psychology, Herrnstein has developed a 'matching law' describing the relation of numbers of feeding acts to numbers of food reinforcements, but its relevance to natural behaviour is questionable since it applies only to variable-interval reinforcement schedules, which may be rare outside operant laboratories (see Kamil and Yoerg, in Bateson and Klopfer, 1982). (Under variable-interval schedules of reinforcements, an animal is reinforced for the first response after a randomly varying interval of time). Food preferences are often transitive (i.e. if food type A is preferred to B, and B to C, then A is preferred to C), and comparison of consumption totals enables tests of quantitative aspects of such transitivity. In centrarchid fish, transitivity can be modelled as the outcome of preferences for individual food types (Colgan and Smith, 1984). The amounts eaten, both when only one type is available and when two are available, can be predicted.

However, as in other areas of motivational analysis, the elucidation of hunger mechanisms is more effectively performed using fine-grained data on the duration and sequencing of individuals' acts and the choices generated by this behavioural flow. Much attention has therefore come to focus on satiation

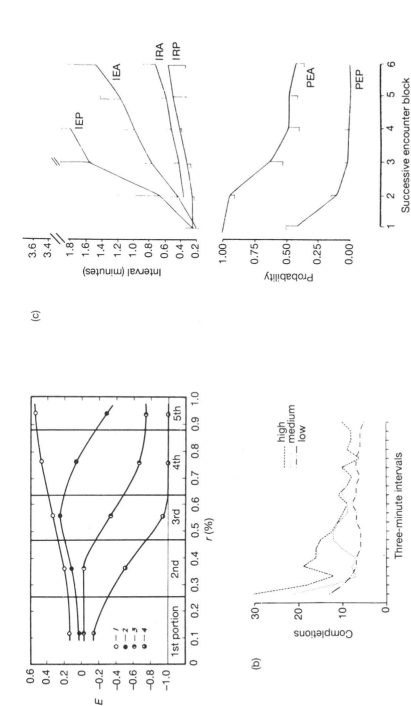

Fig. 2.3 Satiation curves in fish feeding. (a) Carp, *Cyprinus carpio*, feeding electivity (*E*), measuring preference for four prey types as a fraction of the total meal (*r*%) (Ivlev, 1961). (b) Sticklebacks grouped according to initial feeding rates on *Tubifex* worms (Tugendhat, 1960a). (c) Pumpkinseed sunfish, *Lepomis gibbosus*: length of the interval following an encounter in which an adult housefly was eaten (IEA) or refused (IRA) or a pupa was eaten (IEP) or refused (IRP), and probability of the fish eating an adult (PEA) or a pupa (PEP), plotted against successive blocks of about 450 encounters (Colgan, 1973).

curves (Fig. 2.3) generated by fish feeding amid abundant food, where changes in performance reflect changes in internal state only. In general, fish feed more selectively (Fig. 2.3(a)) and slowly (Fig. 2.3(b), (c)) as a meal proceeds, notwithstanding the contrary predictions of classical optimal foraging theory (see Chapter 8 by Hart, this volume). Dill (1983) has summarized the effects of satiation on searching, handling, and ratio measures of behaviour such as attacks per approach. He points out that these effects are adaptive by increasing searching and decreasing handling time when low food availability leads to hunger. Hunger influences the feeding tactics of Australian salmon, *Arripis trutta*, preying on krill of various densities (Morgan and Ritz, 1984). Feeding often occurs in bouts, with periods of ingestion separated by relatively long intervals. Like behaviour in a variety of situations, feeding in pumpkin-seed sunfish is well described by highly skewed probability distributions, such as gamma, whose means are influenced by the individual fish, the food type, and satiation (Colgan and Smith, in Colgan and Zayan, 1986). Goldfish achieve nearly optimal depletion of different food patches by comparing the current feeding rate with that previously experienced elsewhere (Lester, 1984).

An even more detailed investigation of feeding behaviour requires the examination of the behaviour involved in the treatment of each food item. Accepting and rejecting food items during a meal have marked and opposite influences on feeding behaviour in sticklebacks (Thomas, 1974, 1977). After an acceptance, fish search more intensively in the immediate vicinity; such 'area restricted searching' has also been noticed in tetrapods. In contrast, after a rejection a stickleback is more likely to leave the area. It appears that, in addition to the effects of satiation extending over an entire meal, acceptances and rejections result in respective short-term positive and negative changes in feeding motivation. These changes are adaptive if prey are patchily distributed. Further careful scrutiny of such aspects of behaviour is needed for a comprehensive understanding of feeding motivation.

2.4   FEAR AND AVOIDANCE

Fear and avoidance have been intensively examined in some animal groups (e.g. Sluckin, 1979). By way of contrast, these phenomena have been little studied in fish, perhaps because fish ethologists are themselves less motivated by fear than their non-ichthyological counterparts! Research comparing avoidance learning in difference species has investigated, for example, the effect of varying the interval between a conditioned stimulus (generally a light) and a shock using goldfish (Bitterman, 1965). Avoidance conditioning is better in goldfish for which the light has been associated with food or shock, and poorer in goldfish habituated to the light, compared with controls (Braud,

1971). In convict cichlids, fear increases with social isolation (Gallagher *et al.*, 1972) and plays a role in the prior-residence effect in dominance relations (Figler and Einhorn, 1983). The lack of habituation of escape responses to test stimuli in the blue chromis, *Chromis cyaneus* (Hurley and Hartline, 1974) is clearly adaptive. Compared with uninfested fish, sticklebacks infested with cestode larvae recover from a frightening stimulus more quickly (Giles, 1983). This quicker recovery may increase the predation risk of the fish and hence be adaptive to the parasite in reaching its definitive avian host.

2.5   REPRODUCTION

There has been much investigation of functional aspects of mating and parental care in teleosts (reviews: Turner, Chapter 10, and Sargent and Gross, Chapter 11, this volume). In considering underlying motivation, Baerends *et al.* (1955) provided ethology with a classic illustration of the interaction of external stimuli and internal causation in the courtship behaviour of male guppies (Fig. 2.4). Males prefer large females over small ones, which is adaptive since brood size increases with female size. The state of arousal of a courting male is reflected in coloured skin patches, the colour intensity of which is under neural control. Together, female size (the external stimulus) and internal arousal combine to determine the courtship acts of the male, with more intense acts requiring higher combinations. Such coloration patterns are thus most useful in the study of the motivation of courtship and can be examined in many species (e.g. Baerends *et al.*, 1986).

Valuable physiological data are available indicating the mechanisms involved in the motivation of reproductive behaviour. In terms of chemical messages, a variety of hormones operate within individuals, and pheromones between individuals (Liley and Stacey, 1983). Refractory periods (pauses following mating) are known in many species of animals. In the lemon tetra, *Hyphessobrycon pulchripinnis*, males are physiologically limited in their ability to produce sperm and so to fertilize eggs during the daily spawning period, which takes up the first 2 h of the morning (Nakatsuru and Kramer, 1982).

In promiscuous species, the Coolidge effect refers to the reinstatement of copulatory behaviour in a previously satiated male by presenting a new mate (Colgan, 1983). The effect has been most investigated in mammals, and involves dishabituation in the male and differential behaviour by mated and unmated females. In the live-bearer *Poecilia sphenops*, copulation attempts by males are more frequent after a new female is made available (Franck, 1975). Possibly the effect is adaptive for a male by enhancing the number and variety of his offspring. On a longer time span, many species show reproductive rhythms, generally associated with such environmental cues as temperature, tides and rains (Schwassmann, 1980).

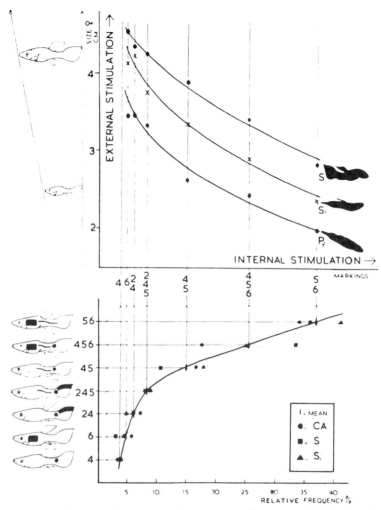

Fig. 2.4 Courtship in male guppies is the outcome of an external stimulus, female size, and internal arousal, as reflected in coloration, calibrated using the relative frequencies of copulation attempts (CA) and sigmoid displays (S, Si). Si, sigmoid intention movement; P_f, posturing behaviour. Reproduced with permission from Baerends *et al.* (1955).

In his classic study of the organization of courtship in the bitterling, *Rhodeus amarus*, Wiepkema (1961) introduced the use of factor analysis for the detection of clusters of activities. Much of the variation in the data can be accounted for by distinguishing sexual responses occurring in courtship, agonistic responses involving aggression and flight, and non-reproductive responses such as feeding and comfort movements. Important external cues include a mating partner and a mussel in which eggs are laid. The number of underlying causal factors may vary among species, however. In the courtship of the blue chromis observed naturally above reefs, Boer (1980) detected only two causal factors, aggression and nesting, which determined both social responses and coloration.

Like a brilliant but brief-lived firecracker, Nelson (1964, 1965) illuminated temporal aspects of reproductive behaviour. In the glandulocaudine fish *Corynopoma riisei*, courting sequences occur randomly in time, with different activities having different durations and taking place with probabilities dependent on the preceding activity. Near fertilization, behavioural sequences become more determinate, as the females are influenced by the cumulative effect of male actions. In stickleback courtship (see Chapter 16 by FitzGerald and Wootton, this volume), stimulation by female dummies affects the occurrence of zigzagging, fanning, and creeping through the nest. These changes can be adequately modelled using two variables, excitation and threshold, which rise during stimulation and creeping through respectively, and fall otherwise. The Markovian nature of courtship sequences in *Barbus nigrofasciatus* led the investigators to conclude that 'barbs do it (almost) randomly' (Colgan, 1978; Putters *et al.*, 1984).

Courtship behaviour is often important for the ethological isolation of closely related species. Motivational analysis indicates how this isolation is achieved proximately. In the closely related *Cichlasoma citrinellum* and *C. zaliosum* of Central America, courtship requires at least 4 days before spawning takes place (Baylis, 1976). Three phases can be distinguished: a phase of pair formation, an intervening phase, and a phase of preparation of the spawning site and spawning itself. Changes in the organization of the behaviour result from differential frequencies of the occurrences of activities. There are also differences between the two species, especially early in courtship, owing to differences in the thresholds for sexual and aggressive behaviour. Males and females of pumpkinseed and bluegill sunfish, and of their hybrids, generally discriminate, among the species and hybrids, the taxon of courting partners (Clarke *et al.*, 1984). The courtships of the parental crosses (pumpkinseed × pumpkinseed and bluegill × bluegill) are least similar, and those of other crosses are intermediate.

In many species of fish, spawning is followed by care of the brood by one or both parents. This may involve fanning the eggs (which require oxygen during their development), consuming diseased eggs, and, especially, defending the eggs and often the brood against predation. In blue gouramis, *Trichogaster trichopterus*, broodiness can be induced by the repeated presentation of conspecific eggs (Kramer, 1973). As in so much of behaviour, there are rhythmic aspects to parental care. For instance, in two species of cichlids, fanning of eggs by the female is greater at night than during the day (Reebs and Colgan, 1990). Experimental manipulation of the light:dark cycle indicates that an endogenous circadian rhythm is involved.

2.6   AGGRESSION

Beyond reproduction, fish are motivated to engage in other social behaviour, both affiliative and agonistic. Indeed, such behaviour may involve other species, as in interspecific shoaling (Ehrlich and Ehrlich, 1973; Chapter 12 by Pitcher, this volume) and territoriality (Thresher, 1978). Just as courting behaviour is the result of internal and external factors, so is agonism. (For the role of hormones see Munro and Pitcher, in Rankin *et al.*, 1983.) In *Haplochromis burtoni*, presentation of a stimulus dummy has been reported to increase by a fixed amount the attack rate against smaller conspecifics (Heiligenberg *et al.*, 1972; Fig. 2.5).

Along the same lines, aggression in male pumpkinseed sunfish defending nests, as reflected in various responses such as approaching stimulus dummies, is the outcome of both motivational state and external stimuli (Colgan and Gross, 1977; Fig. 2.6). The motivational state of the defending fish changes greatly over the four phases of a typical 10 day nesting cycle (Fig. 2.6(a)). Aggressiveness begins low in the initial 2 day nesting phase during which

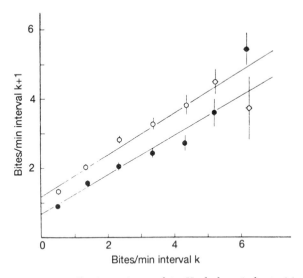

Fig. 2.5 Biting rate in a 5 minute interval in *Haplochromis burtoni* is raised by a fixed amount after a stimulus dummy is presented ($\bigcirc$) compared with baseline measured in the preceding interval ($\bullet$). Reproduced with permission from Heiligenberg *et al.* (1972).

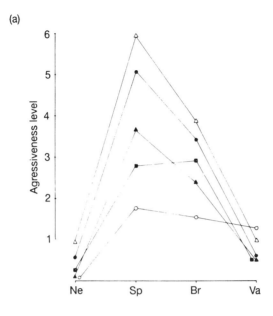

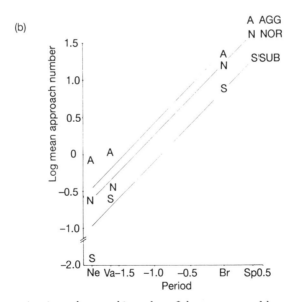

Fig. 2.6 Aggression in male pumpkinseed sunfish, as measured by such responses as approach, is the outcome of an external stimulus, a conspecific wooden dummy, and internal arousal, as reflected in phase of nesting. See text for details. Dummy postures: AGG (△), aggressive; NOR (●), normal; SUB (▲), subordinate. Nesting phases: Ne, nesting; Sp, spawning; Br, brooding; Va, vacating (Colgan and Gross, 1977).

nests are established. Then it rises for the 3 day spawning phase (in which females lay eggs in the nest and the male fertilizes these) and the 4 day brooding phase (in which the male cares for the hatching fry). Finally, it drops in the 1 day vacating phase when nests are abandoned. External stimuli were manipulated by presenting painted plywood dummies of conspecifics in aggressive, normal, and subordinate postures. A contingency-table analysis of the frequencies of approaches (Fig. 2.6(b)) and other responses, with dummy posture and nesting phase as factors, indicated that both external and internal cues play a role in producing behaviour, and enabled a scaling of relative effects of dummies and nesting phases.

Several studies have investigated the motivational factors underlying agonistic behaviour. Using factor analysis, as similarly employed by Wiepkema (1961) and described earlier, Balthazart (1973, 1974) found that two independent factors, attack and flight, accounted for the variation in his data on the frequencies of different acts in *Oreochromis (Tilapia) macrochir*. The outcome of a fight appears to be decided early in the encounter, with subsequent behaviour being an artefact of aquarium confinement. Thresher (1978) has described two components of territorial aggression in the threespot damselfish, *Eupomacentrus planifrons*, with one determining the size of area defended and the second the vigour of defence. Congruent with Thresher's two-component theory are the findings by Colgan *et al.* (1981) that pumpkin-seed sunfish defending nests treat conspecific dummies differently in terms of the frequencies of their aggressive responses but similarly in terms of the spatial occurrence of those responses. The spatial aspects of nest defence are partially determined by the location of the nest rim, and are well modelled using catastrophe theory, with aggressive behaviour viewed as the joint outcome of the reproductive phase of the defending fish and the distance of an intruder from the nest centre. More recently, Putters (in Colgan and Zayan, 1986) has shown that the temporal course of aggressive display in the paradise fish can be modelled as the outcome of two underlying variables.

Whereas the levels of feeding and sexual motivation clearly increase in the absence of the appropriate cues (food and partners), the extent to which aggression similarly changes has been a matter of controversy, both for fish and for other species (especially for the human case, for obviously critical reasons). Like its freshwater cousin the Siamese fighting fish, the yellowtail damselfish, *Microspathodon chrysurus*, seeks opportunities for aggressive display, which can be used as a reward for conditioning (Rasa, 1971). In isolation, individuals show increased comfort behaviour, with each activity following a separate time course: this has been interpreted as a means of achieving behavioural homeostasis in the face of an accumulation of endogenous aggressive energy deprived of its usual releasers. However, based on research on orange chromides, *Etroplus maculatus*, Reyer (1975) concluded that external stimuli, not internal drive, produced aggressive behaviour. Certainly social isolation has been found to decrease aggressiveness in a number of species (e.g. *Haplochromis burtoni*: Heiligenberg and Kramer, 1972).

These conflicting results could be due to differences among species. However, data on the effects of social isolation in the paradise fish could offer a solution (Davis, 1975). By increasing reactivity to both social and non-social stimuli, isolation has general effects on behaviour, and not necessarily specific effects on the readiness for social display. Thus a generalized hyperactivity due to isolation combined with a differential habituation to specific stimuli could account for these apparently contradictory reports. Recent findings with Siamese fighting fish indicate that social isolation decreases the readiness to display to conspecifics but leads to greater rates of display once the fish have been primed (Halperin *et al.*, in press).

It is interesting to note that the boldness of sticklebacks towards pike, *Esox lucius*, covaries with the intensity with which they attack conspecifics when measured in conditions ranging from non-reproduction through nesting to defending newly hatched young (Huntingford, 1976). This covariance suggests that there is a common causal system producing timidity.

2.7 CAUSAL INTERACTIONS

The stream of behaviour constantly reflects choices among options, and therefore the conflict among causal systems for the 'behavioural final common path' (McFarland and Sibly, 1975) demands close attention. The expression of any one causal system is likely to be interfered with by other systems. From a functional viewpoint, animals must trade off priorities. This has been noticed in many studies, such as Wiepkema's (1961) on bitterling reproduction. It has been variously suggested (e.g. McFarland, 1974) that interference may be random, competitive, or time-sharing. In competition, changes in the causal factors of a second activity oust an ongoing activity, whereas in time-sharing, an ongoing activity terminates itself and disinhibits (releases) the second activity. In the most recent of several experiments focusing on how nesting male sticklebacks divide their time between courting females and attending to the nest, Crawford and Colgan (1989) found that most transitions were random and that previous studies suffered from conceptual methodological flaws.

Displacement acts, defined by their irrelevance in the flow of behaviour (Tinbergen, 1952), are a major aspect of conflict behaviour. Their role in the evolution of communicative behaviour has often been invoked as one of the key elements in the phylogeny of motivation. Currently, the entire study of communication – in terms of causal mechanisms, the exchanged information about strength and intentions, and the adaptiveness of signals – is undergoing intense scrutiny. The present discussion will be limited to the motivation of displacement responses. Under the disinhibition hypothesis, displacement responses are viewed as the outcome of balanced conflict between two dominant behavioural tendencies, each cancelling the inhibition exerted by the other on the displacement activity. Other work has suggested that other explanations may also apply. Wilz (1970a, b) conducted experiments on displacement responses in nesting male sticklebacks. In these males the tendencies to lead the female to the nest and to chase her from it are in conflict. The reactions of males to a female dummy were recorded, with the aggressiveness level of the male subjects being manipulated by presenting them with a stimulus male in a bottle. Particular attention was paid to 'dorsal pricking', in which the male nudges the female dummy towards the surface of the water with his dorsum. Wilz showed that dorsal pricking tends to occur in response to an approach by a female when the male is aggressive or when sexual stimulation is lacking. Dorsal pricking causes the female to wait while the male goes off to the nest, performs displacement activities (creeping through and fanning), and returns to court. Females failing to wait are generally attacked; attacks also result if the displacement activities are blocked. Thus dorsal pricking provides the male with an opportunity to perform displacement responses and thereby to switch from primarily aggressive motivation to primarily sexual motivation. Wilz argues that such self-regulation of motivation is the function of these responses. Given these successive motivational changes, the time scale for measurement is critical. Wilz (1972) subsequently showed that aggression inhibits nest gluing, an activity associated with sexual behaviour, in the short term, but over the long term increases it.

The antagonism between sexual and aggressive behaviour has been examined in other species. When approached by a gravid female prepared to spawn, nesting male pumpkinseed and bluegill sunfish gradually shift from chiefly aggressive motivation (attacking the female) to chiefly sexual motivation (courting the female) (Ballantyne and Colgan, 1977). This shift (and the underlying conflict) is indicated by sounds produced by the male rubbing his pharyngeal dental pads together, and hence these sounds play a key communicative role during courtship. (See also Chapter 5 by Hawkins, this volume.) Similarly, by exposing test convict cichlids to stimulus individuals of either sex, and factor analysing the resulting data, Cole *et al.* (1980) showed a mutual inhibition of sex and aggression.

Students of both the motivation of feeding and its functional aspects, often in the context of optimal foraging theory, have been impressed by the manner in which feeding is compromised by other activities. Experimental sticklebacks subjected to electric shocks while feeding show differences in behaviour compared with unshocked control fish (Tugendhat, 1960b). Feeding is interrupted in a manner dependent on the location of the shocks. Fish shocked at the entrance to the feeding areas return less frequently to the shock-free living area, whereas those shocked for taking food items return more frequently. The behavioural conflict is also reflected in changes in the frequency of dorsal spine raising and comfort movements. The total amount of feeding decreases as shock intensity increases. Young, trained carp arrive at a feeding place more quickly when swimming in a shoal habituated to the experimental conditions than in a frightened one (Kohler, 1976). This difference reflects the conflict of a conditioned reaction and mutual attraction among the members of the shoal. In bluntnose minnows, *Pimephales notatus*, foraging latency increases in the presence of a predator and decreases in larger shoals and at higher levels of hunger (Morgan, 1988a). Concomitantly, the cohesiveness of the shoal increases with the size of the shoal and in the presence of a predator, but decreases with greater hunger (Morgan, 1988b). In Siamese fighting fish the interaction of feeding and aggressive display is affected by fear (Hogan, 1974). Fish living in a T-maze offering food or a mirror image prefer food, but those transferred to the maze show a greater tendency to choose the mirror image. This differential effect of fear on feeding and aggression, in response to unfamiliar surroundings, illustrates the need to allow for environmental influences on motivational state in both field and laboratory studies.

The relation between feeding and aggression is of paramount importance in assessing the motivation of predatory species. Likewise, the motivation of cannibals and of feeding specialists such as cleaners removing ectoparasites from hosts, paedophages feeding on young, and lepidophages engulfing mouthfuls of integument are of special interest for the blend of trophic and social motivation (e.g. Sazima, 1983). In sunfish, observations on aggression and predation in socially dominant and subordinate individuals reveal that these two motivational systems are independent (Poulsen and Chiszar, 1975). However, variation in hunger has complex effects on different components of aggressive behaviour, which thus cannot be regarded as a unitary system (Poulsen, 1977). Siamese fighting fish direct more biting attacks toward small guppies and more threat displays towards large ones (Baenninger and Kraus, 1981). Hunger level influences only biting attacks, and mirror exposure only threat displays. Lenke (1982) has argued for the existence of a motivational system for cleaning behaviour separate from feeding in the cleaner wrasse, *Labroides dimidiatus*, since this behaviour is both facultative and unaffected by food deprivation. Various injected amounts of different hormones are reported to have different effects on cleaning behaviour, thus demonstrating the endocrinal background of this motivation. However, more convincing data are needed to establish this system. For their part, hosts are motivated by a conflict between tactile reinforcement and aversive stimulation resulting from the removal of parasites (Losey, 1979). In our laboratory we have studied how male guppies divide their time between feeding and courting females. As expected, males are less attracted to food as satiation proceeds, especially as the number of stimulus females is increased. On interactions between feeding and other activities, females of the cichlid *Pseudocrenilabrus multicolor* care for young in the mouth. In such females, feeding is inhibited by the broodiness of the female and by stimuli from the brood itself (Mrowka, 1986). Hunger inhibits the occurrence of brooding, but not its timing.

The use of stimulus dummies for motivational analysis has been cultivated to a high art under Baerends' leadership at Groningen (e.g. Baerends, 1985). The interactions between attack, escape, and courtship can be manipulated in the jewel cichlid, *Hemichromis bimaculatus*, through the use of such dummies (Rowland, 1975a,b). Different dummies differentially elicit responses associated with each causal system. For instance, red dummies produce strong sexual arousal, which inhibits aggressive activities such as biting. Recent experience also influences behavioural priorities in the fish. An unpaired male courted a dummy, whereas he attacked it after pairing with a female. In a similar vein, communication in *Pseudotropheus zebra*, a cichlid from Lake Malawi, involves the interaction of attack and escape tendencies, as assayed by dummies of various colours and sizes (Vodegel, 1978). A quantitative model accounts for the influence on these tendencies, of dummy size and accessibility, the distance between the fish and dummy, habituation, and which activities have been performed. Subsequent research on three cichlids endemic to Lake George revealed that responsiveness to territorial intruders is the joint outcome of four components: interest in the intruder, avoidance, attraction to the shelter, and persistence (Carlstead, 1983a,b). Attacks and displays are especially affected by avoidance and persistence. The levels of the components vary across species, and change as an encounter with a conspecific or environmental disturbance proceeds. More recently, Rowland and Sevenster (1985) have extended the classical use of stimulus dummies in elicitation of aggressive and sexual responses in threespine sticklebacks.

2.8 CONCLUSIONS

The research reviewed in this chapter indicates a variety of approaches for analysing the causal basis of behaviour into its external and internal components. It is clear that even the foremost analyses, such as those by Baerends (e.g. 1986), have scarcely gained a toehold on the terrain to be investigated. The characteristics of the different causal systems and their interactions in a diversity of teleost species must be investigated. The work of the past two decades has revealed many motivational subsystems within and beyond the 'big four', hunger, fear, aggression and sex. For instance, there appear to be four systems operating in a female Siamese fighting fish as she responds to a conspecific (Robertson, 1979). Surely, as with finding new species and races in an unexplored region, the number of motivational systems discovered will approach an asymptote. The analysis of problems such as individual differences (often found in the research reviewed above (e.g.

Vodegel, 1978, and considered by Magurran in Chapter 13, this volume) requires consideration of their motivational substrates. Beyond research into specific ontogenetic processes such as learning and imprinting, there is a need for examination in teleosts of the ontogeny of motivation, understood as the developmental patterning of causal systems, similar to that carried out for some tetrapod species. Similarly, the phylogeny of motivation merits attention: how does diversification evolve to enable complex interactions such as communication and interspecific cleaning behaviour? In the area of research between the levels of physiology and behaviour (e.g. Bruin, 1983), further studies linking motivation with underlying physiological processes are required to sort through the current conceptual ragbag. At the interface of behaviour and ecology, the adaptiveness of causal mechanisms needs to be assessed. In the past, causation and function have often been confused. Such confusion is understandable: for instance, competition among causal systems and trade-offs among objectives in behavioural ecology are obverse problems. But the separateness of causation and function must be realized: causal analysis does not indicate adaptiveness, and functional analysis does not reveal causation.

Methodologically, the above research reflects the breadth of useful analytical approaches to the problem of motivation: multivariate statistics, information theory, stochastic processes, and specific models. Conversely, motivational differences have direct significance for the methodology of behavioural research. For instance, it is generally hoped that observations from the field and the laboratory are comparable, but this is not always the case. Courtship sequences in sunfish are longer in aquaria than in natural colonies for several reasons: interruptions by other males are generally eliminated; the female cannot entirely escape the attention of the male; and of particular relevance here, in the laboratory the female is placed with the male by the experimenter, hoping she is ready to spawn, whereas in the field she freely enters and leaves the colony (Clarke *et al.*, 1984). Therefore field and small aquarium comparisons must be made with caution.

The available data on the motivation of fish behaviour and the analytical tools for its study provide a guiding framework for research: motivational organization is viewed as a set of clusters of activities causally linked by transition probabilities dependent on both internal and external cues (Itzkowitz, 1979), and durations spent within a particular state are similarly influenced. A steadfast use of this framework will perhaps solve the recondite problems alluded to in Dubos' (1971) dictum, 'Under precisely controlled conditions an animal does as he damn pleases.'

2.9 SUMMARY

The study of the internal proximate causes of responses is a central topic in fish behaviour. The motivation of fish, along with their learning capacities, develops over the lifetime of the individual. Feeding motivation is reflected in satiation over a meal and preferences between available food types. Fear and avoidance operate in learning and social interactions. Motivation for reproductive and agonistic behaviour is dependent on external cues and internal state, especially hormonal levels. Analyses of courting, parental, and aggressive activities indicate the operations of several causal factors. These major causal systems interact mutually in complex patterns to influence behaviour performance. The nature of these systems, the concepts to deal with them, and the relations with physiology and ecology all require much further investigation.

ACKNOWLEDGEMENTS

The writing of this chapter has been supported by Queen's University at Kingston and the Natural Sciences and Engineering Research Council of Canada. For comments and criticism I am grateful to Ian Jamieson, Jim Kieffer, Bob Lavery, and Stephan Reebs.

REFERENCES

Baenninger, R. and Kraus, S. (1981) Some determinants of aggressive and predatory responses in *Betta splendens*. *J. comp. physiol. Psychol.*, **95**, 220–27.

Baerends, G. P. (1985) Do the dummy experiments with sticklebacks support the IRM-concept? *Behaviour*, **93**, 258–77.

Baerends, G. P. (1986) On causation and function of the pre-spawning behaviour of cichlid fish. *J. Fish Biol.*, **29A**, 107–21.

Baerends, G. P., Brower, R. and Waterbolk, H. T. (1955) Ethological studies in *Lebistes reticulatus* (Peters). 1. An analysis of the male courtship pattern. *Behaviour*, **8**, 249–334.

Baerends, G. P., Wanders, J. B. W. and Vodegel, R. (1986) The relationship between marking patterns and motivational state in the pre-spawning behaviour of the cichlid fish *Chromidotilapia guentheri* (Sauvage). *Neth. J. Zool.*, **36**, 88–116.

Ballantyne, P. K. and Colgan, P. W. (1977) Sound production during agonistic and reproductive behaviour in the pumpkinseed (*Leponis gibbosus*), the bluegill (*L. macrochirus*), and their hybrid sunfish. 1–3. *Biol. Behav.*, **3**, 113–135, 207–232.

Balthazart, J. (1973) Analyse factorielle du comportement agonistique chez *Tilapia macrochir* (Boulenger 1912). *Behaviour*, **46**, 37–72.

Balthazart, J. (1974) Non-stationnarité et aspects fonctinnels du comportement agonistique chez *Tilapia macrochir* (Boulenger 1912) (Pisces: Cichlidae). *Acta Zool. Path. Antv.*, **58**, 29–40.

Bateson, P. P. G. (1986) Inprinting as a process of competitive exclusion, in *Imprinting and Cortical Plasticity* (eds J.P. Rauschecker and P. Marler), Wiley, New York, pp. 151–68.

Bateson, P. P. G and Klopfer, P. H. (eds) (1982) *Perspectives in Ethology*, Vol. 5, Plenum, New York, 520 pp.

Baylis, J. R. (1976) A quantitative study of long-term courtship. 2. A comparative study of the dynamics of courtship in two New World cichlid fishes. *Behaviour*, **59**, 117–61.

Bitterman, M. E. (1965) The CS–US interval in classical and avoidance conditioning, in *Classical Conditioning* (ed. W. F. Prokasy), Appleton-Century-Crofts, New York, pp. 1–19.

Boer, B. A. D. (1980) A causal analysis of the territorial and courtship behaviour of *Chromis cyanea* (Pomacentridae, Pisces). *Behaviour*, **73**, 1–50.

Bols, R. J. and Hogan, J. A. (1979) Runway behavior of Siamese fighting fish, *Betta splendens*, for aggressive display and food reinforcement. *Anim. Learn. Behav.*, **7**, 537–42.

Braud, W. G. (1971) Effectiveness of 'neutral' habituated, shock-related, and food-related stimuli as CSs for avoidance learning in goldfish. *Condit. Reflex*, **6**, 153–6.

Brown, D. M. B. and Noakes, D. L. G. (1974) Habituation and recovery of aggressive display in paradise fish (*Macropodus opercularis* (L)). *Behav. Biol.*, **10**, 519–25.

Bruin, J. P. C. D. (1983) Neural correlates of motivated behavior in fish, in *Advances in Vertebrate Neuroethology* (eds J. P. Ewert, R. R. Capranica, and D. J. Ingle), Plenum, New York, pp. 969–95.

Carlstead, K. (1983a) The behavioural organization of responses to territorial intruders and frightening stimuli in cichlid fish (*Haplochromis* spp.). *Behaviour*, **83**, 18–68.

Carlstead, K. (1983b) Influences of motivation on display divergences in three cichlid fish species (*Haplochromis*). *Behaviour*, **83**, 205–28.

Chantrey, D. F. (1978) Short-term changes in responsiveness to models in *Betta splendens*. *Anim. Learn. Behav.*, **6**, 469–71.

Clarke, S. E., Colgan, P. W. and Lester, N. P. (1984) Courtship sequences and ethological isolation in two species of sunfish (*Lepomis* spp.) and their hybrids. *Behaviour*, **91**, 93–114.

Cole, H. W., Figler, M. H., Parente, F. J. and Peeke, H. V. S. (1980) The relationship between sex and aggression in convict cichlids (*Cichlasoma nigrofasciatum* Gunther). *Behaviour*, **75**, 1–21.

Colgan, P. W. (1973) Motivational analysis of fish feeding. *Behaviour*, **45**, 38–66.

Colgan, P. W. (ed.) (1978) *Quantative Ethology*, Wiley, New York, 364 pp.

Colgan, P. W. (1983) *Comparative Social Recognition*, Wiley Interscience, New York, 281 pp.

Colgan, P. W. (1989) *Animal Motivation*, Chapman and Hall, London, 159 pp.

Colgan, P. W. and Gross, M. R. (1977) Dynamics of aggression in male pumpkinseed sunfish (*Lepomis gibbosus*) over the reproductive phase. *Z. Tierpsychol.*, **43**, 139–51.

Colgan, P. W. and Smith, J. T. (1984) Experimental analysis of food preference transitivity in pumpkinseed sunfish (*Lepomis gibbosus*). *Biometrics*, **41**, 227–36.

Colgan, P. W. and Zayan, R. (eds) (1986) *Quantitative Models in Ethology*, Privat, Toulouse, 148 pp.

Colgan, P. W., Nowell, W. A., Gross, M. R. and Grant, J. W. A. (1979) Aggressive habituation and rim circling in the social organization of bluegill sunfish (*Lepomis macrochirus*). *Env. Biol. Fishes*, **4**, 29–36.

Colgan, P. W., Nowell, W. A. and Stokes, N. W. (1981) Nest defence by male pumpkinseed sunfish (*Lepomis gibbosus*): stimulus features and an application of catastrophe theory. *Anim. Behav.*, **29**, 433–42.

Crawford, S. S. and Colgan, P. W. (1989) Motivational models of courtship in male threespine sticklebacks (*Gasterosteus aculeatus*). *Behaviour*, **109**, 285–302.

Davis, R. E. (1975) Readiness to display in the paradise fish *Macropodus opercularis*, L., Belontiidae: the problem of general and specific effects of social isolation. *Behav. Biol.*, **15**, 419–33.

Dill, L. M. (1983) Adaptive flexibility in the foraging behavior of fishes. *Can. J. Fish. Aquat. Sci.*, **40**, 398–408.

Domjan, M. (1983) Biological constraints on instrumental and classical conditioning: implications for general process theory. *Psychol. Learn. Motiv.* **17**, 215–77.

Dubos, R. (1971) In defense of biological freedom, in *The Biopsychology of Development* (eds E. Tobach, L. R. Aronson, and E. Shaw), Academic Press, New York, pp. 553–60.

Ehrlich, P. R. and Ehrlich, A. M. (1973) Coevolution: heterotypic schooling in Caribbean reef fishes. *Am. Nat.*, **107**, 157–60.

Fabricius, E. (1950) Aquarium observations on the spawning of the char (*Salvelinus alpinus*). *Rep. Inst. Freshwat. Res. Drottningholm*, **34**, 14–48.

Figler, M. H. and Einhorn, D. M. (1983) The territorial prior residence effect in convict cichlids (*Cichlasoma nigrofasciatum* Gunther): temporal aspects of establishment and retention, and proximate mechanisms. *Behaviour*, **85**, 157–83.

Franck, D. (1975) Der Anteil des 'Coolidge-Effektes' and der isolationsbedingten Zunahme sexueller Verhaltensweisen von *Poecilia sphenops*. *Z. Tierpsychol.*, **38**, 472–81.

Gallagher, J. E., Herz, M. J. and Peeke, H. V. S. (1972) Habituation of aggression: the effects of visual stimuli on behavior between adjacently territorial convict cichlids (*Cichlasoma nigrofasciatum*). *Behav. Biol.*, **7**, 359–68.

Giles, N. (1983) Behavioural effects of the parasite *Schistocephalus solidus* (Cestoda) on an intermediate host, the three-spined stickleback, *Gasterosteus aculeatus* L. *Anim. Behav.*, **31**, 1192–4.

Halperin, J. R. P., Dunham, D. W. and Ye, S. (in press) Primed aggressiveness increases after social isolation. *Behav. Process.*

Heiligenberg, W. (1976) A probabilistic approach to the motivation of behavior, in *Simpler Networks and Behavior* (ed. J. C. Fentress), Sinauer, Sunderland, MA, pp. 301–13.

Heiligenberg, W. and Kramer, U. (1972) Aggressiveness as a function of external stimulation. *J. comp. Physiol.*, **77**, 332–40.

Heiligenberg, W., Kramer, U. and Schulz, U. (1972) The angular orientation of the black eye-bar in *Haplochromis burtoni* (Cichlidae, Pisces) and its relevance to aggressivity. *Z. vergl. Physiol.*, **76**, 168–76.

Hogan, J. A. (1974) On the choice between eating and aggressive display in the Siamese fighting fish (*Betta splendens*). *Learn. Motiv.*, **5**, 273–87.

Hogan, J. A. and Bols, R. J. (1980) Priming of aggressive motivation in *Betta splendens*. *Anim. Behav.*, **28**, 135–42.

Hogan, J. A., Kleist, S. and Hutchings, C. S. L. (1970) Display and food as reinforcers in the Siamese fighting fish (*Betta splendens*). *J. comp. physiol. Psychol.*, **70**, 351–7.

Huntingford, F. A. (1976) A comparison of the reaction of sticklebacks in different reproductive conditions towards conspecifics and predators. *Anim. Behav.*, **24**, 694–7.

Hurley, A. C. and Hartline, P. H. (1974) Escape response in the damselfish *Chromis cyanea* (Pisces: Pomacentridae): a quantitative study. *Anim. Behav.*, **22**, 430–37.

Itzkowitz, M. (1979) On the organization of courtship sequences in fishes. *J. theor. Biol.*, **78**, 21–8.

Ivlev, V. S. (1961) *Experimental Ecology of the Feeding of Fishes*, Yale University, New Haven, CT, 302 pp.

Klein, R. M., Figler, M. H. and Peeke, H. V. S. (1976) Modification of consummatory (attack) behaviour resulting from prior habituation of appetitive (threat) components of the agonistic sequence in male *Betta splendens* (Pisces, Belontiidae). *Behaviour*, **58**, 1–25.

Kohler, D. (1976) The interaction between conditioned fish and naive schools of juvenile carp (*Cyprinus carpio*, Pisces). *Behav. Process.*, **1**, 267–75.

Kramer, D. L. (1973) Parental behaviour in the blue gourami *Trichogaster trichopterus*

(Pisces, Belontiidae) and its induction during exposure to varying numbers of conspecific eggs. *Behaviour*, **47**, 14–32.

Laming, P. R. and Ebbessan, S. O. E. (1984) Arousal and fright responses and their habituation in the slippery dick, *Halichoeres bivittatus*. *Experientia*, **40**, 767–9.

Laming, P. R. and Ennis, P. (1982) Habituation of fright and arousal responses in the teleosts *Carassius auratus* and *Rutilus rutilus*. *J. comp. physiol. Psychol.*, **96**, 460–66.

Lenke, R. (1982) Hormonal control of cleaning behaviour in *Labroides dimidiatus* (Labridae, Teleostei). *Mar. Ecol.*, **3**, 281–92.

Lester, N. P. (1984) The feed:feed decision: how goldfish solve the patch depletion problem. *Behaviour*, **89**, 175–90.

Liley, N. R. and Stacey, N. E. (1983) Hormones, pheromones and reproductive behavior in fish, in *Fish Physiology*, Vol. 8 (eds W. S. Hoar and D. J. Randall), Academic Press, New York, pp. 1–63.

Losey, G. S. J. (1979) Fish cleaning symbiosis: proximate causes of host behaviour. *Anim. Behav.*, **27**, 669–85.

McFarland, D. J. (ed.) (1974) *Motivational Control Systems Analysis*, Academic Press, London, 523 pp.

McFarland, D. J. and Sibly, R. M. (1975) The behavioural final common path. *Phil. Trans. R. Soc.*, **270B**, 265–93.

Metcalfe, N. B., Huntingford, F. A. and Thorpe, J. E. (1986) Seasonal changes in feeding motivation of juvenile Atlantic salmon (*Salmo salar*). *Can. J. Zool.*, **64**, 2439–46.

Morgan, M. J. (1988a) The influence of hunger, shoal size and predator pressure on foraging in bluntnose minnows. *Anim. Behav.*, **36**, 1317–22.

Morgan, M. J. (1988b) The effect of hunger, shoal size and the presence of predator on shoal cohesiveness in bluntnose minnows, *Pimephales notatus* Rafinesque. *J. Fish. Biol.*, **32**, 963–71.

Morgan, W. L. and Ritz, D. A. (1984) Effects of prey density and hunger state on capture of krill, *Nycitiphanes australis* Sars, by Australian salmon, *Arripis trutta* (Bloch & Schneider). *J. Fish Biol.*, **24**, 51–8.

Mrowka, W. (1986) Satiation restores brood care motivation in the female mouth-brooder *Pseudocrenilabrus multicolor* (Cichlidae). *Physiol. Behav.*, **38**, 153–6.

Nakatsuru, K. and Kramer, D. L. (1982) Is sperm cheap? Limited male fertility and female choice in the lemon tetra (Pisces, Characidae). *Science*, **216**, 753–5.

Nelson, K. (1964) The temporal patterning of courtship behaviour in the glandulo-caudine fishes (Ostariophysi, Characidae). *Behaviour*, **124**, 90–144.

Nelson, K. (1965) After-effects of courtship in the male three-spined stickleback. *Z. vergl. Physiol.*, **50**, 569–97.

Olson, R. D., Kastin, A. J., Mitchell, G. F., Olson, G. A., Coy, D. H. and Montalbane, D. M. (1978) Effects of endorphin and enkephalin analogs on fear habituation in goldfish. *Pharmacol. Biochem. Behav.*, **9**, 111–14.

Peeke, H. V. S. (1983) Habituation sensitization and redirection of aggression and feeding behaviour in the three-spined stickleback (*Gasterosteus aculeatus* L.). *J. comp. Psychol.*, **97**, 43–51.

Poulsen, H. R. (1977) Predation, aggression and activity levels in food-deprived sunfish (*Lepomis macrochirus* and *L. gibbosus*): motivational interactions. *J. comp. physiol. Psychol.*, **91**, 611–28.

Poulsen, H. R. and Chiszar, D. (1975) Interaction of predation and intraspecific aggression in bluegill sunfish *Lepomis macrochirus*. *Behaviour*, **55**, 268–86.

Putters, F. A., Mertz, J. A. J. and Kooijman, S. A. L. M. (1984) The identification of simple function of a Markov chain in a behavioural context: barbs do it (almost) randomly. *Nieuw Arch. Wiskunde*, **2**, 110–23.

Rankin, J. C., Pitcher, T. J. and Duggan, R. T. (eds) (1983) *Control Processes in Fish Physiology*, Croom Helm, London, 298 pp.

Rasa, O. A. E. (1971) Appetence for aggression in juvenile damselfish. *Z. Tierpsychol. Suppl.*, **7**, 70 pp.

Reebs, S. G. and Colgan, P. W. (1990) Nocturnal care of eggs and circadian rhythms of fanning activity in two normally diurnal cichlid fishes, *Cichlasoma nigrofasciatum* and *Herotilapia multispinosa*. *Anim. Behav.*, V. **41**, 303–311.

Reyer, H.-U. (1975) Ursachen und Konsequenzen von Aggressivitat bei *Etroplus maculatus* (Cichlidae, Pisces). *Z. Tierpsychol.*, **39**, 415–54.

Robertson, C. M. (1979) Aspects of sexual discrimination by female Siamese fighting fish (*Betta splendens* Regan). *Behaviour*, **70**, 323–36.

Rowland, W. J. (1975a) The effects of dummy size and color on behavioral interaction in the jewel cichlid, *Hemichromis bimaculatus* Gill. *Behaviour*, **53**, 109–25.

Rowland, W. J. (1975b) System interaction of dummy-elicited behavior in the jewel cichlid, *Hemichromis bimaculatus* Gill. *Behaviour*, **53**, 171–82.

Rowland, W. J. and Sevenster, P. (1985) Sign stimuli in the threespine stickleback (*Gasterosteus aculeatus*): a re-examination and extension of some classic experiments. *Behaviour*, **93**, 241–57.

Sazima, I. (1983) Scale-eating in characoids and other fishes. *Env. Biol. Fishes*, **9**, 87–101.

Schwassmann, H. O. (1980) Biological rhythms: their adaptive significance, in *Environmental Physiology of Fishes* (ed. M. A. Ali), Plenum, New York, pp. 613–30.

Sluckin, W. (ed.) (1979) *Fear in Animals and Man*, Van Nostrand, New York, 317 pp.

Stephens, D. W. and Krebs, J. R. (1986) *Foraging Theory*, Princeton University Press, Princeton, NJ, 247 pp.

Thomas, G. (1974) The influences of encountering a food object on subsequent searching behaviour in *Gasterosteus aculeatus* L. *Anim. Behav.*, **22**, 941–52.

Thomas, G. (1977) The influence of eating and rejecting prey items upon feeding and food searching behaviour in *Gasterosteus aculeatus* L. *Anim. Behav.*, **25**, 52–66.

Thresher, R. E. (1978) Territoriality and aggression in the threespot damselfish (Pisces: Pomacentridae): an experimental study of causation. *Z. Tierpsychol.*, **46**, 401–34.

Tinbergen, N. (1951) *The Study of Instinct*, Clarendon, Oxford, 228 pp.

Tinbergen, N. (1952) 'Derived' activities: their causation, biological significance, origin, and emancipation during evolution. *Q. Rev. Biol.*, **27**, 1–32.

Tinbergen, N. (1963) On aims and methods of ethology. *Z. Tierpsychol.*, **20**, 410–33.

Tugendhat, B. (1960a) The normal feeding behavior of the three-spined stickleback (*Gasterosteus aculeatus* L.). *Behaviour*, **25**, 284–318.

Tugendhat, B. (1960b) The disturbed feeding behavior of the three-spined stickleback: 1. Electric shock administered in the food area. *Behaviour*, **16**, 159–87.

Vodegel, N. (1978) A study of the underlying motivation of some communicative behaviours of *Pseudotropheus zebra* (Pisces, Cichlidae): a mathematical model (1) & (2). *Proc. K. ned. Akad. Wetenschappen*, **81C**, 211–240.

Wapler-Leong, C.-Y. (1974) The attack readiness of male *Haplochromis burtoni* (Cichlidae, Pisces) reared in isolation. *J. comp. Physiol.*, **94**, 219–25.

Warren, E. W. and Callaghan, S. (1976) The response of male guppies (*Poecilia reticulata*, Peters) to repeated exposure to an open field. *Behav. Biol.*, **18**, 499–513.

Wiepkema, P. R. (1961) An ethological analysis of the reproductive behaviour of the bitterling (*Rhodeus amarus* Bloch). *Archs. neerl. Zool.*, **14**, 103–99.

Wilz, K. J. (1970a) Causal and functional analysis of dorsal pricking and nest activity in the courtship of the three-spined stickleback *Gasterosteus aculeatus*. *Anim. Behav.*, **18**, 115–24.

Wilz, K. J. (1970b) The disinhibition interpretation of the 'displacement' activities during courtship in the three-spined stickleback, *Gasterosteus aculeatus*. *Anim. Behav.*, **18**, 682–7.

Wilz, K. J. (1972) Causal relationships between aggression and the sexual and nest behaviours in the three-spined stickleback (*Gasterosteus aculeatus*). *Anim. Behav.*, **20**, 335–40.

Chapter three

Development of behaviour in fish

Felicity A. Huntingford

3.1 QUESTIONS ABOUT THE DEVELOPMENT OF BEHAVIOUR

A newly fertilized egg does not behave. An adult fish responds to its environ-ment with a repertoire of complex, adaptive behaviour patterns. This chapter is about what happens in between. When in the developmental process do co-ordinated behaviour patterns arise, and how does this come about? When and how are social relationships established and what is their role in the life of a young fish? How do external stimuli control behaviour in fish of various ages? When and how do different behaviour patterns take on the motivational relationships that characterize adult behavioural systems? To answer these questions we need to know what young fish actually do at each stage in the development from egg to adult. Equally we might ask how it comes about that different species, sexes and individuals show distinct behavioural responses. This second and very difficult question requires a knowledge of the factors that influence the sequence of developmental events, and how they do so.

3.2 FISH AS SUBJECTS FOR DEVELOPMENTAL RESEARCH

Teleost fish have a number of characteristics that make them suitable subjects for developmental studies. In the first place, although their behaviour is often spectacular, their repertoire of action patterns is not unmanageably complex. It is therefore possible to produce an accurate description of the complete behavioural repertoire of a particular species (an 'ethogram') so that in a number of cases we know where behavioural development is heading. Secondly, the huge number of teleost species (at least 22 000) means that the comparative approach (Huntingford, 1984) can be put to good use; we can, for example, compare the development of behaviour in species with and without parental care (see Chapter 10 by Turner, and Chapter 11 by Sargent

Behaviour of Teleost Fishes 2nd edn. Edited by Tony J. Pitcher. Published in 1993 by Chapman & Hall. ISBN 0 412 42930 6 (HB) and 0 412 42940 3 (PB).

and Gross, this volume) to find out how interactions with parents influence the way behaviour develops. A further advantage of fish as subjects for developmental studies is that they generally produce many offspring whose larval and juvenile stages are free living. This means that fish are potentially accessible for observation and experimental manipulation of the environmental conditions in which they develop. On the other hand, young fish are not easy to rear and their small size makes experimental manipulation of the developing nervous system difficult. Overall, however, the development of behaviour in fish has proved a fruitful area of research, and the aim of this chapter is to describe some of these studies and see if they provide any insights into the process of development. Like other reviews of the subject (Noakes, 1978), this one concentrates on the behavioural ontogeny of the two most intensively studied groups, the salmonids and the cichlids.

3.3   THE EARLY DEVELOPMENT OF BEHAVIOUR IN SALMONIDS

Because of their commercial importance, salmonids have received a lot of attention from scientists. Figure 3.1 summarizes the main behavioural changes that occur in the Atlantic salmon, *Salmo salar*, from fertilization to an age of about 150 days. The first stages of development, up to a few weeks after hatching, normally take place under several centimetres of gravel, where the eggs are buried after fertilization. The first observed movements occur about one-third of the way into embryonic life in the form of weak contractions of the developing heart. These gradually increase in strength and frequency until co-ordinated contractions of the fully formed heart occur. Shortly after heart movement starts, the anterior muscles above the yolk sac being to twitch, causing very slow and irregular bending movements of the trunk, initially with no orderly relationship between different parts of the body. These bending movements become gradually stronger, extending to muscle blocks along the whole of the body, with the two sides co-ordinated to produce S-shaped swimming movements. Trunk movements initially occur spontaneously, but later are elicited by tactile stimulation. Towards the end of embryonic life the jaws, operculae and the pectoral fins begin to move spasmodically, when the axial muscles contract. Shortly before hatching, movements of the jaw and operculae and of the pectoral fins are fully co-ordinated and independent of contractions of the axial musculature. Hatching is the result of swimming movements which are sufficiently frequent and violent to split the egg membrane and allow the young fish to emerge (Abu-Gideiri, 1966).

At first the newly hatched larvae lie mostly on their side, but later they rest upright with the head pointing downwards. As the yolk sac becomes smaller, the body axis gradually becomes horizontal; the young fish initially lie close

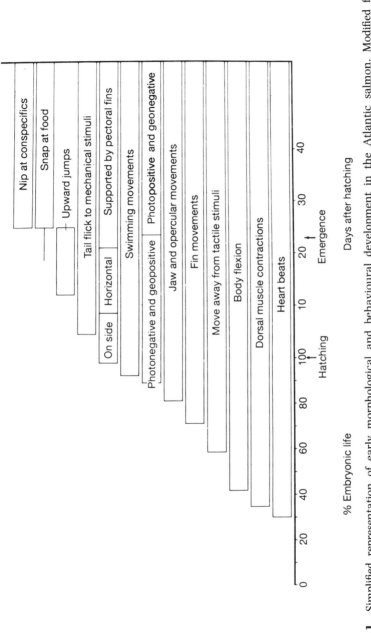

Fig. 3.1 Simplified representation of early morphological and behavioural development in the Atlantic salmon. Modified from Abu-Gideiri (1966) and Dill (1977).

to the ground but later support the front of their body with the pectoral fins. Lateral movements of the body and the tail occur in all these positions, producing forward movement along and finally off the substrate. At this stage swimming may be elicited by mechanical and visual stimuli, and the young fish orientate away from light, towards the ground and into the current. On emergence from the gravel, young salmonids periodically make sudden vertical movements by flexing their tails and pushing against the substrate. These movements, which bring the young fish out of the gravel and into the stream, coincide with a change to positively phototactic behaviour. Emergence takes place just before the yolk supply is exhausted and when development of a functional gut is complete. Atlantic salmon parr tend to remain in contact with the substrate and to swim into the current. Even before emergence, the young fish will dart and snap at a solid object moved in front of its head. Initially bites are poorly co-ordinated, not very successful and directed indiscriminately at any small solid object; later, they become more effective and directed only at likely food items. The young fish begin to chase and nip at their companions at the same time as they start to feed and with similar movements; the fish that are being chased give escape responses (Dill, 1977).

The subsequent behaviour of young salmonids varies from species to species and has been less minutely documented. In the presence of a predator, even very young sockeye, *Oncorhynchus nerka*, and Atlantic salmon parr maintain a minimum distance from the larger fish. Should the predator approach, the parr form schools (see Chapter 12 by Pitcher and Parrish, this volume) and, if it attacks, they escape rapidly and remain motionless either at the water surface or on the bottom (Ginetz and Larkin, 1976; Jacobssen and Jarvi, 1976).

In Atlantic salmon and rainbow trout, *Oncorhynchus mykiss*, agonistic interactions between young fish become more frequent with age, but their form changes: chasing and nipping become less common and are replaced by stereotyped fin displays and head-down postures; these elicit avoidance without overt physical contact (Dill, 1977; Cole and Noakes, 1980). Agonistic behaviour gradually becomes restricted to specific areas as the fish establish and defend feeding territories (Keenleyside and Yamamoto, 1962). As Atlantic salmon parr get older, the size of food they take and the area they need to find it in increases. They leave their territories, move into deeper, more open water, and feed together in dominance-structured groups (Wankowski and Thorpe, 1979a).

This may represent the start of the complex physiological and behavioural changes of smolt transformation that many salmoinds undergo (Hoar, 1988). During smolt transformation, the fish become silvery, change shape and finally abandon their territories. Most smolts then cease to swim into the current and are carried downstream in schools. This takes them to the sea where they feed and grow for a variable period of time before returning to their home stream (which they recognize by its smell) as mature adults ready to breed (Hasler and Scholtz, 1983).

Age of smolting is variable even within a given species; for example, in different populations of Atlantic salmon it ranges from 1 to 7 years in relation to latitude and temperature, which together determine feeding opportunity (Metcalfe and Thorpe, 1990). Within sibling groups kept under hatchery conditions, some individuals maintain feeding and growth through their first winter of life and smolt the following spring, at 18 months; others enter a state of natural anorexia and cease growth from late autumn to early spring and delay smolting for at least 1 year (Metcalfe, *et al.*, 1988; Thorpe, 1989). The autumn drop in appetite and growth is determined during a restricted period in June/July. It seems that fish growing well at this time maintain appetite and smolt early, while their slower-growing siblings cease feeding and delay smolting. The known genetic influences on age of smolting may act by altering the threshold growth rate that differentiates these two developmental patterns (Thorpe, 1989).

In both Atlantic and coho salmon, some male parr mature without migrating at all (see Thorpe, 1989, and Chapter 13 by Magurran, this volume). In contrast to normally-developing parr, these early-maturing males retain their tendency to swim into the current and remain on the spawning grounds. They compete both with other parr and with anadromous males to fertilize eggs, and although some survive to breed in successive years (Garcia de Leaniz, 1990), many do not survive after spawning time (Myers and Hutchings, 1987; Hutchings and Myers, 1987, 1988). As in the case of early smolting, at the population level, the proportion of early-maturing males (which may reach 100%) depends on opportunities for feeding and growth (Myers, 1984). Within a population, male parr with high lipid reserves in the spring are more likely to mature than are more emaciated individuals (Rowe *et al.*, 1991).

3.4 THE DEVELOPMENT OF BEHAVIOUR IN CICHLIDS

Because they are easy to keep in the laboratory and have a complex and variable repertoire of social responses, the behaviour of cichlid fish has been intensively studied. Much of this work has focused on their reproductive and parental behaviour (see Chapter 10 by Turner, this volume), and it was a natural step to extend these investigations beyond the stage at which the young are deserted by their parents into the period from independence to adulthood.

The pattern of development of behaviour after hatching in the substrate-brooding orange chromide, *Etroplus maculatus*, is summarized in Fig. 3.2. In this species, the young remain in a shoal for their first few weeks of life, close

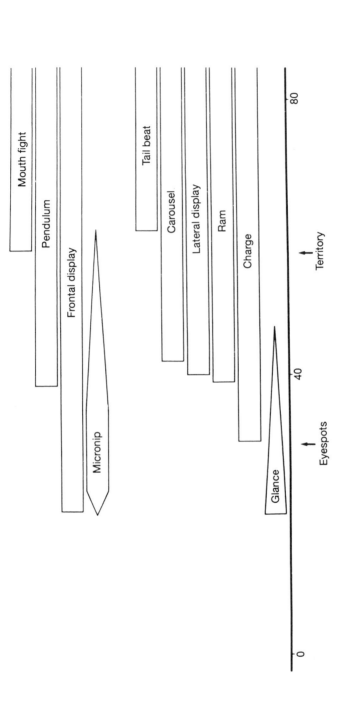

Fig. 3.2 Simplified summary of behavioural development in the orange chromide, *Etroplus maculatus*: the occurrence of ten behaviours is plotted against age. Modified from Wyman and Ward (1973).

to their parents. The young respond to their parents from the moment of hatching and swim towards them, either nipping at their sides (micronip) or curving their body so that their side contacts the parent briefly (glance). When micronipping, the fry are feeding on mucus secreted by their parents' skin; glancing movements probably stimulate mucus production. In another substrate-brooding cichlid, *Cichlasoma citrinellum*, these responses are initially directed towards both parents, but after a few days the young fish make contact preferentially with their father (Noakes and Barlow, 1973). The frequency of interactions between parents and young decreases up to the time the parents desert; the young then remain in the shoal where social encounters occur, initially during competition for food items. To return to the orange chromide, the distance between the young in the shoal gradually increases, and individuals concentrate their activities in a specific area. By 55 days, larger individuals defend small territories within which they feed. Glance and micronip gradually disappear, to be replaced by a variety of other behaviour patterns.

At about 25 days, young orange chromides begin to charge at each other. A charge is like a glance, but involves harder contact with the other fish. Initially the recipient shows no response, but by day 35 frontal display (see below) and the beginning of a charge are enough to elicit retreat. From about 35 days, three new acts are seen. Ramming, a fierce slap with the side of the body, is first shown in encounters in the shoal, but eventually occurs only in a territorial context where it serves to terminate a fight. Lateral display, a static glance, can lead to carouseling, mutual glance-like movements orientated to the opponent's tail resulting in a circular chase. As the fish get older, this complicated movement increases in diameter, becomes more fixed in form, and ceases to involve any physical contact. At 64 days, lateral display in territorial disputes may turn into tail beating; initially the body posture is not maintained as the tail moves, but later, vigorous tail movements are performed on the spot.

Other new acts in the repertoire involve a frontal position with the mouth open and median fins elevated; in other words they look like variations on the micronip theme. Frontal display is the earliest to appear, initially being no more than a brief hesitation with fins raised at the end of a micronip. Later, frontal display is used in sibling interactions as a preliminary to charging, and eventually it elicits retreat even when it is performed alone. If the recipient of a frontal display responds with the same behaviour, the two opponents may perform a back-and-forth pendulum movement. Once territories are established, frontal display becomes an important component of territorial defence, during which mutual frontal display can give rise to mouth fighting.

Thus, in a period of about 70 days, young orange chromides change from living in a closely packed shoal, interacting with their parents by means of just two behaviour patterns, to maintaining feeding territories by an extensive repertoire of discrete and efficient behaviour patterns, finely tuned to the behaviour of other fish. Some 100 days after the agonistic repertoire is complete, fish that have territories begin to breed. During the complex courtship sequence, movements that are very similar to those already described are used in quite different functional contexts. For example, potential mates glance against and nip at each other, skim and dig with similar movements at the spawning site, and may show long bouts of mutual tail beating (Wyman and Ward, 1973).

3.5   WHAT HAVE THESE TWO EXAMPLES SHOWN?

These two studies describe the development of the various functional behavioural systems that we observe in adult fish, namely habitat selection, feeding and antipredator responses, together with agonistic, sexual and parental behaviour. Together, they illustrate a number of general points.

1. Strong, co-ordinated movements emerge gradually from the weak, spasmodic muscle twitches of very young embryos. These early co-ordinated responses themselves change in form and may be replaced by recognizably different acts. In the orange chromide, the component of the glance that is directed towards another fish becomes more marked to produce a charge; holding the movement at the outermost extent of the circle results in lateral display (Fig. 3.3). Thus, as the young fish get older, their repertoire is enlarged by emphasizing or modifying different components of existing actions to form a number of new behaviour patterns. In the development of agonistic behaviour, this often results in overtly aggressive acts with physical contact being replaced by stereotyped movements that act at a distance.

2. In addition to those changes in form, the systems controlling the various behaviour patterns alter as development progresses. A change in internal control mechanisms might show up, for example, as a shift in the sequences in which they are performed (see Colgan, Chapter 2, this volume). The frontal display of the young orange chromide initially occurs at the end of a micronip, later reliably precedes a charge, and finally occurs on its own. Some differences in the way these acts are controlled must underlie such changes. The effect of external stimuli may also change with age, as when the initially photonegative response of Atlantic salmon fry becomes photopositive at emergence. In a social context, charge and frontal display elicit no response in a very young cichlid, but both acts can cause an older opponent to retreat.

3. In both cichlids and salmonids, periods of rapid behavioural reorganization are interspersed with periods when change is more gradual. These times

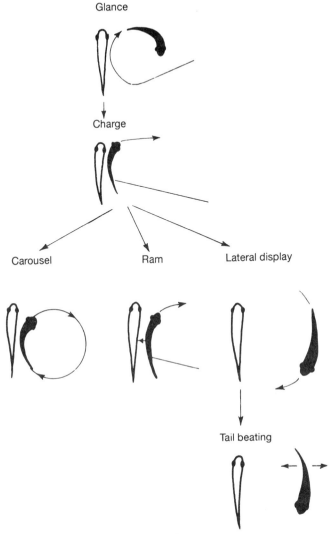

Fig. 3.3 Proposed sequence for the development of adult from juvenile behaviour patterns in the orange chromide. Modified from Wyman and Ward (1973) in Huntingford (1984).

of rapid changes often correspond to important events such as emergence from the redd, migration to a new habitat, or becoming independent of parents.

4. The end result of all these changes is an adult fish with a complete repertoire, but there is more to the process of development than simply

constructing functional adults. Much of the behaviour of a young fish serves an immediate and vital function in survival. For example, micronip and glance collect and maintain a food supply for young cichlids as well as being the precursors of movements used by older fish.

5. The course and timing of developmental events are not identical for all members of a species. Some differences between individuals may be caused by genetic variation. However, different patterns of development may also be the result of condition-dependent switches, as in the cases of early smolting in Atlantic salmon that were growing fast the previous summer and early maturation in parr with good lipid stores in spring.

3.6   WHAT CAUSES BEHAVIOUR TO CHANGE DURING DEVELOPMENT?

The previous sections described a number of behavioural changes that occur during development and have given some answers to the first set of questions spelled out in the introduction. The next stage is to identify and characterize the factors and processes that control the course of development. From the examples given here, a number of possibilities can be suggested. The simplest would be that the behavioural changes come about because the eliciting stimulti in the external environment have altered. On the other hand, behaviour patterns may appear in the repertoire or alter in form because the machinery necessary for their performance reaches a certain stage of maturation, or because physiological conditions within the fish change. Behaviour may change during development because what an animal experiences at one stage alters what happens subsequently. These various possibilities are now discussed in turn. They are not mutually exclusive; indeed, changes in the nervous system or some other aspect of structure or physiology may be the mechanism by which experience modifies the course of development.

Behavioural changes resulting from alteration in external stimuli

The gradual reduction in the frequency of parent-contacting movements that occurs as young cichlids become independent may come about because their parents are producing less mucus. This may actually represent an accelerating process, since parents receiving fewer contacts are stimulated to produce less mucus, hence the rapid nature of the behavioural changes that occur at this time. The appearance of charging movements in cichlids corresponds to the time when their companions start to develop species-typical colour patterns. Here we have behavioural shifts that probably occur because of changes in the external stimuli that the environment provides, although of course the reduction in mucus levels and appearance of colour patterns themselves require explanation.

Behavioural changes accompanying development of the nervous system

Most studies of development of the nervous system in fish are confined to the the embryonic stage, but here clear correlations exist between the state of the nervous system and the behaviour of the developing fish (Noakes and Godin, 1988). For example, the Mauthner cells, a pair of large interneurones that run from the hindbrain to the motor neurones of the spinal cord, are a conspicuous feature of the developing nervous system in fish. When these are functional, the developing fish is able to produce rapid startle movements in response to mechanical stimulation (Abu-Gideiri, 1966; Armstrong and Higgins, 1971). Larval herring show a sharp increase in responsiveness to attacks by predators at a length of about 26 mm, which corresponds with the filling of the otic capsule with gas and the appearance of neuromasts enclosed in the lateral line canal (Blaxter and Fuiman, 1990). Here we have examples of behavioural capacities emerging in parallel with the growth of the neural and sensory structures necessary for their performance.

Behavioural changes resulting from non-neural morphological changes

The gradual appearance of co-ordinated movements of the mouth during feeding and respiration can be related to structural developments in the head. The lower jaw of salmonids can be depressed by a number of separate

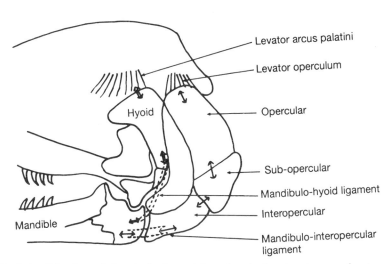

Fig. 3.4 Side view of the skull of a rainbow trout, showing two mechanisms for depressing the lower jaw. Broken lines represent ligaments; arrows represent bone movements. Modified from Verraes (1977).

mechanisms, two of which are shown in Fig. 3.4. When the levator arcus palatini contracts, the roof of the skull swings up and outwards, pulling up the hyoid bone as it does so. The hyoid is attached to the back of the lower jaw (mandible) by a long ligament (the mandibulo-hyoid ligament) so when the hyoid is pulled up, the lower jaw opens. When the levator operculum muscle contracts, this pulls on a series of bones: the opercular, the sub-opercular and the interopercular. This last bone is also attached to the mandible by a ligament (the mandibulo-interopercular ligament), so a pull on the interopercular depresses the lower jaw.

The mandibulo-hyoid ligament is present at hatching, when it makes contact at both ends with cartilage. Weak contraction of the developing levator arcus palatini occurs long before overt jaw movements are observed, and this is thought to stimulate ossification of the cartilage near the points of attachment of the ligament. Once the bones are fully formed, effective lowering of the jaw, as used to generate a respiratory current, is possible. The mandibulo-interopercular ligament and the opercular bones do not develop until much later, and therefore the second mechanism for lowering the jaw, which plays a major role in ingesting food, does not become active until later, at the time the young fish have depleted their yolk sacs. Thus, the two patterns of co-ordinated and effective jaw movement only appear in the repertoire when the relevant system of bone and ligaments is connected up (Verraes, 1977). The subsequent increase in the size of particle that young salmon choose to ingest is paralleled by an increase in the size and suction pressure of their gape (Wankowski and Thorpe, 1979b). In the cichlid *Haplochromis piceatus* the jaw muscles increase isometrically with size, so larger fish are able to eat bigger, more profitable prey. In addition, the retractor dorsalis muscle shifts to a more vertical position, giving larger fish a stronger bite and making it possible for them to crush the cuticle of insect pupae that are inaccessible to smaller fish (Gallis, 1990).

Behavioural changes resulting from alterations in non-neural physiology

Although young salmonids have access to food particles and are capable of ingesting and digesting food as soon as they emerge from the redd, feeding movements occur infrequently until several days later. Up to this time the yolk is not completely used up and so feeding may not occur because the fish have no need for food. One might therefore ascribe this behavioural shift to a deprivation-induced change in feeding motivation.

The autumn anorexia that occurs in Atlantic salmon parr and that will delay smolting seems to be caused by a raising of the threshold lipid reserves below which feeding is activated. Appetite can be reinstated in anorexic fish by a period of food deprivation sufficient to reduce lipid reserves from 2.5% to 1.5%

wet weight. This reinstatement is temporary, being reversed once the fish have made up the loss in reserves (Metcalfe and Thorpe, 1992).

A dramatic surge in circulating levels of the hormone thyroxin has been reported just before smolting in salmon and some of the behavioural and physiological characteristics of smolting salmon can be induced by experimentally raising thyroid hormone levels. However, levels of corticosteroids, prolactin and growth hormone (respectively involved in lipid metabolism, osmoregulatory physiology in fresh water and mineral balance in seawater) are also elevated at the time of smolting. The endocrine basis of smolting is clearly complex, with thyroid hormone interacting with other hormones and with an endogenous cycle to bring about this complex change (Hoar, 1988). The appearance of agonistic behaviour in the paradise fish, *Macropodus opercularis*, parallels histological changes in the gonads (Davis and Kessel, 1975). These authors suggest that the increase in levels of agonistic behaviour and appearance of reproductive movements in the orange chromide may also correlate with increased gonadal activity, and in particular with increased production of gonadal hormones. In these examples, the changes in behaviour that are observed during development relate to, and may well be caused by, alterations in the hormonal state of the young animal.

Behavioural changes caused by experience

Many of the morphological and physiological changes that cause developmental shifts in behaviour are themselves the result of the circumstances impinging on the young animal. These influences may come from within the developing animal; thus the bone growth that allows effective jaw movement in young salmon itself depends on early muscle twitches. Alternatively, developmental changes may be caused by influences external to the animal. Exposure to light is known to influence the way the visual system and visually guided behaviour develop in other groups of animals (mammals, Lund, 1978; amphibians, Copp and McKenzie, 1984). In eyeless cave fish, *Astyanax* spp, the visual pathways develop in the absence of any stimulation and contain a severely reduced number of nerve cells. The optic tectum, which in normal fish processes visual information, is effectively deprived of sensory input. This region of the brain undergoes compensatory innervation by neurones associated with the lateral line system; as a result, the blind fish have particularly well-developed responses to water movements (Kutz *et al.*, 1981; and see Chapter 17 by Parzefall, this volume).

There is a rather special kind of influence impinging on the developing animal from its external environment. This is the experience of particular beneficial or harmful contingencies following performance of a given action or, in other words, learning. Since male *Cichlasoma citrinellum* produce more mucus than females, the development in the fry of a preference for the father may come about because they learn to associate the visual cues provided by the male with a superior food supply. The gradual tightening up in the performance of a number of agonistic acts in young orange chromides may be the result of a general improvement in strength and coordination, but it could also come about as a result of learning that certain acts are more effective in driving off a rival. After a rainbow trout has experienced frontal display and charging in quick succession on a few occasions, frontal display may become a conditioned stimulus which can elicit retreat on its own (Chiszar and Drake, 1975). Here are a number of behavioural changes that may come about through some sort of learning process in the developing animal.

3.7   EXPERIMENTAL STUDIES

Thus we have a number of candidate explanations, which may not be mutually exclusive, for the changes in behaviour that occur as an egg develops into a mature fish. However, since these suggestions are mostly based on correlations, it is not easy to distinguish cause and effect. For example, young *Cichlasoma citrinellum* may develop a preference for their father because they learn that males provide more mucus, but the bias might equally well develop for some other reason and this in turn may stimulate more copious mucus production in the preferred parent. These two possibilities were distinguished by an experiment; young cichlids raised without the opportunity to contact their parents' bodies fail to develop a preference for their father, which supports the learning hypothesis (Noakes and Barlow, 1973). This study illustrates the form that most experimental studies of development take. To find out whether a particular factor is involved in determining the course of behavioural development, we alter the state of that factor in some way; if this changes what the young animals or adults do, then we can conclude that the factor is important and start to investigate the nature of its influence on development. If the course of development proceeds unaltered in spite of the experimental manipulation, we can conclude that the particular factor we manipulated was not essential for the normal process of behavioural development.

This conclusion has to be worded carefully, since negative results of deprivation experiments are notoriously hard to interpret (Bateson, 1981; Huntingford, 1984). Thus it is not easy to be sure that we really have removed a particular environmental influence, and even if fish that really have been so deprived develop normally, this does not prove that the experience in question has no effect in normal development, nor does it show that development is independent of any environmental influences at all. Examples of these kinds of complication are given below. With these provisos in mind, however, such experiments can help us to identify and characterize factors that influence the course of behavioural development.

The developing system can be manipulated in two general ways: by altering the genetic constitution of the zygote or by tampering with the environment in which it develops. It is quite clear that differences in genetic make-up can result in behavioural changes in fish (see Chapter 1 by Danzmann *et al.*, this volume). However, there is very little information available about exactly how genes exert their effects on the development of the nervous system and behaviour in fish, or indeed for most other groups of animals, except possibly fruit flies, nematodes and Siamese cats (Partridge, 1983). Much more is known about the effects of environmental manipulations, and although fewer studies of this sort have been carried out on fish than on birds and mammals, there is still a substantial literature on this subject.

Effects of experience on feeding behaviour

Changes in foraging behaviour with age (for example, shifts from pelagic planktivory to a benthic diet) have been reported for many species of fish (Werner and Gilliam, 1984). Such shifts in diet may be a direct result of morphological changes, such as increased gape size and altered muscle orientation (Gallis, 1990). However, the underlying processes may be more complex. For example, in young cichlids, *Cichlasoma managuense*, reared on flaked food, both the speed of feeding on novel live prey and the number of feeding attempts increase with age over the first month of life. Multiple regression analysis showed this to be an effect of age per se and not due to the associated increase in size (Meyer, 1988). Similarly, although hatchery-reared coho salmon respond enthusiastically to live prey on the first encounter, efficiency of prey capture increases rapidly from an initially low level during the first few hours of the first feeding bout (Paszkowski and Olla, 1985).

Changes in foraging efficiency with experience continue to occur in adult fish. Wild-caught adult fiften-spined sticklebacks, *Spinachia spinachia*, were given a diet of mussel flesh for 2 months before being fed on either *Gammarus* or *Artemia* or both species presented alternately on eight trials. When the trials were separated by 2 days or less, handling time decreased over trials for all prey types, but most particularly for *Artemia* (Fig. 3.5); this effect was accompanied by shorter and simpler foraging sequences. No improvement was found when the trials were separated by more than 2 days, as the fish forgot their learned skills (Croy and Hughes, 1991). Such reversible, experience-induced changes allow fish that naturally experience a range of prey types whose availability fluctuates to maintain their overall foraging efficiency.

Effects of experience on antipredator behaviour

Fish are vulnerable to a range of predators depending on their age and size; development of effective responses to the threat posed by such predators

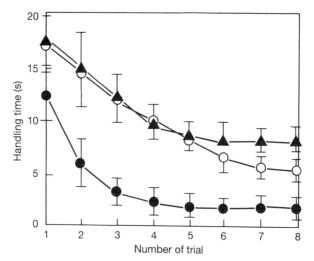

Fig. 3.5 Median handling times (seconds from grasping to swallowing prey; bars are 95% confidence intervals) for the first prey items taken by fifteen spine sticklebacks in successive feeding trials following a 2 month period of being fed on non-living food. ○, Fish tested with living *Gammarus* in all trials; ●, fish tested with living *Artemia* in all trials; ▲, fish tested with *Artemia* and *Gammarus* in alternate trials. Reproduced with permission from Croy and Hughes (1991).

depends on a complex interaction between inherited behavioural tendencies and the effects of experience.

Considering first the issue of how potential predators are recognized. Soon after hatching, jewel cichlids, *Hemichromis bimaculatus*, flee more strongly from simple models bearing two black spots arranged like eyes than they do from models with other arrangements of spots. This discrimination persists in older fish reared without experience of conspecifics, but disappears after some 140 days in fish that had the opportunity to observe the eyes of conspecifics (Coss, 1979). European minnows, *Phoxinus phoxinus*, respond to fright substance (see Chapter 6 by Hara, this volume) on the first occasion that they smell it, even if they have had no prior experience of predators (Magurran, 1986). In contrast, predator-naive minnows do not show fright responses when exposed to the odour of either a natural predator (a pike, *Esox lucius*) or a benign species (the cichlid *Tilapia mariae*). However, if either odour is presented in conjunction with alarm substance on just one occasion, that odour will subsequently elicit a fright response when presented on its own; this effect is particularly marked for the pike odour. The fish form a conditioned response to previously ineffective odours, but appear to do so more readily if the odour comes from a natural predator (Magurran, 1989).

The early development of the Mauthner neurones and their neural connections with the trunk muscles means that young fish respond to direct attack from an early age and without prior experience of predators. Predator-naïve fish of a number of species also show more complex responses to predators, but the strengths of these responses are population-specific. In guppies, *Poecilia reticulata* (Seghers, 1974; Breden *et al.*, 1987; Magurran and Seghers, 1990a), minnow (Magurran, 1986, 1990; Magurran and Pitcher, 1987) and three-spined sticklebacks, *Gasterosteus aculeatus* (Huntingford, 1982; Tulley and Huntingford, 1987a) fish from populations that are naturally exposed to a high level of predation show stronger protective responses than do their counterparts from sites where predators are rare or absent, and these population differences persist in predator-naïve fish. So in fish from sites where there is a serious predation risk, experience of encounters with predators is not necessary for the development of effective antipredator responses.

However, this is not to say that such experience has no effect. In sticklebacks (Benzie, 1965) and in minnows (Magurran, 1990) various aspects of predator avoidance and evasion are significantly enhanced in fish that have been exposed to predatory attacks. Similarly, coho salmon smolts that have experienced two short periods of simulated attack by a piscivorous fish were subsequently more effective at surviving real attacks than were naïve smolts (Olla and Davis, 1989). Conversely, exposure to a potentially dangerous fish that fails to attack can lead to reduced fear responses in subsequent encounters with the same species, by a process of habituation (Csanyi *et al.*, 1989).

Interactions with conspecifics can also influence the development of antipredator responses. Guppies reared in social groups, in which they experience chases (and possibly cannibalistic attacks) by larger fish, subsequently show more effective escape responses when exposed to a predatory fish (Goodey and Liley, 1986). A similar effect is seen in sticklebacks that as fry experienced a period of paternal care (during which fathers chase and retrieve fry when they first leave the nest; Benzie, 1965), but only in fish from sites where the predation risk is high (Tulley and Huntingford, 1987b). Predator-naïve fish from such sites are quicker than fish from safe areas at learning an avoidance conditioning task (involving learning to avoid an area where attacks are experienced; Huntingford and Wright, 1989; Fig. 3.6). Similar site-specific learning has been described for minnows (Magurran, 1990). So one property of fish from sites where predation risk is serious may be a predisposition to learn from adverse experience; this would promote the development of the rich repertoire of antipredator responses that characterize such fish.

Effects of experience on social behaviour

The strong schooling responses that characterize the fry of many species of fish (e.g. jewel cichlids, Chen *et al.*, 1983; guppies, Magurran and Seghers, 1990b) often give way to aggressive interactions as the fish grow larger and

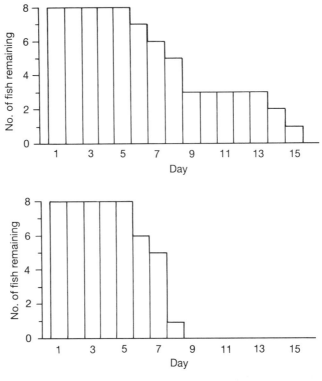

Fig. 3.6   The number of sticklebacks from a low-risk (top) and a high-risk site (bottom) that failed to reach the criterion for learning to avoid a previously favoured feeding patch on successive days of avoidance training. Median number of days to reach criterion = 8 (low-risk site) and 6 (high-risk site); $P < 0.05$, Mann–Whitney test. (Huntingford and Wright, in press).

less vulnerable and as competition for resources increases (Chen *et al.*, 1983). Several studies have shown that most of the major components of the agonistic repertoire appear in isolated fish at the normal age (e.g. Tooker and Miller, 1980). These may be directed at the appropriate stimulus. For example, cichlids, *Haplochromis burtoni*, reared in total isolation with no experience of the normal colour patterns of adults of their own species, preferentially attack models with a black eyebar (which triggers attack in normally reared fish, Hieligenberg, 1976) when tested as adults (Fernald, 1980). In this case, the natural stimulus for attack is simple, and specific experience of the stimulus seems not to be necessary for the normal development of correctly directed attack. However, experience can modify the strength of the elicited response; for example, when given a choice, male *H. burtoni* reared to maturity with either their own or a closely related species spend most of their time attacking the species with which they were reared (Crapon de Caprona, 1982).

Although agonistic actions may develop more or less normally in fish reared in isolation, this does not mean that no practice at all is necessary for the normal development of the movements. Young orange chromides reared in isolation direct their glancing movements against stones and nip at other inanimate objects in the tank; this may provide the experience necessary for development of other behaviour patterns (Wyman and Ward, 1973). Nor does the appearance of a full repertoire of agonistic actions in isolation-reared fish mean that experience has no effect on how these actions are used. A striking and common result is that fish reared in isolation tend to use overtly aggressive acts, rather than non-contact displays. These findings suggest that whereas interaction with other fish is not necessary for the differentiation of agonistic movements during development, it is critically important if the fish are to perform them effectively and in the correct context, and to respond appropriately to them. Experience acquired during fights in older fish can also affect the subsequent nature of agonistic responses. In a number of species, experience of victory or defeat has a profound effect on how animals behave subsequently (Huntingford and Turner, 1987). For example, sticklebacks that have been defeated in one fight are significantly more likely to lose a second fight against a size-matched opponent (Bakker *et al.*, 1989).

As far as sexual behaviour is concerned, male platyfish, *Xiphophorus maculatum*, reared in isolation from soon after birth perform normal sexual movements but fail to associate preferentially with females, unlike males reared with conspecifics. Isolated males, and males reared with both red and black companions, show no preference when offered a choice between red and black females. In contrast, males housed just after sexual maturity with fish of just one colour prefer females of that colour. The females themselves show no preference for black or red males, regardless of their rearing conditions (Ferno and Sjolander, 1973). Here again, whereas experience of conspecifics is not necessary for development of the movements used in sexual interactions, in males it does influence the context in which these movements are used.

Young of some species of cichlid (e.g. *Aspidogramma reitzigi*, Kuenzer, 1962, quoted in Baerends, 1971) reared in isolation from the egg stage prefer models with the colour pattern typical of their species from the moment of hatching, although this preference becomes stronger if they are allowed contact with their parents. Isolated young of other cichlid species hatch without any preference for the normal parental colour; here, a preference develops over a few weeks' association with the parents. For example, adult *Cichlasoma meeki* have grey bodies and red throats. Young reared in isolation show no preference for red or grey, but over the first 3 weeks of their life young kept with their parents gradually become less responsive to all but red and grey models (Baerends and Bearends-van Roon, 1950).

There is some contradictory evidence that these early filial preferences may influence later mate selection; in platyfish, as described earlier, experience at a later stage (closer to maturity) influences choice of mating partners. Young

H. burtoni were reared to maturity and subsequently offered a choice between adults of the two species, with either visual or chemical cues available. Given visual cues, males prefer females of their own species, regardless of previous experience. Males given a choice based on chemical cues, and females given visual cues, prefer potential mates of the species with which they have been reared (Crapon de Caprona, 1982). In that early experience results in filial preferences and later experience may influence mate choice, these examples clearly have something in common with the filial and sexual imprinting for which young precocial birds are famous (Bateson, 1983, and see Chapter 10 by Turner, this volume). However, in the case of the response of cichlids to their parents, there is no clear evidence that learning fails to occur after a certain age.

3.8 WHAT HAVE THESE EXPERIMENTAL STUDIES SHOWN?

The diversity of developmental processes

It is clear from the examples discussed so far that a variety of different factors, both genetic and environmental in origin, determine the course of development of any particular item of behaviour. Their effects may range from the extremely specific (pairing pike odour with alarm substance causes minnows to show one particular response – fright – to one particular smell, that of a pike) to the more general (experience of light alters a range of visually guided behaviour). In addition, these influences, genetic or environmental, with general or specific consequences, can work in all sorts of different ways.

It may be helpful to distinguish between inducing, facilitating, maintaining and predisposing factors in development (Bateson, 1983). **Inducing factors** are drastic influences which switch development from one path to another (e.g. different growth rates in mid-summer in salmon with very different migration patterns); **facilitating factors** speed up processes which will occur in their own time anyway (e.g. naïve minnows school but experience of a predator strengthens this response), and **maintaining factors** preserve the status quo (e.g. once a platyfish develops a preference for females of a particular colour, periodic exposure to such females prevents this preference from waning). Finally, **predisposing** (or **enabling**) **factors** do not themselves bring about a particular developmental change but allow another influence to do so (e.g. thyroid hormone on its own does not cause salmon to smolt, but it is part of the physiological background necessary for this process to occur). These different types of influence grade into one another, and at the end of the day each case has to be analysed in its own right. However, Bateson's taxonomy stresses the diversity of factors controlling behavioural development and may help us to identify features of development shared by different behavioural systems and by different species.

Constraints on learning

The various examples of learning given in this chapter show that, like other animals (Hinde and Stevenson-Hinde, 1973), fishes are not infinitely plastic in what they will learn. Different categories of individual may show different patterns of learning. For example, both sticklebacks (Huntingford and Wright, 1989) and minnows (Magurran, 1990) from sites where predators are abundant alter their behaviour more readily following an adverse experience than do conspecifics from safer sites. Male platyfish learn from the appearance of their companions the characteristics of sexual partners, but similar experience does not influence mate choice in females (Ferno and Sjolander, 1973). *Haplochromis burtoni* learn the smell of potential mates, whereas females learn what they look like (Crapon de Caprona, 1982).

Even for the same individual, certain associations form more readily than others. Rainbow trout may be predisposed to associate the visual cues provided by an approaching conspecific with the painful physical consequences of receiving the nip that normally follows such an approach (Chiszar and Drake, 1975). Similarly, in minnows the association between fish odour and fright induced by alarm substance is stronger if the odour comes from a natural predator rather than from a benign species (Magurran, 1989).

Nor are things learned equally well at all times. There are periods when the developing animal is particularly susceptible to the influence of environmental factors: for example, just after birth when all stimuli are being experienced for the first time in some young cichlids, and just before smolting when the natal environment has proved itself a good one in salmon. Sensitive periods are clearly characteristic of behavioural development in fish just as they are in birds and mammals; in salmon, at least, the time course of the sensitive period has been characterized. In precocial birds, development of a preference for one object biases the animal against making contact with any others, thus bringing the sensitive period to an end (Bateson, 1983). A similar process may well be at work in young cichlids but it is not at all clear what ends the sensitive period for olfactory imprinting in salmon. The developmental processes that underlie such sensitive periods are probably not different from those that act at other times. However, their effects (the formation of preferences which cut the developing animal off from further experience of the same kind) provide particularly dramatic examples of early experiential influences on subsequent development. In addition, just because developmental events occur so rapidly at these times, they may provide valuable insights into the control of behavioural development.

Instincts learning and developmental stability

These various kinds of constraint on learning are part of a broader phenomenon, the stability of many developing systems. Whereas some behaviour patterns are profoundly influenced by specific experiences, others develop to more or less the same end point in a broad range of environmental conditions. Where, how and on what a salmon or a fifteen spine stickleback chooses to feed is influenced throughout its life by continually shifting patterns of food availability; in contrast, the avoidance response of young jewel cichlids to eyespots and the aggressive responses of male *H. burtoni* to black eyebars are altered very little even by the profoundly abnormal experience of complete social isolation. In extreme cases, there almost seems to be an element of self-regulation in the system, with animals seeking out or providing themselves with stimuli when the normal source of these is denied them. Thus isolated orange chromides perform glancing movements against inanimate objects, and do so at a very high frequency (Wyman and Ward, 1973). This may provide enough experience for some degree of differentiation of the agonistic repertoire to occur.

It is the impressive resistance of many behaviour patterns to developmental perturbation that gave rise to the concept of an instinct (or 'innate' behaviour). Thus, in the early ethological literature, a dichotomous classification of behaviour patterns was made into learned responses as opposed to instincts. An instinct develops into its complete, species-specific form under the influence of the genome and without the need for specific environmental contingencies. This dichotomous classification has been criticized on many occasions (Hinde, 1970; Lehrman, 1970; Bateson, 1981) and so far in this chapter the term 'instinct' has been avoided deliberately. When analysis of behavioural development takes the form of careful longitudinal studies of exactly what happens at each stage of development (and the literature on fish behaviour is particularly rich in such accounts), attention is correctly focused on the processes of development as well as its outcomes. It then becomes quite obvious that a particular item of behaviour develops under the continually interacting influence of maturational events initiated by the genome and various aspects of the external environment. The concept of an instinctive behaviour pattern that develops under the influence of the genes gives a misleading impression of the process of behavioural development.

One can trace the movements of ramming in cichlids back to the glancing movements of the newly hatched young, through the body movements of the embryo to the earliest twitches of the trunk muscles. Practising against another fish or an inanimate object may cause the movements of a glance to be differentiated into a ram, and social experience allows ramming to be integrated into an effective agonistic repertoire. Once we know all this, it is clearly meaningless and unnecessary to classify ramming as either learned or instinctive. To use Hebb's analogy (1953), does the area of a rectangle depend on its length or its breadth?

Two useful things remain from the original dichotomy. In the first place, we can assign differences in behaviour to genetic or environmental effects.

Notwithstanding the complex developmental history of ramming, one cichlid may show more of it than another, either because it inherited from its parents certain alleles that enhance aggressive motivation (Huntingford and Turner, 1987) or because by chance it won its first fight; of course, both of these may be the case. Secondly, as I have pointed out, the development of certain behaviour patterns does occur in a wonderfully stable way and it is this that has led some ethologists to retain the concept of an innate behaviour pattern (Alcock, 1984). Such stability may come about for a number of different reasons. For example, development of a particular behaviour may depend on a given environmental influence, but circumstances are such that this influence is reliably available; in real life, most young cichlids grow up in a world containing companions. Even if the environment is unpredictable, development may be stable if the animal is predisposed to respond to, or learn about, a limited range of stimuli. In addition, development may be resistant to disruption of critical influences because other environmental features can produce the same effect. At its most extreme, this may involve the developing animal compensating in some way for deficiencies in its environment. Whichever is the case, the resulting developmental stability is remarkable and requires explanation. However, labelling such behaviour patterns 'instinctive' does not help, and can hinder, progress towards an understanding of how they develop.

3.9   SUMMARY

As a fertilized egg develops into an adult fish, weak, uncoordinated muscle twitches are replaced by efficient, co-ordinated behaviour patterns. These may themselves subsequently be modified in form, in causation, or in function, until they give rise to the full behavioural repertoire of the adult animal. Observational and experimental studies show that these developmental changes are the result of a continuous interaction between maturational processes within the developing animal and various aspects of its external environment. Although complex, this interaction is amenable to experimental investigation. Such studies have shown that many behaviour patterns in fish develop with remarkable stability in spite of environmental perturbation. However, the concept of instinctive behaviour is of limited value in explaining such stability.

ACKNOWLEDGEMENTS

I would like to thank Neil Metcalfe, John Thorpe and an anonymous referee for helpful comments on an earlier draft of this chapter, Tony Pitcher for his patience over the delay in the production of this revised chapter, and Chuck Hollingworth for copy editing a scrappy manuscript efficiently and without complaint.

REFERENCES

Abu-Gideiri, Y. B. (1966) The behaviour and neuroanatomy of some developing teleost fishes. *J. Zool., Lond.*, **149**, 215–41.

Alcock, J. (1984) *Animal Behaviour*, 3rd edn, Sinauer, Sunderland, MA, 596 pp.

Armstrong, P. B. and Higgins, D. C. (1971) Behavioural encephalisation in the bullhead embryo and its neuroanatomy. *J. comp. Neurol.*, **143**, 371–84.

Baerends, G. P. (1971) The ethological analysis of fish behaviour, in *Fish Physiology Vol. VI. Environmental Relations and Behaviour* (eds W. S. Hoar and D. J. Randall), Academic Press, New York, pp. 279–370.

Baerends G. P. and Baerends-van Roon, J. (1950) An introduction to the study of the ethology of cichlid fishes. *Behaviour (Supp.)*, **1**, 1–242.

Bakker, Th. C. M., Feuth-de-Bruijn, E. and Sevenster. P. (1989) Asymmetrical effects of prior winning and losing on dominance in sticklebacks (*Gasterosteus aculeaus*). *Ethology*, **82**, 224–9.

Bateson, P. P. G. (1981) Ontogeny of behavour. *Br. Med. Bull.*, **37**, 159–64.

Bateson, P. P. G. (1983) Genes, environment and the development of behaviour, in *Animal Behaviour*, Vol. 3 (eds T. R. Halliday and P. J. B. Slater), Blackwell, Oxford, pp. 52–81.

Benzie, V. (1965) Some aspects of the anti-predator responses of two species of sticklebacks. DPhil thesis, Oxford University, 150 pp.

Blaxter. J. H. S. and Fuiman, L. A. (1990) The role of the sensory systems of herring larvae in evading predatory fish. *J. mar. biol. Ass., U.K.*, **70**, 413–27.

Breden, F., Scott, M. and Michel, E. (1987) Genetic differentiation for anti-predator behaviour in the Trinidad guppy, *Poecilia reticulata. Anim. Behav.*, **35**, 618–20.

Chen, M. J., Coss, R. G. and Goldthwaite, R. O. (1983) Timing of dispersal in juvenile jewel fish diving development is unaffected by space. *Develop. Psychobiol.*, **16**, 303–10.

Chiszar, D. and Drake, R. W. (1975) Aggressive behaviour in rainbow trout (*Salmo gairdneri* Richardson) of two ages. *Behav. Biol.*, **133**, 425–31.

Cole, K. S. and Noakes, D. L. (1980) Development of early social behaviour of rainbow trout, *Salmo gairdneri. Behav. Process.*, **6**, 97–112.

Copp, S. and McKenzie, R. L. (1984) Effects of light deprivation on development of photo-positive behaviour in *Xenopus laevis* tadpoles. *J. exp. Zool.*, **23**, 219–27.

Coss, R. G. (1979) Delayed plasticity of an instinct: recognition and avoidance of two facing eyes by the jewel fish. *Develop. Psychobiol.*, **12**, 351–4.

Crapon de Caprona, M. D. (1982) The influence of early experience on preferences for optical and chemical cues produced by both sexes in the cichlid fish *Haplochromonis burtoni*. Z. Tierpsychol., **58**, 329–61.

Croy, M. I. and Hughes, R. N. (1991) The role of learning and memory in the feeding behaviour of the fifteen-spined stickleback, *Spinachia spinachia. Anim. Behav.*, **41**, 149–60.

Csanyi, B., Csizmadia, G. and Milosi, A. (1989) Long-term memory and recognition of another species in the paradise fish. *Anim. Behav.*, **37**, 908–11.

Davis, R. E. and Kessel, J. (1975) The ontogeny of agonistic behaviour and the onset of sexual maturation in the paradise fish, *Macropodus opercularis* (Linnaeus). *Behav. Biol.*, **14**, 31–9.

Dill, P. A. (1977) Development of behaviour in alevins of Atlantic salmon, *Salmo salar*, and rainbow trout, *S. gairdneri*, *Anim. Behav.*, **25**, 116–21.

Fernald, R. D. (1980) Response of male cichlid fish, *Haplochromis burtoni*, reared in isolation towards models of conspecifics. *Z. Tierpsychol.*, **54**, 85–93.

Ferno, A. and Sjolander, S. (1973) Some imprinting experiments on sexual preferences for colour variants in the platyfish (*Xiphophorus maculatus*). *Z. Tierpsychol.*, **33**, 418–23.

Gallis, F. (1990) Ecological and morphological aspects of changes in food uptake through the ontogeny of *Haplochromis piceatus*, in *Behavioural Mechanisms of Food Selection* (ed. R. N. Hughes), Springer-Verlag, Berlin, pp. 281–302.

Carcia de Leaniz, C. (1990) Ecology and orientation of juvenile Atlantic salmon. PhD thesis, Aberdeen University, 230 pp.

Ginetz, R. M. and Larkin, P. A. (1976) Factors affecting rainbow trout (*Salmo gairdneri*) predation of migrant fry of sockeye salmon (*Oncorhynchus nerka*) *J. Fish. Res. Bd. Can.*, **33**, 19–24.

Goodey, W. and Liley, N. R. (1986) The influence of early experience on escape behaviour in the guppy, *Poecilia reticulata. Can. J. Zool.*, **64**, 885–8.

Hasler, A. D. and Scholtz, A. T. (1983) *Olfactory Imprinting and Homing in Salmon*, Springer-Verlag, Berlin.

Hebb, D. O. (1953) Heredity and environment in animal behaviour. *Br. J. Anim. Behav.*, **11**, 43–7.

Heiligenberg, W. (1976) A probabilistic approach to the motivation of behaviour, In *Simpler Networks and Behaviour* (ed. J. C. Fentress), Sinauer, Sunderland, MA, pp. 301–13.

Hinde, R. A. (1970) *Animal Behaviour*, McGraw-Hill, New York, 876 pp.

Hinde, R. A. and Stevenson-Hinde, J. (1973) *Constraints on Learning*, Academic Press, London, 488 pp.

Hoar, W. S. (1988) The physiology of smolting salmonids, in *Fish Physiology Vol. XI. The Physiology of Developing Fish, Part B. Viviparity and posthatching juveniles* (eds W. S. Hoar and D. J. Randall), Academic Press, San Diego, pp. 275–343.

Huntingford, F. A. (1982) Do inter and intra-specific aggression vary in relation to predation pressure in sticklebacks? *Anim. Behav.*, **30**, 909–16.

Huntingford, F. A. (1984) *The Study of Animal Behaviour*, Chapman and Hall, London, 412 pp.

Huntingford, F. A. and Turner, A. K. (1987) *Animal Conflict*, Chapman and Hall, London, 448 pp.

Huntingford, F. A. and Wright, P. J. (1989) How sticklebacks learn to avoid dangerous feeding patches. *Behav. Process.*, **35**, 181–9.

Hutchings, J. A. and Myers, R. A. (1987) Escalation of an asymmetric contest: mortality resulting from mate competition in Atlantic salmon, *Salmo salar. Can. J. Zool.*, **65**, 766–8.

Hutchings, J. A. and Myers, R. A. (1988) Mating success of alternative maturation phenotypes in male Atlantic salmon, *Salmo salar. Oecologia*, **75**, 169–74.

Jacobssen, S. and Jarvi, T. (1976) Anti-predator behaviour of two-year-old hatchery-reared Atlantic salmon (*Salmo salar*) and a description of the predatory behaviour of burbot (*Lota lota*). *Zool. Revy.*, **39** (3), 57–70.

Keenleyside, M. H. A. and Yamamoto, F. T. (1962) Territorial behaviour of juvenile Atlantic salmon (*Salmo salar*). *Behaviour*, **119**, 139–69.

Kutz, M. I., Lasek, R. I. and Kaiserman-Abramof, I. R. (1981) Ontophyletics of the nervous system: eyeless mutants illustrate how ontogenetic buffer mechanisms channel evolution. *Proc. natn. Acad. Sci. U.S.A.*, **78**, 397–401.

Lehrman, D. S. (1970) Semantic and conceptual issues in the nature–nurture problem, In *Development and Evolution of Behaviour* (eds L. R. Aronson, E. Tobach, D. S. Lehrman and J. S. Rosenblatt), W. H. Freeman, San Francisco, 656 pp.

Lund, R. D. (1978) *Development and Plasticity of the Brain*, Oxford University Press, New York.

Magurran, A. E. (1986) The development of shoaling behaviour in the European minnow, *Phoxinus phoxinus*. *J. Fish Biol.*, **29**, (supp. A) 159–70.

Magurran, A. E. (1989) Acquired recognition of predator odour in the European minnow (*Phoxinus phoxinus*). *Ethology*, **82**, 216–23.

Magurran, A. E. (1990) The inheritance and development of minnow anti-predator behaviour. *Anim. Behav.*, **39**, 834–42.

Magurran, A. E. and Pitcher, T. J. (1987) Provenance, shoal size and the sociobiology of predator evasion behaviour in minnows. *Proc. R. Soc.*, **229B**, 439–65.

Magurran, A. E. and Seghers, B. H. (1990a) Population differences in predator recognition and attack cone avoidance in the guppy, *Poecilia reticulata*. *Anim. Behav.*, **40**, 443–52.

Magurran, A. E. and Seghers, B. H. (1990b) Population differences in the schooling behaviour of newborn guppies, *Poecilia reticulata*. *Ethology*, **84**, 334–42.

Metcalfe, N. B. and Thorpe, J. E. (1990) Determinant of geographical variation in the age of seaward migrating salmon. *J. Anim. Ecol.*, **59**, 139–43.

Metcalfe, N. B., Huntingford, F. A. and Thorpe, J. E. (1988) Feeding intensity, growth rate and the establishment of life history patterns in juvenile Atlantic salmon. *J. Anim. Ecol.*, **57**, 463–74.

Metcalfe, N. B. and Thorpe, J. E. (1992) Anorexia and defended energy levels in overwintering juvenile salmon. *J. Anim. Ecol.*, **62**, 175–81.

Meyer, A. (1988) Influence of age and size on the response to novel prey of cichlid fish, *Cichlosoma managuense*. *Ethology*, **78**, 199–220.

Myers, R. A. (1984) Demographic consequences of precocious maturation of Atlantic salmon (*Salmo salar*). *Can. J. Fish. Aquat. Sci.*, **41**, 1349–53.

Myers, R. A. and Hutchings, J. A. (1987) Mating of anadromous Atlantic salmon, *Salmo salar*, with mature male parr. *J. Fish Biol.*, **31**, 143–6.

Noakes, D. L. G. (1978) Ontogeny of behaviour in fishes: a survey and suggestions, in *The development of Behaviour* (eds G. M. Burghardt and M. Bekoff), Garland, New York, pp. 103–25.

Noakes, D. L. G. and Barlow, G. W. (1973) Ontogeny of parent-contacting in young *Cichlasoma citrinellum*. *Behaviour*, **46**, 221–57.

Noakes, D. L. G. and Godin, J. G. J. (1988) Ontogeny of behaviour and concurrent development changes in sensory system in teleost fishes, in *Fish Physiology, Vol. XI. The Physiology of Developing Fish, Part B. Viviparity and posthatching juveniles* (eds W. S. Hoar and D. J. Randall), Academic Press, San Diego, pp. 345–96.

Olla, B. L. and Davis, M. W. (1989) The role of learning and stress in predator avoidance of hatchery-reared coho salmon (*Oncorhynchus kisutch*) juveniles. *Aquaculture*, **76**, 209–14.

Partridge, L. (1983) Genetics and behaviour, in *Behaviour*, Vol. 3 (eds T. R. Halliday and P. J. B. Slater), Blackwells, Oxford, pp. 11–57.

Paszowski, C. A. and Olla, B. L. (1985) Foraging behaviour of hatchery-produced coho salmon (*Oncorhynchus kisutch*) smolts on live prey. *Can. J. Fish. Aquat. Sci.*, **42**, 1915–21.

Rowe, D. K., Thorpe, J. E. and Shanks, A. M. (1991) Role of fat stores in the maturation of male Atlantic salmon (*Salmo salar*) parr. *Can. J. Fish. Aquat. Sci.*, **48**, 405–13.

Scholtz, A. T., White, R. J., Muzi, M. and Smith, T. (1985) Uptake of radio tri-iodothyronine in the brain of steelhead trout (*Salmo gairdneri*) during parr-smolt transformation. Implications for the mechanism of thyroid activation on olfactory imprints. *Aquaculture*, **45**, 199–214.

Seghers, B. H. (1974) Schooling behaviour in the guppy *Poecilia reticulata:* an evolutionary response to predation. *Evolution*, **28**, 486–9.

Thorpe, J. E. (1982) Migration in salmonids, with special reference to juvenile movements in freshwater, in *Proc. Salmon Trout Migratory Behav. Symp.* (eds E. L. Brannon and E. O. Salo), School of Fisheries, University of Washington, Seattle, pp. 86–97.

Thorpe, J. E. (1989) Developmental variation in salmonid populations. *J. Fish Biol.*, **35**, 295–303.

Thorpe, J. E., Morgan, R. I. G., Talbot, C. and Miles, M. S. (1983) Inheritance of developmental rates in Atlantic salmon, *Salmo salar* L. *Aquaculture*, **33**, 119–28.

Tooker, C. P. and Miller, R. J. (1980) The ontogeny of agonistic behaviour in the blue gourami (*Trichogaster trichogaster*). *Anim. Behav.*, **28**, 973–88.

Tulley, J. J. and Huntingford, F. A. (1987a) Age, experience and the development of adaptive variation in anti-predator responses in three-spined stickleback, *Gasterosteus aculeatus*. *Ethology*, **75**, 285–90.

Tulley, J. J. and Huntingford, F. A. (1987b) Parental care and the development of adaptive variation in anti-predator responses in sticklebacks. *Animal Behaviour*, **35**, 1570–2.

Verraes, W. (1977) Postembryonic ontogeny and functional anatomy of the ligamentum mandibulo-hyodeum and the ligamentum interoperculo-mandibular with notes on the opercular bones and some other cranial elements in *Salmo gairdneri*. *J. Morph.*, **151**, 111–20.

Wankowski, J. W. J. and Thorpe, J. E. (1979a) Spatial distribution and feeding in Atlantic salmon, *Salmo salar*, juveniles. *J. Fish Biol.*, **14**, 239–47.

Wankowski, J. W. J. and Thorpe, J. E. (1979b) The role of food particle size in the growth of Atlantic salmon. *J. Fish Biol.*, **14**, 351–70.

Wyman, R. L. and Ward, J. A. (1973) The development of behaviour in the cichlid fish *Etroplus maculatus*. *Z. Tierpsychol.*, **33**, 461–91.

Part two

Sensory Modalities

INTRODUCTION

The four chapters in this section of the book cover the major sensory modalities with which teleosts operate, describe how they extract significant cues from the fish's environment, and discuss the role played by information from each type of sense organ in the behaviour and survival of the fish.

Accurate and up-to-date information about the environment is crucial to an animal's success in taking optimal decisions about food, predators or mates. The early ethologists recognized the importance of such sensory inputs to behaviour and carried out extensive investigations to show precisely which stimuli triggered appropriate behaviours. We now know a lot more about how sense organs and sensory analysis work to detect signals from other fish and the fish's habitat. In teleost fish information from the environment about sound, light and water-borne molecules is transduced into nervous messages by various derivatives of the marvellously versatile neuromast cells, evolutionary modifications of cilia. To some extent, information is filtered and analysed in the sense organs, but it is passed to the CNS for more sophisticated signal processing. Behavioural decisions, moderated by motivational state, can then be rapidly made.

Chapter 4 reviews the role of vision, perhaps the major sense of teleost fish. Simon Guthrie and Bill Muntz begin with an analysis of how the characteristics of natural waters affect visual detection of objects both above and in the water. They move on to discuss how surface waves and the dappling pattern help to protect against surface predators, and how the trade-off between conspicuousness and camouflage is affected by spectral shift. A description of the visual hardware and its acuity and sensitivity follows, from the eye and lens, retinal rods and cones to the optic tectum of the mid-brain. The tectum is the site of most visual analysis in fishes, working in a closely analogous fashion to the optic lobe of the forebrain in mammals. Guthrie describes some of his own work on feature-extracting cells in the tectum. The link between ecology and pigments in retinal cones is briefly reviewed, along with work on colour, movement detection and recently discovered ultraviolet-sensitive cones. Visual discrimination abilities are dealt with next; the upper contours of objects turn out to be the most important for shape discrimination. The authors then discuss experiments on visually-mediated behaviours in shoaling, predation in relation to foraging theory, predator recognition, mating and territorial behaviours, including the remarkable 'poster colours' of coral reef fishes. Guthrie and Muntz conclude that greater understanding of the higher levels of image processing in the deeper layers of the tectum, along with elucidation of the precise roles of the elements of visual signals used by fish, are two important goals for the future.

Chapter 5 begins by noting Isaak Walton's advice to anglers to stay silent 'lest they be heard' by their quarry. In this chapter Tony Hawkins covers the role of sound in fish behaviour, a field which until recently suffered from neglect in relation to other sensory modalities. After setting out the basic physics of sound underwater, Hawkins distinguishes between water turbulence and sound, and between effects near and distant from the source. He describes the instruments required to measure sound pressure (volume), frequency (pitch), particle displacement and temporal structure. Sound can travel vast distances underwater, faster than in air, and many underwater habitats are inherently noisy. A salutary warning about the acoustic design of tanks for experiments in fish behaviour is given; it is possible that some fish can detect infrasonic frequencies as low as 1 Hz. Hawkins describes, with the aid of some beautiful drawings, the range of mechanisms by which teleosts make sounds. He then covers acoustic communication in alarm, mating, shoaling behaviour and species and individual recognition. Shoaling fish remain silent to avoid the attention of predators: damselfish and haddock mating calls may attract predators. Hearing ability in teleosts is well developed and, as in birds, temporal structure is better resolved than in humans, although the range of frequencies is lower. Cod, goldfish and salmon can tune an auditory filter to any frequency of interest to help distinguish a signal from background noise. The anatomy and neurophysiology of the sound-reception hardware is described, including details of the hair-cell transducers, the otolith organs, and acoustic amplifiers in the form of swimbladder and bone modifications. Hawkins closes with a discussion of the mechanism of directional hearing in fish, and a plea for more work in this field.

Toshiaki Hara reviews the role of olfaction in fish behaviour in Chapter 6, distinguishing at the outset between senses of taste and smell. Fish are remarkably sensitive to water-borne molecules, but solubility is more significant in aquatic habitats than the volatility important for air-borne olfaction. Hara describes the teleost olfactory system. The paired nasal pits contain folded lamellae bearing olfactory receptor cells on which amino acids bind at specific sites. Likewise derived from cilia, but with fewer associated structures than auditory, visual or lateral line systems, the olfactory neuromasts have direct connections with the forebrain. Hara follows an 'orderly sequence' of molecular, membrane and neural events: coded trains of spike-trains of nervous impulses convey olfactory information to the CNS. Amino acids are detected in tiny quantities as low as $0.1\ \mathrm{nmol\,l^{-1}}$. The chapter continues with a critical account of olfaction and the classic experiments on homing in salmonids. Hara argues that the 'homestream odour' and 'juvenile pheromone' hypotheses are not incompatable, and there are still some questions to be answered about reproductive homing and odour imprinting. Olfaction in feeding behaviour is described, including fish feeding stimulants and attractants. Anglers may note that 'fingerprint' mixtures of straight-chain amino acids are the most effective. In reproductive behaviour, steroid glucuronides released at ovulation act as pheromones keying courtship and mating which are potent at extremely low concentrations. The author discusses the significance of the distinction between odour and taste. Hara concludes with a detailed account of alarm substance in cyprinid fish, leaving us with an intriguing problem in evolution.

In Chapter 7, Horst Bleckmann describes a teleost sense of which humans do not have direct experience: the lateral line, which is sensitive to minute water movements and acts as a close-range object detector. Some lateral lines have become specialized as detectors of electric fields. Bleckmann opens by reviewing the historical controversy about the lateral line, which was variously thought to be for touch, pressure, hearing, balance, taste, mucus secretion and swimbladder gas production. The action of the neuromast transducer system is described next, including specialized versions found in several groups of fishes. Sensitivity is reduced by efferent nerves during the fish's own power strokes. The system responds to near-field water displacements produced by a sound source and to tiny water currents set up by the fish's own motion. Bleckmann considers the controversy over near- and far-field effects. The author's own observations of prey location by fish feeding at the 'deadly clinging trap' of the water surface are covered next: using their lateral lines, fish such as topminnows can accurately compute targets from the curvature of surface-wave fronts and other signal characteristics. The role of the lateral line in reproductive, agnostic and shoaling behaviour is discussed, and Bleckmann moves on to describe lateral line organs which have become specialized for electroreception in at least 5 groups of fishes including much new material published since the first edition of this book. He reviews passive electroreception and geomagnetic navigation, concluding with an overview of the fascinating subject of communication and object location in electric teleosts.

This section of the book does not include the newly-discovered single-cell sense of fishes (Kotrschall, 1991), minor senses of touch and temperature detection (Murray, 1971), or the still-controversial direct magnetic sense of fishes (Hanson and Westerberg, 1987; Walker 1984).

REFERENCES

Hanson, M. and Westerberg, H. (1987) Occurrence of magnetic material in teleosts. *Comp. Biochem. Physiol.*, **86A**, 169–72.

Murray, R. W. (1971) Temperature receptors, in *Fish Physiology, vol 5*, (eds W. S. Hoar and D. J. Randall), Academic Press, New York, Ch 5.

Kotrschal, K. (1991) Solitary chemosensory cells–teste, common chemical sense or what? *Rev. Fish Biol Fish.*, **1**, 1–22.

Walker, M. W. (1984) A candidate magnetic sense organ in yellowfin tuna. *Science*, **224**, 751–2.

Chapter four

Role of vision in fish behaviour

D. M. Guthrie and W. R. A. Muntz

4.1 INTRODUCTION

Despite the generally poor quality of underwater images, fish depend a great deal on vision as a source of sensory information. All but a few (mainly cave-dwelling species) have well-developed eyes, and in those species that inhabit clear-water environments, the variety of colour patterns and specific movements that they display invites comparison between them and the most visually orientated species among birds and mammals. Because of the physical nature of light and its complex interactions with the environment, a variety of different properties of visible objects can be recognized, such as brightness, hue, texture, and contour, as can more subtle differences of degree, such as patch size or pattern grain. Comparative properties such as colour contrast or brightness contrast can also be identified. The extent to which particular visual properties are important depends on (a) the type of visually mediated behaviour, and (b) the restrictions to visual signalling imposed by the aquatic medium.

Visually mediated behaviour ranges in complexity from the simple alerting or attentive state evoked by any novel but non-specific visual event, to the triggering of an elaborate fixed action pattern by means of a highly specific visual signal which consists of a precise grouping of visual properties. The structure of the visual signal enables it to remain effective even when a high concentration of suspended or dissolved matter in the water alters the perceived properties of hue and texture. At the same time, the more complex signals must require more specialized processing by the nervous system and longer reaction times. It is perhaps slightly ironic that most of the waters of very high optical purity inhabited by fishes are so lacking in nutrients that they support only a meagre fauna. Wastwater in Cumbria is perhaps an example of such a water in the UK.

Behaviour of Teleost Fishes 2nd edn. Edited by Tony J. Pitcher. Published in 1993 by Chapman & Hall. ISBN 0 412 42930 6 (HB) and 0 412 42940 3 (PB).

A number of monographs have been devoted exclusively to the vision of fishes, such as Ali (1975), Ali and Anctil (1976), and, appearing since the first edition of this book, Nicol (1989) and Douglas and Djamgoz (1990). These references may be consulted for detailed data on our present knowledge of the visual mechanisms of fishes. Earlier work on the behaviour of fishes and the role of vision is reviewed by Keenleyside (1979).

4.2 THE OPTICAL PROPERTIES OF NATURAL WATERS AND THEIR EFFECTS ON UNDERWATER VISIBILITY

Generalized habitats

In clear oceanic waters and dystrophic lakes there is very little suspended matter, and solar radiation can penetrate to considerable depths. The downwelling surface light in the middle of the day can, for example, be visible to a human observer at a depth of over 800 m, and should be visible to deep-sea fishes at depths of 1000 m (Clarke and Denton, 1962). Ultraviolet and long wavelengths are, however, rapidly attenuated by the water molecules, so that maximum transmission lies between 400 nm and 500 nm. In consequence these waters have a bluish appearance.

Most inshore and inland waters contain appreciable quantities of mineral and organic matter (especially chlorophyll), leading to the rapid attenuation of light with depth, and to the narrowing and shifting of the spectral composition of the light to a band between 500 nm and 600 nm, giving a greenish or yellowish cast to the water. The majority of coastal waters, lowland (oligotrophic and eutrophic) ponds, and rivers fall into this category. Less common are those waters, mainly fresh, carrying high concentrations of silt or plant breakdown products, which give them a brownish or reddish appearance. Here little light penetrates below 3 m, and this is mostly at wavelengths above 600 nm. The so-called 'black' (Muntz, 1973) or 'infra-red' (Levine *et al.*, 1980) waters are typified by some of the tropical fresh waters in South America such as the Rio Negro. These are heavily stained but transparent rivers, usually of high acidity, and are contrasted with uncoloured 'white' waters, which may form part of the same river system. Heavily peat-stained tarns in Northern Britain (e.g. Wise Een Tarn, Cumbria) provide a similar example. Opaque reddish waters are found where red clays form part of the watershed (Pahang and Kalang Rivers, Malaysia). Clearly, fish living in waters that approximate to these types operate under special visual conditions. The very wide variations that occur in the spectral characteristics of fresh waters are well illustrated by Bowling *et al.*'s (1986) study of Tasmanian lakes.

Lines of sight

As Levine *et al.* (1980) have pointed out, the amount and spectral quality of light entering a fish's eye will differ according to the line of sight involved (Fig. 4.1). The overhead view will involve the shortest optical path length and thus luminance attenuation and spectral shift will be least. Further, overhead objects that are above the water surface can be seen if they are within a solid angle of 97° (Snell's window). Horizontal lines of sight will be subject to lower

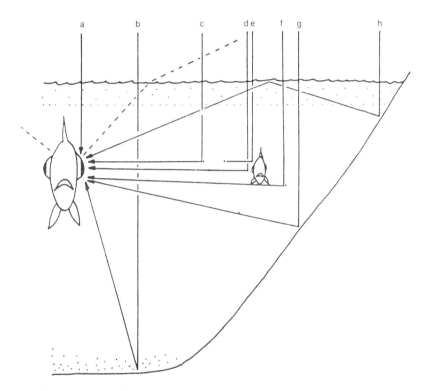

Fig. 4.1 Pathways of light entering a fish's eye. Arrows indicate from the left: a, direct overhead path (high intensity, daylight spectrum); b, long path scattered from bottom (low intensity, modified spectrum); c, light scattered from suspended particles (veiling light); d, light scattered from submerged object; e, path as in d, but light largely absorbed by suspended matter and therefore not reaching the observer; f, as c but longer light path (spectral shift greater, brightness less); g, as b but shorter path reduces the spectral shift produced by differential absorption and reflectance; h, as g and b but total reflectance from water surface has occurred, and spectral shift is greater owing to long path through turbid epilimnion. Stippled areas indicate epilimnion rich in organic particles at surface, and zone of humic particles and 'gelbstoff' near bottom. Dashed line indicates limits of total reflectance (Snell's window).

light levels and greater spectral shifts due to the longer optical path involved. Objects viewed in this direction will also have their apparent contrast reduced by veiling light scattered from particles between eye and object. Furthermore, as the object recedes from the eye, the effect of the veiling light becomes greater, the spectral reflectance of the object is increasingly shifted towards that of the dominant spacelight, and object brightness approximates to background. The downward line of sight from the eye involves the longest optical pathway, with the greatest effects on both the brightness and spectral properties of objects viewed. In shallow water, most of the visual background is provided by the bottom of the pond or river.

Transmission of spatial information

The detection and recognition of objects underwater depends on the rectilinear propagation of light between the object and the observer, and this is affected by the scattering properties of the water body. Very few measurements of image degradation by scattering appear to have been made for natural bodies of water, but it is clear that high spatial frequencies, that is the fine detail of the object being viewed, are particularly strongly degraded (e.g. Wells, 1969; Gazey, 1970).

Temporal and spatial changes in underwater visibility in natural waters

Stratification in optical properties is perhaps most noticeable in fresh waters where a seasonal cycle occurs. The most well known of these changes involves the warm-water layer that forms in lakes and ponds as sunlight hours increase in spring and early summer. The separation of a lighter surface layer (the epilimnion) owing to the action of the sun produces a boundary layer across which the temperature changes abruptly. In large lakes the zone of greatest change (thermocline) occurs between 5 m and 20 m, but in small, shallow ponds the transition zone may be less than 1 m from the surface. The change in temperature between surface waters and the hypolimnion is about 10 °C. In the four examples of Austrian lakes given by Ruttner (1952), the values are 12–14 °C. In eutrophic lakes the hypolimnion becomes depleted of oxygen, relative to the surface layers ($< 50\%$). These differences are due to a rapid growth of planktonic organisms in the surface layers, with a consequent increase in turbidity. As long as wind and wave action are limited, stratification is maintained.

Thermal stratification may also be associated with the segregation of particles that accumulate from other sources. In Green Lake, New York, significant concentrations of particulate calcite ($1.5 \, \mathrm{mg \, l^{-1}}$) are found near the surface (5 m) in June, which later disperse (Brunskill, 1970). At its peak there is six times as much calcite at 5 m as at 45 m.

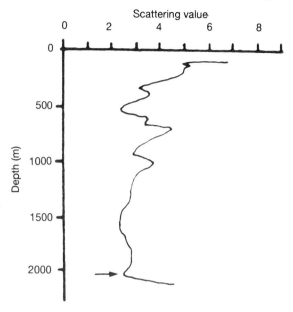

Fig. 4.2 Spatial variations in water turbidity affecting underwater visibility. Turbidity measurements of the deep ocean layers off Vancouver Island, Canada. The light-scattering profile was obtained by means of a nephelometer with an integral light source. Note strongly scattering layers at surface and near the bottom, and distinctive mid-depth layers. Light-scattering values are arbitrary units but 1 unit equals approximately $9\,\mu g\,l^{-1}$ solids. Reproduced with permission from Baker *et al.* (1974).

Although the thermal stratification in marine habitats is less precise than it is in fresh waters during settled midsummer weather, the formation of a layer of less dense water containing higher concentrations of particles is very general. Over the deep bottom formations of the Nittinat Fan off Vancouver Island, high turbidity layers (scattering index over 7 units, range 0–9 units) were found from the surface down to 100–200 m by Baker *et al.* (1974), but the light-scattering depth profile was complex (Fig. 4.2). At one site, for instance, there was an isolated band of light-scattering values at about 650 m. The horizontal light-scattering patterns are also complex because of water movements and underwater contours. At the edge of the Oregon coast, Zaneveld (1974) found that the surface zone of relatively turbid water turned downwards over the shore to a depth of 50–100 m. The clearest water was some way off shore. A turbid shoreline zone caused by water movements can also be observed in freshwater lakes because of wave action.

Striking anomalies of water density are sometimes observed by divers. Limbaugh and Rechnitzer (1951) found 'mirror pools' of cold, high-density water off the Californian coast capable of reflecting surrounding objects, as well as clearly visible moving tongues of cold water extending to a height of 4 ft from the bottom. They point out that these thermal discontinuities affect underwater visibility as a result of (a) differences in turbidity, (b) differences in refractive index, and (c) the formation of 'streamlines' by water movements at the refractive boundaries. 'Streamlines' can often be seen where clear river water empties into a bay.

In fresh water, floating vegetation both shadows-out phytoplankton and reduces the effect of wave action on soft bottom sediments, resulting in greater water clarity. The water column near the bottom of lakes or oceanic depths often contains an accumulation of particulate matter derived from the surface or stirred up from bottom sediments. Two-to-four-fold increases have been recorded by Kullenberg (1974) for North-West African waters, and for North Pacific waters by Baker *et al.* (1974). The concentration of humic substances in Esthwaite Water, UK, is generally highest near the bottom (Tipping and Woof, 1983).

Towards the end of the year in fresh waters much of the marginal and floating vegetation dies back and rots away, and phytoplankton breaks down. If a sample of water is taken in October and November (in the Northern Hemisphere) after the equinoctial gales have caused mixing to take place, a layer of bright yellow material will settle out of the sample. This probably consists of chlorophyll breakdown products and humic substances from the degradation of lignin and cellulose. The early German limnologists referred to this accurately as 'gelbstoff' (yellow substance). Since then the term has often been applied indiscriminately to phytoplankton and other suspended matter containing chlorophyll. Suspended humic matter is particularly associated with bog-lakes and moorland lakes with runoff from sphagnum meadows. Tipping and Woof (1983) describe the autumn (October/November) building up of humic substances in Esthwaite Water, a shallow eutrophic lake in the English Lake District, at depths of 5 m and 14 m. Concentrations are slightly higher at the bottom than near the surface at most times of year, other than the period between November and January. Changes in concentration at the bottom of the water column seem to lead those at the top. These substances appear to absorb strongly in the ultraviolet, and absorbance at 320 nm is maximal in the autumn. Ultraviolet receptors are discussed below. In bog-lakes, Heath and Franks (1982) found that UV-sensitive aggregations containing humic compounds, phosphorus and iron were actively broken down by sunlight, so that accumulations that built up at night were dispersed by noon.

One of the major sources of increased turbidity in waters receiving suspended matter from rivers is the effect of precipitation. In fresh waters, runoff from streams on clays and sedimentary rocks produces high concentrations of coarsely particulate matter with broad effects on visibility, whereas water from old igneous rock formations tends to contain small amounts of mineral particles but may be deeply stained by the decayed sphagnum (peat) characteristic of such areas.

An extreme example of the effects of runoff is provided by the changes in suspended solids for the water in Conowingo Bay near the exit of the Susquehanna River (Maryland) following Hurricane Agnes (Schubel, 1974). The normal level of solids is about $10 \, \text{mg} \, l^{-1}$ (clear oceanic water provides a range of $0.05–1.0 \, \text{mg} \, l^{-1}$ so these waters are usually fairly turbid). On 26 June 1972, the solids concentration was a maximum of $10 \, \text{g} \, l^{-1}$ and remained between 700 and $1400 \, \text{mg} \, l^{-1}$ for some days. Although this is an extreme example, rapid changes in turbidity due to precipitation affect fresh waters and estuaries very generally.

Turbidity changes also accompany freezing and thawing in many lakes and rivers. Stewart and Martin (1982) followed changes in light transmission in a narrow glacial lake with winter ice cover. Hemlock Lake, New York State, is frozen over from mid January to mid March. In June the clearest water in this lake (12 km long, mean depth 78.5 m) is near the bottom (over 50% transmission), but the autumn overturn results in the upper part of the water column becoming most transparent. The transmissivity pattern for December was one of slight turbidity throughout the lake, with high turbidity near the inflow stream as in June. Between February and March the upper part of the water column clears, and a more turbid layer collects near the bottom and near the inflow. Transmissivity ranged from less than 10% to over 50%.

Seasonal changes in the spectral quality of the light may be accompanied by changes in the fishes' visual systems, for example in their visual pigments, and some migrating fishes, such as the salmon, show appropriate changes in their visual pigments when they enter or leave fresh water (Beatty, 1975).

These examples indicate that in many aquatic habitats comparatively large changes in underwater visibility occur, and will significantly modify visual communication by fishes. Some of the ways in which the visual system of fishes takes account of these effects are considered in following sections.

Exposure of fish to surface predators

For many species of teleosts the water film provides a convenient trap for insect prey. At the same time, feeding at the surface is potentially hazardous, mainly because of the attentions of diving birds. Some protection from surface predators is provided by wave action. At reasonable wind speeds, waves result in the distortion, blurring and loss of apparent brightness of objects viewed directly from above. In sunshine the moving pattern of reflected bright sky and dark sky provides a confusing patterned surface, which may be coupled with a complex shadow pattern on the bottom. Convex parts of the water surface will focus light into bright patches, and concave regions will produce darker areas, resulting in the familiar dappled pattern observed from above (e.g. Loew and McFarland, 1990). In addition, the low levels of light reflected from the underwater scene make vision difficult for an overhead observer adapted to the brighter light that prevails above the water surface. Birds coming over still water will remain hidden from fish until they appear overhead and are seen through Snell's window (see Chapter 12 by Pitcher and Parrish, this volume), but where the surface is broken by wavelets, the image of the bird will appear intermittently at angles greater than 45° to the vertical. The thermoclinal surface layer of turbid water and the colouring produced by water runoff also protect fish from surface predators, yet most fish species approach the surface with caution and react with a startle response to the movement of any objects overhead (Eaton and Bombardieri, 1978).

Helfman (1981) has pointed out that many species of temperate lake fish aggregate beneath floating surface objects. He suggests that this allows them to escape detection by an aquatic observer in open water due to a high-increment threshold response for the observer who will be exposed to high ambient light levels. The poor illumination of the target must also be involved. At the same time the sheltering fish have good viewing conditions with a low-increment threshold and reduced effects of glare, while remaining protected from surface predation.

Conspicuousness v. camouflage

For most animals there is a balance of selective forces between signalling effectively to conspecifics, and having these signals intercepted by predators. Endler (1977, 1987) has illustrated this very clearly in his studies on wild populations of the guppy, *Poecilia reticulata*. For example, red or orange spots are much more abundant where the only predator is the freshwater crayfish, which is insensitive to longer wavelengths, and at high light intensities, or high levels of predation, the males court less frequently and when courting use visually conspicuous behavioural elements less often. Stickleback males with brighter scarlet patches are more successful in obtaining females, but are more heavily predated (McPhail, 1969). For many fish species the turbid conditions often found in natural waters offer the opportunity to signal to conspecifics at close range, whereas at a distance they become inconspicuous through loss of contrast and the shift in the spectral properties described on p. 90. Endler (1977) has also pointed out that patch-size contrast alters with distance. That is to say that the dimensions of the contrasting areas or patches presented by the surface of the fish and patterns on it may match those of the background at one distance, but as the distance between fish and background changes, the match is lost due to a difference in apparent patch size.

The nature of fish communities is also of importance. In the clear warm water of a coral reef, several hundred fish species may be present, supported by a large invertebrate fauna (see Chapter 14 by Helfman, this volume).

Visibility is relatively good and territories are often physically identifiable. We find diverse and striking patterns in many species. As Levine *et al.* (1980) point out, there may be trends towards a particular type of colour or pattern within a family. Some groups are relatively sombre in appearance. In the cold, turbid waters of a pond in the UK, only five or six species may be present, and most of these are relatively dull coloured, with few species possessing striking surface patterns. Distinguishing features among these species appear to the human observer to be the dorsal silhouette (when looking upwards), and ventral fin colours (when looking downwards). This signalling repertoire would perhaps be too limited for the competitive conditions of a coral-reef community.

The spectral transmission properties of the water will also affect the conspicuousness of different coloured stimuli, because if a stimulus is to be seen as coloured, its spectral reflectance curve must be changing in a part of the spectrum that concides with the available light. Lythgoe and Northmore (1973) have argued on this basis that in fresh water the most visible colour should be red, whereas in clear oceanic waters yellows, and to a lesser extent blues, will be the most readily visible colours. Surveys of the colours of fishes from different environments give some support to this suggestion. Thus Barlow (1974), discussing the distribution of cichlids in Central America, concluded that blues and greens characterized those species that inhabited the clearest waters, and Levine *et al.* (1980) found that coral fishes, and cichlids from clear waters, were coloured predominantly blue and yellow: both these findings agree with Lythgoe and Northmore's hypothesis. On the other hand Barlow (1974) found that in cichlids from turbid waters, yellow and orange, as well as red, were common, and Levine *et al.* (1980) that cichlids from 'black' waters were often coloured blue, green, or yellow, as well as red. Barlow (1976) also found that the gold morphs of the Midas cichlid, *Cichlasoma citrinellum*, which in spite of the name can in fact be white, yellow, gold, or red, were more common in turbid waters. In clear waters those morphs that occurred were usually yellow or orange, whereas in turbid waters the whole range from white to red occurred. Muntz (1990) has suggested that turbid waters are more 'permissive' than clear waters, allowing a wider range of colours rather than restricting effective colours to the reds alone. He also points out that the Lythgoe and Northmore hypothesis is based on a 'typical' fresh water transmitting maximally at about 570 nm, whereas fresh waters are in fact very variable, often transmitting maximally well above 600 nm.

Another camouflage feature found in many species that live in well-lighted environments or swim close to the water surface is countershading, where the back is dark (melanophores) and the belly lighter (guanin). Furthermore, in many species (like the herring), the scales reflect the prevailing spacelight, thus matching the fish to its background (Denton, 1970).

Finally, it should be remembered that some fishes can change their body coloration to match the background, or can choose their background to match their body colour: these aspects of camouflage are discussed by Muntz (1990).

Predatory fish species like pikes and muskellunge, *Esox* spp., and bass, *Micropterus* spp., appear reasonably well camouflaged to the human observer when seen at intermediate ranges, by virtue of low-contrast flank patterns of spots or patches, and rely for their success on high acceleration and the ability to strike from cover. One of the most interesting examples of camouflage is represented by the angler fishes, *Lophius*, and frog-fishes, *Histrio*, which have to submit to close-range scrutiny by other fishes. In extreme examples the skin texture resembles the surface of a sponge, the outline of the head is broken up by protuberances, and the pupil of the eye (a higher-contrast feature) is reduced to a very small aperture.

4.3 THE VISUAL SYSTEM

The eye

Paired image-forming eyes are found in most species. Benthic species living at depths down to 1000 m tend to have particularly large eyes and luminescent organs, while abyssal forms below this depth have small eyes. Fishes lacking eyes altogether are mostly confined to caves (see Chapter 17 by Parzefall, this volume). In addition the median pineal body is an effective light sensor in many forms.

The teleost eye conforms to the vertebrate type in that it consists of a subspherical chamber containing an inverted retina, and a focusable lens. In compensation for the fact that the cornea/water interface does not contribute significantly to bringing images of objects to a focus, the fish lens is a highly refractive sphere lying far forward in the eyeball. It follows from this that (a) accommodation is by the movement of the lens under the action of external muscles, rather than by changing the shape of the lens, as in mammals, and (b), because of the position of the lens, the iris cannot be 'stopped down' very far. In most teleosts, changes in pupillary diameter under natural conditions are very small and rather sluggish. Thus Charman and Tucker (1973) report that pupil size in the goldfish shows little change with light intensity, and Ali (1959) states that the iris of Pacific salmon is not capable of photomechanical changes under light conditions varied experimentally by a factor of × 4000. Several hours of light or dark adaptation have been found to produce no measurable change in the diameter of the pupil of the perch eye (D. M. Guthrie and J. R. Banks, unpublished observations; measurement accuracy ± 0.3%). However, under the action of drugs a reduction of pupil diameter of 16% occurred after 2 h (= 30% reduction of pupil area).

This lack of pupillary function, even in teleost species which are active under diverse light conditions, may be compensated for by the movement of the outer segments of retinal receptors relative to the pigment layer (retinomotor responses), but as in other vertebrates the major components of light and dark adaptation are achieved by neural and photochemical means, with pupil responses or photomechanical movements having a minor (though no doubt important) effect.

While most teleosts have essentially immobile pupils, there are a few specialized teleosts (eels, flatfish, stargazers) in which the iris is much more mobile, and mobile pupils also occur in selachians. In the eel the pupil diameter can be reduced by 45% by the combined effects of drugs and light (Young, 1950). Even so, these changes are much smaller than those of diurnal mammals and many reptiles (e.g. gekkos). In humans, for example, the pupil diameter is reduced by 75% under strong illumination, equivalent to a 94% reduction in area (de Groot and Gebhard, 1952).

In lower teleosts like the goldfish, pike and trout, accommodation is produced by a single muscle pulling the lens backwards and inwards, but in percoid fishes (bass, perch) a complex set of four or more muscles is present (Schwassmann and Meyer, 1973; Guthrie, 1981). Sivak (1973) measured accommodatory movements in a number of species and found that the transverse excursions were similar in most of them (equivalent to 10–12 dioptres) including the trout, but that rostro-caudal excursions were much greater in the percoid fishes (perch: 40 dioptres). Sivak and Howland (1973) were able to demonstrate by cinematography that quite large accommodatory movements of the lens could occur within 30 ms in the rock bass, *Ambloplites*, in response to natural objects.

Spherical aberration has been measured in the lenses of a number of species, most recently by Sivak and Kreutzer (1981) and by Sroczyński (1977, 1981 and earlier papers). Most lenses are quite well corrected, but there is considerable variation between species, with fish such as the rock bass and the pike having better correction than the goldfish or roach. The better-corrected fishes are more visual than the less-well-corrected fishes. Sivak and Bobier (1978) and Sroczyński (1977, 1981 and earlier papers) have also measured chromatic aberration in a number of species. This, unlike spherical aberration, is found to be substantial. Muntz (1973, 1976) showed that many reef species have a yellow cornea (trigger fish and puffer fish). Wrasse have yellow lenses in addition. Some freshwater species (e.g. cichlids) may possess yellow pigment in the retina as well as the features listed above. Since chromatic aberration is a greater problem at short rather than long wavelengths, such filters may, by filtering out short wavelengths, decrease its adverse effects on vision. Filtering out short wavelength light may also contribute to image sharpness in clear waters where, as in air, scattering is predominantly at short wavelengths. Whereas the benefit of these filters for reef fishes is easy to see, the reason for the very heavy pigmentation found in cichlids living in a long wavelength dominated environment is less clear. There is an extensive older literature on the refractive properties of the fish lens, and the reader should refer to Charman and Tucker's (1973) paper on the goldfish eye for an introduction to it.

Attempts have been made to estimate the visual acuity of the fish eye using minimum cone separation and lens diameter. Tamura (1957) applied this technique to 27 marine species (mostly belonging to percoid families) and obtained values ranging from 4.2' (minutes of arc) for the grouper *Epinephelus*, which is a reef-dwelling predator, to 15.4' in *Chlorophthalmus*, a bottom-dwelling benthic form. Schwassman (1974), using minimal stimuli and electrophysiological methods, obtained a value of 4' for *Lepomis* (a percoid sunfish) and 15' for the goldfish. Behavioural estimates for five teleost species, obtained using black and white gratings, agree well with these findings, varying between 5.5' for the skipjack tuna, *Katsuwonus pelamis* and 10.8' for the minnow, *Phoxinus laevis* (see Muntz, 1974, for a summary). The differences between species may be correlated with the different requirements for mid-depth pursuit predators and planktivores on the one hand, and partly herbivorous bottom feeders like the goldfish on the other.

It is interesting that no species has a visual acuity better than about 4'. Marine mammals have visual acuities of the same order as fishes (8.3' for the harbour seal, *Phoca vitulina*, and 7.1' for the Stellar sea lion, *Eumetopias jubata*: Schusterman and Balliet, 1970), and two species of Australian octopus have been shown to have a visual acuity of 5' (Muntz and Gwyther, 1988). It may be that in the aquatic environment, where high spatial frequencies are heavily attenuated (p. 92), visual acuity better than 4' to 5' serves no useful purpose.

A further specialization related to underwater vision is worth mentioning. Especially in rock pools and other shallow littoral habitats, the cornea of some species (gobies, blennies) may contain oblique layers that reflect away direct rays from the overhead sun while remaining transparent to reflected rays from objects in front of the eye (Lythgoe, 1974).

General structure of the retina

The generalized vertebrate pattern of three tiers of cells – receptors, bipolars and ganglion cells connected perpendicularly by horizontal cells and amacrine cells – is clearly seen in fishes. However, except for one or two species, a pit-like fovea is absent, and instead there are regional variations in the density of retinal cells. Often there is a high density 'area', or area centralis. In the perch, for example, there are about four to five times as many cones per unit area in the upper temporal part of the retina as in anterior regions (Ahlbert, 1968). Sometimes there are also prominent horizontal streaks or bands of specialization in the retina, which may be related to the fish's habitat and behaviour. Munk (1970), discussing the occurrence and significance of such streaks, concludes that they are associated with the detection of objects on the horizon. Collin and Pettigrew (1988a,b) studied the retinas of ten species of coral fish and, in agreement with Munk, concluded that species with marked horizontal streaks inhabited open water where they had an unobstructed view of the sand/water horizon, whereas species with well-developed areae but no streak lived in 'enclosed' spaces without a view of the horizon. Two of the species described by Munk, *Apocheilus lineatus* and *Epiplatys grahami*, had two horizontal streaks of specialization in each eye. Munk suggests the function of the lower streak may be the detection of objects moving into the edge of Snell's window. Some teleost species, such as the blueback herring, *Alosa aestivalis*, feed preferentially on zooplankton just outside Snell's window, where they look bright against the reflection of the dark water beneath them (Janssen, 1981).

The fish retina shows a degree of complexity at least as high as those of higher vertebrates. Three or more cone types and rods, two types of bipolar cell, three to five kinds of horizontal cell, and six or more types of ganglion cell have been observed in different teleost species (Cajal, 1893; Wagner, 1975, 1990). For detailed comparisons between species the reader should refer to Ali and Anctil (1976).

Receptor adaptations

In the typical duplex fish retina, the arrangement of the cones may be irregular, or, as in many of the advanced predatory species, it may form a regular mosaic with rosettes of double cones aligned in rows. The larger cones often form opposed pairs, and according to whether they contain the same pigment or a different one are termed twins or doubles, respectively. The single cones tend either to lie at the centre of a cone rosette (long central singles), or irregularly at the edge of a cone group (short additional or accessory singles). The rods are visible in a tangential section of the retina of a fish like the perch as 10 to 20 small profiles accompanying each cone pair. In some deep-ocean species, clusters of receptors are arranged in groups separated by screening pigment.

The absorption curves of receptor pigments are clearly of considerable interest in relation to colour vision and the spectral quality of the available light. Following the pioneering work of Liebman and Entine (1964) and Marks (1965) in developing the technique of microspectrophotometry, which allows the characterization of the visual pigments of individual receptors, the rod and cone pigments of a very large number of species have been studied. Summaries of the extensive data now available on this topic may be found in Loew and Lythgoe (1978), Levine and MacNichol (1979), Lythgoe and Partridge (1989), and Bowmaker (1990). Since fishes live in a wide range of visual environments, varying particularly in spectral quality and turbidity, many of these studies have attempted to relate the visual pigments of different species to the visual tasks they face in their particular environments. The visual pigments are found to vary between individuals, according to cone type, and according to the part of the retina from which they are sampled; and different fishes occupying the same habitat may also have different pigments. As Levine and MacNichol (1979) point out, given these factors and the complex interactions between the behaviour of animals and the characteristics of their environments, simplistic predictions and most general statements are risky. Nevertheless, some general rules, even if they prove to have many exceptions, appear to be emerging. These are briefly summarized below: references to the original papers may be found in the reviews listed above.

1. Deep-sea species have rod pigments absorbing maximally at relatively short wavelengths (roughly 470 nm to 490 nm) compared with shallow-water and freshwater species. This may be correlated with the predominantly short wavelength light of their environment. In most cases all the receptors have the same visual pigment, but in a few species two pigments are present, segregated into separate rods. These paired pigment species are usually dark in colour as opposed to silvery, and often have specialized bioluminescent organs emitting light of relatively long wavelength, and it has been suggested that the presence of two pigments is related to these factors.

2. Marine fishes living at intermediate depths in coastal waters usually have two cone pigments (apart from the rod pigment) absorbing maximally at around 460 nm and 540 nm respectively. Freshwater fishes living near or on the bottom, particularly if they are crepuscular or nocturnal, also often have two cone pigments, absorbing maximally at around 530 nm and 620 nm. In both cases the long wavelength cones match the spectral quality of the light fairly well. It has been suggested that in this group of fishes the cones sensitive to long wavelengths subserve brightness and sensitivity, and the short wavelength cones add their contribution to subserve colour vision.

3. Fishes living in shallow fresh waters have three cone pigments, two of them similar to those of the deeper-living forms, but with a blue-sensitive cone, absorbing maximally at around 430 nm, in addition. The presence of three as opposed to two cone types, covering a wider range of the spectrum, presumably correlates with the wider spectrum of light available to them: in deeper fresh water no short wavelength light is present and a blue cone serves no purpose. Fishes in this group sometimes have an ultraviolet-sensitive cone as well (p. 103).

4. Fishes inhabiting very shallow fresh waters or tide pools also usually have three cone pigments, but compared with group (3) the 'green' and 'red' cone pigments are shifted towards short wavelengths and are comparable in their spectral positions to the cones of terrestrial animals. The light

environment in very shallow water is of course similar to that of terrestrial animals, since the filtering effect of the water will be small.

5. A number of species have cones sensitive in the ultraviolet. Microspectro-photometers are usually relatively insensitive in the ultraviolet, and the exact forms of the pigments' absorbance curves are consequently uncertain, but the absorption maximum probably lies around 355–360 nm (Harosi and Hashimoto, 1983). Ultraviolet light is heavily absorbed by water and so only present at any intensity in the surface layers, and in some species ultraviolet sensitivity correlates with living near the surface. In the brown trout, *Salmo trutta*, for example, the ultraviolet-sensitive pigment is located in a type of miniature cone that disappears as the animals get older, and older fish live deeper in the water column than juveniles (Bowmaker and Kunz, 1987), and in the pollack, *Pollachius pollachius*, young animals have cones maximally sensitive at 420 nm which change with age to become maximally sensitive at 460 nm: here again the older fish live deeper than juveniles (Shand *et al.*, 1988). It is not known what use fish make of their ultraviolet sensitivity.

The central projection of the visual pathway

The optic nerve contains at least four different kinds of fibre (including efferents) as detected by conduction velocity measurements (Vanegas *et al.*, 1971; Schmidt, 1979). Most of these fibres pass across the midline to the opposite side of the brain, and terminate in the upper layers of the optic tectum. A few fibres are now known to pass ipsilaterally, and some of those that project contralaterally end in anterior nuclei rather than in the tectum. There are some six nuclei lying in the anterior subtectal area, and four of these receive direct retinal input. Pathways from the tectum pass to some of these nuclei, to a median nucleus involved with eye movements (the nucleus isthmi), and to subtectal areas like the torus semicircularis (Fig. 4.3). Several transverse commissures link the two optic tecta, and are believed to be involved in binocular function.

In the goldfish each optic tectum contains about 2 million intrinsic cells (Meek and Schellart, 1978), and these are deployed as 15 or so recognizable and distinct morphological types arranged in successive layers and coincident with the fibrous layers (see Meek and Schellart for details of goldfish and other species). Other species examined possess similar cell types so that the tectum can be regarded as a very conservative structure. The commonest neurone types have their cell bodies within a close-packed layer, the periventricular layer (PVC). Broadly speaking, a point-to-point 'through' system can be seen as separate from a system running parallel to the fibre layers and inter-connecting the through pathways, as in the retina. Degeneration and retrograde-staining experiments indicate that the outer fibre layers contain

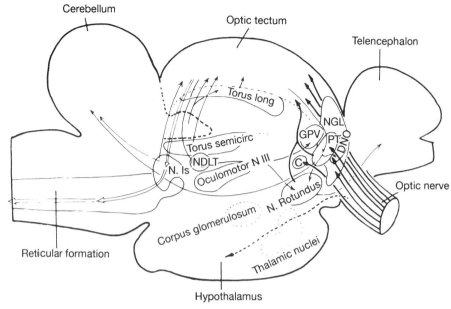

Fig. 4.3   The central projection of the visual pathway in teleosts, based on *Perca* and *Eugerres*. Most retinal afferents project to the contralateral tectum, where information from the retina acquires more complex spatial properties and becomes associated with cerebellar and other brain inputs. Some of the subtectal nuclei are believed to be involved in a number of functions to do with image stabilization. The torus semicir-cularis is a sensory-associated area for visual and auditory mixing. Abbreviations: C, nucleus corticalis; DMO, dorsomedial optic nucleus of thalamus; GPV, geniculatus posterior pars ventralis; NDLT, nucleus dorsolateralis; NGL, geniculatus lateralis ipsum; N. Is, nucleus isthmi; PT, pretectal nuclei; Torus long, torus longitudinalis; Torus semicirc, torus semicircularis.

most of the retinal afferents, and projections from other brain regions enter either in deep layers, or in the most superficial layers (stratum marginale). A recent description of tectal morphology may be found in Meek (1990).

Neural physiology of fish vision

Chromatic processing

Some of the most important early studies on vertebrate retinal physiology were made on the large receptor cells and horizontal cells of fishes. Svaetichin (1953) demonstrated opponent processes recording from what were probably horizontal cells, and Tomita *et al.* (1967) provided evidence of three separate wavelength-sensitive channels from cones. More recent work, summarized for example in Djamgoz and Yamada (1990), has recorded from all the different elements of the retina, confirming the original appearance of chromatic opponent processing at the level of the horizontal cells, and following through the changes in neural response in the different layers until the final output of the retinal ganglion cells. From the behavioural point of view this last level has probably yielded the most significant generalizations (Daw, 1968; Spekreijse *et al.*, 1972). Daw found that in the goldfish 68% of cells are double opponent, that is to say their receptive fields have a roughly circular centre and surround structure, with chromatic opponency at the centre (often red 'on', green 'off'), but also opponency between the centre and the surrounding annulus. Three main types of colour opponent cell were observed by Daw, but Spekreijse *et al.* identified 12 different combinations of red-, blue- and green-sensitive channels, also in the goldfish. Long wavelength 'on' centres were the commonest, and medium- and short-wave-sensitive channels were always complementary. Short-wave-sensitive (blue) centres were also observed. In the tectum of the perch, *Perca fluviatilis*, Guthrie (1981, 1990) found that only 18% of cells had double opponent chromatic properties. The majority were red 'on' centre cells, but blue 'on' centre cells were also seen. A few cells responded best either to wavelengths below 400 nm (380 nm), or to light in the spectral range 650–700 nm. Specific responses to near ultraviolet were occasionally seen, and are interesting in the light of more recent observations on UV-sensitive cones (p. 103). The response to very long wavelengths corresponds quite well to the characteristic curve of perch ventral fins (Guthrie, 1983).

The significance of double opponency is twofold. First, it preserves narrow spectral tuning under different conditions of brightness, and is therefore crucial in hue resolution; and secondly it is effective in mediating responses to colour contrast borders moved across the cell's receptive field. It may also be important in colour constancy under different spacelights: some fish have been shown to have colour constancy (Burkamp, 1923; Ingle, 1985), and it is probably widespread.

Non-chromatic processing

For predation and for obstacle avoidance, the accurate registration of objects in space is clearly necessary, and we find that object position is preserved in local tectal unit responses (the retinotopic map). It is preserved most accurately by the small simple receptive fields of retinal afferents (2.5–12.5° for goldfish: Wartzok and Marks, 1973; Sajovic and Levinthal, 1982). Intrinsic tectal cells tend to have larger receptive fields (30–160°: Sutterlin and Prosser, 1970; O'Benar, 1976; Guthrie and Banks, 1978) and their field structure is irregular, or multicentre (Schellart and Spekreijse, 1976). Even regular annulate receptive fields may have some shape-selective properties (Guthrie, 1981, 1990), but preferences for orientated edges have not been generally observed.

Some degree of directional preference was observed among movement-sensitive afferents in the goldfish by Cronly-Dillon (1964) and Wartzok and Marks (1973). In the upper nasal eye field there was a preference for anteriorward movement, and Zenkin and Pigarev (1969) noted a similar preference in movement detectors in the pike, *Esox*. This coincides with the stimulus most likely to be involved in predation, which also demands accurate coding of object velocity. Information about the velocity of an object can be used in two ways: to drive pursuit movements, or to provide data for object identification. In the first case, direct coding of velocity as impulse frequency may be adequate, but for recognition purposes, tuned units would seem to be required. There is some evidence for tuned units from work on goldfish afferents with reasonably sharp peaks within the range $5-20° \, s^{-1}$ (Wartzok and Marks, 1973). Pike afferents seemed to fall into three response categories corresponding to object velocities of $0.5-5° \, s^{-1}$, $3-40° \, s^{-1}$ and $50° \, s^{-1}$ (Zenkin and Pigarev, 1969).

Certain tectal cells respond, but habituate rapidly, to small movements almost anywhere in the eye field (trout: Galand and Liége, 1975; goldfish: Ormond, 1974; O'Benar, 1976; perch: Guthrie and Banks, 1978). These 'novelty' units, which in the perch lie in the deep tectal layers, are probably associated with the attentive phase of visual behaviour.

Localized electrical stimulation of the tectum causes eye movement and body movements directed towards appropriate parts of visual space. This suggests that heightened local activity in the tectum is sufficient to drive a specific motor programme (Meyer *et al.*, 1970; El-Akell, 1982).

We know very little about the function of the subtectal areas. Although it is now clearly demonstrated that mixed modality units, responding to both visual and auditory stimuli, are found in the torus semicircularis, their significance is unclear (Page and Sutterlin, 1970; Schellart, 1983). In some species (e.g. cichlids), courtship movements are reinforced by specific sounds (Nelissen, 1978).

The pineal body

This structure is a light-sensitive organ that is often well developed in fishes. In most forms it is a sac-like median lobe of the diencephalon projecting obliquely forwards towards the cranial roof. Cells within the pineal secrete the hormone melatonin, an amino acid derivative that causes darkening of the skin by expansion of the melanophores, most usually during the hours of darkness. Melatonin release from the pineal, under the influence of light, is now known to control circadean rhythms in many animals, including fishes, and in some species the pineal is also affected by other factors apart from light, most notably temperature (Falcón and Collin, 1989; Underwood, 1989). Pineal fibres are also believed to project to the hypothalamic–pituitary region and have more general neuroendocrine effects, reducing activity and affecting territorial aggression (Rasa, 1969).

In trout and tuna, the head of the pineal is large and lies below a transparent cartilaginous window in the frontal bone (Rivers, 1953; Morita, 1966), but in other species (most percoid species, for instance) there is no obvious frontal window and the pineal lobe is small and delicate. The electrical activity of the pineal cells, receptors and ganglion cells, has been successfully recorded by a variety of methods (Dodt, 1963; Morita, 1966; Hanyu, 1978). There is general agreement that photic stimuli inhibit ongoing spontaneous activity, and that the degree of inhibition is related to light intensity. Conversely, dimming produces an increase in discharge frequency. In some species (e.g. of *Plecoglossus, Pterophyllum*) there is a phasic response followed by a steady tonic discharge which is related to the maintained light level, although Hanyu found that the quantitative relationship varied for different cells in *Plecoglossus*. The spectral sensitivity of the pineal cells peaks at about 500 nm in the trout (Morita, 1966), and this and the slow response are rod-like characteristics. Hanyu sees the pineal as a dimming or dusk detector in its tonic mode, responding to maintained light levels, and as a detector of shadows by means of the dynamic or phasic response.

4.4 STUDIES ON VISUAL BEHAVIOUR

Training methods for determining aspects of visual discrimination

Spatial proportions

A considerable body of work, mainly provided by the German school of Gestalt behaviourists, has come into existence over the last 50 years (Northmore *et al.*, 1978; Guthrie, 1981). Most of these studies involve training the fish to respond to one of a pair of stimuli, such as geometrical figures, for food reward (operant conditioning). The trained subject (i.e. one which has reached a certain score of correct responses) is then tested with another pair of figures which (or one of which) are slightly altered versions of the two shapes used initially. The ease of transfer, or degree of maintenance of correct choices, measures the extent to which the fish recognizes the alterations that have been made. The problem with many of these transfer experiments is that the precise level of recognition is difficult to assess since two elements of recognition are involved simultaneously in the transfer. If, for example, the transfer is from stimulus pair A/B to stimulus pair A′/B, both A/B differences and A′/A differences are involved. To some extent the ability to generalize opposes the ability to notice that the A′ is different from A. Both processes may be valuable in the wild.

The results suggest that teleosts can discriminate between a variety of simple solid geometrical forms. This can be disrupted by altering the direction of figure/ground contrast in some species like the perch, although such contrast reversal did not affect transfer in the trout and the minnow. Several studies indicate that the upper contour of a figure is more important than the lower in shape recognition. Points, acute angles, or knobs seem to make for easier recognition than small differences in the form of curves or obtuse angles. Interestingly, in relation to spatial frequency, Schulte's carp (Schulte, 1957) found great difficulty in discriminating between four stripes and seven stripes occupying the same area. Most of these tests do not involve fine detail and depend heavily on one group of cyprinoid fish. Herter's (1929) work with the perch, *Perca fluviatilis*, suggests that this species can respond to relatively small differences in test figures (Guthrie, 1981).

Spectral sensitivity and hue discrimination

Spectral sensitivity curves based on the lowest intensities to which teleosts respond at different wavelengths have now been obtained in a number of species, using operant conditioning, classical conditioning, or innate behaviour patterns that are under visual control. Reviews are available in Northmore *et al.* (1978) and Douglas and Hawryshyn (1990). In general the resulting curves show two or more relatively well-defined peaks, indicating the activity of different receptor classes or wavelength dependent channels. Trimodal curves have been obtained even in those species like the shanny, *Blennius pholis*, and the perch that are believed to possess only two classes of cone. In his study on the perch, Cameron (1982) suggests that a 'differencing' channel, comparing outputs from red- and green-sensitive cones, is responsible for the relatively high sensitivity found at 400 nm (third peak). In most cases it is necessary to assume some form of 'differencing' or inhibition between the different spectral channels, reminiscent of the chromatic opponency found at the level of the retina. It is also the case that in many species the form of the spectral sensitivity curve obtained depends on the type of behaviour that was used to obtain it, and hence is presumably related to the visual requirements of that particular task. Overall, the behavioural spectral sensitivity curves that have been obtained appear to reflect considerable neural interaction, and not simply the sensitivity of the animals' receptors.

The question of whether fishes have colour vision, that is whether they can distinguish similar objects of the same brightness differing only in hue, was largely resolved by the work of von Frisch in 1925, working with minnows, and Hurst (1954) using the bluegill, *Lepomis*. Later studies showed that all teleost species tested were capable of making discriminative choices on the basis of hue alone. It has to be emphasized that most behavioural tests have been conducted using visually active freshwater species now known to possess two or more cone types: deep-water rod-dominated species may well be colour blind.

The minimum spectral interval that can be discriminated on the basis of behavioural tests has been determined for the goldfish, and lies between 10 and 20 nm at their point of greatest sensitivity (about 600 nm). This can be compared with intervals of 1–2 nm for Man. Of especial interest is the evidence provided by Burkamp (1923) and Ingle (1985) for colour constancy, for ambient spectral conditions may vary regionally as described on p. 90. Colour constancy is an important property of the visual system which allows hue recognition to remain effective despite widely varying spectral conditions.

Sensitivity to ultraviolet (UV) and polarized light

A number of species have cones sensitive in the UV (p. 103). Behavioural spectral sensitivity curves in many experiments have been found to show exceptionally high sensitivity at short wavelengths. It was originally considered that this was due to some artefact, such as fluorescence, scatter of light on to unadapted areas of the retinas, or the stimulation of the cis-peaks of other receptors, but it is now clear that it is due to the UV cones (Douglas and Hawryshyn, 1990).

Many fishes are also able to respond to polarized light (Waterman, 1975; Loew and McFarland, 1990), and there appears to be a connection between this and the different cone classes, particularly with the UV receptors. Thus Hawryshyn and McFarland (1987), using classical conditioning techniques, showed that the UV, green, and red receptors were all sensitive to polarization, the first being most sensitive to vertical e-vector axis, and the others to horizontal e-vector axis. The blue cones were insensitive to the plane of polarization. Hawryshyn *et al.* (1990) trained rainbow trout, *Oncorhynchus mykiss*, to find a refuge according to the vector of polarized light. Their ability to do this disappeared as the fish grew older, with a time course similar to the loss of the UV cones (p. 103).

Field-based studies

Aggregation responses

A large proportion of fish species form conspecific shoals, pairs or territorial groups (see Chapter 12 by Pitcher and Parrish, this volume), and there is general agreement (Breder, 1951; Hemmings, 1966) that visual signals are important in shoal formation and group cohesion. The elements in the visual appearance of a species that determine aggregation behaviour can be discovered by experimentation using models, or less certainly by observation of the responses of the subject species when confronted by a variety of different species differing in certain details of appearance from its own. The starting point has inevitably to be the appearance of the fish as seen by the human observer, and to this extent it is impossible to avoid a subjective element in many of these studies. Nevertheless, a number of studies support the idea that specific visual features operate to trigger behaviour of this kind and broadly conform to Tinbergen's ideas on releasers and sign stimuli (Tinbergen, 1951).

Keenleyside (1955) found that if the black-and-white tip of the dorsal fin was removed in *Pristella riddlei*, a small characid, the deprived fish was no longer accepted into the shoal. It should be pointed out that several other allied species also have a dorsal fin spot, and that it tends to be flicked in a manner that makes it conspicuous.

Moller *et al.* (1982) suggested that the vertical bars of some weakly electric species (*Gnathonemus petersi, Brienomyrus niger*) were used as a shoal aggregation device. These fish are relatively poor sighted, but Tayssedre and Moller (1983) found that they would exhibit a following response to a striped drum. Patfield (1983), using live fish in adjacent tanks, demonstrated the ability of perch to aggregate with conspecifics, rather than with other local common species (*Rutilus* sp., *Leuciscus* sp.), and studied the effect of the vertical dark bars, dorsal fin spot and red ventral fins on the aggregation responses. The dorsal fin spot appeared to provide an important stimulus for aggregation. In seven out of eight trial series, the 'buttout' aggregation response was much stronger towards normal perch than towards perch from which the spot had been snipped out. Roach with a spot added did not have a significant effect. Results from tests involving ventral fins were not significant. However, weak aggregation responses resulted from exposing perch to roach painted intra-dermally with dark bars.

Katzir (1981) found that the black-and-white striped humbug damselfish, *Dascyllus aruanus*, would aggregate with black-and-white still photographs of conspecifics, rather than with pictures of allied species (*D. marginatus* and *D. trimaculatus*). Experiments with photomontages in which stripes were deleted showed that the long second stripe was the most powerful feature in inducing aggregation.

There is a good deal of anecdotal evidence that species which form stable pairs during the breeding season are able to recognize their mates by visual means. The cichlid *Hemichromis bimaculatus* was shown by Noble and Curtis (1939) to form monogamous pairs capable of individual recognition, and Barlow (1982) was able to demonstrate this in the natural habitat by identifying face patterns of a reef species (*Oxymonacanthus* sp.). Fricke (1973) used a two- or four-choice experiment to demonstrate individual recognition in the laboratory with *Amphiprion bicinctus*. Attack rates of one partner against the other partner, or foreign individuals, in adjacent tanks, were monitored. For the two-choice experiment, discrimination was at the level of 390:4, and where two other foreign individuals were visible in separate compartments (four choices) it was at the level of 615:0. If the partner was dyed with bromocresol green or enclosed in a green plastic jacket, it was unrecognized, and in the event of a new partner being provided, effective recognition was acquired within 24 h.

The responses to visual signals of adult fishes can be regarded as the function of mature nervous and endocrine systems. The responses of fry have to be viewed a little differently. Work in this field has the interest that it may reveal innate elements of recognition-processing dependent on the function of the inexperienced nervous system. Furthermore, the retina is enlarged as the fish grows by marginal accretions of new cells, and the image formed on the retina at this stage is presumably less detailed than that formed by the adult eye.

Baerends and Baerends-van Roon (1950), working with mouthbrooding cichlids (*Tilapia*), found that a variety of simple rules described fry behaviour. They preferred dark objects to light, the lower side of objects, and concave surfaces. No detailed form discrimination was needed for them to locate the mother's mouth. Kuenzer (1968) came to roughly the same conclusion working with the fry of *Nannacara anomala*. These would aggregate with a long oval model of a particular shade of grey more readily than they would with other simple shapes, but a very simplified representation of the mother with its characteristic flank patches was more effective. Precisely detailed models produced no increase in response. When exposed to a variety of geometrical figures, a shape preference diagram could be constructed on the basis of two choice tests. The rules governing some of these choices were obscure, but Kuenzer (1975) expanded his testing scheme with an exhaustive analysis of the visual preferences of the fry of *Hemihaplochromis multicolor*. These tests involved discrimination of colour, grey shades, size, form, dimensionality, contrast with background, spot and annulus features. Responses were categorized as orientation, approach/contact or flight. Curiously perhaps, in view of the bright colours of the parents, grey shades (pale) were preferred to any colour when presented as discs, but blue or green (not red) spot features were preferred over grey. The largest test disc sizes offered were preferred, but preferences for any particular geometrical figure were weak. Face-like configurations were not preferred to the standard stimulus, a disc with a central spot. A noticeable feature of most responses was a decided choice of what might be termed constant contrast. That is to say, if the major feature was made darker, a darker ground was chosen by the fry. Where figure/ground contrast was maintained over a range, medium to high values were chosen. Kuenzer then introduced variations in background brightness, and again values were chosen that suggested that the fish preferred a constant level of background/figure plus ground contrast. There seemed also to be a preferred proportionality between the area of the spot and of the surrounding annulus, in relation to spot contrast. As might be expected, paler spots resulted in larger preferred spot sizes, and darker spots resulted in smaller preferred spot sizes. Sometimes flight reactions were produced by the largest targets (20 mm).

The results of these tests suggest that, as with the experiments of Baerends and Baerends-van Roon (1950), very simple models can trigger the aggregation responses, but that some preferences for levels of contrast, colour, etc. do exist, which probably correspond to those most likely to be encountered under natural conditions. In contradiction to these results are those of Hay (1978) using convict cichlid, *Cichlasoma nigrofasciatum*, fry. He employed both simplified models and precise replicas of fish species. The aggregation preferences evinced by the fry clearly demonstrated an ability to respond positively to both generalized and detailed features possessed by the parent models when compared with models of 'foreign' species. Discrimination was best developed in the very young fry (1–3 days), but as the tendency to flee from larger fish developed, positive responses to detailed models of the parents disappeared. Though these reactions are probably mainly innate, there was a specific enhancement of response with exposure to the appropriate model, suggesting some kind of rather weak imprinting mechanism. Indeed, Hay uses the term 'filial imprinting' to describe it.

Agonistic behaviour and territorial responses

Aggregation responses can be used for discrimination experiments, but much more powerful agents are available to the experimental behaviourists in the territorial behaviour of fishes. In many fishes a nest site provides a localized resource, and intruding male conspecifics are driven off as part of the programme of breeding behaviour. One of the most important studies in this field is that of Stacey and Chiszar (1978) on the breeding male pumpkinseed sunfish, *Lepomis gibbosus*, a centrarchid. They used a variety of cut-out models representing intruding males, with key features deleted. Their general conclusions were that all the seven key features studied had an effect in determining aggression. The most powerful features are the red iris of the eye, and the red patch on the earflap. The blue-speckled body pattern and the yellow chest patches are subsidiary in their ability to promote attacks. Conversely, the black eye and tab features slightly reduce aggression, and the vertical bars displayed by the female strongly reduce aggression. A black dorsal fin spot, characteristic of an allied species, the bluegill, *L. macrochirus*, greatly increases aggression.

Kohda and Watanabe (1982a,b) have analysed the visual signals underlying the agonistic behaviour of a freshwater serranid, *Coreoperca kawamebari*. The most important elements of the body coloration were the 'eyespots' on the posterior edge of the operculae, a dark vertical band through the eye, and vertical stripes on the flanks. During fighting these became more accentuated, and at the end of an encounter were retained by the winner, while the loser fell back to a much less marked pattern. Similarly, if a number of fish were introduced into an aquarium, there was initially considerable fighting. Eventually, however, a dominance order was established, and the fish then retained the colour patterns corresponding to their ranks, even if subsequently placed in isolation. In another example Watanabe *et al.* (1984) report that in the cichlid *Cichlasoma coryphaenoides* the iris darkens instantly when a fish is attacked or chased. This did not happen if the fish was the aggressor. In this case changes in body (as opposed to eye) pattern, while they could occur, played little part over the time course of such encounters.

The importance of the orientation of features has seldom been analysed in detail, but Heiligenberg *et al.* (1972) did examine this, using models to determine the most effective orientation of the black eyebar in male *Haplochromis burtoni*. In territorial males of this species a bar extends downwards from the eye at an angle of 45° to the snout profile, and this triggers attacks in other males. The problem of orientation is complicated by the head-down posture adopted by the fish at close range (the conflict position). Their findings were as follows.

1. Bars at 180° or 0° were more effective than those at 45°
2. The shape of the attack rate/bar angle curve was similar for vertical and horizontal positions of the fish.
3. Vertical positions of the dummy were more effective than horizontal for four out of five bar angles. They noted that even a dummy without an eyebar would provoke attacks if held vertically. Eyebar and body angle behave largely as independent additive features. The finding that there is a more effective signal than the natural one has been noted in herring gulls and fireflies, and must be the result of other constraints. It is clear that within rather wide limits, the eyebar angle is important and attack rate doubles across the bar-angle range. A smaller increment is due to body position. The results indicate that eyebar angle is assessed as a feature related to the geometry of the fish rather than to horizontal and vertical planes. When the male has established a territory, orange shoulder patches appear, and these inhibit attack readiness in other males. Heiligenberg *et al.* (1972) developed a theoretical scheme to explain these interactions (Fig. 4.4). It was suggested earlier (p. 112) that Hay's results with fry support the idea of an acquired element in the recognition process. Fernald (1980) came to the conclusion, from studying *Haplochromis burtoni* reared in isolation, that eyebar recognition was an innate process, but, as discussed by Huntingford (Chapter 3, this volume), Crapon de Caprona (1982), using similar material, found that (at least for the males) early experience was required for the correct interpretation of the complex colour patterns.

Other aspects of vision and visual behaviour in this very interesting fish have also been studied in detail, and are summarized by Fernald (1990).

What might be termed the generalized territorial behaviour of reef fishes has been the subject of enthusiastic study by several authors. The brilliance

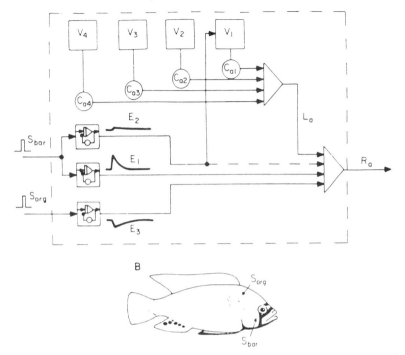

Fig. 4.4 Diagrammatic representation of the quantitative relationships involved in the territorial defence of male *Haplochromis burtoni*. The black eyebars tend to produce a fast (E1) and a slow (E2) effect on attack response, and the orange shoulder patches inhibit (E3) responses. The inputs are seen as acting through feedback-stabilized amplifiers (boxes). The profiles indicate the kind of time-dependent process involved. Slow excitation (E2) has its main effect via an internal motivational system V1–V4. Reproduced with permission from Heiligenberg *et al.* (1972).

and variety of the colour patterns of these species, combined with the high clarity of the reef water, suggest that visual signals are particularly important. Lorenz (1962) put forward the idea that brilliant 'poster' colours found in certain reef fishes were related to their degree of territoriality. For example, the relatively non-competitive trigger fish, *Odonus niger*, is relatively inconspicuous, whereas the bizarrely patterned Picasso fish, *Rhinecanthus aculeatus*, is very competitive and quarrelsome. It also came to be recognized that competition for a resource was likely to be most fierce between related forms with similar requirements. Zumpe (1965), studying three species of butterfly fish, *Chaetodon*, found 192 attacks on conspecifics as compared with 20 attacks on foreign species. The pattern found in these species conforms to a generic type, but is otherwise recognizably different to the human observer. Ehrlich *et al.* (1977), also working with *Chaetodon* (*C. trifascialis*) but using models, came to a similar conclusion; that is to say, the level of aggression displayed towards models of other species was directly proportional to their similarity to conspecifics.

Myrberg and Thresher (1974), using the damselfish, *Eupomacentrus planifrons*, also found that conspecifics were driven off at the greatest range (2.75–4.0 m) whereas an allied species like *E. patilius* could approach to within 1 m. Other genera could approach still closer. The authors point out that *E. planifrons* is a drab species, yet very aggressive, and this conflicts with the advertisement–aggression idea of Lorenz. Other studies confirm the lack of correlation between bright colours and territorial aggression (Fine, 1977).

In contrast, the results of Thresher's later study (1976) on 22 species of intruders into the territory of *E. planifrons* found that only 5% of attacks were on conspecifics. The species attacked at close range belonged to different genera and seemed to be competitive for shelter, nest holes, etc. Another class, attacked at longer range, fed on similar benthic algae, and this resource was presumably more significant.

The visual consequence of the need to recognize conspecifics is that comparatively small differences of pattern and colour have to be recognized rather than large differences of outline, head shape, and the like.

Visual signals and courtship behaviour

General

The more striking aspects of the physical appearance of fishes seem often to be related to communication between conspecifics in the context of breeding behaviour. The stereotyped nature of this kind of behaviour implies highly specific, stable and effective recognition processes which are genetically determined and subject to little modification as a result of experience. The powerful motivational systems involved in this kind of behaviour probably aid experimentation.

Sexual dimorphism

Two kinds of visual signal are involved in courtship, signals usually consisting of movements that initiate courtship responses, and recognition of the sexual status of the partner. Sexual status may be conveyed by morphological differences such as temporary breeding colours, or sexual dimorphism which is a constant feature of the species. Sexual dimorphism of the kind that is obvious to the human observer is rare in most teleost species, and the lack of it may prove a source of difficulty to fish farmers.

In *Megolamphodus swegelsii* the dorsal fin is lance-shaped in the male and triangular in the female; the anal fin is rounded in males of *Nannostomus marginalis* and angular in females. Sexual dimorphism is striking in the dragonet, *Callionymus Iyra*: the male has sail-like fins and turquoise stripes; the female short fins and drab colours. Males of *Puntius arulius* have exposed rays at the margin of the dorsal fin, whereas in the female the membrane extends to the edge. In the spined loach, *Cobitis taenia*, the male fish have much larger pectoral fins than the females, and these fins bear an enlarged scale, the Canestrini scale. Similar morphometrical differences can be observed in certain other species.

During the reproductive season the males of many species exhibit breeding colours. The scarlet belly patch of the male stickleback is the most well known, and other European freshwater and littoral marine species like the minnow, the goby and the charrs, *Salvelinus spp.* exhibit red patches or coloration. As we have seen, these reds and oranges show up well against a greenish spacelight. In both the stickleback and the minnow, the red colour seems to be a signal used to drive off competing males, and it is the bluish-green iridescence of the flanks in both species (Pitcher *et al.*, 1979) that is effective in encouraging the females. In some species the males are much darker than the females: the male black-faced blenny, *Tripterygion tripteronotus*, has a dark head that blackens at spawning, males of the goby, *Gobius fluviatilus*, go entirely dark at spawning, and male surgeonfish (acanthurids) darken in the breeding season. The degree of darkening in *Tilapia* expresses the degree of dominance of that male. Achromatic hues are believed to show up well (p. 97).

Alterations in female coloration are less common. Females of the pumpkin-seed sunfish and those of the blue panchax, *Aplocheilus lineatus*, develop dark vertical bars during the breeding season. In *Lepomis* it is possible using models to show that this feature inhibits male aggression (Stacey and Chiszar, 1978). In *L. cyanellus* and *L. macrochirus* such vertical banding can be produced by focal electrical stimulation of midbrain areas (Bauer and Demski, 1980). It is not known whether the position of these distinctive dimorphic features matters.

Picciolo (1964) showed that the blue anterior-ventral patch of male dwarf gourami, *Colisa lalia*, is critically located for sexual recognition, by displacing it to other sites on models, and Keenleyside (1971) also demonstrated the importance of position with regard to the eye and opercular patch of the long-ear sunfish, *Lepomis megalotis*.

Courtship movements

Courtship displays by males have been described in many different species. Usually it can be seen that there is a repetitive or rhythmic component in the movement. One of the most easily observed among tropical freshwater species is the courtship dance of a common aquarium species, the zebrafish, *Brachydanio rerio*. The male moves through a series of abrupt turns and in an elliptical pattern around the female. The movement is striking and clearly distinguishable from the normal movements. To the human observer it has the effect of emphasizing the flank and fin markings. Minnows perform a low-level waggle dance,hiding the red belly that signals aggressive intent to other males, but exposing the turquoise flank patches to the female. In the labrid *Halichoeres melanochir*, there is a mating system similar to a lek, sexual dimorphism is slight, but there is a complex routine of rhythmic movements that leads to spawning (Moyer and Yogo, 1982).

These dance-like movements are often followed up by the male fish poking or butting the female in the flanks (*Poecilia, Gasterosteus* and *Phoxinus*), and in some species the male eventually curls himself around the body of the female (*Macropodus* sp., *Barbus* sp.).

A specialized sequence occurs in the mouthbrooding cichlid *Haplochromis burtoni*. The male drags his anal fin, which bears egg-like markings, in front of the female; when she comes forward to pick up the eggs, she takes in the sperm which is expelled near the male's anal fin, and fertilization occurs in the mouth. The representation of the egg row on the male's fin is an exact one to the human eye.

To sum up, courtship behaviour often relies chronologically on (i) visual recognition of the opposite sex and of its breeding condition; (ii) the state of readiness conveyed by the male by means of specific movements; (iii) the participatory behaviour of the female. The latter is often less striking than that of the male, but maintains continuity through the female staying near spawning or nesting site. These visual elements may be strongly supported by pheromones or other cues. Sound emission may complement colour changes during courtship or territorial behaviour as, for example, in cichlids. Nelissen (1978) found that among a group of six cichlid species that he studied there was an inverse relationship between the number of colour patterns and the number of sound patterns. Needless to say, visual signals are relatively unimportant in some species, such as the goldfish, where pheromones play the dominant role.

Visual differentiation of social role and developmental status

In many species there is little visible sign of developmental status apart from size. This is true of most European freshwater species. At the other extreme are cichlids, and a number of reef genera (such as *Pomacanthus*) in which juveniles are conspicuously different from adults. There is insufficient space here for detailed descriptions of the range of colour patterns that occur within a species and provide signals to conspecifics, but the reader should refer for an example to the tables of Voss (1977). These depict the appearance of some 20 cichlid species under 20 headings corresponding to differences of sex, age,

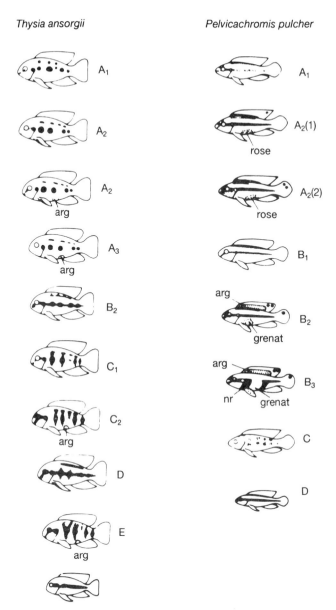

Fig. 4.5 Main colour patterns of two species of cichlids, *Thysia ansorgii* and *Pelvicachromis pulcher*. (i) A_1, neutral livery. (ii) A_2, livery of territorial males (two versions in *P. pulcher*). (iii) Females, early stages of sexual parade: *T. ansorgii*, A_3; *P. pulcher*, B_3. (iv) C_1, *T. ansorgii* alternative to A_2, egg-laying livery (males). (v) Egg-guarding livery: *T. ansorgii*, D, males, C_2, females; *P. pulcher*, B_2 (females). (vi) Fry-guarding livery, *T. ansorgii*, D, males, E, females. (vii) Fear, inferiority: *T. ansorgii*, A_3, B_2; *P. pulcher*, A_1, B_1 followed by C. (viii) Young: special form (*T. ansorgii*); *P. pulcher*, D, colour codes: arg, silver spot in female; grenat, garnet; nr, black; rose, pink. Reproduced with permission from Voss (1977).

breeding role and emotional state. It is worth noting that the transitions from one form to another may be as significant as the patterns themselves (Fig. 4.5).

Although there is some evidence for dominance in shoaling species, colour differences related to this are less well known. However, McKaye and Kocher (1983) found that the dominant individual in a foraging shoal of *Cyrtocara moorei* is a vivid blue colour distinguishable from the others, which have only the basic flank pattern (three dark spots on a pale blue ground).

Visually mediated aspects of predation

A good deal of interest has been focused on prey selection in relation to optimal foraging strategy (see Chapter 8 by Hart, this volume). Hairston *et al.* (1983) showed that as bluegill sunfish, *Lepomis macrochirus*, grow, they are able to detect smaller prey. Cone separation in the retina does not alter, but the angle between adjacent cones becomes smaller. Similar results have been obtained by Wanzenböck and Schiemer (1989) for three species of cyprinid (the roach, *Rutilus rutilus*; the bleak, *Alburnus alburnus*; and the blue bream, *Abramis ballerus*), and by Browman *et al.* (1990) for the white crappie, *Pomoxis annularis*. The improvement in acuity increases the range at which a given prey size can be detected. Gibson *et al.* (1982) point out that this wider range of available prey should allow larger fish to be more selective than smaller ones, with perhaps a bias towards the more valuable, i.e. larger, prey. O'Brien *et al.* (1976) found that large bluegills did select large prey when there was abundant prey. More recently it has been suggested that other factors such as turbidity and effective water volume (shallows) may have a decisive effect on strategy.

Coates (1980a) demonstrated how quickly the approach strategy of predators was altered according to the expected outcome. The five species tested all learned to avoid the non-reward species (*Dascyllus aruanus*). Confer *et al.* (1978) provide some valuable comparative data concerning the distances at which five species of fish (bluegills, *Lepomis macrochirus*; pumpkinseed sunfish, *Lepomis gibbosus*; lake trout, *Salvelinus namaycush*; brook trout, *Salvelinus fontinalis*; and rainbow trout, *Oncorhynchus mykiss*) appear to respond to prey of different sizes. As might be expected, reaction distances (RD) increased as light levels were raised, up to a critical level, but for the pumpkinseed RD was much larger (+ 75% to 115%) than for the brook trout at the two lowest light levels and similar prey size. For lake trout the relationship between turbidity (estimated as extinction coefficient) and RD was a linear one.

Recognition of predators by fish

Many species show a visually mediated startle response to large moving objects. This is probably controlled by the action of the giant Mauthner cells

(Eaton and Bombardieri, 1978). More specific responses to predator species have been studied by Coates (1980b) in the humbug, *Dascyllus aruanus*. Selected predatory species presented in transparent containers had the effect of halving the number of *Dascyllus* swimming above their 'home' coral head, as compared with either an empty container or one containing a non-predatory species presented at the same ranges. Despite the tendencies of the humbugs to retreat, they would also make attempts to drive off the predator, but it was notable that this aggressive behaviour was rarer when the predators were presented at close range (one-fifth the occurrences at long range, for five predator species). The very striking looking *Pterois volitans* was seldom attacked at any range. Hurley and Hartline (1974) in a similar way studied the escape response of the damselfish, *Chromis cyanea*, but using models, and again measuring the escape response of the fish in retreating to their 'home' coral head. They found that the most important elements in releasing the response were size, with large models being more effective than smaller ones, and brightness, with darker models more effective than lighter ones.

One of the most interesting papers in this field resulted from a study by Karplus *et al.* (1982) on the recognition by the reef fish *Chromis caeruleus* of predator's faces when recognition must depend on frontal views of the approaching fish. A large-eyed model with a large upturned mouth (piscivore) was contrasted with a small-eyed model with a small downturned mouth (non-piscivore). Subject position was significantly affected by the piscivore model. If the piscivore model was contrasted with models with small upturned mouths, or large downturned mouths, then the subject moved closer to the model with non-piscivore features. A disruptive pattern of patches applied to the piscivore model significantly reduced the aversive response of the subject. The significance of eye size is also emphasized by the work of Altbäcker and Csányi (1990). In these experiments the responses of the paradise fish *Macropodus opercularis* to pike (which have big eyes) and catfish (which have small eyes) were measured. The former was much more effective in releasing antipredatory behaviour than the latter. A dummy catfish with artificially big eyes was, however, effective. Eyes in a fish-like dummy were most effective when in their normal position and number: that is, two eyes arranged side by side were more effective than four eyes, one eye, or two eyes arranged vertically one above the other. In this context, considerable interest has been aroused among primate ethologists by the discovery of neurones in the monkey temporal lobe specific for face recognition (Perrett and Rolls, 1983; Hasselmo *et al.*, 1989).

The camouflage adopted by would-be predators has two forms: (i) aggressive (Peckhammian) mimicry, or (ii) resemblance to background. In the scale-eater *Proboluchus heterostomus*, the visual appearance of the predator closely resembles that of the harmless species *Astyanax fasciatus*, and this allows it to approach its victims (Sazima, 1977). A more versatile example is provided by the paedophagous cichlid *Cichlasoma orthognathus*, which adopts a stripe when stalking *C. pleurotaenia*, and remains all silver when stalking *C. encinostomus* (McKaye and Kocher, 1983).

4.5 SUMMARY

The physical properties of natural waters produce varying limitations on the visual signalling capabilities of fishes. In fish communities rich in species, especially where the water is reasonably transparent (coral reefs; Lake Malawi) the complex colour patterns, contours and textures distinguish individual species, and it can be clearly demonstrated that species respond differentially to these signals.

Studies on the fish eye show that its ability to resolve detail, while not as good as that of terrestrial animals, is adequate in view of the degradation of edges caused by suspended particles in the water. Contrast direction and hue may therefore have become more important as a means of object recognition. One of the most successful areas of study has been the correlation of receptor pigments of a species with the spectral characteristics of its environment. The bulk of physiological work has so far been centred on the retina, where opponent processes rather similar to those of primates appear to mediate colour contrast and hue discrimination. Our understanding of central visual processing, i.e. beyond the afferent tectal layers, remains fragmentary and offers an interesting though daunting prospect for future work.

Conditioning experiments have provided some useful basic information about shape significance and hue discrimination, but have seldom been related to natural forms or colours. Field and tank trials using pre-existing responses (territorial aggression, shoaling aggregation) and natural stimuli have been among the most exciting studies made in the last 15 years. From the point of view of discriminatory mechanisms, the order of effectiveness under given environmental conditions of hue, texture, contour, etc. is not clearly elucidated for any one example, and remains a goal for the future.

REFERENCES

Ahlbert, I.-B. (1968) The organization of the cone cells in the retinas of four teleosts with different feeding habits. *Ark. Zool.*, **22**, 445–81.

Ali, M. A. (1959) The ocular structure, retinomotor and photo-behavioural responses of juvenile pacific salmon. *Can. J. Zool.*, **37**, 965–96.

Ali, M. A. (ed.) (1975) *Vision in Fishes*, Plenum, New York, 836 pp.

Ali, M. A. and Anctil, M. (1976) *Retinas of Fishes: an Atlas*, Springer-Verlag, Heidelberg, 284 pp.

Altbäcker, V. and Csányi, V. (1990) The role of eyespots in predator recognition and antipredator behaviour in the paradise fish, *Macropodus opercularis* L. *Ethology*, **85**, 51–7.

Baerends, G. P. and Baerends-van Roon, J. M. (1950) An introduction to the study of cichlid fishes. *Behaviour (Supp.)*, **1**, 1–242.

Baker, E. T., Sternberg, R. W. and MacManus, D. A. (1974) Continuous light-scattering profiles and suspended matter over the Nittinat deep sea fan, in *Suspended Solids in Water* (ed. R. J. Gibbs), Plenum, London, pp. 155–72.

Barlow, G. W. (1974) Contrasts in social behaviour between Central American cichlid fishes and coral-reef surgeon fishes. *Am. Zool.*, **14**, 9–34.

Barlow, G. W. (1976) The Midas cichlid in Nicaragua, in *Investigations of the Ichthyofauna of Nicaraguan Lakes* (ed. T. B. Thorson), School of Life Sciences, University of Nebraska, Lincoln, Nebraska, pp. 332–58.

Barlow, G. W. (1982) Monogamy amongst fishes, *Abstract from 4th Congress of European Ichthyologists, Hamburg*.

Bauer, D. H. and Demski, L. S. (1980) Vertical banking evoked by electrical stimulation of the brain in the anaesthetized green sunfish *Lepomis organellus* and bluegills *Lepomis macrochirus*. *J. exp. Biol.*, **84**, 149–60.

Beatty, D. D. (1975) Rhodopsin–porphyropsin changes in paired-pigment fishes, in *Vision in Fishes* (ed. M. A. Ali), Plenum, New York, pp. 635–44.

Bowling, L. C., Steane, M. S. and Tyler, P. A. (1986) The spectral distribution and attenuation of underwater irradiance in Tasmanian inland waters. *Freshwat. Biol.*, **16**, 313–35.

Bowmaker, J. K. (1990) Visual Pigments of Fishes, in *The Visual System of Fish* (eds R. H. Douglas and M. B. A. Djamgoz), Chapman and Hall, London, pp. 81–107.

Bowmaker, J. K. and Kunz, Y. W. (1987) Ultraviolet receptors, tetrachromatic colour vision and retinal mosaics in the brown trout (*Salmo trutta*): age-dependent changes. *Vision Res.*, **27**, 2101–8.

Breder, C. M., jun. (1951) Studies on the structure of fish shoals. *Bull. Am. Mus. Nat. Hist.*, **198**, 1–27.

Browman, H. I., Gordon, W. C., Evans, B. I. and O'Brien, W. J. (1990) Correlation between histological and behavioural measures of visual acuity in a zooplantivorous fish, the white crappie (*Pomoxis annularis*). *Brain Behav. Evol.*, **35**, 85–97.

Brunskill, G. J. (1970) Fayetteville Green Lake, New York, *Limnol. Oceanogr.*, **14**, 133–70.

Burkamp, W. (1923) Versuche über Farbenwiederkennen der Fische, *Z. Sinnesphysiol.*, **55**, 133–70.

Cajal, S. R. (1893) La Rétine des Vertebrés. *Cellule*, **9**, 17–257.

Cameron, N. D. (1982) The photopic spectral sensitivity of a dichromatic teleost fish (*Perca fluviatilis*). *Vision Res.*, **22**, 1341–8.

Charman, W. N. and Tucker, J. (1973) The optical system of the goldfish eye. *Vision Res.*, **13**, 1–8.

Clarke, G. L. and Denton, E. J. (1962) Light and animal life, in *The Sea* (ed. M. N. Hill), Wiley, New York, pp. 456–68.

Coates, D. (1980a) Potential predators and *Dascyllus aruanus*. *Z. Tierpsychol.*, **52**, 285–90.

Coates, D. (1980b) The discrimination of and reactions towards predatory and non-predatory species of fish in the humbug damselfish. *Z. Tierpsychol.*, **52**, 347–54.

Collin, S. P. and Pettigrew, J. D. (1988a) Retinal topography in reef teleosts I. Some species with well developed areae but poorly-developed streaks. *Brain Behav. Evol.*, **31**, 269–82.

Collin, S. P. and Pettigrew, J. D. (1988b) Retinal topography in reef teleosts II. Some species with prominent horizontal streaks and high-density areae. *Brain Behav. Evol.*, **31**, 283–95.

Confer, J. L., Howick, G. L., Corzette, M. H., Framer, S. L., Fitzgibbon, S. and Landesberg, R. (1978) Visual predation by planktivores. *Oikos*, **31**, 27–37.

Crapon de Caprona, M. D. (1982) The influence of early experience on preferences for optical and chemical cues produced by both sexes of a cichlid fish *Haplochromis burtoni*. *Z. Tierpsychol.*, **58**, 329–61.

Cronly-Dillon, J. R. (1964) Units sensitive to direction of movement in goldfish optic tectum. *Nature, Lond.*, **203**, 214–15.

Daw, N. W. (1968) Colour coded ganglion cells in the goldfish retina. *J.Physiol., Lond.*, **197**, 567–92.

de Groot, S. G. and Gebhard, J. W. (1952) Pupil size as determined by adapting luminance. *J. opt. Soc. Am.*, **42**, 492–5.

Denton, E. J. (1970) On the organization of reflecting surfaces in some marine animals. *Phil. Trans. R. Soc.*, **182B**, 154–8.

Djamgoz, M. B. A. and Yamada, M. (1990) Electrophysiological characteristics of retinal neurones: synaptic interactions and functional outputs, in *The Visual System of Fish* (eds R. H. Douglas and M. B. A. Djamgoz, Chapman and Hall, London, pp. 159–210.

Dodt, E. (1963) Photosensitivity of the pineal organ in the teleost *Salmoirideus*. *Experientia*, **19**, 642–3.

Douglas, R. H. and Djamgoz, M. B. A. (eds) (1990) *The Visual System of Fish*. Chapman and Hall, London, 526 pp.

Douglas, R. H. and Hawryshyn, C. W. (1990) Behavioural studies of fish vision: an analysis of visual capabilities, in *The Visual System of Fish* (eds R. H. Douglas and M. B. A. Djamgoz), Chapman and Hall, London, pp. 373–418.

Eaton, R. C. and Bombardieri, R. A. (1978) Behavioural functions of the Mauthner Cell, in *Neurobiology of the Mauthner cell* (eds D. Faber and H. Korn), Raven Press, New York, pp. 221–44.

Ehrlich, P. R., Talbot, F. H., Russell, B. C. and Anderson, G. R. V. (1977) The behaviour of chaetodontid fishes with special reference to Lorenz's poster coloration hypothesis. *J. Zool.*, **183**, 213–22.

El-Akell, A. (1982) The optic tectum of the perch, unpublished PhD thesis, University of Manchester.

Endler, J. A. (1977) A predator's view of animal colour patterns. *Evol. Biol.*, **11**, 319–64.

Endler, J. A. (1987) Predation, light intensity and courtship behaviour in *Poecilia reticulata* (Pisces: Poeciliidae). *Anim. Behav.*, **35**, 1376–85.

Fálcon, J. and Collin, J.-P. (1989) Photoreceptors in the pineal of lower vertebrates: functional aspects. *Experientia*, **45**, 909–13.

Fernald, R. D. (1980) Response of male *Haplochromis burtoni* reared in isolation to models of conspecifics. *Z. Tierpsychol.*, **54**, 850–93.

Fernald, R. D. (1990) *Haplochromis burtoni*: a case study, in *The Visual System of Fish* (eds R. H. Douglas and M. B. A. Djamgoz), Chapman and Hall, London, pp. 443–63.

Fine, M. L. (1977) Communications in selected groups: communication in fishes, in *How Animals Communicate* (ed. T. A. Sebeok), Indiana University Press, Bloomington, pp. 472–518.

Fricke, H. W. (1973) Individual partner recognition in fish: field studies on *Amphiprion bicinctus*. *Naturwissenschaften*, **60**, 104–6.

Frisch, K. von (1925) Farbensinn der Fische und Duplizitat-theorie. *Z. vergl. Physiol.*, **2**, 393–452.

Galand, G. and Liége, G. (1975) Responses visuelles unitaires chez la truite, in *Vision in Fishes* (ed. M. A. Ali), Plenum, New York, pp. 1–14.

Gazey, B. K. (1970) Visibility and resolution in turbid waters. *Underwater Sci. Technol. J.*, June, 105–15.

Gibson, R. M., Li, K. T. and Easter, S. S. (1982) Visual abilities in foraging behaviour of predatory fish. *Trends Neurosci.*, June, 1–3.

Guthrie, D. M. (1981) The properties of the visual pathway of a common freshwater fish (*Perca* fluviatilis) in relation to its visual behaviour, in *Brain Mechanisms of Behaviour in Lower Vertebrates* (ed. P. R. Laming), Cambridge University Press, Cambridge, pp. 79–112.

Guthrie, D. M. (1983) Visual central processes in fish behaviour, in *Recent Advances in vertebrate Neuroethology* (eds J. P. Ewert, R. R. Capranica and D. I. Ingle), Plenum, New York, pp. 381–412.

Guthrie, D. M. (1990) The physiology of the teleostean optic tectum, in *The Visual System of Fish* (eds R. H. Douglas and M. B. A. Djamgoz), Chapman and Hall, London, pp. 279–343.

Guthrie, D. M. and Banks, J. R. (1978) The receptive field structure of visual cells from the optic tectum of the freshwater perch. *Brain Res. Amst.*, **141**, 211–25.

Hairston, N. G., Li, K. T. and Easter, S. S. (1983) Fish vision and the detection of planktonic prey. *Science*, **218**, 1240–42.

Hanyu, I. (1978) Salient features in photosensory function of teleostean pineal organ. *J. comp. Biochem. Physiol.*, **61A**, 49–54.

Harosi, F. I. and Hashimoto, Y. (1983) Ultra-violet visual pigment in a vertebrate: a tetrachromatic cone system in a dace. *Science*, **222**, 1021–3.

Hasselmo, M. E., Rolls, E. T. and Baylis, G. C. (1989) The role of expression and identity in the face-selective responses of neurones in the temporal cortex of the monkey. *Brain Behav. Res.*, **32**, 203–18.

Hawryshyn, C. W. and McFarland, W. N. (1987) Cone photoreceptor mechanisms and the detection of polarised light in fish. *J. comp. Physiol.*, **160A**, 459–65.

Hawryshyn, C. W., Arnold, M. G., Bowering, E. and Cole, R. L. (1990) Spatial organisation of rainbow trout to plane polarised light: the ontogeny of e-vector discrimination and spectral sensitivity characteristics. *J. comp. Physiol.*, **166A**, 565–74.

Hay, T. F. (1978) Filial imprinting in the convict cichlid fish *Cichlasoma nigrofascinatum*. *Behaviour*, **65**, 138–60.

Heath, R. T. and Franks, D. A. (1982) U. V. sensitive phosphorous complexes in association with dissolved humic material and iron in a bog lake. *Limnol. Oceanogr.*, **27**, 564–9.

Heiligenberg, W., Kramer, U. and Schultz, V. (1972) The angular orientation of the black eyebar in *Haplochromis burtoni* and its relevance to aggressivity. *Z. vergl. Physiol.*, **76**, 168–76.

Helfman, G. S. (1981) The advantage to fishes of hovering in the shade. *Copeia*, **2**, 392–400.

Hemmings, C. C. (1966) Factors influencing the visibility of underwater objects, in *Light as an Ecological Factor* (Symp. Br. Ecol. Soc. 6) (eds C. C. Evans, R. Bainbridge and O. Rackham), Blackwell, Oxford, pp. 359–74.

Herter, K. (1929) Dressurversuche an Fischen, *Z. vergl. Physiol.*, **10**, 688–711.

Hurley, A. C. and Hartline, P. H. (1974) Escape response in the damselfish *Chromis cyanea* (Pisces: Pomacentridae): a quantitative study, *Anim. Behav.*, **22**, 430–37.

Hurst, P. M. (1954) Colour discrimination in the bluegill sunfish. *J. comp. Physiol. Psychol.*, **46**, 442–5.

Ingle, D. J. (1985) The goldfish as a retinex animal. *Science*, **227**, 651–4.

Janssen, J. (1981) Searching for zooplankton just outside Snell's window. *Limnol. Oceanogr.*, **26**, 1168–71.

Karplus, I., Goren, M. and Algom, D. (1982) A preliminary experimental analysis of predator face recognition by *Chromis caerulaeus*. *Z. Tierpsychol.*, **61.**, 149–56.

Katzir, G. (1981) Visual aspects of species recognition in the damselfish *Dascyllus aruanus*. *Anim. Behav.*, **29**, 842–9.

Keenleyside, M. H. A. (1955) Some aspects of schooling in fish. *Behaviour*, **8**, 183–249.

Keenleyside, M. H. A. (1971) Aggressive behaviour of the male long ear sunfish (*Lepomis megalotis*). *Z. Tierpsychol.*, **28**, 227–40.

Keenleyside, M. H. A. (1979) *Diversity and Adaptation in Fish Behaviour*, Springer-Verlag, Berlin, 208 pp.

Kohda, Y. and Watanabe, M. (1982a) Agonistic behaviour and color pattern in a Japanese freshwater serranid fish, *Coreoperca kawamebari*. *Zool. Mag.*, **91**, 61–9.

Kohda, Y. and Watanabe, M. (1982b) Relationship of color pattern to dominance order in a freshwater serranid fish *Coreoperca kawamebari*. *Zool. Mag.*, **91**, 140–45.

Kuenzer, P. (1968) Die Auslösung der Nachfolgereaktion bei Erfahrungslosen Jungfischen von *Nannacara anomala*. *Z. Tierpsychol.*, **25**, 257–314.

Kuenzer, P. (1975) Analyse der Auslosende Reizsituationen für die Anschwimm-, Eindring- und Fluchtreaktion Junger *Hemihaplochromis multicolor*. *Z. Tierpsychol.*, **47**, 505–44.

Kullenberg, G. (1974) The distribution of particulate matter in a North-West African coastal upwelling area, in *Suspended Solids in Water*, (ed. R. J. Gibbs), Plenum, London, pp. 95–202.

Levine, J. S. and MacNichol, E. F. (1979) Visual pigments in fishes: effects of habitat, Microhabitat, and behaviour on visual system evolution. *Sens, Process.*, **3**, 95–131.

Levine, J. S., Lobel, P. S. and MacNichol, E. F. (1980) Visual Communication in Fishes, in *Environmental Physiology of Fishes* (ed. M. A. Ali), Plenum, New York, pp. 447–75.

Liebman, P. A. and Entine. G. (1964) Sensitive low light-level microspectrophotometric detection of photosensitive pigments of retinal cones. *J. Opt. Soc. Am.*, **54**, 1451–9.

Limbaugh, C. and Rechnitzer, A. B. (1951) Visual detection of temperature–density discontinuation in water by diving. *Science*, **121**, 395–6.

Loew, E. W. and Lythgoe, J. N. (1978) The ecology of cone pigments in teleost fishes. *Vision Res.*, **18**, 715–22.

Loew, E. W. and McFarland, W. N. (1990) The underwater visual environment, in *The Visual System of Fish* (eds R. H. Douglas and M. B. A. Djamgoz), Chapman and Hall, London, pp. 1–43.

Lorenz, K. (1962) The function of colour in coral reef fishes. *Proc. R. Ins. Great Britain*, **39**, 282–96.

Lythgoe, J. N. (1974) The structure and function of iridescent corneas in fishes, in *Vision in Fishes* (ed. M. A. Ali), Plenum, New York, pp. 253–62.

Lythgoe, J. N. and Northmore, D. P. M. (1973) Problems of seeing colours underwater, in *Colour '73*, Adam Hilger, London, pp. 77–98.

Lythgoe, J. N. and Partridge, J. C. (1989) Visual pigments and the acquisition of visual information. *J. exp. Biol.*, **146**, 1–20.

McKaye, K. R. and Kocher, T. (1983) Head-ramming behaviour by three species of paedophagous cichlids. *Anim. Behav.*, **31**, 206–10.

McPhail, J. D. (1969) Predation and the evolution of a stickleback (*Gasterosteus*). *J. Fish. Res. Bd Can.*, **26**, 3183–208.

Marks, W. B. (1965) Visual pigments of single goldfish cones. *J. Physiol. Lond.*, **178**, 14–32.

Meek, H. (1990) Tectal morphology: connections, neurones and synapses, in *The Visual System of Fish*, (eds R. H. Douglas and M. B. A. Djamgoz), Chapman and Hall, London, pp. 239–277.

Meek, H. and Schellart, N. A. M. (1978) A Golgi study of the goldfish optic tectum. *J. comp. Neurol.*, **182**, 89–122.

Meyer, D. L., Schott, D. and Schaeffer, K. D. (1970) Brain stimulation in the optic tectum of freely swimming codfish *Gadus morhua*. *Pflügers Arch. ges. Physiol.*, **314**, 240–52.

Moller, P., Servier, J., Squire, A. and Boudinot, M. (1982) Role of vision in schooling. *Anim. Behav.*, **30**, 641–50.

Morita, Y. (1966) Enladungsmuster pinealer Neurone der Regenbogenforelle (*Salmo irideus*) by Belichtung des Zwischenhirns. *Pflügers Arch. ges. Physiol.*, **289**, 155–67.

Moyer, J. T. and Yogo, Y. (1982) The lek-like mating system of *Halichoeres melanochir* at Miyake-jina, Japan. *Z. Tierpsychol.*, **60**, 209–26.

Munk, O. (1970) On the occurrence and significance of horizontal band-shaped retinal areae in teleosts. *Vidensk. Meddr. dansk naturh. Foren.*, **133**, 85–120.

Muntz, W. R. A. (1973) Yellow filters and the absorption of light by the visual pigment of some Amazonian fishes. *Vision Res.*, **13**, 2235–54.

Muntz, W. R. A. (1974) Comparative studies in behavioural aspects of vertebrate vision, in *The Eye*, Vol. 6: *Comparative Physiology* (eds H. Davson and L. T. Graham), Academic Press, New York, pp. 155–226.

Muntz, W. R. A. (1976) On yellow lenses in mesopelagic animals. *J. Mar. Biol. Ass. U.K.*, **56**, 963–76.

Muntz, W. R. A. (1990) Stimulus, environment and vision in fishes, in *The Visual System of Fish* (eds R. H. Douglas and M. B. A. Djamgoz), Chapman and Hall, London, pp. 491–511.

Muntz, W. R. A. and Gwyther, J. (1988) Visual acuity of *Octopus pallidus* and *Octopus australis*. *J. exp. Biol.*, **134**, 119–29.

Myrberg, A. A. and Thresher, R. E. (1974) Interspecific aggression and its relevance to the concept of territoriality in reef fishes. *Am. Zool.*, **14**, 81–96.

Nelissen, M. H. J. (1978) Sound production by some Tanganyikan cichlid fishes and a hypothesis for the evolution of their communication. *Behaviour*, **64**, 137–47.

Nicol, J. A. C. (1989) *The Eyes of Fishes*, Clarendon Press, Oxford, 308 pp.

Noble, G. K. and Curtis, B. (1939) The social behaviour of the jewel fish *Hemichromis bimaculatus*. *Bull. Am. Mus. Nat. His*, **76**, 1–46.

Northmore, D., Volkmann, F. C. and Yager, D. (1978) Vision in fishes: colour and pattern, in *Behaviour in Fish and Other Aquatic Animals*, (ed. D. I. Mostofsky), Academic Press, New York, pp. 79–136.

O'Benar, J. D. (1976) Electrophysiology of neural units in the goldfish optic tectum. *Brain Res. Bull.*, **1**, 529–41.

O'Brien, W. J., Slade, N. A. and Vineyard, G. L. (1976) Optimal foraging strategy in fishes. *Ecology*, **57**, 1304–30.

Ormond, R. F. G. (1974) Visual responses in teleost fish, unpublished PhD thesis, University of Cambridge.

Page, C. H. and Sutterlin, A. M. (1970) Visual auditory unit responses in the goldfish tegmentum. *J. Neurophysiol.*, **33**, 129–36.

Patfield, I. (1983) Conspecific recognition in the perch, unpublished PhD thesis, University of Manchester.

Perrett, D. I. and Rolls, E. T. (1983) Neural mechanisms underlying the analysis of faces, in *Advances in Vertebrate Neuroethology* (eds J. P. Ewert, R. R. Capranica and D. I. Ingle), Plenum, New York, pp. 543–68.

Picciolo, A. K. (1964) Sexual and nest discrimination in anabantid fishes of the genera *Colisa* and *Trichogaster*. *Ecol. Monogr.*, **34**, 53–77.

Pitcher, T. J., Kennedy, G. J. A. and Wirjoatmodjo, S. (1979) Links between the behaviour and ecology of freshwater fishes. *Proc. 1st Br. Freshwat. Fish Conf.*, 162–75.

Rasa, O. A. E. (1969) Territoriality and the establishment of dominance by means of visual cues in *Pomacentrus jenkinsii*. *Z. Tierpsychol.*, **26**, 825–45.

Rivers, R. (1953) The pineal apparatus of tunas in relation to phototactic movement. *Bull. Mar. Sci. Gulf Caribb.*, **3**, 168–80.

Ruttner, F. (1952) *Fundaments of Limnology*, University of Toronto Press, Toronto, 242 pp.

Sajovic, P. and Levinthal, C. (1982) Visual cells of the zebra fish optic tectum. Mapping with small spots. *Neuroscience*, **7**, 2407–26.

Sazima, I. (1977) Possible case of aggressive mimicry in a Neotropical scale-eating fish. *Nature, Lond.*, **170**, 510–12.

Schellart, N. A. M. (1983) Acousticolateral and visual processing and their interaction in the torus semicircularis of the trout *Salmo gairdneri*. *Neurosci. Lett.*, **42**, 39–44.

Schellart, N. A. M. and Spekreijse, H. (1976) Shapes of receptive field centers in the optic tectum of the goldfish. *Vision Res.*, **16**, 1018–20.

Schmidt, J. T. (1979) The laminar organization of optic nerve fibres in the tectum of goldfish. *Proc. R. Soc.* **205B**, 287–306.

Schubel, J. R. (1974) Effect of tropical storm Agnes on the suspended solids of the Northern Chesapeake Bay, in *Suspended Solids in Water* (ed. J. R. Gibbs), Plenum, London, pp. 113–32.

Schulte, A. (1957) Transfer- und Transpositions-Versuche mit Monokular Dressierten Fischen. *Z. vergl. Physiol.*, **393**, 432–76.

Schusterman, R. J. and Balliet, R. F. (1970) Visual acuity of the harbour seal and the Stellar sea lion underwater. *Nature, Lond.*, **226**, 563–4.

Schwassman, H. O. (1974) Refractive state. Accommodation and resolving power of the fish eye, in *Vision in Fishes* (ed. M. A. Ali), Plenum, New York, pp. 279–88.

Schwassman, H. O. and Meyer, D. L. (1973) Refractive state and accommodation in three species of *Paralabrax*. *Vidensk. Medd dansk naturh. Foren.*, **134**, 103–8.

Shand, J., Partridge, J. C., Archer, S. N., Potts, G. W. and Lythgoe, J. N. (1988) Spectral absorbance changes in the violet/blue sensitive cones of the pollack, *Pollachius pollachius*. *J. comp. Physiol.*, **164A**, 699–703.

Sivak, J. G. (1973) Accommodation in some species of North American fishes. *J. Fish. Res. Bd Can.*, **30**, 1141–6.

Sivak, J. G. and Bobier, W. R. (1978) Chromatic aberration of the fish eye and the effect of its refractive state. *Vision Res.*, **18**, 453–5.

Sivak, J. G. and Howland, H. C. (1973) Accommodation in the northern rock bass (*Ambloplites rupestris*) in response to natural stimuli. *Vision Res.*, **13**, 2059–64.

Sivak, J. G. and Kreutzer, R. O. (1981) Spherical aberration of crystalline lenses studied by means of a helium–neon laser. *Vision Res.*, **23**, 59–70.

Spekreijse, J., Wagner, H.-J. and Wolbarsht, M. T. (1972) Spectral and spatial coding of ganglion cell responses in goldfish retina. *J. Neurophysiol.*, **35**, 73–86.

Sroczyński, S. (1977) Spherical aberration of the crystalline lens in the roach. *J. comp. Physiol.*, **121**, 135–44.

Sroczyński, S. (1981) Optical system of the eye of the ruff (*Acerina cernua* L.). *Zool. Jb.*, **85**, 316–42.

Stacey, P. B. and Chiszar, D (1978) Body colour pattern and aggressive behaviour of male pumpkinseed sunfish (*Lepomis gibbosus*). *Behaviour*, **64**, 271–304.

Stewart, K. M. and Martin, P. J. H. (1982) Turbidity and its causes in a narrow glacial lake with winter ice cover. *Limnol. Oceanogr.*, **27**, 510–17.

Sutterlin, A. M. and Prosser, C. L. (1970) Electrical properties of the goldfish optic tectum. *J. Neurophysiol.*, **33**, 36–45.

Svaetichin, G. (1953) The cone action potential. *Acta physiol. scand. (Suppl.)*, **106**, 565–600.

Tamura, T. (1957) A study of visual perception in fish, especially on resolving power and accommodation. *Bull. Jap. Soc. Scient. Fish.*, **22**, 536–57.

Tayssedre, C. and Moller, P. (1983) The optomotor response in weakly electrical fish. *Z. Tierpsychol.*, **60**, 265–352.

Thresher, R. E. (1976) Field analysis of the territoriality of the 3-spot damselfish, *Eupomacentrus planifrons*. *Copeia*, **2**, 166–76.

Tinbergen, N. (1951) *The Study of Instinct*, Clarendon Press, Oxford, 228 pp.

Tipping, E. and Woof, C. (1983) Humic substances in Esthwaite Water. *Limnol. Oceanogr.*, **28**, 145–53.

Tomita, T., Kaneko, A., Murakami, M. and Pautler, E. L. (1967) Spatial response curves of single cones in the carp. *Vision Res.*, **7**, 519–31.

Underwood, H. (1989) The pineal and melanin: regulators of circadean function in lower vertebrates. *Experientia*, **45**, 914–22.

Vanegas, H., Essayag-Millan, E. and Laufer, M. (1971) Response of the optic tectum to electrical stimulation of the optic nerve in *Eugerres plumieri*. *Brain Res. Amst.*, **31**, 107–18.

Voss, J. (1977) Les Livrées òn patron de coloration chez les poissons cichlids africains. *Revue fr. Aquarol. Herpetol.*, **4**, 34–80.

Wagner, H.-J. (1975) Patterns of Golgi impregnated neurones in a predator-type fish retina, in *Neural Principles of Vision* (eds F. Zetler and R. Weiler), Springer, Berlin, pp. 7–25.

Wagner, H.-J. (1990) Retinal structure of fishes, in *The Visual System of Fish* (eds R. H. Douglas and M. B. A. Djamgoz), Chapman and Hall, London, pp. 109–157.

Wanzenböck, J. and Schiemer, F. (1989) Prey detection in cyprinids during early development. *Can. J. Fish. Aquat. Sci.*, **46**, 995–1001.

Wartzok, D. and Marks, W. B. (1973) Directionally selective visual units recorded in optic tectum of the goldfish. *J. Neurophysiol.*, **36**, 588–603.

Watanabe, M., Kobayashi, T. and Terami, H. (1984) Changes in eye color during aggressive interaction in the chocolate cichlid, *Cichlasoma coryphaenodes*. *Zool. Sci.*, **1**, 787–93.

Waterman, T. H. (1975) Natural polarized light and e-vector discrimination by vertebrates, in *Light as an Ecological Factor* II (eds G. C. Evans, R. Bainbridge and O. Rackham), Blackwell, Oxford, pp. 305–335.

Wells, W. H. (1969) Loss of resolution in water as a result of multiple small angle scattering. *J. Opt. Soc. Am.*, **59**, 686–91.

Young, J. Z. (1950) *The Life of Vertebrates*, Oxford University Press, Oxford, 767 pp.

Zaneveld, (1974) Spatial distribution of the index of refraction of suspended matter in the ocean, in *Suspended Solids in Water* (ed. R. J. Gibbs), Plenum, London, pp. 87–100.

Zenkin, G. M. and Pigarev, I. N. (1969) Detector properties of the ganglion·cells of the pike retina. *Biofizika*, **14**, 763–72.

Zumpe, D. (1965) Laboratory observations on the aggressive behaviour of some butterfly fishes (Chaetodontidae). *Z. Tierpsychol.*, **22**, 226–36.

Chapter five

Underwater sound and fish behaviour

A. D. Hawkins

5.1   INTRODUCTION

It has long been known that sounds are important to fish. Isaak Walton advised anglers 'to be patient and forbear swearing, lest they be heard'. A wide range of species, including many that are commercially valuable, emit sounds (Tavolga, 1976; Myrberg, 1981), and many species have now been shown to be acutely sensitive to underwater sounds. However, before we consider the acoustic behaviour of fish in more detail, we need to understand what sound is, how sounds are created, and how they are transmitted through water.

5.2   THE NATURE OF UNDERWATER SOUND

Sound is essentially a local mechanical disturbance generated in any material medium, whether a gas or a liquid. It is a remarkably pervasive and ubiquitous form of energy which is often difficult to screen out. Sounds are generated by the movement or vibration of any immersed object, and result from the inherent elasticity of the surrounding medium. As the source moves, kinetic energy is imparted to the medium and is passed on as a travelling elastic wave, within which the component particles of the medium are alternately forced together and then apart (Fig. 5.1). This disturbance propagates away from the source at a high speed, which depends on the density and elasticity of the material.

Fundamentally, sound propagation involves a transfer of energy without any net transport of the medium. Close to a sound source, however, it is not easy to draw a distinction between sound and bulk movements of the medium itself. Local turbulent and hydrodynamic effects occur which involve net motion of the medium, and neither depend upon the elasticity of the medium

Behaviour of Teleost Fishes 2nd edn. Edited by Tony J. Pitcher. Published in 1993
by Chapman & Hall. ISBN   0 412 42930 6 (HB) and 0 412 42940 3 (PB).

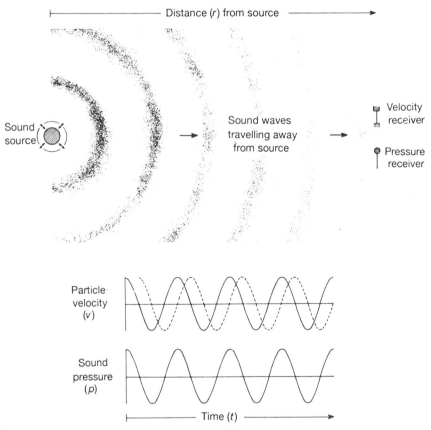

Fig. 5.1 Sound propagation through water. The source sets up a wave which travels through the medium at a constant velocity. The wave can be characterized at a particular point by a to-and-fro motion of the component particles of the medium (the particle velocity, v) or by a variation in pressure above and below the ambient level (the sound pressure, p). If the source moves in a simple harmonic fashion, as here, both parameters vary sinusoidally with time. Distant from the source, v, is in phase with p (solid curve). Close to the source, v may lag p by up to $90°$ (broken curve).

nor propagate at the velocity of sound (see Chapter 7 by Bleckmann, this volume). To a particular sense organ, these hydrodynamic effects may be indistinguishable from sounds.

The to-and-fro displacements that constitute the sound are extremely small, of the order of nanometres. They are accompanied by an oscillatory change in pressure above and below the prevailing hydrostatic pressure, the sound pressure. In a free sound field, where there are no physical obstructions to passage of the sound, and where the advancing wave-front is an almost planar surface, the oscillatory **particle velocity** (v, the first time derivative of the particle displacement) and the **sound pressure** (p) are directly proportional to one another, i.e. $v = p/\rho c$, where c is the propagation velocity (m s^{-1}), and ρ is the density of the medium (kg m^{-3}).

The product (ρc) is termed the **acoustic impedance**, and is a measure of the acoustic properties of the medium (analogous to the resistance of an electrical conductor). The particle velocity is measured in metres per second and the sound pressure in pascals (1 Pa $= 1$ Nm$^{-2} = 10\,\mu$bars $= 10$ dyn cm^{-2}), but because a great range of amplitudes of both quantities are encountered in nature, it has become conventional to express sound levels in terms of a logarithmic measure – the decibel – relative to a reference quantity. Thus:

$$\text{sound pressure level } (SPL) = 20 \log_{10} p/p_{\text{ref}} \text{ dB} \qquad (5.1)$$

where p is the measured sound pressure and p_{ref} is a reference pressure normally taken as 1 μPa for water (1 μPa $= 1$ Pa^{-6}), the SPL being expressed as dB re. 1 Pa). A tenfold increase in sound pressure is equal to 20 dB, a hundredfold increase is 40 dB, and a reduction of one-thousandth is -60 dB.

Many simple sound sources generate regular waves of motion and pressure, where the amplitudes of both pressure and motion vary with time in a sinusoidal manner. Examples are the tuning fork and bell, which generate sounds of a single frequency and wavelength (as in Fig. 5.1), perceived as having a particular pitch. Other sources may generate complex sounds, with much more irregular waveforms, composed of a wide range of sine waves of differing frequency, amplitude and phase.

Sounds inevitably diminish in level as they propagate away from a source. Distant from the source, in a free acoustic field, both pressure and velocity decline with the inverse of the distance (i.e. by a factor of 2, or 6 dB, for a doubling of distance), and both parameters are in phase with one another. Close to a source, however, where the radiating wave-fronts are no longer plane but spherical, the simple plane-wave equation no longer applies. The particle velocity is much higher for a given sound pressure, the so-called **near-field effect** (Fig. 5.2). Within the near field, velocity declines with the inverse square of distance (Harris, 1964), the phase of velocity lagging that of pressure (by 90° close to the source). The extent of the near field depends on the nature of the source. For a simple monopole source (a pulsating sphere), the limit is reached when $r = \lambda/2$, where r is the distance and λ is the wavelength of the sound ($\lambda = c/f$, where f is the frequency). See also Chapter 7 by Bleckmann, this volume.

Sounds also depart from the plane-wave equation close to a reflecting boundary. At a boundary with a 'soft' material, having low acoustic impedance like air, the local amplitude of particle motion is much higher. Conversely, close to a 'hard' sea-bed, the amplitude is much lower. In a small tank in the laboratory, any sound source is completely surrounded by

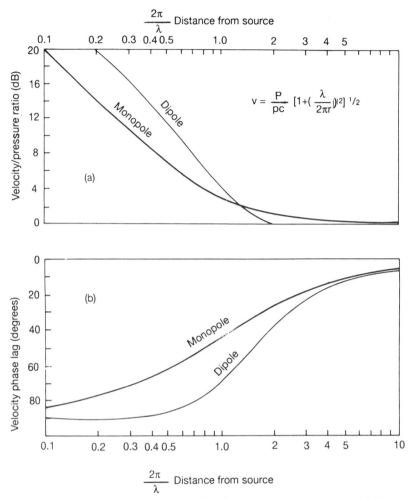

Fig. 5.2 The relationship between particle velocity and pressure varies with distance from a sound source. The velocity/pressure ratio increases closer to the source (a), and shows a phase lag behind pressure (b). Distant from the source (in the far field), velocity always bears the same proportion to pressure, and is in phase with pressure. The near-field effect is illustrated along the major axes of two types of source: a monopole, or pulsating sphere, and a dipole, or oscillating sphere. The spherical wave equation for a monopole is given. After Siler (1969).

reflectors, and the acoustic conditions become very complex. It is no longer possible to predict the particle velocities which accompany a measured sound pressure by simple application of the wave equations. Ideally, experiments on the acoustic behaviour of fish should be performed in large bodies of water, where more predictable acoustic conditions prevail. This is not always possible, and behavioural experiments have often been performed without proper control or measurement of the sound stimulus.

The to-and-fro motion of the particles, whether it is expressed as particle displacement, velocity or acceleration, differs from the sound pressure in that it is inherently directional, usually taking place along the axis of transmission. Thus, the particle velocity, displacement and acceleration are all vector quantities. A single particle motion detector can, if suitably constructed to resolve a signal into its components, detect the axis of propagation (though not necessarily the direction of the source). Sound pressure, on the other hand, is a scalar quantity acting in all directions.

Water has a greater density, lower elasticity and higher sound propagation velocity than air. It has a high acoustic impedance, which essentially means that, for a given sound pressure, the particle motion is smaller (approximately 3500 times less than in air). For a sound of given frequency, the wavelengths (λ) are much longer (by a factor of approximately 4.5), so that the near-field effect extends some distance from the source. In the absence of discontinuities or reflectors, water is a very good medium for the propagation of sound. Underwater explosions may be detected half-way around the world, and the sounds generated by ship's engines and propellers may travel many miles. Indeed, one of the problems in listening to the sounds emitted by particular aquatic organisms is that they are often masked by sounds generated by other sources, some close and others distant.

Any animal moving in water almost inevitably generates sounds and any animal capable of detecting them gains a number of advantages. Because sounds propagate rapidly and effectively through water, the detector is provided with an early notification of the presence of the source, even where there is no direct line of sight to it. Low-frequency sounds, in particular, may propagate around solid objects without being absorbed, and may penetrate dense cover, or propagate around corners, providing an almost instantaneous warning of the movement of something which would otherwise be concealed. Moreover, because sounds can vary in their characteristics, depending on the nature of the source, the detector can potentially gain important information about the object emitting them. By analysing the time structure of the sound or by resolving a complex sound into its component frequencies, the fish may be able to identify precisely the particular source. It is therefore possible to determine whether the source is predator or prey, inert or alive. Perhaps more important, however, a sound receiver is potentially able to determine the direction and even the distance of any source or to determine whether it is coming closer, or moving away. This ability may be especially significant in the sea, where light levels are low and long-distance vision is often impaired.

5.3   DETECTION OF UNDERWATER SOUNDS

Man must detect underwater sounds by means of a hydrophone, an under-water microphone which converts the water-borne sound into an electrical signal which can subsequently be amplified, analysed or broadcast into air. Most hydrophones are sensitive to sound pressure, the pressure-sensitive element being composed of a waterproofed transducer of barium titanate or

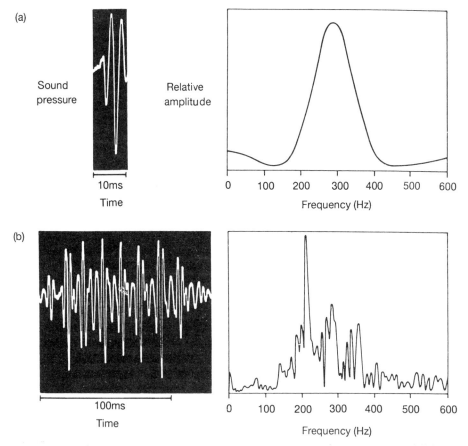

Fig. 5.3   The call of the tadpole fish, *Raniceps raninus* (family Gadidae), shown as the variation in sound pressure with time (the oscillogram, left), and as the spectrum (right), where the sound is broken down into its component frequencies. Here the spectral analysis was performed by mathematical analysis of the oscillogram (by fast Fourier transform). The call is made up of a series of repeated pulses: (a) shows that the spectrum for a single pulse is simple, and typical of that for a heavily damped resonator; (b) shows that repetition of a series of similar pulses gives a spectrum consisting of several more or less regularly spaced frequency bands (the spacing of the bands, in hertz, is the reciprocal of the time interval between the pulses, in seconds).

lead zirconate connected to a high-input impedance preamplifier (Urick, 1983). The detected signals are usually filtered to eliminate noise at extremely low or high frequencies, and then recorded on magnetic tape and monitored on headphones. The temporal structure of the sound can be examined by displaying the sound-pressure waveform on an oscilloscope (Fig. 5.3). Alternatively, a frequency analysis can be prepared, either mathematically (Fig. 5.3) or by passing the signal through a number of parallel narrow-band filters and displaying the spectrum, showing the relative energy at different frequencies. A more elaborate analysis can be performed by means of a sound spectrograph (Pye, 1982), which shows changes in the frequency structure with time (Fig. 5.4).

Sound pressure hydrophones are calibrated in terms of the voltage produced at the output terminals by a sound of a given sound pressure (for example −60 dB re 1 V (i.e. 1 mV) for a sound pressure of 1 Pa). If the particle velocity of the sound wave is required, then it can be calculated using the appropriate wave equation – though only if the sound is measured in a free sound field, which is rarely the case in a small laboratory tank. Direct measurement of particle velocity or particle acceleration is possible by means of a suitably

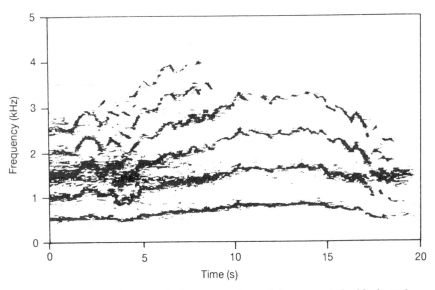

Fig. 5.4 Sound spectrogram of a long spawning call from a male haddock, *Melanogrammus aeglefinus*. The filter bandwidth for analysis was 45 Hz. The spectrogram demonstrates changes in the frequency spectrum of the call with time. The spectrum, like that for the call of the tadpole fish (Fig. 5.3), consists of a series of equally spaced frequency bands and is typical of a call made up from a series of repeated pulses. During spawning, the pulses are produced at varying rates (Fig. 5.7), resulting in changes in the spacing of frequency bands.

mounted seismic accelerometer, though such a sensor is relatively insensitive. Because particle motion is a vector quantity (see above), it is usually necessary to employ three such sensors, directed at mutual right angles.

The first impression gained when a hydrophone is lowered into a large body of water is that the medium is very noisy. Even in a much quieter pond or rock pool, any footsteps or vibrations imparted to the adjacent ground are readily audible. To obtain quiet conditions in an aquarium tank it is necessary to acoustically isolate the tank from the floor, and to switch off any pumps or machinery in the vicinity. Airborne noise is less of a problem because most of it is reflected at the air/water interface. In rivers and lakes, and in the sea, there is a continual and pervasive background noise resulting from turbulence, water flow, breaking waves and spray, together with the sounds of distant storms and precipitation at the surface. Shipping is clearly audible, often at great distances, together with sounds from oil production platforms and drilling rigs, and the explosive sounds of seismic surveying. Superimposed on this background noise are the sounds made by aquatic organisms themselves. Among the latter are the high-frequency calls and sonar signals of aquatic mammals, and the low-frequency sounds of aquatic crustaceans and fish. Many of the sounds of biological origin, like the rasp of a marine snail's radula or the grinding of a sea urchin's teeth, may be purely incidental, but collectively they all contribute to the overall ambient noise. There have been several detailed studies of sea noise, because of its importance for the detection of submarines, surface craft and underwater weapons (Urick, 1983).

5.4   SOUNDS PRODUCED BY FISH

Sound-producing mechanisms

Myrberg (1981) has listed over 50 families of fish containing sound producers. It is questionable whether all these fish really are vocal. Some of them may have emitted the sounds attributed to them only in response to strong stimulation, for example by electric shocks, while others may have produced the sounds incidentally – for example, while consuming crunchy food. Though the latter may be detected, and even acted upon by other listening animals, they do not necessarily imply the production of a call or signal. Nevertheless, many fish do produce calls as part of a particular behavioural repertoire, and the sounds are believed or have been shown to elicit a change in the behaviour of other individuals of the same or different species.

Fish sounds vary in structure, depending on the mechanism used to produce them (Schneider, 1967). Generally, however, they are predominantly composed of low frequencies, with most of their energy lying below 3 kHz. So far, no ultrasonic sounds (above the range of human hearing) have been recorded from fish, though such sounds are produced by marine mammals.

Stridulatory sounds are made by fish rubbing parts of the body together. Characteristically they are rasps and creaks, often made up from a series of very rapidly produced and irregular transient pulses, containing a wide range of frequencies. Members of the grunt family, Pomadasyidae, produce a sharp, vibrant call by grating a dorsal patch of pharyngeal denticles against smaller ventral patches, and in the triggerfish (family Balistidae) the fused anterior spines of the dorsal fin produce a grating sound when moved against their socket. Some catfish of the family Siluridae produce a squeak when the enlarged pectoral spines are moved. Other fish clap or thump different parts of the body together, like the grouper, *Mycteroperca bonaci*, which bangs the opercula or gill covers against the body to produce a low-pitched thump (Tavolga, 1960).

Hydrodynamic sounds are produced by fish that are actively swimming, or rapidly turning. Though much of the disturbance recorded on a hydrophone in the vicinity is generated by water turbulence, a booming or rushing sound is often detectable at several metres. Some elements of the sound may be derived from the internal stresses set up within the body of the animal. Perhaps more important, however, are the pressures and water movements set up by rapid movements of the fish's body. Close examination of the pressure fields around swimming fish (Gray and Denton, 1991) has revealed that fast pressure pulses are generated when fish make rapid swimming movements.

With both stridulatory and hydrodynamic mechanisms, the gas-filled swim bladder may play a subsidiary part, and may impart a hollow resonant quality to the sound. If the swim bladder of a sound-producing white grunt, *Haemulon plumieri*, is deflated, the sound generated by stridulation of the pharyngeal teeth is less loud, and loses its grunt-like quality. The swim bladder is more directly involved in the production of sounds by some physostomatous fish (with open swim bladders), such as the eel, *Anguilla anguilla*, where gas is released from the swim bladder into the oesophagus, giving a sharp pop or squeak.

In perhaps the most specialized and characteristic sound-producing mechanism, paired striated muscles from the body wall compress the swim bladder. These muscles are derived from the trunk musculature, and may simply overlie the swim bladder, as in the Sciaenidae or drums (Schneider, 1967), or they may be attached partly or wholly to the organ as in the Opsanidae or toadfishes (Tavolga, 1960). Within a family, such as the gurnards (Triglidae) or codfishes (Gadidae), there can be substantial differences between the conformation of the muscles in individual species (Fig. 5.5) and even between fish of different sexes, the muscles often being more highly developed in the male (Templeman and Hodder, 1958). The muscle fibres themselves are specialized. They are often red or yellow, with a high myoglobin content and a rich blood supply. Their diameter is thin, they contain a well-developed sarcoplasmic reticulum, and they may show innervation by a large number

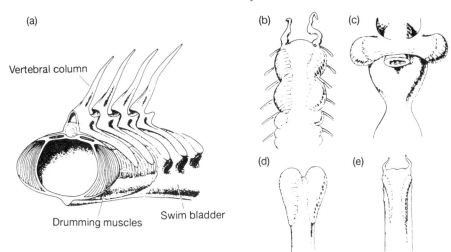

Fig. 5.5 The sound-producing apparatus in gadoid fish. (a) The drumming muscles in the haddock, *Melanogrammus aeglefinus*, are dorsally attached to the wall of the swim bladder overlying strong lateral wings (parapophyses) extending from the anterior vertebrae. Ventrally, the muscle fibres insert upon the tough outer tunic of the gas-filled swim bladder. (b) In the cod, *Gadus morhua*, the muscles are attached dorsally directly to the vertebral parapophyses and ventrally to the swim bladder. (c) In the tadpole fish, *Raniceps raninus*, the muscles are attached dorsally to the sessile ribs of the 2nd vertebra, the fibres running rostro-ventrally to insert on a sheet of connective tissue, continuous with the peritoneum and enclosing the two anterior cornua of the swim bladder, on either side of the oesophagus. (d) In the burbot, *Lota lota*, the only freshwater gadoid, a pair of thin muscle sheets is attached entirely to the swim bladder, as is also the case in (e) the lythe, *Pollachius pollachius*.

of nerve fibres, with many nerve terminals along their length. They contract very rapidly, with a high degree of synchrony, and can be stimulated to contract repeatedly at a very fast rate without going into a sustained state of contraction or tetany. In the haddock each synchronous contraction of the paired muscles results in a brief thump or pulse of sound, the repeated contraction giving rise to longer calls in which the individual pulses may still be detectable or may run together to give a grunt. In different species from the same family, the differing patterns of contraction of the muscles give rise to quite different calls (Fig. 5.6). The presence of this muscular mechanism in a range of unrelated families, and its absence from closely related species, suggests that it has evolved independently in different fish. In the gurnards, the muscles that develop as part of the trunk musculature are innervated by branches of the anterior spinal nerves. This origin suggests that the mechanism evolved from the incidental production of sound through the contraction of originally unspecialized trunk muscles, perhaps during swimming, or accompanying rapid flexure of the body. In more specialized forms, the muscles can

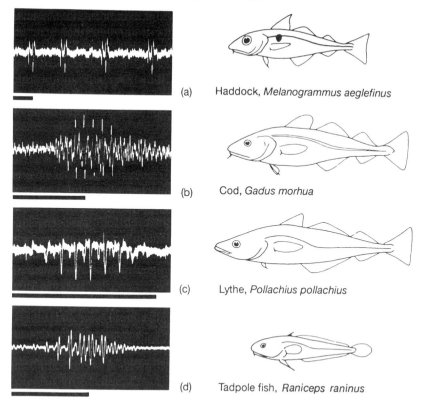

(a)   Haddock, *Melanogrammus aeglefinus*

(b)   Cod, *Gadus morhua*

(c)   Lythe, *Pollachius pollachius*

(d)   Tadpole fish, *Raniceps raninus*

Fig. 5.6 Calls produced by different members of the family Gadidae from the northern North Sea. The time base (black bar) in each case is 100 ms. All these calls were produced during aggressive behaviour outside the spawning season. (a) A short series of closely repeated knocks from the haddock; (b) a grunt from the cod; (c) a grunt from the lythe; (d) a grunt from the tadpole fish. Note that the various grunts, like the longer call of the haddock, are made up of rapidly repeated pulses, the rate of production of pulses rising to over 100 per second in the case of the lythe.

be contracted independently, to produce sound without any accompanying movement of the fish.

The involvement of the swim bladder in sound production by so many fish suggests that the organ plays a key role in the generation of sounds. To generate low-frequency sounds efficiently, it is necessary to move a large body of water. This can most readily be achieved by causing a volume change in the medium (Harris, 1964). Compression of the gas in the swim bladder by contraction of muscles is an effective way of achieving this. We shall see later (p. 000) that the swim bladder is resonant and tends to pulsate at a particular frequency, but the organ is heavily damped and the pulsations rapidly die out. The resultant sound pulse is short, and contains a range of

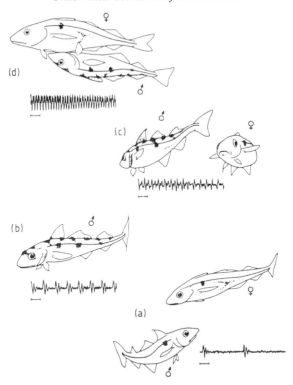

Fig. 5.7 Courtship behaviour and associated calls in the haddock, *Melanogrammus aeglefinus*. The time base (scale bar) for the calls is 50 ms. (a) The male approaches a maturing female, with his fins erect, uttering a short series of repeated knocks. (b) A sexually active male swims along the bottom, in tight circles, with exaggerated body movements and fins erect. A heavy pattern of pigmentation is shown, and long calls of rapidly repeated knocks are uttered. (c) The male leads a ripe female up through the water column, his tail moving from side to side, with all fins extended. A continuous rasping call is produced. (d) The male mounts the female from below, the continuous call reaching a hum. Sound production ceases as eggs and sperm are released into the water. Note that the calls of the male consist of series of repeated pulses, the repetition rate varying in different contexts. The female remains silent throughout courtship.

frequencies centred on the resonant frequency, but a longer call can be generated by repeated contraction of the muscles.

Examples of sounds emitted by fish using this mechanism are illustrated in Figs. 5.3, 5.4, 5.6 and 5.7. In an individual species, like the haddock, different calls may be produced in different contexts (Fig. 5.7), the calls essentially varying in their patterns of pulse modulation. Such variations will be apparent to a receiver capable of resolving the individual pulses in time. However, the variations also result in changes in the frequency structure, as shown by the spectrogram in Fig. 5.4. This analysis, performed with a narrow-tuned filter, is incapable of resolving rapid variations with time but shows a number of spaced frequency bands, the spacing varying with the temporal spacing of the individual pulses. Interestingly, temperature may exert marked effects upon the temporal characteristics of fish sounds. Calls made by the male goby, *Padogobius martensi*, are made up of rapidly repeated pulses, and both the pulse rate and duration of the calls are strongly affected by the environmental temperature (Torricelli *et al.*, 1990).

There is some evidence (review, Hawkins and Myrberg, 1983) that fish can discriminate the calls of their own species and that this is done by recognizing the particular pattern of pulses. For example, members from each of four species of damselfish of the genus *Eupomacentrus* can distinguish their own courtship chirps from those of other species by attending solely to the duration of the interpulse intervals. Indeed, it has been shown that males of the bicolour damselfish, *Pomacentrus partitus*, can recognize the individual sounds of their immediate neighbours (Myrberg and Riggio, 1985).

Communication by sounds

It is evident that the great majority of sounds emitted by fish are produced in a social context, and involve interaction between individuals. There may be some exceptions. There is a bathypelagic fish which is believed to produce a low-frequency echolocation call, perhaps serving to locate the sea-bed, though the evidence for this is unconvincing (Griffin, 1955). In general, however, the calls from fish have been recorded during an encounter between the vocalist and another fish of the same or a different species, and it is usually tacitly assumed that the call involves communication between the animals. Myrberg (1981), in seeking rigorous definition of this term 'communication', decided that it described the transfer of information between individuals, with the functional intent of gaining an adaptive advantage for the sender. This definition accepts that communication may involve individuals of differing species, and does not rule out the possibility that both sender and receiver may obtain mutual benefit.

Sound production commonly occurs in fish when an individual is disturbed by a predator or subjected to a noxious stimulus, as shown by the gurnard, *Trigla lucerna*. When disturbed, this species utters a short grunt while at the same time erecting prominent dorsal-fin spines and unfolding the large brightly marked and coloured pectorals. Generally, calls produced in this context are sharp, with a sudden onset, and they are often accompanied by a strong visual display, perhaps analogous to the 'flash' display shown by some insects. Though there is no documented evidence for fish that such calls drive away predators, or reduce the likelihood of a successful attack, if we argue that these calls are analogous to the well-studied startle displays and calls from insects and mammals (Edmunds, 1974), it is likely that they do serve this function. The same calls may also alert other fish, and help them to escape predators. The advantages to the sender in these circumstances are not clear, unless the fish alerted are kin-related. Myrberg termed this phenomenon 'interception', an attentive listener reacting to the call of another to its own advantage.

One of the most common contexts of sound production is during reproductive activity, where the calls may directly influence the behaviour of prospective mates. Sometimes the sounds accompany complex visual displays (Fig. 5.7), and it is possible that they form only part of a more complex signalling system. Often the calls are produced by the male fish, which in many species shows territorial behaviour. For example, male toadfish, *Opsanus tau*, occupy well-defined sites on the sea-bed and utter long and characteristic 'boatwhistle' calls, even in the apparent absence of prospective mates. It has been established that female fish may approach the sender of the sounds, giving rise to an increased rate of call production by the male (Gray and Winn, 1961). Similarly, male haddock will occupy the floor of an aquarium tank for long periods, swimming in an exaggerated manner and developing a characteristic pattern of pigmentation, while emitting a continual train of sound pulses. A male will often increase the rate of pulse production as another fish approaches, and may subsequently rise from the floor of the tank and lead the fish upwards, flicking the vertical fins while swinging from side to side, and increasing the rate of pulse production still further (Fig. 5.7). If the other fish is a female, this behaviour may lead to a spawning embrace. It is notable that in this species the drumming muscles are more highly developed in the males, especially in the spawning season. The female remains silent throughout courtship, though any males that approach the active male may engage it in an aggressive bout in which both participants produce sounds. From the contexts in which these reproductive calls occur, it seems likely that they serve to advertise the presence and reproductive readiness of the male sender to the females, and may even arouse reproductive activity in the latter. The calls may provide the basis for mate selection by the female. Field evidence has demonstrated that free-ranging female bicolour damselfish use the courtship sounds of conspecific males to locate male nest sites, and that they can distinguish the courtship sounds of different individual males, thus providing a basis for mate assessment (Myrberg *et al.*, 1986). Myrberg has remarked that the preponderance of territorial species producing such calls suggests that this kind of behaviour is characteristic of species whose sexes may be separated by considerable distances, or living in habitats where visual and chemical signals are inadequate.

Several studies, including those of Tavolga (1958) on the frillfin goby, *Bathygobius soporator*, have confirmed by the playback of recorded calls that the sounds may elicit a response even in the absence of the visual and other signals which normally accompany them. However, other workers have emphasized the limited response obtained in playback alone, and have stressed that sound is only part of a more complex assemblage of signals produced by the fish. Though there is evidence from birds that sound production by males may assist in stimulating reproductive maturation in the female, there is only a single observation on a mouthbrooding cichlid, *Oreochromis mossambicus*, that points to this possibility in fishes (Marshall, 1972).

There are several recorded instances of sound production by one male stimulating others to be vocal. Thus, in the haddock, another male may approach a sound-producing male and may engage it in an aggressive display, both fish emitting sounds. Among colonial males of various damselfish, *Eupomacentrus* spp. courtship chirping by an individual male may initiate chirping among neighbouring males on the reef, and some may move towards the territory of the initiator (Myrberg, 1972). These are almost certainly examples of the interception of a call. Since an imitator may lead an approaching female away from the initiator, they point to the adaptive significance of intercepting the calls of others (Krebs and Davies, 1978; and see Chapter 10 by Turner, this volume).

Aggressive fish are often vocal. Male croaking gouramis, *Trichopsis vittatus*, may participate in prolonged bouts, with butting, chasing and lateral displays by one male to another. However, there are numerous examples of fish producing sounds while competing for other resources, for example food or space. Female haddock, though silent during courtship, will readily produce sounds during competitive feeding outside the breeding season. Valinsky and Rigley (1981) have confirmed that sound production can provide significant benefits in such circumstances. Experimentally muted territorial residents of the loach, *Botia horae*, were unable to deter intruders from entering their shelter sites, despite appropriate visual displays. Intact and sham-operated fish were successful in repelling intruders.

It has been suggested that the social aggregation of fish (shoaling: see Chapter 12 by Pitcher and Parrish, this volume) may be facilitated by sounds (Moulton, 1960). Where fish swim in co-ordinated groups or schools, the motions set up in the water by the swimming fish may well be important in maintaining the cohesion of the school under poor visual conditions. It is evident that distinct pressure pulses (and fast water movements) are generated by fish as they make rapid swimming movements, or accelerate from rest. Gray and Denton (1991) have concluded that these pulses can be detected by fish at several fish lengths from the source fish. Furthermore, enough information is provided by the pulse to enable a neighbouring fish to determine the approximate position of the source, though there is no evidence at present that this information is utilized. Saithe, *Pollachius virens*, can school while temporarily blindfolded (Pitcher *et al.*, 1976), and cutting the posterior lateral line nerve indicates that the lateral line system of sense organs may play a significant role during normal schooling. The response in this case is probably to bulk motion of the water, or turbulence, rather than sound per se (see Chapter 7 by Bleckmann, this volume), but sounds may be important in maintaining social cohesion in other species, or over greater distances. So far, however, few sounds have been recorded from schooling species. Indeed, within a family like the Gadidae, sound production is often absent from the more actively schooling species like saithe. One disadvantage in using sound to promote cohesion of the school once assembled is that predators may intercept the sounds, perhaps eliminating any antipredator advantages provided by the shoaling habit. Various studies have shown that predatory fish, and especially sharks, may home in on the incidental sounds produced by struggling or injured prey.

5.5   THE HEARING ABILITIES OF FISH

The realization that many fish were sound producers, and that sound provided an effective channel for communication underwater, prompted an early interest in the hearing abilities of fish. By the end of the 19th century, the morphology of the fish ear had been well described, but critical experimental studies of the hearing characteristics of fish awaited two developments. The first was the application of conditioning techniques, pioneered by von Frisch and his associates, which enabled fish to be trained to respond unambiguously to sound stimuli. The second was the development of controlled electronic means for delivering and measuring underwater sounds. Since the early days, the hearing of a wide range of teleosts and elasmobranchs has been examined, both by means of conditioning experiments under controlled conditions, and by direct observation of the responses of free-ranging and captive fish to sounds. Of particular value have been experiments to determine threshold values for particular sound stimuli, where fish were conditioned to respond to high levels and the stimulus was then progressively reduced to determine the limiting level for the response.

These experiments have not been performed without difficulty. A special problem is the need to perform acoustic studies under suitable conditions, where sounds can be presented without distortion and subsequently measured with precision. Most aquarium tanks are deficient in this respect for the reasons outlined earlier. A range of special tanks have been constructed by different workers, and hydrophones sensitive to particle velocity have been developed to adequately monitor the sound stimuli. Some workers have performed their experimental studies in mid-water in large bodies of water, well away from reflecting boundaries, where measurements of the sound pressure can be used to calculate particle velocities (Hawkins, 1981).

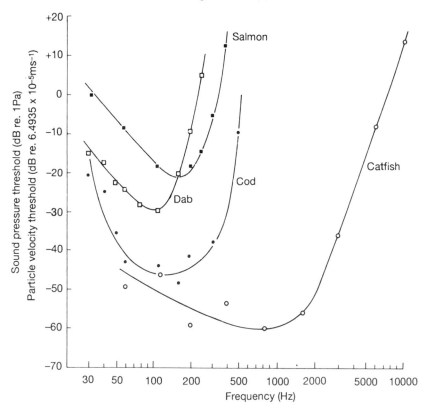

Fig. 5.8 Audiograms for four species of teleost fish, showing auditory thresholds or minimum sound levels detectable by the fish, at a range of different frequencies. The thresholds for the cod, *Gadus morhua* (Chapman and Hawkins, 1973) and freshwater catfish, *Ictalurus nebulosus* (Poggendorf, 1952) are given in terms of sound pressure. Those for the Atlantic Salmon, *Salmo salar* (Hawkins and Johnstone, 1978) and dab, *Limanda limanda* (Chapman and Sand, 1974) are expressed in terms of the particle velocity. Though the thresholds for all four species are directly comparable at a substantial distance from the sound source, in the near field, where the particle velocity increases steeply in relation to sound pressure, the sensitivity of the dab and salmon will increase relative to the other species. The sound pressure thresholds are given in decibels relative to 1 Pa (to convert to dB re. 1 μPa add 120 dB), and particle velocity thresholds are given in decibels relative to $6.4935 \times 10^{-5}\,\mathrm{m\,s^{-1}}$ (the velocity corresponding to a pressure of 1 Pa in the far field).

The sensitivity of fish to sound is conveniently expressed as an audiogram, a curve showing the thresholds or minimum sound levels to which the fish will respond over a range of frequencies. Examples for several species are given in Fig. 5.8. In general, most thresholds for fish have been determined in terms of the measured sound pressures. In some experiments, however, the ratio of sound pressure to particle velocity has been varied. In these circumstances the thresholds for some species follow the sound pressure; but in others the thresholds follow the particle velocity. Thus the units used to express the auditory thresholds in any audiogram differ, depending on the key stimulus. In Fig. 5.8, the audiograms for the cod, *Gadus morhua*, and catfish, *Ictalurus nebulosus*, are given in terms of sound pressure, while the thresholds for the Atlantic salmon, *Salmo salar* and dab, *Limanda limanda*, are given in terms of particle velocity. In an open body of water, distant from the sound source, the two parameters are of course proportional to one another and the audiograms are directly comparable, but within the near field, where particle velocity increases steeply against sound pressure, the sensitivities differ.

Several conclusions can be drawn from the audiograms published for a wide range of fish. In general, fish are sensitive to a rather restricted range of frequencies compared with terrestrial vertebrates, and especially mammals and birds. Even the best fish are relatively insensitive to sound at frequencies above 2 or 3 kHz whereas Man retains a sensitivity above 15 kHz, and some mammals – including aquatic forms – can detect frequencies of over 100 kHz. There is some uncertainty over the lowest frequencies to which fish are sensitive. In many experimental tanks, low-frequency background noise levels are high, limiting the sensitivity that can be shown by fish (see below). It has been suggested, that the cod may show acute sensitivity even to infrasonic frequencies below 1 Hz (Sand and Karlsen, 1986) however this may simply reflect the ability of the fish to detect linear accelerations.

Within their restricted frequency range, many fish are acutely sensitive to sounds, especially those such as the cod and catfish, which respond to sound pressure. Indeed, it has been established that in the sea the cod is not limited by its absolute sensitivity, but by its inability to detect sounds against the background of ambient noise, even under relatively quiet sea conditions (Chapman and Hawkins, 1973). Any increase in the level of ambient noise, either naturally as a result of a storm, or imposed artificially by replaying broad-band white noise, results in an increase in the auditory threshold (a decline in sensitivity), as shown in Fig. 5.9. Many of the differences in sensitivity seen in the audiograms of different species are probably the result of different noise levels prevailing during the experiments. These differences are most evident at low frequencies, where noise levels are much more variable.

Some fish are inherently less sensitive to sounds and are only rarely limited by naturally occurring noise levels. These insensitive species, including the salmon (Fig. 5.8 and 5.9) and flatfish such as the dab and plaice, *Pleuronectes platessa*, have been shown to be sensitive to particle velocity (Fig. 5.10). We have already seen that particle velocity amplitudes in water are very low. Even the best man-made velocity hydrophones are very insensitive compared with their pressure counterparts.

Whether a fish is sensitive to sound pressure or particle velocity depends upon the presence of a gas-filled swim bladder, this organ playing a key role

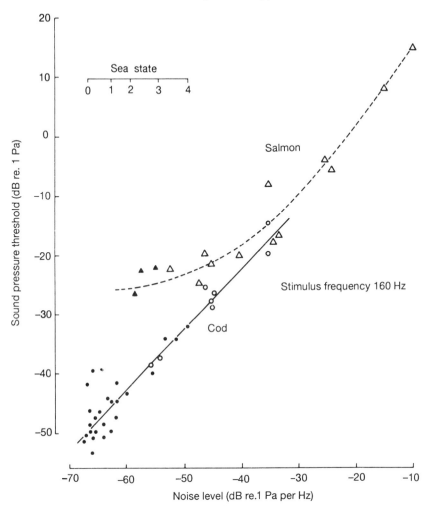

Fig. 5.9 Changes in auditory threshold for cod and Atlantic salmon at 160 Hz, accompanying variations in the level of background noise in the sea. Solid symbols represent measurements made at different natural ambient noise levels; open symbols represent measurements made at higher levels, broadcast from a loudspeaker. The higher the noise level, the higher the threshold through auditory masking. Note that the hearing of the cod is often impaired at quite moderate ambient levels, whereas the hearing of the salmon is only masked at much higher levels (corresponding to sea states greater than 4). The data for the cod, *Gadus morhua*, are taken from Chapman and Hawkins (1973) and those for the salmon, *Salmo salar*, from Hawkins and Johnstone (1978). To convert sound pressure to dB re. 1 μPa add 120 dB.

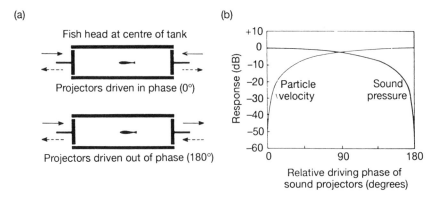

(a)

Fish head at centre of tank

Projectors driven in phase (0°)

Projectors driven out of phase (180°)

(b)

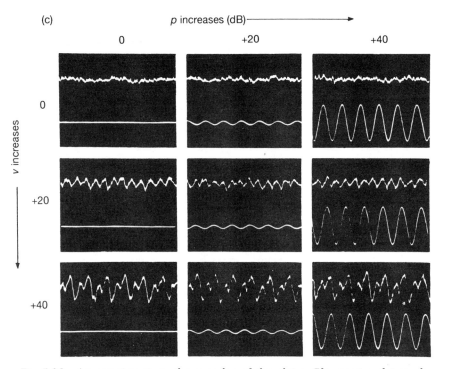

(c)

Fig. 5.10 An experiment on the sacculus of the plaice, *Pleuronectes platessa*, has confirmed that the otolith organ essentially responds to particle motion rather than sound pressure (Hawkins and MacLennan, 1976). The fish was placed at the centre of a standing wave tank (a), where the amplitudes of particle velocity and sound pressure could be varied independently by adjusting the amplitudes and phases of two opposing sound projectors (b). (c) Changes in particle velocity resulted in pronounced changes in the receptor potentials recorded at the saccular macula (the saccular microphonics). Changes in the sound pressure (*p*) had no effect upon the potentials. In these oscillograms the microphonics are on the upper beam and sound pressure is on the lower beam.

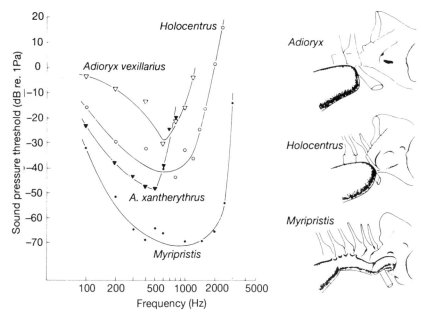

Fig. 5.11 Audiograms of four species of squirrelfish (family Holocentridae). *Myripris-tis kuntee* yields lower sound pressure thresholds and responds over a wider frequency range than other holocentrids. The greater sensitivity of this species is associated with a linkage between the gas-filled swim bladder and the ear. Anteriorly, the swim bladder is divided into two lobes, the medial faces of which make strong contact with a thin membrane in the wall of the auditory bulla, lateral to the sacculus of the ear. In other holocentrids, the anterior of the swim bladder lies close to the skull or in contact with it (as in *Holocentrus ascensionis*). The audiograms for *Adioryx xantherythrus* and *Myripristis kuntee* are taken from Coombs and Popper (1979), and those for *Adioryx vexillarius* and *Holocentrus ascensionis* from Tavolga and Wodinsky (1963). The morphology of the swim bladder and rear part of the skull is redrawn from Nelson (1955).

in hearing as well as in sound production. A linkage between the swim bladder and the ear is characteristic of those fish that are sensitive to sound pressure. Indeed, both the absolute sensitivity of fish and their frequency range appear to depend upon the degree of association between the swim bladder and the ear. Thus, the cypriniform or ostariophysan fish, which have a close connection between the two, show a very acute sensitivity to sounds and have an extended frequency range, as in the catfish, *Ictalurus nebulosus* (Fig. 5.8). In the cod, the anterior portion of the swim bladder is simply placed close to the ear and this species both is less sensitive and has a more restricted frequency range than the catfish. Within a particular family like the Holo-centridae the audiograms may vary, depending on the degree of association between the swim bladder and the ear (Fig. 5.11).

We have seen that the hearing of a fish like the cod may be affected by the prevailing level of ambient noise. Where thresholds are masked in this way, it can be shown that not all frequency components of the background noise are equally effective at impairing detection (Hawkins and Chapman, 1975). If high-level noise is transmitted in a relatively narrow frequency band, successively tuned to different frequencies, the degree of masking for a particular tone is strongly dependent upon the frequency of the noise. Detection of the tone is masked most effectively by noise at the same or immediately adjacent frequencies. This observation provides evidence that interaction between the stimulus and the noise is confined to a narrow range of frequencies on either side of the stimulus. The cod, the goldfish, *Carassius auratus*, and even the salmon appear to possess an auditory filter, capable of being tuned to any frequency of interest, thereby improving the ability of fish to detect signals in the presence of high levels of noise (Tavolga, 1974; Hawkins and Chapman, 1975; Hawkins and Johnstone, 1978). The filter is narrower in the cod than it is in the salmon, the latter (with its reduced sensitivity) being less likely to have its hearing impaired by ambient noise.

The presence of a frequency-selective auditory filter suggests that fish may also be able to distinguish between different frequencies. Behavioural discrimination experiments have confirmed this (Dijkgraaf, 1952; Fay and Popper, 1980). The goldfish can separate tones which differ in frequency by as little as 3–5%, an ability which is poorer than that of Man but broadly comparable to that of many mammals and birds. However, there may be significant differences between fish. Cypriniformes like the goldfish and the European minnow, *Phoxinus phoxinus*, appear to be better than others like the marine goby, *Gobius niger*, and the freshwater bullhead, *Cottus gobio*, at discriminating frequency.

Fish are also able to discriminate signals that differ in amplitude. The goldfish can discriminate 300 Hz tone pulses which differ by 4 dB, and the cod and haddock discriminate 50 Hz tone pulses differing by as little as 1.3 dB. We have seen that many fish calls differ in their patterns of pulse modulation. It is therefore especially interesting to know whether fish can resolve the individual pulses, distinguishing calls by their temporal patterning, or whether the pulses are run together by the auditory system, so that the calls must be distinguished in some other way (perhaps through differences in their frequency spectrum). In fact, the goldfish appears to be able to resolve short-duration pulses much better than Man. Fay (1982) has pointed out that the teleost ear may be well adapted for preserving the fine temporal structure of sounds. The goldfish can also discriminate between sounds that differ in phase; that is, between sounds that begin with a compression and those that begin with a rarefaction, an ability not possessed by Man.

In a medium like water where light levels are low and vision is often impaired, an ability to locate the position of a sound source in space is likely to be especially important. Despite initial doubts, it is now firmly established that some fish have this ability (the evidence is reviewed by Schuijf and Buwalda, 1980). The only species for which extensive experimental data are available, however, is the cod. Field experiments have shown that cod are able to discriminate between spatially separated loudspeakers in both the horizontal and vertical planes, and also that they can orientate towards particular sources. Field observations on predatory fish, and especially sharks, have shown that fish may locate and track down their prey by means of sound, often over large distances (Myrberg *et al.*, 1976). In several respects the auditory localization abilities of fish may exceed those of terrestrial vertebrates. Cod can discriminate between sound sources at different distances (Schuijf and Hawkins, 1983) and can discriminate between diametrically opposed loudspeakers in both the horizontal and vertical planes under circumstances which are ambiguous or confusing for Man (Buwalda *et al.*, 1983). Living in an environment where vision at a distance is often precluded, fish may depend heavily upon the ear for information about their distant surroundings.

5.6   ANATOMY AND ORGANIZATION OF THE AUDITORY SYSTEM

The main sound receptors of fish are the otolith organs of the inner ears. The ears are paired structures embedded in the cranium on either side of the head (Fig. 5.12) close to the midbrain. There are no obvious external structures to indicate their presence, though in the clupeoid fishes there is indirect connection to the exterior via the lateral line system (see below) and in some other fish there is a small opening by way of a narrow endolymphatic duct.

Each ear is a complicated structure of canals, sacs and ducts filled with endolymph, a fluid with a particular ionic composition and special viscous properties. In elasmobranchs and teleosts the ear has three semicircular canals, each incorporating a bulbous expansion, the ampulla, occluded by a jelly-like flap or diaphragm, the cupula. The canals are arranged orthogonally (at mutual right angles). Angular accelerations of the head cause the endolymph to lag behind the movements of the canal, deflecting the cupula and stimulating a population of sensory hair cells mounted on the crest of a saddle-shaped wall extending across the ampulla, the crista.

Three expanded sacs within the ear are linked with one another, and with the semicircular canals (Figs 5.12 and 5.13). The most superior sac, the utriculus, communicates directly with the lumen of the semicircular canals, and with them forms the pars superior. The sacculus communicates with the utriculus by a very small aperture, and also with a posterior diverticulum, the lagena. The sacculus and lagena together constitute the pars inferior. In the teleosts, each of these sacs contains an otolith, a dense mass or stone of calcium carbonate and other inorganic salts within a protein matrix, sitting upon a

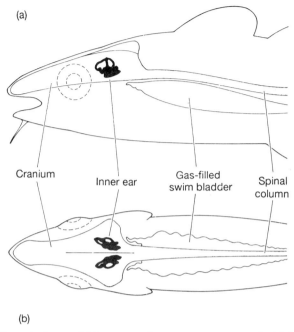

(a)

Cranium

Inner ear

Gas-filled
swim bladder

Spinal
column

(b)

Fig. 5.12   The paired ears of fish are membranous structures embedded in fluid-filled spaces within the cranium, on either side of the midbrain. The position of the two ears is shown for the cod in lateral (a) and dorsal (b) views. The position of the swim bladder in relation to the ears is also shown.

sensory membrane or macula containing many mechanoreceptive hair cells. The body of the otolith is separated from the delicate hair cells by a thin otolithic membrane, which may extend over parts of the macula not covered by the otolith. In elasmobranchs (and in higher vertebrates), the otolith is replaced by a jelly-like cupula containing many small spherules of calcium carbonate, the otoconia. There is an additional macula with an unloaded cupula in elasmobranchs and some teleosts, the macula neglecta, which is often adjacent to the endolymphatic duct.

The various sacs and their enclosed cupulae and otoliths vary in size, orientation and shape from one species to another. The sacculus is generally the largest, but in some catfish, and in the Clupeiformes, the utriculus may exceed the sacculus in size. In the holostean *Amia calva*, and in many Cypriniformes, the lagena is particularly large. The utricular macula and the otolith that surmounts it usually lie predominantly in the horizontal plane, and the sacculus and lagena lie in different vertical planes (Figs 5.12 and 5.13); the maculae are often twisted. The attachment of the otoliths to the macula is only poorly understood. Though the otolithic membrane appears to

(a)

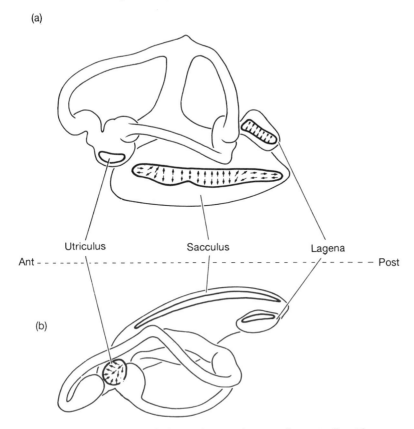

(b)

Fig. 5.13 The main parts of the cod ear, shown schematically. The sensory membranes or maculae of the three otolith organs are shown, with the orientation of the sensory hair cells indicated by arrows. (a) Lateral view of left ear. (b) Dorsal view of left ear.

be thin and fragile, it may serve in some species to suspend or restrain the movements of the otolith. Most otoliths have a complex, sculptured shape, and some appear to have flanges or keels which may be important in influencing their freedom of movement. The otoliths may rotate about their own axes, or move along a curvilinear path, rather than show simple linear translation.

The ears are innervated by the eighth cranial nerve, which sends rami to each of the ampullary organs, and to the various maculae. The conformation of the different nerve branches is often complicated, with the finer ampullary rami combining with the larger rami to the otolith organs. The various branches may also be overlain by other nerve trunks, especially those of the vertical line, which issue from several cranial nerve roots.

Hair cells

The epithelial, mechanoreceptive hair cells of the ear and lateral line neuromasts are found in all vertebrate classes, and show strong structural similarities throughout the group. The physiology of hair cells has been

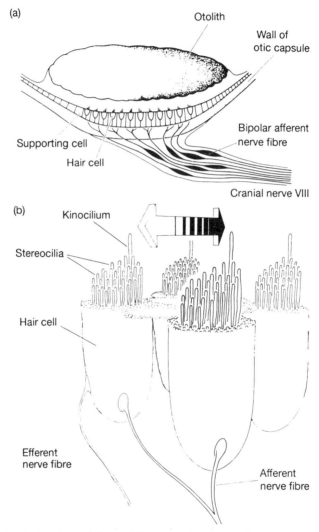

Fig. 5.14 Fine structure of the otolith organ. (a) Schematic cross-section through the utriculus, showing the otolith mounted on the hair cells but separated from the cilia of the cells by an otolithic membrane. Bipolar afferent fibres and finer efferent fibres synapse with the hair cells. (b) The hair cells are directional in their physiological response to stimulation, having a distinct axis defined by the position of the kinocilium. Adjacent hair cells often have a common axis, and many hair cells often synapse with a single afferent fibre.

reviewed by Ashmore and Russell (1983). In fish, the hair cells are typically elongated and cylindrical, surrounded by supporting cells on a firm connective-tissue base. Bipolar afferent nerve fibres synapse with the base of the cells and pass through the basement membrane of connective tissue into the auditory nerve (Fig. 5.14). Finer efferent fibres also terminate on the hair cells. The latter are strengthened at their apical ends by a cuticular plate surmounted by a sensory process made up of cilia embedded within the otolithic membrane. Many stereocilia, packed with microfilaments, are grouped together with steadily increasing length towards a longer, eccentrically placed kinocilium, containing nine double microtubules and two central single tubules. The positioning of the kinocilium towards one side of the cell gives the apical end of the cell a pronounced structural asymmetry.

The hair cell essentially responds to mechanical deflection of its sensory process. Electrical potentials exist at the apical and basolateral ends of the cell, due to differences in ionic content between the extra- and intracellular fluids. Deformation of the cell at a sensitive locus at the apical end results in ionic flow across the cell membrane, producing a progressive depolarization which acts at the base to modulate the release of a chemical transmitter at the afferent synapse. The changing receptor potentials are carried primarily by potassium ions, with calcium as a necessary cofactor.

The transduction properties of the hair cells are influenced by their external mechanical connections, and also by the arrangement of the stereocilia. The stereocilia themselves contain the protein actin, which may influence their mechanical properties. Mechanical constraints and electrical resonances within the cell may tune the cells to particular frequencies. In fish the tuning would appear to be rather broad (see below), but in reptiles, birds and mammals it can be very sharp, with some of the hair cells tuned to high frequencies and others to low. There is evidence that the efferent nerve fibres to the cell may regulate the tuning.

An important feature of the hair cell is that it is directional in its response to mechanical stimulation. Depolarization of the cell and excitation of the afferent fibres are most pronounced when the stereocilia are deflected in the direction of the kinocilium, and hyperpolarization and inhibition of the afferent fibres result when the stereocilia are deflected by shearing forces acting along the same axis in the opposite direction (Fig. 5.14). The response is non-linear and asymmetrical, the positive response resulting from movement towards the kinocilium gradually rising to a higher level than the negative response resulting from movement in the reverse direction. The physiological basis for this directionality is not yet established, but it may be related to the arrangement of the graded stereocilia and filaments that connect them, with the cuticular plate also playing a part.

Scanning electron microscopy has revealed particular patterns in the arrangement of the polarized hair cells in the various sensory epithelia of the octavo-lateralis system. Hair cells in the cristae of the ampullae of the semicircular canals share the same axis and the same polarization. Thus, all the afferent fibres are excited by fluid motion in one direction, and inhibited by motion in the reverse direction. A wider range of hair cell polarization patterns is found in the maculae of the otolith organs. Generally, each macula may be subdivided into a number of regions, within which all the hair cells are morphologically polarized in the same direction, but the pattern may vary greatly between the different maculae (Fig. 5.13). Moreover, between families, and even species, the pattern can vary for any particular macula, reflecting major differences in the shape, size and orientation of the otolith itself. At least five different hair cell patterns have been identified for the saccular macula (Platt and Popper, 1981). Within any macula it is common to find a bidirectional arrangement of hair cells, whereby the presence of a group of hair cells polarized in one direction is usually paralleled by a group of cells polarized in the opposite direction (Fig. 5.13). The pattern may be much more complicated, however. The utriculus, in particular, often has a wide range of hair cell orientations. It must not be forgotten that most diagrams illustrating patterns of hair cell polarization are drawn in two dimensions, whereas the maculae themselves are often curved or twisted.

The functional significances of these different structural patterns in different taxonomic groups of fish are poorly understood.

Within each macula there are different 'types' of hair cell, with varying heights of ciliary bundles and differing relative lengths of the kinocilium and stereocilia. It appears that most afferent nerve fibres synapse only with hair cells of a particular polarization, and respond only to stimulation of the cell in one direction. However, it is clear from the relatively large numbers of hair cells in relation to the few innervating fibres that there is a great convergence upon each fibre, and each fibre may innervate widely separated hair cells. There is much scope for differing patterns of response, depending upon the numbers, types and orientation of the hair cells synapsing with each afferent fibre.

Accessory structures

In many fish the inner ears stand alone, with no ancillary structures or attachments. In others, however, there are well-defined structural linkages with gas-filled cavities. Best known are fish of the order Cypriniformes (the ostariophysan fish), where the anterior end of the swim bladder is coupled to the ear by a chain of movable bones, the Weberian ossicles. Expansion or contraction of the anterior chamber of the bilobed swim bladder results in motion of the ossicles. This motion subsequently causes fluid motion in a small sinus, filled with perilymph, which is then communicated to an endolymphatic transverse canal connecting with the lumen of both saccular chambers. Thus, motion of the anterior wall of the swim bladder results in a deflection of the saccular otolith and stimulation of the hair cells. There is evidence that the wall of the anterior chamber of the swim bladder is kept taut by maintenance of the gas at a slight excess pressure. Small changes in depth by the fish might be expected to move the ossicles to the limit of their range and restrict the functioning of this mechanism, but Alexander (1959) has argued that the high compliance of the swim bladder wall and its high viscosity cause it to act as a high-pass filter, accommodating changes in hydrostatic pressure while still enabling the system of ossicles to respond to rapid variations in pressure.

In the Clupeiformes there is a very different coupling with the ear, the swim bladder entering the cranial cavity. The system in herring, *Clupea harengus*, has been described by Allen *et al.* (1976). The central feature is a pair of pro-otic bullae, each divided into gas-filled and liquid-filled parts by a membrane under tension. The upper part contains perilymph, connected with that of the labyrinth by a fenestra in the upper wall of the bulla. Lateral to the fenestra is a compliant membrane (the lateral recess membrane), positioned in the skull wall at the back of the lateral recess from which all the lateral line canals radiate. The gas-filled part of the bulla is connected to the swim bladder by a long gas-filled duct. When the fish changes depth, the main part of the swim bladder, which has a more compliant wall than the bulla membrane, accommodates the change in hydrostatic pressure by changing volume. Gas may then pass along the duct to equalize pressure between the swim bladder and the bulla. By this means, the volume of the gas-filled part of the bulla remains constant as the fish changes depth. The swim bladder essentially acts as a reservoir of gas for the pro-otic bullae.

Rapid motion of the membrane in the bulla generates motion in the perilymph which is transmitted to the sense organs of the inner ear and lateral line. Thus, changes in pressure in the bulla lead to displacements of the perilymph which are transmitted to the macula of the utriculus, sacculus, and perhaps also the lagena, displacing the sensory processes of the hair cells.

Many other teleosts possess a modification of the anterior end of the swim bladder which may influence the functioning of the ear. In the Mormyridae the swim bladder enters the intracranial space, whereas in others, for example some of the Holocentridae and Sparidae (Perciformes), and the Moridae (Gadiformes), the swim bladder is attached extracranially to the skull, adjacent to the sacculus. In other fishes, alternative gas-filled spaces are utilized, including air-filled branchial cavities in the Anabantidae, and subpharyngeal cavities in the Channidae. Moreover, there is experimental evidence that even in a species like the cod without direct connections between the swim bladder and the ear, the gas contained within the organ may play a part in hearing.

Where a connection exists between the swim bladder and the ear, there may be a further specialization at the macular level. Thus, a highly specialized tripartite utricular macula is found in the Clupeiformes. On the other hand, in the Cypriniformes and Mormyridae, a very simple pattern of hair-cell orientation is encountered in the saccular macula, with two groups of vertically orientated hair cells in opposition to one another.

5.7   HEARING MECHANISMS IN FISH

Role of the otolith organs

The simple otolith organ serves several functions. First, it serves as a gravity receptor, enabling the fish to determine its orientation with respect to the Earth's gravitational field. The heavy otolith tends to shift as the head of the

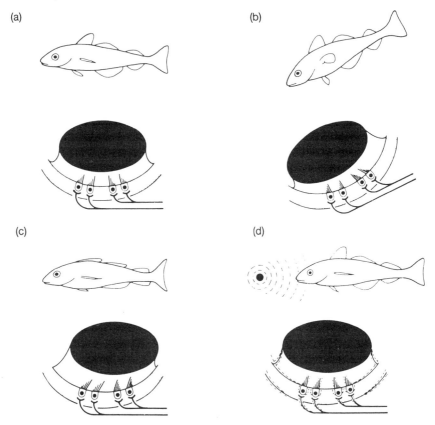

Fig. 5.15   The otolith organs appear to serve several functions. In (a), the fish is stationary and the otolith is in its resting position. In (b) the fish is tilted with respect to the Earth's gravitational field and the otolith also tilts, deflecting the cilia of the hair cells. In (c) the fish is accelerating forwards, the heavy mass of the otolith lagging behind and deflecting the cilia. In (d) the fish is stimulated by a sound wave. The fish flesh is acoustically transparent and moves back and forth with the sound wave. The dense otolith lags behind, creating an oscillatory deflection of the cilia.

fish tilts, deflecting the sensory processes of the hair cells (Fig. 5.15). Such a system is also sensitive to linear acceleration, the otolith tending to lag behind the accelerating fish, or overshooting when the body rapidly comes to rest. In birds and mammals, the detection of these accelerational forces appears to be the main function of the otolith organs. This sensitivity to acceleration may provide a basis for inertial navigation and may explain the apparent sensitivity of the fish to very low frequency sounds (Sand and Karlsen, 1986).

In fish surgical elimination experiments have confirmed that the otolith organs also play a role in sound reception. If a fish without a swim bladder (e.g. the plaice) is placed in a standing wave tank, where the particle motion and sound pressure can be varied independently, the summed extracellular receptor potentials (or microphonics) from the hair cells of the sacculus respond only to changes in particle motion. Variations in the sound pressure have no effect (Fig. 5.10). This evidence that the otolith organ is essentially driven by particle motion (either particle acceleration or particle velocity) is supported by field experiments on the dab, *Limanda limanda*, and the Atlantic salmon, *Salmo salar*. If auditory thresholds are measured for these fish at different distances from the sound source, the sound pressure thresholds measured within the near field are lower, confirming that the fish respond to the greater amplitudes of particle motion close to the source. Pumphrey (1950) suggested that the wave of particle motion passing through the body of these fish moves the tissues of the head, which have a similar acoustic impedance to water, but the dense otoliths or otoconia lag behind, creating a shearing force to stimulate the hair cells (Fig. 5.15).

Particle motion amplitudes in water are very low, especially distant from a sound source, and though the hair cells are believed to be extremely sensitive to mechanical displacement, it is unlikely that such a system can provide for the detection of very weak sounds or operate over a wide frequency range. Certainly, fish like the dab and salmon are relatively insensitive to sounds and have a narrow frequency range. A simple mathematical model of the otolith and its suspension, suggested by de Vries (1956), predicts that the otolith is heavily damped, with a rather low natural frequency of vibration. Its amplitude of motion will progressively decline at frequencies above the natural frequency, which de Vries suggests occurs at a few hundred hertz. In fact, the audiograms of the dab and salmon decline steeply above about 150 kHz.

Role of gas-filled cavities

Those fish having a close association between the swim bladder and the ear are sensitive to sound pressure. If a dwarf catfish is placed in a tank with a strong gradient of particle motion, the same sound pressure audiogram is obtained at all positions. Moreover, in field experiments, the cod – unlike the dab and salmon – shows similar sound pressure thresholds at higher frequencies even within the near field of the source. Surgical experiments on the

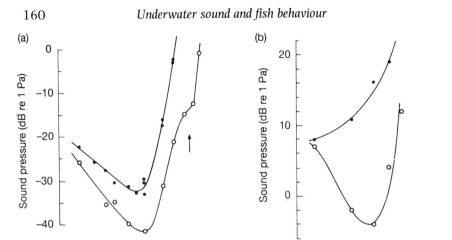

Fig. 5.16 The role of a gas-filled space in hearing. (a) Sound pressure thresholds for two dabs, obtained with ● and without (○) an air-filled balloon beneath the head of the fish. The arrow indicates the resonance frequency of the balloon. Redrawn from Chapman and Sand (1974). (b) Audiograms of two cod with empty (●) and full (○) swim bladders, respectively. Redrawn from Sand and Enger (1973). The audiograms were obtained by measuring the sound pressures necessary to evoke a given amplitude of saccular microphonic potential. They indicate changes in relative sensitivity rather than in the absolute sound pressure thresholds. To convert sound pressures to dB re. 1 μPa add 120 dB.

dwarf catfish have shown that interference with the Weberian ossicles or deflation of the swim bladder results in a decline in sensitivity to sounds. In the cod, which simply has the swim bladder close to the ear, deflation of the swim bladder results in a pronounced drop in the amplitude of microphonic potentials recorded at the saccular macula (Fig. 5.16). Even more remarkable is the observation that placing a small inflated balloon close to the head of a fish lacking a swim bladder (the dab) gives an increased sensitivity, and a more extended frequency range (Chapman and Sand, 1974).

It would seem that the gas-filled cavity acts as an acoustic transformer. Incident sound pressures cause the compressible body of gas within the organ to pulsate, generating a much higher amplitude of particle motion than would otherwise have existed. These locally high particle motions may be coupled directly to the otolith organs of the inner ear, or may simply propagate through the surrounding tissues to stimulate the otolith organs. Various authors have assumed that there is a close correspondence between the behaviour of the swim bladder and that of a free gas bubble in water. The latter can be regarded as a simple mass/spring system, where the spring factor is provided by the low elastic modulus of the contained gas, and the mass results from the high inertia of the surrounding water. If such a bubble is exposed to sound pressures of equivalent amplitude but varying frequency, its mechanical response reaches

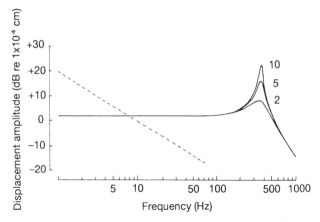

Fig. 5.17 Calculated curves showing the pulsation amplitude at different frequencies of a 1.5 cm radius gas bubble at 20 m depth in a sound field. The sound pressure is kept constant at −20 dB re. Pa. The amplitudes are shown for different degrees of damping of the motion (*Q* values are shown). The water displacements accompanying the same sound pressure in the absence of a gas bubble are shown as a dashed line for comparison. Note that the bubble amplifies the local particle motion over a wide range of frequencies.

a maximum at a particular frequency, the resonant frequency (Fig. 5.17). The sharpness of the resonance depends upon the degree of damping of the bubble, whereas the resonant frequency depends upon the hydrostatic pressure, the volume of gas, and several other factors. If the gas-containing organ behaved acoustically as a free gas bubble, the resonance would provide the animal with a highly sensitive receiver, but there would be other effects. The resonant frequency would depend upon the size of the fish and its depth, and the tuning of the auditory system would vary accordingly. There would be a time delay in detecting sounds, since a lightly damped resonant system takes time to build up to its maximum response. This would have the effect of reducing the time resolution of the system, since it would be unable to preserve rapid amplitude modulations. Moreover, the auditory system would no longer preserve the phase of sounds which differed in frequency, since the phase of response of the gas bubble varies on either side of the resonance.

The actual properties of a gas-filled cavity can be measured directly, either optically or by means of appropriately placed hydrophones. Measurements on the swim bladders of intact, living cod show that the swim bladder is highly damped, and that the resonance frequency is generally well above the hearing range of the fish (Sand and Hawkins, 1973). It provides moderate amplification over a wide frequency range, with little phase distortion, and preserves the ability of the auditory system to respond rapidly to amplitude modulations. If the fish is forced to change depth, the damping at first becomes lighter, but is rapidly restored by the fish. In the herring, with its auditory bullae, the response of the mechanical system extends up to 1 kHz, but falls off rapidly above this frequency and shows an increasing phase lag. Thus, there is evidence that the gas-filled chambers in fish are not highly tuned, with all the attendant disadvantages. Rather, they amplify the particle motion over a wide frequency range with minimal distortion.

Analysis of sound quality

Because the human ear is able to separate sounds into their component frequencies, it is often assumed that other animals distinguish between sounds in a similar fashion. Fish are certainly capable of distinguishing between tones of differing frequency, and masking experiments show that some frequency filtering is performed within the auditory system. However, in fish there is no obvious morphological frequency analyser analogous to the mechanical filter provided by the cochlea of mammals. Behavioural studies of sound communication have indicated that fish discriminate between calls on the basis of differences in repetition rate and duration, rather than frequency or bandwidth (Fine, 1978; Myrberg, 1981). There has therefore been some controversy about whether fish distinguish between sounds through differences in their frequency spectra or their fluctuations in amplitude with time. Though these characteristics are not completely independent, a device for spectral analysis will be quite different from one for temporal analysis. The former will contain narrow frequency filters and show poor time resolution, whereas the latter will have a wide bandwidth but show good temporal resolution.

In the absence of any evidence in fish for a 'place' mechanism for frequency discrimination, analogous to that found in the mammalian ear, it is often assumed that some form of temporal analysis of sounds is carried out by fish. There have been several attempts to examine the mechanisms of analysis through study of the activity of afferent neurones from the otolith organs to the brain in fish. Many, though not all, of the neurones from all three otolith organs are spontaneously active, even under quiet conditions. Some of these increase their discharge rates when stimulated with sound, and most show a high degree of synchrony of firing with the waveform of the sound stimulus, giving one or more spike discharges for every cycle of the stimulus, the spikes being locked to a particular phase of the stimulus cycle. Some neurones respond with a high synchrony to the stimulus, but without showing any change of discharge rate. The frequency response of the individual neurones can be measured by comparing the firing rate, or the degree of synchrony, at different frequencies and amplitudes. The response curves obtained are rather broad, but in several species different neurones are tuned to different frequencies or at least their upper frequency limits vary. However, there is little evidence that the tuning is sufficiently sharp to explain the degree of frequency discrimination obtained from fish in behavioural experiments, and certainly there is no evidence for an array of fibres tuned to different frequencies and innervating different parts of the sensory macula as exists in the cochlea of mammals and to a lesser degree the maculae of amphibians. There is some peripheral filtering of the received signals, but essentially the waveform of the received signal is coded by the discharge rate of the neurones, their firing retaining a high degree of synchrony with the stimulus waveform, often down to very low signal levels. These findings have not entirely ruled out a 'place' mechanism of frequency discrimination. Some authors have pointed to the complex patterns of hair cell orientation in the fish ear, and have suggested that the movement patterns of the otoliths, and the parts of the macula stimulated, may depend upon the frequency. However, closer study of the biophysics of the fish ear is necessary before these speculations can be confirmed.

Fay (1982) has drawn attention to the ability of the goldfish ear to detect rapid amplitude fluctuations in both tonal and noise signals. He suggests that the auditory system of the goldfish is particularly well adapted for temporal resolution, both amplitude modulation of a tonal signal and the frequency of the signal itself being coded by the discharge patterns of the auditory neurones. In this respect, goldfish may differ significantly from Man. Certainly the goldfish is able to discriminate much more rapid amplitude modulation, probably basing the discrimination on temporal variations in the signal rather than spectral cues.

Mechanisms of directional hearing

Van Bergeijk (1964) originally suggested that directional hearing in fish was entirely dependent upon the lateral line system (see Chapter 7 by Bleckmann, this volume). However, surgical elimination experiments on the cod have shown that both the ears are essential for directional detection at a distance from the source, as in terrestrial vertebrates. It is highly unlikely, however, that fish utilize the same direction-finding mechanisms as the latter. In Man the directional cues result from the physical separation of the two ears. Differences in the length of the stimulus path between the source and each of the ears give rise to differences in the time of arrival or phase of the sounds. In addition, at higher frequencies, the head interferes with sound propagation, and has a shadowing effect, causing differences in stimulus intensity at the two ears which are dependent upon direction. For fish, the high velocity of sound in water and the close proximity of the two ears means that differences in stimulus timing are minimal. Moreover, most fish are small relative to the sound wavelength of interest to them, reducing sound shadowing effects, and fish flesh is similar in acoustic properties to the surrounding water, rendering the head effectively transparent. Together, these factors minimize interaural intensity differences. These difficulties, together with the linking of the two ears to the swim bladder, forming a single receptor, led van Bergeijk to assert that directional hearing by means of the ear was not possible.

There is evidence, however, that the otolith organs can provide a basis for the detection of direction. The hair cells of the inner ear have a definite axis of sensitivity, indicated by the position of the eccentrically placed kinocilium. There are orderly patterns of hair cell orientation within each macula, and there is evidence that this segregation is preserved at the level of the primary afferent neurones (Hawkins and Horner, 1981). The movements of the otolith are essentially driven by particle motion, which takes place along a radial axis from the source. It might, therefore, be expected that these movements will stimulate hair cells of differing orientation to a differing degree. By this means the fish should be able to determine the axis of propagation by a process of vector weighing, comparing the outputs of differently orientated groups of hair cells. It has been confirmed by electrophysiological experiments (Fig. 5.18) that particular neurones respond best to stimulation along a particular axis, and that this axis differs from one neurone to another and between otolith organs (Hawkins and Horner, 1981). Given that the fish has two ears, each containing three otolith organs that are potentially sensitive to sound, and each with its own distinctive pattern of hair cell orientation, then a system is available which is potentially capable of determining the axis of propagation in three-dimensional space.

There are two flaws in this model of directional detection. First, detection of the axis of propagation does not in itself indicate the bearing of the source. Particle motion alternately takes place towards and away from the source, and the hair cells are inherently bidirectional, so a simple vector weighing of the kind proposed yields a 180° ambiguity in the detection of direction. Secondly, it is difficult to understand the role of the swim bladder in this process. It would appear that in fish such as the cod and goldfish, in addition to the direct vectorial input from the source, the otolith organs must also receive indirect stimulation from the swim bladder, which carries no directional information per se and which might be expected to dominate the direct signal.

In practice, experiments in mid-water in the sea have shown that the cod can discriminate between opposing sound sources (180° apart) in both the horizontal and vertical planes. Significantly, however, the phase relationship between sound pressure and particle motion is crucial for the fish to perform this discrimination (Buwalda *et al.*, 1983). If the sound is switched to a second opposing source, but with the expected inversion of the phase of sound pressure with respect to particle motion locally abolished (by the addition of a standing wave), the fish is no longer able to discriminate between the two sources. Alternatively, if the sound emanating from a single source is simulated to come from an opposing source by locally inverting the phase relationship, the fish responds as if the sound were coming from the opposing source. Thus, phase comparison between sound pressure and particle motion appears to provide the basis for eliminating the ambiguity in directional detection. This suggests that, far from interfering with the detection of the

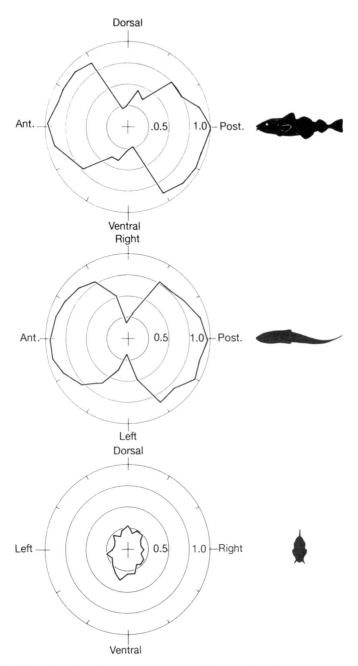

Fig. 5.18 Directionality of a single unit recorded from the left anterior sacculus of the cod, *Gadus morhua*. The polar diagrams represent the relative response recorded from a single afferent nerve fibre (in terms of spikes per hertz), at different angles in three mutually perpendicular planes. The response is highly directional in both the median vertical and horizontal planes.

signals reaching the ear directly from the source, the signal re-radiated from the swim bladder is essential for the elimination of ambiguity.

The precise details of the auditory mechanisms used by the cod to discriminate direction need further examination. It is not yet clear whether all the otolith organs of the ear are implicated, not is it clear whether certain parts of the ear are isolated from stimulation via the swim bladder. It is particularly difficult to determine how far the model developed for the cod can be applied to other species of fish, including those lacking a swim bladder. Progress in achieving a better understanding of hearing mechanisms in fish has now slowed, however, following a period of very active research during the 1970s and early 1980s. The interests of physiologists are now tending to focus on mechanisms operating at the molecular level, and the problems of sensory and neural integrations are largely being neglected. What we have found so far is that the cod and probably many other species of fish are well able to locate sound sources in three dimensions and have a real acoustic sense of space. Their sense of hearing is undoubtedly of great importance in enabling them to find mates, to seek out prey and to avoid predators, often under conditions where other senses cannot operate. Further work is now required, however, both to elucidate the central mechanisms which enable fish to locate and recognize sources of sound, and to understand the role sounds play in their everyday life.

5.8   SUMMARY

Sounds are local mechanical disturbances that propagate rapidly and very effectively through water. Communication by means of sound appears to be widespread in fish, low-frequency calls being produced in a variety of social contexts including competitive and aggressive behaviour and courtship. Fish are acutely sensitive to sounds, though their hearing abilities are confined to low frequencies. They are able to discriminate between sounds of different amplitude and frequency, and between calls that differ in their pulse patterning – an ability that seems to be particularly important in enabling them to distinguish their own calls from those of other species. Fish are also able to determine the direction and even the distance of a sound source. Sounds are important to fish, and may enable them to seek out prey, predators and their own kind, sometimes at great distances, under conditions where other senses may be less effective.

REFERENCES

Alexander, R. McN. (1959) The physical properties of the swim bladders in intact cypriniformes. *J. exp. Biol.*, **36**, 315–32.

Allen, J. M., Blaxter, J. H. S. and Denton, E. J. (1976) The functional anatomy and development of the swimbladder-inner ear-lateral line system in herring and sprat. *J. mar. biol. Ass. U.K.*, **56**, 471–86.

Ashmore, J. F. and Russell, I. J. (1983) The physiology of hair cells, in *Bioacoustics: a Comparative Approach* (ed. B. Lewis), Academic Press, New York, pp. 149–80.

Bergeijk, W. A. van (1964) Directional and nondirectional hearing in fish, *Marine Bio-acoustics*, Vol. 1 (ed. W. N. Tavolga), Pergamon Press, New York, pp. 281–99.

Buwalda, R. J. A., Schuijf, A. and Hawkins, A. D. (1983) Discrimination by the cod of sounds from opposing directions. *J. comp. Physiol.*, **150**, 175–84.

Chapman, C. J. and Hawkins, A. D. (1973) A field study of hearing in the cod, *Gadus morhua*. *J. comp. Physiol.*, **85**, 147–67.

Chapman, C. J. and Sand, O. (1974) Field studies of hearing in two species of flatfish, *Pleuronectes platessa* and *Limanda limanda*. *Comp. Biochem. Physiol.*, **47A**, 371–85.

Coombs, S. and Popper, A. N. (1979) Hearing differences among Hawaiian squirrelfish (family Holocentridae) related to differences in the peripheral auditory system. *J. comp. Physiol.*, **132A**, 203–7.

de Vries, H. (1956) Physical aspects of the sense organs. *Progr. Biophys. biophys. Chem.*, **6**, 207–64.

Dijkgraaf, S. (1952) Uber die Schallwahrnehmung bei Meeresfischerei. *Z. vergl. Physiol.*, **34**, 104–22.

Edmunds, M. (1974) *Defence in Animals*, Longmans, London, 257 pp.

Fay, R. R. (1982) Neural mechanisms of auditory temporal discrimination by the goldfish. *J. comp. Physiol.*, **147**, 201–16.

Fay, R. R. and Popper, A. N. (1980) Structure and function in teleost auditory systems, in *Comparative Studies of Hearing in Vertebrates* (eds A. N. Popper and R. R. Fay), Springer-Verlag, New York, pp. 3–42.

Fine, M. L. (1978) Seasonal and geographical variation of the mating call of the oyster toad-fish *Opsanus tau* L. *Oecologia*, **36**, 45–7.

Gray, G. A. and Winn, H. E. (1961) Reproductive ecology and sound production of the toadfish, *Opsanus tau*. *Ecology*, **42**, 274–82.

Gray, J. A. B. and Denton, E. J. (1991) Fast pressure pulses and communication between fish. *J. mar. biol. Ass. U.K.*, **71**, 83–106.

Griffin, D. R. (1955) Hearing and acoustic orientation in marine animals. Papers on Marine Biology and Oceanography, *Deep Sea Res. (Supp.)*, **3**, 406–17.

Harris, G. G. (1964) Consideration on the physics of sound production by fishes, in *Marine Bio-acoustics* (ed. W. N. Tavolga), Pergamon Press, New York, pp. 233–47.

Hawkins, A. D. (1981) The hearing abilities of fish, in *Hearing and Sound Communication in Fishes* (eds W. N. Tavolga, A. N. Popper and R. R. Fay), Springer-Verlag, New York, pp. 109–33.

Hawkins, A. D. and Chapman, C. J. (1975) Masked auditory thresholds in the cod, *Gadus morhua*. *J. comp. Physiol.*, **103A**, 209–26.

Hawkins, A. D. and Horner, K. (1981) Directional characteristics of primary auditory neurons from the cod ear, in *Hearing and Sound Communication in Fishes* (eds W. N. Tavolga, A. N. Popper and R. R. Fay), Springer-Verlag, New York, pp. 311–28.

Hawkins, A. D. and Johnstone, A. D. F. (1978) The hearing of the Atlantic salmon, *Salmo salar*. *J. Fish Biol.*, **13**, 655–73.

Hawkins, A. D. and MacLennan, D. N. (1976) An acoustic tank for hearing studies on fish, in *Sound Reception in Fish* (eds A. Schuijf and A. D. Hawkins), Elsevier, Amsterdam, pp. 149–69.

Hawkins, A. D. and Myrberg, A. A. (1983) Hearing and sound communication under water, in *Bioacoustics: a Comparative Approach* (ed. B. Lewis), Academic Press, New York, pp. 347–405.

Krebs, J. R. and Davies, N. B. (1978) *Behavioural Ecology*, Blackwell, Oxford, 494 pp.

Marshall, J. A. (1972) Influence of male sound production on oviposition in female *Tilapia mossambica* (Pisces: Cichlidae). *Am. Zool.*, **12**, 633–64.

Moulton, J. M. (1960) Swimming sounds and the schooling of fishes. *Biol. Bull. mar. biol. Lab., Woods Hole*, **119**, 210–23.

Myrberg, A. A. (1972) Using sound to influence the behaviour of free-ranging marine animals, in *Behaviour of Marine Animals*, Vol. 2 (eds J. E. Winn and B. L. Olla), Plenum, New York, pp. 435–68.

Myrberg, A. A. (1981) Sound communication and interception of fishes, in *Hearing and Sound Communication in Fishes* (eds W. N. Tavolga, A. N. Popper and R. R. Fay), Springer-Verlag, New York, pp. 395–452.

Myrberg, A. A., Gordon, C. R. and Klimley, P. (1976) Attraction of free-ranging sharks by low frequency sound, with comments on the biological significance, in *Sound Reception in Fish*, (eds. A. Schuijf and A. D. Hawkins), Elsevier, Amsterdam, pp. 205–28.

Myrberg, A. A., Mohler, M. and Catala, J. D. (1986) Sound production by males of a coral reef fish (*Pomacentrus partitus*): its significance to females. *Anim. Behav.*, **34**, 913–23.

Myrberg, A. A. and Riggio, R. J. (1985) Acoustically mediated individual recognition by a coral reef fish (*Pomacentrus partitus*). *Anim. Behav.*, **33**, 411–16.

Nelson, E. M. (1955) The morphology of the swimbladder and auditory bulla in the Holocentridae. *Fieldiana, Zool.*, **37**, 121–30.

Pitcher, T. J., Partridge, B. L. and Wardle, C. S. (1976) A blind fish can school. *Science*, **194**, 963–5.

Platt, C. and Popper, A. N. (1981) Fine structure and function of the ear, in *Hearing and Sound Communication in Fishes* (eds W. N. Tavolga, A. N. Popper and R. R. Fay), Springer-Verlag, New York, pp. 4–38.

Poggendorf, D. (1952) Die Absoluten Horschwellen des Zwergwelses (*Ameiurus nebulosus*) und Beitrage zur Physik des Weberschen Apparate der Ostariophysen. *Z. vergl. Physiol.*, **34**, 222–57.

Pye, J. D. (1982) Techniques for studying ultrasound, in *Bioacoustics: a Comparative Approach* (ed. B. Lewis), Academic Press, New York, pp. 39–68.

Pumphrey, R. F. (1950) Hearing. *Symp. soc. exp. Biol.*, **4**, 3–18.

Sand, O. and Enger, P. S. (1973) Evidence for an auditory function of the swimbladder in the cod. *J. exp. Biol.*, **59**, 405–14.

Sand, O. and Hawkins, A. D. (1973) Acoustic properties of the cod swimbladder. *J. exp. Biol.*, **58**, 797–820.

Sand, O. and Karlsen, H. E. (1986) Detection of infrasound by the Atlantic cod. *J. exp. Biol.*, **125**, 197–204.

Schneider, H. (1967) Morphology and physiology of sound-producing mechanisms in teleost fishes, *Marine Bio-acoustics*, Vol. 2 (ed. W. N. Tavolga), Pergamon Press, Oxford, pp. 135–58.

Schuijf, A. and Buwalda, R. J. A. (1980) Under-water localisation – a major problem in fish acoustics, in *Comparative Studies of Hearing in Vertebrates* (eds A. N. Popper and R. R. Fay), Springer-Verlag, New York, pp. 43–77.

Schuijf, A. and Hawkins, A. D. (1983) Acoustic distance discrimination by the cod. *Nature, Lond.*, **302**, 143–4.

Siler, W. (1969) Near- and far-fields in a marine environment. *J. acoust. Soc. Am.*, **46**, 483–4.

Tavolga, W. N. (1958) Underwater sounds produced by two species of toadfish, *Opsanus tau* and *Opsanus beta*. *Bull. mar. Sci. Gulf Caribb.*, **8**, 274–84.

Tavolga, W. N. (1960) Underwater communication in fishes, in *Animal Sounds and Communication* (eds W. E. Lanyon and W. N. Tavolga), Am. Inst. Biol. Sci., Washington, DC, pp. 93–136.

Tavolga, W. N. (1974) Signal/noise ratio and the critical band in fishes. *J. Acoust. Soc. Am.*, **55**, 1323–33.

Tavolga, W. N. (ed.) (1976) *Sound Reception in Fishes*, Dowden, Hutchinson and Ross, Stroudsberg, Pennsylvania, 317 pp.

Tavolga, W. N. and Wodinsky, J. (1963) Auditory capacities in fish. Pure tone thresholds in nine species of marine teleosts. *Bull. Am. Mus. nat. Hist.*, **126**, 177–239.

Templeman, W. and Hodder, V. M. (1958) Variation with fish length, sex, stage of sexual maturity, and season in the appearance and volume of the drumming muscles of the swimbladder in the haddock, *Melanogrammus aeglefinus* L. *J. Fish. Res. Bd Can.*, **15**, 355–90.

Torricelli, P., Lugli, M. and Pavan, G. (1990) Analysis of sounds produced by male *Padogobius martensi*, and factors affecting their structural properties. *Bioacoustics*, **2**, 261–75.

Urick, R. J. (1983) *Principles of Underwater Sound for Engineers*, 3rd edn, McGraw-Hill, New York, 342 pp.

Valinsky, W. and Rigley, L. (1981) Function of sound production by the skunk loach *Botia horae* (Pisces: Cobitidae). *Z. Tierpsychol.*, **55**, 161–72.

Chapter six

Role of olfaction in fish behaviour

Toshiaki J. Hara

6.1   INTRODUCTION

The chemical senses of teleosts play a major role in mediating physiological and behavioural responses to the fishes' environment. Chemical stimuli include biochemical products released by conspecifics and other organisms, some of which may reveal the presence and location of food, mates, predators, or spawning sites. Although available information indicates that these chemical signals, including pheromones, are more widespread in the social interactions of fish than might have been suspected, their importance in aspects of fish behaviour is only beginning to be fully appreciated.

Fish detect chemical stimuli through at least two different channels of chemoreception: olfaction (smell) and gustation (taste). The distinction between these two senses in fish is not always as clear as in terrestrial, air-breathing vertebrates, mainly because in fish both olfaction and gustation are mediated by molecules dissolved in water. In terrestrial air-breathers, **olfaction** is defined as the detection via the nose of air-borne molecules emanating from a distance, while **gustation** is the detection of water-soluble chemicals in the mouth. Solubility rather than volatility is more relevant in fish chemoreception. Historically, whether taste and smell in fish are two distinct functions has long been controversial. It was Strieck (1924) who first provided the most convincing experimental evidence indicating that the sense of smell exists in fish. He trained minnows, *Phoxinus phoxinus*, to discriminate between odorous and taste substances. Trained fish were unable to discriminate odorous substances after the forebrain was removed, but they could still perceive taste substances.

Many findings point to olfaction as a general mediator of chemical signals involved in various teleost behaviours. Nevertheless, very little is known about their underlying physiological mechanisms. The main aim of this chapter is

Behaviour of Teleost Fishes 2nd edn. Edited by Tony J. Pitcher. Published in 1993 by Chapman & Hall. ISBN   0 412 42930 6 (HB) and 0 412 42940 3 (PB).

to summarize the basic principles of the olfactory system in teleosts and show how it is used to extract information about the chemical environment in order to control behaviour. For a more detailed description of ultrastructural, electrophysiological, biochemical and behavioural studies on fish chemoreception, the reader is referred to Hara (1982, 1992).

6.2   ANATOMY OF THE OLFACTORY SYSTEM

Peripheral olfactory organ

The olfactory organs of fishes show a considerable diversity, reflecting the degree of development and ecological habits. In the teleost fishes the paired olfactory pits (nasal cavity or olfactory chamber) are usually located on the dorsal side of the head (Fig. 6.1(A)). Each nasal cavity generally has two openings, anterior inlet and posterior outlet. Unlike terrestrial vertebrates, there is no contact between the olfactory and respiratory systems in any teleost species. A current of water enters the anterior and leaves through the posterior naris as the fish swims. In some species (*Zoarces viviparus*, *Gasterosteus aculeatus*, *Spinachia spinachia*, etc.), there is only one opening and water enters the naris and leaves with respiratory movements. The floor of the nasal cavity is lined with the olfactory epithelium or mucosa, which is raised from the floor into a complicated series of folds or lamellae to form a rosette (Fig. 6.1(B)). The arrangement, shape, and degree of development of the lamellae vary considerably among species. In the majority of the teleosts, the lamellae radiate from a central ridge (raphe) arising rostrocaudally from the floor of the cavity. The number of olfactory lamellae also varies greatly among species – from a few in sticklebacks to as many as 120 in eels and morays. The number of lamellae increases to some extent with growth of an individual, but remains relatively constant after the fish reaches a certain stage in development. Additionally, secondary folding of the lamella occurs in some species. Most notable are salmonids (Fig. 6.1(C)); there are between five and ten secondary foldings per lamella in adults but none exists in the parr and younger stages.

No simple correlation has been established between the number of olfactory lamellae and the acuity of the sense of smell. A marked sexual dimorphism has been evolved in deep-sea fishes, ceratoid angler fishes and *Cyclothone* spp. (Marshall, 1967). In these groups the olfactory organ is large in the males but reduced in the females. Marshall speculates that these bathypelagic fishes generally use senses other than olfaction for procurement of food, and that the well-developed olfactory organ in males may be useful in searching for a mate.

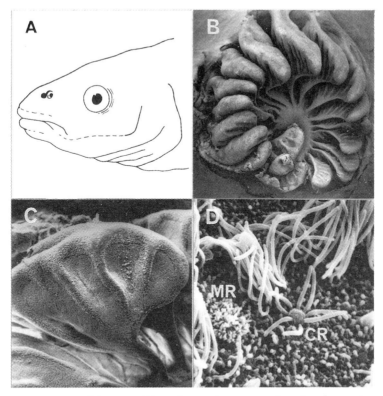

Fig. 6.1 Position of the nose (A) in the rainbow trout, *Oncorhynchus mykiss*, and scanning electron micrographs of (B) an olfactory rosette, (C) lamella, and (D) surface view of the sensory epithelium. CR, ciliated receptor cell; MR, microvillous receptor cell. Widths of photos B, C, and D approximate 4.0 mm, 1.0 mm, and 40 μm, respectively. (Scanning electron micrography courtesy of Dorthy Klaprat, Freshwater Institute, and Dr B. Dronzek and B. Luit, Department of Plant Science, University of Manitoba.)

Ultrastructure of the olfactory epithelium

The olfactory lamella is composed of two layers of epithelium enclosing a thin stromal sheet. The epithelium is separated into two regions, sensory and non-sensory (indifferent). The sensory epithelium shows various distribution patterns among teleosts but may fall into one of the following four types: (1) continuous except for the lamellar margin, (2) separated regularly by the non-sensory epithelium, (3) interspread irregularly with the non-sensory epithelium, and (4) scattered in islets. The sensory epithelium consists of three main cell types: (1) sensory receptor cells (olfactory neurones), (2) supporting or sustentacular cells, and (3) basal cells.

Sensory receptor cells

In teleosts, at least two morphologically distinct receptor cell types generally exist: ciliated and microvillous (Fig. 6.1(D)). The sensory epithelium on each side of the nasal cavity in an average teleost comprises approximately 5–10 million olfactory receptor cells. The receptor cell is a bipolar neurone with a cylindrical dendrite (1.5–2.5 μm in diameter) which terminates at the free surface of the epithelium. This anatomical position contrasts strikingly with visual and auditory receptor cells, which are guarded by membranes, fluid baths, bones and other structures serving to transduce and process the signals before they are preceived (see Chapter 4 by Guthrie and Muntz and Chapter 5 by Hawkins, this volume). The distal end of the dendrite forms a swelling (olfactory knob or vesicle) which protrudes slightly above the epithelial surface. The proximal part of the perikaryon tapers to form an axon. The axons pass through the basement membrance, become grouped in the sub-mucosa, and form the olfactory nerve fascicles, which run posteriorly to end in the olfactory bulb. The ciliated receptor cell has four to eight cilia radiating from an olfactory knob (Fig. 6.1(D)). Each cilium measures 2–7 μm in length and 0.2–0.3 μm in diameter, much shorter than those found in air-breathing vertebrates. Usually the cilia show the 9 + 2 arrangement of microtubules, which is identical with that of common kinocilia (see Chapter 5 by Hawkins and Chapter 7 by Bleckmann, this volume). The microvilli number from 30 to 80 depending upon the species, and are 2–5 μm long and about 0.1 μm wide. The free ending of the receptor cells, whether ciliated or microvillous, is the only bare portion of the dendrite and is exposed directly to stimulant molecules.

A unique feature of the olfactory system is the regenerative capacity of the receptor neurones. The receptor cells are continually renewed in normal adults. Experimental severance of the olfactory nerve or treatment of the olfactory mucosa with toxicants (e.g. heavy metals) causes degeneration of the sensory neurones followed by the reconstitution of a new population of functional neurones (Evans *et al.*, 1982). This renewal process of the olfactory receptor cells may be considered to be an adaptation of the system to impairment caused by enviromental hazards during the normal life of the animal.

Non-sensory cells

The supporting cells are columnar epithelial cells extending vertically from the epithelial surface to the basal lamina, forming a mosaic interspersed with receptor cells. The free surface is flat, with relatively few irregular microvilli. Supporting cells adjoin receptor cells, ciliated non-sensory cells, and other supporting cells at the free surface by means of a junctional complex consisting mainly of an apical tight junction. Ciliated non-sensory cells are typical columnar epithelial cells with a wide flat surface (4–7 μm), from which a number of long kinocilia (20–30 μm in salmon) extend. The beating of these cilia creates weak currents over the lamellae, presumably assisting in water renewal and the transport of stimulant molecules in the olfactory organ. The basal cells are small and undifferentiated cells lying adjacent to the basal lamina and having no cytoplasmic processes reaching the free surface. The basal cell in the sensory epithelium is assumed to be the progenitor of the receptor or supporting cells. In addition to the cell types described, mucous (goblet) cells are abundant in the indifferent epithelium.

In summary, the olfactory epithelium consists of three principal cell types: receptor, supporting and basal cells. The main function of the receptor cells is to detect, encode, and transmit information about the chemical environment to the olfactory bulb and higher brain centres. The basal cells appear to be stem cells and become active during cell turnover and reconstitution of the olfactory epithelium. The function of the supporting cells in olfactory perception is not defined, but they are likely to have a significance beyond mere mechanical support. Some olfactants cause the release of secretory products from supporting cells, changing the nature of the mucus bathing the epithelial surface.

Olfactory bulb and tract

The olfactory nerve fibres, unmyelinated axons of the receptor neurones, course to the olfactory bulb where they make a synaptic contact with the second-order bulbar neurones in the form of glomeruli (Fig. 6.2). Some olfactory fibres of *G. aculeatus* collect into fascicles that course through the bulbs into the telencephalon without forming the glomeruli. The axons within the olfactory nerve are arranged into several groups of bundles. In the carp, *Cyprinus carpio*, for example, the olfactory nerve consists of two main bundles, medial and lateral. The former is derived from the more rostral lamellae, and the latter from the more caudal. The fibres do not branch until they terminate.

The olfactory bulb in fishes is poorly differentiated and the lamination is not so distinct as in higher vertebrates. The mitral cell, the most characteristic cell found in the bulb, has a relatively large cell body and usually more than one dendrite ending in different glomeruli (Fig. 6.2). This is in marked contrast to mammalian mitral cells, in which only a single main dendrite ends in each glomerulus. It is significant that the axon of a receptor cell does not terminate in more than one glomerulus, and that each glomerulus receives neural inputs only from a limited group of several olfactory receptor cells (Fig. 6.2). The glomerular synapses are always directed from the olfactory nerve to the mitral-cell dendrites.

The axons of the mitral cells form the majority of the centripetal fibres of

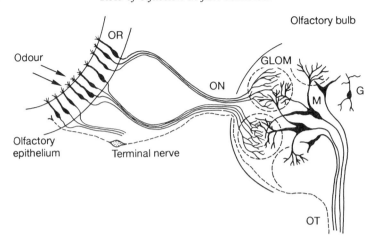

Fig. 6.2 Schematic representation of the cellular anatomy of the peripheral olfactory system and the neural organization in the olfactory bulb of teleosts. Olfactory receptor cells (OR) lie in the olfactory epithelium. Bundles of receptor-cell axons form the fila olfactoria, which coalesce as the olfactory nerve (ON). Axon terminals branch at the olfactory bulb in the form of glomeruli (GLOM), where they synapse with processes of second-order neurones, mitral cells (M). Mitral-cell axons project centrally as the olfactory tract (OT). G, granule cells.

the olfactory tract through which information from the olfactory bulb is conveyed to the telencephalic hemispheres (Fig. 6.3). The olfactory tract consists of two main bundles, lateral and medial. Both bundles are further subdivided into several small bundles. Some fibres run directly to the hypothalamus, and some cross in the anterior commissure. The centrifugal fibres originating in the telencephalon run backwards through the medial portion of the tract, terminate in the granule-cell layer, and form synapses with granule-cell dendrites of the bulb (Oka *et al.*, 1982). The mitral cells are thus under the influence of higher centres through the centrifugal fibres. The olfactory tract fibres number about 10^4. The convergence of the primary olfactory nerve fibres (about 5–10 million) upon secondary neurones would therefore be about 1000:1. This ratio is approximately the order that is estimated for the mammalian olfactory system. Such a high convergence ratio is implicated in the high sensitivities found in the olfactory system.

The olfactory tract fibres appear to terminate bilaterally in the following telencephalic hemispheres (Fig. 6.3): (1) medial terminal field in the area ventralis telencephali, (2) lateral terminal field in the ventrolateral part of the area dorsalis telencephali, and (3) posterior terminal field in the central part of the area dorsalis telencephali (Oka *et al.*, 1982). Recent studies support the view that the olfactory system is composed of two separate systems: (1) the lateral olfactory system, in which the lateral part of the olfactory bulb receives

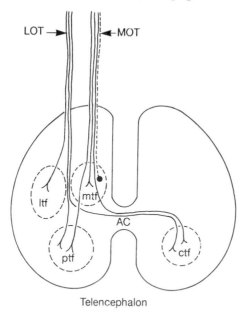

Telencephalon

Fig. 6.3 Diagram of the central olfactory pathway in teleosts (horizontal view). AC, anterior commissure; LOT, lateral olfactory tract; MOT, medial olfactory tract; ctf, contralateral terminal field; ltf, lateral terminal field; mtf, medial terminal field; ptf, posterior terminal field; dotted line in MOT represents a centrifugal fibre. Modified from Oka *et al.* (1982).

inputs mainly from the lateral bundle of the olfactory nerve and sends outputs to the lateral olfactory tract, and (2) the medial olfactory system, in which the medial part of the olfactory bulb receives inputs mainly from the medial bundle of the olfactory nerve and sends outputs to the medial olfactory tract (Satou *et al.*, 1983; Satou, 1990). The segregation of the teleost olfactory system may be homologous to two functionally distinct subsystems in higher vertebrates, i.e. the main and accessory olfactory systems.

Terminal nerve

The axons of another cranial nerve, the terminal nerve, course centrally into the telencephalon in association with the medial olfactory tract in a wide range of vertebrates including teleosts (Fig. 6.2; Demski and Schwanzel-Fukuda, 1987). In goldfish, large neuronal cell bodies occur within the medial half of the olfactory nerves and the rostral part of the olfactory bulb (Demski and Northcutt, 1983). The peripheral process of some ganglion cells of the terminal nerve end in the olfactory epithelium, and the central processes project to various areas in the telencephalon, diencephalon and mesencephalon as well as to the retina.

6.3   FUNCTION OF THE OLFACTORY SYSTEM

Interactions of odorants with receptors

Sensory information about the chemical environment is transmitted to the brain by olfactory receptor neurones through an orderly sequence of molecular, membranous and neural events. The processes are initiated by the impingement of odorous molecules upon the surface of the olfactory mucosa. Biochemical studies using a preparation enriched in plasma membranes from the olfactory mucosa have shown that binding of odorant amino acids is a physiologically relevant measurement of an early event in olfactory transduction (Brown and Hara, 1982; Fesenko *et al.*, 1983; Brand and Bruch, 1992). In these studies a sedimentable membrane – mitochondria fraction derived from olfactory rosettes is incubated with radiolabelled amino acids, and specific binding activities are assayed by either a rapid filtration or a centrifugation method. Binding meets many of the basic receptor criteria such as saturability, reversibility, affinity, stereospecificity, and quantity of sites. The extent of binding of a series of amino acids parallels their relative electrophysiological effectiveness. Studies using a cilia-enriched preparation have provided evidence indicating that at least two separate populations of odorant binding sites, one for neutral and the other for basic amino acids, exist in the rainbow trout olfactory system (Rhein and Cagan, 1983). This supports the earlier hypothesis that odorant recognition sites are proteins and integral parts of the cilia. Amino acid binding sites have also been reported in isolated membrane preparations from the olfactory epithelium of coho salmon, *Onocorhynchus kitsutch*, carp, channel catfish, *Ictalurus punctatus*, and skate, *Dasyatis pastinaca*. All the data favour the hypothesis that specific classes of olfactory receptor binding sites exist for these stimuli.

Neurophysiology of olfactory reception

It is now generally accepted that odorant molecules bind to receptor proteins in the ciliary plasma membrane, enabling them to activate a G protein. The activated G protein then activates adenylyl cyclase. The resulting increase in cAMP concentration opens ion channels in the membrane, causing membrane depolarization. This eventually leads to the generation of spike potentials by which the sensory information is transmitted to the higher-order olfactory brain. These electrical signals can be tapped, amplified and displayed at various levels of the olfactory system.

Receptor activity

When the nares of a fish are infused with water containing an odorous chemical, three types of response may be recorded from the olfactory epithelium using appropriate electrophysiological techniques. The first type is the electro-olfactogram (EOG), a slow negative voltage change recorded from the epithelial surface (Ottoson, 1971). This is a summated receptor potential and represents an excitatory response generated by many receptor cells in response to odour stimulation. The second type is a unitary action potential recorded extracellularly from an olfactory neurone intraepithelially, or from olfactory nerve twigs at some distance from the nares. Many receptor neurones exhibit spontaneous activity, i.e. impulse discharges without odour stimuli, and respond to odour stimulation by, in the majority of cases, increasing impulse discharges. Generally, receptor cells are broadly tuned, and each cell may respond to many odorants differing in quality. The third type is a transmembrane voltage change recorded using an intracellular microelectrode. This provides insight into transmembrane mechanisms associated with cellular activation by odour; however, it is extremely difficult to obtain such recordings, mainly because of the small size of the receptor cell. A high correlation exists among these three types of response when examined in the same species.

Olfactory bulbar activity

Infusion of odorous chemicals into the nares induces rhythmic, oscillatory responses (induced waves) in the olfactory bulb. The olfactory bulb also develops a slow potential shift superimposed with regular oscillation when recorded through a DC-coupled preamplifier. The slow potential is produced in the dendritic network within the glomeruli, and the induced waves are the result of sychronous activity of the secondary bulbar neurones.

Sensitivity and specificity of olfaction

The aquatic environment surrounding fishes makes their olfaction unique; the entire process takes place in water, and therefore volatility of odorants is less relevant than for those in air. Traditionally, however, volatile chemicals primarily odorous to humans had been widely utilized as stimuli for studying fish olfaction. Many of the early published detection thresholds such as $3.5 \times 10^{-18}\,\mathrm{mol\,l^{-1}}$ β-phenethyl alcohol for *Anguilla anguilla*, 1×10^{-15} $\mathrm{mol\,l^{-1}}$ butyl alcohol for *Ictalurus catus*, and $5.7 \times 10^{-10}\,\mathrm{mol\,l^{-1}}$ morpholine for salmonids are not supported by studies employing modern electrophysiological techniques (Hara, 1975).

Since Sutterlin and Sutterlin (1971) and Suzuki and Tucker (1971) independently provided electrophysiological evidence that olfactory receptors of Atlantic salmon, *Salmo salar*, and white catfish, *Ictalurus catus*, respectively, are highly sensitive to amino acids, research on fish olfaction has centred on recordings of electrical responses to amino acids at different levels of the olfactory system. Amino acids are potent olfactory stimuli in a wide variety of fish species, and the threshold concentrations lie as low as 0.1 nmoll^{-1} (Hara, 1975, 1982). Generally, unsubstituted L-α-amino acids containing unbranched and uncharged side chains are the most effective olfactory stimuli. Ionically charged α-amino and α-carboxyl groups appear essential; acylation of the former or esterification of the latter results in reduced activity. The amino acids effective as olfactory stimuli are thus characterized by being simple, short and straight-chained, with only certain attached groups. Although some species specificities exist, several receptor site types or transduction mechanisms are likely to be involved in amino acid detection: (1) two or three neutral amino acid receptors, and (2) basic amino acid receptors (Caprio and Byrd, 1984; Sveinsson and Hara, 1990). Except for the basic amino acid receptor, these receptor sites are not mutually exclusive. It is not clear whether these distinct receptor sites occur in separate receptor neurones. The amino acid specificities found for fish olfaction are similar to those for bacterial chemotaxis and a neutral amino acid transport system in the mammalian ileum. Although peptides are normally nonstimulatory, salmon gonadotrophin releasing hormone (GnRH) and its analogues are potent olfactory stimulants for rainbow trout, with threshold concentrations ranging from 10^{-16} to 10^{-14} moll^{-1} (Andersen and Døving, 1991).

Bile acids, especially taurine and glycine conjugates, are also potent olfactory stimulants for salmonid fishes, with a threshold concentration ranging between 0.1 and 10 nmoll^{-1} (Døving *et al.*, 1980; Hara *et al.*, 1984). Ciliated receptor cells which are distributed more distally on each lamella are primarily responsible for bile acid stimulation, and more proximally distributed microvillous receptor cells are for amino acid stimulation (Thommesen, 1983). Responses induced by bile salts project to the medial part of the olfactory bulb, and amino acids elicit responses in the lateral part of the bulb (Døving *et al.*, 1980). However, this hypothesis is not necessarily supported by recent studies with rainbow trout embryos, in which ciliated receptor cells respond to both amino acids and bile acids prior to the maturation of microvillous receptor cells (Zielinski and Hara, 1988). Bile salts are similarly stimulatory for goldfish olfactory systems (Sorensen *et al.*, 1987), but they are only slightly stimulatory in channel catfish at high concentrations, 10^{-4} moll^{-1} and higher (Erickson and Caprio, 1984). High sensitivity of the olfactory receptor of salmonids to bile acids is implicated in their role as specific chemical signals (pheromones) in homing migration.

It has been established that two pheromones, sex steroids 17α,20β-dihydroxy-4-pregnen-3-one (17, 20P) and F-series prostaglandins (PGFs) are involved in spawning of goldfish (Sorensen *et al.*, 1987, 1988). The steroid 17, 20P is an extremely potent olfactory stimulant and may be the most stimulatory odorant described in fish. It has a detection threshold of approximately 10^{-12} moll^{-1}, and at a concentration of 10^{-8} moll^{-1} elicits an EOG response three times the size of that elicited by $10^{-5} \, \text{mol} \, l^{-1}$ L-serine. Some of its precursors and metabolites are also stimulatory, but to lesser degrees. Recent EOG studies have shown that testosterone is a potent odorant in precocious male Atlantic salmon parr, with a threshold concentration of $10^{-14} \, \text{mol} \, l^{-1}$, for a limited period of the year (October) (Moore and Scott, 1991).

The olfactory epithelium of goldfish is also acutely sensitive to prostaglandins, especially $PGF_2\alpha$ and its metabolite 15-keto-$PGF_2\alpha$ (Sorensen *et al.*, 1988). Prostglandin $F_2\alpha$ is detected at a threshold of approximately $10^{-10} \, \text{mol} \, l^{-1}$. Cross-adaptation experiments indicate that olfactory receptor sites for each of these odorant groups are separate and mutually exclusive from those for amino acids and bile salts. 5β-Pregnane-3α,17α-diol-20-one-3α-glucuronide and other related steroids, believed to be the sex pheromones for African catfish, *Clarias gariepinus*, is detected at concentrations between 10^{-9} and $10^{-11} \, \text{mol} \, l^{-1}$ (Resink *et al.*, 1989b).

Olfaction – taste distinction and overlap

In fish, as discussed earlier, amino acids are important chemical signals eliciting various physiological and behavioural reactions through both the olfactory and gustatory systems. In the majority of species studied, olfactory receptors are stimulated by a wide spectrum of amino acids, with a tendency to tune more to their neutral portion. The taste system demonstrates more variabilities. At one extreme, taste receptors of species such as catfish generally respond to a wide variety of amino acids; at the other extreme, species such as rainbow trout respond to only a limited selection of amino acids; and the rest fall between these two extremes (Yoshii *et al.*, 1979; Goh and Tamura, 1980; Kiyohara *et al.*, 1981; Caprio, 1982; Marui *et al.*, 1983). Thus functional separation between olfaction and taste is obvious for this group of chemicals, though their specific roles in fish behaviour are not always clearly defined. As the number of stimulant chemical groups specific for the olfactory and gustatory systems increases, functional separation between the two systems becomes clearer. Recently defined reproductive pheromones in goldfish, sex steroids and prostaglandins, are detected only by olfactory receptors and mediate spawning behaviour (Sorensen *et al.*, 1987, 1988), while carbon dioxide and marine toxins are detected through the gustatory system and seem to play a significant role in avoiding noxious substances in the environment (Yamamori *et al.*, 1988). There are, however, other groups of chemicals such as bile salts which stimulate both olfactory and gustatory systems with equivalent thresholds (Hara *et al.*, 1984), though their functional significance remains to be determined. There has long been a belief that olfaction is the immediate mediator of chemical communication in fishes, simply because of its lower active thresholds. However, because the fish gustatory system is as sensitive to as many chemical stimuli as the olfactory system, we can no longer consider olfaction to be the sensory system for detection of aquatic chemicals only on the basis of its lower effective concentration.

6.4   OLFACTION AND HOMING MIGRATION IN SALMON

Homing migration is displayed by many fish species. These fish may spend their early life in a home territory, migrate to an ecologically entirely different area, then return to spawn in their home territory (for a general review of fish migration see McKeown, 1984). The salmon are best known for their spectacular migrations. This impressive feat comprises three migratory phases: (1) downstream journey of the young to the ocean, (2) return of spawning adults to the coastal area near the entrance to their home stream, and (3) the upstream migration. Phases (1) and (2) are thought to rely primarily on non-chemosensory mechanisms such as drift, random movement, and celestial, magnetic and sun-compass orientation, and will not be discussed here. Homing of salmon is well documented (e.g. Hasler, 1966; Harden Jones, 1968; Thorpe, 1988), and a number of studies on salmonid migration have demonstrated the importance of chemical information in home-stream discrimination (reviews Hara, 1970; Cooper and Hirsch, 1982; Hasler, 1983). Hasler and Scholz (1983) describe the historical development of the olfactory hypothesis for salmonid homing, endocrine control of the olfactory imprinting process, and other factors controlling migration. Despite considerable research effort, many aspects of homing are still poorly understood and remain controversial (Quinn, 1990).

Olfactory hypothesis

Over a hundred years ago, Buckland (1880) postulated an olfactory basis for salmonid homing, and his statement is often quoted today:

> When the salmon is coming in from the sea he smells about till he scents the water of his own river. This guides him in the right direction, and he has only to follow up the scent, in other words, to 'follow his nose', to get up into fresh water, i.e., if he is in a travelling humour. Thus a salmon coming up from the sea into the Bristol Channel would get a smell of water meeting him. 'I am a Wye salmon', he would say to himself. 'This is not the Wye water; it's the wrong tap, it's the Usk. I must go a few miles further on', and he gets up stream again.

The olfactory hypothesis, cast into modern terms by Hasler and his students, proposes that salmonids 'imprint' to certain distinctive odours of the home stream during the early period of residence, and as adults they use this information to locate the home stream, at least during the last stages of the homing migration. The results of olfactory impairment experiments repeated over 20 times using seven different salmonid species are remarkably consistent; in 16 experiments the olfactory sense appeared to be necessary for correct homing, and, in addition, two studies demonstrated that blind fish homed nearly as well as control fish. Thus vision is not essential for relocating the original stream, at least during upstream migration. At present, little is known about how olfactory imprinting might operate in salmon homing. Greater olfactory learning activity concomitant with elevated levels of thyroid hormones during smolting has been implicated in olfactory imprinting (Morin *et al.*, 1989). Whatever the underlying mechanism of imprinting, the smolt stage is of critical importance, because it is during this period that salmon become indelibly imprinted to distinctive odours of their natal tributary. This ability appears to be acquired rather than inherited; young salmon transplanted from their natal tributary into a second stream before smolt transformation occurs will return to that second stream. There is evidence that this process is rapid. Less than 10 days, or perhaps only several hours, appears sufficient for imprinting to take place (Hasler and Scholz, 1983). The nature of the olfactory cues in home streams has been variously characterized as volatile, non-volatile, organic and inorganic. If odours distinctive to streams are often-suggested mixtures, or 'bouquets', as opposed to a single chemical, and fish discriminate subtle differences of their chemical composites, the identification of the active ingredient by conventional methods appears extremely difficult.

Homing of artificially imprinted salmon

Studies on artificial imprinting using coho salmon, *Oncorhynchus kisutch*, in Lake Michigan have been well described (Cooper and Hirsch, 1982; Hasler, 1983; Hasler and Scholz, 1983). The basic principle was to expose salmon to low levels of synthetic chemicals (either morpholine or phenethyl alcohol) in place of, or in addition to, natural home-stream odours, to determine whether as adults they could be attracted to a stream scented with that chemical. The fish were exposed to the chemicals in a fish hatchery during the smolt stage and then released directly into the lake. During the spawning migration, two separate streams were scented respectively with morpholine and phenethyl alcohol to simulate home streams for the experimental fish. Approximately 95% of the morpholine fish were captured in the morpholine-scented stream and 92% of the phenethyl alcohol fish were recovered in the phenethyl-alcohol-scented stream. The general conclusions drawn are that the fish are decoyed to the streams by the synthetic chemicals, able to learn or imprint to them during a brief period of the smolt stage, and retain these cues without being again exposed to the chemicals. However, two fundamental questions remain unanswered in these studies. Do salmonids really smell morpholine and phenethyl alcohol? How do synthetic chemicals make stream water specific for fish? Studies with modern techniques (electrophysiological, cardiac conditioning, and behavioural) have been unable to demonstrate that salmon detect these chemicals at the concentrations used. Alternatively, morpholine would alter the chemistry of the whole stream water to create a novel condition to which salmon imprint. Dodson and Bitterman (1989), examining the uniqueness of a chemosensory compound in goldfish, suggest that the interaction of the component stimuli produces a new stimulus that acquires associative strength independently of the separate components.

Do the imprinting odours serve as a sign stimulus to release a stereotyped behaviour pattern in fish? Migrating salmon are known to make a wrong choice at a stream junction, or bypass the natal tributary (over-shooting). These fish rectify their errors through 'backtracking', that is, they eventually swim back downstream and reach their natal tributary. Fish deprived of their sense of smell swim downstream but do not swim back upstream. Thus the presence of the home-stream odour evokes a positive rheotaxis, and absence of the odour evokes a negative rheotaxis. This is demonstrated in the following experiments in which salmon smolts were exposed to either morpholine or phenethyl alcohol, stocked in Lake Michigan, and captured as adults in streams scented with the chemicals (Hasler and Scholz, 1983). The fish were then transported to and released in a different section of the river located upstream from the capture site. Morpholine-treated salmon migrated up-stream when morpholine was present and downstream when it was absent. Phenethyl-alcohol-exposed fish, on the other hand, swam downstream in both cases.

Continuous excitation of the olfactory system induces sensory adaptation or fatigue so that a given odour is no longer perceived. The behaviour of a salmon ascending the river system seems analogous to that of a dog following an odour track; the dog does not stay exactly on the track but progresses along it criss-crossing. When morpholine was introduced on either the right or left side of a stream, movements of morpholine-treated fish were confined to the morpholine sides. If a fish swam out of the odour trail, it swam downstream until encountering the odour again (Hasler and Scholz, 1983).

Pheromone hypothesis

An idea that pheromones emitted by juvenile fish attract migrating adult salmon into their natal streams has existed for some time. For example, G. H. Parker suggested, "It is barely possible that a certain race of fish may give off emanations that differ chemically from those of other races, and that one might attribute the return of individual races to home streams to their power to sense the familiar emanation" (Chidester, 1924). This hypothesis has been further formulated by Nordeng (1977), who proposes that (1) populations or races of salmon in different streams emit pheromones that serve to identify fish distinctly from each other, (2) the memory of this population-specific pheromone is inherited, and (3) homing adults follow pheromone trails released by juveniles residing in the stream, that is, the juveniles provide a constant source of population odour. The pheromones are thought to be released from the skin mucus. The olfactory bulbar neurones of the Arctic charr, *Salvelinus alpinus*, responded differentially to mucus emanating from different populations of the same species (Døving *et al.*, 1975). Further electrophysiological and chemical analyses have shown that the skin mucus of salmonids contains species-specific compositions of amino acids responsible for olfactory stimulation (Hara *et al.*, 1984). Some species of fish are known to recognize other members of their own species by mucus. However, the functional significance of chemosensory responses to conspecifics is not fully understood, and they may play a major role in individual or sexual recognition.

Hasler's experiments on homing of salmon treated with synthetic chemicals cannot, however, be explained by the pheromone hypothesis. In all these experiments, there were no young salmon present in the scented streams at the time the adults were attracted to them. In addition, all the fish used were from the same spawning stock and randomly separated into groups for exposing to different odours. Therefore, the fact that different experimental groups behaved distinctively would seem to rule out the possibility that the result of artificial imprinting experiments could be explained by the pheromone hypothesis. The imprinting and pheromone hypotheses may not necessarily be mutually exclusive; under natural conditions population-specific odour may be just one component in the chemical environment to which a salmon imprints as a juvenile and responds as an adult.

Hormonal regulation of homing and olfactory imprinting

Evidently, sex hormones play important roles in regulating reproductive behaviour in a number of teleost species (Stacey, 1983). Oestradiol-17β and testosterone increase dramatically during the course of the spawning season in salmonids, reaching a peak at early stages (Scott and Sumpter, 1983; Ueda *et al.*, 1984). Increased sex-hormone levels in both males and females of migrating salmon are correlated with high sensitivity to or discrimination of home-stream water (Hasler and Scholz, 1983). Spawned-out salmon with low levels of sex hormones no longer respond to their home-stream odour. Administration of gonadotrophin into salmon at a non-migratory stage increases locomotor activity and upstream movement, as the levels of sex hormones increase. Generally, a relationship between nasal and genital function exists in vertebrates. In fish, injections of oestradiol or testosterone modify olfactory bulbar activity by altering thresholds of responses to chemical and electrical stimulation (reviews Hara, 1970; Demski and Hornby, 1982). However, its underlying physiological mechanisms have not been well understood.

Juvenile salmon undergo marked transitions in morphology, physiology and behaviour just prior to their seaward migration (smolt transformation). This process is complex, and appears to be under the control of the endocrine system, particularly thyroid hormones and cortisol (Hoar, 1988). Hasler and Scholz (1983) demonstrated that (1) serum concentrations of these hormones rise just before smolting occurs, and (2) injection of thyroid-stimulating hormone (TSH) or adrenocorticotrophic hormone (ACTH) into presmolts induces these transitions. There is also evidence that olfactory imprinting occurred owing to TSH treatment. Presmolts receiving TSH or TSH plus ACTH and exposed to either morpholine or phenethyl alcohol demonstrated the ability to track their respective odour upstream, whereas those receiving ACTH, saline or no injections did not (Hasler and Scholz, 1983). Only gonadotrophin-injected smolts, treated with TSH or TSH plus ACTH, displayed this attraction. Therefore TSH injections mimic the events that activate olfactory imprinting in natural smolts (see also Chapter 3 by Huntingford, this volume). Together these findings are taken to show that episodic peaks in plasma concentration are associated with olfactory imprinting. However, Bern (1982) warns, in his closing remarks at a symposium on salmon smolting, that "We do not have information on blood levels of thyroid hormones and cortisol at different stages in various salmonids. We still need information on the kinetics of hormone production and degradation . . . we really do not know the physiological roles of thyroid hormones in salmonids (indeed in fish generally)".

6.5   OLFACTION AND FEEDING

The majority of species rely upon information received by all their senses for food detection, recognition and selection. Whatever the senses involved, feeding behaviour shows a stereotyped sequence of behavioural components (Atema, 1980, and see Chapter 8 by Hart, this volume). The first step in the feeding sequence is arousal to the presence of food (alert or arousal stage). Animals become alerted to the presence of food by changing their respiratory and swimming patterns or activity levels. This process is primarily mediated by olfaction. The ensuing locating or searching phase shows a wide diversity and is under the control of various sensory activities, including chemical senses, depending upon the ecological niche of the animals involved. Feeding behaviour is completed by the food uptake and ingestion phase, which is triggered by gustatory stimuli. Many of the earlier studies dealt with general behavioural reactions to prey organisms and food extracts, and consequently the distinction of senses involved is not always clear. The literature on the role of chemoreception in feeding behaviour has been reviewed on several occasions (e.g. Kleerekoper, 1969; Hara 1971, 1975, 1982; Atema, 1980, Jones, 1992).

Detection of food

The significance of olfaction in the procurement of food by fishes was established experimentally by Bateson as early as 1890. He lists more than 15 freshwater and marine species that seek and recognize their food by the sense of smell alone. One of the interesting features common to these species is, Bateson notes, that they are all more or less nocturnal and remain in hiding by day. Catfish, for example, populate the bottoms of ponds, lakes and stagnant parts of rivers, and feed mostly at night. Detection of live prey (concealed earthworms) by ictalurid catfish, *Ictalurus nebulosus*, was abolished when olfaction was blocked by sectioning the olfactory tract, but not when barbels were removed. The yellow bullhead, *I. natalis*, on the contrary, shows essentially normal feeding behaviour with its olfactory epithelium cauterized, and even feeds normally with its forebrain removed. In the latter species external taste serves to locate dead bait at a distance and to trigger the pick-up reflex, whereas internal taste serves as a second, more restrictive screen for food intake by controlling swallowing (Atema, 1980). Thus, in the yellow bullhead, the entire feeding behavioural sequence from alert to swallowing is controlled by taste. Unlike catfish, yellowfin tuna, *Thunnus albacares*, which are highly visual predators, constantly cruise in large schools in the upper water layers in search of prey such as anchovy and squid. When a prey school is encountered, hungry tuna go into a feeding frenzy. They rely on vision in distant-prey detection under normal circumstances. However, under laboratory conditions, yellowfin tuna can detect and distinguish between odours of intact prey organisms, responding to odour quality of various prey rinses (Atema, 1980). Feeding experience with prey causes a gradual shift in responses from one to another prey odour or artificial mixture of odours. Furthermore, the hunger stage of tuna strongly influences intensity of responses. Localization of the odour source does not usually occur. Thus, in nature, tuna may use their olfactory sense when they cross odour trails of prey. Responses to the odour appear to be followed by orientation using other, probably visual, cues (Atema *et al.*, 1980) Under a two-choice situation, whitefish, *Coregonus clupeaformis*, which are midwater plankton-feeding coregonids (Salmoniformes), show preference (attraction) for food extract. The behavioural reaction is eliminated when the nares of whitefish are cauterized. The preference reaction is characterized by a nosing and exploratory behaviour followed by a stationary posture in the side containing food extract, consequently resulting in a gradual decrease in crossing a boundary (Hara *et al.*, 1983).

Feeding stimulants

Although behavioural studies to identify the active ingredients in food materials are fraught with inconsistencies in the response criteria, preparations of food materials, and chemical senses involved, there is now evidence that feeding behaviour in different fish species is stimulated by somewhat different chemical substances (Hara, 1982, 1992).

The following generalization regarding the chemical nature of feeding stimulants can be made. To date, all feeding stimulants identified for teleosts are (1) of low molecular weight (< 1000), (2) non-volatile, (3) nitrogenous, and (4) amphoteric. The generalization applies to most of the cases in which amino acids, betaine, other amino acid-like substances and nucleotides, particularly inosine, have been implicated. The high sensitivity to a relatively broad spectrum of amino acids and the apparent species-specific arrays of relative acuities suggest that biologically meaningful feeding stimulants may consist of fingerprint-like mixtures or chemical images (Atema, 1980). For most species studied, single compounds such as proline, betaine, glycine and alanine have been identified as representing major effectiveness in these mixtures, although mixtures are nearly always more effective than single compounds. Because L-proline is dominant among the gustatory amino acids in most species tested, almost exclusively in some salmonids, and because L-proline is one of the most abundant amino acids in invertebrate tissues, it may play an important role in some aspects of feeding.

Anatomical separation of feeding behaviour

In the cod, *Gadus morhua*, electrical stimulation of the lateral part of the lateral olfactory tract elicits the characteristic feeding behaviour (Døving and Selset, 1980). In goldfish the feeding response induced by a food odour is abolished by sectioning of the lateral part of the olfactory tract, but sectioning of the medial part has little effect (Stacey and Kyle, 1983). The lateral part of the olfactory bulb of salmonids responds to feeding stimulant amino acids, whereas the medial part responds to bile acids (Thommesen, 1978). These functional differentiations are consistent with the structural organization of the teleost olfactory fibre connections, and suggest that olfactory neural inputs relevant to feeding may be processed through a channel relatively independent of those for other behaviour patterns such as reproduction (Satou, 1992).

6.6   OLFACTION AND REPRODUCTION

Olfaction exerts a functional role in every aspect of the reproductive process, from initial attraction and recognition of sexual status to sexual development of the young. Considerable work on the role of chemical signals, or pheromones, in reproductive behaviour of fishes has been conducted; however, as is the case with feeding behaviour, the sensory system involved and the nature of pheromones have not been rigorously investigated (Kleerekoper, 1969; Hara, 1975; Colombo *et al.*, 1982; Liley, 1982; Pfeiffer, 1982; Liley and Stacey, 1983; Stacey, 1983). Consequently, most of what is known of the interface between olfaction and reproduction is derived from studies on laboratory animals such as rodents (Stoddart, 1980). Recently, however, it has been demonstrated that hormones and their metabolites may commonly serve as reproductive pheromones in fish (Sorensen, 1992). In this Section, some of the major developments emphasizing interactions between olfaction and reproductive behaviour currently in progress will be discussed.

The most convincing demonstration of the role of olfaction in mating is that exposure of males to water holding a gravid female elicits courtship behaviour in the goby, *Bathygobius soporator* (Tavolga, 1956). Of various internal body fluids from gravid females, only ovarian fluid elicits the behaviour. However, anosmic males do not respond to the female odour. Whether the stimulating ovarian substance is secreted by the eggs or by the ovary itself is not clear from this experiment. Many studies point to the gonads as a source of sex pheromones (Colombo *et al.*, 1982). In *Gobius jozo*, urine is more effective than ovarian washing in triggering courtship behaviour, and fluids collected from gravid females are more effective than those of postvitellogenic ones (Colombo *et al.*, 1982). This contrasts with the observations of Tavolga (1956) that urine is inactive in *B. soporator*. Studies on *G. jozo* indicate that the steroid glucuronides are likely candidates. 5β-Reduced androgen conjugates, particularly etiocholanolone glucuronide produced by a male mesorchial gland, attract gravid females (Colombo *et al.*, 1982). The sex attractant secreted by the ovaries of female zebrafish, *Brachydanio rerio*, also consists of steroid glucuronides, most likely a mixture containing oestradiol-17β and testosterone glucuronides (van den Hurk and Lambert, 1983). Overian pheromones in both species not only elicit courtship behaviour, but also inhibit fighting; anosmic males show extremely aggressive behaviour towards females. Some of the sex steroid hormones are known to influence olfactory mechanisms, probably mediated by steroid feedback on brain areas involved in centrifugal control of the olfactory bulb (Demski and Hornby, 1982).

Shortly after ovulation, the female goldfish releases a pheromone from the ovary which attracts the male and elicits the persistent courtship that accompanies spawning (Partridge *et al.*, 1976). It is now apparent that periovulatory female goldfish sequentially release two hormonal pheromones, 17, 20P and PGFs. Environmental cues trigger an ovulatory surge in gonado-trophic hormone (GtH) in a vitellogenic female, which sequentially stimulate 17, 20P synthesis by the ovary. The latter hormone induces female oocyte maturation and is released to the water where it functions as a preovulatory priming pheromone. This pheromone evokes a surge in circulating GtH in males, which stimulates the synthesis of testicular 17, 20P. This in turn evokes an increase in milt production by the time of ovulation and spawning. At the time of ovulation, females produce PGFs to mediate follicular rupture and to trigger female spawning behaviour. Circulating PGFs are subsequently meta-bolized and released into the water, where they function as a postovulatory pheromone that stimulates male sexual arousal, thus resulting in spawning synchrony (Dulka *et al.*, 1987; Sorensen *et al.*, 1987, 1988, Sorensen, 1992). In African catfish, on the other hand, steroid glucuronides of the seminal vesicle such as 5β-pregnane-3α,17α-diol-20-one-3α-glucuronide play an important role as sex pheromone (Resink *et al.*, 1989a).

Although the ovarian sex pheromones are essential for triggering a series of courtship actions by the males, a complete courtship repertoire is expressed only in the presence of other sensory cues (see Chapter 16 by FitzGerald and Wootton, this volume). In the male threespine stickleback, *Gasterosteus aculeatus*, for example, the functioning of the olfactory system during repro-ductive behaviour (nest building, zigzag dance, and fanning) is supplemented by the gustatory system (Segaar *et al.*, 1983). The olfactory neural inputs, albeit not absolutely necessary, support nest building and are essential for subsequent development of courtship behaviour as expressed by the zigzag dance. General olfactory stimuli, rather than a specific pheromone, are assumed to influence the development of nest building by inducing hormonal changes necessary to start the reproductive cycle. The olfactory inputs are essential for promoting the sexual tendency during the early reproductive cycle, and once the zigzag scores have reached an essential level, sectioning the olfactory nerve no longer affects normal sexual activities, probably because the loss of chemical information via the olfactory system is compensated by functioning of the gustatory nerves (ixth and xth). Stimuli from eggs and embryos are perceived by receptors located in the area of the pharyngeal roof between the first and second gill arches and excite sexual behaviour during most of the reproductive cycle.

Most experiments dealing with the role of olfaction in reproductive behaviour in teleosts have generally considered the olfactory system as a uniform whole. However, as discussed earlier, both anatomical and electrophysiological studies support the contention that the olfactory system may be composed of two relatively independent parts (Satou *et al.*, 1983; Satou, 1990, 1992). Sectioning of the medial olfactory tract of goldfish reduces male sexual behaviour to the level of fish with complete tract section, whereas cutting the lateral olfactory tract has no effect (Stacey and Kyle, 1983). Only the medial olfactory tract responds to pheromones and both tracts respond to amino acids and food odour (Sorensen *et al.*, 1991). Similarly, feeding stimulant amino acids to salmonids elicits responses more in the lateral part of the olfactory bulb, and responses induced by steroid bile acids project to the medial part of the bulb (Døving *et al.*, 1980). The medial olfactory tract contains fibres of the terminal nerve, a cranial nerve containing luteinizing hormone-releasing hormone-immunoreactive material (Münz *et al.*, 1981; Kah *et al.*, 1984). It is suggested that this system may play an integrative function in processing olfactory and optic information in teleosts and is an essential element of the anatomical substrates for the role of olfaction and vision in homing and reproduction. A recent study, however, failed to demonstrate that the terminal nerve mediates responses to pheromones in goldfish (Fujita *et al.*, 1991).

6.7 OLFACTION AND THE FRIGHT REACTION

Von Frisch (1938) accidentally discovered that when an injured European minnow, *Phoxinus phoxinus*, was introduced into a shoal, they became frightened and dispersed. "In fact, the incident that first drew the attention of von Frisch to this problem was his observation of a kingfisher swooping down and attacking a school [= shoal, Ed.] of minnows. After successfully capturing one, as the bird was flying away it dropped the minnow and von Frisch noticed that the other minnows, which were still schooling, suddenly dispersed". (Quoted from Hasler and Scholz, 1983.) The general characteristics of the phenomenon is that when the skin of a fish is damaged (e.g. by a predator), alarm substance cells are broken and release alarm substance (Schreckstoff). Nearby conspecifics smell the alarm substance and show a fright reaction (Schreckreaktion). Their fright reaction may then be treated as a visual signal by other conspecifics, leading to rapid transmission of the signal through a group of fish. This alarm-signal system consists of two main components: a behavioural component, the fright reaction, and a morphological component, the alarm substance (cells). In view of the comprehensive reviews by Pfeiffer (1982) and Smith (1982, 1992), the following will focus on recent developments, together with a brief summary of earlier reviews.

Fright reaction behaviours

The fright reaction is not stereotyped and differs considerably from species to species, involving cover seeking, closer crowding, rapid swimming or immobility. Thus seven intensities of response are arbitrarily established when a skin extract of the minnow is introduced into an aquarium containing minnows. The most intense reaction involves sudden flight into the hiding place, emergence, rapid swimming around the tank, and avoiding the feeding place for a considerable length of time. In a typical experiment, an extract of 0.002 mg of chopped fish skin was sufficient to elicit fright reaction in minnows in a 14-litre aquarium. In the common shiner, *Notropis cornutus*, a skin preparation increases shoal cohesion and polarization and decreases the variability in overall shoal dimensions (Heczko and Seghers, 1981). The reaction is generally accepted to be mediated by olfaction; however, as von Frisch (1941) himself noticed, involvement of gustatory or other senses is not entirely ruled out. The extract of the skin and accompanying mucus contains various organic compounds such as fatty acids, carbohydrates, proteins, phospholipids and free amino acids, many of which stimulate gustatory as well as olfactory systems. Although the fright reaction may be initiated by a chemical stimulus, it can be transmitted through a school of fish by vision. When two aquaria with a group of fish in each are placed close together, an alarm-substance-induced fright reaction in one tank would trigger a fright reaction in the fish in the adjacent tank without exchanging water or chemical stimuli.

Alarm substance

The alarm substance is contained in a large epidermal cell, the 'club cell'. Alarm substance cells can be distinguished from mucous cells, the other main fish epidermal secretory cells, by their negative reaction to periodic acid Schiff's (PAS) reagent. The mucous cells secrete the mucus at the skin surface through a pore, but the alarm substance cells lack an opening at the skin surface and release their contents only when the skin is injured. Alarm substance cells are usually found closer to the basal layer than the mucous cells and autofluoresce under ultraviolet light. Skin extract from breeding males or testosterone-treated fathead minnows, *Pimephales promelas*, which lose their alarm substance cells, does not induce a fright reaction. Skin extract from females or non-breeding males, which retain alarm substance cells, does induce a fright reaction (Smith, 1982). The timing of the loss of alarm substance cells in nature corresponds with the development of androgen-induced secondary sexual characters and with higher levels of testosterone. Seasonal loss of alarm substance in male fatheads is interpreted as an adaptation to their abrasive spawning habits, reducing the chance of the alarm system 'misfiring' and interfering with spawning or parental care. The seasonal loss of alarm substance is found in males of seven cyprinid species, six of which have abrasive spawning behaviour.

Identification of the alarm substance

The alarm substance was variously characterized as being purine- or a pterine-like, small-ringed or double-ringed, polypeptide-like, or histamine-like. A mixture of several compounds including amino acids and peptides was also suspected for channel catfish, and the skin extract of Atlantic salmon, which lack the alarm substance, was about equally effective (Tucker and Suzuki, 1972). Recent chemical analyses suggest that hypoxanthine-3(N)-oxide is most likely to be the alarm substance identified in the minnows. This substance is colourless, non-flourescent, slightly water soluble, and unstable in solution (Argentini, 1976, cited in Pfeiffer, 1982). Behavioural experiments have confirmed that hypoxanthine-3(N)-oxide has full biological activity. However, its chemostimulatory effectiveness has not been determined.

Taxonomic distribution of the alarm substance – fright reaction

The alarm substance – fright reaction system described above appears to be restricted to one phylogenetically related group of fish, the Ostariophysi. However, possession of alarm substance and fright reaction is not universal within the group. Several groups lack one or both components of the alarm system. The reaction is not associated with any particular habit or type of social behaviour within the groups. The fright reaction is not species specific, fish react to the alarm substance emitted by others, but the intensity of the response is related to the phylogenetic proximity of the species. The fright reaction is genetically determined, and appears at a certain stage of development regardless of prior experience. The reaction does not develop until some time after the alarm substance is formed in the skin. In minnows, it appears at 51 days after hatching. Shoaling behaviour, by contrast, appears much earlier than the fright reaction.

Biological significance of the fright reaction

It is conceivable that individuals are likely to gain protection as a result of their responses to alarm substance released by a conspecific upon attack by a predator. It is, however, difficult to understand how the release of alarm substance can provide protection or other benefits to the damaged individual, and how such a signalling system might have evolved. Smith (1982) discusses some of the hypotheses, such as kin selection and chemical repellent, regarding the adaptive significance of fish alarm substance. Further research is required to answer this intriguing question properly.

6.8 SUMMARY

Fish detect chemical stimuli through two major chemosensory channels, olfaction and taste. Unlike terrestrial animals, in fish all chemical activators are mediated by water, and chemoreception occurs entirely in the aquatic environment. Therefore, a distinction between olfaction and taste is tenuous, and volatility of olfactory stimulants (odours) is less relevant. Because the fish gustatory system is as sensitive as the olfactory system, we can no longer consider olfactory to be the sensory modality for detection of aquatic chemicals only on the basis of its lower effective concentration. The olfactory system is not a uniform whole; at least two types of olfactory receptor cell detect and encode chemical signals. The encoded olfactory information is transmitted and integrated into behavioural patterns through spatially separated neuro-anatomical substrates within the olfactory centre. Thus feeding behaviour elicited by substances such as amino acids may be processed via the **lateral olfactory system**, whereas other behaviour such as reproduction may be processed through the **medial olfactory system**.

Four principal areas of interface between olfaction and fish behaviour are described.

1. Homing migration in salmon. The olfactory sense seems essential for homing migration. Salmonids imprint to certain distinct odours of the home stream during the early period of residence, and as adults they use this olfactory information to locate the home stream – the 'olfactory hypothesis'. Experiments in which salmon artificially imprinted to morpho-line homed to a stream scented with the chemical verified the olfactory hypothesis. However, the physiological basis for olfactory detection of morpholine has not been established. A role of thyroid hormones in imprinting during smolting is implicated. Homing may be an inherited response to population-specific pheromone trails released by descending smolts – the 'pheromone hypothesis'.

2. Feeding. Feeding behaviour, a stereotyped sequence of behavioural components, is diversely controlled by various sensory modalities, depending upon the ecological niche of the animal. The initial arousal or alert process is primarily mediated by olfaction. Feeding stimulants, fingerprint-like mixtures or chemical images, seem species specific, and have the following chemical nature: (a) low molecular weight, (b) non-volatile, (c) nitrogenous, and (d) amphoteric. L-Proline and related amino acids may have the dominant role in some fish species.

3. Reproduction. Olfaction exerts a functional role in every aspect of the reproductive process. Recent studies point to the gonads as a source of fish reproductive pheromones, and indicate that steroids are likely candidates. Periovulatory female goldfish sequentially release two hormonal phero-mones, 17,20P and PGFs. Male African catfish release steroid glucuronide pheromones from the seminal vesicle. Hormones and their metabolites may commonly serve as reproductive pheromones in fish. Evidence is also presented that loss of chemical information via the olfactory system is compensated by functioning of the gustatory system.

4. Fright reaction. When the skin of an ostariophysid fish is damaged, alarm substance cells are broken and release alarm substance. Nearby conspecifics smell the alarm substance and show a fright reaction. Hypoxanthine-3(N)-oxide is likely to be the alarm substance identified in minnows. The adaptive significance of the fright reaction still remains an intriguing question.

ACKNOWLEDGEMENTS

I thank Dr Tony J. Pitcher for providing the opportunity to write this chapter. I also thank Scott Brown, Robert Evans, Dorthy Klaprat, Torarinn Sveinsson, and Chunbo Zhang for technical and Carol Catt for secretarial assistance. Research discussed here was supported by a Natural Sciences and Engineering Research Council of Canada grant (A7576).

REFERENCES

Andersen, Ø. and Døving, K. B. (1991) Gonadotropin releasing hormone (GnRH) – A novel olfactory stimulant in fish. *NeuroReport*, **2**, 458–460.

Atema, J. (1980) Chemical sense, chemical signals, and feeding behaviour in fishes, in *Fish Behaviour and its Use in the Capture and Culture of Fishes* (eds J. E. Bardach, J. J. Magnuson, R. C. May and J. M. Reinhart), International Center for Living Aquatic Resources Management, Manila, pp. 57–101.

Atema, J., Holland, K. and Ikehara, W. (1980) Olfactory responses of yellowfin tuna (*Thunnus albacares*) to prey odors: chemical search image. *J. Chem. Ecol.*, **6**, 457–65.

Bateson, W. (1890) The sense-organs and perceptions of fishes; with remarks on the supply of bait. *J. Mar. Biol. Ass. U.K.*, **11**, 225–56.

Bern, H. A. (1982) Epilog, in *Salmonid Smoltification* (eds H. A. Bern and C. V. W. Mahnken). *Aquaculture* (Special Issue), **28**, v–x and 1–270.

Brand, J. G. and Bruch, R. C. (1992) Molecular mechanisms of chemosensory transduction: gustation and olfaction, in *Fish Chemoreception* (ed. T. J. Hara), Chapman and Hall, London, pp. 126–149.

Brown, S. B. and Hara, T. J. (1982) Biochemical aspects of amino acid receptors in olfaction and taste, in *Chemorception in Fishes* (ed. T. J. Hara), Elsevier, Amsterdam, pp. 159–80.

Buckland, J. (1880) *Natural History of British Fishes*, Unwin, London.

Caprio, J. (1982) High sensitivity and specificity of olfactory and gustatory receptors of catfish to amino acids, in *Chemoreception in Fishes* (ed. T. J. Hara), Elsevier, Amsterdam, pp. 109–34.

Caprio, J. and Byrd, R. P., jun (1984) Electrophysiological evidence for acidic, basic, and neutral amino acid olfactory receptor sites in the catfish. *J. gen. Physiol.*, **84**, 403–22.

Chidester, F. E. (1924) A critical examination of the evidence for physical and chemical influences of fish migration. *J. exp. Biol.*, **2**, 79–118.

Colombo, L., Belvedere, P. C., Marconato, A. and Bentivegna, F. (1982) Pheromones in teleost fish, in *Reproductive Physiology in Fish* (eds C. J. J. Richter and H. J. Th. Goos), Pudoc, Wageningen, pp. 84–94.

Cooper, J. C. and Hirsch, P. J. (1982) The role of chemoreception in salmonid homing, in *Chemoreception in Fishes* (ed. T. J. Hara), Elsevier, Amsterdam, pp. 343–62.

Demski, L. S. and Hornby, P. J. (1982) Hormonal control of fish reproductive behaviour: brain–gonadal steroid interactions. *Can. J. Fish. Aquat. Sci.* **39**, 36–47.

Demski, L. S. and Northcutt, R. G. (1983) The terminal nerve: a new chemosensory system in vertebrates? *Science, Wash. D.C.*, **220**, 435–7.

Demski, L. S. and Schwanzel-Fukuda, M. (eds) (1987) *The Terminal Nerve (Nervus Terminalis): Structure, Function and Evolution. Ann. N.Y. Acad. Sci.*, **519**, 1–469.

Dodson, J. J. and Bitterman, M. E. (1989) Compound uniqueness and the interactive role of morpholine in fish chemoreception. *Biol. Behav.*, **14**, 13–27.

Døving, K. B. and Selset, R. (1980) Behavior patterns in cod released by electrical stimulation of olfactory tract bundles. *Science, Wash. D.C.*, **207**, 559–60.

Døving, K. B., Nordeng, H. and Oakley, B. (1975) Single unit discrimination of fish odours released by char (*Salmo alpinus*) polulations. *Comp. Biochem. Physiol.*, **47A**, 1051–63.

Døving, K. B., Selset, R. and Thommesen, G. (1980) Olfactory sensitivity to bile acids in salmonid fishes. *Acta physiol. scand.*, **108**, 123–31.

Dulka, J. G., Stacey, N. E., Sorensen, P. W. and Van Der Kraak, G. J. (1987) A steroid sex pheromone synchronizes male–female spawning readiness in goldfish. *Nature, Lond.*, **325**, 251–3.

Erickson, J. R. and Caprio, J. (1984) The spatial distribution of ciliated and microvillous olfactory receptor neurons in the channel catfish is not matched by a differential specificity to amino acid and bile salt stimuli. *Chem. Senses*, **9**, 127–41.

Evans, R. E., Zielinski, B. and Hara, T. J. (1982) Development and regeneration of the olfactory organ in rainbow trout, in *Chemoreception in Fishes* (ed. T. J. Hara), Elsevier, Amsterdam, pp. 15–37.

Fesenko, E. E., Novoselov, V. I., Krapivinskaya, L. D., Mjasoedov, N. F. and Zolotarev, J. A. (1983) Molecular mechanisms of odor sensing VI. Some biochemical characteristics of a possible receptor for amino acids from the olfactory epithelium of the skate *Dasyatis pastinaca* and carp *Cyprinus carpio. Biochim. biophys. Acta*, **759**, 250–56.

Frisch, K. von (1938) Zur Psychologie des Fisch-Schwarmes. *Naturwissenschaften*, **26**, 601–6.

Frisch, K. von (1941) Ueber einen Schreckstoff der Fischhaut und seine biologische Bedeutung. *Z. vergl. Physiol.*, **29**, 46–145.

Fujita, I., Sorensen, P. W., Stacey, N. E. and Hara, T. J. (1991) The olfactory system, not the terminal nerve, functions as the primary chemosensory pathway mediating responses to sex pheromones in goldfish. *Brain Behav. Evol.*, **38**, 313–321.

Goh, Y. and Tamura, T. (1980) Olfactory and gustatory responses to amino acids in two marine teleosts – red sea bream and mullet. *Comp. Biochem. Physiol.*, **66C**, 217–24.

Hara, T. J. (1970) An electrophysiological basis for olfactory discrimination in homing salmon: a review. *J. Fish. Res. Bd Can.*, **27**, 565–86.

Hara, T. J. (1971) Chemoreception, in *Fish Physiology*, Vol. 5 (eds W. S. Hoar and D. J. Randall), Academic Press, New York, pp. 79–120.

Hara, T. J. (1975) Olfaction in fishes. *Progr. Neurobiol.*, **5**, 271–335.

Hara, T. J. (ed.) (1982) *Chemoreception in Fishes*, Elsevier, Amsterdam, 433 pp.

Hara, T. J. (ed.) (1992) *Fish Chemoreception*, Chapman and Hall, London, 373 pp.

Hara, T. J., Brown, S. B. and Evans, R. E. (1983) Pollutants and chemoreception in aquatic organisms, in *Aquatic Toxicology* (ed. J. O. Nriagu), Wiley, New York, pp. 247–306.

Hara, T. J., Macdonald, S., Evans, R. E., Marui, T. and Arai, S. (1984) Morpholine, bile acids and skin mucus as possible chemical cues in salmonid homing: electrophysiological re-evaluation, in *Mechanisms of Migration in Fishes* (eds J. D. McCleave, G. P. Arnold, J. J. Dodson and W. H. Neill), Plenum, New York, pp. 363–78.

Harden Jones, F. R. (1968) *Fish Migration*, Edward Arnold, London, 325 pp.

Hasler, A. D. (1966) *Underwater Guideposts – Homing of Salmon*, University of Wisconsin Press, Madison, 155 pp.

Hasler, A. D. (1983) Synthetic chemicals and pheromones in homing salmon, in *Control Processes in Fish Physiology* (eds J. C. Rankin, T. J. Pitcher and R. T. Duggan), Croom Helm, Lndon, pp. 103–16.

Hasler, A. D. and Scholz, A. T. (1983) *Olfactory Imprinting and Homing in Salmon. Investigations into the Mechanism of the Imprinting Process*, Springer-Verlag, Berlin, 134 pp.

Heczko, E. J. and Seghers, B. H. (1981) Effects of alarm substance on schooling in the common shiner (*Notropis cornutus*, Cyprinidae). *Env. Biol. Fishes*, **6**, 25–9.

Hoar, W. S. (1988) The physiology of smolting salmonids, in *Fish Physiology*, Vol. 11, Part B (eds W. S. Hoar and D. J. Randall), Academic Press, New York, pp. 275–343.

Hurk, R. van den and Lambert, J. G. D. (1983) Ovarian steroid glucuronides function as sex pheromones for male zebrafish, *Brachydanio rerio. Can. J. Zool.*, **61**, 2281–7.

Jones, K. A. (1992) Food search behaviour in fish and the use of chemical lures in commercial and sports fishing, in *Fish Chemoreception* (ed. T. J. Hara), Chapman and Hall, London, pp. 288–319.

Kah, O., Chambolle, P., Dubourg, P. and Dubois, M. P. (1984) Immunocytochemical localization of luteinizing hormone-releasing hormone in the brain of the goldfish *Carassias auratus. Gen. comp. Endocrinol.*, **53**, 107–15.

Kiyohara, S., Yamashita, S. and Harada, S. (1981) High sensitivity of minnow gustatory receptors to amino acids. *Physiol. Behav.*, **26**, 1103–8.

Kleerekoper, H. (1969) *Olfaction in Fishes*, Indiana University Press, Bloomington, 222 pp.

Liley, N. R. (1982) Chemical communication in fish. *Can. J. Fish. Aquat. Sci.*, **39**, 22–35.

Liley, N. R. and Stacey, N. E. (1983) Hormones, pheromones, and reproductive behavior in fish, in *Fish Physiology*, Vol. IX (eds W. S. Hoar, D. J. Randall and E. M. Donaldson), Academic Press, New York, pp. 1–63.

McKeown, B. A. (1984) *Fish Migration*, Croom Helm, London, 224 pp.

Marshall, N. B. (1967) The olfactory organs of bathypelagic fishes, *Symp. Zool. Soc. Lond.*, **19**, 57–70.

Marui, T., Evans, R. E., Zielinski, B. and Hara, T. J. (1983) Gustatory responses of the rainbow trout (*Salmo gairdneri*) palate to amino acids and derivatives, *J. Comp. Physiol.*, **153A**, 423–33.

Moore, A. and Scott, A. P. (1991) Testosterone is a potent odorant in precocious male Atlantic salmon (*Salmo salar* L.) parr. *Phil. Trans. R. Soc. Lond.* B, **332**, 241–244.

Morin, P.-P., Dodson, J. J. and Doré, F. Y. (1989) Thyroid activity concomitant with olfactory learning and heart rate changes in Atlantic salmon, *Salmo salar*, during smoltification, *Can. J. Fish. Aquat. Sci.*, **46**, 131–6.

Münz, H., Stumpf, W. E. and Jennes, L. (1981) LHRH systems in the brain of platyfish. *Brain Res. Amst.*, **221**, 1–13.

Nordeng, H. (1977) A pheromone hypothesis for homeward migration in anadromous salmonids, *Oikos*, **28**, 155–9.

Oka, Y., Ichikawa, M. and Ueda, K. (1982) Synaptic organization of the olfactory bulb and central projection of the olfactory tract, in *Chemoreception in Fishes* (ed. T. J. Hara), Elsevier, Amsterdam, pp. 61–75.

Ottoson, D. (1971) The electro-olfactogram, in *Handbook of Sensory Physiology*, Vol. 4, Part 1 (ed. L. M. Beidler), Springer-Verlag, Berlin, pp. 95–131.

Partridge, B. L., Liley, N. R. and Stacey, N. E. (1976) The role of pheromones in the sexual behaviour of the goldfish. *Anim. Behav.*, **24**, 291–9.

Pfeiffer, W. (1982) Chemical signals in communication, in *Chemoreception in Fishes* (ed. T. J. Hara), Elsevier, Amsterdam, pp. 307–326.

Quinn, T. P. (1990) Current controversies in the study of salmon homing. *Ethol. Ecol. Evol.*, **2**, 49–63.

Resink, J. W., Schoonen, W. G. E. K., Albers, P. C. H., File, D. M., Notenboom, C. D., van den Hurk, R. and van Oordt, P. G. W. J. (1989a) The chemical nature of sex attracting pheromones from the seminal vesicle of the African catfish, *Clarias gariepinus. Aquaculture*, **83**, 137–51.

Resink, J. W., Voorthuis, P. K., van den Hurk, R., Peters, R. C. and van Oordt, P. G. W. J. (1989b) Steroid glucuronides of the seminal vesicle as olfactory stimuli in African catfish, *Clarias gariepinus. Aquaculture*, **83**, 153–66.

Rhein, L. D. and Cagan, R. H. (1983) Biochemical studies of olfaction: binding specificity of odorants to a cilia preparation from rainbow trout olfactory rosettes. *J. Neurochem.*, **41**, 569–77.

Satou, M. (1990) Synaptic organization, local neuronal circuitry, and functional segregation of the teleost olfactory bulb. *Progr. Neurobiol.*, **34**, 115–42.

Satou, M. (1992) Synaptic organization of the olfactory bulb and its central projection, in *Fish Chemoreception* (ed. T. J. Hara), Chapman and Hall, London, pp. 40–59.

Satou, M., Fujita, I., Ichikawa, M., Yamaguchi, K. and Ueda, K. (1983) Field potential and intracellular potential studies of the olfactory bulb in the carp: evidence for a functional separation of the olfactory bulb into lateral and medial subdivisions. *J. comp. Physiol.*, **152A**, 319–33.

Scott, A. P. and Sumpter, J. P. (1983) The control of trout reproduction: basic and applied research on hormones, in *Control Processes in Fish Physiology* (eds J. C. Rankin, T. J. Pitcher and R. T. Duggan), Croom Helm, London, pp. 200–220.

Segaar, J., de Bruin, J. P. C. and van der Meché-Jacobi, M. E. (1983) Influence of chemical receptivity on reproductive behaviour of the male three-spined stickleback (*Gasterosteus aculeatus* L.). *Behaviour*, **86**, 100–166.

Smith, R. J. F. (1992) Alarm signals in fishes. *Rev. Fish Biol. Fish.*, **2**, 33–63.

Smith, R. J. F. (1982) The adaptive significance of the alarm substance–fright reaction system, in *Chemoreception in Fishes* (ed. T. J. Hara), Elsevier, Amsterdam, pp. 327–342.

Sorensen, P. W. (1992) Hormones, pheromones and chemoreception, in *Fish Chemoreception* (ed. T. J. Hara), Chapman and Hall, London, pp. 199–228.

Sorensen, P. W., Hara, T. J. and Stacey, N. E. (1991) Sex pheromones selectively stimulate the medial olfactory tracts of male goldfish. *Brain Res. Arist.*, **558**, 343–347.

Sorensen, P. W., Hara, T. J. and Stacey, N. E. (1987) Extreme olfactory sensitivity of mature and gonodally-regressed goldfish to a potent steroidal pheromone, 17α, 20β-dihydroxy-4-pregnen-3-one. *J. comp. Physiol.*, **160A**, 305–13.

Sorensen, P. W., Hara, T. J., Stacey, N. E. and Goetz, F. Wm. (1988) F prostaglandins function as potent olfactory stimulants that comprise the postovulatory female sex pheromone in goldfish. *Biol. Reprod.*, **39**, 1039–50.

Stacey, N. E. (1983) Hormones and Reproductive Behaviour in Teleosts, in *Control Processes in Fish Physiology* (eds J. C. Rankin, T. J. Pitcher and R. T. Duggan), Croom Helm, London, pp. 117–129.

Stacey, N. E. and Kyle, A. L. (1983) Effects of olfactory tract lesions on sexual and feeding behavior in the goldfish. *Physiol. Behav.*, **30**, 621–8.

Stoddart, D. M. (1980) *The Ecology of Vertebrate Olfaction*, Chapman and Hall, London, 234 pp.

Strieck, F. (1924) Untersuchungen über den Geruchs- und Geschmackssinn der Elvitzen. *Z. vergl. Physiol.*, **2**, 122–54.

Sutterlin, A. M. and Sutterlin, N. (1971) Electrical responses of the olfactory epithelium of Atlantic salmon (*Salmo salar*). *J. Fish. Res. Bd Can.*, **28**, 565–72.

Suzuki, N. and Tucker, D. (1971) Amino acids as olfactory stimuli in freshwater catfish, *Ictalurus catus* (Linn.). *Comp. Biochem. Physiol.*, **40A**, 399–404.

Sveinsson, T. and Hara, T. J. (1990) Analysis of olfactory responses to amino acids in Arctic char (*Salvelinus alpinus*) using a linear multiple-receptor model. *Comp. Biochem. Physiol.*, **97A**, 279–87.

Tavolga, W. N. (1956) Visual, chemical and sound stimuli as cues in the sex discriminatory behavior of the gobiid fish, *Bathygobius soporator*, *Zoologica*, **41**, 49–64.

Thommesen, G. (1978) The spatial distribution of odour induced potentials in the olfactory bulb of char and trout (Salmonidae). *Acta physiol. scand.*, **102**, 205–17.

Thommesen, G. (1983) Morphology, distribution, and specificity of olfactory receptor cells in salmonid fishes. *Acta physiol. scand.*, **117**, 241–9.

Thorpe, J. E. (1988) Salmon migration. *Sci. Prog.*, *Oxf.*, **72**, 345–70.

Tucker, D. and Suzuki, N. (1972) Olfactory responses to Schreckstoff of catfish, in *Olfaction and Taste* IV (ed. D. Schneider), Wissenschaftliche Verlagsgesellschaft, Stuttgart, pp. 121–7.

Ueda, H., Hiroi, O., Hara, A., Yamauchi, K. and Nagahama, Y. (1984) Changes in serum concentrations of steroid hormones, thyroxine, and vitellogenin during spawning migration of the chum salmon, *Oncorhynchus keta*. *Gen. comp. Endocrinol.*, **53**, 203–11.

Yamamori, K., Nakamura, M., Matsui, T. and Hara, T. J. (1988) Gustatory responses to tetrodotoxin and saxitoxin in fish: a possible mechanism for avoiding marine toxins. *Can. J. Fish. Aquat. Sci.*, **45**, 2182–6.

Yoshii, K., Kamo, N., Kurihara, K. and Kobatake, Y. (1979) Gustatory responses of eel palatine receptors to amino acids and carboxylic acids. *J. gen. Physiol.*, **74**, 301–17.

Zielinski, B. and Hara, T. J. (1988) Morphological and physiological development of olfactory receptor cells in rainbow trout (*Salmo gairdneri*) embryos. *J. comp. Neurol.*, **271**, 300–311.

Chapter seven

Role of the lateral line in fish behaviour

Horst Bleckmann

7.1 INTRODUCTION

The lateral line system is nearly ubiquitous among anamniotic vertebrates. It was already present in the earliest heterostracans and probably has a monophyletic origin (Northcutt, 1989). The lateral line provides petromyzontid agnathans, all extant cartilaginous and bony fishes, most larval and even some postmetamorphic amphibians with a hydrodynamic sense of which humans do not have direct experience. In teleost fish the lateral line is visible externally as rows of small pores found on the trunk and the head. These pores, which – according to Parker (1904) – were first described by Stenon in 1664, connect the outside medium with a sophisticated canal system formed by bone or scales. Lateral line canals contain numerous sensory hillocks or neuromasts. Together with freestanding neuromasts these canals constitute the mechano-sensory lateral line system of teleost fish (Fig. 7.1(A), (B)).

Early investigators considered the lateral line as a system of glands (Fuchs, 1895). It was Jakobson (1813) who first concluded from observing an extensive nerve supply that the lateral line is a sensory system which is probably stimulated mechanically. Knox (1825) and Leydig (1850, 1851) supported this idea and the lateral line was soon accepted as a mechano-receptive organ especially adapted for aquatic life.

Despite these early investigations the mechanically relevant stimuli remained disputed. Some investigators assumed the lateral line to be an organ of touch (Merkel, 1880; de Sede, 1884), to have an intermediate function between touch and hearing (Leydig, 1850, 1851; Schulze, 1870; Dercum, 1880), to be an accessory auditory organ (Emery, 1880; Mayser, 1881; Bodenstein, 1882), an equilibrium organ (Lee, 1898) or a pressure receptor (Fuchs, 1895). Richard (1896, quoted by Parker, 1902) even believed that the lateral line was connected with the production of gas in the swim bladder.

Behaviour of Teleost Fishes 2nd edn. Edited by Tony J. Pitcher. Published in 1993 by Chapman & Hall. ISBN   0 412 42930 6 (HB) and 0 412 42940 3 (PB).

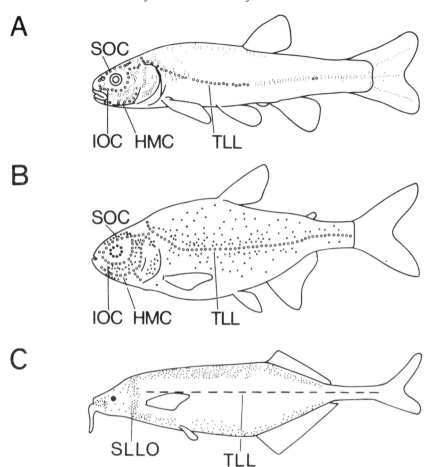

Fig. 7.1 (A) and (B) Distribution of ordinary lateral line organs (A) the European minnow, *Phoxinus phoxinus*, and (B) the blind cave fish *Astyanax mexicanus* ●, free neuromasts; ○, canal pores (enlarged for clarity). (C) Distribution of specialized lateral line organs in the weakly electric fish *Gnathonemus petersi*. The ordinary lateral line organ on the trunk is indicated by a dashed line. HMC, hyomandibular canal; IOC, infra-orbital canal; SLLO, specialized lateral line organs; SOC, supra-orbital canal; TLL, trunk lateral line; Redrawn from Dijkgraaf (1934), Grobbel and Hahn (1958), and Szabo (1974).

In 1908 Hofer convincingly demonstrated that the lateral line functions as a hydrodynamic receptor. He observed distinctive fin movements when weak water currents were impinging locally on the head or trunk of a pike, *Esox lucius*, and that unilateral cauterization of the lateral line rendered the operated part of the body insensitive. From these experiments Hofer concluded that fish can 'feel at a distance'. Dijkgraaf (1963) demonstrated that fish can use self-induced water currents for close-range obstacle detection and Schwartz (1965, 1971) was the first to show that surface-feeding fish detect and locate sources of surface waves with parts of the cephalic lateral line.

In cartilaginous fish two more kinds of end organs have been classified as mechanosensory lateral line: the spiracular organs (also found in most non-teleost bony fishes), which are suspected to function as proprioceptors monitoring the position of the hyomandibula, and the vesicles of Savi, which may function in substrata vibration detection (Barry and Bennet, 1989). The spiracular organs and the vesicles of Savi will not be discussed in this review.

Physiological experiments seem to indicate that the lateral line serves a dual function, as both a mechano- and a chemoreceptor (Katsuki and Yamashita, 1969; Kawamura and Yamashita, 1983). In amphibians chemical stimulation of the lateral line, however, does not appear to be a behaviourally relevant stimulus (Görner and Mohr, 1988). Most likely the chemosensitivity of the lateral line is an incidental by-product of normal receptor physiology and thus will not be considered in this chapter. Recently the lateral line has also become a subject of developmental studies. This topic also will not be discussed, but the reader who is interested in this theme is referred to Blaxter (1987).

In 1917 Parker and van Heusen discovered that catfish respond to weak galvanic currents. Since that time it became evident that the lateral line of many fishes consists of 'ordinary' mechanoreceptive neuromasts (hereafter referred to as lateral line) and of 'specialized' organs (electroreceptors), stimulated by weak electric fields (Fig. 7.1(C)).

7.2 PHYSICAL PROPERTIES OF MIDWATER WAVES

Dipole field equations

One physical definition of sound is the collection of disturbances set up in a medium by a source of movement. Complex acoustic stimuli can be described as an infinite series of terms, each series consisting of a monopole (pulsating sphere), a dipole (constant-volume vibrating sphere), a quadrupole, and so forth. True monopole lateral line stimuli are probably rare and will not be discussed here. Because contributions of successive terms fall off with increasing powers of the distance, sufficiently far from the source the dipole moment (assuming that a monopole moment is missing) of a series dominates all higher terms (for a thorough discussion of monopole and dipole fields see Kalmijn, 1988a). In front, above, below and to the sides a swimming fish generates dipole-like water motions. A dipole, like all other sound sources, simultaneously generates both pressure waves and particle displacements (Fig. 7.2, inset). The acoustic pressure, p, caused by a dipole of the radius r is given by

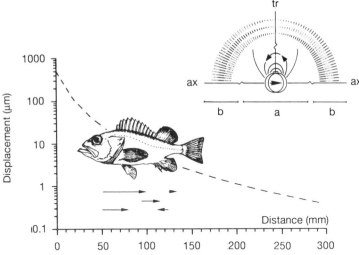

Fig. 7.2 Diagram showing differential movement between fish and surrounding water in the near field. A dipolar sound source of 30 mm diameter is assumed to vibrate in front of the fish with a peak-to-peak displacement of 500 μm. The broken curve gives the calculated decrease of near-field water displacement over distance. The arrows symbolize the displacement amplitude of the water at the source distances 50 and 120 mm (top pair), the horizontal displacement of the fish caused by the vibrating sphere (middle arrow), and the relative motion between fish and water (bottom pair). Note the phase reversal which occurs between the head and tail region. Adapted from Denton and Gray (1983). Inset: Dipole field of a vibrating sphere. a, Dipolar local flow near the source; arrows along the axis (ax) show the direction of sphere displacement and of axial flow. b, Far-field flow region where the propagated sound pressure waves dominate. Arrowheads suggest particle velocities, heavy shading denotes compression, light shading rarefaction. tr, Transverse plain. The transition zone between the dipolar and radial regions is not shown. Redrawn from Kalmijn (1988a).

(Equations 7.1 to 7.3 have been taken from Kalmijn, 1988a):

$$p = -\frac{\rho c k^2 r^2}{2D_r} U \cos\theta \cos(\omega t - kr) - \frac{\rho c k r^2}{2D_r^2} U \cos\theta \sin(\omega t - kr), \quad (7.1)$$

where U is the amplitude of axial source velocity, D_r is the radial distance from the centre of the sphere to the point of interest, θ is the angle of radiation, c is the speed of sound, k, is the wavenumber $= \omega/c = 2\pi/\lambda$ (with $\omega = 2\pi f$, λ is the wavelength, f is frequency in Hz), ρ is the density of water, and t is time. (A list of symbols is given on pages 236–7.) The first term of Equation (7.1) represents what we usually think of as the propagating sound wave. It starts right at the source and is proportional to $1/D_r$ and to k^2. The second term of the equation describes the near-field pressure which is proportional to $1/D_r^2$

and to k. Due to the difference in attenuation, the near-field pressure (Equation 7.1) dominates only close to the sound source. Note that the acoustic pressure is proportional to the cosine of the angle of radiation, θ (Fig. 7.2, inset).

Harris and van Bergeijk (1962) demonstrated that the lateral line is not sensitive to sound pressure waves but to near-field water motions. The water velocity caused by a small (if compared with λ) constant-volume vibrating sphere can be subdivided into a radial, v_r, and a tangential, v_θ, particle velocity component:

$$v_r = -\frac{k^2 r^3}{2D_r} U \cos\theta \cos(\omega t - kD_r) - \frac{kr^3}{D_r^2} U \cos\theta \sin(\omega t - kD_r)$$

$$+ \frac{r^3}{D_r^3} U \cos\theta \cos(\omega t - kD_r). \tag{7.2}$$

The first term of Equation 7.2 describes the particle velocity caused by the propagating sound wave. Here v, like p, is proportional to $1/D_r$ and to k^2. The second term represents the intermediate flow, where v is proportional to $1/D_r^2$ and to k. The third term of Equation 7.2 describes the local flow; here v is proportional to $1/D_r^3$ and independent of k (and thus independent of stimulus frequency). Due to the differences in attenuation, close to the source the local and intermediate flow dominate and define the dipole near field. Note that the radial particle velocities of the dipole field, like p, are proportional to the cosine of the angle of radiation, θ. The tangential velocity component, v_θ, is given by:

$$v_\theta = -\frac{kr^3}{2D_r^2} U \sin\theta \sin(\omega t - kD_r) + \frac{r^3}{2D_r^3} U \sin\theta \cos(\omega t - kD_r). \tag{7.3}$$

The first term corresponds to the intermediate flow, and the second term to the local flow. The tangential velocities of the local and intermediate flow are proportional to the sine of the angle of radiation. The actual water velocity at any given point is the vector sum of v_r and v_θ:

$$v = \sqrt{(v_r^2 + v_\theta^2)} \tag{7.4}$$

The motion of water particles can not only be described in terms of velocity, but also in terms of displacement, d, or acceleration, a, where:

$$d = \int v \, dt \quad \text{and} \quad a = \frac{\partial v}{\partial t} \tag{7.5a and 7.5b}$$

Boundary layer

The velocity of a fluid at the interface of a solid is always the same as that of the solid. This no-slip condition implies that there is a gradient in the speed of flow near every surface and that the velocity gradients are developed entirely within fluids rather than between fluids and solids. If the fluid is homogeneous the velocity decreases smoothly to zero as the surface is approached. The velocity gradient region is associated with the term 'boundary layer'. For d.c. (direct) and a.c. (alternating) flow, respectively, the thicknesses of the boundary layer, i.e. the distance normal to the surface of a plate (fish) where the local velocity has risen to 0.99% of the free stream velocity, is approximated by (Lighthill, 1980):

$$\delta = 5\sqrt{[(l\mu)/(pU)]} \quad \text{or} \quad \delta = \sqrt{[(2\mu)/(\rho\omega)]} \qquad \text{(7.6a, and 7.6b)}$$

where δ is a measure of the thickness of a laminar boundary layer, l is the distance downstream from the leading edge of the object (e.g. a fish or a plate), μ is the dynamic viscosity of fluid, and ρ is fluid density. To give numerical examples: for fresh water of $20\,°C$ the thickness of the boundary layer measured 3 cm downstream is 5 mm for a water (or fish) velocity of $3\,\text{cm s}^{-1}$ and 1.6 mm for a velocity of $30\,\text{cm s}^{-1}$. In case of a.c. flow the boundary layer thickness is $540\,\mu\text{m}$ at 1 Hz and $54\,\mu\text{m}$ at 100 Hz.

Midwater lateral line stimuli

The biologist who wants to study the biological significance of the lateral line faces the problem that little is known about natural lateral line stimuli. This especially holds true for midwater and bottom-dwelling fishes. In most electrophysiological experiments and theoretical calculations concerned with lateral line function, a vibrating sphere has been used as a stimulus source. This is so because the stimuli produced are easily understood and characterized (Equations 7.1–7.4). Dipole-like wave stimuli, however, do not consider important natural stimulus properties like vorticity and turbulence.

In the time domain, wave stimuli caused by aquatic animals (fish, frogs, and crustaceans) can range from transient to sinusoidal; frequency components above about 30 Hz are rare in these waves (Kirk, 1985; Montgomery *et al.*, 1988; Enger *et al.*, 1989). Transient wave stimuli caused by a rapid onset of motion (e.g. a startle response), however, can have frequency components of up to at least 100 Hz (Bleckmann *et al.*, 1991). What, up to now, has been little appreciated in lateral line research is the fact that even a slowly swimming fish generates a trail of vortices (photo, Fig. 12.3). Similar to the wake behind a boat, this vortex trail – which has a complex three-dimensional ladder-like arrangement (Blickhan *et al.*, 1990, and Fig. 7.3) – can outlast its creator by at least several seconds. These vortices, which represent separated, rolled-up parts of the boundary layer, can contain frequency components up to at least 100 Hz (Fig. 7.3, inset).

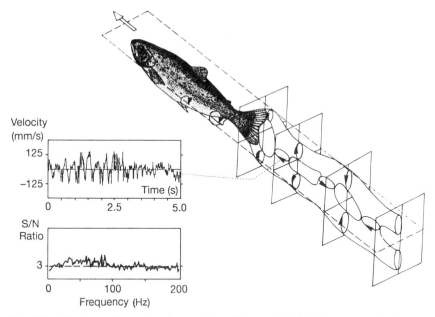

Fig. 7.3 The three-dimensional extension of the wake behind a trout swimming at 48 cms^{-1} in a water tunnel. Arrows indicate flow direction of individual water particles (after Blickhan, unpublished). Inset: Water velocity recorded in the wake 3–5 cm behind a swimming trout (top) and spectral distribution of the signal-to-noise ratio (S/N ratio) (bottom). Values above the broken line are assumed to override the noise level. After Bleckman *et al.* (1991).

Midwater noise

Every animal faces the problem that it must be able to extract meaningful stimuli from background noise. Although direct measurements are lacking, it is safe to assume that fishes that live in different habitats often face different hydrodynamic conditions. Whereas ponds, lakes, and the open ocean mostly provide quiet-water habitats, the ocean shoreline or a rapidly running river may provide turbulent, high-frequency water conditions. Most likely the morphological variability of the peripheral lateral line (p. 208) at least in part reflects the hydrodynamic environmental factors which might have shaped this sensory system during the course of evolution.

Effective lateral line stimuli

The active space of a sensory system covers the area over which a stimulus can be detected. It depends on the initial stimulus amplitude, the damping properties of the medium through which the stimulus propagates, the noise level, the threshold sensitivity of the sensory system in question, and the ability of an animal to extract a meaningful stimulus from background noise. Fishes are nearly neutrally buoyant. For this reason they tend to move with the water mass. Thus lateral line organs can only detect the spatial derivative of the imposed local flow field integrated along the length of the sensory arrays. The spatial derivative of the imposed local and intermediate flow (Equations 7.2, 7.3) falls off as approximately one over the fourth power of distance (Kalmijn, 1988a). Owing to this strong attenuation, the movement of the medium adjacent to the wave source is greater than that of the fish itself. For those parts of the fish which are furthest away from the source the reverse is true (Denton and Gray, 1983, 1988). From the physical principles given above it becomes apparent that even under low noise conditions the lateral line usually will be stimulated by dipole-like water waves only if the wave source is very near (cm to dm range). This is due to (1) the strong attenuation of the local and intermediate flow (Equation 7.2) and (2) the fact that at a source distance which is large compared with the size of the fish, the whole animal moves with the water mass. As a result, net movements between sound-induced particle motions and a potential hydrodynamic receiver do not occur.

Behind a swimming fish, non-dipole-like, circular and turbulent water motions (Fig. 7.3) can be observed. For this reason some body parts of a fish that swims or hovers in the wake caused by another fish should experience some kind of net hydrodynamic stimulation, even at distances of several metres.

7.3 MORPHOLOGY OF THE ORDINARY LATERAL LINE

Lateral line canals

The most striking part of the lateral line is a sophisticated canal system recessed in the skin or in the bony tissue of the skull or scales. The head lateral line of most teleost fish comprises at least three canals, one of which passes forwards and above the eye (supra-orbital canal), another forward and immediately below the eye (infra-orbital canal), and a third downwards and over the lower jaw (hyomandibular canal) (Fig. 7.1 (A), (B)). Examples of cephalic lateral line canal loss and replacement by superficial neuromasts are known, however. Head canal diameter varies widely, from less than 200 μm up to several mm. Wider canals may be covered by an elastic connective tissue, reinforced partially by bone. The trunk of the earliest bony fishes was characterized by dorsal, medial, and ventral lateral line canals (Northcutt, 1989). Most recent fish, however, have only one trunk canal, but in many species multiple trunk lateral lines still occur (Coombs *et al.*, 1988, 1992; Webb, 1989). In most fishes, lateral line canals open to the environment through a series of pores, which, in different species, vary in number, shape and size. Pores may be flush with the surface of the canal or sit at the end of tubules or canaliculi extending from the canal. Secondary and even tertiary branching of tubules may occur.

There are indications that teleost fish that live in turbulent water have a higher canal specialization (i.e. secondary and tertiary branching of the canals), an increased number of pores and only a few superficial neuromasts. In contrast, teleost fish that live in quiet waters and which swim slowly tend to have a reduced, simple canal system and an increased number of superficial neuromasts (Dijkgraaf, 1963; Bleckmann and Münz, 1990; Vischer, 1990) Model experiments show that lateral line canals are accessory structures that form a mechanical filter which determines the stimulus which finally reaches a neuromast (Denton and Gray, 1988; Bleckmann and Münz, 1988, 1989).

Neuromast

The basic unit of the mechanosensitive lateral line is the neuromast. It consists of sensory hair cells, supporting cells and mantle cells which sit above the basement membrane in the epidermis (Fig. 7.4(A)). The ciliary bundles of hair cells project into a jelly-like substance, the cupula. Cupulae are composed of an outer sheet and a central core. Lateral line cupulae can have a flat, ribbonlike, or rodike shape. Depending on the species investigated and the position on the body, cupular length of freestanding neuromasts can vary at least between 40 and 300 μm (Blaxter and Fuiman, 1989; Teyke, 1990). It was Schulze (1861) who first observed that slight water motions made lateral line cupulae move and bend. Superficial neuromasts, which are often found at the bottom of a shallow pit or groove in the skin, are distributed in a definite arrangement on the head and the trunk of teleost fish (Fig. 7.1(A), (B)). Superficial neuromasts are often smaller ($< 100\,\mu$m in diameter) than canal neuromasts, which can have a diameter (length) of more than 600 μm (Münz, 1989). Consequently superficial neuromasts have fewer hair and support cells than canal neuromasts. Canal neuromasts (Fig. 7.4(B)) and their cupula vary in shape, topography, and orientation with respect to canal axis. Neuromast number, placement, and design may reflect certain evolutionary adaptations. For instance, in the blind cave fish *Astyanax hubbsi*, which heavily depends on its lateral line, the freestanding neuromasts on average are twice as large and have longer cupulae than the neuromasts of the sighted river fish, *A. mexicanus* (Teyke, 1990).

Lateral line hair cell

The hair cells of the lateral line (Fig. 7.4(C)) are similar to those in the auditory and vestibular organs of vertebrates. The bundle of 30 to 150 hairs on the hair cells, called stereovilli, were first recognized by Schulze (1861). The stereovilli (diameter 0.2–0.8 μm) grow longer from one edge of the hair bundle

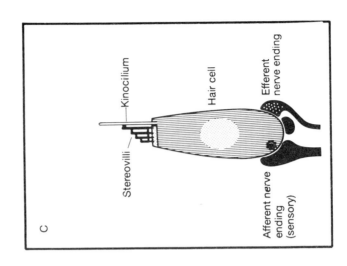

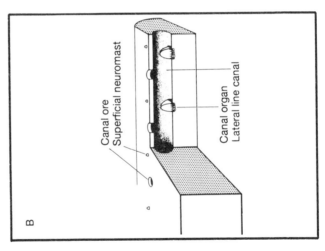

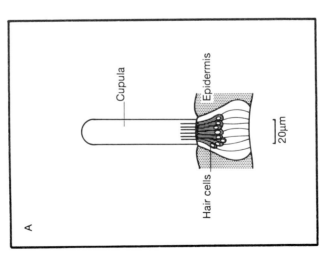

A

Cupula

Epidermis

Hair cells

20μm

B

Canal ore
Superficial neuromast

Canal organ
Lateral line canal

C

Kinocilium

Hair cell

Efferent
nerve ending

Stereovilli

Afferent nerve
ending
(sensory)

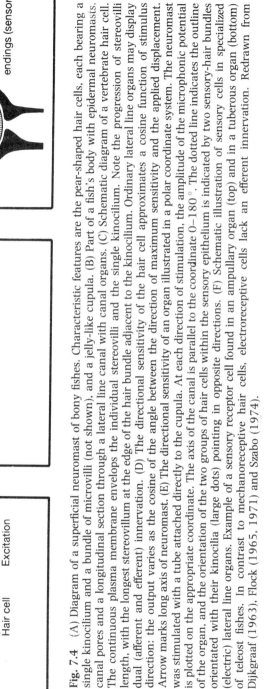

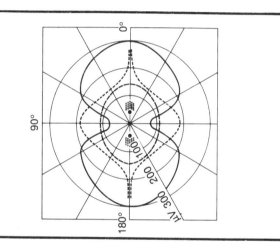

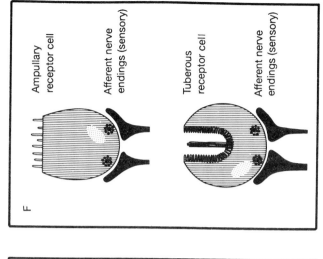

Fig. 7.4 (A) Diagram of a superficial neuromast of bony fishes. Characteristic features are the pear-shaped hair cells, each bearing a single kinocilium and a bundle of microvilli (not shown), and a jelly-like cupula. (B) Part of a fish's body with epidermal neuromasts, canal pores and a longitudinal section through a lateral line canal with canal organs. (C) Schematic diagram of a vertebrate hair cell. The continuous plasma membrane envelops the individual stereovilli and the single kinocilium. Note the progression of stereovilli length, with the longest stereovillum at the edge of the hair bundle adjacent to the kinocilium. Ordinary lateral line organs may display dual (afferent and efferent) innervation. (D) The directional sensitivity of the hair cell approximates a cosine function of stimulus direction: the output varies as the cosine of the angle between the direction of maximum sensitivity and the applied displacement. Arrow marks long axis of neuromast. (E) The directional sensitivity of an organ illustrated in a polar coordinate system. The neuromast was stimulated with a tube attached directly to the cupula. At each direction of stimulation, the amplitude of the microphonic potential is plotted on the appropriate coordinate. The axis of the canal is parallel to the coordinate 0–180°. The dotted line indicates the outline of the organ, and the orientation of the two groups of hair cells within the sensory epithelium is indicated by two sensory-hair bundles orientated with their kinocilia (large dots) pointing in opposite directions. (F) Schematic illustration of sensory cells in specialized (electric) lateral line organs. Example of a sensory receptor cell found in an ampullary organ (top) and in a tuberous organ (bottom) of teleost fishes. In contrast to mechanoreceptive hair cells, electroreceptive cells lack an efferent innervation. Redrawn from Dijkgraaf (1963), Flock (1965, 1971) and Szabo (1974).

to the other. Lateral line hair cells have a single membrane-bound kinocilium (diameter about $0.3 \, \mu m$) with the true $9 + 2$ ciliary pattern. The kinocilium always occurs eccentrically at the tall edge of the bundle, thus all hair cells display a morphological polarization. Within a neuromast, hair cells are usually orientated such that the kinocilium is pointing in the direction of the long axis of the neuromast. Hair cells are mechanoelectrical transducers (Roberts *et al.*, 1988). The appropriate stimulus is a mechanical shearing force applied to the distal end of the hair bundle. An individual hair cell is connected to the CNS by afferent (sensory) nerve fibres whose cell bodies are within the lateral line nerve ganglion, and by efferent nerve fibres whose cell bodies are within the brain (Fig. 7.4(C)).

Central pathways

Defined on the basis of the possession of a separate ganglion and distinct areas of peripheral innervation, up to four lateral line nerves can be distinguished: the dorsal and ventral anterior lateral line nerves, the middle lateral line nerve, and the posterior lateral line nerve (Northcutt, 1989). Roots of the anterior, middle, and posterior lateral line nerves enter the ipsilateral brainstem and terminate in the medial and caudal nucleus of the lateralis area and in the eminentia granularis of the cerebellum (McCormick, 1989; Puzdrowski, 1989). In fish studied so far, information from canal and superficial neuromasts is carried by separate afferent fibres (Münz, 1985). The axons of the secondary cells of the primary lateral line nuclei collectively course bilaterally, with contralateral dominance. The largest portion of this pathway synapses in the midbrain torus semicircularis. The final ascending pathway for mechano-reception involves the relay of information from midbrain to thalamus, and finally to telencephalon (McCormick, 1989).

7.4   PHYSIOLOGY OF THE ORDINARY LATERAL LINE

Transfer properties

Every displacement of the cupula results in a motion of the hair bundle. Physical parameters which determine the movement of the cupula are a frequency-dependent combination of viscous and inertial fluid forces and the cupula sliding stiffness (van Netten and Kroese, 1987; van Netten, 1990). As a result of these combined forces the deflection of the cupula of the ruff, *Acerina cernua*, is proportional to the velocity (at frequencies below 10 Hz), the acceleration (at intermediate frequencies), or the displacement (at frequencies above 300 Hz) of the water flow around the cupula (van Netten, 1990). The relationship between cupula displacement and electrical response of a neuromast is such that the hair cell potential changes linearly with the deflection of the cupula. Displacement of the stereovilli in the direction of the kinocilium causes a depolarization, displacement in the opposite direction a hyperpolarization of the hair cell membrane potential (Flock, 1971; Kroese and van Netten, 1989). Consequently, the response of a single hair cell varies with the stimulus angle in a cosine fashion (Fig. 7.4(D)). In a neuromast, adjacent hair cells are orientated with their kinocilia facing in opposite directions, i.e. a displacement of the cupula will cause responses of opposite polarities from the two sets of cells, which work $180°$ out of phase. As a consequence, lateral line neuromasts, just like individual hair cells, are directionally sensitive (Fig. 7.4(E)).

Lateral line afferent fibres are spontaneously active. A single afferent couples only to hair cells that have the same orientation. An afferent fibre therefore responds best (with a decrease or increase of the spontaneous discharge frequency) if the cupula is bent in one of the two possible directions with respect to the most sensitive axis of the neuromast (Münz, 1985). Canal organs are most sensitive to water currents parallel to the axis of the canal, i.e. in this case the amplitude of the displacement of the canal fluid and the electrophysiological responses of single neuromasts are maximal (Sand, 1981; Denton and Gray, 1983; Gray, 1984). Primary lateral line afferents react to sinusoidal wave stimuli with phase coupling (Coombs and Janssen, 1990). Stimulus intensity is encoded by the degree of phase coupling and by the fibre's maximal firing rate. Saturation of firing rate may be reached 80 dB above threshold (Münz, 1985; Elepfandt and Wiedemer 1987; Bleckmann *et al.*, 1989b). Measurements show that in the trout *Salmo gairdneri* the mean diameter of the afferent fibre population in the tail region is larger than half-way down the trunk. The corresponding conduction velocities decrease from $31 \pm 8\,\mathrm{m\,s}^{-1}$ to $19 \pm 4\,\mathrm{m\,s}^{-1}$ (Kroese *et al.*, 1989). The regional variation in conduction velocity probably enhances the accuracy with which temporal aspects of hydrodynamic stimuli can be decoded. This may be a prerequisite to analyse complex three-dimensional wave patterns such as the vortex trail caused by another fish (Fig. 7.3).

Sensitivity of the lateral line

The response of an individual hair cell depends on the water flow around the cupula. It should be noted, however, that the hydrodynamic stimulus which drives the cupula is different from the stimulus the animal detects. The reason for this is that the velocity of the water at the surface of the animal is proportional to a fractional derivative of the velocity of the water volume outside the boundary layer. The influence of lateral line canals is such that the velocity of the water in the canal is even more proportional to the first full derivative of the velocity of the water outside the boundary layer (Kalmijn,

1989). As a consequence free neuromasts rank between velocity- and acceleration-sensitive detectors whereas canal neuromasts usually function as acceleration detectors (Münz, 1985; Kroese and Schellart, 1987; Coombs and Janssen, 1990). Neuromast placement as well as cupula and canal dimensions are thus ways to influence the frequency response and the sensitivity of lateral line neuromasts.

The absolute sensitivity of the lateral line is very high. For freestanding neuromasts, minimal velocity thresholds are less than $0.03 \, \text{mm s}^{-1}$ (Münz, 1985; Coombs and Janssen, 1990), for canal neuromasts acceleration thresholds of 0.3 to $20 \, \text{mm s}^{-2}$ have been reported (e.g. Münz, 1985; Bleckmann and Münz, 1990; Coombs and Janssen, 1990).

Response to pressure waves

Equations 7.1 to 7.3 show that a vibrating sphere simultaneously generates both pressure waves and particle displacements. The lateral line usually is not stimulated by the pressure component of a sound wave, nevertheless some fish have accessory structures which serve this function. For example, in herring, *Clupea harengus*, the swim bladder, inner ear, and parts of the lateral line are linked. The central feature is a pair of pro-otic auditory bullae, which act as pressure–displacement converters. Each bulla is divided by a membrane into a gas-filled lower part and a liquid-filled upper part. Whereas the liquid-filled part is hydrodynamically connected via a fenestra in the upper wall of the bulla with the ear and the head lateral line, the gas-filled lower part is air-connected with the swim bladder via a duct. The auditory bullae allow sound pressure changes to generate flows of liquid which stimulate the sense organs of both the inner ear (mainly those of the utriculus) and the lateral line (Blaxter *et al.*, 1981).

Efferent fibres

Efferent fibres provide the opportunity for a dynamic control of sense organ properties. From physiological experiments we know that the sensitivity of lateral line organs can actively be reduced by means of efferent fibres. This especially may occur immediately before and during the power strokes of swimming movements associated with the contraction of white 'anaerobic' swimming muscles, i.e. those associated with escape, turning or chasing (Russell, 1974). However, in some fish there is also evidence for an excitatory efferent action. Thus, it seems that efferent nerve stimulation can result either in inhibition or in excitation of lateral line sensory activity (review, Roberts and Meredith, 1989). Physiological studies further show that the electrical stimulation of efferent fibres can cause changes in amplitude, shape, and phase of microphonic and summating potentials (Russell and Lowe, 1983). Efferent activity may also influence the mechanical properties of hair cells (Ashmore, 1984), i.e. this could be one of the mechanisms used to alter the response properties of hair cell receptor systems. In sum the impact of the efferent system on the lateral line is still not very well understood and may be far more complicated than was previously thought (see also Tricas and Highstein, 1990).

7.5   BEHAVIOUR

Identification and localization of stationary objects

The biological significance of the lateral line system can only be recognized by using behavioural endpoints. In behavioural tests one has to make sure, however, that the response to a hydrodynamic stimulus was not mediated in part or completely by other sensory systems. Potential candidates are the inner ear, which may detect both pressure waves and whole body motions (Fig. 7.3), and the common cutaneous sense.

Some fish use self-induced water motions to detect stationary objects with the aid of the lateral line. This ability has been thoroughly investigated in the blind Mexican cave fish *Anoptichthys jordani* (now *Astyanax mexicanus*). If *Anoptichthys* is confronted with a new object, it swims around restlessly, avoiding collision with and maintaining a narrow gap between itself and the object. On approaching the object, the fish accelerates and then glides past it in close proximity. Behavioural experiments show that *A. jordani* gains information about its environment by analysing distortions of the self-induced flow field with the aid of the lateral line. Weissert and von Compenhausen (1981), von Campenhausen *et al.* (1981), and Hassan (1986) provided evidence for this when they trained *A. jordani* to differentiate either between tunnels that contained vertical bars in various combinations of size, number and positions, or between pairs of grids which consisted of equidistant vertical bars. To give some examples: *A. jordani* can be trained to discriminate between two 'fences', each of which has six bars differing only slightly in respect to their relative position. If pairs of grids are presented, the fish are able to discriminate between grids if the difference between the bar intervals is at least 1.25 mm (e.g. 10 mm v. 8.75 or 11.25 mm bar intervals). Ablation experiments indicate that the canal neuromasts may play a crucial role in analysing self-induced flow fields (Abdel-Latif *et al.*, 1990).

Theoretical calculations show that the magnitude of a self-induced flow field increases with both the fish's cross section and its swimming speed (Hassan, 1985). All else being equal, a small, slowly swimming fish will experience a weaker lateral line stimulus than a large and/or fast-swimming fish. Smaller cave fish on average swim faster than larger fish and in general tend to increase their swimming speed when placed in unfamiliar surroundings or when confronted with a new object (Teyke, 1988). The increase in swimming speed also leads to a decrease in boundary layer thickness (Equation 7.6a) which at least in theory should enhance lateral line perception. Blind cave fish can even use lateral line information to develop an inner map of their environment (Teyke, 1989).

Lateral line mediated responses to moving objects

Fish with an intact lateral line learn to associate an approaching disc of varying diameter either with punishment or with food. For instance, a food-conditioned blinded minnow, *Phoxinus phoxinus*, can locate a glass filament with a diameter of only 0.25 mm at a distance of up to 10 mm (Dijkgraaf, 1963). In trained fish, tests with weak local water currents flowing from a pipette nozzle tend to produce the same reactions.

In teleost larvae, free neuromasts play an important role in sensory awareness. If a probe approaches a larva it will respond with escape swimming. Experiments show that the observed flight responses are at least in part mediated by the lateral line but not by inner ear or touch receptors (Blaxter and Fuiman, 1989).

A swimming fish generates a complex three-dimensional vortex trail (p. 207) which may provide other fish with useful information. Such information could include the time elapsed since a fish swam by, its size, swimming speed, and swimming direction. It may even be that the wake behind a swimming fish contains species-specific information. Most likely the wake generated by a swimming fish can be sensed by another fish at long distances (several metres) and times (several seconds) relative to the location and presence, respectively, of the sender. Detection of the 'historical' turbulent stimulus can occur in two ways. First, a potential receiver may swim through the wake caused by a sender. When this happens, different parts of the fish – even at large distances – should receive differential water motions. Second, in a river the water current carries the turbulent historical stimulus to a downstream receiver at the speed of the water current. We do not know whether the lateral line can evaluate such spatially non-uniform complex hydrodynamic events. But if so, such a task should be facilitated by the dispersed arrangement of lateral line neuromasts.

Prey detection

Some fish are capable of using the lateral line to feed on zooplankton. This ability enables them to feed at night or through the long periods of darkness at high latitude. Behavioural evidence for lateral line detection of planktonic prey has been obtained in the mottled sculpin, *Cottus bairdi* (Hoekstra and Janssen, 1985). Blinded sculpin will take live prey and react to other moving objects but ignore dead prey. Inactivation of the lateral line by external application of streptomycin, or mechanical blocking of canal pores, eliminates their feeding response. Behavioural experiments have shown that other fish species which use the lateral line to detect planktonic prey include the amblyopsid blind cave fish (Poulson, 1963), the piper, *Hyporhamphus ihi*, and the Antarctic fish *Pagothenia borchgrevinki* (Montgomery, 1989).

Blinded pike, *E. lucius*, attack live fish from distances of 5 to 10 cm. The animals fail to attack after permanent extirpation of the lateral line (Wunder, 1927). Enger *et al.* (1989) conducted similar experiments with the bluegill, *Lepomis macrochirus*. Intact bluegills attack live fish in daylight and under covert infrared illumination. If the lateral line is blocked by cobalt ions (Karlsen and Sand, 1987), the bluegills only attack the fish under infrared light after direct touch. Controls show that the cobalt treatment does not noticeably affect the fish's feeding motivation (Enger *et al.*, 1989).

Lateral line and obstacle entrainment

Many river fish live in a noisy environment of rapid water flow with much turbulence created by submerged rocks, branches, or roots. As fishermen know, trout prefer specific locations from which they only venture to seize pieces of drifting debris. In a fast stream trout often orientate upstream behind rocks and remain there, having no contact with the bottom and no regular or rhythmic swimming movements. Behavioural experiments indicate that brook trout, *Salvelinus fontinalis*, use submerged objects for their own hydro-dynamic advantage. Terminal speed, i.e. the speed where the fish can no longer maintain a stationary position, is $47 \, \mathrm{cm \, s^{-1}}$ if no obstacles are present but $86 \, \mathrm{cm \, s^{-1}}$ if the fish can entrain behind small objects (Sutterlin and Waddy, 1975). The locations chosen by the trout are those where the fish needs to expend a minimum of energy to maintain its position. Under dark conditions bilateral denervation of the posterior lateral line reduces the degree to which trout can entrain an object, i.e. in this case the terminal speed is only $51 \, \mathrm{cm \, s^{-1}}$. These results suggest that the posterior lateral line is involved in detecting flow discontinuities, enabling a trout to maintain position with minimum energy expenditure (Sutterlin and Waddy, 1975).

Lateral line and intraspecific communication

The use of 'water touch' in sexual display behaviour was first reported by Stahr (1897, quoted by Hofer, 1908). Commonly two fish keep alongside each other at a distance of only a few centimeters, their heads often pointing in opposite directions. Although direct experimental evidence is not at hand, sexual stimulation may be done by pushing movements that create small flows of water against a companion (e.g. Webb, 1982; Enquist *et al.*, 1990). In the hime salmon, *Oncorhynchus nerka*, the lateral line may be involved in the detection of water waves caused by vibrational body movements which these animals perform in order to synchronize spawning (Satou *et al.*, 1987, 1991).

Lateral line function and shoaling

Another striking behaviour which at least partly depends on lateral line function is schooling (shoal is a general term for a social group of fish; the term 'school' is restricted to a synchronously turning and accelerating fish group) (Pitcher, 1979). Fish schools are well known for their remarkable synchrony, which even persists if the school executes complicated evasive manoeuvres. This requires individual fish to respond quickly to short-term changes in the velocity and direction of their neighbours. In normal saithe, *Pollachius virens*, schools there is a significant clustering of neighbours parallel to each other. This is the position in which changes in velocity of the nearest fishes are most apparent visually. However, even temporarily blindfolded fish are capable of matching changes in velocity and heading with at least their first two neighbours. Only when the fish are both blinded and have their posterior lateral line nerves cut at the opercula are they then unable to school (Pitcher *et al.*, 1976; Partridge and Pitcher, 1980).

Recently it has been shown that a fish which makes a rapid swimming movement generates a pressure pulse whose major frequency components are above 300 Hz (Gray and Denton, 1991). The amplitude of such a pulse can reach several Pascal which is more than 40 dB above the acoustic threshold of many fish species (Fay, 1988). As has already been mentioned (page 214), lateral line perception of a pressure pulse is only possible if the receiving fish has pressure-displacement transducers which indirectly stimulate lateral line neuromasts and/or if the receiving fish is moved in a pressure gradient such that a suprathreshold net-movement between water and receiver occurs. Gray and Denton (1991) have shown that this can be the case if the distance between the fish which emits the pressure pulse and the receiving fish is less than one body length. Such short distances may occur in tightly packed fish schools.

In the bream, *Abramis brama*, both lateral line and vision are employed to maintain individual distance (Pitcher, 1979). Bream with an intact lateral line cover more ground (as measured by swimming behaviour) and make more periodic forays, spending about 20% of their time up to 12 body lengths from their fellows: the converse was observed when there was no lateral line information. Thus it seems possible that deprivation of lateral line information about neighbouring fish that could nevertheless be seen makes bream more 'anxious' and hence increases their motivation to be part of the group (Pitcher, 1979).

7.6   PERCEPTION OF WATER SURFACE WAVES

For terrestrial insects the water/air interface often becomes a deadly clinging trap. No wonder that some teleost species are specialized to prey on insects fallen into the water. In doing so they make use of capillary surface waves which are generated whenever the prey is struggling. The physical properties of water surface waves differ from those of midwater waves, nevertheless the cephalic lateral line of surface-feeding fish is adapted to detect and encode surface wave information (Schwartz, 1965, 1971).

Physical properties of the water surface

The particular properties of the water surface are created by the asymmetry of intermolecular forces between water and air. Water surface waves are boundary waves caused by moving a solid through the water, by shock, wind, and all similar forces that generate and maintain waves. Water waves transport energy, but not mass.

Depth impact

The maximal horizontal and vertical displacement components A_x and A_z caused by a surface wave can be calculated according to (Lighthill, 1980):

$$A_x = A_0/[\sinh(kh)]\cosh[k(h-z)] \tag{7.7a}$$

$$A_z = A_0/[\sinh(kh)]\sinh[k(h-z)] \tag{7.7b}$$

where A_0 is the maximal displacement amplitude of the particles at the water surface above the observation point, z is vertical distance below the water surface, and h is water depth. For $h > \lambda$ and $z < h$, i.e. in deep water not too close to the bottom, A_x equals A_z; hence the particles describe a circular path (Fig. 7.5(A)). For $h < \lambda$ or z about h, i.e. in shallow water or in deep water very close to the bottom, particles move in an elliptical path (Fig. 7.5(B)). In general, the depth impact of surface waves is small and decreases with decreasing λ. For instance, at a depth of only one λ (e.g. 23.6 mm at 10 Hz or 2.9 mm at 140 Hz) the vertical movement of water particles is already less than one five-hundredth of that observed at the surface.

Dispersion characteristics

Each short, local disturbance of the water surface generates a single wave train (click) that has many frequencies. During horizontal propagation, such a click deceases in amplitude and changes in frequency composition, duration, and time course (Fig. 7.5(C)). There are two reasons for this: (1) gravity waves

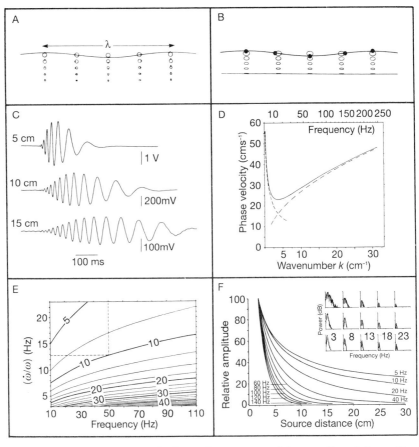

Fig. 7.5 (A) and (B) Motion of fluid particles in a sinusoidal wave of length λ travelling from left to right on deep water (A) and shallow water (B). In (A) the maximum surface elevation is $0.02\,\lambda$, and particles on the surface describe circles of this radius. In each case the particle's instantaneous position on its circular path is shown. In (B) the maximum surface elevation again is $0.02\,\lambda$. Water depth is $0.16\,\lambda$. A particle's instantaneous position on its elliptical path is shown only for those in the top row, but the motion of every particle in the same vertical line again is in phase. In (A) and (B) the movement of particles is to scale in relation to wave amplitude and depth of observation point, but particle size is exaggerated for legibility. (Redrawn from Lighthill, 1980). (C) The nature of a click stimulus as it passes points 5, 10, and 15 cm. (D) The wave speed C_{ph} for ripples on deep water (solid curve). Asymptotes for pure capillary waves (dashed-dotted curve) and gravity waves (broken curve) are also shown. Transition occurs around a frequency of 13 Hz, where phase velocity is minimal. (E) Range contours as function of relative frequency change and local frequency. For example, a stimulus with a local frequency of 50 Hz and a relative frequency change of 12.8 Hz has travelled a distance of 10 cm as indicated by the intersection of dashed lines. (F) Calculated attenuation of water surface waves (Equation 7.10) in dependence on stimulus frequency and distance from vibration source. Inset: change of spectral composition of water surface waves during stimulus propagation (numbers 3, 8 etc. denote cm). Modified from Bleckmann *et al.* (1989).

($\ll 13$ Hz) have normal dispersion characteristics, i.e., the longer waves travel faster than the shorter waves; (2) capillary waves ($\gg 13$ Hz) are abnormally dispersive, i.e. shorter waves travel faster than the longer waves (Sommerfeld, 1970). For $h \gg \lambda$ the following equation describes the dispersion relationship of surface waves (Lighthill, 1980):

$$C_{ph} = \frac{\omega^2}{k^2}\left[g + \frac{Tk^2}{\rho}\right]k^{-1} \tag{7.8}$$

where g is gravitational acceleration, T is a coefficient of surface tension, and C_{ph} is phase velocity in $cm\,s^{-1}$. The minimal phase velocity of water surface waves is $23\,cm\,s^{-1}$ at a wavelength of about $1.7\,cm$ (corresponding to a frequency of 13 Hz) (Fig. 7.5(D)).

After travelling a certain distance, a dispersive wave group consists of a band of different wavelength and frequencies, respectively. Locally λ can be defined by the crest-to-crest distance, but owing to dispersion (Equation 7.8) the distance of successive crests differs by a small amount (Fig. 7.5(C)). These differences reflect both the distance-dependent frequency modulation of the wave group and the speed with which the wave packet spreads in space. Theoretical calculations show that the source distance can be calculated unequivocally if only two parameters, namely the local frequency ω and the frequency modulation $\dot\omega$ around this local frequency, are known. The exact relation is (Käse and Bleckmann, 1987):

$$D = -4.5C_{ph}(\omega/\dot\omega) \tag{7.9}$$

where D is source distance and the dotted variable denotes the derivative with respect to time. Figure 7.5(E) displays the range contours calculated according to Equation 7.9.

Attenuation

Concentric surface waves are strongly attenuated during propagation. The slope of attenuation depends on both stimulus frequency and the distance the stimulus has travelled. For a distance $\geqslant 2$ cm the damping of water waves can be approximated by (Bleckmann, 1988):

$$A = A_i[\sqrt{(R_i/R)}]\exp[(16v\pi^2/\lambda^2 C_{ph})(R_i - R)] \tag{7.10}$$

where A_i is the displacement amplitude of the wave stimulus at distance R_i, A is the actual wave amplitude at distance R, and v is the kinematic viscosity of water. Due to geometrical spreading, attenuation occurs mostly in the vicinity of a wave source. In addition, the water surface behaves like a low-pass filter, i.e. attenuation strongly increases with frequency (decreasing λ) (Fig. 7.5(F)).

Biologically relevant stimuli and noise

A common source of prey-borne vibrations is aquatic, semiaquatic and terrestrial insects that have fallen into the water (Lang, 1980). 'Noise' waves are caused by wind or water currents, by falling leaves, seeds, and twigs and by fish which contact the water surface in order to feed. The frequency content of prey (insect) stimuli and background noise (i.e. all other wave types) that potentially interferes with the relevant messages, differ close to the source. The spectrum of background noise is typically of narrow bandwidth below 20–50 Hz, while insect stimuli are broad-band and extend to frequencies above 50 Hz (Fig. 7.6).

Water surface disturbances may differ not only in frequency, but also in time structure and duration (Fig. 7.6). Insect stimuli are more irregular in the time course than other wave types; their duration often exceeds 1.5 s, with values of 60 s or more being common. Wind waves can last for hours or even days. However, the duration of all other wave types usually does not exceed 1.5 s (Bleckmann, 1988). Surface waves relevant in the present context are in the mm (insect waves) to cm range (wind waves, wind speed 6–12 km h^{-1}) (Lang, 1980; Bleckmann and Rovner, 1984). In contrast to frequency content, time structure, and duration, wave amplitude alone is not sufficient for distinguishing between different stimuli.

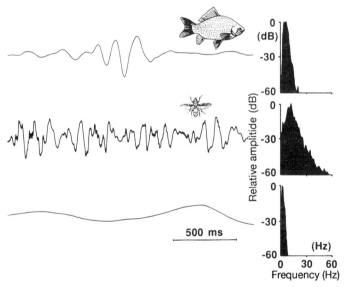

Fig. 7.6 Examples of water surface waves caused by the goldfish *Carassius auratus* and the fly *Calliphora vicina*. The bottom trace shows wind generated waves recorded in the natural biotope of the fishing spider *Dolomedes triton*. Amplitudes are not to scale. Right: relative amplitude spectra of corresponding wave stimuli. (From Bleckmann *et al.*, 1989.)

Behaviour

Threshold sensitivity

In terms of displacement the topminnow, *Aplocheilus lineatus*, and the African butterfly fish, *Pantodon buchholzi*, are most sensitive in the frequency range of 50–140 Hz. Within this range they respond to surface wave stimuli which have a peak-to-peak (p-p) amplitude of less than 0.02 μm (Bleckmann *et al.*, 1989). Thus surface-feeding fish can perceive the low-amplitude high-frequency displacement components that characterize prey stimuli. The signal-to-noise ratio may be further improved by the rise in the fish's threshold displacement for frequencies below 14 Hz, i.e. for frequencies which under natural conditions are mainly caused by wind (Fig. 7.6).

Determination of target angle

In surface-feeding fish the determination of wave direction is highly accurate ($\pm 5°$). Behavioural experiments indicate that arrival time and/or phase differences between different organs are the cues used for target angle determination (Bleckmann *et al.*, 1989a). Owing to the low propagation speed of surface waves (p. 221), this probably reflects a good localization strategy.

Distance determination

If one knows the frequency-dependent attenuation and phase velocity of surface waves, the distance to a wave source can be determined by measuring (1) the curvature of the wave front, (2) the relative amplitude decrease per unit of distance (Equation 7.10), and (3) the local frequency modulation of the initial part of the wave stimulus (Equation 7.9). In addition the amplitude spectrum of a wave stimulus can give some information about source distance if compared with a commonly experienced standard.

Surface-feeding fish use at least three of the four information sources just mentioned. This can be deduced from behavioural experiments: if clicks of different amplitudes and thus of different upper frequency limits are presented, the localization errors are small (less than 10%) and independent of stimulus amplitude, frequency content, and source distance (Fig. 7.7(A) to (C)). Thus the determination of source distance cannot be based solely on the evaluation of the wave spectrum. As has been shown (cf. Fig. 7.5(E)), decoding of local frequency and frequency modulation of a click is one way to determine the distance to the wave source. Consequently the presentation of single-frequency stimuli, i.e. of stimuli which do not contain any frequency modulation at all, leads to an impaired distance determination. With increasing stimulus frequency surface-feeding fish increasingly underestimate the source

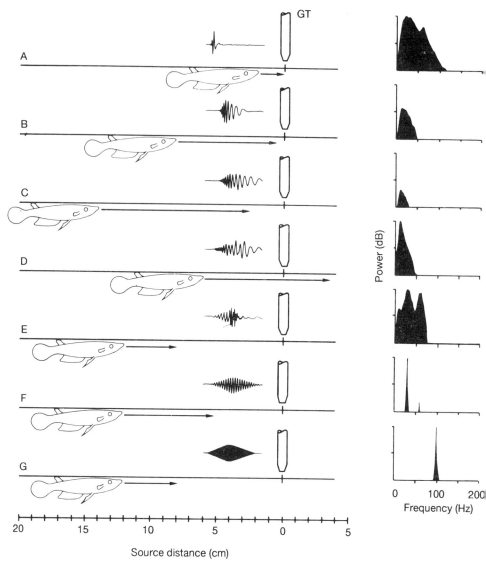

Fig. 7.7 (Left) Mean swimming distance (arrow length) of *Aplocheilus lineatus* towards the wave source (small vertical lines drawn below the glass tube, GT) which is dependent on stimulus (insets) and source distance (x-axis). The stimulus amplitudes are not to scale. In all cases the surface waves were produced by blowing a defined air stream through a glass tube onto the water surface. (Right) Spectra of the corresponding wave stimuli. The y-axis in all cases is 0 to −60 dB. In (A), (B) and (C) the highest amplitude value of (A) was set equal to 0 dB. In (D) to (G) the highest amplitude value of the corresponding spectrum was defined as 0 dB. The stimuli were a click at 3 cm (A), 10 cm (B), and 15 cm (C) source distance, a simulated click (D), an upward frequency-modulated wave stimulus (E), a 35 Hz wave stimulus (F), and a 100 Hz wave stimulus (G).

distance if it exceeds 6–8 cm (Fig. 7.7(F) and (G)) (Hoin-Radkovski *et al.*, 1984; Bleckmann, 1988). This may indicate that the amplitude spectrum is also evaluated, i.e. if no other cues are available, a high-frequency stimulus is 'expected' to have travelled a shorter distance than a low-frequency one. That surface-feeding fish do follow this strategy is also supported in experiments with artificial frequency-upward modulated stimuli containing high-amplitude, high-frequency components. These stimuli also cause an underestimate of source distance (Fig. 7.7(E)). Considering the filter properties of the water surface (Equation 7.10), this probably is a good localization strategy. When surface-feeding fish are confronted with a computer-controlled wave stimulus which is generated at 7 cm but simulates the frequency modulation of a click at 15 cm, they swim beyond the wave source (Fig. 7.7(D)). This demonstrates again that the frequency modulation indeed is an important cue used for distance determination. The use of frequency modulation for distance determination is further supported in experiments with the topminnow *A. lineatus*, in which all but one cephalic neuromast have been destroyed. Although these animals have no way to determine the curvature of the wave-front or the decrease in amplitude per unit of distance, they still show an increase of swimming distance with source distance if clicks are presented (Müller and Schwartz, 1982).

Stimulus discrimination and communication

That the evaluation of wave stimuli is based on frequency and amplitude discrimination is supported in experiments in which fish were trained to differentiate between stimuli that differed either in one or the other of these two parameters. For instance, in the range 10–150 Hz, *A. lineatus* can discriminate water wave frequencies to a resolution of 10% (Bleckmann *et al.*, 1981). Amplitude discrimination is also well developed: *A. lineatus* can distinguish a 70 Hz wave stimulus of 2.3 μm p-p amplitude from one of the same frequency but with an amplitude of only 0.14 μm (Bleckmann *et al.*, 1989a). Most likely surface-feeding fish use their wave-discriminating ability to differentiate between more complex prey and non-prey waves, but this has not been tested.

In fighting fish, *Betta splendens*, both nest building and the hatching of the juveniles are done by the male. At an age of about 3 days, up to 150 juveniles leave the air-bubble nest and begin to investigate their surroundings. In doing so they always keep in contact with the surface film of the water since as anabantid fish they depend on breathing air. In the event of danger the male takes an oblique position with the head close to the water surface and begins to produce surface waves by trembling movements of his pectoral fins. Up to a distance of 40 cm, the juveniles, even in complete darkness, orientate themselves to the vibration source and then intermittently swim in the wave-source direction. Reaching the signalling male they are sucked up by him and transported back to the nesting site (Kühme, 1961). The response of juvenile *B. splendens* to water surface waves can also be elicited with artificial wave stimuli, probably perceived with head neuromasts which are already well developed at that time. Single-frequency signals between 8 and 10 Hz and with an amplitude of at least 13 μm turned out to be the most effective stimulus. Only one or two minutes after stimulus onset, more than 70% of all fish had accumulated close to the wave centre (Kaus and Schwartz, 1986).

7.7 SPECIALIZED LATERAL LINE ORGANS (ELECTRORECEPTORS)

Electroreception is one of the most recently discovered sensory modalities. In 1917 Parker and van Heusen noted that the catfish *Ictalurus nebulosus* responds to galvanic currents of less than 1 μA. Parker and van Heusen did not, however, consider the electrical sensitivity to be of any biological significance. Hatai *et al.* (1932) give some indications that the unusual vibratory sensitivity of catfish observed prior to earthquakes is partly due to terrestrial electricity which precedes seismic events. Roth (1968) finally demonstrated that catfish not only can detect electric fields but in addition are able to sense the polarity of such a field. The ancient Egyptians and Romans discovered that some fish are capable of producing strong electric fields (Wu, 1984). At that time the physical concept of electricity was, however, unknown.

Lissmann (1958) was the first to give behavioural evidence that the South American knife fish *Gymnarchus niloticus* uses weak self-generated electric fields for object identification. And Kalmijn and Adelmann (in Kalmijn, 1974) finally demonstrated that some electroreceptive fish locate prey by the bioelectric fields (p. 227) emanating from the prey. Since these early studies, numerous aspects of the anatomy, peripheral and central physiology, evolution, ecology, and behaviour of the electrosensory system have been investigated (recent reviews, Bullock and Heiligenberg, 1986; Kalmijn, 1988b; Bell, 1989; Heiligenberg, 1989; Kramer, 1990a). For lack of space this chapter can only summarize the most basic issues.

Natural electric stimuli

The natural electric fields to which animals may be exposed are of four types (data are mostly taken from Kalmijn 1974).

1. Inanimate electric fields due to geomagnetic variations, tidal forces, tectonic processes, the contact of chemically dissimilar media, or lightning. With the exception of nearby lightning these fields are usually weak and rarely exceed some μV cm^{-1}.

2. Motional electric fields, induced whenever water is moving or a fish is swimming through the Earth's magnetic field. Depending on flow and swimming speed, motional electric fields can have a magnitude of up to $0.5\,\mu\text{V cm}^{-1}$.

3. Animate electric fields, produced incidentally by electrochemical disparities in the internal and external milieu of an animal or a plant. These d.c. (direct current) bioelectric fields, which in injured animals can reach more than $1\,\text{mV cm}^{-1}$, are modulated by respiratory gill movements of fish and other relative movements of body parts (Roth, 1972).

4. Animate electric fields produced by specialized electric organs, which are derived from muscle or nervous tissue. It had long been known that a few species, such as the electric eel, *Electrophorus electricus*, can produce strong stunning electric discharges of up to several hundred volts (Cavallo, 1786). Only 40 years ago Lissmann (1951) discovered that some fish constantly surround themselves with a weak electric field whose amplitude – if measured in water whose conductivity is similar to that of the natural habitat – rarely exceeds a few mV cm^{-1}. Within the teleosts, electric fish are mostly found in the knife fishes (Gymnotiformes) of South America and in the elephant-nose fishes (Mormyriformes) of Africa. In addition some Perciformes (Uranoscopidae) have electric organs. Wave-type weakly electric fish, i.e. most knife fishes, fire their electric organ in a rather sinusoidal manner; in these fish the individual and species-specific discharge

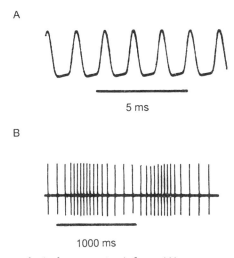

A

5 ms

B

1000 ms

Fig. 7.8 Electric signals (voltage v. time) from (A) a wave species, *Apteronotus albifrons*, and (B) a pulse species, *Brienomyrus brachyistius*. Note in (B) the variable pattern of pulse intervals, which is controlled by a pacemaker in the medulla. Changes in pulse intervals are used to generate different kinds of social signals. After Hopkins (1986).

frequency may be as low as 20 Hz or as high as 1800 Hz. In wave species the discharge frequency often is stable over hours or even days. In pulse species, i.e. in all but one mormyriform (*Gymnarchus niloticus*), but also in many gymnotiforms, the pulse interval is long compared with pulse duration and pulses may occur at a highly regular or irregular rate (e.g. Fig. 7.8B). Recently it has been discovered that some synodontid catfish also produce weak electric discharges in either continuous or burst-like fashion (Hagedorn *et al.*, 1990).

Electroreceptors

With the exception of the Uranoscopidae (Bullock *et al.*, 1983) all electric fish, but also many fish that lack electric organs, have low-frequency electroreceptors. Electroreceptors were identified as such in the search for the basis for an established behavioural response (Bullock, 1986). Most likely, the electroreceptors of teleosts phylogenetically developed from the same primordium as mechanosensitive hair cells of the ordinary lateral line. Electroreceptors are ordinal characteristics of all non-teleost fish, except the Holosteans (Bodznick and Boord, 1986; Northcutt, 1986). Known electroreceptive teleosts constitute less than 1% of all teleost species and belong to two distantly related groups, the Ostariophysi and the Osteoglossiformes (McCormic and Braford, 1988). Within the Ostariophysi, electroreceptors are found in the South American order Gymnotiformes (Gymnotidae: many genera and species), and in the large and world-wide order Siluriformes (many families and species). Electroreceptive Osteoglossiformes include the African order Mormyriformes (Gymnarchidae: one species, and Mormyridae: many genera and species) and the featherback fish, *Xenomystus nigri* (Notopteridae) (Bullock *et al.*, 1983). Electroreceptors are also found in many amphibians (Fritzsch and Münz, 1986), in the Australian water mammal *Platypus* (Scheich *et al.*, 1986) and in the spiny anteater. *Tachyglossus* (Gregory *et al.*, 1989). It should be noted, however, that *Platypus* and *Tachyglossus* are three to four orders of magnitude less sensitive to electric stimuli than electroreceptive fish and amphibians.

With one possible exception (Andres *et al.*, 1988) catfish, like all electroreceptive non-teleost fish, have only ampullary low-frequency (passband about 0.1–50 Hz) electroreceptors. Ampullary electroreceptors are either within the epidermis (many freshwater species) or are located in a deep invagination of the epidermal basement membrane (most saltwater species). The apical membrane of the sensory cells bears microvilli and/or a single kinocilium which protrudes into a flask-shaped canal, the ampulla. The ampulla is connected to the epidermal surface either by a jelly-filled canal or by way of specialized cells (Szabo, 1974). Since their discovery, ampullary organs were regarded as pressure-, mechano-, thermo-, chemo- and electroreceptors (for a complete review of the history see Bullock, 1986). Finally the evidence that electroreception is the biological function was provided by experiments based on behaviour (Dijkgraaf, 1968). In freshwater teleosts the number of receptor cells in each ampullary organ may vary between one in gymnarchids and as many as 20 in some catfish. In most teleosts, each organ is innervated by a single nerve fibre that contacts every receptor cell (Zakon, 1986). The ampullary organs of teleosts are excited by anodal stimuli at the canal pore, while those of non-teleosts are excited by cathodal stimuli (Zakon, 1986). Electrophysiological and behavioural experiments show that the ampullary system is as sensitive as $0.3 \, \mu V \, cm^{-1}$ in some freshwater catfish (Knudsen, 1974); for saltwater cartilaginous fish, *Raja clavata*, detection thresholds of $0.005 \, \mu V \, cm^{-1}$ have been reported (Kalmijn, 1988b). Ampullary organs are used in passive electrolocation, i.e. they are used to detect motional electric fields (Peters and Wijland, 1974) and low-frequency electric events caused incidentally by all aquatic organisms (Roth, 1972).

In the first studies of electric sense in gymnotiforms it became clear that there must be a second type of electroreceptor. Since the physiological study of Bullock *et al.* (1961) we know that gymnotiform fish, besides having ampullary organs, also possess high-frequency (depending on the species, 50 up to 1800 Hz) electroreceptors, called 'tuberous organs'. In gymnotiforms a tuberous receptor organ has 20–30 sensory cells, in specialized receptors this value can be even higher. Tuberous receptor cells have no kinocilium. Their apical surfaces have many microvilli which are far more densely packed than those on ampullary receptor cells (Fig. 7.4(F)). Tuberous electroreceptors are also present in the primitive African mormyriform fishes, which probably have evolved an electrosensory system independently from the gymnotiforms (Bullock *et al.*, 1975). In mormyriforms two types of high-frequency electro-receptors can be distinguished: mormyromasts and Knollenorgans. With rare exceptions Knollenorgans have no more than 10 receptor cells (Zakon, 1986). Mormyromasts have two morphologically distinct populations of sensory cells. The deeper part of a mormyromast is composed of three to five B cells which are innervated by a single nerve fibre. The more superficial A cells are typically innervated by two nerve fibres which are separate from the single fibre which innervates the B cells (Bell *et al.*, 1989). A and B cells terminate in separate brain areas (Bell and Szabo, 1986).

Tuberous organs of wave- and pulse-type gymnotiform fish are used in active electrolocation, i.e. with the aid of these organs the fish obtain sensory information from distortions of their own electric field, caused by various objects with conductivities different from that of the surrounding water (Fig. 7.9). Physiologically the tuberous organs of wave-type gymnotiform fish can be subdivided into two types. P-type receptors or probability coders change their rate of firing in accordance with modulations in the local amplitude of an electric signal. In contrast, T-type receptors or time coders fire only one spike in each electric organ discharge cycle, at a fixed phase in reference to

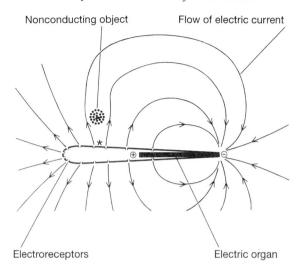

Fig. 7.9 The principle of active electrolocation. The location of the electric organ is indicated by the black bar. The solid lines give the current flow associated with electric organ discharge. The electroreceptors, which monitor voltage changes across the epidermis, are found in pores of the anterior body surface. Each object with conductivity different from that of the surrounding water distorts the current pattern and thus alters transepidermal voltage in the area of skin (*) nearest to the object Redrawn from Heiligenberg (1977).

the zero-crossing of an electric field (Heiligenberg, 1989). Phase locking (1:1) thresholds for T-units range from $300\,\mu\mathrm{V\,cm^{-1}}$ to $1.5\,\mathrm{mV\,cm^{-1}}$. Thresholds of P-units vary from $1.8\,\mathrm{mV\,cm^{-1}}$ to $6.0\,\mathrm{mV\,cm^{-1}}$.

As has already been mentioned Mormyriforms also have two types of tuberous receptors, the Knollenorgans and the mormyromasts. Knollenorgans have a threshold of about 0.1 mV across the skin and – independent of stimulus amplitude – fire one time-locked spike to transient electric signals. Knollenorgans have their afferent inputs centrally suppressed during the fish's own electric organ discharge and thus inform only about electric signals caused by other fish, i.e. these organs are used for electric communication. Mormyromasts have a higher threshold ($10–120\,\mathrm{mV\,cm^{-1}}$) than the Knollen-organs and are mainly used for active electrolocation. Both A and B cells of mormyromasts encode amplitude modulations of the animal's own electric organ discharge caused, for instance, by objects submerged in the water. In general an amplitude increase of the EOD causes a decrease in spike latency and an increase in the number of spikes elicited (Bell, 1990). B cells, in addition, are extremely sensitive to distortions of the EOD. If the amplitude and the duration of a $250\,\mu s$ lasting EOD is held constant a distortion which corresponds to a one degree phase shift is already sufficient to cause a significant decrease in first-spike latencies (Emde and Bleckmann, 1992, 1993). Under natural conditions pulse distortions are caused by capacitive objects like other animals and plants. Mormyrids may use the differences in A and B cell sensory output to distinguish between purely ohmic and capacitive objects (see below).

Tuberous receptors (especially the T-units) of gymnotiform wave species are often sharply tuned with a distinct best frequency which is matched to the peak power of the species-specific electric organ discharge. In pulse gymnotiform fish, tuberous receptors are usually more broadly tuned, but in many species the frequency sensitivity is not well matched to the peak power of the species-specific electric organ discharge. The tuning curves of the tuberous receptors of mormyrids are broad and often have no distinct best frequency (review, Zakon, 1986).

Central pathways

Ampullary and tuberous electroreceptors are innervated by lateral line nerve fibres. In all teleosts primary electroreceptive afferents terminate in the ipsilateral electrosensory lateral line lobe of the medulla. This lobe lies topographically lateral to the primary mechanosensory lateral line centre (medial octavolateral nucleus). In species with tuberous receptors the electro-sensory lateral line lobe has distinct ampullary and tuberous recipient zones. Furthermore, in gymnotiform fish the tuberous receptors that code for timing terminate separately from those that code for amplitude (Heiligenberg, 1989). This provides the basis for independent initial processing of stimulus amplitude and phase by higher-order neurones. In mormyriforms the two types of tuberous receptors, Knollenorgans and mormyromasts, also project to different central targets (Bell, 1989). From the electrosensory lateral line lobe most ascending fibres target the contralateral metencephalic nucleus praeeminen-tialis and the midbrain electrosensory nucleus in the torus semicircularis, which is laminated in gymnotiforms but has a nuclear organization in mormyriforms. The electrosensory torus has both ascending and descending connections. Ascending connections reach the tectum and the diencephalon or pretectum (McCormick and Braford, 1988). Descending connections target the nucleus praeeminentialis and the reticular formation. In contrast to all other teleosts, the valvula cerebelli of mormyriforms is an enormous structure that overshadows the rest of the brain in size. Mormyriforms are unique in that the valvula cerebelli receives substantial input from the electroreceptive part of the torus. For a thorough comparison of the central electroreceptive pathways among different teleosts see Finger *et al.* (1986).

Decoding of electrical stimuli including the consequences for motor output provides some of the best-understood cases in neuroethology. The reader who is interested in further anatomical and physiological details is referred to Bastian (1986), Carr and Maler (1986), Bell (1989) and Heiligenberg (1991).

Passive mode of electroreception

Prey detection

Catfish, *Ictalurus*, and some weakly electric fish (for example *Gymnotus carapo* and *Apteronotus albifrons*) locate the position of a prey fish, even when all but the bioelectric stimuli emanating from the prey are attenuated by a layer of agar (Kalmijn, 1974). *Ictalurus* and weakly electric fish no longer respond to the prey, however, when a thin, electrically insulating film of polyethylene is added to the agar. That the ampullary electroreceptors and not visual or chemical cues are involved in this behaviour can be seen in experiments in which the presence of a prey fish is simulated by passing low-frequency current between two electrodes buried in sand. In this case, *Ictalurus*, *Gymnotus* and *Apteronotus* display the same characteristic feeding response to the electrodes as they do to actual prey (Kalmijn, 1974).

Long-distance orientation

The possible navigation of migratory fish by using the electric fields they induce when swimming through the Earth's magnetic field (p. 227) is a fascinating idea. Since, under natural conditions, the effective low-frequency ampullary stimulus depends on the direction in which the animal is moving, the Earth's magnetic field indirectly may provide electrosensitive fish with orientational cues, i.e. these cues may indicate the fish's heading relative to the Earth, its velocity and latitude. Magnetic orientation is well established in some elasmobranch fishes which not only detect the direction, but also the polarity of a magnetic field (Kalmijn, 1988b). The possibility of using magnetic and electric cues, respectively, for orientation has been of interest in studies of long-distance migrating eels, *Anguilla rostrata*, and Atlantic salmon, *Salmo salar*. Rommel and McCleave (1972) have claimed electrosensitivity for *S. salar* and *A. rostrata*, but repeated efforts to confirm a sufficient electrosensitivity for these species have failed (e.g. Enger *et al.*, 1976; Bullock *et al.*, 1983). Furthermore there are no anatomical specializations in the hindbrain as always seen in electroreceptive teleosts (Meredith *et al.*, 1987). Some catfish, however, can orientate in weak electric fields. This can be inferred from their ability to select the correct (rewarded) direction from among several electric fields arranged in a circle (Peters and van Wijland, 1974; Kalmijn *et al.*, 1976, quoted by Bullock, 1982).

Active mode of electroreception

Object identification

Lissmann (1958) was the first to propose the theory that weakly electric fish may locate nearby objects by detecting the distortions which these produce in the fish's own electric organ field (Fig. 7.9). Indeed, the weakly electric fish *Gymnarchus niloticus* could be trained to respond to the presence of a glass rod 2 mm in diameter, even though the rod was placed in a porous pot which excluded mechanical and visual detection (Lissmann and Machin, 1958). Some weakly electric fish, *Gnathonemus petersii*, can even utilize capacitive and resistive characteristics for object discrimination (von der Emde, 1990). Although the threshold of tuberous receptors may be less than 1 mV, the active range, i.e. the range at which a 1 cm object can be detected and discriminated, covers only a few cm (Bastian, 1986). This is due to the strong attenuation of dipole-like electric fields. For electric fish, the active range is typically of ellipsoid shape, with the greatest sensitivity parallel to the long axis of the fish's body. According to Knudsen (1975), the field potential generated by gymnotiform fish falls off with the inverse square of the distance for all distances beyond 10 cm and according to the cosine of the angle between the main body axis of the fish and the recording electrode. The magnitude of the electric field vectors falls off according to the inverse cube of the distance.

Weakly electric fish depend upon the active electric sense in manoeuvring nocturnally or in muddy environments. This can be shown by the following experiment: if a hole, through which fish were previously trained to pass, is mechanically blocked by electrically 'transparent' agar, they will persist in attempting to enter the hole (Kalmijn, 1974). Through the use of electro-location, blinded electric fish are also able to maintain postural control near a substratum (Meyer *et al.*, 1976).

Communication

A fascinating aspect of electric fish behaviour is social communication via the unique electric modality. Electric communication occurs when one fish, the sender, emits an electric signal that evokes a behavioural response from another fish, the receiver. Electric fish can detect each other's presence by using their high-frequency tuberous receptors to sense the electric-organ discharge of a conspecific. In a low-noise laboratory environment, the range at which signals can be detected by a recipient fish may cover in a small mormyrid as much as 135 cm (Squire and Moller, 1982).

All else being equal, the communication range depends on water resistivity. The net effect is such that in water of high resistivity, i.e. in water which is similar to that found in the natural habitat of weakly electric fish, there is an increase in communication range (Knudsen, 1974). Electric communication is used in courtship, aggression, appeasement, and sometimes in identifying the species, sex, and even the individual (Hagedorn, 1986; Hopkins, 1986; Kramer, 1990a). In pulse-type electric fish, males and females may generate electric signals that differ in waveform (Shumway and Zelick, 1988; Hopkins *et al.*, 1990). Especially in mormyrids, single pulses may be separated by a highly variable sequence of time intervals (Fig. 7.8(B)) which are under central control. Behavioural experiments indicate that in some pulse species the electric organ discharge waveform may play a role in species and sex recognition (Hopkins, 1986; Shumway and Zelick, 1988). In other fish and especially in fish with identical electric organ discharges, species recognition is accomplished by reliance upon interpulse intervals (Kramer, 1990b). In addition to the recognition function of electric communication – at the level either of species or of sex – weakly electric fish may generate signals to warn of impending attack, to signal submission or for courtship (Hagedorn, 1986). Possible displays include discharge cessations, brief accelerations, bursts, buzzes, and rasps. A sharp increase in discharge frequency, for example, may indicate threat, whereas discharge arrests or cessations may indicate submission in the losing animal. Some weakly electric fish can discriminate interpulse time intervals that differ by only 2% (Kramer, 1990a). In some mormyrid species electrical signalling serves schooling behaviour (Moller, 1976).

Russell *et al.* (1974) and Bauer and Kramer (1974) were the first to describe echo responses in mormyrid fish. An echo response is characterized by an extremely short-latency firing (depending on species, 12–22 ms) of the electric organ discharge (EOD) of one animal in response to the EOD of a conspecific. The echo response may function to avoid jamming of the EOD used in electrolocation and/or it might serve social communication. A second response type, similar to the echo, is the preferred avoidance response (Heiligenberg, 1977; Lücker and Kramer, 1981). This response is a specific avoidance of a discharge at a fixed latency following an electric stimulus from a conspecific EOD. Lücker and Kramer (1981) noted that in *Pollimyrus isidori* the preferred avoidance response is characteristic of sexually mature females, but not of males, which produce a typical 10–12 ms echo. Hence a male *Pollimyrus* may be able to probe other fish with bursts of pulses in order to determine their sex.

In wave-type electric fish, sexually mature males and females also may possess characteristic EOD waveforms or discharge frequencies (Kramer, 1985; Kramer and Otto, 1988). For instance, adult females and juveniles of *Eigenmannia virescens* have an EOD waveform that resembles a sinusoid much more closely than does that of adult males. *Eigenmannia* either spontaneously or after conditioning may distinguish between male and female waveforms (Kramer and Zupanc, 1986; Kramer and Otto, 1988). In wave-type electric fish, daily injections of androgens may elicit frequency decreases of the EOD (Meyer, 1984). The frequency tuning of electroreceptors follows hormone-induced frequency shifts even if the receptors do not experience the animal's electric organ discharge (Meyer *et al.*, 1984). At least approximate matching of EOD peak power frequency and electroreceptor tuning is necessary for electroreception.

Wave-type electric fish have an individual EOD rate which usually is constant over hours but may change over the course of weeks and months (Heiligenberg and Bastian, 1984). If it happens that the EOD rate of a neighbouring fish is similar, electrolocating performance may be impaired. This could be the reason why wave species, in experiments in which a clamp device which holds the frequency difference Δf between the fish's EOD and an artificial electric signal constant is used, usually shift their EOD rate away from that of the interfering signal. In a clamped-stimulus situation a Δf of 3 Hz elicits the strongest responses (Bullock *et al.*, 1972). In a more natural, unclamped situation, however, the response of a jammed fish can be highly variable, i.e. depending on its sex or age a fish may not respond at all or it may raise or lower its EOD frequency in a given situation (Kramer, 1987). The described change of frequency, first discovered by Watanabe and Takeda (1963), is called 'jamming avoidance response'. The jamming avoidance response is thought of as a mechanism which minimizes the effect of interfering signals (Heiligenberg, 1989) or, alternatively, as a mechanism to maximize the effect of the other fish's EOD on the fish's own 'carrier frequency' by shifting the frequency by a few Hz to an optimal beat frequency for better signal discrimination (Kramer, 1990b). Studies have shown that the amplitude and phase of the signal at the animal's body surface are modulated when the electric organ discharges of a neighbour mix in the water with the animal's own electric organ discharge. The modulation depth in amplitude and phase varies along the fish's body, and these differences are detected by the P- and T-type electroreceptor system to a remarkable precision. *E. virescens*, for example, can detect phase differences as small as $0.5 \mu s$ (Heiligenberg, 1989).

7.8 SUMMARY

The lateral line organs of teleost fishes can be divided into 'ordinary' (mechanosensitive) organs and 'specialized' (electrosensitive) organs. The mechanosensitive lateral line is found in all teleosts and comprises superficial neuromasts which in many fish are partly transformed into canal organs. All neuromasts are covered by a gelatinous cupula which encloses the sensory hairs from the underlying mechanosensitive hair cells. The appropriate stimuli for the mechanosensitive lateral line are minute net water displacements which shear the cupula parallel to its base, causing a gliding movement over the sensory hair bundle.

The ordinary lateral line organs are used as 'distant touch' receptors. They serve mainly to detect and locate moving animals, as well as inanimate mobile and immobile objects at short range on the basis of current-like water disturbances. The ordinary lateral line of teleosts is of special importance for the detection and localization of prey, for enemy avoidance, obstacle entrainment, schooling, and intraspecific communication.

The specialized electrosensitive lateral line organs, found in the African freshwater order Mormyriformes, the South American order Gymnotiformes, and in the world-wide order Siluriformes, comprise two broad classes of receptors: the ampullary organs, which are sensitive to weak low-frequency voltage changes, and the tuberous organs, which are sensitive to weak high-frequency voltage changes. These specialized lateral line organs are used to detect electric stimuli produced by inanimate sources (e.g. electric currents indirectly caused by the Earth's magnetic field) or incidentally produced stimuli generated by other aquatic organisms. The former serves an orientation function, and the latter is engaged in prey-capture behaviour.

Weakly electric fish produce their own electric fields by means of electric organs. Along with specialized tuberous receptors, the electric field forms part of an active sensory system for object location and short-range intra- and interspecific communication.

LIST OF SYMBOLS

a Acceleration
A Displacement amplitude at distance R
A_o Maximal displacement amplitude of the particles at the water surface above the observation point
A_x Horizontal displacement component
A_z Vertical displacement component
c Speed of sound
C_{ph} Phase velocity
d Displacement
D Source distance
D_r Radial distance from the centre of the sphere to the point of interest
f Frequency
g Gravitational acceleration
h Water depth
k Wave number $= \omega/c = 2\pi/\lambda$
I Distance downstream
p Acoustic pressure
r Radius of dipole
R Wave stimulus distance
t Time

T	Coefficient of surface tension
U	Amplitude of axial source velocity
v	Particle velocity
v_r	Radial particle velocity component
v_θ	Tangential particle velocity component
z	Vertical distance below the water surface
λ	Wavelength
δ	Thickness of boundary layer
θ	Angle of radiation
μ	Dynamic viscosity of fluid
ρ	Fluid (water) density
v	Kinematic viscosity
ω	$2\pi f$

ACKNOWLEDGEMENTS

I thank Drs B. Fritzsch, P. Görner, B. Kramer, H. Münz, and R. Zelick for comments on early versions of the manuscript and K. Grommet and Dr M.-D. Crapon de Caprona for help with some figures. I am indebted to Dr R. Blickhan who generously gave me the permission to publish the drawing of the fish-generated wake shown in Fig. 7.3(B). My original research was generously supported by grants of the Deutsche Forschungsgemeinschaft (Bl 242) and by the Bennigsen-Foerder-Preis of Nordrhein-Westfalen.

REFERENCES

Abdel-Latif, H., Hassan, E. S. and Campenhausen, C. von (1990) Sensory performance of blind Mexican cave fish after destruction of the canal neuromasts. *Naturwissenschaften*, **77**, 237–9.

Andres, K. H., Düring, M. von and Petrasch, E. (1988) The fine structure of ampullary and tuberous electroreceptors in the South American blind catfish *Pseudocetopsis* spec. *Anat. Embryol.*, **177**, 523–35.

Ashmore, J. F. (1984) The stiffness of the sensory hair bundle of frog saccular hair cells. *J. Physiol., Lond.*, **350**, 20P.

Barry, M. A. and Bennet, M. V. L. (1989) Specialized lateral line receptor systems in elasmobranchs: the spiracular organs and vesicles of Savi, in *The Mechanosensory Lateral Line. Neurobiology and Evolution* (eds C. Coombs, P. Görner and H. Münz), Springer, New York, pp. 591–606.

Bastian, J. (1986) Electrolocation. Behavior, Anatomy, and Physiology, in *Electroreception* (eds T. H. Bullock and W. Heiligenberg), John Wiley and Sons, New York, pp. 577–612.

Bauer, R. and Kramer, B. (1974) Agonistic behaviour in mormyrid fish: latency relationship between electric discharges of *Gnathonemus petersii* and *Mormyrus rume*. *Experientia*, **30**, 51–2.

Bell, C. C. (1989) Sensory coding and corollary discharge effects in mormyrid electric fish. *J. exp. Biol.*, **146**, 229–53.

Bell, C. C. (1990) Mormyromast electroreceptor organs and their afferent fibers in mormyrid fish. III. Physiological differences between two morphological types of fibers. *J. Neurophysiol.*, **63**: 319–32.

Bell, C. C. and Szabo, T. (1986) Electroreception in mormyrid fish. Central anatomy, in *Electroreception* (eds T. H. Bullock and W. Heiligenberg), New York, John Wiley & Sons, pp. 375–464.

Bell, C. C., Zakon, H. and Finger, T. E. (1989) Mormyromast electroreceptor organs and their afferent fibers in mormyrid fish: I. Morphology. *J. comp. Neurol.*, **286**: 391–407.

Blaxter, J. H. S. (1987) Structure and development of the lateral line. *Biol. Rev.*, **62**, 471–514.

Blaxter, J. H. S. and Fuiman, L. A. (1989) Function of free neuromast of marine teleost larvae, in *The Mechanosensory Lateral Line. Neurobiology and Evolution* (eds S. Coombs, P. Görner and H. Münz), Springer, New York, pp. 481–99.

Blaxter, J. H. S., Denton, E. J. and Gray, J. A. B. (1981) Acousticolateralis system in clupeid fishes, in *Hearing and Sound Communication in Fishes* (eds W. N. Tavolga, A. N. Popper and R. R. Fay), Springer, Heidelberg, pp. 39–59.

Bleckmann, H. (1988) Prey identification and prey localization in surface feeding fish and fishing spiders, in *Sensory Biology of Aquatic Animals* (eds J. Atema, R. R. Fay, A. N. Popper and W. N. Tavolga), Springer, New York, pp. 619–41.

Bleckmann, H. (1991) Orientation in the aquatic environment with aid of hydro-dynamic stimuli. *Verh. dt. zool. Ges.*, **84**, 105–24.

Bleckmann, H. and Münz, H. (1988) The anatomy and physiology of lateral line mechanoreceptors in teleosts with multiple lateral lines. *Verh. dt. zool. Ges. 81* (ed. F. G. Barth), Gustav Fischer, Stuttgart, p. 288.

Bleckmann, H. and Münz, H. (1989) Mechanical factors in the excitation of highly branched lateral line canals. In *Proc. 17th Göttingen Neurobiol. Conf.* (eds N. Elsner and W. Singer), Stuttgart, Thieme, p. 270.

Bleckmann, H. and Münz, H. (1990) Physiology of lateral line mechanoreceptors in a teleost with highly branched, multiple lateral lines. *Brain, Behav. Evol.*, **35**, 240–50.

Bleckmann, H. and Rovner, J. (1984) Sensory ecology of the semiaquatic spider *Dolomedes triton*. I. Roles of vegetation and wind-generated waves in site selection. *Behav. Ecol. Sociobiol.*, **14**, 297–301.

Bleckmann, H., Breithaupt, T., Blickhan, R. and Tautz, J. (1991) The time course and frequency content of hydrodynamic events caused by moving fish, frogs, and crustaceans. *J. comp. Physiol.*, **168A**: 749–757.

Bleckmann, H., Tittel, G. and Blübaum-Gronau, E. (1989a) The lateral line system of surface-feeding fish: anatomy, physiology, and behavior, in *The Mechanosensory Lateral Line. Neurobiology and Evolution* (eds S. Coombs, P. Görner and H. Münz), Springer, New York, pp. 501–26.

Bleckmann, H., Waldner, I. and Schwartz, E. (1981) Frequency discrimination of the surface-feeding fish *Aplocheilus lineatus* – a prerequisite for prey localization? *J. comp. Physiol.*, **143A**, 485–90.

Bleckmann, H., Weiss, O. and Bullock, T. H. (1989b) The physiology of mechanosensory lateral line areas in the thornback guitarfish, *Platyrhinoidis triseriata* (Elasmobranchii). *J. comp. Physiol.*, **164A**, 67–84.

Blickhan, R., Krick, C. and Nachtigall, W. (1990) Flow in the vicinity of swimming fish. In *Verh. dt. zool. Ges.* (ed. H. D. Pfannenstiel), Gustav Fischer, Stuttgart, New York, p. 630.

Bodenstein, E. (1882) Der Seitenkanal von *Cottus gobio. Z. wiss. Zool.,* **37**, 121–45.

Bodznick, D. and Boord, L. (1986) Electroreception in chondrichthyes: central anatomy and physiology, in *Electroreception* (eds T. H. Bullock and W. Heiligenberg), John Wiley and Sons, New York, pp. 483–98.

Bullock, T. H. (1982) Electroreception. *A. Rev. Neurosci.,* **5**, 121–70.

Bullock, T. H. (1986) Introduction, in *Electroreception* (eds T. H. Bullock and W. Heiligenberg), John Wiley and Sons, New York, pp. 1–12.

Bullock, T. H. and Heiligenberg, W. (1986) *Electroreception,* John Wiley and Sons, New York, 722 pp.

Bullock, T. H., Behrend, K. and Heiligenberg, W. (1975) Comparison of the jamming avoidance response in gymnotoid and gymnarchid electric fish: a case of convergent evolution of behavior and its sensory basis. *J. comp. Physiol.,* **103A**, 97–121.

Bullock, T. H., Bodznick, D. A. and Northcutt, G. (1983) The phylogenetic distribution of electroreception: evidence for convergent evolution of a primitive vertebrate sense modality. *Brain Res. Rev.,* **6**, 25–46.

Bullock, T. H., Hagiwara, S., Kusano, K. and Negishi, K. (1961) Evidence for a category of electroreceptors in the lateral line of gymnotid fishes. *Science,* **134**, 1426–7.

Bullock, T. H., Hamstra, R. H. and Scheich, H. (1972) The jamming avoidance response of high-frequency electric fish. II. Quantitative aspects. *J. comp. Physiol.,* **77A**, 23–48.

Campenhausen, C. von, Riess, I. and Weissert, R. (1981) Detection of stationary objects by the blind cave fish *Anoptichthys jordani* (Characidae). *J. comp. Physiol.,* **143A**, 369–74.

Carr, C.C. and Maler, L. (1986) Electroreception in gymnotiform fish, in *Electroreception* (eds T. H. Bullock and W. Heiligenberg), John Wiley and Sons, New York, pp. 319–73.

Cavallo, T. (1786) A complete treatise on electricity, in *Theorie and Practice,* Vol. 2, C. Dilly, London, pp. 309–11.

Coombs, S. and Janssen, J. (1990) Behavioral and neurophysiological assessment of lateral line sensitivity in the mottled sculpin, *Cottus bairdi. J. comp. Physiol.,* **167A**, 557–67.

Coombs, S., Janssen, J. and Webb, J. (1988) Diversity of lateral line systems: evolutionary and functional considerations, in *Sensory Biology of Aquatic Animals* (eds J. Atema, R. R. Fay, A. N. Popper and W. N. Tavolga), Springer, New York, pp. 553–93.

Coombs, S., Janssen, J., Montgomery, J. (1992) Functional and evolutionary implications of peripheral diversity in lateral line systems, in *The Evolutionary Evolution of Hearing* (eds. D. B. Webster, R. R. Fay and A. N. Popper) Springer, New York, pp. 267–94.

Denton, E. J. and Gray, J. A. B. (1983) Mechanical factors in the excitation of clupeid lateral lines. *Proc. R. Soc.,* **218B**, 1–26.

Denton, E. J. and Gray, J. A. B. (1988) Mechanical factors in the excitation of the lateral lines of fishes, in *Sensory Biology of Aquatic Animals* (eds J. Atema, R. R. Fay, A. N. Popper and W. N. Tavolga), Springer, New York, pp. 595–617.

Dercum, F. (1880) The lateral sensory apparatus of fishes. *Proc. Acad. nat. Sci. Philad.,* 152–4.

Dijkgraaf, S. (1934) Untersuchungen über die Funktion der Seitenorgane an Fischen. *Z. vergl. Physiol.,* **20**, 162–214.

Dijkgraaf, S. (1963) The functioning and significance of the lateral line organs. *Biol. Rev.,* **38**, 51–106.

Dijkgraaf, S. (1968) Electroreception in the catfish, *Amiurus nebulosus. Experientia,* **24**, 187–8.

Elepfandt, A. and Wiedemer, L. (1987) Lateral-line responses to water surface waves in the clawed frog, *Xenopus laevis. J. comp. Physiol.*, **160A**, 667–82.

Emde, G. von der (1990) Discrimination of objects through electrolocation in the weakly electric fish, *Gnathonemus petersii. J. comp. Physiol.*, **167A**, 413–21.

Emde, G. von der and Bleckmann, H. (1992) Extreme phase-sensitivity of afferents which innervate mormyromast electroreceptors. *Naturwi.* **79**, 131–3.

Emde, G. von der and Bleckmann, H. (1993) Differential responses of two types of electroreceptive afferents to signal distortions may permit capacitance measurement in a weakly electric fish, *Gnathonemus petersii. J. comp. Physiol. A* (in press).

Emery, C. (1880) Le specie del genera fierasfer nel golfo di napoli e regioni limitrofe, *Fauna and Flora des Golfes von Neapel.* 2 Monographie, Leipzig, pp. 1–76.

Enger, P. S., Kalmijn, A. J. and Sand, O. (1989) Behavioral investigations of the functions of the lateral line and inner ear in predation, in *The Mechanosensory Lateral Line. Neurobiology and Evolution* (eds S. Coombs, P. Görner and H. Münz), Springer, New York, pp. 575–87.

Enger, P. S., Kristensen, L. and Sand, O. (1976) The perception of weak electric D. C. currents by the European eel (*Anguilla anguilla*). *Comp. Biochem. Physiol.*, **54A**, 101–3.

Enquist, M., Leimar, O., Ljungberg, T., Mallner, Y. and Segerdahl, N. (1990) A test of sequential assessment game: fighting in the cichlid fish *Nannacara anomala. Anim. Behav.*, **40**, 1–14.

Fay, R. R. (1988) *Hearing in vertebrates: a psychophysics databook.* Hill-Fay Associates, Winnetka, Illinois, pp. 1–621.

Finger, T. E., Bell, C. C. and Carr, C. E. (1986) Comparison among electroreceptive teleosts, in *Electroreception* (eds T. H. Bullock and W. Heiligenberg), John Wiley and Sons, New York, pp. 465–81.

Flock, A. (1965) Electron microscopic and electrophysiological studies on the lateral line canal organ. *Acta Oto-lar.*, **199**, 1–90.

Flock, A. (1971) Sensory transduction in hair cells, in *Handbook of Sensory Physiology*, Vol. 1, *Principles of Receptor Physiology* (ed. W. R. Loewenstein), Springer, New York, pp. 396–441.

Fritzsch, B. and Münz, H. (1986) Electroreception in amphibians, in *Electroreception* (eds T. H. Bullock and W. Heiligenberg), John Wiley and Sons, New York, pp. 483–98.

Fuchs, S. (1895) Über die Funktion der unter der Haut liegenden Canalsysteme bei den Selachiern. *Arch. gesamte Physiol.* **59**, 454–78.

Görner, P. and Mohr, C. (1988) The lateral line organ, a mechano- and chemoreceptor? in *Sense Organs. Interfaces between Environment and Behavior* (eds N. Elsner and F. G. Barth), Stuttgart, Thieme, p. 160.

Gray, J. (1984) Interaction of sound pressure and particle acceleration in the excitation of the lateral-line neuromasts of sprats. *Proc. R. Soc.*, **220B**, 299–325.

Gray, J. A. B. and Denton, E. J. (1991) Fast pressure pulses and communication between fish. *J. Mar. Biol. U.K.*, **71**, 83–106.

Gregory, J. E., Iggo, A., McIntyre, A. K. and Proske, U. (1988) Responses of electroreceptors in the snout of echidna, *J. Physiol.*, **414**, 521–38.

Grobbel, G. and Hahn, G. (1958) Morphologie und Histologie der Seitenorgane des augenlosen Höhlenfischs *Anoptichthys jordani* im Vergleich zu anderen Teleostiern. *Z. Morph. Ökol. Tiere*, **47**, 240–63.

Hagedorn, M. (1986) The ecology, courtship, and mating of gymnotiform electric fish, in *Electroreception* (eds T. H. Bullock and W. Heiligenberg), John Wiley and Sons, New York, pp. 497–525.

Hagedorn, M., Womble, M. and Finger, T. E. (1990) Synodontid catfish: a new group of weakly electric fish. *Brain, Behav. Evol.*, **35**, 268–77.

Harris, G. G. and Bergeijk, W. A. van (1962) Evidence that the lateral-line organ responds to near-field displacements of sound sources in water. *J. acoust. Soc. Am.*, **34**, 1831–41.

Hassan, E. S. (1985) Mathematical analysis of the stimulus for the lateral line organ. *Biol. Cybern.*, **52**, 23–36.

Hassan, E. S. (1986) On the discrimination of spatial intervals by the blind cave fish (*Anoptichthys jordani*). *J. comp. Physiol.*, **159A**, 701–10.

Hatai, S., Kokubo, S. and Abe, N. (1932) The earth currents in relation to the responses of catfish. *Proc. imp. Acad. Japan*, **8**, 478–81.

Heiligenberg, W. (1977) Principles of electrolocation and jamming avoidance in electric fish. A neuroethological approach, in *Studies of Brain Function*, Vol. 1 (ed. V. Braitenberg), Springer, New York, pp. 1–85.

Heiligenberg, W. (1989) Coding and processing of electrosensory information in gymnotiform fish. *J. exp. Biol.*, **146**, 255–75.

Heiligenberg, W. (1991) *Neural nets in electric fish*. Cambridge, Massachusetts, MIT Press, pp. 1–179.

Heiligenberg, W. and Bastian, J. (1984) The electric sense of weakly electric fish. *A. Rev. Physiol.*, **46**, 561–83.

Hoekstra, D. and Janssen, J. (1985) Non-visual feeding behavior of the mottled sculpin, *Cottus bairdi*, in Lake Michigan. *Env. Biol. Fishes*, **12**, 111–17.

Hofer, B. (1908) Studien über die Hautsinnesorgane der Fische. I. Die Funktion der Seitenorgane bei den Fischen. *Ber. Kgl. Bayer. Biol. Versuchsstation München*, **1**, 115–68.

Hoin-Radkovski, I., Bleckmann, H. and Schwartz, E. (1984) Determination of source distance in the surface-feeding fish *Pantodon buchholzi* (Pantodontidae). *Anim. Behav.* **32**, 840–51.

Hopkins, C. D. (1986) Behavior of Mormyridae, in *Electroreception* (eds T. H. Bullock and W. Heiligenberg), John Wiley and Sons, New York, pp. 527–76.

Hopkins, C. D., Comfort, N. C., Bastian, J. and Bass, A. H. (1990) Functional analysis of sexual dimorphism in an electric fish, *Hypopomus pinnicaudatus*, order Gymnotiformes. *Brain, Behav. Evol.*, **35**, 350–67.

Jakobson, L. (1813) Extrait d'un mémoire sur un organe particulière de sens dans les raies et les squales. *Nouveau Bull. Sci. Soc. philomatique, Paris*, **3**, 332–7.

Kalmijn, A. J. (1974) The detection of electric fields from inanimate and animate sources other than electric organs, in *Handbook of Sensory Physiology*. Vol. III/3. *Electroreceptors and Other Specialized Receptors in Lower Vertebrates* (eds H. Autrum, R. Jung, W. R. Loewenstein and D. M. MacKay), Springer, New York, pp. 147–200.

Kalmijn, A. J. (1988a) Hydrodynamic and acoustic field detection, in *Sensory Biology of Aquatic Animals* (eds J. Atema, R. R. Fay, A. N. Popper and W. N. Tavolga), Springer, New York, pp. 83–130.

Kalmijn, A. J. (1988b) Detection of weak electric fields, in *Sensory Biology of Aquatic Animals* (eds J. Atema, R. R. Fay, A. N. Popper and W. N. Tavolga), Springer, New York, pp. 151–86.

Kalmijn, A. J. (1989) Functional evolution of lateral line and inner ear sensory systems, in *The mechanosensory lateral line. Neurobiology and evolution* (eds S. Coombs, P. Görner and H. Münz), New York, Springer, pp. 187–216.

Karlsen, H. E. and Sand, O. (1987) Selective and reversible blocking of the lateral line in freshwater fishes. *J. exp. Biol.*, **133**, 249–62.

Käse, R. and Bleckmann, H. (1987) Prey localization by surface wave-ray tracing – fish track bugs like oceanographers track storm. *Experientia*, **43**, 290–93.

Katsuki, Y. and Yamashita, S. (1969) Chemical sensitivity of lateral line organs in the goby, *Gobius giurinus. Proc. Japan Acad.*, **45**, 209–14.

Kaus, S. and Schwartz, E. (1986) Reaction of young *Betta splendens* to surface waves of the water, in *Verh. dt. zool. Ges.* (eds F. G. Barth and E. A. Seyfarth), Gustav Fischer, Stuttgart, pp. 218–19.

Kawamura, T. and Yamashita, S. (1983) Chemical sensitivity of lateral line organs in the goby, *Gobius giurinus. Comp. Biochem. Physiol.*, **74A**, 253–257.

Kirk, K. L. (1985) Water flows produced by *Daphnia* and *Diaptomus*: implications for prey selection by mechanosensory predators. *Limnol. Oceanogr.*, **30**, 679–86.

Knox, R. (1825) On the theory of the existence of a sixth sense in fishes; supposed to reside in certain peculiar tubular organs, found immediately under integuments of the head in sharks and rays. *Edinburgh J. Sci.*, **2**, 12–16.

Knudsen, E. I. (1974) Behavioral thresholds to electric signals in high frequency electric fish. *J. comp. Physiol.* **91A**, 333–53.

Knudsen, E. I. (1975) Spatial aspects of the electric fields generated by weakly electric fish. *J. comp. Physiol.*, **99A**, 103–18.

Kramer, B. (1985) Jamming avoidance in the electric fish *Eigenmannia*: harmonic analysis of sexually dimorphic waves. *J. exp. Biol.*, **119**, 41–69.

Kramer, B. (1987) The sexually dimorphic jamming avoidance response in the electric fish *Eigenmannia* (Teleostei, Gymnotiformes). *J. exp. Biol.*, **130**, 39–62.

Kramer, B. (1990a) *Electrocommunication in Teleost Fishes. Behavior and Experiments*, Springer, New York, 240 pp.

Kramer, B. (1990b) Sexual signals in electric fishes. *Trends Ecol. Evol.*, **5**, 247–50.

Kramer, B. and Otto, B. (1988) Female discharges are more electrifying: spontaneous preference in the electric fish, *Eigenmannia* (Gymnotiformes, Teleostei). *Behav. Ecol. Sociobiol.*, **23**, 55–60.

Kramer, B. and Zupanc, G. K. H. (1986) Conditioned discrimination of electric waves differing only in form and harmonic content in the electric fish *Eigenmannia. Naturwissenschaften*, **73**, 679–80.

Kroese, A. B. A. and Netten, S. M. van (1989) Sensory transduction in lateral line hair cells, in *The Mechanosensory Lateral Line. Neurobiology and Evolution* (eds S. Coombs, P. Görner and H. Münz), Springer, New York, pp. 205–84.

Kroese, A. B. A. and Schellart, N. A. M. (1987) Evidence for velocity- and acceleration-sensitive units in the trunk lateral line of the trout. *J. Physiol.*, **394**, 13.

Kroese, A. B. A., Prins, M. and Schellart, N. A. M. (1989) Regional differences in conduction velocity and fibre diameter in posterior lateral line nerve axons in the trout. *J. Physiol.*, **418**, 136.

Kühme, W. (1961) Verhaltensstudien am maulbrütenden (*Betta anabatoides* Bleeker) und nestbauenden Kampffisch (*B. splendens* Regan). *Z. Tierpsychol.*, **18**, 629–76.

Lang, H. H. (1980) Surface wave discrimination between prey and nonprey by the backswimmer *Notonecta glauca* L. (Hemiptera, Heteroptera). *Behav. Ecol. Sociobiol.*, **6**, 233–46.

Lee, F. S. (1898) The function of the ear and lateral line in fishes. *Am. J. Physiol.*, **1**, 128–44.

Leydig, F. (1850) Über die Schleimkanäle der Knochenfische. *Müll. Arch. Anat. Physiol.*, 170–81.

Leydig, F. (1851) Über die Nervenköpfe in den Schleimkanälen von *Lepidoleprus umbrina* und *corvina. Müll. Arch. Anat. Physiol.*, 235–40.

Lighthill, J. (1980) *Waves in Fluids*, Cambridge University Press, Cambridge, 504 pp.

Lissmann, H. W. (1951) Continuous electrical signals from the tail of a fish, *Gymnarchus niloticus* Cuv. *Nature, Lond.*, **167**, 201–2.

Lissmann, H. W. (1958) On the function and evolution of electric organs in fish. *J. exp. Biol.*, **35**, 156–91.

Lissmann, H. W. and Machin, K. E. (1958) Electric receptors in a non-electric fish (*Clarias*). *Nature, Lond.*, **199**, 88–9.

Lücker, H. and Kramer, B. (1981) Development of a sex difference in the preferred latency response in the weakly electric fish, *Pollimyrus isidori* (Cuvier et Valenciennes) (Mormyridae, Teleostei). *Behav. Ecol. Sociobiol.*, **9**, 103–9.

McCormick, C. A. (1989) Central lateral line mechanosensory pathways in bony fish, in *The Mechanosensory Lateral Line. Neurobiology and Evolution* (eds S. Coombs, P. Görner, and H. Münz), Springer, New York, pp. 341–64.

McCormick, C. A. and Braford, M. R. (1988) Central connections of the octavolateralis system: evolutionary considerations, in *Sensory Biology of Aquatic Animals* (eds J. Atema, R. R. Fay, A. N. Popper and W. N. Tavolga), Springer, New York, pp. 733–56.

Mayser, M. (1881) Vergleichende anatomische Studien über das Gehirn der Knochenfische mit besonderer Berücksichtigung der Cyprinoiden. *Z. wiss. Zool.*, **36**, 259–364.

Meredith, G. E., Roberts, B. L. and Maslam, S. (1987) Distribution of afferent fibers in the brainstem from end organs in the ear and lateral line in the European eel. *J. comp. Neurol.*, **265**, 507–20.

Merkel, F. (1880) *Über die Endigungen der sensiblen Nerven in der Haut der Wirbeltiere*, Verlag der Stiller'schen Hof- und Universitätsbuchhandlung, Rostock, 214 pp.

Meyer, D. L., Heiligenberg, W. and Bullock, T. H. (1976) The ventral substrate response of fishes: a new postural control mechanism in fishes. *J. comp. Physiol.*, **109A**, 59–68.

Meyer, J. H. (1984) Steroid influences upon discharge frequencies of intact and isolated pacemakers of weakly electric fish. *J. comp. Physiol.*, **154A**, 659–68.

Meyer, J. H., Zakon, H. H. and Heiligenberg, W. (1984) Steroid influences upon the electrosensory system of weakly electric fish: direct effects upon discharge frequencies with indirect effects upon electroreceptor tuning. *J. comp. Physiol.* **154A**, 625–31.

Moller, P. (1976) Electric signals and schooling behavior in a weakly electric fish, *Marcusenius cyprinoides* L. (Mormyriformes). *Science*, **193**, 697–9.

Montgomery, J. C. (1989) Lateral line detection of planktonic prey, in *The Mechanosensory Lateral Line. Neurobiology and Evolution* (eds S. Coombs, P. Görner and H. Münz), Springer, New York, pp. 561–74.

Montgomery, J. C., Macdonald, J. A. and Housley, G. D. (1988) Lateral line function in an antarctic fish related to the signals produced by planktonic prey. *J. comp. Physiol.*, **163A**, 827–33.

Müller, U. and Schwartz, E. (1982) Influence of single neuromasts on prey-localizing behavior of surface-feeding fish *Aplocheilus lineatus*. *J. comp. Physiol.*, **149A**, 399–408.

Münz, H. (1985) Single unit activity in the peripheral lateral line system of the cichlid fish *Sarotherodon niloticus* L. *J. comp. Physiol.*, **157A**, 555–68.

Münz, H. (1989) Functional organization of the lateral line periphery, in *The Mechanosensory Lateral Line. Neurobiology and Evolution* (eds S. Coombs, P. Görner and H. Münz) Springer, New York, pp. 285–98.

Netten, S. M. van (1990) Hydrodynamics of the excitation of the cupula in the fish canal lateral line. *J. acoust. Soc. Am.*, **89**: 310–319.

Netten, S. M. van and Kroese, A. B. A. (1987) Laser interferometric measurement on the dynamic behavior of the cupula in the fish lateral line. *Hearing Res.*, **29**, 55–61.

Northcutt, G. (1986) Electroreception in nonteleost bony fishes, in *Electroreception* (eds T. H. Bullock and W. Heiligenberg), John Wiley and Sons, New York, pp. 257–86.

Northcutt, R. G. (1989) The phylogenetic distribution and innervation of craniate mechanoreceptive lateral lines, in *The Mechanosensory Lateral Line. Neurobiology and Evolution* (eds S. Coombs, P. Görner and H. Münz) Springer, New York, pp. 17–78.

Parker, G. H. (1902) Hearing and allied senses in fishes. *Bull. U.S. Bur. Fish.*, **22**, 45–64.

Parker, G. H. (1904) The function of the lateral-line organs in fishes. *Bull. U.S. Bur. Fish.*, **24**, 185–207.

Parker, G. H. and Heusen, A. P. van (1917) The responses of the catfish, *Amiurus nebulosus*, to metallic and non-metallic rods. *Am. J. Physiol.*, **44**, 405–20.

Partridge, B. L. and Pitcher, T. J. (1980) The sensory basis of fish schools: relative roles of lateral line and vision. *J. comp. Physiol.*, **135A**, 315–25.

Peters, R. C. and Wijland, F. van (1974) Electro-orientation in the passive electric catfish, *Ictalurus nebulosus* LeS. *J. comp. Physiol.*, **92A**, 273–80.

Poulson, T. L. (1963) Cave adaptations in amblyopsid fishes. *Am. Midl. Nat.*, **70**, 257–90.

Pitcher, T. J. (1979) Sensory information and the organisation of behaviour in a shoaling fish. *Anim. Behav.*, **27**, 126–49.

Pitcher, T. J., Partridge, B. L. and Wardle, C. S. (1976) A blind fish can school. *Science*, **194**, 963–5.

Puzdrowski, R. L. (1989) Peripheral distribution and central projections of the lateral line nerves in goldfish, *Carassius auratus*. *Brain, Behav. Evol.*, **34**, 110–31.

Roberts, B. L. and Meredith, G. E. (1989) The efferent system, in *The Mechanosensory Lateral Line. Neurobiology and Evolution* (eds S. Coombs, P. Görner and H. Münz), Springer, New York, pp. 445–59.

Roberts, W. M., Howard, J. and Hudspeth, A. J. (1988) Hair cells: transduction, tuning, and transmission in the inner ear. *A. Rev. Cell Biol.*, **4**, 63–92.

Rommel, S. A. and McCleave, J. D. (1972) Sensitivity of American eel (*Anguilla rostrata*) and Atlantic salmon (*Salmo salar*) to weak electric and magnetic field. *J. Fish. Res. Bd Can.*, **30**, 657–63.

Roth, A. (1968) Electroreception in the catfish, *Amiurus nebulosus* Z. vergl. Physiol., **61**, 196–202.

Roth, A. (1972) Wozu dienen die Elektrorezeptoren der Welse? *J. comp. Physiol.*, **79A**, 113–35.

Russell, C. J., Myers, J. P. and Bell, C. C. (1974) The echo response in *Gnathonemus petersii* (Mormyridae). *J. comp. Physiol.*, **92A**, 181–200.

Russell, I. J. (1974) Central and peripheral inhibition of lateral line input during the startle response in goldfish. *Brain Res., Amst.*, **80**, 517–22.

Russell, I. J. and Lowe, D. A. (1983) The effect of efferent stimulation on the phase and amplitude of extracellular receptor potentials in the lateral line system of the perch (*Perca fluviatilis*). *J. exp. Biol.*, **102**, 223–38.

Sand, O. (1981) The lateral line and sound reception, in *Hearing and Sound Communication in Fishes* (eds W. N. Tavolga, A. N. Popper and R. R. Fay), Springer, Heidelberg, pp. 459–78.

Satou, M., Takeuchi, H., Takei, K., Hasegawa, T., Okumoto, N. and Ueda, K. (1987) Involvement of vibrational and visual cues in eliciting spawning behaviour in male hime salmon (landlocked red salmon, *Oncorhynchus nerka*). *Anim. Behav.*, **35**, 1556–8.

Satou, M., Shiraishi, A., Matsushima, T. and Okumoto, N. (1991) Vibrational communication during spawning behavior in the hime salmon (landlocked red salmon, *Oncorhynchus nerka*). *J. comp. Physiol. A*, **168**, 417–28.

Scheich, H., Langner, G., Tidemann, C., Coles, R. B. and Guppy, A. (1986) Electroreception and electrolocation in platypus. *Nature, Lond.*, **319**, 401–2.

Schulze, F. E. (1861) Über die Nervenendigung in den sogenannten Schleimkanälen der Fische und über entsprechende Organe der durch Kiemen atmenden Amphibien. *Arch. Anat. Physiol. Wiss. Med. Lpz.*, 759–769.

Schulze, F. E. (1870) Über die Sinnesorgane der Seitenlinie bei Fischen und Amphibien. *Arch. mikrosk. Anat.*, **6**, 62–88.

Schwartz, E. (1965) Bau und Funktion der Seitenlinie des Streifenhechtlings *Aplocheilus lineatus*. *Z. vergl. Physiol.*, **50**, 55–87.

Schwartz, E. (1971) Die Ortung von Wasserwellen durch Oberflächenfische. *Z. vergl. Physiol.*, **74**, 64–80.

Sede, P. de (1884) La ligne latérale des poissons osseux. *Revue Scientifique Série de Tom*, **7**, 467–70.

Shumway, C. A. and Zelick, R. D. (1988) Sex recognition and neural coding of electric organ discharge waveform in the pulse-type weakly electric fish, *Hypopomus occidentalis*. *J. comp. Physiol.*, **163A**, 465–78.

Sommerfeld, A. (1970) *Vorlesung über theoretische Physik*. Band 11. *Mechanik der deformierbaren Medien*, Akademische Verlagsgesellschaft, Leipzig, 446 pp.

Squire, A. and Moller, P. (1982) Effects of water conductivity on electrocommunication in the weak electric fish *Brienomyrus niger* (Mormyriformes). *Anim. Behav.*, **30**, 375–82.

Sutterlin, A. M. and Waddy, S. (1975) Possible role of the posterior lateral line in obstacle entrainment by brook trout (*Salvelinus fontinalis*). *J. Fish. Res. Bd Can.*, **32**, 2441–6.

Szabo, T. (1974) Anatomy of the specialized lateral line organs of electroreception, in *Handbook of Sensory Physiology*, Vol. 111/3 (eds H. Autrum, R. Jung, W. R. Loewenstein and D. M. MacKay), Springer, New York, pp. 13–58.

Teyke, T. (1988) Flow field, swimming velocity and boundary layer: parameters which affect the stimulus for the lateral line organ in blind fish. *J. comp. Physiol.*, **163A**, 53–61.

Teyke, T. (1989) Learning and remembering the environment in the blind cave fish *Anoptichthys jordani*. *J. comp. Physiol.*, **164A**, 655–62.

Teyke, T. (1990) Morphological differences in neuromasts of the blind cave fish *Astyanax hubbsi* and the sighted river fish *Astyanax mexicanus*. *Brain, Behav. Evol.*, **35**, 23–30.

Tricas, T. C. and Highstein, S. M. (1990) Visually mediated inhibition of lateral line primary afferent activity by the octavolateralis efferent system during predation in the free-swimming toadfish, *Opsanus tau*. *Exp. Brain Res.*, **83**, 233–6.

Vischer, H. A. (1990) The morphology of the lateral line system in 3 species of Pacific cottoid fishes occupying disparate habitats. *Experientia*, **46**, 244–50.

Watanabe, A. and Takeda, K. (1963) The change of discharge frequency by a.c. stimulus in a weak electric fish. *J. exp. Biol.*, **40**, 57–66.

Webb, J. F. (1989) Gross morphology and evolution of the mechanoreceptive lateral-line system in teleost fishes. *Brain, Behav. Evol.*, **33**, 34–53.

Webb, P. W. (1982) Avoidance response of fathead minnow to strikes by four teleost predators. *J. comp. Physiol.*, **147A**, 371–8.

Weissert, R. and Campenhausen, C. von (1981) Discrimination between stationary objects by the blind cave fish *Anoptichthys jordani* (Characidae). *J. comp. Physiol.*, **143A**, 378–81.

Wu, C. H. (1984) Electric fish and the discovery of animal electricity. *Am. Scient.*, **72**, 598–607.

Wunder, W. (1927) Sinnesphysiologische Untersuchungen über die Nahrungsaufnahme bei verschiedenen Knochenfischarten. *Z. vergl. Physiol.*, **6**, 67–98.

Zakon, H. H. (1986) The electroreceptive periphery, in *Electroreception* (eds T. H. Bullock and W. Heiligenberg), John Wiley and Sons, New York, pp. 103–56.

Part three

Behavioural Ecology

INTRODUCTION

Four major topics are reviewed in this behavioural ecology section of the book: feeding, mating, social behaviour and the impact of environmental variation in space and time. In addition, two case studies conclude the section. In the first three pairs of chapters, the first member of each pair gives a synoptic review of the topic, while the second highlights one aspect in detail. A review of foraging is followed by how feeding decisions are constrained by predators; mating is followed by the evolutionary basis of parental care; social behaviour (shoaling) by the importance of individual differences. The next pair of chapters, on the diurnal light regime and on littoral fishes, centre on spatio-temporal patterning by the evolutionary template of habitat. The behavioural ecology section closes with two case studies on cave dwelling fishes and on sticklebacks.

Behavioural ecology is the study of the ways in which behaviour is influenced by natural selection in relation to ecological conditions. Along with its sister discipline of sociobiology, defined as the study of factors shaping the evolution of social behaviours, this approach has stimulated an explosion of fresh insights since the mid 1970s. A recent review of the field is given by Krebs and Davies (1991). The search for optimal evolutionarily stable solutions to animals' problems in search, choice and conflict entails attempts to match theory with experimental evidence, so that the testing of alternative theories forms a core in the discipline. Since the first edition, a great deal of fundamental new work in this relatively new discipline has focussed on fish.

In Chapter 8, Paul Hart reviews our understanding of the decision rules about how fishes feed by examining the relationship between the now-extensive body of foraging theory and the vast amount of empirical work on how fishes actually feed. Theory is the 'skeleton on which to hang the flesh of fact'. He points out that the global problem is the optimal allocation of time and resources to feeding, reproduction and defence which will maximize the individual's lifetime reproductive success. Hart discusses how, once fishes are in feeding mode, they deal with patchy and variable food, alternative prey types and sizes, the presence of other fishes, predators, morphological constraints and learning. Recent advances in diet selection models are described. Observed behaviour generally represents trade-offs between the benefits from nutritional energy gained with the costs and risks involved. Many early theories failed to account for real feeding behaviour, but recent advances in theory have provided us with considerable insight of these trade-offs. Hart sees the future of foraging work in developing dynamic optimization models which can account for changes in the major driving factors for 'each turn of the feeding cycle' as appetite returns.

Milinski, in Chapter 9, analyses the trade-offs between feeding and predator risk in detail, using many results from his own extensive series of elegant experiments on sticklebacks. Feeding fish take greater risks when hungry or parasitized. Smaller vulnerable fishes will settle for sub-optimal food. Milinski concludes by noting that the subtlety of fish responses may make it difficult to predict the precise details of behaviour, although some workers have managed to equate perceived risk to quantity of food in elegant experiments.

Chapter 10, by George Turner, reviews mating behaviour in teleosts as analysed using ESS models. A conflict of interest between the sexes is reflected in the number of sexual partners and degree of parental care. There are many solutions to the conflict so that the stable breeding systems which have evolved vary greatly between species. Turner discusses the evidence for mate choice and sexual selection in fishes, including the arguments about the genetic foundation for choosing mates. This is followed by an analysis of breeding strategies. The chapter encompasses polygamy, alternative male strategies, monogamy, sex change, hermaphroditism, polygyny, polyandry, leks and hotspots, broodcare helpers, asexual reproduction in clones of poeciliids, and sex reversal in wrasse and others, some of which defy explanation by current theory. The new edition of Turner's chapter places fish mating firmly in the mainstream of current behavioural theory.

Craig Sargent and Mart Gross extend their analysis of the evolution of parental care in Chapter 11, first presented in the first edition of this book. The concept that present reproduction bears a cost limiting future reproduction is termed the 'Williams Principle' after G. C. Williams. Evidence from fishes is presented that there is such a cost. The authors develop a quantitative model of this trade-off between breeding now or later, and use it to analyse the prevalence of male broodcare, changes in parental behaviour with age, brood size, and territory defence using examples from sticklebacks, medaka, sunfish, guppies, sculpins and cichlids. Sargent and Gross review a number of tests of the Williams Principle using the recently-introduced technique of dynamic programming optimization. Sargent and Gross put forward three areas for the next phase of research, the effect of social status on optimal reproductive effort; competion for resources which are not divisible; conflict between the sexes about the degree of parental care; and a re-visitation of Zahavi's handicap principle.

In Chapter 12, Julia Parrish and myself review shoaling behaviour. This new edition, as well as covering laboratory tests of the functions of shoaling, places a greater emphasis on recent quantitative field measurements than in the first edition. The main theme remains that the most useful view of shoaling pivots on the rules governing an individual fish's decisions to join, leave or stay with the social group. An older conceptual framework, emphasizing group behaviour and impressive coordination has been less profitable in providing insight of this fascinating behaviour. After outlining heuristic definitions of shoaling, a brief perspective on structure and function in fish shoals concludes with their impact on the management of commercial fisheries. Next, the attractive possibility of hydrodynamic advantage is reviewed, but there is still little direct evidence. Two major sections cover the costs and benefits of predator defence and foraging in shoals, in both bases shoal size being a critical factor in determining fish's decisions. Next, the trade off between feeding and predator risk is reviewed. The chapter proceeds with discussions of position, size-sorting, migration, the genetic basis of shoaling, mixed-species shoals and concludes with the search for rules governing optimal shoal size.

In Chapter 13, Anne Magurran focuses on individual differences in fish behaviour, reinforcing the points made by Turner, Hart, and Pitcher and Parrish in previous chapters and centred on recent theoretical advances in understanding the evolution of behaviour. At one time, animals that failed to perform what was considered specie-specific behaviour, for example standing idly by during courtship, were labelled as aberrant or sick, or were even omitted from data altogether. This chapter shows just how misguided that view was since such individual differences are now seen as critical to our understanding of behavioural evolution. Individual differences in foraging, antipredator responses, mating, sex change, aggression, and habitat use are reviewed including evidence from her own work on guppies and minnows. Distinct alternative behavioural strategies, far from being an aberration, may represent an evolutionarily stable solution to the problem of optimizing individual reproductive success. Furthermore, appropriate behaviour may be contingent upon what others are doing, or upon the patchiness of resources in space and time. In some cases the different behaviours may reflect genetic differences, while in others fish are endowed with the behavioural plasticity to opt for one solution or the other. Magurran concludes by warning against describing behaviour in terms of the average individual.

We now come to two chapters on spatial and temporal patterning. In Chapter 14, Gene Helfman shows how the behaviour of coral-reef, kelp forest and lake fishes is shaped by the daily cycle of light, twilight and darkness. Few fishes can be optimally adapted to all these light conditions, and so in each of these communities, daytime species of zooplanktivores, invertebrate eaters and piscivores are replaced by sets of nocturnal species, although herbivores and cleaners tend to be diurnal. At night, killer molluscs and killer isopods threaten refuging fishes. Twilight has a profound effect on the evolution, physiology and behaviour of fishes in both of these sets, since many ambush predators, striking from below, have the advantage at low light levels during the 'changeover' period. Helfman finishes with a discussion of differences between the roles of twilight periods in the tropics and temperate habitats.

Robin Gibson, in Chapter 15, reviews the behavioural ecology of intertidal fishes, where the normal medium for locomotion is removed for hours at a time. About five families are specialized for sea shore life, including residents, transients and visitors. Some intertidal fishes avoid the problem by migrating to and fro, but residents shelter, burrow or have evolved amphibious traits. Gibson describes the physical changes which shore fishes experience, along with the morphological, physiological and behavioural adaptations associated with habitat selection, maintenance of position, site fidelity, territoriality, reproduction and the shore fishes' impressive homing ability. Gibson warns against the lack of appropriate tidal cues in laboratory experiments on these fishes. He continues with an account of how fish synchronize behaviour with the tidal cycle, and concludes with an account of fish which make amphibious excursions onto land.

The final two chapters in this section are case studies revisiting the conceptual themes of earlier chapters. Gerard FitzGerald and Bob Wootton review the behavioural ecology of the sticklebacks in Chapter 16, since Tinbergen's pioneering experiments one of the most popular fish for behavioural work and one which has been the subject of significant advances in our understanding of foraging and mating behaviour. Sticklebacks bear the double advantage of being easy to maintain in the laboratory and easy to observe in the wild. In the wild, insight of the forces shaping behaviour have come from comparison of stickleback communities with differing habitats, competition, food, predation and parasitism, mate choice, parental behaviour and cannibalism. FitzGerald and Wootton leave us with the prediction that stickleback studies will contribute as much to behavioural ecology as they did to the classic study of ethology.

Jakob Parzefall reviews the behavioural ecology of cave dwelling fishes in Chapter 17, which closes this section. Members of 14 teleost families have become such troglobionts. Feeding and reproductive behaviour in cave-dwellers are compared with their epigean relatives, including experiments that reveal a genetic basis to some of these differences. He discusses how cave fishes, most of which are blind, deal with food, mates and rivals; in many cases behaviour appears to be generalized and reduced. Competing theories to account for this regressive evolution are analysed. Pheromone-mediated sexual behaviour, nocturnal traits and the circadian clock may be pre-adaptations for cave life. The absence of visual displays may also be necessary for the adoption of cave living, there are no examples of the gradual reduction of such behaviours. Parzefall argues that cave fishes could be important for insight of behavioural evolution, since he feels that changes in many of their traits owe more to the accumulation of neutral mutations than to stabilizing natural selection.

One major topic in fish behaviour not covered by this book is that of migration (Smith, 1985).

REFERENCES

Smith, R. J. F. (1985) *The control of fish migration.* Springer-Verlag, Berlin, 243 pp.
Krebs, J. R. and Davies, N. B. (1991) *Behavioural Ecology,* 3rd edition, Blackwell, Oxford.

Chapter eight

Teleost foraging: facts and theories

Paul J. B. Hart

8.1 INTRODUCTION

A short while observing a feeding fish might give the impression that it would be impossible to understand why behavioural changes occur as they do. A foraging roach, *Rutilus rutilus*, for example, cruises about its habitat digging into the bottom, snapping at a drifting insect in mid-water and plucking at a piece of leaf at the water's edge. How can these apparently disparate actions be explained by a unified foraging theory? The introduction of evolutionary concepts has helped to provide theories that predict decision rules for foraging animals, the principal concept being that efficient feeders will be favoured by natural selection. The art of constructing evolutionary foraging models is to choose the right decision criterion by which different foraging strategies can be judged. In this chapter I do not have the space to review all of fish foraging behaviour or all the foraging models that exist. My intention is to show how foraging models are being used to understand fish foraging. The ultimate test of a theoretical interpretation of a natural process is that it can be used to make predictions that work: Newton's theory of gravitation makes it possible to get spacecraft to the Moon and back. At present foraging theory comes nowhere near that goal. Its main function just now is to explain in an imperfect way why foraging fish do what they do.

Foraging behaviour involves a hierarchical set of decisions. Once a fish has decided to feed (see Chapter 2 by Colgan, this volume), it is faced with a number of further choices. Very often it will be necessary to look for food, and the fish might choose a search path to maximize the energy gain per unit of search time. Food is usually patchily distributed, with different patches having higher or lower profitabilities. The fish has to decide how long to stay feeding in a particular patch, and, within the patch, which size or species of prey it would be appropriate to take.

Behaviour of Teleost Fishes 2nd edn. Edited by Tony J. Pitcher. Published in 1993 by Chapman & Hall. ISBN 0 412 42930 6 (HB) and 0 412 42940 3 (PB).

The principal consideration in this chapter will be with foraging behaviour, but as this is constrained by the physiology and morphology of the fish (Brett, 1979; Brett and Groves, 1979; Bone and Marshall, 1982), they too should be mentioned. Similarly the behaviour of an individual will be influenced by the presence of conspecifics and by predators. The resulting strategies of food gathering are a compromise between the benefits derived from food gathered and the costs associated with the strategy. Costs and benefits need to be expressed in terms of a common currency. As a fish gathers food, it takes in and expends energy and nutrients. Most models assume that costs and benefits can be expressed in units of energy, which is regarded as the common currency. It is often assumed further that maximizing net energy gain is an optimal strategy, but this is not so unless the animal is also maximizing its fitness (Sih, 1982), defined as the animal's lifetime reproductive success. This has never been measured in a fish and for only a select range of other species (Clutton Brock, 1988; Newton, 1989). Very often, however, energy and fitness maximization are correlated (Hart and Connellan, 1984).

The concept of optimization as applied to animal behaviour has been extensively criticized (Gould and Lewontin, 1979; Kitcher, 1985; Gray, 1987; Ollason, 1987). A defence is given by Stephens and Krebs (1986) and a philosophical analysis by Mitchell and Valone (1990). A solution to a problem is only optimal under a particular set of assumptions (Hart, 1989): change the assumptions and the optimal solution changes. For a detailed discussion see Ollason (1987). Despite the criticisms I will continue to use optimality thinking in this chapter as an organizing principle but with an awareness of its limitations.

There are two ways in which efficient feeding strategies can originate; they can be acquired over evolutionary time, the results of natural selection being inbuilt into each individual, or they can be learned during the lifetime of the fish. All vertebrates have complex neurophysiology and learning capacities (MacPhail, 1982); consequently one would expect learning to have an effect on feeding behaviour, modifying the simple strategies characteristic of most of optimal foraging theory (OFT) (Dill, 1983; Milinski, 1984a). This chapter will therefore also examine how learning can modify feeding strategies, and how foraging fish use their learning abilities to respond to environmental variability.

8.2   BEHAVIOURAL FEEDING STRATEGIES

When to start feeding

Modern approaches to the study of motivation, which are discussed in detail by Colgan in Chapter 2 of this volume (and see also Colgan, 1989), argue that motivational state is determined by the combined effects of internal and external

causal factors. The degree of stomach fullness and other physiological parameters (Holmgren *et al.*, 1983) are internal factors and the presence of prey
of the appropriate size and species is an external factor. The internal state
influences the way in which a fish responds to external stimuli. In pike, *Esox
lucius*, Clark (1986) has studied the response of semi-satiated and hungry
individuals to a range of prey-like stimuli. A series of trials were carried out
in which a recently fed or a hungry fish was presented with a silhouette of a
minnow model which moved only in the horizontal plane. The recently fed
pike was unresponsive to the prey for a median of 211.1 s each trial, whereas
the hungry pike was unresponsive for a median of only 23.1 s. Further evidence
shows that the way a pike responds to a prey-like stimulus is strongly influenced
by its internal state.

A very hungry fish will increase its rate of food intake over the first few
minutes of a feeding bout. For example threespine sticklebacks, *Gasterosteus
aculeatus*, that have been starved for 72 h feed faster when given access to food
than do fish starved for 24 h. At the end of a 40 minute feeding session all fish
will have consumed the same number of prey (Beukema, 1964, 1968). A very
hungry stickleback completes more prey capture attempts than after a few
minutes' feeding (Tugendhat, 1960), although the number of completions
rises again before declining slowly towards the end of the feeding session.
Tugendhat (1960) proposed that the shape of the curve was a sign that the
stretch receptors in the stomach changed their response as the stomach filled.
The controlling role of stretch receptors is also discussed by Beukema (1964),
although he suggests in addition that a systemic need has an influence, as
illustrated by the higher feeding rate seen in fish that have been deprived of
food for a long time (see also Chapter 2 by Colgan, this volume).

As the stomach fills during a feeding bout, the fish gradually reduces the
readiness with which it attacks and ingests prey. For sticklebacks the rate of
prey capture falls off exponentially, but the fish continues to initiative prey
capture at a steady rate, even when nearly full (Tugendhat, 1960). The
difference between a hungry and a full fish is that the latter completes fewer
feeding attempts and spends more time moving away from areas where food
is available.

Feeding takes second place to vigilance and reproduction under certain
conditions (Heller and Milinski, 1979; Milinski, 1984b; Noakes, 1986; Godin
and Sproul, 1988; Hart and Gill, 1993). When threat of predation is greater
than the threat of starvation, the fish will suppress feeding even when hungry.
Likewise, during the breeding season, a male threespine stickleback will defend
its nest at the expense of feeding opportunities.

The motivation to feed can also change through time independently of
external factors. Juvenile Atlantic salmon, *Salmo salar*, in their first year are
of two types (Metcalfe *et al.*, 1986): those that continue to grow through their
first winter and those that stop growing. The non-growers orientate less to

food, attack food less, and fewer attacks lead to capture as winter approaches. These changes occur in the absence of competitors or predators and must therefore be the result of internal changes which switch off feeding. These behavioural changes are correlated with a change in habitat use as winter advances. In rivers, juvenile salmon change from holding feeding stations above stones on the stream bed to sheltering beneath stones. In this position the fish adopt a low-intake feeding strategy by which they minimize the cost of obtaining a maintenance ration.

Searching for food

Once a fish has decided to start feeding, its first task is to find suitable food items. Many will be clumped and occur in patches so that the foraging fish has to travel. Assuming that fish are maximizing their fitness, one is led to two considerations; the first concerns the route the fish should take to maximize the net gain of food, the second is concerned with the speed at which and the pattern with which the animal should move. Optimizing the route has been studied quite extensively (e.g. Smith, 1974; Pyke, 1978; Jansen, 1982). Optimal speed has been studied mainly theoretically by Ware (1975, 1978, 1982) and Pyke (1981).

The search paths of the threespine sticklebacks looking for *Tubifex* worms on a grid of containers on the floor of a 9 cm × 27 cm aquarium showed marked responses to the presence of prey (Thomas, 1974). If food was found and eaten, the linear distance travelled after discovery decreased significantly (Fig. 8.1). Also, after a fish had approached and rejected a food item, the linear distance travelled increased significantly. This 'area-avoiding search' moves the fish fast and directly away from an area where a rejected item has been found, so increasing the likelihood of discovering a new area with desirable food.

In many theoretical studies it has been assumed that fish either cruise steadily through the water in search of food, or they sit and wait (O'Brien *et al.*, 1990). Recent work has shown that many fish that feed on plankton are more likely to move in jerks, stopping to search (Evans and O'Brien, 1988; O'Brien *et al.*, 1989, 1990). This stop–start type of search has been called saltatory search. It has been found that white crappies, *Pomoxis annularis*, alter their speed of movement and the distance travelled between stops in response to changes in prey size (O'Brien *et al.*, 1989). The fish travel further and move faster away from an unsuccessful search stop when feeding on large prey. They also pause for a shorter time which may be related to the ease with which larger prey items can be detected (O'Brien *et al.*, 1990). A simulation model was developed by O'Brien *et al.* (1989) to evaluate the efficiency of the observed changes.

The model is a derivative of the simple diet choice model which is described

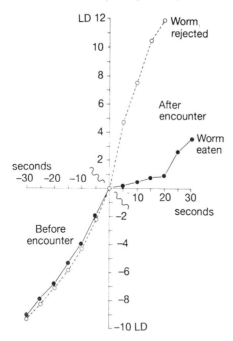

Fig. 8.1 Sticklebacks searching for *Tubifex* set in a rectangular grid on the floor of an aquarium. Each point on the curves shows the linear distance (LD) travelled between 5 s intervals both before and after discovering a worm (signified by the squiggle at the intersection of the two axes). Broken curve shows the resulting LDs before and after fish had found and rejected a worm. The solid curve is for fish that found and ate the worm. Reproduced with permission from Thomas (1974).

in detail later (p. 261). This assumes that the foraging animal is maximizing the long-term rate of net energy gain (R). This is the currency the fish uses to evaluate the fitness-yielding properties of alternative strategies. In their model O'Brien *et al.* (1989) put $R = (E_I - E_E)/T_T$. E_I is the energy input from prey captured, E_E is the energy expended in capturing prey, including search energy, and T_T is the total time taken to search and capture prey. The model predicts that white crappies feeding on small prey should travel 5.5 cm between stops to maximize R. Observed distances were 5.2 cm. For large prey the model predicts travel distances of 13.5 cm as opposed to the observed mean of 9.8 cm. The curve of R against distance travelled has a flatter peak for large prey. If the fish is faced with two sizes of prey and search times are long, then the theory predicts that the fish should drop the small prey from the diet. This is a similar prediction to one derived from a discrete version of the marginal value theorem which will be discussed later (p. 268) (Stephens *et al.*, 1986).

Maximizing R is assumed by O'Brien *et al.* (1989) to be the criterion determining the search strategy chosen. In moving from one point to another, a fish might be approaching food, but at the same time it might be exposing itself to predators (Pyke, 1981). Consequently it is not always easy to choose the optimality criterion. Ware (1978) made the assumption for pelagic fish that an appropriate criterion would be growth-rate maximization. Growth rate was made a function of the net food intake per unit time, the standard metabolic rate and the cost of swimming, each being expressed as functions of body weight. A simulation of the energy budget of fish of different sizes showed that the optimal swimming speed during foraging was a function of food concentration and fish size (Fig. 8.2).

The searching behaviour of a foraging fish is constrained by the structure of the habitat. The pressure to be efficient can be so great that within-species changes in morphology have been developed in threespine sticklebacks to cope with the demands of different parts of the habitat (Bentzen and McPhail, 1984; McPhail, 1993). In the Cowichan Lake drainage system, British Columbia, Lavin and McPhail (1986) found three morphotypes, littoral, limnetic and intermediate. Littoral types feed mostly on the lake bottom and have a longer upper jaw than do fish in the other two categories. They can also ingest larger prey, which they handle faster than do limnetic types. The limnetic and intermediate types are more adept at taking prey from the water column, a skill which is a function of their greater gill raker density, number and length. A similar dimorphism has been found for bluegill sunfish, *Lepomis macrochirus*, individuals which live either in the weed or in the open water (Ehlinger and Wilson, 1988)

The limnetic and benthic habitats differ most in their structural complexity. Within the littoral, threespine sticklebacks still use the water column to search in (Ibrahim and Huntingford, 1989), but they must search for food through a maze of plants. It is likely that sticklebacks will have their encounter rates with prey strongly influenced by the density of vegetation or the need to search through sediment and sort prey from detritus. For example, bluegill sunfish

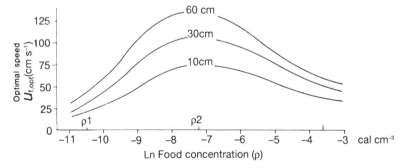

Fig. 8.2 Theoretical relation between the optimal foraging speed ($U_{f, opt}$) and food concentration (ρ) at 15 °C for three different sizes of fish, 10, 30 and 60 cm. Reproduced with permission from Ware (1978).

searching for prey at similar densities in open water, on bare sediment and in vegetation had encounter rates of 0.77, 0.01, and 0.02 prey s^{-1}, respectively (Mittelbach, 1981). In addition, bluegills hover for longer when searching in weed than they do in open water. The hover time is likely to be partly determined by genetic differences (Ehlinger and Wilson, 1988). Roach were more efficient than perch, *Perca fluviatilis*, at feeding in open water, the performance being reversed at high stem densities (Winfield, 1986). Similar findings were reported by Diehl (1988) for perch, bream, *Abramis brama*, and roach. Attack frequency and number of captured chironomid larvae decreased for all three species with increasing complexity of the artificial vegetation. Perch were the least affected. Opposite effects were obtained through a reduction of light intensity. The capture efficiency of roach and bream was uninfluenced by darkness whereas perch efficiency was significantly reduced.

Juvenile Atlantic salmon have a sit-and-wait search strategy. They find a suitable waiting place and dart out to capture prey items that drift by in the current (Metcalfe *et al.*, 1986, 1987). The path that the fish takes from shelter to prey is not always the shortest possible and can be influenced by internal factors (Metcalfe *et al.*, 1986) and the presence of a predator (Metcalfe *et al.*, 1987). As autumn advances, juvenile fish that will stop growing in winter become less willing to travel to a passing food particle, taking only those that pass very close. In the summer the same fish will often swim much further than they need to reach food. Likewise, juveniles that have recently seen a predator are less willing to move far to get food.

Choosing prey and exploiting patches

Maximizing net energy gain – general development of theory

The metaphor of animals as cost–benefit analysers was first employed to develop foraging theory by MacArthur and Pianka (1966) and Emlen (1966). Thereafter the theory was developed and diversified by numerous authors (Stephens and Krebs, 1986; Schoener, 1987) but most were concerned with two problems: diet or patch choice. The theory behind both is mathematically the same, as pointed out by Stephens and Krebs (1986). Whether exploiting a patchily distributed prey or a range of prey sizes or species, the predator employs the same sequence of 'search–encounter–decide'. The simplest theory can be derived from the **Holling disc equation** (Holling, 1959), which makes a useful link between functional responses and diet or patch choice. The Holling disc equation treats all encountered prey as identical, whilst diet or patch theory adds the complication of prey types or patches with different properties. These affect their worth as items of choice. Although the theory of patch and diet selection is mathematically the same, the behavioural decision taken by the forager is different. In the patch model the critical decision is how long the forager should stay in the encountered patch, whilst for the diet selection model the forager has to decide which prey types to include in the diet.

Both models use Holling's disc equation to calculate the average rate of energy intake. They then partition the average to take account of differences between prey types or patches. A word is necessary here on the restricted meaning of the term 'type'. It is used to mean any prey entity that can be recognized by the predator as being distinct. In this sense two size classes of the same species would be two prey types.

To start the derivation, all the time spent searching is labelled as T_s and all the time spent handling is labelled as T_h. This means that all the time spent foraging (T_f) is the sum of the two, or $T_s + T_h$. If we call the energy gained from foraging E_f, then we can express the rate of net energy gained as

$$R = E_f/(T_s + T_h) \tag{8.1}$$

To turn this into a useful equation its components need to be given more detail. The encounter rate with prey per unit time can be called λ, while s can be used to represent the cost per unit time. The total search costs then become sT_s. The average energy gained per encounter is labelled by $\bar{e}$. The energy gained, E_f can now be decomposed and expressed as $\lambda T_s \bar{e}$. If $\bar{h}$ is used to label the time spent handling a prey, then $T_h = \lambda T_s \bar{h}$. As encounters are linearly related to T_s we can now write R in a new way as

$$R = (\lambda T_s \bar{e} - sT_s)/(T_s + \lambda T_s \bar{h}) \tag{8.2}$$

Note that all times are now expressed in terms of search time. T_s can be cancelled throughout to give

$$R = (\lambda \bar{e} - s)/(1 + \lambda \bar{h}) \tag{8.3}$$

which is Holling's disc equation.

In the form so far developed R is a function only of the number of prey encountered, their average energy content and average handling time. As $\bar{e}$ and $\bar{h}$ will be fixed for a particular habitat, R is really a function only of λ. In nature prey may come in different-sized packets which contain different amounts of energy and take different amounts of time to handle. Alternatively, patches may have greater or fewer prey which affect the yield the predator can expect from a given searching time in the patch. These refinements will be discussed in the next two sections together with data testing the theory.

Before moving to the next section, a word needs to be said about the nature of R. In the context of decision making, R is the currency of the foraging process and allows the animal to assess the worth of different options. In the form developed so far, R is given as if it were deterministic. In fact the energy gained (the top line of the disc equation) is a random variable, G_i, which describes the net gain from the *i*th encounter. A particular realization is g_i. In the same way the time taken to gain g_i, which can be called T_i, is also a random variable of which t_i is a realization. The disc equation (Equation 8.3) should be written as

$$E(G)/E(T) = (\lambda\bar{e} - s)/(1 + \lambda\bar{h}) \qquad (8.4)$$

where $E(g)$ is the expected value of G, or its mean in normal parlance, and $E(T)$ is the expected value of T. Does $E(G)/E(T) = E_f/T_f$? Yes, but only when n, the number of prey items in the sequence encountered, approaches infinity (Stephens and Krebs, 1986).

Throughout most of what follows we assume that foragers are maximizing $E(G)/E(T)$, the long-term average rate. In some situations foragers might be maximizing $E(G/T)$, or the expected rate per encounter (Stephens and Krebs, 1986). Other models use different decision variables. For example some models of risk-sensitive foraging propose that the animal is minimizing the risk of starvation (Stephens and Paton, 1986). The stochastic dynamic models of Mangel and Clark (1988) assume that animals are maximizing fitness to the end of some period T. These alternatives will be discussed later.

Theory of diet choice and some tests

The derivation of the disc equation can now be extended to make it possible to predict which of a set of prey types should be included in a fish's diet. This theoretical extension will be followed by a description of some experimental tests of the theory. The simplest model has been called the **basic prey model** or BPM for short.

The disc equation provides a prediction of how much energy would be gained by a forager taking a series of identical prey types. In nature prey encountered will vary in size and species, each type having more or less energy and costing more or less time to handle. Handling time is what is called an 'opportunity cost'. Whilst handling a prey type, the forager cannot search for and handle another prey type which might have yielded a higher return.

The way in which the disc equation is modified to deal with different prey types is best understood when presented as a two-prey system. Imagine a pond with one prey species divided into two discrete size groups, P_1 and P_2. A single size group of predators searches for the prey and a rule is required to determine the conditions under which each prey size will be taken, assuming that the predator is maximizing its long-term energy gain. In the pond the two sizes of prey contain different amounts of energy, which can be labelled as e_1 and e_2 (joules). These values are the benefit of catching each prey type. It will also be the case that each prey size takes different amounts of time to pursue and handle, which can be labelled h_1 and h_2 (seconds). Dividing energy gained by the cost gives the profitability of each prey type. In the pond let the relationship between the *e*s and the *h*s be such that $e_1/h_1 > e_2/h_2$. That is, prey type 1 is more profitable than is prey type 2. The prey sizes might also occur at different densities, which will be signalled to the searching predator by the encounter rates λ_1 and λ_2. When a prey size is at a high density the predator will encounter it frequently, and vice versa for low prey densities.

It is now possible to modify Equation 8.3. First of all we assume that the search costs are constant and can be left out of any further calculation. It is then possible to partition the gains and costs between the two prey types so that $\bar{e} = \lambda_1 e_1 + \lambda_2 e_2$ and $\bar{h} = \lambda_1 h_1 + \lambda_2 h_2$. Equation 8.3 then becomes

$$R = (\lambda_1 e_1 + \lambda_2 e_2)/(1 + \lambda_1 h_1 + \lambda_2 h_2) \tag{8.5}$$

This would be the net energy gain if both prey were taken. What is required is a rule allowing the forager to 'know' whether it should take only P_1, the most profitable prey type, or both prey types. The net gain from taking P_1 alone would be $R_1 = \lambda_1 e_1/(1 + \lambda_1 h_1)$. Clearly, if this is bigger than the R obtained from taking both prey types, the fish should concentrate on P_1. At what point should the fish switch from taking only the first prey type to taking both? The only variable that can change from one foraging occasion to another is λ, as e and h are assumed to be fixed for each prey type. It can then be asked at what value of λ_1 should the fish switch from taking P_1 to taking both types? An expression for the switchover point can be arrived at by writing

$$\lambda_1 e_1/(1 + \lambda_1 h_1) > (\lambda_1 e_1 + \lambda_2 e_2)/(1 + \lambda_1 h_1 + \lambda_2 h_2) \tag{8.6}$$

In words, when the net energy gain from prey type 1 alone exceeds the net gain that would accrue from taking both types, the fish should eat only prey type 1. Inequality 8.6 can be rearranged and terms cancelled to give $1/\lambda_1 > (h_2 e_1/e_2) - h_1$; $1/\lambda_1$ is the inverse of the encounter rate and is the average time the predator has to search between prey of type 1. When this search time increases so that its value exceeds the right-hand side of the last inequality, the fish should switch to taking both prey types.

The analysis for just two prey makes it plain that the forager should take either the most profitable type or both. The magnitude of the decision variable p_i, which is the probability of taking prey type i, is consequently either 0 or 1.

This simple derivation for just two prey types can be extended to many prey types, each with their own values of pe, h and λ. A full derivation of the multiprey case is given by Stephens and Krebs (1986).

Three predictions about prey choice flow from the theory.

1. **The zero–one rule**. Prey types should be in the diet or not, with no intermediate states or partial preferences, i.e. $p_i = 1$ or 0.
2. **Ranking by prey profitability**. The predator should rank prey types by their profitability and then add them to the diet until the decision criterion is fulfilled.
3. **Taking a prey type is independent of its encounter rate**. As was shown by the decision inequality, it was the encounter rate with prey type

1 that determined whether prey type 2 should be included in the diet. Even if extremely abundant, a less profitable prey type should not be included in the diet unless the encounter rate with more profitable types falls below the critical level.

Like all mathematical models, the predictions depend heavily on the assumptions (Hart, 1989). For the BPM these are as follows.

1. Searching for and handling prey are mutually exclusive.
2. Encounter with prey is sequential and the intervals between encounters are described by a Poisson distribution.
3. The es, hs and λs for each foraging occasion are fixed and do not change with p_i.
4. Encounter without attack incurs no costs.
5. The predator has complete information about all the parameters, recognizes prey types instantaneously and does not improve its performance through learning.

This model of diet selection has been tested on a wide range of animals, from ants and crabs to sunfish and howler monkeys (Stephens and Krebs, 1986). Many of the tests show qualitative, if not quantitative agreement with predictions, although there is one deviation which is common. In many instances the zero–one rule is not fulfilled, so that foragers eat unprofitable prey even when conditions predict that they should not. This disagreement is consistent with the idea that as animals cannot have complete knowledge of their environment they must continue to sample non-profitable elements, so updating what they know (see also p. 273).

A detailed study of diet selection and its consequences by the sunfishes, *Lepomis* spp., has been made by Werner and his co-workers (Werner, 1974, 1977, 1984; Werner and Hall, 1974; Werner and Mittelbach, 1981; Werner *et al.*, 1981, 1983a,b). The times taken to chew and swallow natural and artificial prey of known size (handling time) by the bluegill sunfish and green sunfish, *L. cyanellus*, were determined by Werner (1974). Energetic costs of handling were assumed to be proportional to handling time. As prey size increased, handling time increased, slowly at first and then exponentially until an upper limit determined by morphology was reached. Werner showed that larger fish have greater latitude in the sizes of prey they can take profitably.

A multiprey version of the BPM (Equation 8.5) was used by Werner and Hall (1974) to make predictions about the diet composition of bluegills which were then tested by experiment. Bluegills were allowed to feed for a very short time on *Daphnia magna*, 3.6, 2.5, and 1.4 mm in length at six different densities. The densities were adjusted to take account of the different availabilities of each size, small *Daphnia* being less visible than large ones. At the lowest prey density (20 of the largest prey with the abundances of the other two adjusted accordingly), all three prey sizes were taken with equal frequency. As abundance increased to a maximum of 350 per size class, the bluegills concentrated more and more on the large prey (Fig. 8.3). The main effect of increasing prey density was to reduce the search time and thereby increase the encounter rate. Using the model, Werner and Hall (1974) calculated the search time after which the fish should switch from one of

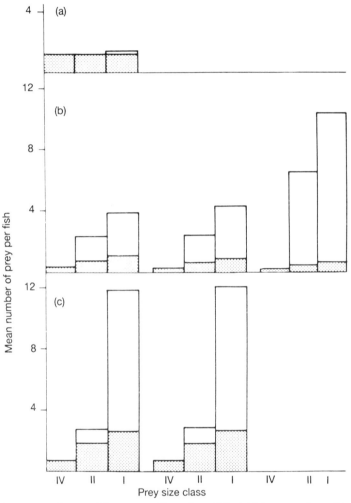

Fig. 8.3 A test of optimal foraging in bluegill sunfish. (a) Mean number of *Daphnia* of three different size classes eaten per fish at 20 prey per size class (ɪ, largest; ɪᴠ, smallest). (b) Number of *Daphnia* of different sizes eaten per fish at (from left to right), densities of 50, 75 and 200 prey per size class. (c) Numbers eaten with prey at 300 and 350 per size class. In each histogram the stippled area represents the expected numbers in the stomachs if items were eaten as encountered (taking into account differential visibility). Reproduced with permission from Werner and Hall (1974).

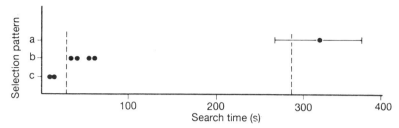

Fig. 8.4 The way in which bluegill sunfish switch their feeding habits as prey density changes: predictions and tests. The selection patterns for different prey sizes, as shown in Figs. 8.3(a), (b) and (c), are plotted against search time in seconds. A mean ($\pm$ SE) is plotted for the experiments at the lowest density, otherwise individual experiments are plotted. The dashed lines are the points at which the selection pattern should change as determined theoretically. If $T_s < 29$ s, only the largest prey (size 1) should be taken, and if $T_s > 29$ s, the two largest prey types should be taken as encountered.

the feeding patterns in Fig. 8.3 to another. The result is reproduced in Fig. 8.4, showing that the theory predicts well what the fish actually do.

Having tested diet theory with bluegills, Werner went on to use the theory to help understand habitat use and competition, both intra- and interspecific (Werner, 1984). The success of his enterprise throws up an interesting problem with the BPM. When testing the theory, Werner and Hall (1974) presented prey in swarms so that the assumption of sequential encounter was not satisfied. In addition, the fish were only allowed to feed for a very short time, only taking about four prey at the lowest densities; this scheme prevented hunger or learning from influencing the results. Most of the tests of the BPM using fish as subjects, and discussed by Stephens and Krebs (1986) in their survey of tests of the theory, violated some of the assumptions.

A test of the BPM, fulfilling all but the fifth assumption, was carried out by Hart and Ison (1991). Threespine sticklebacks were offered two sequences of seven, millimetre size groups of the Isopod crustacean *Asellus*, ranging between 3 and 9 mm. In the first sequence the optimal diet, as calculated from the BPM, consisted of 4, 5 and 6 mm *Asellus* only, whilst the optimal diet from the second sequence included 7 mm prey as well. The results (Fig. 8.5) failed to agree with the theoretical prediction as the fish never took a significant proportion of large prey and always took 3 mm *Asellus*. Prey under 6 mm were taken with almost 100% certainty. The 7 mm prey were taken more often under the second sequence than the first; 8 and 9 mm prey were nearly always rejected. A finer analysis showed that hungry fish were more ready to take 7 and 8 mm prey than were those that had already consumed a number of prey. The detail of this analysis has been examined further by Hart and Gill (1992).

The sequence of events modelled by the BPM can be visualized as the flow

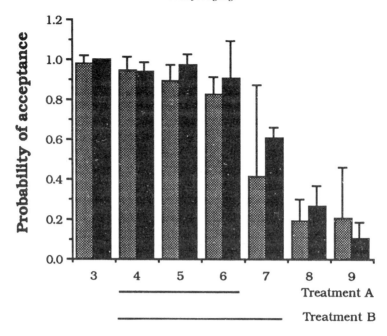

Prey size group (mm)

Fig. 8.5 The probability of acceptance of seven size groups of *Asellus* when offered to threespine sticklebacks for treatments A and B. The vertical bars at the top of each column represent one standard deviation. Stippled columns denote results for treatment A and black for B. Under treatment A, the encounter rate with the 5, 6 and 7 mm *Asellus* was higher than for treatment B. The bars under the horizontal scale show the prey sizes that should have been included in the optimal diet under the two treatments. Reproduced with permission from Hart and Ison (1991).

diagram shown in Fig. 8.6(a). The forager only makes one decision, which is whether or not the perceived prey is in the optimal diet. The experiment with sticklebacks (Hart and Ison, 1991) shows that the flow diagram in Fig. 8.6(b) would be a more realistic representation. The fish must now make three decisions after the prey has been perceived: is the gut empty, is the prey less than 7 mm, and if so is there room to fit the current prey into the gut? In this way the fish's response to prey is a dynamic process through which the fish relates the decisions it makes to its internal needs and the properties of the prey encountered.

Other evidence, also from sticklebacks, shows that static optimization models ignore the importance of the dynamic nature of decision making (Hart, 1989; Hart and Gill, 1993). One example comes from the study of Heller and Milinski (1979), who showed that a hungry stickleback preferred to feed on the densest region of a swarm of *Daphnia*. As they filled up and

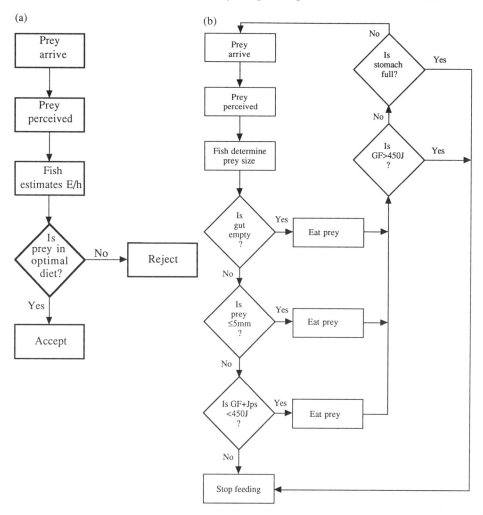

Fig. 8.6 (a) The sequence of events and decisions faced by a foraging fish as encapsulated by the basic prey model (BPM). The rectangles represent events or processes; the diamond represents a decision. (b) A more detailed representation of events and decisions derived from experiments by Hart and Gill (1992) on threespine sticklebacks. Symbolism as in (a). GF = gut fullness, J_{ps} = joules contained in the prey size on offer, J = joules. Reproduced with permission from Hart and Ison (1991) and Hart and Gill (1992).

hunger dropped off, the fish transferred their attention to peripheral areas of the swarm. This change of preference was interpreted as reflecting the changing balance of net gain. A very hungry fish judges it most important to avoid starvation and feeds at the highest rate it can. High rates are best achieved in the densest parts of the swarm. As hunger wanes, starvation becomes less of a threat and risk of predation is more critical. When feeding on the densest region of the prey swarm, the fish is not able to watch as effectively for predators (Milinski, 1984b; Godin and Sproul, 1988). As the fish fills with food, it transfers to the less dense margins of the swarm, where it can be more effective in watching for predators.

The dynamic nature of decision making can be modelled using dynamic optimization (Stephens and Krebs, 1986). Very often the number of factors that should be included make analytical methods unusable. Recently (Houston *et al.*, 1988; Mangel and Clark, 1988) stochastic dynamic programming methods have been developed which model behaviour through discrete time periods. The decision criterion, or currency, used to decide between options is survival to the end of the chosen time horizon followed by a state-dependent fitness gain. The methods make it possible to incorporate the effects of changing internal state on decisions and are likely to become more widespread in the next few years.

Exploiting patchy resources

The Holling disc equation can also be modified to account for how fish (and other foragers) might exploit patchily distributed resources. The simplest theory produces rules that apply only to solitary foragers. Most studies of fish foraging on patchy resources have been done with species that most often feed in the company of conspecifics. As a result the **basic patch model** (Stephens and Krebs, 1986) does not apply. When intraspecific interactions are present it is more useful to use game theory (see later and Maynard Smith, 1982). In this section I will briefly outline the patch model and two tests of it, then describe a special case and a test of it, after which I will discuss some work where game theory is used to understand fish interactions.

In the basic prey model the forager encounters different types of prey and has to decide whether they should be included in the diet. At a higher level in the hierarchy of behaviours employed during foraging, the forager will encounter clumps of prey. The basic patch theory predicts the time to be spent in a patch once encountered. It is assumed that an encountered patch will always be entered. Before modifying the disc equation, the principal new element to introduce is a gain function which relates cumulative gain to residence time. It is a reasonable assumption to suppose that at first the cumulative gain will increase fast, but after a while the items in the patch will become depleted so that the rate of gain falls off and the cumulative gain reaches an asymptote (Fig. 8.7).

The disc equation can now be modified to describe the rate of gain expected when a forager is exploiting an environment where prey are found in i patch types, each with its own food abundance, and characterized by a particular gain function $g_i(t_i)$, where t_i is the time spent foraging in patch i. As before

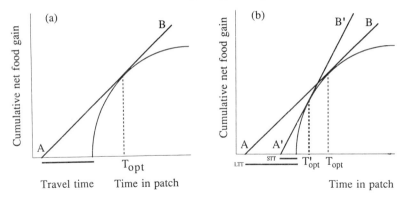

Fig. 8.7 (a) A graphical interpretation of the Marginal Value Theorem, true only when all patches have the same gain function (see text for details). The curve shows the cumulative net gain for different times spent in the patch. The horizontal axis also shows the travel time between patches. The line AB gives the rate of energy gain per unit travel time + time in the patch. Where AB grazes the cumulative gain curve the maximum achievable rate is given. A perpendicular dropped from this point to the horizontal axis gives the optimal time (T_{opt}) to spend in the patch. (b) The way in which a change in travel time alters the optimal time to spend in the patch. LTT, long travel time; STT, short travel time; T'_{opt}, optimal time in the patch when travel time is short; T_{opt}, the optimal time in the patch when travel time is long. See text for further details of the model.

λ_i is called the encounter rate, but now as patches per unit time. In this development t_i is the decision variable (cf. p_i, the probability of taking prey size i, in the prey model). The Holling disc equation now becomes

$$R = \left[\sum_{i=1}^{i=j} \lambda_i g_i(t_i) - s \right] \bigg/ \left(1 + \sum_{i=1}^{i=j} \lambda_i t_i \right). \tag{8.7}$$

As with the prey model, Equation 8.7 is only valid if five assumptions are true. The assumptions are identical in form to those for the patch model with some changes. For example assumption 3 (p. 263) requires that encounter rates when searching are independent of the residence times chosen. Assumption 4 is changed to define the gain function, $g_i(t_i)$, which must be, in a mathematical sense, well defined, continuous, deterministic and negatively accelerated (Stephens and Krebs, 1986).

In a particular habitat the animal needs to choose a set of t_i values such that R is maximized. This occurs when the forager stays in a patch until the rate of gain on leaving, $[g_i'(t_i)]$, is equal to the average rate of energy intake for the habitat. The rate of gain on leaving the patch has been called the marginal rate, explaining why the theory is sometimes called the **marginal value theorem**, or MVT for short (Charnov, 1976; Stephens and Krebs, 1986), but not to be confused with MTV (Knopfler and Sting, 1985).

When all patches have the same gain function and $s = 0$, the MVT can be interpreted graphically as shown in Fig. 8.7. This follows when Equation 8.7, with $g_1(t_1) = g_2(t_2) = \cdots = g_n(t_n)$, is differentiated with respect to t_i and then set to zero. The derivative is

$$g'(t_{opt}) = \lambda g(t_{opt})/(1 + \lambda t_{opt}) \qquad (8.8)$$

where t_{opt} is the residence time yielding the maximum R. As the patches all have the same gain function, only one optimum residence time is required.

Bluegill sunfish foraging for patches of invertebrate prey in weed patches in aquaria used a rule of constant residence time to decide when to leave a patch (Marschall *et al.*, 1989). Although this rule derived from the MVT was closest to the observed behaviour, the fish's behaviour was better accounted for by a model of random residence times. DeVries *et al.* (1989) used the MVT to predict the patch leaving time and found that bluegills appeared to be using a rule based on the rate of food gain at the point of leaving. It was also found that fish stayed longer in a patch and took more prey than theory predicted. Staying longer than predicted has also been found for starlings, *Sturnus vulgaris*, by Kacelnik (1984), for humans by Hart and Jackson (1986) and is predicted by McNamara's (1982) study of optimal patch use in a stochastic environment. In a stochastic environment, uncertainty about reward rates may make it worth staying longer in a patch than the MVT predicts to gain more information about potential future gains.

A discrete version of the MVT (Stephens *et al.*, 1986) has been tested with pike (Hart and Hamrin, 1990). The MVT and BPM require that patches or prey are encountered sequentially. For many predators, different prey or patch types are encountered simultaneously. This is likely to be the case for pike, which will encounter shoals of prey composed of several sizes and species (Pitcher, 1980). In addition, the pike will most likely have only one chance to capture prey, as its attack will cause the shoal to either break up (and the individual fish go into hiding) or move away. This means that as the pike attacks it will have to opt for one particular prey type. Which one should it take?

Hart and Hamrin (1988) found that pike with continuous access to two sizes of rudd, *Scardinius erythrophthalmus*, preferred the smaller size. The discrete version of the MVT with simultaneous encounter, predicted that as the search time for the prey school increased beyond a critical time the pike should switch to the larger rudd. The predictions were not upheld by experiment (Hart and Hamrin, 1990). Experiments with black-capped chickadees, *Parus atricapillus*, also found that the simultaneous encounter model failed to account for the bird's prey choice (Barkan and Withiam, 1989).

In a fish shoal, behaviour of an individual depends on what others in the group are doing (see Chapter 12 by Pitcher and Parrish, this volume); a single foraging fish would be constrained only by the properties of the food distribution in ways that have just been discussed. The way in which groups of threespine sticklebacks exploit patches of food was investigated by Milinski (1979) in a game theory context. Two experiments were done: in the first, prey were delivered every 2 s at one end of the tank and every 10 s at the other, producing a ratio of inputs of 5:1. In the second experiment, the time intervals were 2 s and 4 s, giving a ratio of 2:1. After 3–4 minutes' feeding the six sticklebacks introduced into the tank had divided themselves into two groups, the size of each having the same ratio to each other as did the food input rates, 2:1 and 5:1. In the second experiment the prey inputs were swapped from one end to the other so that the low-input end became the high-input end and vice versa. As a result some of the fish also changed ends so that a new distribution was established, but still in proportion to prey abundance. Matching between fish group ratios and input ratios was not as exact as before the switch.

These data are consistent with the predictions of the **ideal free distribution** (IFD) (Fretwell and Lucas, 1970). This theory predicts that animals exploiting patches will distribute themselves over patches in proportion to the profitability of the patches. Once achieved, the distribution is an evolutionarily stable strategy (ESS) (Parker, 1984). The IFD is contingent upon the following conditions: (1) patches have different profitabilities, (2) increasing competition decreases the patch profitability, (3) there is no resource guarding so predators are free to leave or enter as they wish (hence the name of the distribution), (4) all individuals choose the patch where their expected gains will be the highest, and (5) all individuals have the same ability to gather food. In the case of Milinski's results, this last assumption was shown to be untrue.

A more detailed experiment used six marked fish (Milinski, 1984a). During the course of a trial, the number of *Daphnia* caught by the six fish varied between individuals from more than 30 to fewer than 10. This difference between individuals was consistent over time. The better competitors sampled the patches at the start of the trial and then settled down to exploit the one chosen. The less skilful fish continued throughout the trial to switch from one end of the tank to the other. Despite these differences in competitive ability the overall distribution of individuals mimicked an IFD. A theoretical analysis by Sutherland and Parker (1985) showed that several distributions of good and bad competitors could produce the same sums of competitive abilities in the two patches. In only one would the densities of the competitors be in proportion to the input rates, so mimicking an ideal free distribution (Milinski, 1986).

The idea that some individuals have a lower gain from patch exploitation is not compatible with the idea of an ESS, where all individuals are supposed to gain the same. The sorts of distributions predicted by Sutherland and Parker (1985) and shown by threespine sticklebacks indicate that when competitors differ in abilities, then each will behave so as to maximize its rewards, within the constraints determined by its capacities. In this sense, the strategy that evolves is dependent on the phenotypic limitations of individuals. This type of strategy was called a phenotype-limited ESS by Sutherland and Parker (1985). Although pay-offs to individuals differ, each is doing the best it can given its particular abilities. In the next section constraints will be examined in more detail.

8.3   CONSTRAINTS ON FEEDING STRATEGIES

Behavioural strategies for foraging are constrained by morphology. The largemouth bass, *Micropterus salmoides*, the green sunfish and the bluegill, living in lakes in Michigan, USA, exploit different size ranges of prey, which are related to mouth shapes and positions and to body form. The bluegill is very good at locating and snapping up small prey requiring fine manoeuvrability for their capture, whereas largemouth bass are best at capturing faster-moving prey which require rapid pursuit (Werner, 1977). Their differences in diet are reflected in the relation between gain per unit effort and prey size. The effect body shape has on manoeuvrability throughout the teleosts is discussed in detail by Webb (1982).

Capture of fast-moving prey requires a well-integrated sequence of jaw movements, which have been closely studied in several species (Rand and Lauder, 1981, pickerel; Wainwright, 1986, pumpkinseed sunfish, *Lepomis gibbosus*; Wainwright and Lauder, 1986, sunfishes in general; Chu, 1989, surfperches; Muller, 1989, teleosts in general). When chain pickerel, *Esox niger*, attack prey they employ two types of fast start and two types of jaw movement (Rand and Lauder, 1981). The start used depended on how far the pickerel was from the prey, and the jaw movements varied with the position of the prey. Prey in corners were caught with a technique that increased the suction created by the opening mouth, so making it harder for the prey to escape. The results show how intimate is the relation between the behaviour employed and the morphology of the prey-catching equipment. They also illustrate the ways in which the behavioural repertoire, and as a consequence the prey types taken, will be limited by the structure of the fish. Morphology alone confines the diet to a subset of what is available; behaviour allows the fish to further refine its selection in response to features that will change from day to day.

8.4   LEARNING AND FORAGING BEHAVIOUR

The models that have been described assume that a foraging fish responds in a constant way to certain variables of the environment. Few foraging environments are simple enough to make it possible for a predator to survive for long by using unvarying rules. Within the domain frequented by a foraging fish, the positions of patches and the types of prey will change with the seasons, so a predator must be constantly modifying its behaviour if it is to stay alive. An example would be the way the size composition of a river-fish community changes over a year. In early summer there will be large numbers of fish fry available, which will disappear as the season advances. As a result a pike, for example, would eat different sizes and species of fish at different times of the year (Mann, 1982).

Learning makes it possible for the forager to adapt to systematic variation in its environment, and can take two forms (Krebs *et al.*, 1983): 'learning how' and 'learning about'. In both cases the evolutionary strategy has been to invest in learning capacity which will cost the developing animal both time and energy. The capacity will only evolve if the benefits to be gained outweigh the costs (Orians, 1981), and in many organisms they are too small or life is too short for the benefit of learning to be large (Staddon, 1983), a restriction that does not apply to most fish.

'Learning how' mostly changes the costs or benefits of foraging. For example Werner *et al.* (1981) found that the capture rate of bluegills feeding on *Chironomus plumosus* increased with successive experiments so that at the end of a series the rate was four times greater. Bluegills are also able to improve their habitat-specific foraging efficiency through learning, although not all individuals improve equally (Ehlinger, 1989).

In the fifteenspine stickleback, *Spinachia spinachia*, the mean frequency of attack and the mean attack efficiency (proportion of attacks leading to capture) both increased with experience (Croy and Hughes, 1990, 1991). These changes were observed when fish were catching *Gammarus*, *Artemia* or a mixture of the two. The handling time for the first prey item in a sequence decreased with experience when the sticklebacks were eating either of the two prey types. The degree of change observed was a function of the prey species offered. Hungry fish learnt faster than those that were 50% and 95% satiated although there was no interdependence between hunger and experience. Several other fish species have been shown to increase their feeding efficiency with experience (summary, Dill, 1983). The important point is that changes in components of efficiency with time could differentially alter the profitabilities of different foods (Hughes, 1979; Croy and Hughes, 1990). If experience of a prey lowers pursuit and handling costs, the net return will increase and the prey could change its rank in the diet, leading to switching.

Assessing profitabilities of food sources comes under the heading of 'learning about'. The basic prey and patch models assume that prey or patches are divided into categories or types that are recognized instantaneously. In fact prey may be divided into subtypes that are ambiguous (Stephens and Krebs, 1986). Two subtypes for a pike might be threespine and ninespine sticklebacks measuring 40 mm, and recognizing the difference could be crucial to the pike (Hoogland *et al.*, 1957). Stephens and Krebs (1986) ask how a long-term rate maximizer should treat types when it knows they are divided into ambiguous subtypes. They divide the problem into three parts: the value of information about subtypes, tracking a changing environment, and patch sampling. Because of space limitations I will discuss only tracking a changing environment and patch sampling, both of which have been studied with fish.

Over time patches may change quality and it may not be possible for a forager to track the changes without sampling. Threespine sticklebacks spend some time sampling when confronted with two sources of prey (Milinski, 1979, 1984a). How should the forager allocate time to available patches so as to keep track of changes? Where there are just two patches the problem of how to sample so as to maximize pay-off has been called the two-armed bandit problem (Krebs *et al.*, 1978) and has been applied to sticklebacks by Thomas *et al.* (1985). The term 'two-armed bandit' derives from the analogy with a slot machine, normally called a 'one-armed bandit'. In the problem tackled, there are two sources of reward rather than one.

In the experiments carried out by Thomas *et al.* (1985) threespine stickle-backs were trained to choose between one of two adjacent compartments in an aquarium. Each compartment had an associated pay-off which in the experiments was the ubiquitous *Tubifex* worm. One worm was given as a reward with a probability that depended on the schedule being used. One compartment always had a higher probability (p) of reward than did the other (q). It was argued that the fish might use three methods to choose a compartment, given that its goal was to maximize pay-off over a fixed number of trials. It could choose exclusively the side with the highest probability of a pay-off, and this would yield the maximum pay-off. Alternatively it could use a probability-matching rule in which it devoted time to compartments in proportion to the probability of getting a pay-off. Finally, the fish could choose a compartment at random. Predicted pay-offs from these three alternatives were calculated for varying values of q, p being held constant at 0.9. By the end of the 11 day experiment, all fish were choosing exclusively the compartment which gave the highest pay-off. Over the first 6 days, the fish could have been either choosing at random or using the probability-matching rule. After day 6, fish steadily increased the proportion of visits to the most rewarding compartment until they visited it all the time. Further analysis showed that the bigger the difference between p and q the less time it look the fish to learn which was the best side. It has been found in a number of species that the time taken to learn the differences in gain from two patches is inversely related to the difference between them (Kamil and Yoerg, 1982).

A second experiment examined the effects of hunger on the fish's sampling behaviour. Hungry fish mostly maximized their pay-off over the last half of the experiment whilst satiated fish behaved as if they were probability matching. This difference may relate to the need a fish has to balance competing requirements. For the hungry fish the most important task is to reduce the risk of starvation, whilst the replete fish must also explore the environment to find new food sources. Probability matching may be a compromise between the needs of optimally exploiting the environment and gaining new information from it.

When a forager exploits a patch in the company of conspecifics it will be necessary to take account of what they are doing, too (see also Chapter 12 by Pitcher and Parrish, this volume). In these circumstances it is assumed that natural selection will push individuals towards a distribution that is an ESS (Parker, 1984). When resources have variable pay-offs, for example when food patches change in quality over time, a genetically determined ESS would be too inflexible. An alternative is to have an ESS achieved through learning, as in the example discussed earlier of threespine sticklebacks exploiting two patches (Milinski, 1984a). A learning rule for several conspecifics foraging from a patchily distributed resource was devised by Harley (1981) and is called the **relative pay-off sum** (RPS) learning rule. This rule, once adopted by all the population, cannot be bettered by any other rule, so leading to an ESS.

On first encounter with the two patches it is assumed by the RPS model that one patch is chosen arbitrarily. In response to the reward received from the patch, the individual reassesses its choice, staying with the patch chosen if it proved profitable and moving to the next if it did not. At each return the fish updates its assessment and compares it with its memory of past profitabilities. The model allows for varying rates of memory decay but only applies to fish selecting prey from two separate sources.

Regelmann (1984) modified this RPS learning rule to allow fish to have unequal foraging abilities and to account for travel time between food sources. The new theory was tested by Milinski (1984a) and his results have already been discussed on p. 271. The RPS rule predicts that the fish will distribute themselves according to the patch profitabilities, that good competitors will decide earlier where to feed and will switch less between patches, that the distribution of bad competitors will be determined by patch profitabilities and by the distribution of good competitors, and that travel costs will decrease the amount of switching. The predictions accurately described the outcome of Milinski's experiment, indicating that the sticklebacks were using some form of RPS learning rule to regulate their behaviour. The fish all showed some sampling behaviour in the early part of each trial, but soon settled to a steady exploitation frequency.

The degree to which learning is involved in foraging is only now being investigated extensively and there is much work to be done. Fish are well able to learn, so there is every reason to expect significant changes in our understanding of the rules governing their behaviour as new results become available.

8.5 RISK AND FORAGING BEHAVIOUR

I discussed earlier (p. 261) that $E(G)/E(T) = E_f/T_f$ in the BPM only when the sequence of prey encountered is very long. In a mathematical context the equality is only true when n, the sequence of prey, approaches infinity. A foraging animal is never in a position to fulfil this requirement, so the shorter runs of prey it encounters could well have considerable variability in E (e.g. Hart and Ison, 1991). The BPM and other models derived from it assume that variability in reward is not important to the animal. For an individual close to starvation, variability in E could be critical. For many diurnal birds it is necessary for sufficient food to be gathered through the day to last the animal overnight. If it is an hour before dusk and the bird has some way to go before it has accumulated sufficient energy for the night, then the variability of the food resources being collected is critical. For reviews of foraging under conditions where variability matters, often known as risk-sensitive foraging, see Stephens and Krebs (1986) and Real and Caraco (1986).

Modellers of risk-sensitive foragers have proposed two features that the animal might be using as a decision criterion: either they could be trying to maximize a linear combination of mean and variance or they could be trying to minimize the risk of starvation (Stephens and Paton, 1986). The first has been called 'variance discounting' by Real (1980) and the second 'shortfall minimizing' by various authors. In variance discounting, the forager is maximizing $\mu\text{-}k\sigma^2$, where μ is the mean food reward, σ^2 is its variance and k is a constant which expresses the degree to which the forager discounts the variance of reward. Using the rufous hummingbird, *Selaphoruous rufus*, Stephens and Paton (1986) found that the shortfall minimization model best accounted for observed behaviour.

Most of the work on risk-sensitive foraging has been done with birds or mammals, both with high metabolic rates and therefore more vulnerable to starvation. Despite their lower metabolic rates, fish can be sensitive to starvation: Ivlev (1961) found that bleak, *Alburnus alburnus*, die after a few days without food whereas pike can survive months of deprivation. These differences relate to the type of diet eaten by the two species, with bleak eating small surface insects and planktonic organisms in an analogous manner to small passerines.

For shortfall minimizers, theory predicts a switch from risk aversion (avoiding a variable food source) when reserves are above those required to risk proneness (preferring a more variable food source) when reserves are lower than required. An experimental study of risk-sensitive foraging in the bitterling, *Rhodeus sericus*, showed that a group of seven fish were all risk averse when their reserves were higher than required. Only one out of a group of six fish switched to being risk prone when energy reserves fell below those required (Young *et al.*, 1990). Increasing the deficit in fish running below energy requirements forced all individuals to become risk prone. This is another illustration of the dynamic properties of fish foraging behaviour which are probably best modelled by the stochastic dynamic programming methods of Mangel and Clark (1988).

8.6   SUMMARY AND CONCLUDING REMARKS

I have chosen to use foraging theory as a skeleton on which to hang the flesh of fact. As Einstein remarked to Heisenberg (reported in Heisenberg, 1971) "It is the theory which decides what we can observe." Likewise Carr (1987), writing about history, claims that "...not all facts about the past are historical facts..." the meaning of which is made clear when he writes further that "It is the historian who decided for his own reasons that Caesar's crossing of that pretty stream, the Rubicon, is a fact of history, whereas the crossing of the Rubicon by millions of other people before or since interests nobody at all." So, the facts that I have chosen to review are largely determined by the theory I discuss, but this is not a bad thing so long as one is aware of the temporary nature of much theory and of its limitations.

All the theory used in this chapter has three common elements (Stephens and Krebs, 1986): (1) the foraging problem to be analysed (e.g. which prey type to take), (2) the decision criterion to be used to choose between options (e.g. shortfall minimization), and (3) the assumptions which constrain the forager's choices (e.g. prey are encountered sequentially). A taxonomy of the theories discussed is shown in Fig. 8.8. It may well be that static optimization models, the line that has been most fecund over the past twenty years, are soon to become extinct, mainly because of the increasing need to address the dynamic nature of foraging.

A recent review of foraging behaviour in the threespine stickleback (Hart and Gill, 1993), and material covered in this chapter illustrate this dynamic quality of foraging behaviour (Fig. 8.9). The static models of foraging behaviour, starting with MacArthur and Pianka (1966) and Emlen (1966), have concentrated on the regular features of the feeding cycle shown in Fig. 8.9. MacArthur was a pioneer in developing ecological models that concentrated on the repeatable features of ecological processes and down-played the significance of history (MacArthur, 1972). It is now possible to see that for a complete understanding of foraging, the variability which accompanies each turn of the feeding cycle must be accounted for. For each

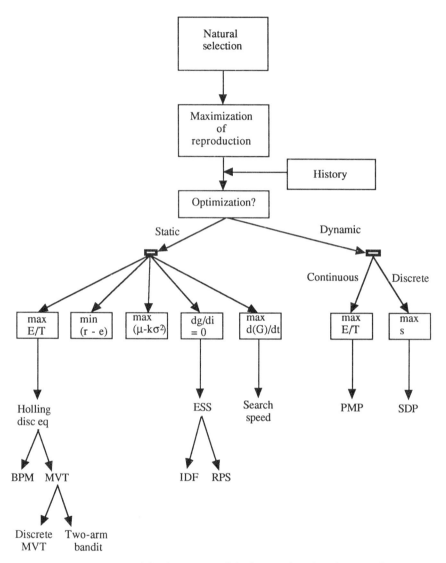

Fig. 8.8 A taxonomy of the foraging models discussed in this chapter. The top two rectangles show the assumptions underlying optimization models. The rectangle marked 'History' is a reminder of the role of the unique sequence of events each organism experiences. Models are divided into static, which exclude time, and dynamic. The seven small rectangles define currencies used in various models. Abbreviations: max E/T, maximization of long-term net rate of energy intake; min $(r - e)$, minimization of the difference between energy requirements (r) and energy intake (e); $\max(\mu - k\sigma^2)$, maximization of the mean reward devalued by a proportion of its variance; dg/di = 0, rate of gain to each individual in the group is the same; max d(G)/dt, maximization of growth rate; max s, maximization of survival; ESS, evolutionarily stable strategy; PMP, Pontryagin's maximum principle; SDP, stochastic dynamic programming; BPM, basic prey model; MVT, marginal value theorem; IDF, ideal free theory; RPS, relative pay-off sum. (© 1992 Paul J. B. Hart.)

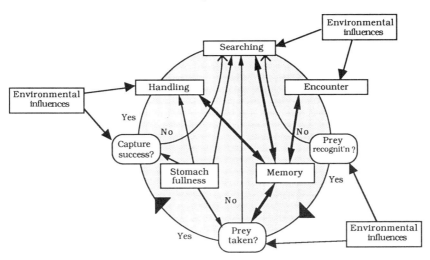

Fig. 8.9 The cycle of feeding events and decisions for a foraging fish together with the internal and external factors influencing it. Each turn of the cycle will be made unique by the particular combinations of the environmental influences in combination with the fish's internal state. (© 1992 Paul J. B. Hart.)

turn of the cycle in Fig. 8.9, the factors in the boxes can take different values. In addition the internal state of the fish will alter, either through changes in hunger or through learning. Static optimization models cannot handle this and we now need to develop new models that can.

ACKNOWLEDGEMENTS

I thank Michael Blake and Andrew Gill for comments on an earlier version of the manuscript.

REFERENCES

Barkan, C. P. L. and Withiam, M. L. (1989) Profitability, rate maximization, and reward delay: a test of the simultaneous-encounter model of prey choice with *Parus atricapillus. Am. Nat.,* **134**, 254–72.

Bentzen, P. and McPhail, J. D. (1984) Ecology and evolution of sympatric sticklebacks (*Gasterosteus*): specialization for alternative trophic niches in the Enos Lake species pair. *Can. J. Zool.,* **62**, 2280–86.

Beukema, J. J. (1964) A study of the time pattern of food intake in the three-spined stickleback (*Gasterosteus aculeatus*) by means of a semi-automatic recording apparatus. *Archs néerl. Zool.,* **16**, 167–8.

Beukema, J. J. (1968) Predation by the threespine stickleback (*Gasterosteus aculeatus* L.). *Behaviour*, **31**, 1–126.

Bone, Q. and Marshall, N. B. (1982) *Biology of Fishes*, Blackie, Glasgow, 253 pp.

Brett, J. R. (1979) Environmental factors in growth, in *Fish Physiology*, Vol. VIII: *Bioenergetics and Growth* (eds W. S. Hoar, D. J. Randall and J. R. Brett), Academic Press, New York, pp. 599–675.

Brett, J. R. and Groves, T. D. D. (1979) Physiological energetics, in Fish Physiology, Vol. VIII: *Bioenergetics and Growth* (eds W. S. Hoar, D. J. Randall and J. R. Brett), Academic Press, New York, pp. 279–352.

Carr, E. H. (1987) *What is History?* 2nd edn, Penguin, London, 188 pp.

Charnov, E. L. (1976) Optimal foraging theory: the Marginal Value Theorem. *Theoret. Popul. Biol.*, **9**, 129–36.

Chu, C. T. (1989) Functional design and prey capture dynamics in an ecologically generalized surfperch (Embiotocidae). *J. Zool., Lond.*, **217**, 417–40.

Clark, I. S. (1986) The interaction of factors influencing the predatory behaviour of the pike *Esox lucius* L. Unpublished PhD thesis, University of Leicester, 176 pp.

Clutton Brock, T. H. (ed.) (1988) *Reproductive Success. Studies of Individual Variation in Contrasting Breeding Systems*, University of Chicago Press, Chicago, 538 pp.

Colgan, P. W. (1989) *Animal Motivation*, Chapman and Hall, London, 168 pp.

Croy, M. I. and Hughes, R. N. (1990) The combined effects of learning and hunger in the feeding behaviour of the fifteen-spined stickleback (*Spinachia spinachia* L.), in *Behavioural Mechanisms of Food Selection* (ed. R. N. Hughes) (NATO ASI Series, Vol. G20), Springer, Heidelberg, pp. 215–33.

Croy, M. I. and Hughes, R. N. (1991) The role of learning and memory in the feeding behaviour of the fifteen-spined stickleback (*Spinachia spinachia* L.). *Anim. Behav.*, **41**, 161–170.

DeVries, D. R., Stein, R. A. and Chesson, P. L. (1989) Sunfish foraging among patches: the patch-departure decision. *Anim. Behav.*, **37**, 455–64.

Diehl, S. (1988) Foraging efficiency of three freshwater fishes: effects of structural complexity and light. *Oikos*, **53**, 207–14.

Dill, L. M. (1983) Adaptive flexibility in the foraging behaviour of fishes. *Can. J. Fish. Aquat. Sci.*, **40**, 398–408.

Ehlinger, J. T. (1989) Learning and individual variation in bluegill foraging: habitat-specific techniques. *Anim. Behav.*, **38**, 643–58.

Ehlinger, J. T. and Wilson, D. S. (1988) A complex foraging polymorphism in bluegill sunfish. *Proc. natn. Acad. Sci. U.S.A.*, **85**, 1878–82.

Emlen, J. M. (1966) The role of time and energy in food preference. *Am. Nat.*, **100**, 611–17.

Evans, B. I. and O'Brien, W. J. (1988) A re-analysis of the search cycle of a planktivorous salmonid. *Can. J. Fish. Aquat. Sci.*, **45**, 187–92.

Fretwell, S. D. and Lucas, H. L. (1970) On territorial behaviour and other factors influencing habitat distribution in birds. *Acta Biotheor.*, **19**, 16–36.

Godin, J.-G. J. and Sproul, C. D. (1988) Risk taking in parasitized sticklebacks under threat of predation: effects of energetic need and food availability. *Can. J. Zool.*, **66**, 2360–67.

Gould, S. J. and Lewontin, R. C. (1979) The spandrels of San Marco and the Panglossian paradigm: a critique of the adaptationist programme. *Proc. R. Soc.*, **205B**, 581–98.

Gray, R. (1987) Faith and foraging, in *Foraging Behaviour* (eds A. C. Kamil, J. R. Krebs and H. R. Pullium), Plenum Press, New York, pp. 69–140.

Harley, C. B. (1981) Learning the evolutionarily stable strategy. *J. theor. Biol.*, **89**, 611–33.

Hart, P. J. B. (1989) Predicting resource utilization: the utility of optimal foraging models. *J. Fish Biol.*, **35** (Supp. A), 271–7.

Hart, P. J. B. and Connellan, B. (1984) The cost of prey capture, growth rate and ration size in pike (*Esox lucius*) as functions of prey weight. *J. Fish Biol.*, **25**, 279–91.

Hart, P. J. B. and Gill, A. B. (1992) Constraints on prey size selection by the three-spined stickleback: energy requirements and the capacity and fullness of the gut. *J. Fish Biol.*, **40**, 205–218.

Hart, P. J. B. and Gill, A. B. (1993) Evolution of foraging behaviour in threespine stickleback, in *The Evolution of the Threespine Stickleback* (eds M. A. Bell and S. A. Foster), Oxford University Press, Oxford.

Hart, P. J. B. and Hamrin, S. F. (1988) Pike as a selective predator. Effects of prey size, availability, cover and jaw dimensions. *Oikos*, **51**, 220–26.

Hart, P. J. B. and Hamrin, S. F. (1990) The role of behaviour and morphology in the selection of prey by pike, in *Behavioural Mechanisms of Food Selection* (ed. R. N. Hughes) (NATO ASI Series, Volume G20) Springer, Heidelberg, pp. 235–54.

Hart, P. J. B. and Ison, S. (1991) The influence of prey size and abundance, and predator phenotype, on prey choice by threespine stickleback. *J. Fish Biol.*, **38**, 359–372.

Hart, P. J. B. and Jackson, P. H. (1986) The influence of sex, patch quality, and travel time on foraging decisions by young adult *Homo sapiens* L. *Ethol. Sociobiol.*, **7**, 71–89.

Heisenberg, W. (1971) *Physics and Beyond*. Harper, New York. xviii. 247 pp.

Heller, R. and Milinski, M. (1979) Optimal foraging of sticklebacks on swarming prey. *Anim. Behav.*, **27**, 1127–41.

Holling, C. S. (1959) The components of predation as revealed by a study of small-mammal predation on the European sawfly. *Can. Ent.*, **91**, 293–320.

Holmgren, S., Grove, D. J. and Fletcher, D. J. (1983) Digestion and the control of gastro-intestinal motility, in *Control Processes in Fish Physiology* (eds J. C. Rankin, T. J. Pitcher and R. Duggan), Croom Helm, London, pp. 23–40.

Hoogland, R. D., Morris, D. and Tinbergen, N. (1957) The spines of sticklebacks (*Gasterosteus* and *Pungitius*) as a means of defence against predators (*Perca* and *Esox*). *Behaviour*, **10**, 205–37.

Houston, A., Clark, C., McNamara, J. and Mangel, M. (1988) Dynamic models in behavioural and evolutionary ecology. *Nature, Lond.*, **332**, 29–34.

Hughes, R. N. (1979) Optimal diets under the energy maximisation premise: the effects of recognition time and learning. *Am. Nat.*, **113**, 209–21.

Ibrahim, A. A. and Huntingford, F. A. (1989) Laboratory and field studies of the effect of predation risk on foraging in three-spined sticklebacks (*Gasterosteus aculeatus*). *Behaviour*, **109**, 46–57.

Ivlev, V. S. (1961) *Experimental Ecology of the Feeding of Fishes*, Yale University Press, New Haven, CT, 302 pp.

Jansen, J. (1982) Comparison of searching behaviour for zooplankton in an obligate planktivore, blueback herring (*Alosa aestivalis*) and a facultative planktivore, bluegill (*Lepomis macrochirus*). *Can. J. Fish. Aquat. Sci.*, **39**, 1649–54.

Kacelnik, A. (1984) Central place foraging in starlings (*Sturnus vulgaris*) I. Patch residence time. *J. Anim. Ecol.*, **53**, 283–99.

Kamil, A. C. and Yoerg, S. J. (1982) Learning and foraging behaviour, in *Perspectives in Ethology*, Vol. 5 (eds P. P. G. Bateson and P. H. Klopfer), Plenum Press, New York, pp. 325–46.

Kitcher, P. (1985) *Vaulting Ambition. Sociobiology and the Quest for Human Nature*, MIT Press, Cambridge, MA, 456 pp.

Knopfler, M. and Sting (1985) Money for nothing, in *Brothers in Arms* (eds Dire Straits), Phonogram Ltd, London.

Krebs, J. R., Kacelnik, A. and Taylor, P. (1978) Test of optimal sampling by foraging great tits. *Nature, Lond.*, **275**, 27–31.

Krebs, J. R., Stephens, D. W. and Sutherland, W. J. (1983) Perspectives in optimal foraging theory, in *Perspectives in Ornithology* (eds G. H. Clark and A. H. Bush), Cambridge University Press, New York, pp. 165–221.

Lavin, P. A. and McPhail, J. D. (1986) Adaptive divergence of trophic phenotype among freshwater populations of threespine stickleback (*Gasterosteus aculeatus*). *Can. J. Fish. Aquat. Sci.*, **43**, 2455–63.

MacArthur, R. H. (1972) *Geographical Ecology. Patterns in the Distribution of Species*, Harper and Row, New York, 269 pp.

MacArthur, R. H. and Pianka, E. R. (1966) On optimal use of a patchy environment. *Am. Nat.*, **100**, 603–9.

McNamara, J. M. (1982) Optimal patch use in a stochastic environment. *Theoret. Popul. Biol.*, **21**, 269–88.

MacPhail, E. M. (1982) *Brain and Intelligence in Vertebrates*, Oxford University Press, Oxford, 423 pp.

McPhail, J. D. (1993) Speciation and the evolution of reproductive isolation in the sticklebacks (*Gasterosteus*), in *The Evolution of the Threespine Stickleback* (eds M. A. Bell and S. A. Foster), Oxford University Press, Oxford.

Mangel, M. and Clark, C. W. (1988) *Dynamic Modeling in Behavioral Ecology*, Princeton University Press, Princeton, NJ, 308 pp.

Mann, R. H. K. (1982) The annual food consumption and prey preference of pike (*Esox lucius*) in the River Frome, Dorset. *J. Anim. Ecol.*, **51**, 81–95.

Marschall, E. A., Chesson, P. L. and Stein, R. A. (1989) Foraging in a patchy environment: prey-encounter rate and residence time distributions. *Anim. Behav.*, **37**, 444–54.

Maynard Smith, J. (1982) *Evolution and the Theory of Games*, Cambridge University Press, Cambridge, 224 pp.

Metcalfe, N. B., Huntingford, F. A. and Thorpe, J. E. (1986) Seasonal changes in feeding motivation of juvenile Atlantic salmon (*Salmo salar*). *Can. J. Zool.*, **64**, 2439–46.

Metcalfe, N. B., Huntingford, F. A. and Thorpe, J. E. (1987) The influence of predation risk on the feeding motivation and foraging strategy of juvenile Atlantic salmon. *Anim. Behav.*, **35**, 901–11.

Milinski, M. (1979) An evolutionarily stable feeding strategy in sticklebacks. *Z. Tierpsychol.*, **51**, 36–40.

Milinski, M. (1984a) Competitive resource sharing: an experimental test of a learning rule for ESSs. *Anim. Behav.*, **32**, 233–42.

Milinski, M. (1984b) A predator's cost of overcoming the confusion-effect of swarming prey. *Anim. Behav.*, **32**, 1157–62.

Milinski, M. (1986) A review of competitive resource sharing under constraints in sticklebacks. *J. Fish Biol.*, **29** (Supp. A), 1–14.

Mitchell, W. A. and Valone, T. J. (1990) The optimization research program: studying adaptations by their function. *Q. Rev. Biol.*, **65**, 43–52.

Mittelbach, G. (1981) Foraging efficiency and body size: a study of optimal diet and habitat use by bluegills. *Ecology*, **62**, 1370–86.

Muller, M. (1989) A quantitative theory of expected volume changes of the mouth during feeding in teleost fishes. *J. Zool., Lond.*, **217**, 639–62.

Newton, I. (ed.) (1989) *Lifetime Reproduction in Birds*, Academic Press, London, 479 pp.

Noakes, D. L. G. (1986) When to feed: decision making in sticklebacks, *Gasterosteus aculeatus*. *Env. Biol. Fishes*, **16**, 95–104.

O'Brien, W. J., Browman, H. I. and Evans, B. I. (1990) Search strategies of foraging animals. *Am. Scient.*, **78**, 152–60.

O'Brien, W. J., Evans, B. I. and Browman, H. I. (1989) Flexible search tactics and efficient foraging in saltatory searching animals. *Oecologia*, **80**, 100–110.

Ollason, J. G. (1987) Artificial design in natural history: why it is so easy to understand animal behaviour, in *Alternatives, Perspectives in Ethology*, Vol. 5 (eds P. P. G. Bateson and P. H. Klopfer), Plenum Press, New York, pp. 233–57.

Orians, G. H. (1981) Foraging behaviour and the evolution of discriminatory abilities, in *Foraging Behaviour, Ecological, Ethological and Psychological Approaches* (eds A. C. Kamil and T. D. Sargent), Garland STPM Press, New York, pp. 389–405.

Parker, G. A. (1984) Evolutionarily stable strategies, in *Behavioural Ecology. An Evolutionary Approach*, 2nd edn (eds J. R. Krebs and N. B. Davies), Blackwell, Oxford, pp. 30–61.

Pitcher, T. J. (1980) Some ecological consequences of fish school volumes. *Freshwat. Biol.*, **10**, 539–44.

Pyke, G. H. (1978) Are animals efficient foragers? *Anim. Behav.*, **26**, 241–50.

Pyke, G. H. (1981) Optimal travel speeds of animals. *Am. Nat.*, **118**, 475–87.

Rand, D. M. and Lauder, G. V. (1981) Prey capture in the chain pickerel, *Esox niger*. Correlations between feeding and locomotion behaviour. *Can. J. Zool.*, **59**, 1072–8.

Real, L. A. (1980) Fitness and uncertainty, and the role of diversification in evolution and behaviour. *Am. Nat.*, **115**, 623–38.

Real, L. A. and Caraco, T. (1986) Risk sensitive foraging in stochastic environments. *A. Rev. Ecol. Syst.*, **17**, 371–90.

Regelemann, K. (1984) Competitive resource sharing: a simulation model. *Anim. Behav.*, **32**, 226–32.

Schoener, T. W. (1987) A brief history of optimal foraging theory, in *Foraging Behaviour* (eds A. C. Kamil, J. R. Krebs and H. R. Pullium), Plenum Press, New York, pp. 5–67.

Sih, A. (1982) Optimal patch use: variation in selective pressure for efficient foraging. *Am. Nat.*, **120**, 666–85.

Smith, J. N. M. (1974) The food searching behaviour of two European thrushes, I. Description and analysis of search paths. *Behaviour*, **48**, 276–302.

Staddon, J. E. R. (1983) *Adaptive Behaviour and Learning*, Cambridge University Press, Cambridge, 555 pp.

Stephens, D. W. and Krebs, J. R. (1986) *Foraging Theory*, Princeton University Press, Princeton, NJ 247 pp.

Stephens, D. W. and Paton, S. R. (1986) How constant is the constant of risk-aversion? *Anim. Behav.*, **34**, 1659–67.

Stephens, D. W., Lynch, J. F., Sorenson, A. E. and Gordon, C. (1986) Preference and profitability: theory and experiment. *Am. Nat.*, **127**, 533–53.

Sutherland, W. J. and Parker, G. A. (1985) Distribution of unequal competitors, in *Behavioural Ecology. Ecological Consequences of Adaptive Behaviour* (eds R. M. Sibly and R. H. Smith), Blackwell, Oxford, pp. 255–73.

Thomas, G. (1974) The influence of encountering a food object on the subsequent searching behaviour in *Gasterosteus aculeatus* L. *Anim. Behav.*, **22**, 941–52.

Thomas, G., Kacelnik, A. and van der Meulen, J. (1985) The three-spined stickleback and the two-armed bandit. *Behaviour*, **93**, 227–40.

Tugendhat, B. (1960) The normal feeding behaviour of the three-spined stickleback (*Gasterosteus aculeatus* L.). *Behaviour*, **15**, 284–318.

Wainwright, P. C. (1986) Motor correlates of learning behaviour: feeding on novel prey by pumpkinseed sunfish (*Lepomis gibbosus*). *J. exp. Biol.*, **126**, 237–47.

Wainwright, P. C. and Lauder, G. V. (1986) Feeding biology of sunfishes: patterns of variation in the feeding mechanism. *Zool. J. Linn. Soc.*, **88**, 217–28.

Ware, D. M. (1975) Growth, metabolism and optimal swimming speed of a pelagic fish. *J. Fish. Res. Bd Can.*, **32**, 33–41.

Ware, D. M. (1978) Bioenergetics of pelagic fish: theoretical change in swimming speed and relation with body size. *J. Fish. Res. Bd Can.*, **35**, 220–28.

Ware, D. M. (1982) Power and evolutionary fitness of teleosts. *Can. J. Fish. Aquat. Sci.*, **39**, 3–13.

Webb, P. W. (1982) Locomotor patterns in the evolution of Actinopterygian fishes. *Am. Zool.*, **22**, 329–42.

Werner, E. E. (1974) The fish size, prey size, handling time relation in several sunfishes and some implications. *J. Fish. Res. Bd Can.*, **31**, 1531–6.

Werner, E. E. (1977) Species packing and niche complementarity in three sunfishes. *Am. Nat.*, **111**, 553–78.

Werner, E. E. (1984) The mechanisms of species interactions and community organization in fish, in *Ecological Communities. Conceptual Issues and the Evidence* (eds D. R. Strong, jun., D. Simberloff, L. G. Abele and A. B. Thistle), Princeton University Press, Princeton, NJ, pp. 360–82.

Werner, E. E. and Hall, D. J. (1974) Optimal foraging and the size selection of prey by the bluegill sunfish (*Lepomis macrochirus*). *Ecology*, **55**, 1042–52.

Werner, E. E. and Mittelbach, G. G. (1981) Optimal foraging: field tests of diet choice and habitat switching. *Am. Zool.*, **21**, 813–29.

Werner, E. E., Gilliam, J. F., Hall, D. J. and Mittelbach, G. G. (1983a) An experimental test of the effects of predation risk on habitat use in fish. *Ecology*, **64**, 1540–48.

Werner, E. E., Mittelbach, G. G. and Hall, D. J. (1981) The role of foraging profitability and experience in habitat use by the bluegill sunfish. *Ecology*, **62**, 116–25.

Werner, E. E., Mittelbach, G. G., Hall, D. J. and Gilliam, J. F. (1983b) Experimental tests of optimal habitat use in fish: the role of relative habitat profitability. *Ecology*, **64**, 1523–39.

Winfield, I. J. (1986) The influence of simulated aquatic macrophytes on the zooplankton consumption rate of juvenile roach, *Rutilus rutilus*, rudd, *Scardinius erythrophthalmus*, and perch, *Perca fluviatilis*. J. Fish Biol., **29** (Supp. A), 37–48.

Young, R. J., Clayton, H. and Barnard, C. J. (1990) Risk-sensitive foraging in bitterlings, *Rhodeus sericus*: effects of food requirements and breeding site quality. *Anim. Behav.*, **40**, 288–97.

Chapter nine

Predation risk and feeding behaviour

Manfred Milinski

9.1 INTRODUCTION

Teleost fish are frequently in danger of being preyed upon. Predators of many taxa are specialized piscivores and there are more of them that attack small fish. Therefore, young fish normally live under a high risk of predation, which decreases as they grow older and bigger. Growing fast is not only a good strategy for escaping the prey spectrum of many predators, but also for increasing reproductive success (e.g. Werner and Gilliam, 1984). Bigger teleosts are generally able to produce more numerous offspring than smaller ones because bigger females produce more eggs, and bigger males can defend breeding sites better and have a higher social rank; for example, dominant male threespine sticklebacks, *Gasterosteus aculeatus*, maintain larger territories and have priority of access to females (Li and Owings, 1978). Furthermore, larger males have a higher fitness because they are preferred by females (e.g. Downhower and Brown, 1980; Rowland, 1989). Thus, there is a high selection pressure to feed most efficiently in order to grow quickly.

Optimal foraging theory (e.g. Stephens and Krebs, 1987) predicts how an animal should proceed to achieve a maximum rate of energy intake, given its potential types of food and their distribution. Several investigations have shown that animals probably forage using the optimality rules predicted. These include decisions on where to forage, how long to sample, which food items to select and when to leave a food patch in order to find another one (see Chapter 8 by Hart, this volume). Finding food in the most economic way is only one of the many subgoals for maximizing fitness. Another one is the avoidance of being preyed upon. Unfortunately, an animal is mostly not able to fulfil the tasks sequentially, for example foraging when hungry and avoiding predators when satiated. Often there is some interaction between foraging and risk of predation: (a) the animal is more conspicuous by its inevitable

Behaviour of Teleost Fishes 2nd edn. Edited by Tony J. Pitcher. Published in 1993 by Chapman & Hall. ISBN 0 412 42930 6 (HB) and 0 412 42940 3 (PB).

movements while foraging; (b) it has to concentrate on finding food at the expense of vigilance for predators; (c) food can be plentiful in places of high risk of predation. Therefore we expect to find a trade-off between energetic returns from foraging and the risk of predation accepted by the animal.

9.2   BALANCING FEEDING AND PREDATION RISK

Many predators adopt an ambush mode of hunting; they wait motionless and often concealed until the prey is within range and then strike suddenly. Several birds, such as kingfishers and herons (Fig. 9.1), and some fish, for example the pike, *Esox lucius*, adopt this strategy. So also during foraging a small fish has to be vigilant in case of a sudden attack. If in some places the risk of predation is higher than in others, the fish could avoid these places, especially when the predator is known to be present.

A problem arises when the predator is detected in the place where food is most plentiful. By feeding there, the fish could ostensibly maximize its rate of food intake, but with a low probability of surviving the foraging trip. Here the best strategy depends on the specific circumstances of the situation and on the fish's behavioural options. Let us assume that it does not pay the fish to wait and come out a night (but see Chapter 14 by Helfman, this volume) or even to starve for a while (see Chapter 8 by Hart, this volume). Hence the best tactic might be a compromise, i.e. feeding near the predator but not closer than at a point where the risk of predation would override the pay-off for feeding

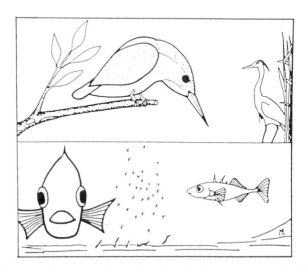

Fig. 9.1   When feeding, a small fish takes the risk of being attacked by birds and fish predators.

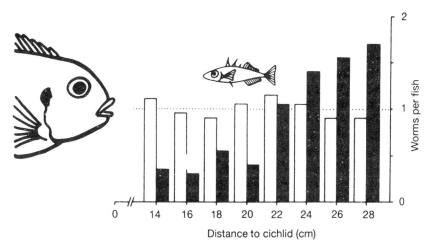

Fig. 9.2 Mean number of *Tubifex* worms on which sticklebacks fed at different distances from a fish predator; each fish was allowed to consume eight worms. Light columns, no cichlid; dark columns, cichlid present. After Milinski (1985a).

most efficiently. The stickleback in Fig. 9.1 has to approach the big perch when it tries to pick up the row of *Tubifex* worms, which are a prey of high quality. How close would the stickleback dare to approach the predator in order to have a good meal? In an experiment (Milinski, 1985a), threespine sticklebacks were allowed singly to feed upon *Tubifex* worms which lay on the bottom of plexiglass cylinders. These were of a height such that the fish while feeding lost sight of a big cichlid, *Oreochromis mariae*, which observed each movement of the stickleback from behind a glass partition. If the sticklebacks dared to feed at all in the presence of the predator, we would expect them to avoid the prey next to the cichlid. The fish indeed preferred the worms that were furthest away from the predator (Fig. 9.2), and they fed more slowly than without the cichlid nearby. Thus, the sticklebacks arrived at a compromise between feeding efficiently and avoiding the predator. We cannot decide whether this compromise is quantitatively the best solution.

Under natural conditions there are many different influences possibly determining the optimum of the compromise. If we could detect and measure all of them, including the animal's constraints, optimality theory would be a suitable tool to predict the best compromise. A great problem for this approach consists in the need to sum up different short-term currencies of fitness, i.e. we have to convert net energy gain from a foraging strategy into probability of death from the risk of predation incurred. In principle, this is possible to solve (McNamara and Houston, 1986; Mangel and Clark, 1986; Abrahams and Dill, 1989). As a first step we should determine qualitatively which influences affect the balance between foraging and avoiding the risk of predation.

1. The animal should be more disposed to take the risk of predation with decreasing value of alternative food in a safer place.
2. If the risk of predation is increased, the animal's behaviour should change to compensate for this; for example, it should reduce its feeding rate in order to become more vigilant and ready to take evasive actions, or it should simply avoid the risky food.
3. An increased need for food should render an animal more willing to accept risky food. The probability of starvation should be proportional to the risk of predation taken in order to feed efficiently. Such a motivational influence is a dynamic one because the state of hunger changes during feeding, and so does the amount of risk which should be accepted.

In the following I will consider the evidence for these influences that is available from studies on teleost fish (general review, Lima and Dill, 1990).

9.3   VALUE OF ALTERNATIVE FOOD AND THE RISK OF PREDATION

Whether fish are willing to run a higher risk of predation if the patch containing a predator provides much more food than an alternative patch without a predator was investigated by Cerri and Fraser (1983). Minnows, *Rhinichthys atratulus*, were confined to an artificial stream system consisting of several compartments, some of which contained predatory fish, *Semotilus atromaculatus*. The prey fish could easily switch between compartments whereas the larger predators could not. The availability of food was low in all compartments in one treatment of the experiment, but there was additional food in the predator compartments in another treatment. Prey fish were found in predator compartments more frequently when there was additional food. This result appears to show that the minnows spent more time close to the predators and therefore accepted a higher risk when patches with predators allowed a higher feeding rate than alternative places.

Cerri and Fraser (1983) suggested that the response of the minnows to predation risk was constant and therefore independent of food reward, which means that the fish did not accept more risk in order to have a better feeding opportunity (but see Milinski, 1985b). To solve this problem, Holbrook and Schmitt (1988) undertook a number of experiments with young black surfperch, *Embiotoca jacksoni*, which could choose simultaneously among four food patches. One experiment exposed foragers to a complete range of variation in predation risk from a kelp bass, *Paralabrax clathratus*, and food density in the four quadrants of a tank: high food and no predator, low food and no predator, high food plus predator, low food plus predator. The experiment involved three treatments where the relative reward between high-food and low-food patches was varied. When little difference in density of food existed between quadrants, the only effect predators had was to lower absolute use of both risky food patches (Fig. 9.3(a)), as predicted by Cerri and Fraser. However, as the comparative value of the more food-rich patch increased, there was a proportionately greater decline in absolute use of the risky low-food patch compared with the risky higher-food patch (Fig. 9.3(b)). A further increase in the comparative value of the more food-rich quadrants led to a near abandonment of the risky food-poor patch (Fig. 9.3(c)). Here predation risk heightened selectivity, whereas in other experiments that Holbrook and Schmitt (1988) undertook, risk either did not influence selectivity or even lowered it.

These results indicate that fish possess an array of responses to food and predators (see also Fraser and Huntingford, 1986). Which strategy should a fish optimally adopt? Gilliam and Fraser (1987) offer a simple rule, 'minimize the ratio of mortality rate to foraging rate' (or 'minimize the risk of death per unit energy consumed'). This rule was used successfully to predict the food density in the riskier patch that should induce a shift of the foragers from the safer (one predator) low-food to the more hazardous (two predators) high-food site in an experiment with minnows, *Semotilus atromaculatus*. Gilliam and Fraser found that the minnows chose the riskier site when feeding there was optimal. In a similar experiment with bluegill sunfish, *Lepomis macrochirus*, as foragers and largemouth bass, *Micropterus salmoides*, as predator Gotceitas (1990) also found the patch preference switch predicted by Gilliam and Fraser's (1987) rule. All fish in Gilliam and Fraser's and in Gotceitas' experiments had a similar hunger state. This is important because with increasing hunger the fish requires a higher feeding rate. Here the fish should apply the rule only to those sites which allow at least for "a target feeding rate" (Gilliam and Fraser, 1987).

A key to understand the function of the fishes' variable tactics is therefore the fish's hunger state, which determines the relative fitness consequences from feeding and avoiding predation. The optimal combination of risk and food reward is state-dependent (e.g. McNamara and Houston, 1990). Less hungry fish should shift the compromise between feeding and avoiding predation towards avoiding predators. For example, juvenile pink salmon, *Oncorhynchus gorbuscha*, were given a choice of two habitats under predation risk, open water with a high prey density and vegetation with a low prey density. When hungry the salmon occupied the open area to a greater extend than when satiated (Magnhagen, 1988a).

When the value of the alternative food patches is very low, the fish probably have to forage in the risky patches. This was different in a study of several species of armoured catfish (Loricariidae) by Power (1984). The larger size

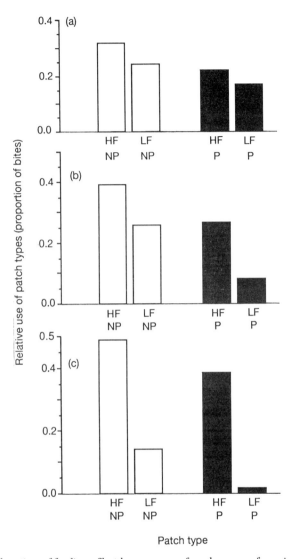

Fig. 9.3 Allocation of feeding effort by young surfperch among four simultaneously available patch types representing the complete combination of variation in food density (HF, high food; LF, low food) and risk (NP, no predator; P, predator present). Results for three levels of disparity in food density between available (HF and LF) patches are plotted separately: (a) low food disparity; (b) moderate food disparity; (c) high food disparity. Data are the proportion (mean) of bites taken on each of the four concurrently available patch types. Modified after Holbrook and Schmitt (1988).

classes of these fish, having outgrown their vulnerability to piscivorous fishes, restricted their foraging to deeper water and did not exploit standing crops of algae in shallow water, even in the dry season when food was most limited. Only in shallow water were they susceptible to several species of bird predators. The large loricariids could afford to avoid the risky habitat, because there were no seasonal changes in mortality rates although somatic growth rates decreased when food was scarce.

9.4 INCREASED RISK OF PREDATION

We might expect compromise between feeding and avoiding predation to be shifted to more pronounced predator avoidance when the risk of predation is increased. Werner *et al.* (1983) investigated whether three size classes of bluegill sunfish, of which the smallest class was very vulnerable to predation by largemouth bass, responded to both increased predation risk and habitat profitability in choosing habitats in which to feed. The bluegills were stocked on both sides of a divided pond which contained largemouth bass on one side only. Each side of the pond consisted of three different habitats, of which only one offered refuges from predation (dense vegetation) but had much less food than the others. It turned out that only fish of the smallest size were preyed upon, although bluegills of the medium size class were assumed to be chased more often than the largest fish. The smallest fish therefore had an increased risk of predation as compared with the other size classes. Without predators, bluegills of all three size classes preferred the benthos, where food was most plentiful (Fig. 9.4). With predators present, only the smallest fish had a less pronounced preference for the benthos, and foraged nearly as often in the less profitable vegetation. Thus, the size class that had an increased risk of predation balanced the two conflicting demands in that they suffered only a little from predation, but grew more slowly than the other size classes. It could not be decided whether all small fish divided their time between different habitats in the same way, or whether some individuals were more willing to take the risk of foraging in the benthos than others. By studying five different natural lakes, Werner and Hall (1988) found that the size at which bluegills shifted to the riskier but more profitable habitat was directly correlated with the density of the largemouth bass in these lakes. Gilliam (1982; see also Werner and Gilliam, 1984) has provided an explicit model of how predation risk and growth rates in different habitats interact to affect lifetime fitness.

Can bluegills actually sense changes in food availability or predation risk in different habitats? Ehlinger (1986) showed experimentally that these fish are very adept at assessing differences in foraging rates between open water and vegetation habitats and switch accordingly. Similarly, when a large-mouth bass was tethered in a pool, algae-grazing minnows, *Campostoma*

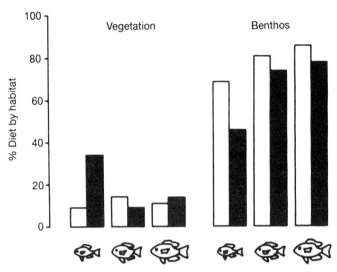

Fig. 9.4   Average percentage composition of the diet in two of three habitats for three size classes of bluegill sunfish. Light columns, no predator; dark columns, predator present. Modified after Werner *et al.* (1983).

anomalum, only fed at a distance of at least 1.3 m from the predator (Power and Matthews, 1983). When structural complexity providing refuge from predation is varied independently of food density and predation risk in experimental studies, each of these variables usually influences the fishes' preference significantly (Schmitt and Holbrook, 1985; Savino and Stein, 1989; Gotceitas and Colgan, 1990).

We have seen that increased risk of predation in one habitat can alter the choice of places in which to forage. It is also possible that within a patch the risk of predation is different for feeding upon one type of food as compared with others; this may alter the choice of items by a forager within a given habitat. In a study of juvenile coho salmon, *Oncorhynchus kisutch*, Dill (1983) investigated the influence of a predator on the distance a coho would swim upstream from its holding station to intercept drifting prey. Bigger prey items were detected and attacked from a greater distance than smaller ones, suggesting that the fish maximized their rate of success for each prey size. On the other hand, by swimming a greater distance the fish becomes more visible, and hence presumably more vulnerable, to predators. Thus, the fish should swim shorter distances for a given prey item when it perceives risk of predation to be high. Dill tested this hypothesis by measuring the attack distance of the fish in response to three sizes of surface-drifting flies in an artificial stream channel. In some trials fish were presented singly with a photograph of a big rainbow trout, resembling a predator for young coho, and in other trials the

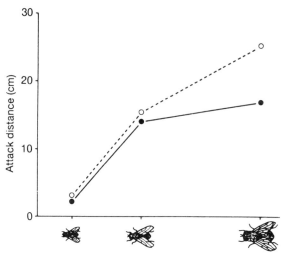

Fig. 9.5   The effect of the presentation of a model predator on the distance swum to attack flies of different size by juvenile coho salmon. Broken curve, no predator; solid curve, predator present. Modified after Dill (1983).

coho could feed undisturbed. Attack distances were reduced in the presence of the 'predator', particularly in response to the largest prey (Fig. 9.5). These results support the hypothesis that juvenile coho salmon are less willing to travel long distances, or to spend long periods of time moving, to obtain prey when risk of predation is increased. The risk seems to grow exponentially with distance travelled, as can be seen from the marked effect when the prey were large. The extent of the reduction in attack distance was greater at higher frequencies of predator presentation, suggesting that coho are able to judge the level of risk and adjust their behaviour accordingly (Dill and Fraser, 1984).

Increased risk of predation reduced the attack distance most for the large, profitable prey. Thus, in the presence of predators, coho should capture fewer and smaller prey items than in their absence (as has been shown by Ibrahim and Huntingford, 1989, for sticklebacks and by Godin, 1990, for guppies) and therefore suffer from a reduced energy intake, as did the small bluegills when choosing to forage in a safer but less profitable habitat (Werner *et al.*, 1983). In the next section we shall see that fish can change their social behaviour to compensate for increased risk of predation.

9.5   PREDATOR DETECTABILITY AND FEEDING EFFICIENCY

When foraging, a fish is at risk – especially from a suddenly attacking predator – because handling food can impair a full view around the fish for some seconds. If there were a way of improving the probability of detecting a predator so that it could be observed from a greater distance, the fish could evaluate the actual risk and sometimes continue feeding for longer. There might be no net benefit if predator detection were improved by investing more time in scanning than in feeding. A more profitable strategy consists in enlarging the number of eyes looking around, which has been shown to work in bird flocks (e.g. Powell, 1974; Lazarus, 1979). Whether the members of a fish shoal have a similar advantage has been investigated by Magurran *et al.* (1985). They observed shoals of 3, 6, 12 or 20 European minnows, *Phoxinus phoxinus*, foraging on an artificial food patch during the simulated stalking approach of a model pike. Minnows in large shoals reduced their foraging sooner but remained feeding on the patch for longer (Fig. 9.6). The relatively late reaction of small shoals to the model and the rapid cessation of feeding provide good evidence that they detected the predator later than larger shoals.

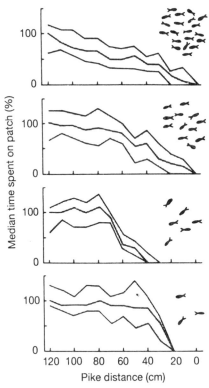

Fig. 9.6 Percentage time (median and interquartile range) that shoals of 20, 12, 6 or 3 minnows spent on a food patch during the approach of a model pike. Modified after Magurran *et al.* (1985).

Several specific antipredator behaviours were observed which substantiated this point (see also Chapter 12 by Pitcher and Parrish, this volume).

Why did smaller shoals leave the patch earlier than larger ones? There could be several functional explanations for this result.

1. Fish in small shoals could not observe the approaching pike and therefore had no possibility of calculating the actual risk. A good strategy would then be to suppose the highest risk and flee immediately upon detection of the predator.
2. Fish in small shoals must be more certain that the predator will successfully attack because they cannot rely on the confusion effect (see Chapter 12).
3. If the predator strikes successfully, a member of a small shoal has a higher chance of being the victim than each member of a large shoal (but see Chapter 12).

Although these effects reduce the risk for members of a large shoal, they do not provide complete safety. By continuously reducing their feeding effort during the predator's approach, the fish in large shoals shifted the compromise more and more from feeding to vigilance in monitoring the pike's approach. Again we would expect that this compromise is also affected by the fishes' hunger level. When shoals of 3, 5, 7, 10, 15 and 20 bluntnose minnows, *Pimephales notatus*, were presented with food after 5, 24 or 72 h of deprivation and eventually with a predator, foraging was affected significantly by shoal size, predator presence and hunger (Morgan, 1988). As hunger level increased, fish in the larger shoals increased their foraging even when a predator was present (although their level of foraging was less than when the predator was absent at that hunger level). However, fish in the smallest shoals (three and five minnows) did not show an increase in foraging with hunger when a predator was present.

Magurran *et al.* (1985) showed that the minnows in larger shoals did not suffer from increased competition for food, which could play a role under different conditions (Street *et al.*, 1984) and might also influence the compromise achieved. For example, when seeming competition was increased for young coho salmon, they reduced the weighting given to risk of predation and travelled further to prey than when they were tested in visual isolation (Dill and Fraser, 1984). There should be an optimal shoal size minimizing the cost/benefit ratio caused by competition and risk of predation. (See Chapter 12 by Pitcher and Parrish, this volume.)

9.6 HUNGER AND AVOIDANCE OF PREDATION

In this section we will see how altering the need for food, i.e. the risk of starvation, changes the compromise between feeding and antipredator behaviour.

Since hunger decreases during feeding, we would expect that the compromise would change as a function of satiation. This will be illustrated by a study on sticklebacks which have to scan for predators while feeding upon dense swarms of water fleas.

Aggregations of tiny prey animals in the zooplankton are a valuable food source, especially for small fish. Big fish with large mouths can suck in many items from the swarm of prey in one gulp if the swarm density is high enough (McNaught and Hasler, 1961). From less-dense swarms, fish have to catch one item after another. Small fish can economically adopt this latter tactic (Janssen, 1976) even if prey items are very small. Such particulate feeders, too, can increase their feeding rate when they choose a swarm of a high density (Heller and Milinski, 1979) because the distances between prey items are shorter there. Many predators, however, suffer from a so-called confusion effect when they attack dense swarms of prey (see Chapter 12 by Pitcher and Parrish, this volume). I will show in the following that a stickleback's decision to attack in a high-density swarm of water fleas not only increases the fish's hunting success, but also affects its own risk of predation.

During the summer, threespine sticklebacks live mostly on planktonic prey (Wootton, 1976, and Chapter 16 by FitzGerald and Wootton, this volume). Whether these fish preferentially attack swarms of a high density was investigated in the laboratory (Milinski, 1977a). The fish were given singly the choice between two test tubes, one containing a swarm of 40 water fleas, *Daphnia magna*, the other containing only two water fleas. One group of sticklebacks was hungry whereas another had become nearly satiated before the trial by feeding upon *Tubifex* worms. The results were rather puzzling because most of the hungry fish attacked the swarm, whereas most less-hungry fish chose the single items. Further experiments confirmed that the hungry fish did not simply overlook the singletons but really chose to attack the swarm (Milinski, 1977b; Heller and Milinski, 1979). Similarly, Morgan and Ritz (1984) found that hungry Australian salmon, *Arripis trutta*, attacked higher densities of swarming krill, *Nyctiphanes australis*, and less-hungry fish preferred to attack lower swarm densities. Does this mean that hunting in a high density is most profitable for hungry fish, and attacking a low-density swarm provides the highest energy intake for more-satiated fish? This was tested in another series of experiments in which hungry sticklebacks and more-satiated ones were allowed to feed upon either a high or a low prey density kept constant during a trial (Heller and Milinski, 1979). The hungry fish fed significantly faster in the high prey density whereas the less-hungry ones achieved a higher feeding rate in the low density. Therefore, it pays a hungry fish to choose a high-density swarm and a less-hungry one to attack a lower density.

If only the risk of starvation is important, optimal foraging theory alone does not explain these results. If hunting in place A is more profitable than in

B, there is no reason for a less-hungry predator to hunt in B. Thus, there must be other costs at stake, which increase with the density of the prey swarm attacked. A candidate for this could be the confusion effect, from which threespine sticklebacks have been shown to suffer (Ohguchi, 1981). Hungry sticklebacks, however, must be able and willing to overcome the confusion imposed by a high-density swarm, otherwise they could not achieve their highest feeding rate there. Since it seems to be far-fetched to assume that the ability to overcome the confusion is dependent on hunger, it must be concluded that the less-hungry fish were not willing to pay the costs for overcoming the confusion.

What could these costs consist of? If the fish has to concentrate on the difficult task of tracking one out of many similar-looking targets in a dense swarm, it probably cannot pay sufficient attention to other events, e.g. a suddenly approaching predator of its own (Fig. 9.1). There is always a certain probability of a sudden attack by an ambush predator. If this risk is lower than the costs of not reducing its energy deficit as quickly as possible, the stickleback should overcome the confusion and attack the high-density swarm.

To test this hypothesis, the stickleback's expected risk of predation was increased. At the new level it should prefer a lower prey density than with the baseline risk level (Milinski and Heller, 1978). A model of a European kingfisher, *Alcedo atthis*, an efficient predator of sticklebacks (Kniprath, 1965), was moved over the tank before the stickleback was allowed to choose between different prey densities. Frightened fish attacked the lowest prey density whereas unfrightened ones preferred the highest density (Fig. 9.7). Similar

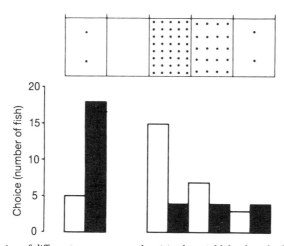

Fig. 9.7 Choice of different prey swarm densities by sticklebacks which were either undisturbed (light columns) or frightened by a model kingfisher (dark columns). The box at top shows the different prey densities available simultaneously. After Milinski and Heller (1978).

results were obtained by Jakobsen and Johnsen (1989) when they tested the effects of alarm substance on feeding behaviour of zebra danio fish, *Brachydanio rerio*. Normally these fish attacked high densities of water fleas, but when exposed to alarm substance (which reveals the presence of a predator) they preferred lower and less confusing prey densities, also lowering their feeding rate.

Now, one question remains to be answered. Is a hungry stickleback attacking a high-density swarm less likely to detect an approaching predator than a similarly hungry one attacking a low density? To test this, hungry sticklebacks were allowed to prey upon either a high density of water fleas or a low one. After they had consumed the fourth water flea, a conspicuous model of a kingfisher, or a cryptic model, or no model, was flown over the tank (Milinski, 1984a). If the fish detected the predator, they should hesitate before attacking the fifth water flea. Both quick and slow fish (as determined by the time they needed for catching the fourth water flea) reacted similarly to the cryptic model and the conspicuous one in the low prey density (Fig. 9.8). In the high prey density, however, the fish which fed at a higher rate reacted only to the conspicuous model. Therefore they must have overlooked the cryptic predator. Some of the slower fish must have seen the cryptic model and some must have overlooked it in the high prey density. Thus, the costs of overcoming the confusion effect increase with feeding rate and prey density. In a similar experiment guppies, *Poecilia reticulata*, in varying hunger states were presented singly with one of five different prey densities, either without a predator or with a predator, a jewel cichlid fish, *Hemichromis bimaculatus*, which was allowed to attack after the guppy had consumed the second water flea (Godin and Smith, 1988). Guppies which attacked the higher prey densities and those feeding at a higher rate (i.e. those having shorter inter-capture intervals before the predator attacked) were caught more often by the cichlid than those feeding more slowly on low prey densities. Hungrier guppies had a higher feeding rate and therefore took a higher risk of predation.

In conclusion, when the sticklebacks were very hungry, they took the risk of overlooking a suddenly approaching predator in order to feed most efficiently. When the risk of predation was increased, or when the fish were less hungry, they gave higher priority to predator detection, thereby sacrificing feeding efficiency. The nervous system is limited in its ability to process sensory data (i.e. in its channel capacity), which makes it increasingly difficult to fulfil two different tasks at the same time as each single task gets harder (Milinski, 1990a). If there is information overload, the animal has to decide which task, either foraging or avoiding predation, to fulfill less efficiently. Evidence for such a trade-off between foraging and avoiding predation is the common observation that fish reduce their feeding rate in the presence of a predator (Milinski, 1984a, 1985a; Fraser and Huntingford, 1986; Giles, 1987; Metcalfe *et al.*, 1987a; Prejs, 1987; Holbrook and Schmitt, 1988; Magnhagen, 1988b;

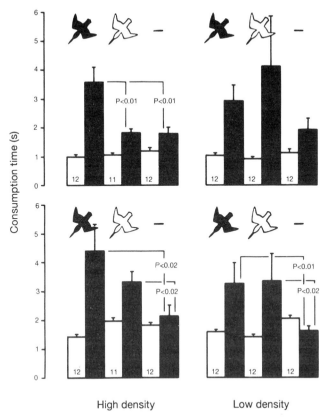

Fig. 9.8 Mean (+ SE) consumption time for the fourth water flea (light columns) and for the fifth one (dark columns) of fast fish (upper part) and of slow fish (lower part) feeding either upon a high or a low prey density. After the fourth water flea had been caught, sticklebacks were shown either a conspicuous model predator (dark bird), or a cryptic one (light bird), or no predator (dash). Number of fish (trials is given in light columns; *P* values are from Mann-Whitney *U*-test. Modified after Milinski (1984a).

Morgan, 1988; Ibrahim and Huntingford, 1989; Jakobsen and Johnsen, 1989; Gotceitas and Colgan, 1990). This saves a larger fraction of the channel capacity available to pay attention to the predator. Both sticklebacks and guppies overlooked the predator when they fed at a high rate. If, however, an Atlantic salmon, *Salmo salar*, has no time to decide more slowly e.g. between suitable and unsuitable items because it waits for food passing quickly in the water current, it makes more mistakes in its choice of diet with a predator near by (Metcalfe *et al.*, 1987b). Also threespine sticklebacks ceased discrimination in favour of the more profitable food items after a model kingfisher had been moved over the tank (Ibrahim and Huntingford, 1989).

Another dynamic motivational influence on the trade-off between risk of predation and food intake was found in coho salmon (Dill and Fraser, 1984). With increasing satiation the fishes' attack distance was progressively reduced, but only when predation risk was high. Thus, hungry coho did not appear to weight risk as heavily as they did when more satiated.

9.7　PARASITISM AND THE RISK OF PREDATION

We have seen that when fish have to balance two conflicting demands, i.e. to feed efficiently and to avoid a predator of their own, they arrive at a compromise which is influenced by their state of hunger and by the actual risk of predation. Hunger can be increased either by starvation or by enhanced needs, e.g. because parasites use up a great proportion of the energy the food contains (Milinski, 1990b). Threespine sticklebacks are frequently parasitized by the cestode *Schistocephalus solidus*, which can exceed the weight of its host (Arme and Owen, 1967). Parasitized sticklebacks are hungrier than uninfested ones, not only because they have to nourish their parasites and swimming is energetically more demanding for them (Lester, 1971), but also because they are poor competitors and have therefore to feed upon less-profitable prey (Milinski, 1984b).

We would expect that parasitized fish give less weight to predator avoidance in order to find enough food. Threespine sticklebacks infested by *S. solidus* and unparasitized ones were allowed to settle singly in a tank before they were frightened by a model heron's head lowered suddenly over the test tank, so that the tip of the bill hit the water surface (Giles, 1983; see also Godin and Sproul, 1988). The fish either stopped swimming immediately or jumped away from the model in alarm to remain still somewhere else. The mean recovery time, however, was twice as long for unparasitized fish as for parasitized ones (Giles, 1983). Moreover, the recovery time was significantly negatively correlated with parasitic load. Since feeding occurred mostly after the recovery period, the parasitized sticklebacks resumed feeding much earlier and most probably at a higher risk of being preyed upon than the healthy ones. Only after the healthy sticklebacks had been starved for 24 h did they have a similar foraging activity after predator stimulus to that of the unstarved parasitized fish (Giles, 1987).

We would predict that heavily parasitized sticklebacks should approach a predator less cautiously than healthy ones in order to find food. When the experiment shown in Fig. 9.2 was repeated with sticklebacks infested by *S. solidus* (Milinski, 1985a), there was no obvious influence of the cichlid's presence on the sticklebacks' foraging behaviour. They fed as quickly and as close to the cichlid's position as with no predator present. This risky behaviour was especially profitable concerning energy gain when a parasitized stickleback was competing for food with a healthy one in the presence of the predator. The parasitized fish clearly outcompeted the healthy fish by consuming three times as many worms as the latter dared to catch. Also under natural conditions parasitized sticklebacks competing for food with healthy ones accept a higher risk of predation and indeed are preyed upon at a higher rate than healthy ones (Jakobsen, *et al.*, 1988).

Dace, *Leuciscus leuciscus*, parasitized by the eyefluke *Diplostomum spathaceum* have impaired vision and therefore forage less efficiently, preferentially close to the water surface where they are highly susceptible to bird predators (Crowden and Broom, 1980). Parasites influence their teleost hosts in various ways that increase the fishes' risk of being preyed upon. The higher need for food forces the infested fish to give less weight to avoidance of predation than to feeding, although they are slower when fleeing.

9.8 THE PREY'S WEIGHTED RESPONSE TO PREDATION RISK

The prey's response to the presence of a predator is not expected to be either all or nothing. Predators vary in the amount of risk they impose on a given prey (e.g. Wahl and Stein, 1988). The response of a given prey is therefore expected to covary with the amount of danger it expects from a given predator (Webb, 1982). The decision of whether, when and how to flee from a predator is an economic one based on the relative costs of fleeing and staying (Ydenberg and Dill, 1986), and it has been shown that the prey's predator avoidance behaviour is threat-sensitive (Helfman, 1989). I have already discussed the success of Gilliam and Fraser's (1987) rule ('minimize the ratio of mortality rate to foraging rate'), examples of minnows changing from feeding to vigilance as a pike model approaches (Magurran *et al.*, 1985), and of bluegill sunfish and armoured catfish changing their habitat preference as they grow larger and become safer from predation (Werner *et al.*, 1983; Power, 1984).

Also after the predator has disappeared we should not expect the prey to resume foraging immediately by 100%, but instead a gradual increase of its foraging activity as the risk of the predator's return decreases. Juvenile Atlantic salmon changed their foraging strategy markedly after a brief exposure to a model trout predator. The relative priority given to foraging and predator avoidance varied with time elapsed since the predator was last sighted; as a consequence, intake rates in the 20 min following the predator presentation averaged only 33% of the pre-predator level, but had increased to 57% 20 min later, and had recovered completely within 2 h (Metcalfe *et al.*, 1987a). There was some indication that the fish fed at a greater-than-normal rate thereafter, probably owing to the energy deficit which had increased during the period of reduced foraging.

Is it possible to measure how the fish rates the actual risk of predation in terms of units of energy it is prepared to miss from reduced feeding? Recently, the ideal free distribution of fish between two feeding sites (Milinski, 1979) has been used to investigate the trade-off between foraging and predator avoidance quantitatively. When the risk of predation is increased in one of the food patches, both minnows (Pitcher *et al.*, 1988) and guppies (Abrahams and Dill, 1989) balance the risk of being eaten against the benefit of getting food. When both patches had equal food but one contained a predator of guppies, there was a certain preference for the predator-free site. Then Abrahams and Dill titrated risk against energy by adding food to the risky patch until it had equal value from the guppies' point of view, so that they returned to a 1:1 distribution. The amount of food added to the predator site was the energetic equivalent of the risk of predation. By this method different currencies of fitness become quantitatively comparable.

9.9   SUMMARY

Often fish must avoid being preyed upon when they are feeding. When both demands are conflicting (i.e. maximizing food intake is achieved only at the expense of efficiently avoiding predation and vice versa), it has been shown experimentally that fish make a compromise by fulfilling either or both needs less efficiently. They take a greater risk in order to feed more efficiently when the need for food is increased by starvation or by parasites, or when feeding is much more rewarding in places with predators. Smaller fish which have an increased risk of predation accept less rewarding food in order to avoid the predator than do bigger fish. Diet selection within a patch can be altered if one type of food is riskier to feed upon than others. By foraging in large shoals, fish can detect and monitor an approaching predator more easily and can continue feeding for longer than fish in smaller shoals. The examples discussed provide qualitative and quantitative evidence for teleosts changing their behaviour adaptively when costs and benefits of feeding and avoidance of predation vary. Because of the difficulty in measuring all costs and benefits and knowing how accurately the fish may do this sum, it is difficult to predict the compromise behaviour quantitatively. Successful attempts to solve this problem are discussed.

REFERENCES

Abrahams, M. V. and Dill, L. M. (1989) A determination of the energetic equivalence of the risk of predation. *Ecology*, **70**, 999–1007.

Arme, C. and Owen, R. W. (1967) Infections of the three-spined stickleback, *Gasterosteus aculeatus* L., with the plerocercoid larvae of *Schistocephalus solidus* (Müller, 1776), with special reference to pathological effects. *Parasitology*, **57**, 301–14.

Cerri, R. D. and Fraser, D. F. (1983) Predation and risk in foraging minnows: balancing conflicting demands. *Am. Nat.*, **121**, 552–61.

Crowden, A. E. and Broom, D. M. (1980) Effects of the eyefluke, *Diplostomum spathaceum*, on the behaviour of dace (*Leuciscus leuciscus*). *Anim. Behav.*, **28**, 287–94.

Dill, L. M. (1983) Adaptive flexibility in the foraging behavior of fishes. *Can. J. Fish. Aquat. Sci.*, **40**, 398–408.

Dill, L. M. and Fraser, A. H. G. (1984) Risk of predation and the feeding behavior of juvenile coho salmon (*Oncorhynchus kisutch*). *Behav. Ecol. Sociobiol.*, **16**, 65–71.

Downhower, J. F. and Brown, L. (1980) Mate preferences of female mottled sculpins, *Cottus bairdi*. *Anim. Behav.*, **28**, 728–34.

Ehlinger, T. J. (1986) Learning, sampling and the role of individual variability in the foraging behavior of bluegill sunfish (*Lepomis macrochirus*). PhD thesis, Michigan State University.

Fraser, D. F. and Huntingford, F. A. (1986) Feeding and avoiding predation hazard: the behavioral response of the prey. *Ethology*, **73**, 56–68.

Giles, N. (1983) Behavioural effects of the parasite *Schistocephalus solidus* (Cestoda) on an intermediate host, the three-spined stickleback, *Gasterosteus aculeatus* L. *Anim. Behav.*, **31**, 1192–4.

Giles, N. (1987) Predation risk and reduced foraging activity in fish: experiments with parasitized and non-parasitized three-spined sticklebacks, *Gasterosteus aculeatus* L., *J. Fish Biol.*, **31**, 37–44.

Gilliam, J. F. (1982) Habitat use and competitive bottlenecks in size-structured fish populations. PhD thesis, Michigan State University.

Gilliam, J. F. and Fraser, D. F. (1987) Habitat selection under predation hazard: test of a model with foraging minnows. *Ecology*, **68**, 1856–62.

Godin, J.-G. J. (1990) Diet selection under predation risk, in *Behavioural Mechanisms of Food Selection* (ed. R. N. Hughes) (NATO ASI Series G, Vol. 20), Springer, Berlin, pp. 739–70.

Godin, J.-G. and Smith, S. A. (1988) A fitness cost of foraging in the guppy. *Nature, Lond.*, **333**, 69–71.

Godin, J.-G. J. and Sproul, C. D. (1988) Risk taking in parasitized sticklebacks under threat of predation. *Can. J. Zool.*, **66**, 2360–67.

Gotceitas, V. (1990) Foraging and predator avoidance: a test of a patch choice model with juvenile bluegill sunfish. *Oecologia*, **83**, 346–51.

Gotceitas, V. and Colgan, P. (1990) Behavioural response of juvenile bluegill sunfish to variation in predation risk and food level. *Ethology*, **85**, 247–55.

Helfman, G. S. (1989) Threat-sensitive predator-avoidance in damselfish–trumpetfish interactions. *Behav. Ecol. Sociobiol.*, **24**, 47–58.

Heller, R. and Milinski, M. (1979) Optimal foraging of sticklebacks on swarming prey. *Anim. Behav.*, **27**, 1127–41.

Holbrook, S. J. and Schmitt, R. J. (1988) The combined effects of predation risk and food reward on patch selection. *Ecology*, **69**, 125–34.

Ibrahim, A. A. and Huntingford, F. A. (1989) Laboratory and field studies of the effect of predation risk on foraging in three-spined sticklebacks (*Gasterosteus aculeatus*). *Behaviour*, **109**, 46–57.

Jakobsen, P. J. and Johnsen, G. H. (1989) The influence of alarm substance on feeding in zebra danio fish (*Brachydanio rerio*). *Ethology*, **82**, 325–7.

Jakobsen, P. J., Johnsen, G. H. and Larsson, P. (1988) Effects of predation risk and parasitism on the feeding ecology, habitat use, and abundance of lacustrine threespine stickleback (*Gasterosteus aculeatus*). *Can. J. Fish. Aquat. Sci.*, **45**, 426–31.

Janssen, J. (1976) Feeding modes and prey size selection in the alewife (*Alosa pseudoharengus*). *J. Fish. Res. Bd Can.*, **33**, 1972–5.

Kniprath, E. (1965) Stichlinge als Nahrung des Eisvogels und des Teichhuhns. *Orn. Beob.,* **62**, 190–92.

Lazarus, J. (1979) The early warning function of flocking in birds: an experimental study with captive *quelea. Anim. Behav.,* **27**, 855–65.

Lester, R. J. G. (1971) The influence of *Schistocephalus* plerocercoids on the respiration of *Gasterosteus* and a possible resulting effect on the behavior of the fish. *Can. J. Zool.,* **49**, 361–6.

Li, S. K. and Owings, D. H. (1978) Sexual selection in the three-spined stickleback: I. Normative observations. *Z. Tierpsychol.,* **46**, 359–71.

Lima, S. L. and Dill, L. M. (1990) Behavioral decisions made under the risk of predation: a review and prospectus. *Can. J. Zool.,* **68**, 619–40.

McNamara, J. M. and Houston, A. I. (1986) The common currency for behavioral decisions. *Am. Nat.,* **127**, 358–78.

McNamara, J. M. and Houston, A. I. (1990) State-dependent ideal free distributions. *Evol. Ecol.,* **4**, 298–311.

McNaught, D. C. and Hasler, A. D. (1961) Surface schooling and feeding behavior in the white bass, *Roccus chrysops* (Rafinesque), in Lake Mendota. *Limnol. Oceanogr.,* **6**, 53–60.

Magnhagen, C. (1988a) Predation risk and foraging in juvenile pink (*Oncorhynchus gorbusha*) and chum salmon (*O. keta*). *Can. J. Fish. Aquat. Sci.,* **45**, 592–6.

Magnhagen, C. (1988b) Changes in foraging as a response to predation risk in two gobiid fish species, *Pomatoschistus minutus* and *Gobius niger. Mar. Ecol. Progr. Ser.,* **49**, 21–6.

Magurran, A. E., Oulton, W. J. and Pitcher, T. J. (1985) Vigilant behaviour and shoal size in minnows. *Z. Tierpsychol.,* **67**, 167–78.

Mangel, M. and Clark, C. W. (1986) Towards a unified foraging theory. *Ecology,* **67**, 1127–38.

Metcalfe, N. B., Huntingford, F. A. and Thorpe, J. E. (1987a) The influence of predation risk on the feeding motivation and foraging strategy of juvenile Atlantic salmon. *Anim. Behav.,* **35**, 901–11.

Metcalfe, N. B., Huntingford, F. A. and Thorpe, J. E. (1987b) Predation risk impairs diet selection in juvenile salmon. *Anim. Behav.,* **35**, 931–3.

Milinski, M. (1977a) Do all members of a swarm suffer the same predation? *Z. Tierpsychol.,* **45**, 373–88.

Milinski, M. (1977b) Experiments on the selection by predators against spatial oddity of their prey. *Z. Tierpsychol.,* **43**, 311–25.

Milinski, M. (1979) An evolutionarily stable feeding strategy in sticklebacks. *Z. Tierpsychol.,* **51**, 36–40.

Milinski, M. (1984a) A predator's costs of overcoming the confusion-effect of swarming prey. *Anim. Behav.,* **32**, 1157–62.

Milinski, M. (1984b) Parasites determine a predator's optimal feeding strategy. *Behav. Ecol. Sociobiol.,* **15**, 35–7.

Milinski, M. (1985a) Risk of predation of parasitized sticklebacks (*Gasterosteus aculeatus* L.) under competition for food. *Behaviour,* **93**, 203–16.

Milinski, M. (1985b) The patch choice model: no alternative to balancing. *Am. Nat.,* **125**, 317–20.

Milinski, M. (1990a) Information overload and food selection, in *Behavioural Mechanisms of Food Selection* (ed. R. N. Hughes) (NATO ASI Series G, Vol. 20), Springer, Berlin, pp. 721–38.

Milinski, M. (1990b) Parasites and host decision-making, in *Parasitism and Host Behaviour* (eds C. J. Barnard and J. M. Behnke), Taylor and Francis, London, pp. 95–116.

Milinski, M. and Heller, R. (1978) Influence of a predator on the optimal foraging behaviour of sticklebacks (*Gasterosteus aculeatus* L.). *Nature, Lond.*, **275**, 642–4.

Morgan, M. J. (1988) The influence of hunger, shoal size and predator presence on foraging in bluntnose minnows. *Anim. Behav.*, **36**, 1317–22.

Morgan, W. L. and Ritz, D. A. (1984) Effect of prey density and hunger state on capture of krill, *Nyctiphanes australis* Sars, by Australian salmon, *Arripis trutta* (Bloch & Schneider). *J. Fish Biol.*, **24**, 51–8.

Ohguchi, O. (1981) Prey density and selection against oddity by three-spined sticklebacks. *Adv. Ethol.*, **23**, 1–79.

Pitcher, T. J., Lang, S. H. and Turner, J. R. (1988) A risk-balancing trade-off between foraging rewards and predation risk in shoaling fish. *Behav. Ecol. Sociobiol.*, **22**, 225–8.

Powell, G. V. N. (1974) Experimental analysis of the social value of flocking by starlings (*Sturnus vulgaris*) in relation to predation and foraging. *Anim. Behav.*, **22**, 501–5.

Power, M. E. (1984) Depth distribution of armored catfish: predator-induced resource avoidance? *Ecology*, **65**, 523–8.

Power, M. E. and Matthews, W. J. (1983) Algae-grazing minnows (*Campostoma anomalum*), piscivorous bass (*Micropterus* spp.), and the distribution of attached algae in a small prairie-margin stream. *Oecologia*, **60**, 328–32.

Prejs, A. (1987) Risk of predation and feeding rate in tropical freshwater fishes: field evidence. *Oecologia*, **72**, 259–62.

Rowland, W. J. (1989) Mate choice and the supernormality effect in female sticklebacks (*Gasterosteus aculeatus*). *Behav. Ecol. Sociobiol.*, **24**, 433–8.

Savino, J. F. and Stein, R. A. (1989) Behavioural interactions between fish predators and their prey: effects of plant density. *Anim. Behav.*, **37**, 311–21.

Schmitt, R. J. and Holbrook, S. J. (1985) Patch selection by juvenile black surfperch (Embiotocidae) under variable risk: interactive influence of food quality and structural complexity. *J. exp. Mar. Biol. Ecol.*, **85**, 269–85.

Stephens, D. W. and Krebs, J. R. (1987) *Foraging Theory*, Princeton University Press, Princeton, NJ, 247 pp.

Street, N. E., Magurran, A. E. and Pitcher, T. J. (1984) The effects of increasing shoal size on handling time in goldfish *Carassius auratus* L. *J. Fish Biol.*, **25**, 561–6.

Wahl, D. H. and Stein, R. A. (1988) Selective predation by three esocids: the role of prey behavior and morphology. *Trans. Am. Fish. Soc.*, **117**, 142–51.

Webb, P. W. (1982) Avoidance responses of fathead minnow to strikes by four teleost predators. *J. comp. Physiol.*, **147**, 371–8.

Werner, E. E. and Gilliam, J. F. (1984) The ontogenetic niche and species interactions in size-structured populations. *A. Rev. Ecol. Syst.*, **15**, 393–425.

Werner, E. E. and Hall, D. J. (1988) Ontogenetic habitat shifts in bluegill: the foraging rate–predation risk trade-off. *Ecology*, **69**, 1352–66.

Werner, E. E., Gilliam, J. F., Hall, D. J. and Mittelbach, G. G. (1983) An experimental test of the effects of predation risk on habitat use in fish. *Ecology*, **64**, 1540–48.

Wootton, R. J. (1976) *The Biology of the Sticklebacks*, Academic Press, London, 387 pp.

Ydenberg, R. C. and Dill, L. M. (1986) The economics of fleeing from predators. *Adv. Study Behav.*, **16**, 229–49.

Chapter ten

Teleost mating behaviour

George F. Turner

10.1   INTRODUCTION

In this chapter, I review the nature and function of teleost mating behaviour, using two basic premises. (1) The force shaping adaptation is natural selection acting through the differential survival and elimination of competing genes (Dawkins, 1976). This does not deny that other factors, such as developmental and genetic constraints may limit and channel the direction of this process, or even fundamentally alter the conditions under which it operates. (2) The appropriate procedure for analysing social behaviour is the logic of game theory in which an animal's behaviour is modelled to determine the evolutionarily stable strategy (ESS). A pure ESS is a strategy which yields higher benefits (pay-offs) in a population of like individuals than any other biologically realistic alternative. A mixed ESS (the second 'S' now standing for 'state'!) can occur when, at a particular ratio in the population, two strategies have equal pay-offs (see Maynard Smith, 1982, and Parker, 1984, for reviews and more precise definitions).

The mating behaviour of an animal will inevitably be related to the extent and type of parental care practised, as care-giving may influence opportunities for further mating. Parental care strategies are explained elsewhere in this volume (Chapter 11 by Sargent and Gross), and I will cover only those aspects essential to the understanding of mating behaviour.

Strategies and tactics

A mating **strategy** is a genetically coded set of rules by which an animal's lifetime reproductive behaviour is guided. A mating **tactic** is a set of activities, which if successful lead to reproduction. A strategy may consist of one or more tactics.

If one sex includes individuals that have different types of mating behaviour, these are known as alternative mating tactics. Alternative mating tactics are produced in two ways. A strategy may allow for changes according to

Behaviour of Teleost Fishes 2nd edn. Edited by Tony J. Pitcher. Published in 1993 by Chapman & Hall. ISBN   0 412 42930 6 (HB) and 0 412 42940 3 (PB).

circumstances. A male, for example, if relatively large, may defend a territory, whereas if smaller, he may mimic a female and try to fertilize the eggs of females spawning with a larger male. This type of strategy is known as a conditional strategy. In some cases, an individual may use both tactics, either alternating in quick succession, or switching from one to another at a later time of life or larger size. Alternatively, two tactics may be equally successful in a population. Usually there will be a frequency-dependent equilibrium: at a certain ratio of one tactic to the other, all individuals will have equal fitness, i.e. individuals adopting each tactic leave the same number of offspring. This is known as a mixed ESS. If one tactic becomes more frequent, it will be less successful, and the population will tend to return to the equilibrium. This mixture of tactics may be achieved either through each individual using a strategy of performing each tactic with a probability equal to the equilibrium proportions, or else each individual performs one tactic only and natural selection adjusts the relative numbers of the two types.

Sex roles

A fundamental influence on the mating systems of animals is the difference between the sexes in their contribution to the fertilized egg. Sperm are small and are probably less energetically expensive to produce than eggs. Thus it is often possible for a male to fertilize a female's entire batch of eggs at little cost, and consequently still be able to remate in a relatively short time. Thus, while a female's reproductive output is limited by her own fecundity, a male often has the opportunity to increase his reproductive output through mating with many females. Polygamous mating is thus generally beneficial to males, but rarely so for females. Consequently, the principal way in which a female can increase her reproductive success is through the exercise of mate choice. Both male polygamy and female choosiness mean that, while almost all females can find a mate, many males are unable to breed. Where there is a genetic basis for differential reproductive success of individuals, sexual selection operates, and it generally operates more on males than on females. It should be stressed that these are overall trends and that there are exceptions: a male's supply of sperm is not inexhaustible (Nakatsaru and Kramer, 1982; Baird, 1988); male parental care or female behaviour may limit the possibility of polygyny; females may be polygamous and may benefit by increasing the genetic variability of their offspring; males may exercise mate choice.

10.2   MATE CHOICE

Not all fish exercise mate choice, other than to choose a partner of the correct species and sex. In many species, males appear to accept any female willing to mate – even males of species practising long-term biparental care may show no evidence of choosiness (Barlow, 1986). In the lumpsucker, *Cyclopterus lumpus*, females spawn only once per season, males are conspicuously coloured and perform parental care. Goulet and Green (1988) have shown that despite the apparent opportunity for female choice, neither the number of eggs laid in a nest, nor their survival, is dependent on male size, nest locality or nest quality. It appears that females are not choosy in this species, and that it is possible that there is no real basis on which to choose between males.

Where mate choice occurs, it can be based either on immediate or material benefits or on the delayed benefits of a partner's genetic composition – choice for 'good genes' (review, Halliday, 1983). Into the former category would come characteristics such as mate size, parental experience or nest-site quality which would enhance the survival of its mate or progeny. For males, selection for female fecundity or increased probability of paternity are also immediate benefits. There is no doubting the selective benefits of choice on these bases.

Choice for immediate benefits

Male preference for large, fecund females has been demonstrated in a number of fish species (Table 10.1). When given a choice of two females, male sticklebacks allocate their courtship effort in direct proportion to the relative fecundities of the females (Sargent *et al.*, 1986), but male mosquito fish devote more effort to smaller females than expected (Bisazza *et al.*, 1989) (Fig. 10.1). Size-assortative mating has been demonstrated in biparental cichlids, *Cichlasoma* spp. (Perrone, 1978; Barlow, 1986; McKaye, 1986), but this is not necessarily due to male preference for fecund females: both sexes may choose partners which are more efficient protectors of young. Laboratory experiments have shown that male *Cichlasoma citrinellum* do not appear to express a preference for large females, and the observed pattern may be a result of female preference for large males, coupled with female competition (McKaye, 1986). In the livebearing fish *Poeciliopsis lucida* there is apparently no preference for large females (Keegan-Rogers and Schultz, 1988).

Sperm competition between males may lead to a preference by male livebearing fishes for virgin females or those that have recently given birth (Table 10.1). In some species, females that have recently given birth produce hormones which have the effect of stimulating male mating attempts.

Females of many species are known to discriminate between nest sites of different quality. Often competition between males leads to good nests being occupied by larger males. Laboratory experiments have indicated that, in some species, females choose on the basis of nest quality and others choose larger males directly (Table 10.1).

Where parental care is practised, choice of larger mates may help in increasing offspring survival (Perrone, 1978; Downhower *et al.*, 1983;

Table 10.1 Examples of mate choice for immediate benefits

Species	Family	Source
Choice for large fecund female		
Mottled scupin, *Cottus bairdi*	Cottidae	Downhower and Brown (1980)
Pupfish, *Cyprinodon macularis*	Cyprindontidae	Loiselle (1982)
Goby, *Chaenogobius isaza*	Gobiidae	Hidaka and Takahashi (1987)
Mosquito fish, *Gambusia holbrooki*	Poeciliidae	Bisazza *et al.* (1989)
Threespine stickleback, *Gasterosteus aculeatus*	Gasterosteidae	Sargent *et al.* (1986)
Coho salmon, *Oncorhynchus kisutch*	Salmonidae	Sargent *et al.* (1986)
Choice for virgin or recently spent female		
Molly, *Poecilia latipinna*	Poeciliidae	Farr and Travis (1986)
Mosquito fish, *Gambusia holbrooki*	Poeciliidae	Bisazza *et al.* (1989)
Choice for large males controlling better nest sites		
Goby, *Padogobius martensi*	Gobiidae	Bisazza and Marconato (1988b)
Sculpin, *Cottus hangiongensis*	Cottidae	Goto (1987)
Choice for better nest sites controlled by large males		
Mottled sculpin, *Cottus bairdi*	Cottidae	Downhower *et al.* (1983)
Bullhead, *Cottus gobio*	Cottidae	Bisazza and Marconato (1988a)
Choice for males guarding eggs		
Threespine stickleback, *Gasterosteus aculeatus*	Gasterosteidae	Ridley and Rechten (1981) Belles-Isles *et al.* (1990)
Bullhead, *Cottus gobio*	Cottidae	Marconato and Bisazza (1986)
Painted greenling, *Oxylebius pictus*	Hexagrammidae	De Martini (1985)
Angel blenny *Coralliozetus angelica*	Chaenopsidae	Hastings (1988)
Garibaldi, *Hypsypops rubicundus*	Pomacentridae	Sikkel (1988)
Wrasse, *Symphodus ocellatus*	Labridae	Wernerus *et al.* (1989)
Carmine triplefin, *Axoclinus carminalis*	Tripterygiidae	Petersen (1989)
Damselfishes, *Stegastes dorsopunicans and Microspathodon chrysurus*	Pomacentridae	Petersen (1990)
Fathead minnow, *Pimephales promelas*	Cyprinidae	Unger and Sargent (1988)
Tesselated darter, *Etheostoma olmstedi*	Percidae	Constantz (1985)

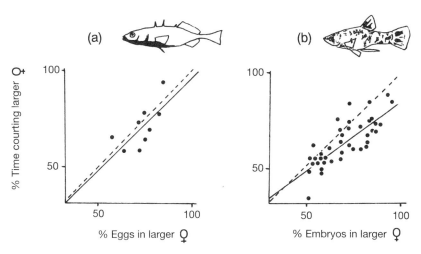

Fig. 10.1 Optimal and non-optimal allocation of courtship time by male fishes. In each experiment males were presented with a simultaneous choice of two females. Solid curves, fit to observed results; broken curves, expected relationship. (a) Male threespine sticklebacks allocated courtship time to females in proportion to the number of eggs they contained. (b) The proportion of time spent courting large female mosquitofish was less than the proportion of embryos they contained. Redrawn from Sargent *et al.* (1986) and Bisazza *et al.* (1989).

Keenleyside *et al.*, 1985; McKaye, 1986). More experienced partners may be better at caring (Colgan and Salmon, 1986) and female *Cichlasoma citrinellum* prefer to pair with more experienced males (Barlow, 1986), although assessing experience of a prospective partner may be difficult – if experience is correlated with size, choosing mates on this basis would seem easier and would often lead to the same end.

Often, females spawn more frequently with males that are already guarding eggs (Table 10.1). Guarding parents may not only protect young: often parental duties are energetically expensive and a parent may compensate for this by eating some of its own offspring. This is particularly likely where a polygamous male guards a number of clutches. He has little to lose, and a little filial cannibalism may increase the likelihood of a male's survival to another season, or allow him to effectively protect the remaining broods. In a recent study (Petersen, 1990) on the damselfish *Segastes dorsipunicans*, it was estimated that 21% of all clutches are eaten by parental males, compared with 20% consumed by predators. This is obviously not in the interests of the female that produces the eggs. Laying eggs in a nest which already contains a number of eggs reduces the probability that a female's brood will be consumed, provided that the male tends to eat a constant number of eggs rather than a fixed proportion (attack abatement effect – see Chapter 12 by Pitcher and Parrish, this volume). Jamieson and Colgan (1989) found that although female sticklebacks spawn more often in nests with eggs than in those without eggs, they generally make their decision before they even see inside the nest (Fig. 10.2). Males with eggs are brighter, court more and females are more likely to follow them to their nests. Male sticklebacks care for free-swimming

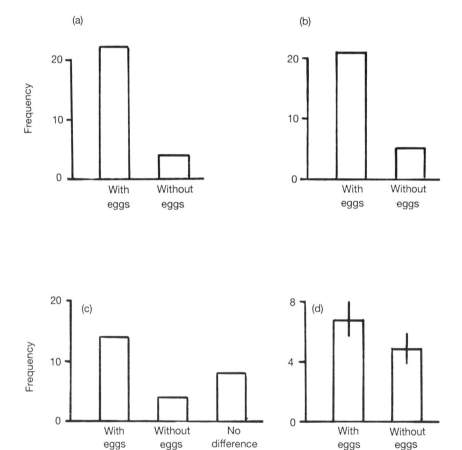

Fig. 10.2 Female sticklebacks prefer males with eggs in their nests because of the males' behaviour and coloration. In a simultaneous choice of males, females significantly preferred to spawn with males whose nests contained eggs (a). However, females generally responded first to males whose nests contained eggs (b): these males were usually brighter (c) and performed more zigzag displays (shown are means and standard errors) (d). Females only entered nests of males without eggs on four occasions, and rejected them. Data from Jamieson and Colgan (1989).

fry and cease courtship 4 days after their first spawning. Males without eggs have no such time constraint and can afford less costly courtship tactics. In other species, males do not care for fry, and may continue to mate throughout the breeding season: thus males might not increase their courtship intensity, and active female choice for nests with eggs is likely to be more important than it is in sticklebacks.

In many species, a male is more likely to eat the last batch of eggs laid. He benefits by getting his meal when his reserves are lowest, and shortens his guarding period. To reduce the possibility of being the female to lay the last egg batch with a particular male, female painted greenling, Garibaldi, and *Stegastes dorsopunicans* (Table 10.1) prefer nests with recently laid eggs.

Choice for good genes

In many fish species, males develop conspicuous colours or extravagant structures when breeding (Fig. 10.3). Many of these 'epigamic' features appear to serve no function in parental care or in fighting between males. Bright colours may make a male more conspicuous and more easily noticed by a female, even if there is no real active preference. However, in laboratory studies, where two or more males are within sight of a female, preference for bright males has been demonstrated in sticklebacks (McPhail, 1969; McLennan and McPhail, 1990), guppies (Breden and Stoner, 1987; Houde, 1987) and cichlids (Hert, 1989).

Epigamic characters may be energetically expensive to make, hamper swimming efficiency, or increase conspicuousness to predators. Sex and species recognition could be achieved through much less costly means. The occurrence of epigamic features demands an explanation not based on immediate benefits. Two main explanations have been proposed: Fisher's runaway sexual selection hypothesis (1930) and Zahavi's handicap principle (1975). Fisher contends that males with characteristics which are attractive to females will have higher fitness through their high number of matings, and thus selection will favour females mating with such males, since their sons will also be attractive. The initial female preference may be for male conspicuousness, viability or some completely randomly chosen feature. To produce really exaggerated male characteristics, Fisher suggests that female preference may be open-ended. For example, if a female always chose to mate with the brighter of two males, the brightest males in the population would mate most often, and any new mutation for even brighter males would be even more popular. Thus, males will tend to become brighter and brighter, until their conspicuousness to predators puts a brake on the process. Theoretical studies have given conflicting assessments of the feasibility of this process (Lande, 1980; O'Donald, 1983), and there are inadequate data available to assess whether Fisher's process occurs (Huntingford, 1984).

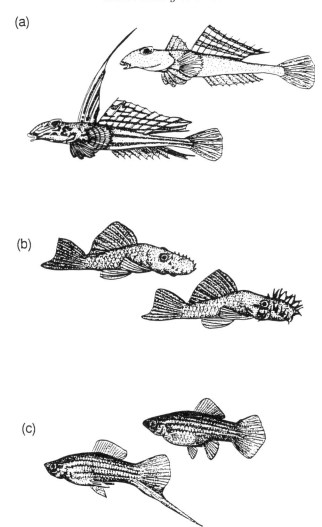

Fig. 10.3 Examples of costly male courtship structures. Males are shown ahead of, and below, females. (a) Bright colours and extended fins in the dragonet, *Callionymus lyra*. (b) Head ornamentation in the catfish, *Ancistrus dolichopterus*. (c) Caudal fin extension in the swordtail, *Xiphophorus helleri*.

Zahavi contends that costly epigamic characters are selected because they are costly, that they are indicative that a male has been able to stay alive and healthy despite the handicap of these features, and that he must therefore possess other good genetic qualities, which though they may be cancelled out by the handicap in his sons, will be entirely beneficial in daughters. As a result of a number of theoretical objections to the workability of the handicap principle (e.g. Davis and O'Donald, 1976; Maynard Smith, 1976), Zahavi (1977) restricted the principle to cases where the quality of male adornment is phenotypically plastic and dependent on male condition as a prerequisite for its manufacture. A male will use as much of his energy reserves as are available to construct courtship structures, which will thus provide an honest signal of his quality. This modified hypothesis appears plausible.

Hamilton and Zuk (1982) have proposed that selection for exaggerated male traits provides females with a way of assessing their partner's susceptibility to parasites. This could work either through a modified handicap principle, where males in worse condition were incapable of producing showy structures, or by means of the structures themselves being directly indicative of parasitic infection, for example a male peacock with intestinal parasites could find it difficult to keep his feathers clean. Indirect evidence supporting this explanation is provided by Ward (1989), who showed that the degree of sexual dichromatism in British freshwater fishes is correlated with the number of parasitic genera reported to infest the species. However, it is not clear that using the number of parasitic genera provides a meaningful measure of a species' history of parasitism.

Milinski and Bakker (1990) have shown that female sticklebacks spend more time near the brighter of two males. Under green light, when the red courting colours cannot be seen, the preference disappears (Fig. 10.4). When the brighter males were infected with white-spot disease (*Ichthyophthirius multifiliis*) and allowed to recover for a few days, their colours were less intense, and they were no longer preferred by females. Thus, female preference for bright colours results in choice for males in good condition. While Milinski and Bakker's results indicate that the handicap principle may be operating in sticklebacks, a more parsimonious explanation is that females are choosing males that will be better able to defend their offspring (a choice for immediate benefits).

While the Fisherian or Zahavian processes are required to explain the evolution of costly courtship characteristics, selection for good genes could occur in other ways. Viability of mates (which may be heritable) can be indicated by size, age and status. In cases where mates provide parental care, these characters are the same as those which aid survival of offspring, and it is not possible to distinguish the two mechanisms. Where no paternal care is provided, it might seem that active female choice can only be for good genes. In many fish species with no parental care or female care, males aggregate in communal display areas, and some males receive more matings than others – are they being selected for good genes?

In the utaka, *Copadichromis eucinostomus*, females brood the eggs in their mouths, and the eggs are only in the open for a few seconds at the most. Again, males aggregate on spawning arenas, and some are favoured by females

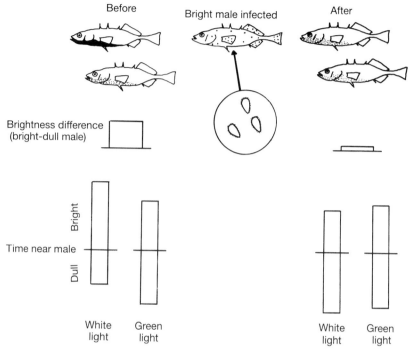

Fig. 10.4 Preference for brightly coloured male sticklebacks leads to females mating with males in better condition. When given a simultaneous choice of two male sticklebacks, females spent more time courting the brighter male. Under green light, when the red colour of the courting male could not be seen, the preference disappeared. After the brighter male had been parasitized and allowed to recover, it was no longer clearly brighter than the other male. Females now showed no preference for either male under white or green light. Data from Milinski and Bakker (1990).

(McKaye, 1984). However, predation on eggs is intense. On one arena alone, six different species of specialized egg predators were recorded, and conspecifics will also eat eggs when possible. Defence of the spawning territory by males appears to be important in ensuring egg survival, and could form a basis for female preference. The wrasse *Halichoeres melanochir* (Moyer and Yogo, 1982) produces planktonic eggs which are not cared for by either parent. Males on central territories are more favoured by females. Central territories may be safer for spawning females as a result of the selfish herd effect (see Pitcher and Parrish, Chapter 12, this volume). Although this may lead to male competition and thus to superior males mating more often, this may be a side-effect of females' concern for their own immediate safety.

Is there any evidence that females force competition on males by randomly choosing a favourite spawning site? In the bluehead wrasse, *Thalassoma*

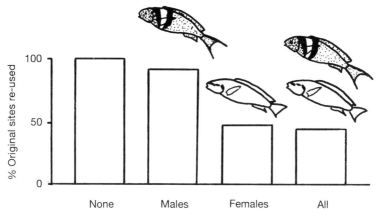

Fig. 10.5 Females choose spawning sites in the bluehead wrasse. Territorial male blue-headed wrasse aggregate on a few spawning sites on a reef. When all fish were removed from a reef and replaced, all the spawning areas previously used were re-used by the fish. When all territorial males were removed and replaced by those from another reef, most sites are again re-used. Removal of all resident females or of all fish led to the abandonment of most of the old nest sites, indicating that females choose spawning sites. Data from Warner (1990).

bifasciatum, some areas are strongly preferred for spawning, while other seemingly similar sites are ignored. In an ingenious set of experiments, Warner (1988, 1990) exchanged entire populations of the wrasse between reefs, and found that breeding sites which had been used continually for 12 years were now ignored (Fig. 10.5), and new sites were apparently chosen at random from within the set of suitable sites. He found that when only females were exchanged between reefs, even though males initially remained at their usual display areas, females chose new spawning sites and the males were forced to move. Young females breeding for the first time apparently learn the position of the new sites from older females. In this way, males have to compete for the females' favoured breeding areas.

These examples show that, although some males get more matings than others, there is only a little circumstantial evidence that female fishes directly choose males on the basis of their genetic qualities.

Constraints on mate choice

It has been demonstrated that choosiness may often be beneficial and that it sometimes occurs, but it must be remembered that there are costs involved (see Parker, 1983, for a theoretical treatment). Risk of predation entailed by leaving the normal home range, loss of foraging opportunities and risk of a potential mate reproducing with a competitor are all good reasons for not being too fussy. Downhhower and Brown (1980) modelled mate choice in female mottled sculpins, which are known to prefer to mate with larger males. They found that the model that best agreed with field data on the distribution of eggs in nests was one where females chose the larger of two males, indicating that very little effort is expended in mate choice in this species. In the topminnow, *Poeciliopsis lucida*, males mating with asexual hybrids do not pass their genes onto future generations. While the males learn to avoid the more abundant hybrid forms, rare female hybrids are not so easily distinguished, and up to ten different sympatric asexual hybrid forms are found in the field (Keegan-Rogers and Schultz, 1988).

Mate choice may also be limited by the behaviour of members of the other sex. Female sticklebacks prefer brighter-coloured males, but dull males can increase their reproductive success by interfering with their more attractive neighbours (McLennan and McPhail, 1990). Where females are aggregated to exploit some resource, a dominant male may control a large area containing several females. This kind of resource-defence polygyny occurs commonly and limits female choice.

While it can be argued that harem polygyny ensures that females mate with the strongest males, one example shows clearly that aggressive dominance by one sex can occur in conflict with the interests of the other sex. Shpigel and Fishelson (1983) studied two species of damselfish, *Dascyllus aruanus* and *D. marginatus*, which live in groups within the protection of branching corals. In the Gulf of Aqaba, in the Red Sea, many of the suitable corals are populated by both species. In any given coral, except for very large ones, there is only one dominant male which is able to breed. Consequently, only females of the same species are able to breed. As predation on fish moving away from the coral is intense, the females of the other species must wait until the dominant male dies, whereupon the largest female (of either species) in the group becomes the dominant male.

10.3   SEXUAL SELECTION IN FEMALE FISHES

While it is to be expected that, in general, sexual selection will be more strongly felt by males, there is growing evidence of its importance in female fishes. Aggressive competition between female fish has been observed in a number of cases. Subordinate female pipefish (Rosenqvist, 1990) are inhibited from developing brightly coloured skin folds which are attractive to males. Female pufferfish, *Canthigaster valentini*, live within the territory of a polygynous male and compete for his attention using a sigmoid abdomen-flexing display

Table 10.2 Colour patterns and structures found in ripe adult female fishes, but not in quiescent females, males or juveniles

Family, species	Female pattern	Breeding behaviour	Source
Pipefish (Syngnathidae)			
Nerophis ophidion	Reddish skin on abdomen	Monogamy; Male care	Rosenqvist (1990)
Wrasse (Labridae)			
Xyrichthys martinicensis	Transparent window over bright ovary	Polygyny; no care	Baird (1988)
Gobies (Gobiidae)			
Chaenogobius isaza	Reddish abdomen	Polygyny; male care	Hidaka and Takahashi (1987)
Cichlids (Cichlidae)			
Chromidotilapia guentheri	Pink abdomen silver dorsal	Monogamy; biparental care	Myrberg (1965)
Cichlasoma festae	Orange with black bars	Monogamy; biparental care	Turner (unpublished)
Cichlasoma friedrichstahli	Yellow and Black	Monogamy; biparental care	Turner (unpublished)
Cichlasoma nicaraguense	Red abdomen, blue cheeks	Monogamy; biparental care	Konings (1989)
Cichlasoma nigrofasciatum	Gold abdomen, turquoise fins	Monogamy; biparental care	Meral (1973)
Cichlasoma salvini	Red abdomen	Monogamy; biparental care	Turner (unpublished)
Crenicichla spp.	Pink abdomen dorsal ocelli	Monogamy; biparental care	Turner (unpublished)
Laetacara curviceps	Blue and black	Monogamy; biparental care	Turner (unpublished)
Microgeophagus ramirezi	Pink abdomen	Monogamy; biparental care	Turner (unpublished)
Pelvicachromis spp.	Bright abdomen, bright dorsal	Monogamy; biparental care	Voss (1980)
Thysia angsorgii	Silver spot above vent	Monogamy; biparental care	Voss (1980)

(Gladstone, 1987). Females of the biparental convict cichlid, *Cichlasoma nigrofasciatum*, exclude rivals from the vicinity of their partner (Meral, 1973). Conspicuous colours are found in courting females, but not in males or juveniles, of a number of taxa (Table 10.2). Limited male fertility could lead to female competition in haremic species such as the straight-tailed razor fish (Baird, 1988) and the pufferfish. Where males are able to care for a single brood at one time, and the quality of care is related to male size, females should compete for the largest male. In pipefishes a small male may not be able to accommodate a full batch of eggs from a large female in his pouch. Many of the conspicuous colours adopted by females are located on the abdomen (Table 10.2), suggesting that an exaggeration of fecundity increases their chance of being chosen by a male.

10.4 POLYGAMY

Since males are generally able to produce enough sperm to mate more frequently than females, polygyny should be frequent. Where females occupy restricted ranges, it is often possible for a male to maintain a territory encompassing those of several females. Harems of this kind are found, for example, in species of wrasse (Robertson and Hoffman, 1977; Thresher, 1979), pufferfish (Gladstone, 1987), razorfish (Baird, 1988), damselfish (Shpigel and Fishelson, 1983), sea basses (Petersen and Fischer, 1986) and cichlids (Kuwamura, 1986; Kuwamura *et al.*, 1989). Perrone and Zaret (1979) suggest that harem polygyny occurs in small species which require refuges from predators, while Breitburg (1987) believes that only larger species which are able to drive off interspecific competitors can control the resources necessary for harem formation, and Moyer and Yogo (1982) suggest that low population density prevents harem formation. A thorough comparative study is lacking.

In cases where the male protects eggs on the substrate, he may easily care for a number of batches at little or no extra cost: polygamy through control of spawning sites has been suggested as a major determinant of the predominance of paternal care in fishes (Baylis, 1981). This strategy is used by sticklebacks (see Chapter 16 by FitzGerald and Wootton, this volume), sculpins (Downhower *et al.*, 1983; Marconato and Bisazza, 1986), gobies (Bisazza and Marconato, 1988b) and some damselfish (Petersen, 1990).

Where males do not provide parental care, arena polygamy is found in some mouthbrooding cichlids (McKaye, 1984), and pelagic spawning reef fishes. In bluegill sunfish, *Lepomis macrochirus*, arenas are formed by guarding males, probably for the protection of their eggs (Gross and MacMillan, 1981). In both arena spawners and paternal guarders, unlike in haremic species, the mating success of territorial males is dependent on female choice.

10.5 ALTERNATIVE MALE STRATEGIES

If large, dominant or attractive males are successful at obtaining a large number of mates, what is the best strategy for a small, subordinate or unattractive male? One answer is to be a female, but this option does not appear to be available in all taxa. In many species, however, smaller males do not establish territories or develop breeding colours, but remain at the breeding grounds and attempt to fertilize the eggs of females spawning with territorial males (see also Chapter 13 by Magurran, this volume). These 'sneakers' are reported from a variety of taxa, including wrasse (Warner and Hoffman, 1980; Taborsky *et al*, 1987), sunfish (Gross, 1984), salmonids (Gross, 1984; Sigurjonsdottir and Gunnarson, 1989) and cichlids (Chan and Ribbink, 1990).

In the bluegill sunfish (Gross 1984) and the wrasse, *Symphodus ocellatus* (Taborsky *et al.*, 1987; van den Berghe *et al.*, 1989), there are two types of non-territorial males: sneakers and satellites. Sneakers are small, cryptic and resemble females. They rush into the nest during spawning: this tactic achieves few fertilizations. Satellite sunfish mimic females. Satellite wrasse are localized around the territory of a single dominant male. They defend the territory against other small males, adopt some bright colours, and actively court females, although most of their reproduction appears to be through sneaking. In these species a sneaker may graduate into satellite, which is a more successful tactic. Gross (1984) found evidence that, in sunfish, neither sneakers nor satellites grow into territorial males, and suggested that this may be a genetically based mixed ESS, in which the lifetime fitness of sneaker/satellite males and territorial males may be equal. Sneaker sunfish are on average 2–3 years old, satellites 4, and territorials 8–9. Thus, the higher success of the territorial tactic may be offset by the fact that reproduction is delayed. Taborsky *et al.* (1987) found that when a territorial male wrasse was removed, it was never replaced by a satellite, so it is possible that there is also a genetic basis to alternative male mating tactics in this species. However, Chan and Ribbink (1990) have shown that sneakers of the cichlid *Pseudocrenilabrus philander* can become territorial males, and so this species follows a flexible conditional strategy. This may be more appropriate for a short-lived species, such as *P. philander*, which lives in fluctuating temporary pools which may contain a small number of adults. However, it is difficult to see why wrasse and sunfish lack such flexibility.

10.6 MONOGAMY

It has been stressed that there is a strong selection pressure on males to be polygamous. However, apparently monogamous pairing is not unknown in fishes. Where biparental care increases fry survival appreciably, long-term pair bonds and care of fry may reduce the potential for polygamy: this occurs in many substrate-brooding cichlids (Barlow, 1986; Rogers, 1988). In a Tanganyikan mouthbrooder, *Xenotilapia flavipinnis*, a pair stays together to raise a succession of broods in the same territory (Yanagisawa, 1986). Competition for territories may favour monogamy in this and other species (McKaye, 1984; Kuwamura *et al.*, 1989) and defence of feeding territories or refugia, even when no parental care is given, can lead to pairing and thence to monogamy (Fricke and Fricke, 1977; Perrone and Zaret, 1979; Baylis, 1981; Pressley, 1981; Driscoll and Driscoll, 1988; Donaldson, 1989; Lobel, 1989). In the predatory harlequin bass, *Serranus tigrinus*, pairs of fish hunt co-operatively, and generally spawn together (Pressley, 1981).

The occurrence of monogamous pairing does not mean that there is no selection pressure on males to be polygamous. The benefits of biparental care may be offset for males if there is sufficient opportunity for polygamy or if male care provides little extra benefit to fry. In normally biparental cichlids, a female-biased sex ratio can encourage desertion (Keenleyside, 1983) or bigamy by males (Limberger, 1983; Keenleyside, 1985). Such bigamous males provide some care for their offspring, but even monogamous males of these species invest relatively little in parental care. Males of monogamously paired butterfly fishes often attempt to mate with females occupying adjacent territories (Lobel, 1989). In Lake Tanganyika cichlids, fry that feed in the water column require more protection from predators, and monogamous pairs form, whereas in species with benthic fry, protection is provided by the female alone, freeing the males to establish harems (Kuwamura, 1986; Nagoshi and Gashagaza, 1988). Females of another Lake Tanganyika species, *Perissodus microlepis*, when deserted by their mates, may dump their fry into the protection of other parental cichlids (Yanagisawa, 1985), thus lowering the cost to males that desert. Monogamy is thus a rare system in fishes: it can be essentially regarded as harem polygyny except that unusual circumstances generally limit the harem to one female.

10.7   SEX CHANGE

If there is a greater advantage to being large in one sex than in the other, selection would favour sex change provided that the cost of the change was not too high. This is known as the size-advantage hypothesis (Ghiselin, 1969; Warner, 1975). One consequence of sex change is that there is no selection for equality of sex ratios, and in polygynous species adult sex ratios are generally female biased. This will lead to a higher population fecundity than in normal dioecious species.

There is a great variety of sex-determining mechanisms in fish (e.g.

Yamamoto, 1969), and the structural differences associated with reproduction are less in externally fertilizing species than in internally fertilizing animals such as birds and mammals. Thus there may be less phylogenetic or ontogenetic constraint on sex change in fishes than other taxa. Sex changing in fishes in widely distributed, being found in at least four orders including seven families of Perciformes (Shapiro, 1984), and has probably evolved on at least 12 separate occasions (Charnov, 1982). The size-advantage hypothesis predicts that where large males have an opportunity to achieve many matings, individuals should first be females and change to males at larger sizes (protogyny), but that otherwise the benefits of increasing fecundity in larger females should favour change from male to female (protandry). Anemone fishes, *Amphiprion* spp., live in the shelter of a sea anemone, and as predation on individuals crossing between anemones is high, there is little opportunity for polygamy and the fish live as monogamous pairs. These fishes are protandrous. In polygamous species, large males can either control a social group of several females, as in the cleaner wrasse, *Labroides dimidiatus* (Robertson and Hoffman, 1977), or exclude other males from favoured territories, as in the rainbow wrasse (Warner and Hoffman, 1980). These species are protogynous.

However, once these systems have evolved, they exhibit varying degrees of stability. In a population of the anemone fish *Amphiprion clarkii* living in an area with a high number of hosts and low predation pressure, males did not exploit the potential for polygamy or grow to larger sizes (Ochi, 1989). In fact females were sometimes able to control two territories, each with a resident male, although apparently they only chose to spawn with one partner. Perhaps polygyny is prevented by the fact that males care for the eggs, and that females aggressively exclude others from their territory, preventing the smaller male from attracting further mates. Presumably male care which did not interfere with polygyny had already developed in the ancestors of the anemone fishes, as it is found in most of the other pomacentrids.

In fishes, the timing of sex change is generally environmentally induced, and depends on social factors. Removal of a male from populations of protogynous species results in a large female becoming a male in *Anthias squamipinnis*, *Thalassoma duperrey*, cleaner wrasse (Shapiro, 1984) and some damselfish (*Dascyllus aruanus* and *D. marginatus*) (Shpigel and Fishelson, 1983). In rainbow wrasse, there are both primary males and protogynous hermaphrodites. At high population density, the success of large territorial males is reduced by interference by the small female-coloured primary males, and fewer females change sex (Warner and Hoffman, 1980). In *T. duperrey*, if a large female is confined in an aquarium with a small primary male, which does not court, the female will change sex (Shapiro, 1984). Thus the mechanism of sex change in *Thalassoma* would appear to involve assessment of the bahaviour of the territorial secondary males by the females.

Changing sex takes time, but in a small social group where there is little immigration of competitors, such as those of the cleaner wrasse and the damselfishes at Aqaba, when the single dominant male is lost, there is no possibility of breeding unless one fish changes sex anyway. In the anemone fish, loss of a partner rarely leads to sex change by a resident male (Ochi, 1989). In the parrotfish *Sparisoma aurofrenatum* (Shapiro, 1984) and the angelfish *Centropyge bicolor* (Aldenhoven, 1986), larger females may change sex and become subordinate (and non-breeding) males: these can more rapidly assume the role of a dominant male than females that wait for a vacancy before changing sex. In the angelfish, this subordinate male strategy is favoured by high territory density, which inceases the number of vacancies, and thus reduces the loss of breeding time in fish that change sex early.

10.8   SIMULTANEOUS HERMAPHRODITISM

As well as sex-changing species, there are also simultaneous hermaphrodites among the reef fish, and several recent studies have investigated the factors which stabilize such systems. The monogamous harlequin bass is a simultaneous hermaphrodite. In *Serranus baldwini* (Petersen and Fischer, 1986) and *S. fasciatus* (Fischer and Petersen, 1987), dominant males control a harem of simultaneous hermaphrodites, which reproduce almost entirely as females. Why do they retain male function at all? Petersen and Fischer have considered several explanations. The opportunity for sneaking, while quite high (9%) in *S. fasciatus*, is much lower in *S. baldwini*, and is unlikely to be of much significance. The sex allocation pattern in *S. baldwini* may be advantageous in other habitats within its range. The possession of male tissue may shorten the delay in switching to a dominant male, which as we have already seen may be a significant factor in protogynous species. Finally, it may be that the development of males may be recent and the maintenance of hermaphrodites is a historical legacy likely to be eliminated by selection in time.

Black hamlets, *Hypoplectrus nigricans*, have no males and are pure simultaneous hermaphrodites. Each defends its own feeding territory, which it leaves to travel to a breeding territory in the late evening. As fertilizing eggs is less costly than laying them, it might be expected that individuals should try to behave as males wherever possible and lead other individuals into behaving as females (Fischer and Petersen, 1987, and references therein). This is avoided through egg-trading. Each individual adopting the female role will only lay a small number of eggs, and not its full batch. It then fertilizes a batch of eggs laid by its partner. In this way there is little opportunity for specialization as a male, because an individual that finds a partner which does not lay eggs in turn can abandon a male specialist and look for another, more co-operative mate. As mating occurs late in the evening, there may be few opportunities for further mating in the day. If the population consisted initially of egg traders, a male specialist would only be able to fertilize a single batch with each mate, while its egg-trading partner would have little difficulty in finding another egg-trader to co-operate in producing two full spawns. If egg-traders can recognize reliable trading partners and spawn repeatedly with them, they would waste less time than those that mate randomly and thus might mate with male specialists. In the black hamlet, most individuals spawn each day with the same partner. Thus mating in this species depends on cooperation through reciprocation, perhaps with individually recognizable partners. The chalk bass, *Serranus tortugarum*, has a similar mating system (Fisher, 1984).

10.9   SEXUAL AND ASEXUAL REPRODUCTION

The evolution and maintenance of sexual reproduction is a complex and controversial subject. The problem is that an asexually reproducing individual should produce twice as many descendants as one that produces males and females, since in the latter case it takes two individuals to produce each batch of fertilized eggs, while each asexual individual can do this on its own. One popular explanation (Fischer, 1930) for the advantage of sexuality is that it enables rapid adaptation within a population through the mixing up of genes between different individuals. This ensures that the genes for sexual reproduction are present in a wide variety of genotypes, of which, in a rapidly changing environment, some are likely to be well adapted to any eventuality. Thus, while an asexual population may prosper through its twofold fecundity advantage, it will eventually be extinguished by an environmental change with which it is not equipped to deal.

In fishes, one well-documented case of self-fertilization (i.e. simultaneous hemaphroditism) occurs in *Rivulus marmoratus* (Harrington, 1961). In poeciliid fishes, parthenogenetic clones have been produced through natural hybridization between female *Poeciliopsis monacha* and male *P. lucida*. The hybrid offspring, termed *P. monacha-lucida*, are all females, which produce eggs containing almost entirely maternal genes, and thus if they mate with *P. monacha* males, they produce normal *P. monacha* offspring. Matings with *P. lucida* produce more of the hybrid 'hemi-clone', in which the maternal genes are passed down unchanged as in a clonal or asexual organism, but while the paternal genes are expressed in the offspring, they are not incorporated into the gametes. Occasionally the hybrids produce diploid eggs containing paternal as well as maternal genes. When fertilized, these become triploid individuals, designated *P. monacha 2-lucida*. These produce triploid eggs which require sperm for stimulation to develop, but not for fertilization, and are thus true clones (Schultz, 1969). How do the clonal and sexual populations compare in terms of their ability to cope with a changeable environment? A particularly rapidly changing component of the environment is provided by parasites. With their large fecundities and short generation times, parasites are capable of quickly developing resistance to host defences. Lively *et al.* (1990) showed that the sexual species, *P. monacha*, had much lower parasite infestation than members of the most abundant sympatric clone. However, when the genetic variation in the sexual species was reduced through a temporary reduction in population size, or bottleneck, it was also highly susceptible, but recovered when new genetic material was introduced from another area.

10.10   SUMMARY

Teleost fishes exhibit a tremendous variety of reproductive strategies. Sexual selection acts mainly, but not entirely, on males. Females rarely gain much from polyandry, but profit by choice of males for immediate and perhaps genetic qualities. However, mate choice can carry a cost and may be constrained by the behaviour of members of the other sex. Males principally increase their fitness through multiple matings, although this may be limited by the need for parental care and joint territory defence. Alternative reproductive tactics – sneakers and satellites – exist as phenotype-limited conditional strategies and probably as part of a genetically determined mixed ESS. Gender may be considered as a behavioural strategy, and the stability of sex-changing and simultaneously hermaphroditic mating systems are discussed. Asexual populations of fishes are known, but despite local success there is evidence that these suffer from an inability to cope with fluctuating environments.

ACKNOWLEDGEMENTS

I thank Rosanna Robinson, particularly for helping to sort out my collection of reprints, and the staff of the Monkey Bay Fisheries Research Station (Malawi) and of the JLB Smith Institute of Ichthyology (South Africa) for assisting my searches through their libraries. Jay Stauffer and Dennis Tweddle kindly commented on an earlier draft of this chapter.

REFERENCES

Aldenhoven, J. (1986) Different reproductive strategies in a sex-changing coral reef fish *Centropyge bicolor* (Pomacanthidae). *Aust. J. mar. Freshwat. Res.*, **37**, 353–60.
Baird, T. A. (1988) Abdominal windows in straight-tailed razor-fish, *Xyrichtys martinicensis*: an unusual female sex character in a polygynous fish. *Copeia*, **1988(2)**, 496–9.

Barlow, G. W. (1986) Mate choice in the monogamous and polychromatic Midas cichlid, *Cichlasoma citrinellum. J. Fish Biol.,* **29** (Supp. A), 123–33.

Baylis, J. R. (1981) The evolution of parental care in fishes, with particular reference to Darwin's rule of male sexual selection. *Env. Biol. Fishes,* **6**, 233–51.

Belles-Isles, J.-C., Cloutier, D. and FitzGerald, G. J. (1990) Female cannibalism and male courtship tactics in threespine sticklebacks. *Behav. Ecol. Sociobiol.,* **26**, 363–8.

van den Berghe, E. P., Wernerus, F. and Warner, R. R. (1989) Female choice and the mating cost of peripheral males. *Anim. Behav.,* **38**, 875–84.

Bisazza, A. and Marconato, A. (1988a) Female mate choice, male–male competition and parental care in the river bullhead, *Cottus gobio* L. (Pisces: Cottidae). *Anim. Behav.,* **36**, 1352–60.

Bisazza, A. and Marconato, A. (1988b) Reproductive strategies in fish with parental care. *Monitore zool. ital.,* **22**, 497–8.

Bisazza, A., Marconato, A. and Marin, G. (1989) Male mate preferences in the mosquitofish *Gambusia holbrooki. Ethology,* **83**, 335–43.

Breden, F. and Stoner, G. (1987) Male predation risk determines female preference in the Trinidad guppy. *Nature, Lond.,* **329**, 831–3.

Breitburg, D. L. (1987) Interspecific competition and the abundance of nest sites: factors affecting sexual selection. *Ecology,* **68**, 1844–55.

Chan, T.-Y. and Ribbink, A. J. (1990) Alternative reproductive behaviour in fishes, with particular reference to *Lepomis macrochira* and *Pseudocrenilabrus philander. Env. Biol. Fishes,* **28**, 249–56.

Charnov, E. L. (1982) *The Theory of Sex Allocation* (Monographs in Population Biology, Vol. 18), Princeton University Press, Princeton, NJ.

Colgan, P. W. and Salmon, A. B. (1986) Breeding experience and parental behaviour in convict cichlids (*Cichlasoma nigrofasciatum*). *Behav. Process.,* **13**, 101–18.

Constantz, G. D. (1985) Allopaternal care in the tesselated darter, *Etheostoma olmstedi* (Pisces: Percidae). *Env. Biol. Fishes,* **14**, 175–83.

Davis, J. W. F. and O'Donald, P. (1976) Sexual selection for a handicap, a critical analysis of Zahavi's model. *J. theor. Biol.* **57**, 345–54.

Dawkins, R. (1976) *The Selfish Gene,* Oxford University Press, Oxford.

De Martini, E. E. (1985) Social behaviour and coloration changes in painted greenling, *Oxylebius pictus* (Pisces: Hexagrammidae). *Copeia,* **1985(4)**, 966–75.

Donaldson, T. J. (1989) Facultative monogamy in obligate coral-dwelling hawkfishes (Cirrhitidae). *Env. Biol. Fishes,* **26**, 295–302.

Downhower, J. F. and Brown, L. (1980) Mate preferences of female mottled sculpins, *Cottus bairdi. Anim. Behav.,* **28**, 728–34.

Downhower, J. F., Brown, L., Pederson, R. and Staples, G. (1983) Sexual selection and sexual dimorphism in sculpins. *Evolution,* **37**, 96–103.

Driscoll, J. W. and Driscoll, J. L. (1988) Pair behaviour and spacing in butterflyfishes (Chaetodontidae). *Env. Biol. Fishes,* **22**, 29–37.

Farr, J. A. and Travis, J. (1986) Fertility advertisement by female sailfin mollies, *Poecilia latipinna* (Pisces: Poeciliidae). *Copeia,* **1986(2)**, 467–72.

Fischer, E. A. (1984) Egg trading in the chalk bass, *Serranus tortugarum,* a simultaneous hermaphrodite. *Z. Tierpsychol.,* **66**, 143–51.

Fischer, E. A. and Petersen, C. W. (1987) The evolution of sexual patterns in the seabasses. *Bioscience,* **37**, 482–9.

Fisher, R. A. (1930) *The Genetical Theory of Natural Selection,* Clarendon, Oxford.

Fricke, H. W. and Fricke, S. (1977) Monogamy and sex-change by aggressive dominance in coral reef fish. *Nature, Lond.,* **266**, 830–32.

Ghiselin, M. T. (1969) The evolution of hermaphrodism among animals. *Q. Rev. Biol.,* **44**, 180–208.

Gladstone, W. (1987) The courtship and spawning behaviours of *Canthigaster valentini* (Tetraodontidae). *Env. Biol. Fishes*, **20**, 255–61.

Goto, A. (1987) Polygyny in the river sculpin, *Cottus hangionensis* (Pisces: Cottidae), with special reference to male mating success. *Copeia*, **1987(1)**, 32–40.

Goulet, D. and Green, J. M. (1988) Reproductive success of the male lumpfish (*Cyclopterus lumpus* L.) (Pisces: Cyclopteridae): evidence against female mate choice. *Can. J. Zool.*, **66**, 2513–19.

Gross, M. R. (1984) Sunfish, salmon and the evolution of alternative reproductive strategies and tactics in fishes, in *Fish Reproduction: Strategies and Tactics* (eds G. W. Potts and R. J. Wootton), Academic Press, London, pp. 55–75.

Gross, M. R. and MacMillan, A. M. (1981) Predation and the evolution of colonial nesting in bluegill sunfish (*Lepomis macrochirus*). *Behav. Ecol. Sociobiol.*, **8**, 163–74.

Halliday, T. R. (1983) The study of mate choice, in *Mate Choice* (ed. P. P. G. Bateson), Cambridge University Press, Cambridge, pp. 3–32.

Hamilton, W. D. and Zuk, M. (1982) Heritable true fitness and bright birds. A role for parasites? *Science*, **218**, 384–7.

Harrington, R. W. (1961) Oviparous hermaphrodite fish with internal fertilisation. *Science*, **134**, 1749–50.

Hastings, P. A. (1988) Female choice and male reproductive success in the angel blenny, *Corralliozetus angelica* (Blennoidea: Chaenopsidae). *Anim. Behav.*, **36**, 115–24.

Hert. E. (1989) The function of egg-spots in an African mouthbrooding cichlid fish. *Anim. Behav.*, **37**, 726–32.

Hidaka, T. and Takahashi, S. (1987) Reproducive strategy and interspecific competition in the lake-living gobiid fish isaza, *Chaenogobius isaza*. *J. Ethol.*, **5**, 185–96.

Houde, A. E. (1987) Mate choice based on naturally occurring color-pattern variation in a guppy population. *Evolution*, **41**, 1–10.

Huntingford, F. A. (1984) *The Study of Animal Behaviour*, Chapman and Hall, London.

Jamieson, I. G. and Colgan, P. W. (1989) Eggs in the nests of males and their effect on mate choice in the three-spined stickleback. *Anim. Behav.*, **38**, 859–65.

Keegan-Rogers, V. and Schultz, R. J. (1988) Sexual selection among clones of unisexual fish (*Poeciliopsis*, Poecilidae): genetic factors and rare female advantage. *Am. Nat.*, **132**, 846–68.

Keenleyside, M. H. A. (1983) Mate desertion in relation to adult sex ratio in the biparental cichlid fish *Herotilapia multispinosa*. *Anim. Behav.*, **31**, 683–8.

Keenleyside, M. H. A. (1985) Bigamy and mate choice in the biparental cichlid fish *Cichlasoma nigrofasciatum*. *Behav. Ecol. Sociobiol.*, **17**, 285–90.

Keenleyside, M. H. A., Rangeley, R. W. and Kuppers, B.U. (1985) Female mate choice and male parental defense behaviour in the cichlid fish *Cichlasoma nigrofasciatum*. *Can. J. Zool.*, **63**, 2489–93.

Konings, A. (1989) *Cichlids from Central America*, TFH Publications, Neptune City, NJ.

Kuwamura, T. (1986) Parental care and mating systems of cichlid fishes in Lake Tanganyika: a preliminary field survey. *J. Ethol*, **4**, 129–46.

Kuwamura, T., Nagoshi, M., and Sato, T. (1989) Female-to-male shift in mouthbrooding in a cichlid fish, *Tanganicodus irsacae*, with notes on breeding habits of two related species in Lake Tanganyika. *Env. Biol. Fishes*, **24**, 187–98.

Lande, R. (1980) Sexual dimorphism, sexual selection and adaptation in polygenic characters. *Evolution*, **34**, 292–305.

Limberger, D. (1983) Pairs and harems in a cichlid fish, *Lamprologus brichardi*. *Z. Tierpsychol.*, **62**, 115–44.

Lively, C. M., Craddock, C. and Vrijenhoek, R. C. (1990) Red queen hypothesis supported by parasitism in sexual and clonal fish. *Nature, Lond.*, **344**, 846–66.

Lobel, P. S. (1989) Spawning behaviour of *Chaetodon multicinctus*: pairs and intruders. *Env. Biol. Fishes*, **24**, 125–130.

Loiselle, P. V. (1982) Male spawning-partner preference in an arena breeding teleost, *Cyprinodon macularis californiensis* Girard (Atherinomorpha: Cyprinodontidae). *Am. Nat.*, **120**, 721–32.

McKaye, K. R. (1984) Behavioural aspects of cichlid reproductive strategies: patterns of territoriality and brood defence in Central American substratum spawners and African mouthbrooders, in *Fish Reproduction: Strategies and Tactics* (eds G. W. Potts and R. J. Wootton), Academic Press, London, pp. 243–73.

McKaye, K. R. (1986) Mate choice and size assortative pairing by the cichlid fishes of Lake Jiloa, Nicaragua. *J. Fish Biol.*, **29** (Supp. A), 135–50.

McLennan, D. A. and McPhail, J. D. (1990) Experimental investigations of the evolutionary significance of sexually dimorphic nuptial coloration in *Gasterosteus aculeatus* (L.): the relationship between male colour and female behaviour. *Can. J. Zool.*, **68**, 482–92.

McPhail, J. D. (1969) Predation and the evolution of a stickleback (*Gasterosteus*). *J. Fish. Res. Bd Can.*, **26**, 3183–208.

Marconato, A. and Bisazza, A. (1986) Males whose nests contain eggs are preferred by female *Cottus gobio* L. (Pisces, Cottidae). *Anim. Behav.*, **34**, 1580–82.

Maynard Smith, J. (1976) Sexual selection and the handicap principle. *J. theor. Biol.*, **57**, 230–42.

Maynard Smith, J. (1982) *Evolution and the Theory of Games*, Cambridge University Press, Cambridge.

Meral, G. H. (1973) The adaptive significance of territoriality in new world cichlidae, PhD thesis, University of California, Berkeley, 394 pp.

Milinski, M. and Bakker, T. C. M. (1990) Female sticklebacks use male coloration in mate choice and hence avoid parasitised males. *Nature, Lond.*, **344**, 330–33.

Moyer, J. T. and Yogo, Y. (1982) The lek-like mating system of *Halichoeres melanochir* (Pisces: Labridae) at Miyake-Jima, Japan. *Z. Tierpsychol*, **60**, 209–26.

Myrberg, A. A. (1965) A descriptive analysis of the behaviour of the African cichlid fish *Pelmatochromis guentheri*. *Anim. Behav.*, **13**, 312–29.

Nagoshi, M. and Gashagaza, M. M. (1988) Growth of larvae of a Tanganyikan cichlid, *Lamprologus attenuatus*, under parental care. *Jap. J. Ichthyol.*, **35**, 392–5.

Nakatsaru, K. and Kramer, D. L. (1982) Is sperm cheap? Limited male fertility and female choice in the lemon tetra (Pisces, Characidae). *Science*, **216**, 753–5.

Ochi, H. (1989) Mating behaviour and sex-change of the anemone fish *Amphiprion clarkii*, in the temperate waters of southern Japan. *Env. Biol. Fishes*, **26**, 257–75.

O'Donald, P. (1983) Sexual selection by female choice, in *Mate Choice* (ed. P. P. G. Bateson), Cambridge University Press, Cambridge, pp. 53–66.

Parker, G. A. (1983) Mate quality and mating decisions, in *Mate Choice* (ed. P. P. G. Bateson), Cambridge University Press, Cambridge, pp. 141–66.

Parker, G. A. (1984) Evolutionarily stable strategies, in *Behavioural Ecology*, 2nd edn (eds J. R. Krebs and N. B. Davies), Blackwell Scientific Publications, Oxford, pp. 30–61.

Perrone, M. (1978) Mate size and breeding success in a monogamous cichlid fish. *Env. Biol. Fishes*, **3**, 193–201.

Perrone, M. and Zaret, T. M. (1979) Parental care patterns of fishes. *Am. Nat.*, **113**, 351–63.

Petersen, C. W. (1989) Females prefer mating males in the carmine triplefin, *Axoclinus carminalis*, a paternal brood guarder. *Env. Biol. Fishes*, **26**, 213–21.

Petersen, C. W. (1990) The occurrence and dynamics of clutch loss and filial cannibalism in two Caribbean damselfishes. *J. exp. Mar. Biol. Ecol.*, **135**, 117–33.

Petersen, C. W. and Fischer, E. A. (1986) Mating system of the hermaphroditic coral-reef fish, *Serranus baldwini*. *Behav. Ecol. Sociobiol.*, **19**, 171–8.

Pressley, P. H. (1981) Pair formation and joint territoriality in a simultaneous hermaphrodite: the coral reef fish *Serranus tigrinus*. *Z. Tierpsychol.*, **56**, 33–46.

Ridley, M. and Rechten, C. (1981) Female sticklebacks prefer to spawn with males whose nests contain eggs. *Behaviour*, **76**, 152–61.

Robertson, D. R. and Hoffman, S. G. (1977) The role of female mate choice and predation in the mating systems of tropical labroid fishes. *Z. Tierpsychol.*, **45**, 298–320.

Rogers, W. (1988) Parental investment and division of labor in the Midas Cichlid (*Cichlasoma citrinellum*). *Ethology*, **79**, 126–42.

Rosenqvist, G. (1990) Male mate choice and female–female competition for mates in the pipefish *Nerophis ophidion*. *Anim. Behav.*, **39**, 1110–15.

Sargent, R. C., Gross, M. and van den Berghe, E. P. (1986) Male mate choice in fishes. *Anim. Behav.*, **34**, 545–60.

Schultz, R. J. (1969) Hybridisation, unisexuality and polyploidy in the teleost *Poeciliopsis* (Poeciliidae) and other vertebrates. *Am. Nat.*, **103**, 605–19.

Shapiro, D. Y. (1984) Sex reversal and sociodemographic processes in coral reef fishes, in *Fish Reproduction: Strategies and Tactics* (eds G. W. Potts and R. J. Wootton), Academic Press, London, pp. 103–18.

Shpigel, M. and Fishelson, L. (1983) Ecology and sociobiology of coexistence in two species of *Dascyllus* (Pomacentridae: Teleostei). *Bull. Inst. Oceanog. Fish.*, **9**, 207–24.

Sigurjonsdottir, H. and Gunnarsson, K. (1989) Alternative mating tactics of Arctic Charr, *Salvelinus alpinus*, in Thingvallavatin, Iceland. *Env. Biol. Fishes*, **26**, 159–76.

Sikkel, P. (1988) Factors influencing spawning site choice by female Garibaldi, *Hypsypops rubicundus* (Pisces: Pomacentridae). *Copeia*, **1988(3)**, 710–18.

Taborsky, M., Hudde, B. and Wirtz, P. (1987) Reproductive ecology of *Symphodus* (*Crenilabrus*) *ocellatus* – a European wrasse with four types of male behaviour. *Behaviour*, **102**, 82–118.

Thresher, R. E. (1979) Social behaviour and ecology of two sympatric wrasses (Labridae: *Halichoeres* spp.) off the coast of Florida. *Mar. Biol.*, **53**, 161–72.

Unger, L. M. and Sargent, R. C. (1988) Allopaternal care in the fathead minnow, *Pimephales promelas*: females prefer males with eggs. *Behav. Ecol. Sociobiol.*, **23**, 27–32.

Voss, J. (1980) *Colour Patterns of African Cichlids*, T. F. H. Publications, Neptune City, NJ.

Ward, P. (1989) Sexual showiness and parasitism in freshwater fish: combined data from several isolated water systems. *Oikos*, **55**, 428–9.

Warner, R. R. (1975) The adaptive significance of sequential hermaphrodism in animals. *Am. Nat.*, **109**, 61–82.

Warner, R. R. (1988) Sex-change and the size advantage models. *Trends Ecol. Evol.*, **3**, 133–6.

Warner, R. R. (1990) Male versus female influences on mating-site determination in a coral reef fish. *Anim. Behav.*, **39**, 540–48.

Warner, R. R. and Hoffman, S. G. (1980) Local population size as a determinant of mating systems and social composition in two tropical marine fishes, *Thalassoma* sp. *Evolution*, **34**, 508–18.

Wernerus, F. M., Lejeune, P. and van den Berghe, E. P. (1989) Transmission of mating success among neighbouring males in the Mediterranean labrid fish *Symphodus ocellatus*. *Biol. Behav.*, **14**, 195–206.

Yamamoto, T. (1969) Sex differentiation, in *Fish Physiology*, Vol. III (eds W. S. Hoar and D. J. Randall), Academic Press, New York, pp. 117–75.

Yanagisawa, Y. (1985) Parental strategy of the cichlid fish *Perissodus microlepis*, with special reference to intraspecific brood 'farming out'. *Env. Biol. Fishes*, **12**, 241–9.

Yanagisawa, Y. (1986) Parental care in a monogamous mouthbrooding cichlid *Xenotilapia flavipinnis*. Jap. J. Ichthyol., **33**, 249–61.

Zahavi, A. (1975) Mate selection – a selection for a handicap, *J. theor. Biol.*, **53**, 205–14.

Zahavi, A. (1977) The cost of honesty. (Further remarks on the handicap principle). *J. theor. Biol.*, **67**, 603–5.

Chapter eleven

Williams' principle: an explanation of parental care in teleost fishes

Robert Craig Sargent and Mart R. Gross

11.1   INTRODUCTION

Parental care may be defined as an association between parent and offspring after fertilization that enhances offspring survivorship. This phenomenon has attracted the attention of evolutionary biologists since Darwin; however, it was not until the recent development of behavioural ecology and sociobiology (e.g. Williams, 1966a; Trivers, 1972; Alexander, 1974; Wilson, 1975) that the variety of parental care patterns in animals has attracted such rigorous study. Perhaps because we are mammals, we tend to think of parental care as being the principal occupation of females, possibly with some help from males. A survey of the vertebrates, however, reveals that mammals are merely at one end of the spectrum, with predominantly female care, and fishes are at the other end, with predominantly male care (Table 11.1). Within teleost fishes with external fertilization (about 85% of all teleost families), one finds that the four states of parental care, ranked in descending order of their frequencies, are: no care, male care, biparental care, and female care. This seemingly peculiar trend has attracted considerable attention from evolutionary biologists, who have proposed several hypotheses about the origins of parental care in fishes (reviews, Maynard Smith, 1977; Blumer, 1979; Perrone and Zaret, 1979; Baylis, 1981; Gross and Shine, 1981; Gross and Sargent, 1985).

Fishes have several attributes that make them ideal subjects for the study of the behavioural dynamics of parental care within the context of life-history evolution. First, they exhibit considerable phylogenetic diversity of the states of parental care. Secondly, many species are easily studied in the field. Thirdly,

Behaviour of Teleost Fishes 2nd edn. Edited by Tony J. Pitcher. Published in 1993 by Chapman & Hall. ISBN   0 412 42930 6 (HB) and 0 412 42940 3 (PB).

Table 11.1 Approximate phylogenetic distribution of the states of parental care (expressed in percentage terms), by family, over the major vertebrate classes. Within each class, if no species within a family is known to exhibit parental care, then that family is classified as 'no care'. If at least one species within a family exhibits one of the other states of parental care, then that family is classified under that state. Families that exhibit more than one of the other states of parental care are counted more than once. The data are presented at the family level to increase the likelihood that the states of parental care in the table are independent of phylogenetic relationships. The data were compiled from the following sources: mammals, Kleiman (1977), Kleiman and Malcolm (1981); birds, Lack (1968); reptiles, Porter (1972); amphibians, Salthe and Meacham (1974), Gross and Shine (1981); and fishes, Breder and Rosen (1966), Blumer (1982), Gross and Sargent (1985)

Taxon	Male care	Female care	Biparental care	No care
Mammals	0	90	10	0
Birds	2	8	90	0
Reptiles	0	15	Rare	85
Amphibians	24	32	14	30
Non-teleost fishes	6	66	0	28
Teleost fishes	11	7	4	78

many species adapt readily to the laboratory where one can conveniently control or manipulate variables of interest. Fourthly, guarding the eggs or brood on the substratum is the predominant form of parental care in fishes (Table 11.2). Guarding, unlike feeding, is a divisible resource (Wittenberger, 1981), in the sense that a unit of parental resource may be given to one or several offspring. Thus, a fish may be able to guard large clutches as easily as small ones (Williams, 1975: p. 135). Put another way, with respect to guarding, offspring survival is independent of offspring density, and dependent only upon the amount of parental care. This assumption makes the dynamics of parental behaviour in fishes relatively easy to model mathematically.

Rather than catalogue the remarkable diversity of kinds of parental care exhibited by fishes (review, Breder and Rosen, 1966), we focus in this chapter on an important evolutionary principle and show how it leads to a better understanding of parental behaviour. We refer to this principle after its major progenitor, G. C. Williams (1966a, b). Williams' principle is that natural selection favours maximizing remaining lifetime reproductive success, subject to trade-offs among fitness components, such as present versus future reproduction. Williams' assumption that reproduction has a cost can be modelled into a general framework for predicting and studying a variety of problems on the evolution and behavioural ecology of fish parental care. We illustrate the model with selected examples from fishes, and discuss its implications for future research.

Table 11.2 Approximate distribution of teleost families with external fertilization over the four most common forms of parental care as classified by Blumer (1982). These categories are not necessarily mutually exclusive; some families contain species with more than one form of parental care. The data were compiled from Blumer (1982), with additions noted in Gross and Sargent (1985)

				State of parental care			
Form of parental care	Male	Male or biparental	Male or female	Biparental	Male, female or biparental	Biparental or female	Female
Substratum guarding	28	6	2	10	5	2	1
Mouthbrooding	2	1	1	0	2	0	1
External egg-carrying	3	1	0	0	0	0	2
Brood pouch	1	0	0	0	0	0	1

11.2   THE COST OF REPRODUCTION – WILLIAMS' PRINCIPLE

Williams (1966a, b) proposed that a parent that continues to invest in its offspring does so at the expense of its potential future reproduction. This assumption can be easily accommodated in a static resource-allocation model; Pressley (1976) and Carlisle (1982) were among the first to apply such models to parental care in fishes. We begin with the simplest model possible (e.g. Coleman *et al.*, 1985; Sargent and Gross, 1985). If we assume that a parent, at any point in time, is selected to maximize its remaining lifetime reproductive success, we can model the trade-off between present and future reproduction as follows:

$$RS = P(RE) + F(SE) \qquad (11.1)$$

where RS is the parent's expected remaining lifetime reproductive success; P is present reproduction (the number of offspring at stake times their expected survivorship); F is the parent's expected future reproduction; RE is reproductive effort (*sensu* Williams, 1966a, b) – the proportion of parental resources (energy, time) devoted to fecundity and offspring survivorship; and SE is somatic effort (*sensu* Williams, 1966b) – the proportion of resources devoted to the parent's own growth, maintenance and survival. $SE = 1 - RE$.

The sum of RE and SE represents the total resources available to a parent ($RE + SE = 1$); therefore, reproductive effort (e.g. parental care) is at the expense of somatic effort and potential future reproduction of the parent. Thus, Williams' principle is that natural selection favours animals that maximize their remaining lifetime reproductive success, subject to this constraint. We make the common biological assumption that P and F both increase with diminishing returns (Real, 1980) with increasing RE and SE, respectively (Fig. 11.1(a)). In other words, we assume that small investments into P or F yield higher returns per investment than do large investments into P or F. The optimal reproductive effort (RE^*), which maximizes RS, will occur where $dR/dRE = dP/dRE + dF/dRE = 0$, or:

$$dP/dRE = -dF/dRE \qquad (11.2)$$

The assumption of diminishing returns (Fig. 11.1(a)) means that d^2P/dRE^2, d^2F/dRE^2, and thus d^2RS/dRE^2 are always negative. Because d^2RS/dRE^2 is negative for all values of RE, and because we assume that there exists an RE^* for which $dRS/dRE = 0$, then this RE^* must maximize RS. At this level of RE, the parent obtains equal rates of return on its investments into present and future reproduction (Equation 11.2; Fig. 11.1(b)). In our model the cost of reproduction (CR) is the loss in expected future reproduction (F) that is attributable to present reproduction (P). We will use our CR model to analyse parental investment; however, first we discuss some controversy surrounding Williams' principle, and present evidence of its applicability for fishes.

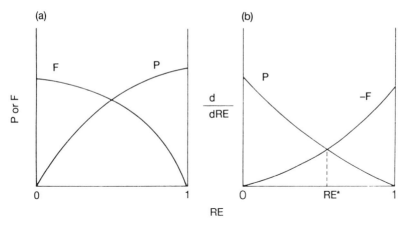

Fig. 11.1 (a) Present reproduction (P) is assumed to increase with diminishing returns with increasing reproductive effort (RE), and future reproduction (F) is assumed to increase with diminishing returns with increasing somatic effort ($1 - RE$). (b) The optimal reproductive effort (RE^*) occurs where the rates of return on investment into present and future reproduction are equal ($dP/dRE = -dF/dRE$).

There has been considerable discussion as to how to measure reproductive effort (RE) (Hirshfield and Tinkle, 1975; Pianka and Parker, 1975; Hirshfield, 1980), and whether or not a cost of reproduction (CR) even exists (Lynch, 1980; Bell, 1984a,b). Bell, for example, has suggested that a zero or positive correlation between present and future reproduction (P and F) among individuals within a population refutes the idea of a CR. We believe that such controversy stems from a misunderstanding of Williams' original model. In particular, it is important to note that the cost of reproduction is based on RE being a proportion of an individual's resource budget, and that the dynamics of the trade-off are assumed to be independent of resource-budget size. Although two adults may have the same optimal RE (RE^*), one may have higher fecundity and higher growth because it has a larger resource budget. Thus, lack of an observed negative correlation between P and F among individuals within a population cannot by itself refute the idea of a CR. See van Noordwijk and de Jong (1986), and Grafen (1988) for further discussion of the difficulties of interpreting correlations among fitness components, among individuals within the same population. We agree with Grafen (1988) that demonstrating trade-offs among fitness components is best accomplished with manipulative experiments in which correlations are computed among experimental treatments.

Several studies with fishes indicate that a CR trade-off does in fact exist. Perhaps the best-known experiment is that of Hirshfield (1980), who studied the Japanese medaka, *Oryzias latipes*, among several temperature- and

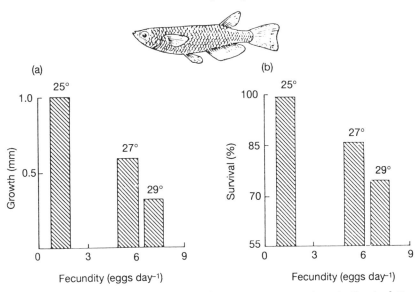

Fig. 11.2   The cost of reproduction over three temperature regimes in *Oryzias latipes* (Hirshfield, 1980). (a) Adult female growth decreases as fecundity increases. (b) Adult female survival (percentage not sick or dead) decreases as fecundity increases.

ration-controlled environments. When ration was held constant, total energy devoted to eggs and to growth (total production) did not differ among temperatures; however, as temperature increased among treatments, fecundity increased, and growth and survival both decreased. Although we do not know how temperature affected RE^*, these results indicate a CR (Fig. 11.2), because of the negative correlation between P and F among treatments, while total production remained constant. Further evidence is supplied by Reznick (1983), who found a similar negative correlation between reproduction and growth among five stocks of guppies, *Poecilia reticulata*, all of which had similar values for total production. In species with male parental care, it is known that males lose weight with increasing residence time on the territory, and males suffer heavy mortality at the end of a reproductive bout or breeding season (*Lepomis gibbosus*: Gross, 1980; *Pimephales promelas*: Unger, 1983, Sargent unpublished; *Gasterosteus aculeatus*: Sargent, 1985, FitzGerald *et al.*, 1989). Thus Williams' assumption of a cost of reproduction seems to hold for fishes.

If optimal RE^* is independent of resource-budget size, as assumed by our model, then P and F should both increase as ration increases. This pattern has been reported for several species of fishes (e.g. *Oryzias latipes*: Hirshfield, 1980; *Poecilia reticulata*: Reznick, 1983; *Cichlasoma nigrofasciatum*: Townshend and Wootton, 1984).

11.3   PARENTAL CARE EVOLUTION IN FISHES: WHY IS MALE CARE SO COMMON?

Traditionally, models for the evolution of parental care have contained a trade-off between parental care and the number of offspring in the brood (e.g. Trivers, 1972; Maynard Smith, 1977); however, there may be a more important trade-off for fishes. Unlike some other vertebrates, fishes continue to grow after they have become reproductive. In fishes, both female fecundity and number of eggs fertilized by males tend to increase with body size (Gross and Sargent, 1985). Any energy spent on reproduction is at the expense of growth; thus, a fish's expected future reproduction depends on how much it invests in growth, and on how big it gets (e.g. Reznick, 1983). Our model assumes that parental care in fishes will be sensitive to this trade-off between reproduction and growth.

How might the costs of parental care to growth affect which sex is most likely to evolve parental care? Imagine a species of fish with external fertilization, no parental care, and males that hold territories where they obtain all of their spawnings. Assume that if either sex shows parental care, it loses growth. Then the sex most likely to evolve care is the one with the lower relative costs to future fertility, because it will receive the higher relative benefits from care. It is well known in fishes that female fecundity increases with accelerating returns with increasing body length (Bagenal and Braum, 1978). If a male's ability to obtain matings were to increase linearly, or with diminishing returns with body length, then females would have the higher costs to future fertility, and thus would be less likely to evolve parental care. We illustrate this below.

Consider the following example from Gross and Sargent (1985) (Fig. 11.3). Assume that two fish have just spawned and that each one has present length s. If a parent guards its offspring, it can expect to be length g in the next breeding season. If it deserts, however, it will attain a larger size, d, because it has not invested energy into parental care. A fish's relative cost of parental care to future fertility is the ratio of its expected gain in fertility if it deserts to its expected gain in fertility if it guards (D/G). This ratio, which is always greater than one, is ranked in descending order of F against body-length curve types: accelerating, linear, and diminishing (Fig. 11.3). Gross and Sargent (1985) examined the shapes of these curves for five species of fishes with male care: the bluegill sunfish, *Lepomis macrochirus*, the pumpkinseed sunfish, *L. gibbosus*, the rock bass, *Ambloplites rupestris*, the three-spine stickleback, *Gasterosteus aculeatus*, and the mottled sculpin, *Cottus bairdi*. In each species the female curves increased with accelerating returns with body length, but the male curves increased either linearly, or with diminishing returns. Thus, for these five species, females would have the higher relative costs to their future fertility if they showed parental care. This difference between the sexes

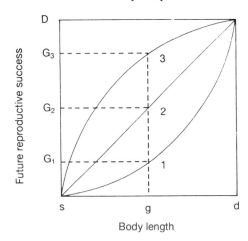

Fig. 11.3 The relationship between expected future reproductive success and body length may be accelerating (1), linear (2), or diminishing (3). At the next breeding season, a fish that is presently of length *s* will attain length *g* if it guards, or *d* if it deserts. If the expected gain of a fish that deserts is the same for each type of curve, then the relative cost of guarding, D/G, to expected future fertility is ranked in descending order among curve types: accelerating (1), linear (2), and diminishing (3).

may be why male care rather than female care evolved in these species, and may provide a general explanation for the prevalence of male parental care in fishes. Gross and Sargent (1985) review this and competing hypotheses for the evolution of male parental care in fishes.

An additional hypothesis worth mentioning is sexual selection via female choice (e.g. Peterson and Marchetti, 1989). Theory suggests that female choice should be based on male resources that contribute to the survival of her offspring (e.g. Trivers, 1972; Williams, 1975; Kirkpatrick, 1985). Conceivably, such a process could lead to the evolution of male parental care.

For example, Kirkpatrick (1985) examined the role of male parental investment on the evolution of female mating preference in his study of Weatherhead and Robertson's (1979) 'sexy son' hypothesis. He assumed that polygynous males trade off parental investment, and thus short-term offspring survival, against long-term survival of a male's sons through the superior genes of the 'sexy father'. Females are assumed to be able to discriminate among the whole range of male phenotypes. Females that choose 'sexy males' as mates suffer low fecundity owing to their mate's low parental investment, but have sons with high viability. Kirkpatrick then let female preferences evolve, and found that at equilibrium, female fecundity is maximized, which occurs where male parental investment is maximized. Under some conditions, sexual selection leads to higher levels of male parental care than would have been favoured by natural selection alone (Kirkpatrick, 1985).

Recently, Curtsinger and Heisler (1988) found an interesting counter-example to Kirkpatrick's general result of female mating preference evolving toward maximizing female fecundity and male parental investment. Whereas Kirkpatrick (1985) used haploid and polygenic models, Curtsinger and Heisler (1988) used a diploid model so that they could explore the effects of dominance. By assuming overdominance in female mating preference in favour of the most extreme 'sexy' males with low parental investment, they were able to produce a stable polymorphism in both female mating preference and male trait. At the equilibrium, female fecundity and male parental investment are not maximized. Curtsinger and Heisler suggest that this kind of counterexample supports the 'sexy son' hypothesis. However, as Kirkpatrick (1988) points out, Curtsinger and Heisler's (1988) assumption of over-dominance imposes a genetic constraint that prevents female mating preference evolving to maximizing female fecundity and male parental investment. At equilibrium, the additive genetic variance for female preference is zero. The assumption of overdominance in female mating preference seems somewhat restrictive, and perhaps unrealistic. It will be interesting to see what other counterexamples such models produce in the future.

For the present, however, it appears that natural selection favours females choosing mates on the basis of male parental investment. Several empirical studies on fishes with paternal care have found a positive association between female preference and offspring survival. The following examples indicate species and cues for female choice: *Cottus bairdi*, male size (Downhower and Brown, 1980, 1981); *Cottus gobio*, male size, eggs in nest (Marconato and Bisazza, 1986, 1988; Bisazza and Marconato, 1988); *Gasterosteus aculeatus*, territory quality – nest site concealment (Sargent and Gebler, 1980; Sargent, 1982); *Lepomis macrochirus*, male size, territory quality – centre position (Gross and McMillan, 1981); *Pimephales promelas*, eggs in nest (Sargent, 1988, 1989; Unger and Sargent, 1988). Female preference for males with eggs may have led to the evolution of males adopting unrelated eggs as part of their mating strategy (e.g. Rohwer, 1978; Ridley and Rechten, 1981; Marconato and Bisazza, 1986; Unger and Sargent, 1988). However, whether or not sexual selection led to the evolution of male parental care from a state of no care is still an open question.

11.4 BEHAVIOURAL ECOLOGY OF PARENTAL CARE: THE COST OF REPRODUCTION AND THE DYNAMICS OF PARENTAL BEHAVIOUR

Our CR model suggests that levels of parental care are unlikely to be static within an individual over time, or among individuals under different conditions.

We look in particular at four phenomena that are common to species with parental care and ask:

1. How should parental behaviour vary with age of the offspring?
2. How should parental behaviour vary with brood size?
3. How should parental behaviour vary with the probability of breeding again?
4. How should male parental care vary with increased investment in male–male competition?

Our approach to answering these questions is through a simple graphical analysis of the model. We consider parents that have already allocated energy to mating and that are now defending eggs or offspring; thus, any pending expenditures of RE are in the form of parental care. Recall from Equation 11.2 that the optimal reproductive effort, RE^*, is the one that yields equal rates of return on investment into present and future reproduction (P versus F), which is a very useful property of the model. For example, if the rate of return on investment into P, dP/dRE, were raised, then RE^* would increase so as to balance the two rates of return, $dP/dRE = -dF/dRE$ (Fig. 11.4). Similarly, if the rate of return on investment into future reproduction, dF/dSE ($= -dF/dRE$), were raised, then the optimal somatic effort would increase and RE^* would decrease (Fig. 11.5). Thus, we can study the dynamics of parental care by perturbing (i.e. manipulating) the trade-off between present and future reproduction, and looking for evidence of the predicted shift in RE^*. Because

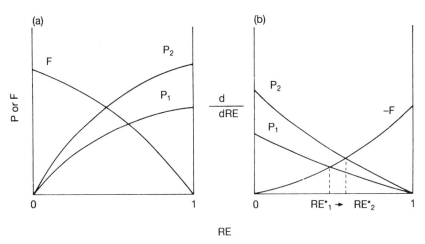

Fig. 11.4 Raising the rate of return of investment into present reproduction (e.g. by increasing brood size) raises the optimal reproductive effort. (a) A parent with a large brood has a higher expected present reproduction than a parent with a small brood ($P_2 > P_1$). (b) As dP/dRE increases, RE^* increases.

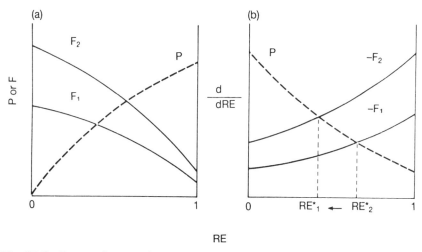

Fig. 11.5 Raising the rate of return on investment into future reproduction (e.g. by increasing the likelihood of remating) raises the optimal somatic effort, which lowers the optimal reproductive effort. (a) A parent that experiences F_2 has a higher expected future reproduction than a parent that experiences F_1. (b) As $-dF/dRE$ increases, RE^* decreases.

we cannot measure RE^* directly, we look for a negative correlation between present and future reproduction among treatments.

Parental behaviour and offspring age

In fishes, it has been observed that, at first, parental care increases as the offspring get older, but later, parental care decreases as the offspring approach independence (e.g. van Iersel, 1953; Barlow, 1964). Although these parental-behaviour dynamics can be explained as reflecting the physical needs of the offspring, they may instead reflect the parent's resolution of a trade-off between P and F. We can describe this process with a progression of CR models whose trade-offs change with time, or, more precisely, with offspring age. Let us consider four developmental stages of fish offspring: (1) newly fertilized eggs (i.e. zygotes); (2) eggs that are about to hatch; (3) wriggler; and (4) free-swimming fry (Keenleyside, 1979). The hypothetical P versus F trade-offs for these developmental states are presented in Fig. 11.6. We assume that during a brood cycle a parent's resource budget remains roughly constant, and that changes in RE^* over time produce changes in the net levels of observed parental behaviour. We also assume that, as the brood cycle progresses, changes in a parent's ability to improve offspring survival far outweigh any decreases in its ability to affect its own survival (and thus F).

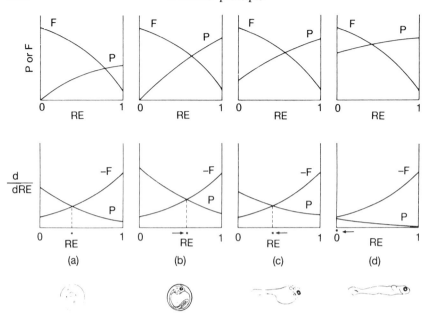

Fig. 11.6 Hypothetical parental reproductive trade-offs for four different stages of offspring development: (a) zygote, (b) late embryo, (c) wriggler, and (d) free-swimming fry. The top row shows present and future reproduction versus reproductive effort; the lower row shows the rates of return on investment into present and future reproduction versus reproductive effort. Under this model, RE^* increases between stages a and b, and decreases between stages b and c and between stages c and d. See text for assumptions and explanation.

Therefore, for simplicity, we treat the $F(RE)$ curve as not changing with offspring age; however, the $P(RE)$ curve depends on offspring age (Fig. 11.6). Let

$$P(RE) = bl(RE) \tag{11.3}$$

where b is brood size, which is assumed to be a constant, and l is the probability of the offspring surviving to reproduce, which increases with diminishing returns with increasing RE. At spawning (Fig. 11.6(a)), offspring survivorship is zero if the parent offers no care, but improves somewhat with maximum parental care ($RE = 1$). Just before hatching (Fig. 11.6(b)), the offspring are still completely dependent on parental care ($P = 0$ if $RE = 0$), but now the parent can achieve higher juvenile survivorship with maximum parental care ($RE = 1$), simply because the offspring are older and thus closer to independence. Because the rate of return on investment into $P(dP/dRE)$ has increased over that for newly spawned eggs, RE^* also increases. When the offspring are at the wriggler stage (Fig. 11.6(c)), they have some finite probability of surviving on their own ($P > 0$ if $RE = 0$), and the difference in offspring survivorship between $RE = 0$ and $RE = 1$ is now less than when the offspring were just hatching. Thus the parent now enjoys a lower dP/dRE than for hatching eggs, and so should lower its RE. Finally, when the fry are free-swimming (Fig. 11.6(d)), the difference between offspring survivorship when $RE = 0$ and $RE = 1$ is very small, and is always less than $-dF/dRE$. Because the parent has higher expected future reproduction when $RE = 0$ than it can gain if it invested anything in its offspring ($RE > 0$), the optimal reproductive effort is zero ($RE^* = 0$), and the parent deserts its offspring.

Thus, we have extended our basic CR model to one that changes with time over a brood cycle, and with it we can explain why the dynamics of parental care change with offspring age. Although the dynamics we have described in Fig. 11.6 will vary among species, they are likely to represent the general pattern in fishes with parental care. To illustrate these dynamics we consider the threespine stickleback, *Gasterosteus aculeatus*.

The threespine stickleback is a species characterized by males that aggregate in shallow water, establish territories, build nests, court females, and care for the developing eggs and newly hatched fry (Wootton, 1976). A single male may undergo several brood cycles during the breeding season. It is well known for the threespine stickleback that paternal fanning increases with age of the eggs until about a day before hatching, and then continues to decline as the offspring approach independence (van Iersel, 1953; van den Assem, 1967). Although this result is predicted by our model, the pattern in fanning over the brood cycle might also be explained by the eggs requiring more oxygen as they get older, and less oxygen after they hatch. Thus fanning may not corroborate our model. Another form of parental care in the threespine stickleback is defence of the brood from predators. Unlike fanning, a brood's requirements in defence from predators are likely to be independent of age. In sticklebacks the major brood predators are conspecifics of either sex (Kynard, 1978). Motivation for brood defence can be measured by presenting a persistent threat to the brood on a male's territory, and then recording the resident male's behaviour. One such threat is another male stickleback in a glass cylinder next to a resident male's nest (Wootton, 1976). Over a brood cycle such tests show that as the eggs approach hatching, brood defence increases (defence is measured as bites per minute towards an intruder). After hatching, however, brood defence wanes as the fry approach independence (Huntingford, 1977; Sargent and Gebler, 1980; Sargent, 1981). Similar brood defence dynamics were found in the pumpkinseed sunfish, *Lepomis gibbosus* (Colgan and Gross, 1977). Thus, both fanning and brood defence exhibit the kinds of dynamics described by our model (Fig. 11.7(a)).

Parental behaviour and brood size

Several laboratory and field studies have shown considerable variation in the levels of parental behaviour among parents, and much of this variation can

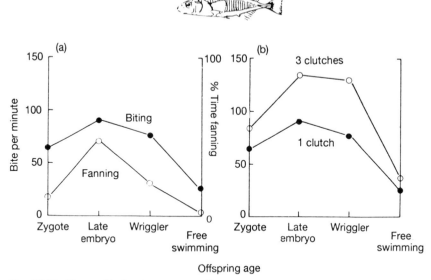

Fig. 11.7 Parental behaviour versus offspring age in *Gasterosteus aculeatus* (Sargent 1981). (a) Both fanning and brood defence parallel the *RE**s given in Fig. 11.6. Each data point represents the mean for 8 fish. Fanning and biting both increase significantly up to the late embryo stage, and then decrease significantly afterward (*P* < 0.05) (b) Males with three clutches show higher brood defence than males with one clutch at the zygote, late embryo and wriggler stages (*P* < 0.05).

be explained by a positive correlation between parental behaviour and brood size (e.g. van Iersel, 1953; Pressley, 1981). Under our model, there may be two components to this correlation.

1. Bigger parents may have larger resource budgets, which allow them to accumulate bigger broods and thus to show higher brood defence. Several studies have documented a positive correlation between parental body size and parental behaviour (Downhower and Brown, 1980; Gross, 1980).
2. Parents with larger broods experience a higher rate of return on investment into brood defence (d*P*/d*RE*, Fig. 11.4), and thus should show higher brood defence in order to maximize remaining lifetime reproductive success.

To test these hypotheses, Sargent (1981) randomly assigned competing male threespine sticklebacks either one or three clutches. Males with three clutches consistently showed higher brood defence than males with one clutch (Fig. 11.7(b)), during each stage of offspring development. Larger males may have had higher brood defence than smaller males; however, this difference was not statistically significant. Thus, although resource-budget size may affect the level of parental behaviour, brood size has an additional effect, apparently because parents with larger broods enjoy a higher rate of return on brood defence. Ridgway (1989) found similar results in his field study on the smallmouth bass, *Micropterus dolomieui*.

Parental behaviour and expected future reproduction

A parent's expected future reproduction will vary with a number of factors, both intrinsic and extrinsic. Examples of intrinsic factors may include age, health, and social status; extrinsic factors may include sex ratio, food availability, and predators. Our model predicts that changes in expected future reproduction will produce changes in RE^*. For example, raising the rate of return on investment into future reproduction $(-dF/dRE)$ lowers RE^* (Fig. 11.5), or the optimal investment in P.

An example of this is found in the parental behaviour of males of the bi-parental cichlid *Herotilapia multispinosa* (Keenleyside, 1983). In this species both parents guard the eggs on the substratum, and guard the fry after they hatch. Each guarding male is defending only one clutch, and either must complete the brood cycle or desert his offspring before he can mate again. Keenleyside studied these fish in ponds in which he manipulated the adult sex ratio (5:7, 1:1, and 7:5, females to males). Note that as the proportion of females increases among ponds, so does a guarding male's expected future reproduction (but not that of the female). According to our model, as a male's chance of remating increases, his RE should decrease. One would expect, therefore, that as the proportion of females increases among ponds, so should the proportion of guarding males who desert their offspring. This is precisely what Keenleyside found (Fig. 11.8). His experiment corroborates our model and demonstrates that the optimal reproductive effort is sensitive not only to a change in the rate of return on investment into the offspring at stake, but also to a change in the rate of return on investment into future reproduction.

Simultaneous trade-offs: paternal behaviour and male–male competition

Almost all species of fishes that guard their eggs and fry exhibit territorial defence of the spawning site. For many species with paternal care, males establish nests in dense aggregations. Males on adjacent territories routinely fight among themselves, presumably for access to females. Time and energy spent on intrasexual aggression may be at the expense of paternal care, and at the expense of future reproduction.

To test this hypothesis, Sargent (1985) extended the basic CR model to include two simultaneous trade-offs: (1) present versus future reproduction; and (2) within present reproduction, mating versus parental care. In simplest

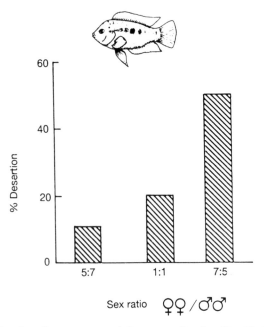

Fig. 11.8 Male desertion versus adult sex ratio in *Herotilapia multispinosa* (Keenleyside, 1983). As the adult sex ratio (proportion of females) increases, the proportion of males that desert their offspring increases.

form, the model is as follows.

$$RS = L(PE) + c(ME) + F(SE), \tag{11.4}$$

where L is survival of the brood at stake; c is reproduction that the male may gain in the present brood cycle by acquiring additional clutches; PE is parental effort – the proportion of parental resources devoted to improving the survival of the offspring at stake; ME is mating effort – the proportion of parental resources devoted to increasing clutch size during the present brood cycle; F and SE are as in Equation 11.1. $RE = PE + SE$; $PE + ME + SE = 1$. As in the simpler model, L, c, and F are all assumed to increase with diminishing returns with increasing PE, ME, and SE, respectively. Assuming that there exists an optimal allocation of effort such that $0 < PE^*$, ME^*, $SE^* < 1$, then at the optimum the parent enjoys equal rates of return on investment into each of its options.

$$dL/dPE = dc/dME = dF/dSE \tag{11.5}$$

Similar to the analysis of Equation 11.2, if dc/dME were raised by a constant (and dL/dPE and dF/dSE were both unaffected), then ME^* would increase and PE^* and SE^* would both decrease.

If dense male nesting aggregations are attractive to females, and if intrasexual territorial aggression functions in obtaining mates, then it may be possible to experimentally manipulate a male's perception of his rate of return on investment into mating by manipulating the density of nesting males. Sargent (1985) conducted such an experiment with threespine sticklebacks.

Male sticklebacks were divided into two treatments: competitive and solitary. Replicate aquaria were divided in half with a partition, and one male nested in each compartment. Competitive males were separated by transparent partitions; solitary males by opaque partitions. Each male was allowed to spawn with one female per brood cycle, and was allowed as many brood cycles as he could achieve within a breeding season. Assuming that competitive males perceived a higher rate of return on mating effort than solitary males, Sargent predicted that relative to solitary males, competitive males would show higher motivation to court females, less parental care, more time and energy spent per brood cycle, and fewer total brood cycles before the end of the season. Each prediction was corroborated.

The competitive males spent considerable time fighting and displaying at the transparent partition, which resulted in their spending less time in nest tending and parental care than solitary males. Late in the first brood cycle, competitive males were more likely to court a test female than were solitary males. Competitive males lost weight faster than solitary males, thus intrasexual aggression imposed an energetic cost on top of parental care. Competitive males took longer to build their nests because they spent time fighting, and also took longer to hatch their eggs because they spent less time fanning. (The length of time between spawning and hatching increases as fanning decreases (van Iersel, 1953); the difference in hatching time between competitive and solitary males has been reported previously by van den Assem (1967)). These temporal costs of territoriality resulted in competitive males having longer brood cycles than solitary males. Both the energetic and temporal costs of territoriality contributed to solitary males averaging more brood cycles per breeding season than the competitive males.

In summary, territorial males may incur higher temporal and energetic costs than solitary males. These higher costs probably reflect a higher rate of return on investment into mating, and can be measured in terms of more rapid weight loss, and lower expectation of future brood cycles.

11.5 DYNAMIC MODELS

Our section on paternal care and intrasexual aggression illustrates how time may be an important component of the cost of reproduction, and our section on parental behaviour and offspring age illustrates how behavioural trade-offs may change over time. Thus time appears to be an important variable in understanding the dynamics of parental care, yet it is lacking in static models. The approach of dynamic optimization explicitly incorporates time. A relatively easy, numerical approach to dynamic modelling is dynamic programming (reviews, Houston and McNamara, 1988; Mangel and Clark, 1988). A dynamic programming model for reproductive success in general form can be written as follows:

$$RS(t, T) = RS(t, t + 1) + RS(t + 1, T) \qquad (11.6)$$

where *RS* is reproductive success, which is a function of the current instant in time (*t*), and the time horizon (*T*), or the end of the time course over which reproductive success is measured. Taking time in units of days, and *T* as the end of the animal's life, then Equation 11.6 states that reproductive success between today and the end of the animal's life is equal to the sum of two components: (1) reproductive success between today and tomorrow; and (2) cumulative reproductive success between tomorrow and the end of the animal's life. The first component includes all offspring that become independent 'today', and survival of dependent offspring should the parent die (Sargent, 1990). The second term is expected residual lifetime reproductive success, assuming the parent survives until 'tomorrow'.

A parent is assumed to be able to choose a behaviour so as to maximize its expectation of remaining lifetime reproductive success, *RS(t, T)*. The optimal behaviour is likely to depend on two state variables: (1) parental state (e.g. energy reserve); and (2) offspring state (e.g. a combination of offspring number and age – Sargent, 1990).

Dynamic programming models for single-parent care

To investigate parental investment in fishes as a dynamic optimization problem, Sargent (1990) constructed a series of dynamic programming models. These models explore care by one parent only, where parental care is assumed to be a divisible resource (*sensu* Wittenberger, 1981); that is, one unit of care may be given to one or several offspring. Because guarding is likely a divisible resource (Williams, 1975), these models are particularly suited to fishes. There were two state variables: parental energy reserve, and offspring number or age. To simplify the presentation, we leave the state variables out of our dynamic programming equations here, and refer the reader to Sargent (1990) for a more thorough treatment of the state variables and the state dynamics.

Feed or care?

Here we model a parent's ability to keep itself and its offspring alive until the offspring are independent; thus, our time horizon is the end of a brood cycle.

We assume that our parent has all of the eggs it will get during this brood cycle, and that further mating is impossible until after the time horizon; thus, this model is a dynamic version of our four-stage static model, *Parental behaviour and offspring age* (p. 343). We began by making several simplifying assumptions.

1. All offspring are the same age.
2. Clutches, or discrete fractions of a clutch, live or die as a unit within a time step. This assumption was made to keep the offspring state space and state dynamics tractable.
3. The strategy set was to feed (itself), care (for its offspring), or both feed and care.
4. The trade-offs are built in through behaviour-dependent effects on parent and offspring survival, and through behaviour-dependent effects on parental energy reserve. Feed reduces both parent and offspring survival, but increases parental energy reserve. Care reduces parental survival and energy reserve, but improves offspring survival.

The dynamic programming equation is of the following general form:

$$RS(t, T) = (1 - s_p(t))s_o(t, T) + s_p(t)RS(t + 1, T) \qquad (11.7)$$

where $s_p(t)$ is parent survival during t and $s_o(t, T)$ is offspring survival without care from now (t) until hatching (T). Thus the first term on the right-hand side of Equation 11.7 tallies offspring survival if the parent dies, whereas the second term tallies all expected future reproductive success if the parent survives until $t + 1$. To solve the model, we initialize terminal fitnesses at T. Then starting at $T - 1$, we iterate backward over all combinations of the state variables (i.e. parental energy reserve and offspring number), and solve for the behaviour that maximizes RS for each state–variable combination in each time step. Thus we find expected behavioural trajectories through time and through the state space.

The following general patterns were found (Sargent, 1990), and are presented graphically in Fig. 11.9.

1. As energy reserve increases, feed decreases, and care increases.
2. As offspring age increases, feed decreases, and care increases.
3. As offspring number increases, feed decreases and care increases.
4. Care does not decrease to zero at the end of the brood cycle, unless offspring survival without care increases with increasing offspring age.

These results agree well with those for our static models, *Parental behaviour and offspring age*, and *Parental behaviour and brood size*, above. Moreover, the dynamic model has the added advantage of an explicit separation of energy and time as components of the cost of reproduction.

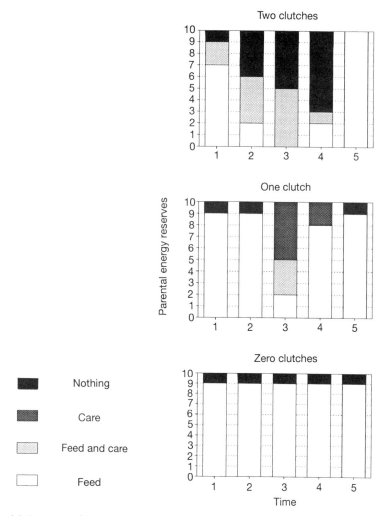

Fig. 11.9 A graphical representation of the model, Feed or Care?, modified from Sargent (1990). Along the horizontal axis is time, or offspring age; along the vertical axis is parental energy reserve. The three possible nest states are shown. A parent is assumed to have either 0, 1 or 2 clutches of eggs in its nest, and these clutches are assumed to survive or die as whole units. The parent's strategy set is assumed to consist of three behaviours: Feed (itself), Care (for eggs in the nest), or a combination of Feed and Care. The per-time-step clutch survival with care is 0.9. Clutch survival without care is 0.3 up until hatching (at $t = 3$), after which clutch survival without care increases with time until it equals clutch survival with care (i.e. 0.9), at $t = 5$. It is informative to examine how parental care is distributed among the levels of energy reserve, nest states, and time steps. Parental care is more likely the more energy a parent has, and the more clutches it has. Parental care increases with clutch age up until hatching, and decreases with increasing age as the offspring approach independence. Note the similarity between these predictions and those of the static model in Fig. 11.6, and the observed pattern in sticklebacks in Fig. 11.7.

Feed, mate or care?

Sargent (1990) constructed a second version of the dynamic programming model in order to understand the simultaneous trade-offs among feeding, mating and providing parental care. Basically, this model is a simple extension *Feed or care* (above), with the major exception that a male with eggs may continue to mate with females and add more eggs to his nest. If a male with eggs continues to mate, then the offspring in his nest may be of different ages; thus, the end of the breeding season was chosen to be the time horizon. As in the previous model, there are two state variables: parental energy reserve, and offspring number/age. There are several basic assumptions.

1. It takes one time step for a male to mate and acquire one clutch, and it takes a clutch three time steps to hatch. Thus, there are eight possible levels of nest state: zero clutches, one clutch age 1, one clutch age 2, one clutch age 3, two clutches ages 1 and 2, two clutches ages 1 and 3, two clutches ages 2 and 3, and three clutches ages 1, 2 and 3.
2. Clutches live or die as units within a time step.
3. The strategy set was to feed, mate, care, or any combination of these behaviours.
4. The trade-offs are built in through behaviour-dependent effects on parent and offspring survival, and through behaviour-dependent effects on parental energy reserve. Feed reduces parent and offspring survival, but increases parent energy reserve. Mate reduces parent and offspring survival, reduces parental energy reserve, but increases offspring number. Care reduces parent survival and energy reserve, but increases offspring survival.

The dynamic programming equation has the following general form:

$$RS(t, T) = s_3(t, H) + (1 - s_p(t))s_{1,2}(t, H) + s_p(t)RS(t + 1, T) \qquad (11.8)$$

where $s_3(t)$ is the probability that an age 3 clutch hatches in t, $s_{1,2}(t, H)$ is the probability that age 1 or 2 clutches survive until hatching if the parent dies, and $s_p(t)$ is the probability that the parent survives t. Thus the first term on the right-hand side of Equation 11.8 tallies any clutches that hatch during t, the second term tallies any clutches that hatch after t if the parent dies, and the third term tallies the parent's expected future reproductive success if it survives to $t + 1$. To solve Equation 11.8, we begin by initializing terminal fitnesses at T. Then starting at $T - 1$, we iterate backward in time over all combinations of the two state variables, and solve for the optimal behaviour for each state–variable combination in each time step.

The following general patterns were found (Sargent, 1990), and are presented graphically in Fig. 11.10.

1. As energy reserve increases, feed decreases, and mate and care both increase.

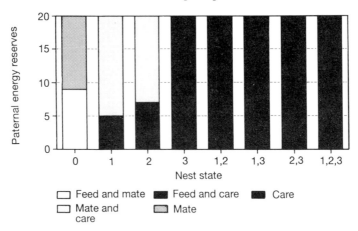

Fig. 11.10 A graphical representation of the dynamic programming model, *Feed, mate or Care?*, modified from Sargent (1990). Along the horizontal axis are the eight levels of nest state (i.e. combinations of clutch age/number); along the vertical axis are the levels of paternal energy reserves. For explanation see text. This graph represents a stationary distribution; i.e. the results are constant in time for the first 16 (out of 30) time steps in the breeding season.

2. As clutch age increases, feed and mate both decrease, and care increases.
3. As clutch number increases, feed and mate both decrease, and care increases.
4. If courtship behaviour increases the mortality of the eggs already in a male's nest, then after some combination of clutch age and number, a male will cease courting females, and switch into a 'parental phase' (*sensu* van Iersel, 1953). This produces the phenomenon of brood cycling (e.g. Peterson and Marchetti, 1989).

The 'parental phase' can be visualized by following male behaviour versus nest state in Fig. 11.10. If we assume that a male enters the breeding season with maximal energy reserves and zero clutches (i.e. nest state = 0), then his optimal behaviour is to mate. Assuming that this first mating attempt was successful, he next has one clutch age one (i.e. nest state = 1); whereupon his optimal behaviour is to mate and care. If this second mating attempt was also successful, he next has two clutches, aged one and two (i.e. nest state = 1, 2), whereupon his optimal behaviour is care. Care remains the optimal behaviour until both clutches hatch; thus, the male enters the parental phase after nest state = 1, 2. On the other hand, if the male's second mating attempt is unsuccessful and his first clutch survives, he then enters nest state = 2, whereupon his optimal behaviour is to mate and care. Males with one clutch do not enter the 'parental phase' until that clutch is age three (i.e. nest state = 3). Thus, this model predicts that the fewer clutches a male has, the later he enters the parental phase, which agrees qualitatively with van Iersel's (1953) results for threespine sticklebacks.

This model illustrates that with respect to energy reserves, feeding is traded off against mating and parental care; however, with respect to clutch number/age, mating and parental care are traded off against each other. Moreover, this model illustrates how the parental phase and brood cycling in nature may be part of a parent's optimal decision, rather than a constraint placed on the system, as in the model *Feed or care?* Although this model has not yet been tested explicitly, it does agree well with known fish behaviour (van Iersel, 1953; Sargent, 1988; Peterson and Marchetti, 1989).

11.6   SUMMARY AND SUGGESTIONS FOR RESEARCH

Williams' principle assumes that reproduction always has a cost. As animals attempt to maximize their remaining lifetime reproductive success, the cost of reproduction will limit the amount that they can invest into future reproduction. Animals that maximize this trade-off obtain equal rates of return on investments into present and future reproduction. This trade-off, when modelled, provides a powerful technique for analysing parental behaviour in fishes. With it we can explain the prevalence of male parental care, the changes in parental behaviour over a brood cycle, the positive correlation between parental behaviour and brood size, the negative correlation between parental behaviour and expected future reproduction, and the dynamics of paternal behaviour with intrasexual competition. Future research might well be directed into the following areas.

1. The effects of social status on optimal reproductive effort. Throughout we have assumed that the optimal reproductive effort (RE^*) is independent of resource-budget size, which we supported with four species of fishes in which both present and future reproduction increase with the size of the resource budget. There may be an interesting exception to this assumption, however. Specifically, in many species of fishes, dominant individuals, which presumably have the largest resource budgets, can inhibit or even prevent the reproduction of subordinate individuals (e.g. Jones and Thompson, 1980). Thus an individual's rate of return on investment into present reproduction may depend on social status, which in turn will affect RE^*.

2. Indivisible resources. For mathematical ease, we limited our model and its interpretations to those parental behaviours that represent divisible resources to the offspring (Wittenberger, 1981). Guarding, a divisible resource, is the most common parental behaviour in fishes; however, other parental behaviours, such as aeration of the eggs and feeding the young,

may represent indivisible resources (i.e. what is given to one offspring cannot be shared by other offspring). If the eggs or fry are under competition for parental care, then our model would have to be made more complex to describe parental behaviour. Unfortunately, there has been little research on competition among siblings for parental resources in fishes.

3. Biparental care and games between the sexes. Because each sex benefits if the other provides care, it may be informative to return to Maynard Smith's (1977) formulation of parental care as a game between the sexes. A game-theory approach is essential to understanding biparental care. Moreover, regardless of which sex provides care, a game-theory approach should also provide insights into the interactions between parental care in one sex and mate choice in the opposite sex. Hammerstein and Parker (1987) review several games between the sexes, and like Maynard Smith (1977) find that many of these games have multiple evolutionarily stable strategies (ESSs), depending on the parameters. To our knowledge, no one has yet attempted to test these models for fishes.

The optimal level of parental care for either sex will likely depend in part on offspring number, age, and energy reserves of one or both parents, and these variables are likely to change over time, depending on each parent's behaviour. Therefore, we suggest that it may be useful to model parental care as a dynamic game between the sexes. To our knowledge, there are no dynamic models that explore the interaction between biparental care and mate choice by both sexes; however, Crowley *et al.* (1991) have taken a step in this direction, by formulating simultaneous mate choice by both sexes as a dynamic game. They assume that each sex benefits more by having a large mate than a small one, and that under certain conditions, both sexes will be choosy. Their general results can be summarized as follows: (1) the limiting sex (i.e. the sex that is capable of mating least frequently, usually females) is more likely to be choosy; (2) choosiness increases as the encounter rate with prospective mates increases; and (3) choosiness increases as predation risk decreases. Unfortunately, in the model of Crowley *et al.* (1991), parental care (or more precisely, reproductive investment) is constant, and is not a 'decision variable' under parental control. We suggest that dynamic models that combine parental care (e.g. Sargent, 1990) and mate choice (e.g. Crowley *et al.*, 1991) will provide additional insights into the ecology and evolution of parental care in fishes.

4. Handicaps and honest signalling. During the mid 1970s, Zahavi (1975, 1977) introduced a novel hypothesis, the 'handicap principle'. Zahavi postulated that male epigamic characters represent handicaps (in the sense that they reduce male survival) that honestly signal male quality to prospective females. This hypothesis has generated substantial controversy

(Maynard Smith, 1976, 1985; Kirkpatrick, 1986; Pomiankowski, 1987) as to whether Zahavi's mechanism could work at all, and if so under what conditions. Recent models, however (e.g. Hasson, 1989; Grafen, 1990a, b), demonstrate that not only can the handicap principle work, but that it can work under a broad range of mathematical assumptions.

Grafen's (1990b) basic ESS model of the handicap principle assumes the following: (1) males vary in quality of interest to females (environmental and/or genetic quality); (2) males also vary in level of advertisement to females (e.g. courtship behaviour, expression of an epigamic character); (3) females base their mate choice on male advertisement; and (4) for a given level of advertisement, the relative cost to a low-quality male is greater than to a high-quality male. Grafen found that the ESS is for male advertisement to be positively correlated with male quality, and for females to prefer males with the highest levels of advertisement.

Although most students of sexual selection are interested in female choice based on male genetic quality, Grafen's analysis has profound implications for fishes with male parental care. If the handicap principle operates in these species, then we should find that male nuptial coloration, courtship behaviour, and degree of development of epigamic characters (e.g. enlarged fins) should be positively correlated with quality of paternal care, and ultimately, with offspring survival. Similar arguments were made by Hoelzer (1989). Future research in this area looks most promising.

ACKNOWLEDGMENTS

The first edition of this chapter was funded by a NATO Postdoctoral Fellowship (National Science Foundation, USA) to R.C.S., and by an NSERC of Canada Operating Grant (UO244) and an NSERC of Canada University Research Fellowship to M.R.G. The second edition was supported by NSF grants (BSR-8614640, BSR-8918871) to R.C.S., and by an NSERC of Canada Operating Grant to M.R.G.

REFERENCES

Alexander, R. D. (1974) The Evolution of Social Behaviour. *A. Rev. Ecol. Syst.*, **5**, 325–83.

Assem, J. van den (1967) Territory in the three-spined stickleback, *Gasterosteus aculeatus* L. An experimental study in intra-specific competition. *Behaviour (Supp.)*, **16**, 1–167.

Bagenal, T. B. and Braum, E. (1978) Eggs and early life history, in *Fish Production in Fresh Waters* (ed. T. B. Bagenal), Blackwell, Oxford, pp. 165–201.

Barlow, G. W. (1964) Ethology of the Asian teleost *Badis badis*. V. Dynamics of fanning and other parental activities, with comments on the behaviour of the larvae and postlarvae. *Z. Tierpsychol.*, **21**, 99–123.

Baylis, J. R. (1981) The evolution of parental care in fishes, with reference to Darwin's rule of male sexual selection. *Env. Biol. Fishes*, **6**, 223–51.

Bell, G. (1984a) Measuring the cost of reproduction. I. The correlation structure of the life table of a plankton rotifer. *Evolution*, **38**, 300–13.

Bell, G. (1984b) Measuring the cost of reproduction. II. The correlation structure of the life tables of five freshwater invertebrates. *Evolution*, **38**, 314–26.

Bisazza, A. and Marconato, A. (1988) Female mate choice, male–male competition and parental care in the river bullhead, *Cottus gobio* L. (Pisces, Cottidae). *Anim. Behav.*, **36**, 1352–60.

Blumer, L. S. (1979) Male parental care in the bony fishes. *Q. Rev. Biol.*, **54**, 149–61.

Blumer, L. S. (1982) A bibliography and categorization of bony fishes exhibiting parental care. *Zoo. J. Linn. Soc.*, **76**, 1–22.

Breder, C. M., Jun. and Rosen, D. E. (1966) *Modes of Reproduction in Fishes*, T.F.H. Publications, Neptune City, NJ, 942 pp.

Carlisle, T. R. (1982) Brood success in variable environments: implications for parental care allocation. *Anim. Behav.*, **30**, 824–36.

Coleman, R. M., Gross, M. R. and Sargent, R. C. (1985) Parental investment decision rules: a test with bluegill sunfish. *Behav. Ecol. Sociobiol.*, **18**, 59–66.

Colgan, P. W. and Gross, M. R. (1977) Dynamics of aggression in male pumpkinseed sunfish (*Lepomis gibbosus*) over the reproductive phase. *Z. Tierpsychol.*, **43**, 139–51.

Crowley, P. H., Travers, S. E., Linton, M. C., Cohn, S. L., Sih, A. and Sargent, R. C. (1991) Mate density, predation risk, and the seasonal sequence of mate choices: a dynamic game. *Am. Nat.*, **137**, 567–96.

Curtsinger, J. W. and Heisler, I. L. (1988) A diploid "sexy son" model. *Am. Nat.*, **132**, 437–53.

Downhower, J. F. and Brown, L. (1980) Mate preferences of female mottled sculpins, *Cottus bairdi. Anim. Behav.*, **28**, 728–34.

Downhower, J. F. and Brown, L. (1981) The timing of reproduction and its behavioral consequences for mottled sculpins, *Cottus bairdi*, in *Natural Selection and Social Behaviour* (eds R. D. Alexander and D. W. Tinkle), Chiron Press, New York and Oxford, pp. 78–95.

FitzGerald, G. J., Guderley, H. and Picard, P. (1989) Hidden reproductive costs in the threespine stickleback (*Gasterosteus aculeatus*). *Exp. Biol.*, **48**, 295–300.

Grafen, A. (1988) On the uses of data on lifetime reproductive success, in *Reproductive Success* (ed. T. H. Clutton-Brock), The University of Chicago Press, Chicago, pp. 454–71.

Grafen, A. (1990a) Sexual selection unhandicapped by the Fisher process. *J. theor. Biol.*, **144**, 473–516.

Grafen, A. (1990b) Biological signals as handicaps. *J. theor. Biol.*, **144**, 517–46.

Gross, M. R. (1980) Sexual selection and the evolution of reproductive strategies in sunfishes (*Lepomis:* Centrarchidae), PhD thesis, University of Utah, University Microfilms International No. 8017132, Ann Arbor, Michigan, pp. 1–139.

Gross, M. R. and McMillan, A. M. (1981) Predation and evolution of colonial nesting in bluegill sunfish (*Lepomis macrochirus*). *Behav. Ecol. Sociobiol.*, **8**, 163–74.

Gross, M. R. and Sargent, R. C. (1985) The evolution of male and female parental care in fishes. *Am. Zool.*, **25**, 807–22.

Gross, M. R. and Shine, R. (1981) Parental care and mode of fertilization in ectothermic vertebrates. *Evolution*, **35**, 775–93.

Hammerstein, P. and Parker, G. A. (1987) Sexual selection: games between the sexes, in *Sexual Selection: Testing The Alternatives* (eds. J. W. Bradbury and M. B. Andersson), John Wiley and Sons, Chichester, pp. 119–42.

Hasson, O. (1989) Amplifiers and the handicap principle in sexual selection: a different emphasis. *Proc. R. Soc.*, **235B**, 383–406.

Hirshfield, M. F. (1980) An experimental analysis of reproductive effort and cost in the Japanese medaka, *Oryzias latipes*. *Ecology*, **61**, 282–92.

Hirshfield, M. F. and Tinkle, D. W. (1975) Natural selection and the evolution of reproductive effort. *Proc. nat. Acad. Sci. U.S.A.*, **72**, 2227–31.

Hoelzer, G. (1989) The good parent process of sexual selection. *Anim. Behav.*, **38**, 1067–78.

Houston, A. I. and McNamara, J. M. (1988) A framework for the functional analysis of behaviour. *Behav. Brain Sci.*, **11**, 117–63.

Huntingford, F. A. (1977) Inter- and intraspecific aggression in male sticklebacks. *Copeia*, **1977**, 158–9.

Iersel, J. J. A. van (1953) An analysis of parental behaviour of the male three-spined stickleback (*Gasterosteus aculeatus* L.). *Behaviour (Supp.)*, **3**, 1–159.

Jones, G. P. and Thompson, S. M. (1980) Social inhibition of maturation in females of the temperate wrasse *Pseudolabrus celidotus* and a comparison with the blennoid *Tripterygion varium*. *Mar. Biol.*, **59**, 247–56.

Keenleyside, M. H. A. (1979) *Diversity and Adaptation in Fish Behaviour*, Springer-Verlag, Berlin, 208 pp.

Keenleyside, M. H. A. (1983) Mate desertion in relation to adult sex ratio in the biparental cichlid fish *Herotilapia multispinosa*. *Anim. Behav.*, **31**, 683–8.

Kirkpatrick, M. (1985) Evolution of female choice and male parental investment in polygynous species: the demise of the "sexy son". *Am. Nat.*, **125**, 788–810.

Kirkpatrick, M. (1986) The handicap mechanism of sexual selection does not work. *Am. Nat.*, **127**, 220–40.

Kirkpatrick, M. (1988) Consistency in genetic models of the sexy son: a reply to Curtsinger and Heisler. *Am. Nat.*, **132**, 609–10.

Kleiman, D. G. (1977) Monogamy in mammals. *Q. Rev. Biol.*, **52**, 39–69.

Kleiman, D. G. and Malcolm, J. R. (1981) The evolution of male parental investment in mammals, in *Parental Care in Mammals* (eds D. J. Gubernick and P. H. Klopfer), Plenum, New York, pp. 347–87.

Kynard, B. E. (1978) Breeding behavior of a lacustrine population of threespine sticklebacks (*Gasterosteus aculeatus* L.). *Behaviour*, **67**, 178–207.

Lack, D. (1968) *Ecological Adaptations for Breeding in Birds*, Methuen, London, 410 pp.

Lynch, M. (1980) The evolution of cladoceran life histories. *Q. Rev. Biol.*, **55**, 23–42.

Mangel, M. and Clark, C. W. (1988) *Dynamic Modeling in Behavioral Ecology*, Princeton University Press, Princeton, NJ, 308 pp.

Marconato, A. and Bisazza, A. (1986) Males whose nests contain eggs are preferred by female *Cottus gobio* L. (Pisces, Cottidae). *Anim. Behav.*, **34**, 292–305.

Marconato, A. and Bisazza, A. (1988) Mate choice, egg cannibalism and reproductive success in the river bullhead, *Cottus gobio* L. *J. Fish Biol.*, **33**, 905–16.

Maynard Smith, J. (1976) Sexual selection and the handicap principle. *J. theor. Biol.*, **57**, 239–42.

Maynard Smith, J. (1977) Parental investment: a prospective analysis. *Anim. Behav.*, **25**, 1–9.

Maynard Smith, J. (1985) Mini review. Sexual selection, handicaps, and true fitness. *J. theor. Biol.*, **115**, 1–8.

Noordwijk, A. J. van and de Jong, G. (1986) Acquisition and allocation of resources: their influence on variation in life history traits. *Am. Nat.*, **128**, 137–42.

Perrone, M., jun and Zaret, T. M. (1979) Parental care patterns of fishes. *Am. Nat.*, **85**, 493–506.

Peterson, C. W. and Marchetti, K. (1989) Filial cannibalism in the Cortez damselfish *Stegastes rectifraenum. Evolution*, **43**, 158–68.

Pianka, E. and Parker, W. (1975) Age-specific reproductive tactics. *Am. Nat.*, **109**, 453–64.

Pomiankowski, A. N. (1987) Sexual selection: the handicap principle does work – sometimes. *Proc. R. Soc.*, **231B**, 123–45.

Porter, K. R. (1972) *Herpetology*, W.B. Saunders, Toronto, 524 pp.

Pressley, P. H. (1976) Parental investment in the threespine stickleback, *Gasterosteus aculeatus*, MSc thesis, University of British Columbia, Vancouver, 78 pp.

Pressley, P. H. (1981) Parental effort and the evolution of nest-guarding tactics in the threespine stickleback, *Gasterosteus aculeatus* L. *Evolution*, **35**, 282–95.

Real, L. A. (1980) On uncertainty and the law of diminishing returns in evolution and behavior, in *Limits to Action: the Allocation of Individual Behavior* (ed. J. E. R. Staddon), Academic Press, New York, pp. 37–64.

Reznick, D. (1983) The structure of guppy life histories: the tradeoff between growth and reproduction. *Ecology*, **64**, 862–73.

Ridgway, M. S. (1989) The parental response to brood size manipulation in smallmouth bass (*Micropterus dolomieui*). *Ethology*, **80**, 47–54.

Ridley, M. and Rechten, C. (1981) Female sticklebacks prefer to spawn with males whose nests contain eggs. *Behaviour*, **76**, 152–61.

Rohwer, S. (1978) Parental cannibalism of offspring and egg raiding as a courtship strategy. *Am. Nat.*, **112**, 429–40.

Salthe, S. N. and Meacham, J. S. (1974) Reproductive and courtship patterns, in *Physiology of the Amphibia*, Vol. 2 (ed. B. Lofts), Academic Press, New York, pp. 307–521.

Sargent, R. C. (1981) Sexual selection and reproductive effort in the threespine stickleback, *Gasterosteus aculeatus*, PhD thesis, State University of New York, Stony Brook, University Microfilms International No DA 8211237, Ann Arbor, Michigan, pp. 1–85.

Sargent, R. C. (1982) Territory quality, male quality, courtship intrusions and female nest-choice in the threespine stickleback, *Gasterosteus aculeatus*. Anim. Behav., **30**, 364–74.

Sargent, R. C. (1985) Territoriality and reproductive tradeoffs in the threespine stickleback, *Gasterosteus aculeatus*. *Behaviour*, **93**, 217–26.

Sargent, R. C. (1988) Paternal care and egg survival both increase with clutch size in the fathead minnow, *Pimephales promelas*. *Behav. Ecol. Sociobiol.*, **23**, 33–8.

Sargent, R. C. (1989) Allopaternal care in the fathead minnow, *Pimephales promelas*: setpfathers discriminate against their adopted eggs. *Behav. Ecol., Sociobiol.*, **25**, 379–85.

Sargent, R. C. (1990) Behavioural and evolutionary ecology of fishes: conflicting demands during the breeding season. *Ann. zool. fenni.*, **27**, 101–18.

Sargent, R. C. and Gebler, J. B. (1980) Effects of nest site concealment on hatching success, reproductive success, and parental behaviour of the threespine stickleback, *Gasterosteus aculeatus*. *Behav. Ecol. Sociobiol.*, **71**, 137–42.

Sargent, R. C. and Gross, M. R. (1985) Parental investment decision rules and the Concorde fallacy. *Behav. Ecol. Sociobiol.*, **17**, 43–5.

Townshend, T. J. and Wootton, R. J. (1984) Effects of food supply on the reproduction of the convict cichlid, *Cichlasoma nigrofasciatum*. *J. Fish Biol.*, **23**, 91–104.

Trivers, R. L. (1972) Parental investment and sexual selection, in *Sexual Selection and the Descent of Man* (ed. B. Campbell), Aldine-Atherton, Chicago, pp. 136–79.

Unger, L. M. (1983) Nest defense by deceit in the fathead minnow, *Pimephales promelas*. *Behav. Ecol. Sociobiol.*, **13**, 125–30.

Unger, L. M. and Sargent, R. C. (1988) Allopaternal care in the fathead minnow, *Pimephales promelas:* females prefer males with eggs. *Behav. Ecol. Sociobiol.*, **23**, 27–32.

Weatherhead, P. J. and Robertson, R. R. (1979) Offspring quality and the polygyny threshold: the "sexy son" hypothesis. *Am. Nat.*, **113**, 201–8.

Williams, G. C. (1966a) Natural selection, the costs of reproduction, and a refinement of Lack's principle. *Am. Nat.*, **100**, 687–90.

Williams, G. C. (1966b) *Adaptation and Natural Selection*, Princeton University Press, Princeton, NJ, 308 pp.

Williams, G. C. (1975) *Sex and Evolution*, Princeton University Press, Princeton, NJ, 200 pp.

Wilson, E. O. (1975) *Sociobiology: the New Synthesis*, Harvard University Press, Cambridge, MA, 698 pp.

Wittenberger, J. F. (1981) *Animal Social Behaviour*, Duxbury Press, Boston, 722 pp.

Wootton, R. J. (1976) *The Biology of the Sticklebacks*, Academic Press, London, 388 pp.

Zahavi, A. (1975) Mate selection–selection for a handicap. *J. theor. Biol.*, **53**, 205–14.

Zahavi, A. (1977) The cost of honesty (further remarks on the handicap principle). *J. theor. Biol.*, **67**, 603–5.

Chapter twelve

Functions of shoaling behaviour in teleosts

Tony J. Pitcher and Julia K. Parrish

12.1 INTRODUCTION

Predators and food are the proximal keys to understanding fish shoals; on a second-to-second basis, synchronized cooperation defeats predators, while food-gathering in shoals reflects a shifting balance between joining, competing in, or leaving the group. In the wild, predators may arrive and attack while shoaling fish are feeding, and so vigilance and other means of gathering information about predators are crucial behaviours. Once a predator is detected, defence generally takes precedence over feeding since an animal's life is worth more than its dinner, but the presence of social companions in a shoal allows individuals to choose among an effective range of trade-offs between feeding, attack mitigation and escape.

Travelling fish schools display impressive coordination, and were once viewed as egalitarian, leaderless societies in which cooperation preserved the species (Breder, 1954; Shaw, 1962; Radakov, 1973). In contrast to such classical group-selectionist views, behavioural ecology reveals social behaviour to be no more glamorous than animals cooperating only when it pays. Distinct coexisting behavioural strategies of sneaking or scrounging are often evolutionarily stable (Barnard, 1984b; Parker, 1984). In fish shoals, homogeneity, synchrony and three-dimensional structure were over-emphasized in the 1960s and 1970s; experimental work from the early 1980s onwards began to reveal that individuals constantly reappraise the costs and benefits of being social and belonging to a shoal. Such reappraisal is reflected in decisions to join, stay with or leave the group. Observed behaviour allows us a glimpse of these underlying tensions. In teleosts, a major constraint is swimming; fish that are physiologically and morphologically adapted to cruise fast, such as mackerel, break ranks less often to avoid the alternative of rapid dispersal. Under some circumstances, however, the underlying tensions between individuals may be uncovered in even the most phalanx-like of cruising fish schools.

Behaviour of Teleost Fishes 2nd edn. Edited by Tony J. Pitcher. Published in 1993 by Chapman & Hall. ISBN 0 412 42930 6 (HB) and 0 412 42940 3 (PB).

But what we, and the fishes' predators, perceive is impressive synchrony, and so we are led to ask what selective forces have shaped this attribute. Although we can demonstrate that school members act as selfish individuals, moment-to-moment decisions about whether and how to respond to a given stimulus are influenced by the behaviour of close neighbours, and possibly by fish further away in the group. Especially in open-water species, where there are no spatial refuges other than the school, how the school responds as a whole can have a drastic influence on the fate of individual members. While individuals will always attempt to maximize their survival by exploiting the vulnerability of others, in fish schools under attack by a predator it is usually in the interest of individuals to coordinate behaviour. Although once considered the evidence of egalitarian cooperation, we no longer need to invoke altruism to account for group-wide coordination in the face of a predator. Individual members of a school responding incorrectly bring immediate attention to themselves by becoming spatially separated or creating a localized zone of confusion (p. 388) and, as revealed by recent experiments, such incorrect responses can be fatal. Risk of predation as a selective force of evolution should drive the stereotypy of individual responses, and so produce coordinated, highly synchronous manoeuvres by fish schools. Many other factors also contribute to the uniformity of schools, such as hunger (Robinson and Pitcher, 1989a, b), asymmetrical pay-offs in feeding competition (Pitcher *et al.*, 1986a), and competition for spawning site (Magurran and Bendelow, 1990).

Shoaling behaviour has attracted much speculation about function (e.g. Shaw, 1978; Partridge, 1982a), but until the early 1980s few critical experiments were performed. The aim of this chapter is to review in the light of current theory the areas where such evidence of function has been gathered. The arguments presented in this chapter do not support the view that shoaling is primarily a matter of cover-seeking (Williams, 1964; Hamilton, 1971), or the idea that hydrodynamics is a major factor (e.g. Breder, 1976). Furthermore, the evolution of fish shoals is unlikely to have been driven by simple attack avoidance and attack dilution.

Our presentation of the functions of fish shoals is set in the context of the thesis that fish shoals are social groups brought about by individual fish which choose to join and stay with their fellows. The rules of assembly of fish in shoals are governed on a second-to-second time scale by an interacting complex of environmental factors among which food and predation risk are paramount.

12.2   DEFINITIONS OF SHOALING

A clarification of terms will facilitate discussion of function, since shoaling behaviour continues to suffer from the semantic confusion of the 1960s

(Keenleyside, 1955; Hemmings, 1966; Radakov, 1973). Groups of fish that remain together for social reasons are here termed **shoals** (Kennedy and Pitcher, 1975; Pitcher, 1983), in an analogous way to the term 'flock' for birds. Defined as a social group of fish 'shoal' has no implications for structure or function.

Synchronized and polarized swimming groups are termed **schools**. Schooling is therefore one of the behaviours exhibited by fish in shoals (Fig. 12.1), and schools have a structure measured in polarity and synchrony. Shoals that travel are almost bound to school on the way. Other action patterns of fish in social groups, such as foraging or antipredator manoeuvres, can be accommodated within the taxonomy outlined in Fig. 12.1.

In North America the term 'school' may still sometimes be used to cover both phenomena distinguished in this chapter. However, the 'shoal' and 'school' definition appear synonymous with 'social aggregation' and 'polarized school' definitions introduced by Norris and Schilt (1988). The dichotomy

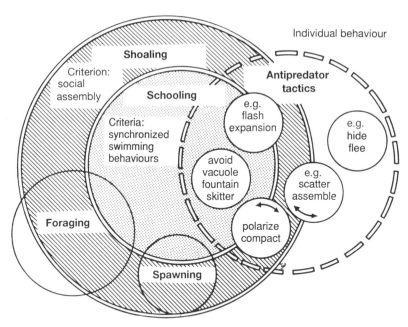

Fig. 12.1 Relationship of shoaling and schooling behaviour illustrated by a Venn set diagram. Criteria for the two behaviours are indicated. Three other behaviours are superimposed as examples of how other functional categories relate to these terms. Some further specific examples of antipredator tactics are shown; spawning and foraging behaviours might be similarly elaborated. Unlike earlier definitions, this scheme can augment descriptions of observed behaviours. Reproduced with permission from Pitcher (1983).

between 'facultative' and 'obligate' schooling fish introduced by Breder (1959) has, however, been largely rejected in favour of direct estimates of the proportion of time which shoaling fish spend travelling in polarized groups (Partridge, 1982b; Pitcher, 1983). For a full discussion of these definitions see Pitcher (1983).

12.3   PERSPECTIVE ON STRUCTURE AND FUNCTION IN SHOALS

This chapter concentrates on function, and does not attempt a full review of recent work on the sensory basis or structural dynamics of schooling. Nevertheless some consideration of what we have learned from structure is necessary for a balanced consideration of the field. In addition, we will consider briefly how shoaling behaviour impinges upon the capture of fish in human fisheries (further behavioural details are given by Wardle, Chapter 18, this volume).

Much of the long search for structure in fish schools has been sterile for the explanation of function, not least because of the failure to distinguish between shoaling and schooling. Measurements of position and movements for individual fish, crucial to tests of current theory, have generally been lost in pooled values for the structural parameters of nearest neighbour distances, bearings and polarity (e.g. Cullen *et al.*, 1965; Hunter, 1966; Pitcher, 1973; Inagaki *et al.*, 1976; Partridge *et al.*, 1980).

Individuals have been distinguished only rarely; very few rigorous accounts of individual positions have been published. Healey and Prieston (1973) identified peripheral and central fish in schools of coho salmon, *Oncorhynchus kisutch*. Pitcher *et al.* (1982b) measured individual position preferences in mackerel, *Scomber scombrus*, schools. Muscialwicz and Cullen (unpublished experiments) demonstrated consistent individual 'initiators' and 'followers' in zebra danio, *Brachydanio rerio*, shoals, and Pitcher (1979b) presented similar evidence in bream, *Abramis brama*, shoals, but did not recognize consistent individuals. Robinson (1991) showed that preferences for lead and side positions in schools of herring, *Clupea harengus*, depended upon the hunger state of the individual. Experiments also suggested that different join, leave and stay tendencies among individuals with hunger might select for homo-geneity in fish schools (Robinson and Pitcher, 1989b).

Work on school structure has found difficulty in assessing the costs and benefits to individuals because they are difficult to translate into the measures of lifetime fitness which strict interpretations of theory demand (Krebs and Davies, 1978). This is not least because of the difficulty of expressing antipredator, feeding and hydrodynamic advantages in equivalent units.

Based on data and behaviour in bird flocks, Mangel (1990) has examined the costs and benefits of group formation with a unified theory employing general units of fitness in a dynamic programming model. Metabolic reserves of the individual at the end of the modelled period are considered to be related to its probability of reproducing. Separate foraging and predation risk factors are incorporated in the overall model. In detail, bird flocks subsume different social structures from fish shoals, for example individual birds have a wide range of vocal calls for food assembly, alarm etc., and so it would be most valuable to restructure Mangel's model with the particular characteristics of fish shoals. Potentially this type of model could be used to predict how changes in food availability, predation risk and other factors will affect the observed group size. The model could be tested by comparing simulated results with observed changes in shoal size. This type of approach is a relatively new perspective on behavioural ecology, and we can expect developments in this area in the near future.

Elective group size in shoals

In the looser organization of the shoal, the precise spatial positioning and coordination with swimming neighbours or schooling is not needed, but we still require a measure of cohesion (packing) and synchrony since, as we will see, these are adjusted according to circumstances. Cohesion in terms of the distances and synchrony between fish could be measured using the same three-dimensional methods used for schools (Pitcher, 1975; Pitcher and Lawrence, 1985), but usually can be scored adequately by eye, as in the aggregation scores by John (1964), Andorfer (1980) and Cerri (1983). A more quantitative measure was devised by Helfman (1984) and Wolf (1983), who based their work on measuring the distribution of fish groups of different sizes.

Using a stricter criterion for a group, including only fish considered to be behaving together within 4 body lengths, Pitcher *et al.* (1983) introduced the term **elective group size (EGS)**; 'elective' distinguishes the measure from group sizes deliberately created by the experimenter. EGS turns out to be a sensitive measure of the fish's perception of the current balance of costs and benefits of shoaling, and can be measured either for individuals or for groups. For example, Fig. 12.2 shows the distribution of EGS values in undisturbed shoals of minnows, *Phoxinus phoxinus*, and dace, *Leuciscus leuciscus*, foraging in semi-natural conditions in a fluvarium. In these shoals, many small fish find it of benefit to forage individually or in small groups, wheareas larger fish are in somewhat larger shoals.

Similarly, McNicol and Noakes (1984) demonstrated a very sensitive response of juvenile brook trout, *Salvelinus fontinalis*, to crowding, group numbers, food and water flow. Behaviour changed rapidly from shoaling through individual territory-holding to dominance hierarchies.

EGS tells us something about the forces acting on the shoal: variance in EGS

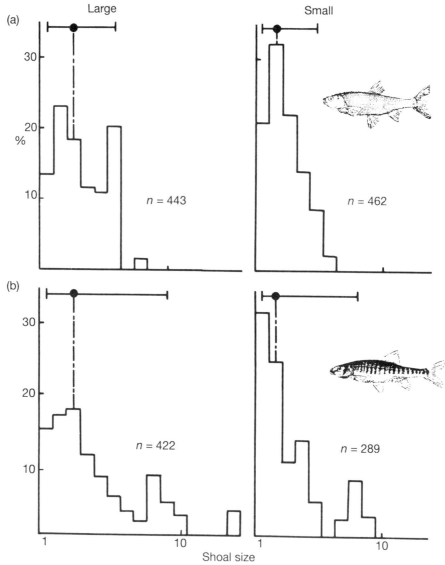

Fig. 12.2 Elective group size (EGS) for two length classes of (a) dace and (b) minnows shoaling in a large 10 m fluvarium; 25 large (left) and small (right) fish of each species were present (mean lengths: dace, 175 and 97 mm; minnows, 56 and 39 mm). Histograms show the percentage of fish observations for each observed group size: totals given as *n*. Groups were scored as fish within 4 body lengths, recorded at a standard time once per day over a 10-day period. Median group sizes shown by solid circles, 90 percentiles by bars. For further details see Pitcher *et al.* (1983).

reflects the stability of those forces. In a constant environment, with no changes in predation risk, food availability or migratory tendencies, EGS would converge on a single value. Changing environments require compensatory adjustments in the EGS. Therefore fish with broad distributions of group sizes probably experience volatile conditions in time, requiring constant adjustment, or in space, such that a wide range of group size is observed. Tracking these changes in relation to environmental factors is a major challenge for the next phase of research on fish shoals.

Hager and Helfman (1991) presented single minnows with a series of choices of shoal sizes to join, ranging from one to 28 fish. In the absence of external stimuli, minnows exhibiting a preference chose the larger of the two shoals. However, discrimination between shoal sizes weakened as overall size increased and as the difference between shoal-pairs presented decreased. In the presence of a predator, however, minnows chose large shoal sizes, made their choices more quickly, and showed a heightened degree of ability to discriminate between shoal size.

EGS varies greatly among species according to life-history traits. In stream-dwelling fishes, small group sizes may result from the availability of refuges other than the shoal. However, open-water fishes, with no structural refuge but the group, can occur in extremely large groups. Such groups can become even larger during periods of migration.

Questions of scale in fish shoals

At first sight, it may seem surprising that the same terminology can deal adequately with very large marine shoals, perhaps several kilometres long, of millions of herring, and small freshwater shoals of a few hundred minnows. Accurate acoustic measurements of fish densities (e.g. Misund, 1991) show that large marine shoals often consist of subgroups on at least one (metres), and possibly two (tens of metres), size scales. Moreover, although we tend to see them in small groups, freshwater shoals in lakes and rivers are likely to have a larger-scale organization analogous to the more dramatic large marine shoals, albeit more constrained by available habitat.

While in this chapter we can show that the terms and definitions of shoaling set out above provide powerful tools for investigating individual behaviour in the subunits in relation to proximal foraging and predator factors, we are only just beginning to understand the rules governing the assembly of subunits on larger scales of organization. The reason for this is largely methodological: at the scale of the subgroups, effective experimental protocols to test meaningful hypotheses can be performed in large laboratory aquariums. On larger scales, even field experiments become more expensive and difficult, and we have to rely largely on measurements taken in the wild.

Shoals and schools in the wild

Individual fish are difficult to visualize on all but the most recent acoustic instruments. Traditional oceanographic measurements are integrated over increments of distance and time that are too large to answer questions about shoaling. The necessary scale and precision in space and time of underwater measurements of fish and their dynamic environment has only recently become feasible.

Evidence suggests that there is large seasonal variability in shoal sizes in the wild, although the precise environmental factors driving such dynamic changes are not yet clear. For example, using purse seine catches, Freon (1991) analysed large seasonal variations and a steady decrease in school size of sardines with stock depletion in the Senegal fishery. Working with modern acoustic instruments from the Fisheries Institute in Bergen, Norway, Misund (1991) has recently demonstrated similar changes in shoal size with area and season in Norwegian herring (Fig. 12.3). Misund's work generally confirms the prediction of Pitcher and Partridge (1979) that the density of fish in schools is around one fish per cube of body length. Misund has also visualized subgroup structure within herring schools using an acoustic 'cell integration' technique (Fig. 12.4). One problem with this visualization is that the boundaries of the school may be blurred because of the method of the acoustic scanning. The resolution is still not quite sufficient to 'see' the definite edges of the schools and their subgroups as we observe by eye. Nevertheless, this type of work will become increasingly important in shoaling research.

A purse seine fishery for horse mackerel, *Trachurus murphyi,* off Chile exhibits fascinating dynamics with season and time of day. Changes in shoaling behaviour appear to track food availability (T. Antezana, 1991, pers. comm.). The fish constitute a viable commercial purse seine catch only when shoaling densely, and so the fishermen's behaviour tracks that of the fish. In winter the fishery is at night over oceanic water as the fish migrate from depth to feed in the pelagic zone on myctophids and euphausids. In the austral summer, the fishery is in the daytime over the continental shelf, where the fish feed on copepods and squid. The purse seine fleet often follows the fish out to sea over 4 to 5 days along with a plume of upwelling nutrient-rich water which drives the plankton production. When the plankton patch on which the horse mackerel are feeding diminishes (or disperses), and the fish are eating less food, the fish switch from dense schooling to looser shoals and begin to disperse. This behavioural decision is matched by the fishermen's decision to stop fishing and look for another area of *Trachurus* shoals (Hancock *et al.,* in press).

It is increasingly important to understand and to forecast these larger-scale phenomena in fish shoaling since it is at this level that shoaling behaviour has the greatest impact on commercial fisheries for shoaling species.

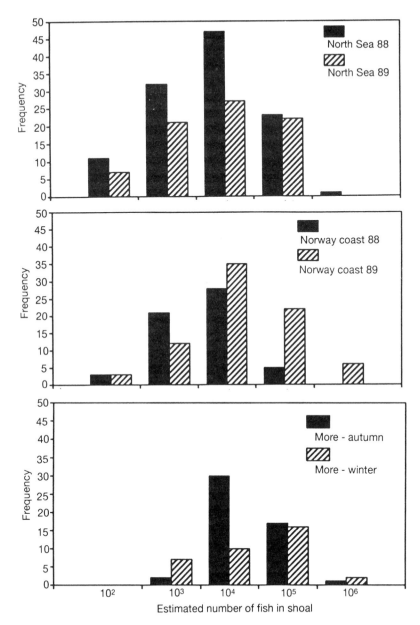

Fig. 12.3 Estimated numbers of fish in discrete Atlantic herring shoals estimated using sonar techniques. Top: North Sea; middle: Norway coast (Lofoten and Gratangen); bottom: Norwegian fjord, More, autumn and winter. Average school sizes differed significantly among regions and seasons. Redrawn from Misund (1991).

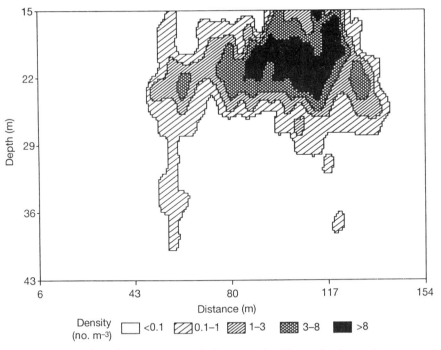

Fig. 12.4 Packing density stucture of a herring school from a fjord near Grantangen, Norway, in 1989, as estimated by cell integration acoustics. It is possible that the gradual edges to the school may partly be artefacts of the acoustic scanning method. The herring density in the centre of the school is approximately one fish per body length cubed. Reproduced with permission from Misund (1991).

Instability of commercial fisheries for shoaling species

It is easy to show that fisheries for shoaling species are inherently unstable (Murphy, 1980) since humans acting as predators have a range of sophisticated devices (e.g. spotter planes, sonar) which render fish shoals detectable at large ranges.

The simplest production models for assessing fisheries depend upon the logistic population growth model:

$$dB/dt = rB[1 - (B/B_\infty)] - qEB \qquad (12.1)$$

where B is population biomass, B_∞ is carrying capacity for biomass of this species, r is rate of increase of biomass, E is fishing effort, t is time, and q is a catchability coefficient for this species in this fishery. The exact form of the equation does not especially concern us here as essentially similar results derive from the various modifications to this model which are used in practice

(a)

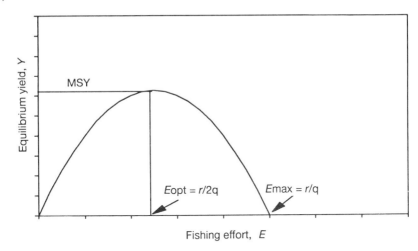

(b)

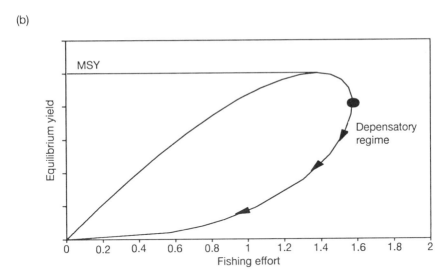

Fig. 12.5 (a) Sustainable yield (catch in weight) from a fishery (single species unit-stock) under equilibrium conditions plotted against fishing effort as predicted from the classic Schaefer surplus production fishery model. MSY denotes maximum sustainable yield. Arrows indicate optimum and maximum effort. Catchability, q, is constant. An equivalent graph could be drawn for fish population biomass. (b) In shoaling species, fish catchability, q, is not constant and increases as population size is diminished by fishing because the capture probability per shoal remains almost the same. A depensatory backward-bending yield-against-effort curve may result. Oval blob denotes point of inflection. For further details see text.

in fishery management. The equilibrium yield (y) is a parabola:

$$Y = qEB_\infty[1 - (qE/r)] \tag{12.2}$$

which forms the familiar equilibrium fishery yield curve of Schaefer, shown in Fig. 12.5(a) (Pitcher and Hart, 1982). Note that catch per unit effort, U, declines as E increases and tracks the decline in population biomass as the population is depleted, $U = Y/E = qB$. The essential feature of this theory is that the catchability of the fish, q, is a constant, U/B. This quantity is an important output estimated by many of the modern fishery assessment techniques.

But serious problems arise in shoaling species where purse seiners armed with sonar can locate and catch a whole shoal almost irrespective of total stock density. Catch per unit effort (measured as catch per shot of the net) can be virtually constant. The serious corollary is that catchability can increase in inverse proportion to abundance:

$$q = aB^b \tag{12.3}$$

and the standard assessment models cannot easily be applied (a is a proportionality constant). For clupeids, b has been estimated at $-.4$ (MacCall, 1976). If we now substitute this relationship instead of constant catchability, q, in the standard Schaefer model (Equation 12.2), then:

$$Y = aB^b EB_\infty[1 - (aB^b E/r)] \tag{12.4}$$

giving a backward-bending curve which is unstable beyond the inflection, as shown in Fig. 12.5b (Csirke, 1988). Beyond the inflection point the stock may be forced to collapse even if effort decreases. This effect is known as inversely density-dependent or 'depensatory'. Clark (1974) analysed an analogous depensatory collapse based on a stock recruitment relationship for shoaling species derived from encounter theory. Both Csirke's and Clark's analyses alert us to a real risk of collapse happening in pelagic shoaling stocks. In fact, increases in catchability with large decreases in stock size have been reported in fisheries for Californian sardine, Atlantic menhaden, Norwegian herring and South African sardine.

Understanding how changes in the shoaling dynamics of important commercial species impacts on the stability of fisheries currently presents a major challenge to behavioural and fishery scientists.

12.4 HYDRODYNAMIC ADVANTAGE IN SHOALING

The beating tail of a swimming fish generates a wake of counter-rotating spinning vortices (Fig. 12.6), which can be visualized using special techniques. It is reasonable to assume that schooling fishes avoid swimming into the wake of those in front and that this constrains their positions in the school. Teleost fishes exhibit a wide diversity of tail shape, including tails specialized for acceleration (e.g. barracuda), fast cruising (e.g. tuna), or manoeuvrability (e.g. perch) (Yates, 1983), but it is not yet known whether these differences influence fish positions in schools as would be expected.

The idea of positive hydrodynamic advantage, rather than passive avoidance of wakes, has been put forward several times (Breder, 1965; Belyayev and Zuyev, 1969; Weihs, 1973, 1975). Zuyev and Belyayev (1970) filmed 20 to 30 *Trachurus* cruising in a flume, and found that fish within stable subgroups exhibited tailbeat frequencies proportional to the distance from the front of the group. This was taken as evidence that fish in front were having to work harder.

Weihs (1973, 1975) formalized wake theory, making five specific predictions about spacing and behaviour in travelling schools which would enable fish to make energy savings up to 65%. Fish gaining benefits 1 to 3 and 5 should swim in a diamond lattice.

1. Lateral neighbours should be 0.4 body lengths apart.
2. Their tailbeats should be in antiphase.
3. Fish in the next row should be centred between the leaders, 6.7 body lengths behind.
4. Lateral neighbours should swim 0.3 body lengths apart to maximize benefit from a lateral 'push-off' effect.
5. Uplift from rigid pectorals in negatively buoyant fish could aid fish behind.

None of the extensive work on the three-dimensional structure of fish schools in the 1970s supported the rigid diamond lattice envisaged by Weihs, but there remains the possibility that advantage could be gained through a statistical tendency to adopt the advantageous positions. Using a very large amount of three-dimensional data on cruising schools of herring, *Clupea harengus*, cod, *Gadus morhua*, and saithe, *Pollachius virens*, (more than 20000 frames at 3 per second), Partridge and Pitcher (1979) demonstrated that predictions 1 to 3 were not met. For prediction 3, even when four out of only 659 fish found centred behind two leaders were 5 body lengths behind, they did not maintain these putatively beneficial positions as might be expected . Indeed, the snouts of following fish were very often ahead of the region where the swimming vortices stabilize, and often ahead of the tail itself, so no hydrodynamic benefit could accrue on this theory. In fact, Partridge and Pitcher made a small error in the lateral spacing predicted by Weihs' theory (Parrish, pers. comm.), but recalculating the figures does not alter the conclusion that the majority of fish were not found near to the predicted positions.

When fish swim faster, tailbeat frequency increases but amplitude, 'stride length' and vortex spacing stay the same (Videler and Wardle, 1991). Weihs' hydrodynamic model therefore predicts no change in relative positions with swimming speed. But in fact school structure alters and becomes more compact with speed (Pitcher and Partridge, 1979). MacCullum and Cullen (pers. comm.) have also shown that the tailbeat prediction, number 2 above, does not hold for schooling yellowtail scad, *Trachurus mccullochi*. Water flow in the wake of swimming fish is currently being described with much greater precision in three dimensions (J. J. Videler, unpublished data), as shown in Fig. 12.6, (see also Chapter 7 by Bleckmann, this volume), and it is possible that this will improve our understanding.

For prediction 5, teleost fish do not hold their pectorals rigidly out when swimming. Indeed the fast-cruising, negatively buoyant scombrids, tunas and istiophorids have evolved precise grooves in which to tuck the pectorals away. Most elasmobranchs which swim with rigid pectorals to generate lift do not school, but a few, such as cownose rays, *Rhinoptera bonasus*, school in the same cohesive way as teleosts (Rogers *et al.*, 1990), and so pectoral lift is a possibility here (Klimley, 1985).

A much more serious criticism of the hydrodynamic hypothesis can be put forward on theoretical grounds. Benefits from Weihs' predictions 1 to 3 accrue only to alternate rows in the school, and therefore are evolutionarily unstable in the absence of altruism. There is no hint of the likely scramble for good positions by fish finding themselves in non-benefit rows. It has been suggested that if advantageous fish positions are shuffled by turns, hydrodynamic benefit might be shared in time (Blake, 1983), but this does not overcome the fundamental objection that no evidence of instability in poor positions is seen.

Benefit from prediction 4, the lateral 'push-off' effect, is fundamentally different in this respect since individual fish gain from their own behaviour, but no unequivocal evidence in support has been produced yet. The saithe studied by Partridge and Pitcher (1979) swam about 0.9 body lengths apart and would have gained about 35% of the maximum benefit from this lateral effect had their neighbours been on the same horizontal level, but saithe neighbours tended to be about 10 degrees above or below. For the lateral advantage to be maximized, fish should choose to swim next to neighbours of the same size. Evidence of a significant tendency towards such choice of neighbour size has been found in herring and mackerel schools (Pitcher *et al.*, 1985). Additional evidence supporting the lateral-neighbour effect has been put forward for bluefin tuna schools (Partridge *et al.*, 1983).

Shiner, *Notropis heterodon*, shoals in a small aquarium became deeper vertically and more compact when attacked by largemouth bass (Abrahams and Colgan, 1985). It was argued that the fish sacrificed a flat, hydrodynamically efficient structure in favour of an unobstructed view of the predator. Abrahams and Colgan (1987) argued forcefully for a trade-off of hydrodynamic advantage against reduction of predator risk. These authors' conclusion rests on two untested assumptions. First, a flat school of six fish was assumed to be

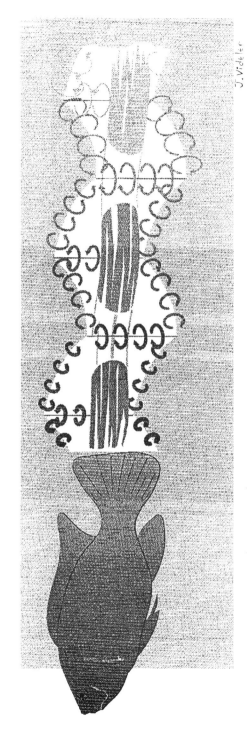

Fig. 12.6 Representation of the water flow and vortices in the wake of a swimming fish. Illustration courtesy of J. Videler.

more hydrodynamically efficient than a vertically stacked school. This prediction is not easily obtained from Weihs' mainly two-dimensional theory. Secondly, they assume that the field of view is obscured by neighbours in the flat school because the predator attacks on the same horizontal level as the shiners. Although this may have been how the experiment was arranged in the aquarium tank, it is difficult to see how the field of view argument could apply in the wild. In other cyprinids, small changes in nervousness which alter shoal size, structure and compaction can be explained without invoking hydrodynamics.

There have been many reports of fish using less oxygen in groups (Abrahams and Colgan, 1985; see Parker, 1973, for earlier references), and hydrodynamic advantage is often claimed as the cause of the lower energy demands. Unfortunately, none of these experiments has contained a satisfactory control for the motivational effects of group size. Group size has a profound effect upon the behaviour of individual fish. As will be discussed later, fish in smaller groups are more timid and nervous and consequently have higher respiratory rates (Itazawa *et al.*, 1978). Since small shoals are less likely to detect a predator early and are less effective in escaping from an attack, the higher respiratory rate of nervous fish might be interpreted functionally as a greater readiness for careful, considered evasion manoeuvres using red muscles. (Emergency escape in teleosts employs the white muscle, which is anaerobically fuelled and engenders recuperation costs.) Long-term growth and energy budgets of minnows are, however, not affected by group size, so fish must compensate for differences in the behaviour in groups (Cui and Wootton, 1989).

Drag-reducing fish mucus has sometimes been linked to hydrodynamic advantage in schools. By adding Polyox, a non-toxic high-molecular-weight polymer of ethylene oxide, to the water, Breder (1976) showed that menhaden, *Brevoortia patronus*, a fast-cruising schooling species, beat their tails almost twice as fast as controls and swam proportionally faster. Breder argued that the Polyox simulated the effect of fish mucus in reducing drag in schools. In fact, there was no evidence for drag reduction because swimming speed increased proportionally with tailbeat frequency. Furthermore, the experiment did not control for slime production, nor for group size, which could have been different between the Polyox and control treatments. Parrish and Kroen (1988) tested the hypothesis again using silversides, *Menidia menidia.* Tailbeats were proportional to swimming speed, but there was no discernible drag reduction measured as a decrease in tailbeat frequency of amplitude for a given swimming speed. Moreover, the measured sloughing rate of mucus from a fish was so small that millions of individuals would be needed to generate concentrations approaching even the lowest Polyox concentration tested.

In conclusion, no valid evidence of hydrodynamic advantage in travelling schools has yet been produced; the existing evidence contradicts most aspects of the only quantitatively testable hydrodynamic theory published. It is certainly surprising that no evidence has been yet been put forward that fish positions in schools are, at the very least, constrained by the thrust vortices shown in Fig. 12.6.

Nevertheless, since fish in the wild often exhibit remarkably regular three-dimensional spacing, the idea of hydrodynamic advantage remains attractive and plausible. Minnows maintaining station in a standing wave in a flume tank adopted more regular school spacing than those in normal flow (unpublished observations associated with the experiments described in Pitcher, 1973). It seems very possible that the hydrodynamic theory needs more attention as the exact dimensions of the wake are quantified, especially as Weihs' theory contains a number of internal inconsistencies. In compact, highly polarized schools, perhaps pockets of water travel with the group of fish, making inside positions in this slipstream less costly. On the other hand, hydrodynamics could not help fish in unstructured shoaling, and is therefore unlikely to have been a primary reason for the evolution of the social groups we term shoals.

12.5 ANTIPREDATOR FUNCTIONS OF SHOALING

Behaviour in fish shoals has been shaped by an evolutionary arms race (Dawkins and Krebs, 1979) with predators; what we currently observe is behaviour reflecting the shoaling fish's tendency to stay one fin ahead in this race. The behaviour has, of course, also been shaped by foraging, as discussed later in this chapter: predator attack must have evolutionary priority, though (Dawkins and Krebs' life/dinner principle). Since the first edition of this book, recent reviews of antipredator functions in fish shoals have been given by Godin (1986a), Magurran (1990) and Parrish (1992).

This section aims to review how shoaling fish may counter predator attack by avoidance, dilution, abatement, evasion, detection, mitigation, inspection, inhibition, prediction and confusion. These terms are defined in Table 12.1 and must be regarded as candidate principles because not all have yet been demonstrated experimentally.

A quantitative conflation of these factors over the whole spectrum of predators suffered by our shoaling prey will shift the delicate balance between the 'join, leave and stay' rules of assembly. The time scale of such shifts has a profound effect on behaviour. Where they occur on a second-to-second basis, fish generally try to trade off the costs and benefits through bet-hedging behaviour. Fish may be able to track seasonal shifts by altering the shoaling rules on a seasonal basis: this often occurs during spawning. Where shifts happen over the life span of an individual, evolutionary changes in shoaling behaviour are registered. Fish in shoals do not necessarily behave homo-

Table 12.1 Logical categories defining the ways in which fish shoals may counter attacks of predators

Strategy	Definition
Avoidance	Avoiding coming into attack range of predator. Predator may or may not be detected.
Dilution	Reduction of risk for an individual member of a group as group size increases because predator is attacking only one of the group (or, strictly, less than the total number). Predator is detected.
Abatement	Reduction of risk with group size for an individual member of a population because of search and dilution. Predator is detected.
Evasion	Reducing the success of an attack by moving out of strike range of a detected predator or by beating the predator's manoeuvrability during a strike. May apply to individual behaviours (e.g. skittering) or to the group as a whole (e.g. flash expansion).
Detection	An individual becoming aware of the presence of a predator, usually (but not always) denoted by some small behavioural cue signalling alertness. Sensory cues from the predator may be direct (visual, auditory, chemosensory), or indirect, mediated via changes in neighbour fish's behaviour signalling alertness.
Mitigation	Reducing the probability of success of an attack which has already been launched by a detected predator.
Inspection	Gaining information about a potential predator while approaching it, and then returning to the group.
Inhibition	Reducing the likelihood of a detected and attacking predator launching a strike.
Confusion	Reducing the success of a attack that has been launched, by beating the predator's sensory (or cognitive) capacity.

geneously – different individuals may adopt alternative trade-offs, both on a moment-to-moment time scale and over a lifetime.

Predator avoidance

It has often been implied that avoidance is automatically enhanced by grouping in shoals. When visual range is low in relation to the relative speed of predator and prey, grouping of prey has been considered to be favoured over dispersal, since the travelling predator has less chance of coming within the detection envelope of a group than the many such envelopes of scattered fish. Although implicitly group-selectionist, since the detection probability for any particular fish is identical in both the dispersed and grouped cases, this idea has been re-invented several times (Brock and Riffenburgh, 1960; Olsen, 1964; Cushing and Harden-Jones, 1968; Wilson, 1975; Partridge, 1982a) and, as first pointed out by Williams (1964), is mistaken. In one variant of the theory (Taylor, 1984), predators that encounter fewer prey as a consequence of overlapping detection envelopes of grouped prey are supposed to go away somewhere else to eat. It is important to realize that the optical qualities of water (see Chapter 4 by Guthrie and Muntz, this volume) render most shoals barely more detectable than a single individual (Murphy, 1980).

The simple avoidance argument confuses the probability of detection of a shoal with the fitness of individual fish. The average group member is protected only if the consumption rate of the predator is lower when feeding on groups. Treisman (1975) shows that if the predator can only eat one individual (or some small proportion of the group) while the rest flee, grouping may be favoured. As our hypothetical predator's total kill per group increases, the loner strategy will be favoured. If this were a major factor in the evolution of fish shoals, most surviving predator species would be expected to be 'single prey eaters'.

Two additional arguments lead us to believe that the passive-avoidance hypothesis is not well supported by evidence, and therefore has probably not had a major influence on the evolution of shoaling.

First, like herds of mammals, many shoals have been shown to be intimately accompanied by their fish predators. For example, roach, *Rutilus rutilus*, in a British river were never out of range of predatory pike, *Esox lucius* (Pitcher, 1980); barracuda (Sphyraenidae) accompany shoals of grunts (Pomadasyidae) on coral reefs. For fish living near cover in some form, such as weeds or substrate, dispersal and refuging would meet the aim of predator avoidance more effectively than grouping. When there are few conspecifics, this is precisely what shoaling minnows do (Magurran and Pitcher, 1983, 1987). Similarly, Savino and Stein (1982) found that bluegill sunfish, *Lepomis macrochirus*, exposed to largemouth bass, *Micropterus salmoides*, predators schooled less when cover was increased.

Secondly, fish shoals near the surface are more visible from the air than individuals. This takes care of the argument about pelagic shoalers, since their diving avian predators (gannets, terns, pelicans, auks) certainly detect shoals at long range. Avian predation is a large source of mortality in such fish, and so, by virtue of this conspicuousness, shoaling cannot have been selected to minimize detection. Again dispersal would be more effective than huge marine shoals for simple avoidance of these predators.

Helfman (see Chapter 14, this volume) shows how the behaviour of shoals on coral reefs may be temporally and spatially patterned by minimizing contact with predators, and this could be taken in support of the avoidance hypothesis. Daily migration routes are perpetuated culturally in grunt shoals (Helfman and Schultz, 1984), but such regular highways tend to concentrate predators (Hobson, 1968). In fact, many fish shoals seem to act as an aggregation source for their predators, supporting our view that avoidance is not a major factor in the evolution of shoaling.

In conclusion, like social groups in other vertebrates, fish shoals are, we think, unlikely to have evolved primarily to avoid detection by predators. This end is better served by crypsis or refuging, an evolutionary route taken by the many teleosts that rarely shoal.

Shoal adaptations to minimize detection by predators

Although avoidance may not be a primary reason behind the evolution of shoaling, there is nevertheless evidence that minimizing detection is important in shaping and maintaining shoal behaviour in fish that have already evolved the trait. Under good visibility (clear environment and/or high visual acuity), groups of animals may themselves become more detectable than individuals. For example, flocks of birds are generally more conspicuous than individuals. Although this situation will rarely apply in the restricted visibility underwater (Chapter 4), we may expect visibility to be minimized where possible.

An oblate spheroid shoal in open water (e.g. herring: Pitcher and Allan, unpublished) could be an adaptation in this respect. This shape minimizes the detection envelope of the shoal underwater (Pitcher and Partridge, 1979) because elongated and flattened shoals produce a spherical envelope of visibility (see Chapter 4 by Guthrie and Muntz, this volume), a smaller detection volume than the ellipsoid visibility envelope of a spherical object. But using accurate acoustics, Misund (1991) reported oblate spheroid herring shoals only near the surface and bottom; elsewhere in the water column shoals tended towards spheres, so this issue remains unclear, especially as many other factors probably affect shoal shape.

Dilution of attack and the abatement effect

It has often been stated that the probability of capture by a predator is diluted by grouping per se. It seems that this 'dilution effect' has often been misunderstood. The theory states that an individual group member gains advantage simply through a reduced probability of being the one attacked in an encounter with a predator (Bertram, 1978; Foster and Treherne, 1981). This probability is the reciprocal of group size, giving the prediction that the log attack rate per individual will plot linearly with a slope of -1 against log group size. Such relationships have been unequivocally demonstrated in several experiments (e.g. Godin 1986b) and in Foster and Treherne's paper. The point, however, is that simple dilution of attack in a group cannot be a selection pressure favouring grouping because the effect applies only to members of a particular group being attacked and not to the rest of the population outwith this group.

Assume, for example, that a predator has a fixed attack rate and picks an individual prey at random, all prey are within detection range in groups of equal numbers. The analogy is with a predator with perfect vision armed with a rifle. For a fixed total number of individuals distributed among several groups, the average probability of being the one attacked is the same whatever the size of the group.

Foster and Treherne (1981) claim to have distinguished a dilution effect from predator confusion in clupeids attacking flotillas of surface-living, constantly moving insects. Fish are not detected by the insects until an attack occurs, and so the picture is not confounded by vigilance or social alarm signals. Almost exactly as predicted by the dilution law, individual insects in larger groups suffered a lower attack rate. The confusion effect was distinguished because attacks launched on larger groups of the insects were less successful. The results support dilution of attack on particular groups, but, as argued above, this does not prove dilution to be a selection pressure leading to grouping.

Foster and Treherne's results could be explained by an alternative hypothesis since they ignore possible reluctance by the fish to attack larger groups, a behaviour which can be seen in many fish predators. The decision not to launch an attack is not the same as the confusion effect, which reduces accuracy and strike success. In fact, larger insect groups suffered significantly fewer attacks than individuals, an observation supporting predator reluctance. In support of the dilution theory, these authors cite Hamilton's (1971) theoretical analysis of an unseen snake attacking frogs in a pond, but in fact Hamilton's argument is not concerned with dilution in the sense used by these authors. (His argument rests on the assumption that the snake eats a single frog, the prerequisite for dilution.) Hamilton's main point is to analyse the dangers of individuals finding themselves on the margins of groups, leading to behaviour in which animals seek cover behind others. Marginal individuals are proportionally fewer in larger groups (for a given shape of group). Hamilton's snake is very like our predator with a rifle because all frogs are detected and within range.

A form of attack dilution can operate, provided that attack is considered along with search or encounter. Previously we assumed that all prey were in equal-sized groups. But we will now allow groups of different sizes to coexist, and endow individuals with the freedom to join or leave any group. A searching predator encounters groups of prey rather than individuals. When a prey group is encountered, one individual is singled out for attack. Unlike our predator with perfect vision and a rifle, this time we draw an analogy with a nocturnal predator searching through the habitat equipped with low-resolution radar and an old firearm accurate only at close range. Under these circumstances, grouping pays: for example, an individual leaving a group of 100 reduces its chance of encounter from one to a half, but its chance of death given an attack goes from one-hundredth to one. Put another way, this is simply that it will be a good idea to be a member of a group if a predator comes across you, predators being unpredictable and nasty things. Turner and

Pitcher (1986) termed this the 'attack abatement' effect to distinguish it from previous analyses, although the basic arguments are very similar to those of Vine (1971). Turner and Pitcher (1986) showed in addition that attack abatement is an evolutionarily stable strategy (ESS) for all prey to group, even though this may not maximize benefit for all individuals. But many still refer to the 'abatement' grouping benefit as 'dilution' (e.g. Inman and Krebs, 1987).

Cognitive dilution benefit? Attack dilution in a larger group could confer benefits indirectly in a more subtle way, through the wider spread of information that a hungry predator was in the vicinity. In larger groups, more fish gain this useful information once an attack is launched. We can draw an analogy with ships under radio silence in a naval convoy knowing of the vicinity of an energy submarine. Because of dilution, individuals are able to experience an attack, meanwhile gaining knowledge about the attacking predator at no extra cost. The benefits of this knowledge would not always be immediate, but may be delayed until fish were able to join in group evasion manoeuvres, as described below (p.387). Most individuals in small scattered groups would not have this advantage, although the precise level at which the effect operated might vary with the nature of the predator and the scale of the grouping. The value of knowledge about the proximity of a predator could help select for grouping behaviour, but it is difficult to see how this 'cognitive dilution benefit' could be tested directly by experiment.

Dilution benefit by time-wasting. Barnard (1984a) suggests a related benefit through time-wasting: once a kill is made, the rest of a group can escape while the predator handles the prey. This benefit would help any group animals being attacked by a predator which can consume only a small part of the group, so that, balanced against the probability of being the one attacked, a clear benefit of attack dilution emerges. Taylor (1984) derives a mathematical model equivalent to this 'time-wasting' hypothesis.

The problem in advocating this as a general reason for shoaling to evolve is partly that many predators eat many fish from the shoal, and partly because many shoals are accompanied by a plethora of individual predators of many species (p. 411). Furthermore, handling time is not likely to be an important issue where predators eat whole prey rapidly and are not satiated by one fish, for instance in fast-moving pelagic predators like tuna. It could be more significant for ambush predators like pike, which may refrain from feeding for a while after capturing a large prey.

Predator evasion

Experiments have confirmed that the attack success of aquatic piscivores declines with prey group size. This has been shown for cephalopod, pike and perch predators (Neill and Cullen, 1974), sticklebacks (Milinski, 1979), piranhas (Tremblay and FitzGerald, 1979), and mink (Poole and Dunstone, 1975), and seems to be widely applicable. Fish that are separated from a group are more likely to be eaten by a predator (minnows: Magurran and Pitcher, 1987; guppies: Godin and Smith, 1988; silversides: Parrish, 1989a): pike for example have reached a stage in the arms race where their attack behaviour aims to split individuals from the shoal. How, then, does shoaling protect fish under attack?

Compaction. When a potential underwater predator is detected, shoals become more compact and cohesive (e.g. Ruppell and Gosswein, 1972; Andorfer, 1980). The dramatic effect of a predator on the EGS of a minnow shoal is illustrated in Fig. 12.7; compaction persists for some considerable time after the predator has left.

Small groups which become separated from a larger shoal may rejoin through a narrow 'neck' or 'pseudopodium' of rapidly moving fish, two or three fish wide: the smaller shoal appears to deflate like a balloon as its members transfer to the main body (Fig. 12.8; Radakov, 1973: Pitcher and Wyche, 1983). Clearly, behaviour has evolved that puts a premium on preserving the integrity of the group.

A major reason for compaction (i.e. shoaling fish adhering more closely to conspecifics when alarmed) is to enable the fish to take advantage of cooperative escape tactics, such as those illustrated in Fig. 12.8. Some tactics with a wide occurrence are discussed in detail below. Individuals cooperate because these behaviours only work when fish behave as a coordinated group.

Why does compaction help the performance of synchronized manoeuvres? It may be to take advantage of the potential for very rapid communication using pressure waves recently discovered by Gray and Denton (1991) in Atlantic herring, *Clupea harengus*, sprat, *Sprattus sprattus*, and whiting, *Merlangius merlangus*. These rapid pressure waves can be detected up to one fish length away using otoliths and the lateral line organs, but their intensity falls off rapidly at greater interfish distances. Gray and Denton suggest that rapid turns in fish schools may be synchronized by very rapid communication of velocity and bearing changes using this system.

Evade and the minimum approach distance. Through judiciously unhurried movements, many fish attempt to maintain a distance of more than 15 body lengths between themselves and the predator (e.g. Jakobsson and Järvi, 1978; Pitcher and Wyche, 1983; Magurran and Pitcher, 1987). Weihs and Webb (1983) have shown a clear benefit in maintaining such a minimum approach distance in order to preserve an advantage in relative manoeuvrabilitry and acceleration by the smaller prey. There may be a trade-off between approach

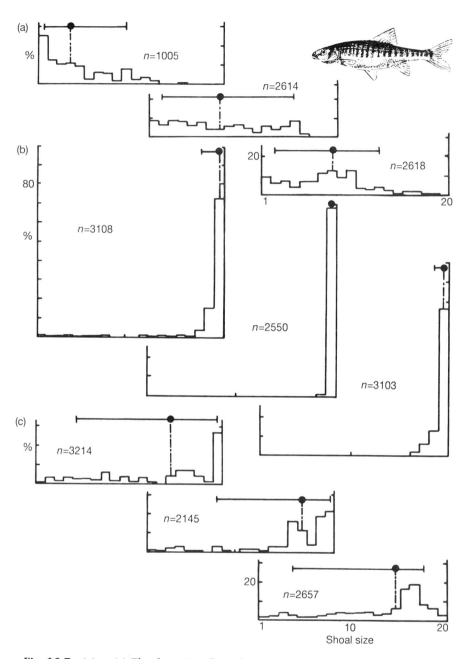

Fig. 12.7 (a) to (c) The dramatic effect of exposure to a predator on the elective group size (EGS) of shoals of 20 minnows. Measures were made in a large arena tank. Three separate replicates are shown from left to right. Each histogram gives the relative frequency of records of groups within 4 body lengths scored under standard conditions. Medians, quartiles and numbers of observations are shown above each histogram. (a) Baseline measurements for undisturbed minnows; (b) 2 hours after a 1-hour exposure to a hunting pike, which was then removed; (c) 1 day after the pike exposure. Reproduced with permission from Pitcher *et al.* (1983).

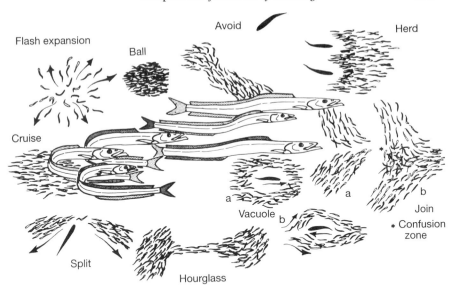

Fig. 12.8 Schematic diagram of the repertoire of antipredator tactics in schools of sandeels (US, sand lance; Ammodytidae) under threat from hunting mackerel. Tactics are drawn in plan view: the small turning school of sandeels is drawn from the side in the foreground. Observations were made in a 12 m arena tank. Further details are given by Pitcher and Wyche (1983).

distance and risk in shoals if we consider the balance between the predation risk and the cost of flight (Ydenberg and Dill, 1986).

Fountain effect. This manoeuvre, in which fish initially fleeing from the predator turn and pass by its sides in the opposite direction, has been widely reported (Potts, 1970; Nursall, 1973; Radakov, 1973; Pitcher and Wyche, 1983; Magurran and Pitcher, 1987). Velocity relative to the predator is maximized and the shoal may reassemble behind the stalking predator. Hall *et al.* (1986) show how this behaviour, which is also preserved around divers or fishing gear, can come about through a simple visual response of individuals in groups of fish (see Wardle, Chapter 18, this volume).

Trafalgar effect. Webb (1980) showed that the average latency between sudden stimulus and fast acceleration to escape was less in shoals than for individuals. Webb's result was probably brought about by the rapid transfer of information across a shoal by observation of and reaction to the startles of fellow fish, as much as to external frightening stimuli. Treherne and Foster (1981) likened this to the rapid transfer by flags of battle signals through Nelson's fleet at Trafalgar, and estimated that a wave of reaction to an attack could propagate through a flotilla of surface-living insects seven times faster than the approach speed of the predator. Evasion turns are thought to spread through bird flocks in similar fashion (Potts, 1984), although in this case birds synchronized their turns by anticipating the approaching wave of reaction. Unrehearsed high kicks in lines of human chorus girls spread faster than the human reaction time through similar anticipation. Potts' 'chorus-line' hypothesis is therefore a subset of the Trafalgar effect. There is also evidence for an effect of this kind in saithe schools (Pitcher, 1979a). Godin and Morgan (1985) investigated the Trafalgar effect by approaching killifish shoals with model predators. The speed of transmission of information across the shoal was roughly twice the predator's approach speed and was independent of shoal size.

Ball. Pelagic shoalers, out of reach of cover, may take refuge in a tightly packed ball (Breder, 1951; Pitcher and Wyche, 1983). For example, over 500 sandeels (mean length 140 mm) threatened by auks were concentrated into a 50 cm ball measured by Grover and Olla (1983). For fish shoals, such jostling throngs provide an example of Hamilton's (1971) 'selfish-herd' effect, in which animals avoid dangerous margins by seeking shelter in the centre of a group (see also Wilson, 1975). In such a case, average individuals would probably be better off dispersed than in a conspicuous ball, but perhaps the benefits of getting the central locations outweigh the costs of dispersal behaviour.

Predator confusion

Overt predator attack may be made less likely or less effective through confusion generated in the predator by multiple potential targets. Landeau and Terborgh (1986) showed that the median time to first successful prey capture increased with shoal size as a consequence of predator confusion. Individuals who join a group all benefit equally from the confusion effect in its simple form, where predator attacks are spoilt by the multiple targets. As may be expected, however, detailed examination of the behaviour reveals conflicts as some individuals attempt to gain more of the benefit.

The neurophysiological basis of the confusion effect is through overloading of the visual analysis channel, either peripherally or centrally (Broadbent, 1965). Alternatively, confusion could be cognitive, the 'embarrassment of riches' effect, like a dog unable to choose between several juicy bones.

Humans watching radar screens (or trying to catch small fish from a large tank) suffer from perceptual confusion, and Milinski (1990b) demonstrates that humans suffer from inability to hit targets as density and distraction increases just as sticklebacks do (see Milinski, Chapter 9, this volume). Because visual predators that target individuals rely on complex movement and trajectory analysis (Guthrie, 1980), it is difficult for them to overcome visual channel confusion. The most thorough experimental analysis of the confusion effect is on sticklebacks feeding on swarming invertebrate prey (Ohguchi, 1981).

Large piscivores (like tuna) which lunge and gobble rapidly and randomly are one evolutionary attempt to beat the confusion effect. Smaller piscivores (like trumpet fish and cornet fish) appear to have evolved a visual 'lock-on' in an attempt to minimize confusion, and, as we shall see, this too has had consequences in the shoaling arms race. Many shoaling fish are thin and silvery. Synchronous turns in a school may cause fish to disappear momentarily and confuse pursuing predators.

Whatever the mechanism, several tactics performed by shoaling fish appear to increase predator confusion beyond that engendered merely by multiple targets. This occurs mainly through increasing the relative movements between individuals. It is here that the opportunity to cheat on the equitably shared standard confusion arises. Two analyses based on minnow, *Phoxinus phoxinus*, shoals are described below.

In European minnows a sequence of behaviours based on **'skittering'**, a Mauthner-driven startle acceleration (Webb, 1976; Eaton *et al.*, 1977; Guthrie, 1980), may reflect an evolutionary race between those who dodge and those who preserve shoal integrity (Magurran and Pitcher, 1987). Minnows that skitter accelerate rapidly for 5 to 10 body lengths, brake, rise in the water 3 or 4 body widths, and then resume coordinated behaviour with the group, rapidly slotting into a new position in the shoal. Although skittering fish may increase the confusion effect, thereby benefiting all shoal members, the important point is that an individual that skitters may gain greater benefit by confounding a predator which may have attempted to 'lock on' to that individual. Neighbours to the original location are then placed at greater risk. This could explain why the frequency of skittering increases as an attack progresses (Magurran and Pitcher, 1987), and why the behaviour often occurs in bursts, appearing to spread through adjacent fish.

Frequent slow and deliberate passings of one fish by another seen in polarized minnow schools under attack (Fig. 12.9) may represent attempts to get neighbours to optomotor smoothly (Shaw, 1965; Shaw and Tucker, 1965) and inhibit skittering so that a neighbour may remain as a target when the test fish itself skitters. (The optomotor response is to maintain station with a moving stimulus using visual cues; see Wardle, Chapter 18, this volume.)

Synchrony of skittering avoids the penalty while maintaining predator confusion. Two behaviours achieve this end. The 'group jump' is a synchronized and polarized skitter by the majority of shoal members, which in effect relocates the whole shoal rapidly. In large schools of flat-iron herring, *Harengula thrissina*, subschools near an attacking predator would frequently break away from the main school and then immediately re-integrate in a new location (Parrish, unpublished data). In minnows, the group jump is not

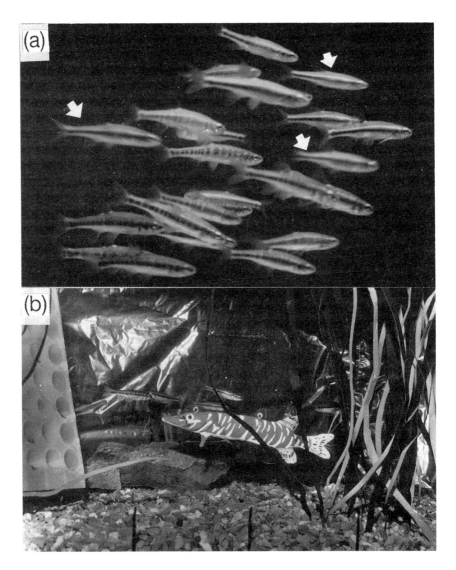

Fig. 12.9 (a) Polarized school of minnows under threat from a pike. Fish in the foreground are out of focus, but other images (arrows) are blurred by movement during passing manoeuvres within the school. Passing probably enhances predator confusion, and may represent a conflict between school neighbours (see text). Photograph taken by flash in an arena tank with the aid of S. Smythe and B. Partridge. (b) Minnows leaving a feeding patch when approached by a model pike. Ice-cube tray feeding patch at left, pike model approaching from behind weeds. The model predator is detected further away by minnows in larger shoals, but the fish allow it to approach closer before leaving the patch. For more details see Magurran *et al.* (1985). Photograph taken by E. Pritchard.

performed very often relative to normal skitters, perhaps because a predator could learn to counter by anticipating the new location of the group.

The second behaviour is 'flash expansion', a behaviour which has been widely reported from teleost shoals (Potts, 1970; Nursall, 1973; Pitcher, 1979b; Pitcher and Wyche, 1983). Here, a polarized group of fish, often stationary, explodes in all directions like a grenade, the fish rapidly reassembling after swimming 10 to 15 body lengths. In effect, the predator's group of potential targets vanishes, to reappear somewhere else. Although a most effective confusion manoeuvre, the danger in this behaviour lies in the fact that school structure is momentarily broken down, because after the school explodes, individual fish are scattered and the predator can concentrate on individual targets. For this reason, and because Mauthner startles are energetically expensive, the tactic is relatively rare. There is clearly a premium on reassembly; subgroups exhibiting flash expansions within a school of flat-iron herring, *Harengula thrissina*, rejoined the main school within seconds. However, in minnows, reassembly can take as long as 30s if the predator presses home an attack. Rapid reassembly is best; the details of such group startles give clues as to how this is done.

Blaxter *et al.* (1981) showed that fish must retain an awareness of their relative positions while accelerating because they do not collide. The lateral line may mediate this effect, because fish with severed lateral line nerves did occasionally collide when startled (Partridge and Pitcher, 1980). Blaxter's research also showed that most startles were in a direction away from a sound (non-visual) stimulus, although the physiological basis of the directionality is not yet understood.

The dangers of joining: confusion zone in schools

Poor coordination in travelling subschools which join may provide evolutionary pressure to avoid splitting in schooling pelagic fish. Pitcher and Wyche (1983) described a zone of confusion at the rear of coalescing sandeel (Ammodytidae) schools, where the lack of coordination may attract predator attack (Hobson, 1968). Selection may favour individuals that minimize the chances of being caught in a confusion zone. This could work by enhancing synchronous turning, which shuffles fish positions. Furthermore, the precise location of the danger zone varies greatly with the angle of incidence of the joining subschools, and is not predictable, so inhibiting a melee to get out of the dangerous rear stations. Confusion zones therefore provide selection pressure for high coordination and synchrony but, paradoxically, this behaviour itself generates the zones. The cost to individuals of occasionally being caught in a confusion zone could actually be greater than for less cohesive shoaling, but all would bear this cost and the situation could be an ESS.

It is evident that not all joining subgroups result in confusion zones as seen

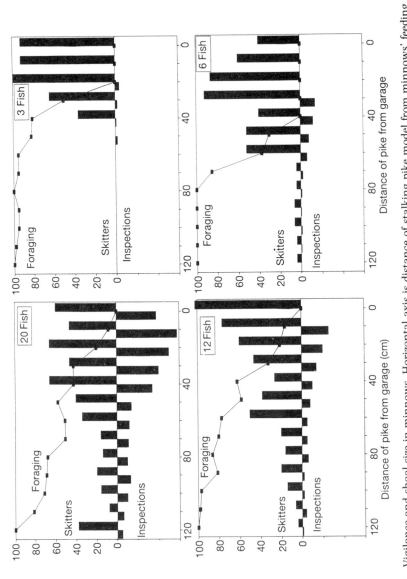

Fig. 12.10 Vigilance and shoal size in minnows. Horizontal axis is distance of stalking pike model from minnows' feeding patch, see Fig. 12.9. Vertical axis is percentage of fish foraging (line), and numbers for inspections and skitters. Histograms show relative values of skitters per fish (above line) and inspections per fish (below line). Minnows cease foraging later and more abruptly in small shoals, while skitters and inspections suggest earlier detection of the threat in larger shoals. Redrawn from values in Magurran *et al.* (1985).

in the sandeels. The phenomenon was never observed in flat-iron herring schools in Baja California (Parrish, unpublished), although this may have been because small subgroups were re-integrating back into a stationary larger school, that is, both subgroups were not moving.

Early predator detection

Minnows in larger shoals detect an approaching predator sooner (Magurran *et al.*, 1985) (Fig. 12.10 and see Chapter 9 by Milinski, this volume); birds in flocks have a similar advantage (Powell, 1974; Kenward, 1978; Lazarus, 1979).

There are two reports of an apparent converse of the early warning effect. First, Seghers (1981) reported that spottail shiner, *Notropis hudsonius*, shoals reacted to an approaching model later than individuals, but he failed to detect any consistent effect with shoal numbers. This may have been because alarm and flight were recorded and the behaviour of individual fish was not seen in sufficient detail to show when the fish first detected the predator. Seghers may have scored the low-intensity warning skitters of a group (see below). Secondly, Goodey and Liley (1985) reported that escape responses of guppies, *Poecilia reticulata*, to a swooping aerial predator model were no faster in groups than for solitary fish, perhaps because all fish need to respond to such a rapidly appearing threat without reference to other group members. This is not the same situation as detection of a gradually approaching threat from a stalking predator.

In fish shoals, detection does not necessarily bring about flight, as is usual in bird flocks. Equally, detection of a predator does not necessarily imply any overt behavioural response. Unequivocal experiments on detection would probably have to be neurological. Fortunately, however, in many fish species we can detect small changes in behaviour which mean that the fish has seen a potential threat. In minnows the discovery of two behaviours was the key to showing how vigilance increased with shoal size. These were skittering and predator inspection (Fig. 12.10), and both are described below.

Godin *et al.* (1988) performed an elegant experiment in which shoals of a characin, the glowlight tetra, *Hemigrammus erythrozonus*, in mid-water were exposed to a light flash at randomly allocated positions around their periphery. The light-flash technique gets around the problem of distinguishing between detection and response. The results showed that as group size increased, the tetras were more likely to detect the stimulus. The shape of this vigilance relationship (Fig. 12.11a) with group size increased in a very similar way to a theoretical line predicted from signal detection theory (Lazarus, 1979), most benefit occuring as group size increased up to 20 fish. But in larger groups more fish responded to the alarming stimulus than predicted (Fig. 12.11b), probably because of social transmission of the alarm through the Trafalgar effect.

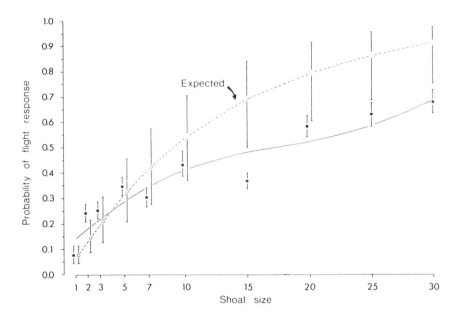

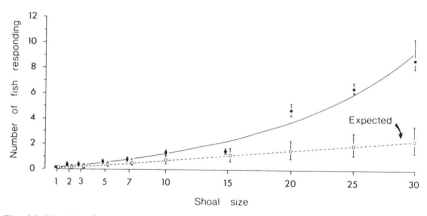

Fig. 12.11 Startle responses of glowlight tetras in a range of shoal sizes to a randomly-located light flash. 10 replicate trials. Observed data (solid line and circles) is fitted by a polynomial: expected relationship (broken line, open circles) was predicted from a signal detection model. For further details see Godin *et al.* (1988). (a-top) Group vigilance: probability of observing at least one response. Means and SEs from arcsine transform. (b-bottom) Numbers of individual fish responding to stimulus. Means and SEs from log transforms.

The two experiments by Godin *et al.* and Magurran *et al.*, provide pretty solid support for the 'many eyes' hypothesis (Bertram, 1978; Barnard, 1983) operating in fish shoals; the average probability among all individuals of detecting a predator is higher in larger groups. However, in fish shoals detection will be mediated by the degree of polarity and the degree of compaction within a given shoal shape. Provided that the information about a putative predator is shared, as in bird alarm calls, an individual in the group clearly gains an advantage through shoaling. The implications of this greater vigilance in the group for the optimal allocation of time to foraging and other behaviours is discussed in Section 12.6 below. But note that the evidence in fish shoals does not show whether the increased group vigilance allows individuals to gain a double benefit by reducing their costs by being less vigilant. This effect, which would produce strong selection pressure favouring grouping (Lazarus, 1985), was demonstrated in Bertram's (1980) original work on ostrich groups, and in da Silva and Terhune's (1988) experiments on harbour seals.

Warning skitters – an alarm signal?

Minnows that detect an approaching predator (real or model) perform a low-intensity skitter, often returning to near their original position in a curved path (Magurran and Pitcher, 1987). The movement does not appear to be fast enough to be a Mauthner fast-start response, unlike most skitters at later stages of attack (see p. 389), although the difference is probably only one of degree. The behaviour is probably still fast enough to minimize the chance of a predator 'locking-on' to the performer.

The skitter may act as a warning to shoal fellows, but is not precisely analogous to warning calls in bird flocks since it may be repeated by increasing numbers of individuals as the predator approaches, and indeed several fish may perform the behaviour together at a later stage.

The cost in lost feeding of a warning skitter in the minnow experiments was very small and many fish performed the behaviour; the fish ceased to forage for a few seconds, although this might not be the case in the wild. Performing a skitter might also cost by attracting the predator's attention, but this would be countered by the confusion effect, and by rapidly rejoining the group. By making sure that the group knows of the danger, the warner's chances of survival are enchanced as the group can now perform antipredator mano-euvres. From these pressures it is easy to see how a warning signal beneficial to all individuals in the group could evolve.

Skittering has not been seen in flat-iron herring shoals, perhaps because predators are always either in plain view or attack so rapidly that there is no time for a warning signal.

Attack inhibition

Breder (1959), Hobson (1968), and others have described several instances of apparent predator inhibition by mass displays of shoaling fish. Motta (1983) describes several instances of displays by coral-reef prey to their predators which appeared to be analogous to mobbing by birds (Bertram, 1978), although some of these instances may have been unrecognized inspection visits. One of us (TJP) has once seen a group of freshwater bream, *Abramis brama*, push a predatory pike to the surface of a large arena tank by swimming underneath it. Since the behaviour is rare, it has not yet been investigated experimentally.

Synchronously turning silvery fish shoals have frightened scuba divers (Pitcher, 1979a) and there are anecdotal accounts of such 'flash displays' scaring predators (Springer, 1957; Hobson, 1968), but predator inhibition by 'flashing' also remains to be investigated by actual experiment. Flashing may act as a social signal within the shoal (E. J. Denton, 1991, pers. comm.).

Attack inhibition has been suggested as a function for predator inspection, as discussed below.

Predator inspection behaviour

During the early stages of an attack by a stalking pike, individual minnows leave the shoal and approach to within 4 to 6 body lengths of the predator. There they wait for a second, turn, and return to the shoal. Pitcher *et al.* (1986b) termed this behaviour **predator inspection** (Fig. 12.12). The behaviour may also be performed more rapidly, with no obvious pause (e.g. Allan and Pitcher, 1986). The same behaviour is shown to models and, initially, to other intruders such as large non-predatory fish. Inspection has now been described for a wide range of shoaling fish species which are attacked by stalking or ambush predators, but has not been seen in pelagic schooling species. The evolution and function of predator inspection has recently been reviewed by Pitcher (1992).

Although individuals will inspect alone, small groups will perform the behaviour together, and sometimes almost the whole group will inspect *en masse*. The probability of inspecting in a group near to the maximum size was about one-third in the live predation experiment performed by Magurran and Pitcher (1987). Group sizes are the key to understanding inspection behaviour, but any candidate explanations of the function and evolution of inspection behaviour must account for the range of group sizes seen (Fig. 12.12). As we will see, no one explanation has yet achieved this (Pitcher, 1992).

There are large individual differences in tendency to inspect, and shoals may consist of individuals with strategies of behaviour leading to differing trade-offs between risk and information (Murphy and Pitcher, 1991). Approach is often in a 'leisurely' manner, perhaps reducing the risk of attracting attention by a sudden movement. Approaching a predator as an individual clearly carries a

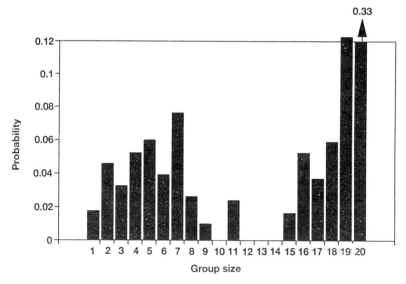

Fig. 12.12 Predator inspections of a hunting pike by minnows in shoals of 20. Vertical axis shows the probability of individual fish inspecting the predator in groups of different sizes (data from experiments reported by Magurran and Pitcher, 1987).

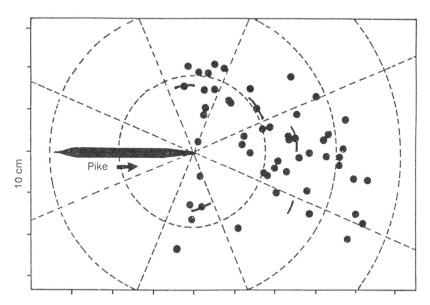

Fig. 12.13 Mapped positions at which inspecting minnows halted and turned away from a pike during predator inspection. Bars indicate median positions, which despite large variation, are significantly further away for sectors in front of the pike's jaws.

cost, inspectors tending to avoid the most dangerous 'attack cone' area proximal to the jaws of the predator as illustrated in Fig. 12.13 (George, 1960; Pitcher and Graham, unpublished data; Magurran and Seghers, 1990a). In some experiments, risk was assessed directly when predator strikes were made only towards inspectors (Magurran, 1990; Pitcher and Murphy, unpublished data). Experiments show that risk of strike is diluted by inspection group size, so the large numbers of inspections in small groups or singletons are an enigma.

Predator inspection increases in frequency as a pike comes closer, largely through recruitment of new fish to the behaviour (Pitcher, 1992), but the behaviour decreases rapidly when a serious attack is launched. Experiments have provided evidence that inspection gives the fish information about the status and attack readiness of the predator and allows them to shift feeding to safer locations (Pitcher, 1992). Inspection habituates more slowly to realistic predator models (Csanyi, 1985; Magurran and Girling, 1986), so it is possible that inspection allows fish to tell friend from foe. Furthermore, fish that perform predator inspection visits gather knowledge about the predator's precise location and current state, which may be of advantage if sudden flight is needed from a predator that had previously only been in the vicinity but begins to launch an overt attack. It is interesting that individuals frequently repeat inspections, giving the shoal a second-by-second assessment of likely attack.

Several experiments have demonstrated that information is transferred from inspectors to other fish in the shoal through behavioural changes in the tendencies to join, leave and stay (Pitcher *et al.*, 1986b, Magurran and Higham, 1988; Croy, 1990). Most of the information transfer seems to be visual, because in these experiments 'receiver' fish behind a one-way mirror saw only 'transmitter' fish which in turn saw the predator approaching.

Attack anticipation. Magurran and Pitcher (1987) noted that minnows were able to anticipate a strike by a stalking pike, thereby taking swift synchronized evasion. Human observers are able to make similar predictions, either live or when watching videotape. The precise cues from the predator remain unclear, although it may relate to an 'unnatural-looking' stillness and indication of tension in the pike's body. Sometimes pike quivered before a strike, but the effect was not consistently linked to this. Prediction was almost certainly achieved through inspection: attack anticipation has been measured quantitatively in minnows (Pitcher, 1992; Pitcher and Murphy, unpublished data).

Attack inhibition and attack invitation. Approaches to predators reported in some mammals and birds have been interpreted as attack inhibition (= pursuit deterrence, e.g. Kruuk, 1976), whereby the prey informs the predator that it has been identified, located and is currently monitored. A subsequent attack by the predator is less likely to be a success, so the behaviour of both predator and prey can be shaped by natural selection; Thompson's gazelles have evolved displays (stotting) which analysis has suggested perform this function (Caro, 1986). The element of surprise by the predator is consequently lost, so the predator redirects its attention to prey that have not signalled that detection. An alternative hypothesis is attack invitation (= pursuit invitation, Smythe, 1977), in which the attack is provoked by the prey, thereby gaining the initiative in an incident which the predator may lose, thus gaining time to get on with other important business to hand, such as feeding or mating. Predator invitation is inevitably highly risky and there is not much evidence for it.

Attack inhibition is an attractive possibility for inspection, but the experimental evidence is equivocal. Two experiments have employed a protocol in which a pike behind a one-way mirror is either hidden from or visible to a feeding minnow shoal (Magurran, 1990; Pitcher and Murphy, unpublished data), but have provided opposite results, probably because of differences in the experimental protocol (Pitcher, 1992). At present we can only state that it is possible, but not yet proven, that inspection may partially inhibit attack. Attack invitation is not supported by any of the experimental results.

Tit for tat. Several experiments have now provided support for Milinski's (1987) suggestion that predator inspection behaviour is driven by a 'tit-for-tat' (TFT) strategy of alternating cooperation between pairs of fish (Dugatkin, 1988; Milinski *et al.*, 1990a,b, Barrie *et al.*, 1992), although the concept has generated a lot of controversy (Dugatkin, 1990, 1991; Lazarus and Metcalfe, 1990; Masters and Waite, 1990; Milinski, 1990a; Dugatkin and Alfieri, 1991a; Pitcher, 1992).

Dugatkin and Alfieri (1991b) have shown that there are persistent differences in inspection behaviour between individuals, and that individuals remember the behaviour of past partners and gauge new bouts accordingly. Thus 'cooperate' and 'defect' may not be rigid behaviours as much as relative terms. TFT is a strong candidate explanation for inspection, but frequent observations of single fish and large groups inspecting seem to confound TFT's predictions. Dugatkin (1990) has suggested that the evolution of cooperation may not be evolutionarily stable such that metapopulations may have subgroups (i.e. individual shoals) with widely ranging ratios of cooperators to defectors.

Mitigation of inspection costs. Pitcher (1992) suggests that inspection is opportunistic; depending on the circumstances, individuals which inspect may choose from a range of mechanisms that mitigate risk of attack. Risk dilution (safely in numbers) operates to form large inspecting groups when perceived risk is high (Murphy and Pitcher, unpublished data). Furthermore, dilution cannot be entirely ruled out as an explanation for behaviour in other sizes of shoals, including all experiments on tit-for-tat. The high number of singleton inspectors remains unexplained: perhaps they bear the costs of inspection alone in order to manipulate the rest of the group to their advantage. Through rapid serial repetition of a standard act at a rate proportional to perceived danger, inspection may itself be an easily detected, discriminated and remembered signal (Guilford and Dawkins, 1991).

Relative predator and prey size

For small pike, only careful unobtrusive stalking had a pay-off in prey capture, Magurran and Pitcher's (1987) results suggesting that shoaling cyprinids are currently ahead in the arms race. This effect was, however, strongly dependent upon size; for most experiments pike encountered minnows of approximately optimal size (55 mm long; based on handling time in relation to energy value; Hart and Connellan, 1984). But when larger pike were employed in the laboratory, shoaling was little protection against rapid predator capture. It is interesting that minnows would be sub-optimal-sized prey for such larger pike in the wild, and an optimally foraging pike might ignore them.

Shoaling and parasitism

It has been suggested that increased parasitism may be a cost of shoaling, but, at least in freshwater fish, a survey of 60 species shows that this is not the case (Poulin, 1991). Indeed, elegant experiments by Poulin and FitzGerald (1989a) demonstrated that juvenile sticklebacks in larger shoals bore a lower individual risk of infection by ectoparasitic crustaceans, *Argulus canadiensis*. In fact, the fish tended to form larger shoals in the presence of the ectoparasites, and altered their microhabitat (Poulin and FitzGerald 1989b).

12.6 FORAGING FUNCTIONS OF SHOALING

Visibility is restricted underwater; even in the clearest water, absorption and scattering of light mean that objects are visible only over a few tens of metres. The visual range in many waters extends only a few metres (see Guthrie and Muntz, Chapter 4, this volume). Under these conditions, foraging for patchy food in a social group has immense benefits, although this aspect of shoaling behaviour was virtually ignored in the classical literature. Radakov (1973) reviews many anecdotal Russian observations which, unlike much contemporary western research at that time, considered enhancements to foraging through social observation.

Many costs and benefits in foraging shoals were explored quantitatively during the 1980s, and are reviewed briefly below. Classical **optimal foraging theory (OFT)** assumes that animals will maximize their energy intake through diet and food-patch selection (see Hart, Chapter 8, this volume; Krebs, 1978). Similarly, the marginal-value theorem predicts when foragers should leave a food patch (Charnov, 1976; Hart, Chapter 8, this volume). Social behaviour was not taken into account in the simple classical versions of these theories; the only group-related prediction of classical OFT or the marginal-value theorem was that individuals will tend to congregate on good patches (Krebs, 1978) until depletion makes alternative sites more attractive. Unfortunately, this is not how foraging groups of fish behave (Pitcher *et al.*, 1983). Recent developments in foraging theory have taken a wider range of real behaviours into account. There is now considerable support for the idea that fish feed to maximize their net energy intake under prevailing environmental conditions, subject to constraints imposed by predation risk and social benefits (see Chapter 8 by Hart, this volume; Stephens and Krebs, 1986). One problem has been that the 'constraints' often seem to be as important in shaping behaviour as the rule of energy maximization. Nevertheless many of the factors driving foraging have been clarified using the new generation of foraging models.

Fish foraging in shoals gain benefits through faster location of food, more time for feeding, more effective sampling, information transfer and opportunity for copying. Increased costs with group size are perceived through greater competition in various forms. Experiments which have investigated these costs and benefits, usually through manipulation of shoal size, are discussed next.

Individuals finding food faster

Faster location of patchy food in larger shoals has been experimentally demonstrated by manipulating the shoal size of marked individuals in two cyprinid species, minnows and goldfish, *Carassius auratus* (Fig. 12.14; Pitcher *et al.*, 1982a) and in stone loach, *Noemacheilus barbatulus* (Street and Hart, 1985). A simple model based on a random food searcher that can detect when other searchers find food can explain this result. As the density of searchers rises, randomly located food is found sooner by one fish or another because the aggregate search rate is higher for each additional fish searching. Non-finders then benefit by moving to the food.

Bergelson *et al.* (1986) reported the exact opposite to these results using bloodfin tetras, *Aphyocharax anisitsi*, where although larger groups found food more quickly, individual fish in shoals appeared to take longer. The reason for this anomaly is not clear, but the tetras were not timed individually, individual performances being calculated from a theoretical equation.

Faster food location by individual fish in groups suggests strong selection pressure for the ability to tell when others have found food. Goldfish searchers

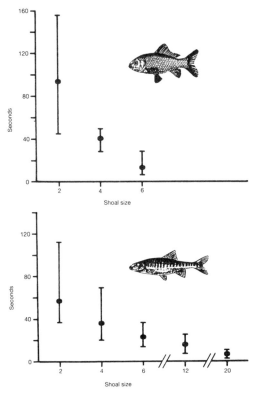

Fig. 12.14 Goldfish (top) and minnows (bottom) in larger shoals find food faster.
Vertical axis is the time spent by a focal individual before locating a randomly placed
food item. Points show the medians and quartiles. Reproduced with permission from
Pitcher *et al.* (1982a).

watch for the 'nose-down' feeding posture of a food finder (Magurran, 1984),
minnows react to 'wriggling' foragers, and mackerel also appear to use visual
cues from excited food finders, Krebs *et al.* (1972) reported a similar finding
for flocking tits in experiments in which flock size was manipulated. Faster
food location in larger groups of stone loach, a nocturnally feeding cobitid
(Street and Hart, 1985), is more likely to be the result of water turbulence set
up by food locators than vision or olfaction. In bird flocks, some individuals
are consistently better food locators (Barnard, 1984b), but such differences
have not yet been investigated for fish.

   At first sight it is surprising that behaviours like 'nose down' or 'wriggling'
which enable food locators to be detected are not strongly countered by
behaviour to conceal or even bluff a find. Concealment would keep a valuable
patch to the finder, whereas bluff might throw others off the scent while the

bluffer moves off to fresh search areas. Bluffing would be a viable strategy only at low frequencies, but it is difficult to see why concealment behaviour is not more obvious when foraging fish groups are observed. A possible reason may be that the morphophysiological constraints on capturing and subduing prey make eating difficult to conceal: the 'nose-down' goldfish is using efficient suction from its protrusible cyprinid jaw mechanism (Lauder, 1983), and piscivores have to shake prey in their jaws to incapacitate and swallow them.

Time allocated to feeding

In larger groups, cyprinids feeding on patchy food allocate more of their time budget to foraging (Magurran and Pitcher, 1983) because they are less timid, spending a smaller proportion of their time in cover (Fig. 12.15). Even when an approaching predator is detected, fish in larger shoals may allow it to approach more closely before stopping feeding (Fig. 12.9(b); Magurran *et al.*, 1985), will approach a dangerous area away from cover more readily (Milinski, 1985a, and Chapter 9, this volume), or may resume feeding more readily after fright (FitzGerald and van Havre, 1985). Individuals are less

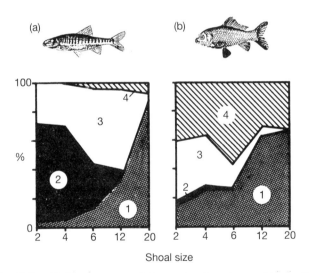

Fig. 12.15 Shifts of behaviour towards increasing foraging and lower timidity with shoal numbers in (a) minnows and (b) goldfish. Shaded areas represent proportion of time spent in the behavioural categories (medians for eight replicates; categories grouped for figure from 15 original scores). Categories are: 1, forage; 2, in weeds; 3, swim bottom; 4, swim mid-water. Both species of cyprinid shoal, but only minnows spend an appreciable time in polarized synchronized schools. Although goldfish spend less time in weed, and minnows spend less time in mid-water, the trends in the two species are remarkably similar (data from Magurran and Pitcher, 1983).

timid, spend less time in exclusively vigilant behaviour, and forage longer in larger shoals, all of which are direct benefits to fish that join small groups of conspecifics. Much of this particular benefit may accrue to fish in multispecies shoals, even where the fish are not feeding on precisely the same food (Allan, 1986).

The conflicting demands of vigilance and foraging, exemplified by the work of Caraco (1980), Sih (1980) and Lendrem (1983), are described by Hart in Chapter 8 (this volume) and, for fish shoals, are dealt with below.

Sampling feeding sites

Experiments have confirmed that fish in larger shoals gain greater foraging benefit through sampling behaviour. When feeding in an environment containing several food patches of different quality, although feeding most on the best patch, fish make visits to feed on poorer patches. Such sampling continues in goldfish even when the location of the best patch is completely stable over many trials (Pitcher and Magurran, 1983). In larger goldfish shoals, individuals make more visits to poor patches, partly because more time is budgeted to foraging. Pitcher and Magurran demonstrated that information gathered on visits to other patches was used to change foraging tactics when the location of the best feeding patch was switched: the visits therefore represented genuine sampling behaviour. More sampling took place in larger shoals (note that all experiments controlled for patch depletion).

Cowie and Krebs (1979) reported sampling of poor feeding sites in great tits. In goldfish, Lester (1981, 1984) showed that foraging time on two food patches departed from OFT predictions the more the location of the better patch was altered. This could be interpreted as reflecting greater sampling behaviour in the face of more unpredictability of patch quality.

For any given food distribution varying in space and time, there is likely to be an optimal sampling regime which produces maximum food intake while allowing for unpredictability. Stephens and Krebs (1986) present a lucid analysis of recent models representing foraging strategies designed to track enviromental changes, but very few have been applied to fish shoals.

A simple reason for observing more sampling in larger shoals is that individuals have more foraging time at their disposal. Theory, however, may in the future be able to show how individuals benefit from doing more sampling rather than just spending their larger foraging budget on the best patch. One novel approach to modelling social foraging is by Clark and Mangel (1984), and is based on the value of information rather than classical OFT's emphasis on energy/time budgets. Clark and Mangel's model predicts that individuals should sample before taking decisions to stay with, leave or join foragers on a patch, precisely the kind of behaviour which has been clearly observed in cyprinid shoals.

Passive information transfer

Active information transfer (AIT) (Pitcher *et al.*, 1982a) occurs when distinct overt behaviour is employed to communicate information such as the presence and location of food, as in the famous honey-bee dance or in flocking nectar bats (Howell, 1979). **Passive information transfer (PIT)** (Magurran, 1984) occurs when inadvertent behavioural cues about food are used by other animals. Insight of the manipulative nature of animal signals (Krebs and Dawkins, 1984) suggests that, in evolutionary terms, there may be a continuous spectrum between these two extremes, but when observing animals now, behaviour is likely to be optimized for either AIT or PIT. AIT may be found when a signaller benefits from giving information about food to others through kinship or manipulation; PIT may be seen where behavioural detection of feeding is currently the winning strategy in the evolutionary race, perhaps because feeding is hard to conceal.

In fish shoals, PIT allows the faster food-finding effect discussed above. In addition, by reciprocal transfer of single goldfish that had knowledge about the location of good food patches ('informed' fish) between goldfish shoals with different food patch arrangements, Pitcher and Magurran (1983) demonstrated that PIT enabled fish in larger shoals to recover more rapidly from changes in the location of the good patch. In these experiments, the effect occurred even when 'misinformed' fish were in the majority, although the precise relative values of the patches and shoal size might influence this point. Using evidence largely from bird flocks, Giraldeau (1984) has suggested that a 'skill pool' effect enhances foraging when individual differences in foraging specializations are shared through social learning. In a broad sense, the differences in information about good foraging sites in this experiment fit the 'skill pool' definition.

There is as yet no evidence for AIT in fish shoals, nor for Ward and Zahavi's (1973) 'information centre' hypothesis in which details of the location and quality of food are shared among groups. Information centres exist in the social insects and in many mammals, and have been experimentally demonstrated in quail (de Groot, 1980).

Competition for food

As the benefits of being in group go up, costs due to increased intraspecific competition for food also increase (Bertram, 1978). In goldfish shoals this increase in competition is reflected in a decrease in handling time for each food item (Street *et al.*, 1984). Presumably, goldfish speeded up handling to avoid interference from others. Japanese medaka, *Oryzias latipes*, ate faster in larger groups (Uematsu and Takamori, 1976). A similar effect was noted for house sparrows by Barnard (1984). Bolting food in this way may bear digestive costs in both animals.

Eggers (1976) demonstrated that competition costs are reduced if prey densities are high or if the distance between school members is large. The effects of competition do not appear to become pronounced until a group size of about a dozen individuals is reached (Morgan and Colgan, 1987). The growth rate of Japanese medaka under standard densities and feeding regimes giving equal food intake increased with group size from 2 to 6 fish, but did not increase further as shoal numbers went up to 12 fish (Kanda and Itazawa, 1978).

Competition between two sizes of fish was scored in minnow shoals (Pitcher *et al.*, 1986a) and might be expected to be more severe in larger groups. Small minnows were forced out of feeding pots by large intruders, so small fish were forced to move about on the patch more. Milinski (1984) has demonstrated alternative coexisting feeding strategies depending upon competitive ability in sticklebacks. Fish that switched feeding patches more performed poorly in a test of competitive ability.

Such movers and stayers do not seem to be equivalent to the 'producers' and scrounging kleptoparasites seen in bird flocks (Brockmann and Barnard 1979). Giraldeau (1984) considers 'joiners' and 'discoverers' in relation to new food sites. Barnard and Sibly's model (1981) predicts that group size will have a strong effect upon the stable ratio of producers and scroungers, but despite attempts, consistent individual strategies like these have not yet been observed in fish shoals.

The ideal free distribution (IFD). IFD predicts that individuals in foraging populations will distribute themselves among available resources, such as food patches, in proportion to the rewards encountered (Fretwell and Lucas, 1970; Parker, 1978; Milinski, 1979; Sutherland, 1983). In this way each individual will gain the same food intake; optimal foraging for all in an alternative sense of OFT. Godin and Keenleyside (1984) timed six cichlids feeding on two patches as appearing to fit the ideal free distribution. Milinski (1979) reported an apparently similar finding for sticklebacks foraging on water fleas. However, more detailed investigation by Milinski (1984) included careful measurements of individual food intake and behaviour and revealed large consistent differences between fish. The basic IFD was not an appropriate model in these circumstances, although an alternative optimality theory could be modified *post hoc* to take account of the social interaction (the relative pay-off sum rule; see Hart, Chapter 8, this volume). 'Switchers' were poorer competitors who changed patches more frequently and had lower food intakes. Milinski's work shows that even though the gross picture may ostensibly fit an IFD prediction, the precise behaviour of individuals depends upon the amount of social competition for food. For example, Godin and Keenleyside may have obtained a different result by using a different group size in relation to the availability of food.

The IFD's prediction of equality of food intake is not likely to be met in practice because individuals vary in their ability to compete or scrounge food located by others, as has been clearly demonstrated in fish shoals (Wolf, 1987; Milinski, 1988) and in bird flocks (Barnard *et al.*, 1982). Theoretical IFD individuals are 'ideal' and 'free' because of the absence of constraints on where they should go or how they should perform. In the real world, shoaling fish are not free of such constraints and therefore basic IFD theory does not make such precise predictions as might be hoped. Not surprisingly, information also alters the fishes' decisions as to where to forage (Abrahams, 1989). A realistic range of factors governing distributions of fish found on patches are discussed comprehensively by Milinski (1988).

Forage area copying

Before individuals can benefit from foraging in a group, they have to join others feeding. Fish may join from a distance, but the same behaviour can be observed on a smaller scale within loosely structured shoals. In minnow and goldfish shoals, small groups with low EGS remain in visual contact. Individuals will soon join a group in which much 'nose-down' foraging is observed. This joining behaviour has been termed **forage area copying (FAC)** by Barnard and Sibly (1981), following Krebs *et al.* (1972), who first observed the phenomenon in bird flocks.

FAC does not necessarily entail moving to an actual food find: joining a group of foragers is sufficient. Moreover, sometimes this behaviour may be difficult to distinguish from aggregation driven by other motivations such as antipredator defence. For example, herons joined polystyrene models erected on mudflats (Krebs, 1974), and rudd joined conspecifics held in jars (Keenleyside, 1955), but food or defence reasons for joining were not disentangled at that time.

FAC has been tested experimentally for goldfish by Pitcher and House (1987). A shoal of test goldfish was placed next to a trained shoal feeding on the better of two patches behind a transparent barrier. The test fish had a choice between two food patches of equal food density. FAC was demonstrated when the test fish fed on the patch next to the trained shoal. In contrast, classical OFT predicts an equal distribution of test fish on the patches.

FAC depends critically upon individual reward rate. Minnows in shoals of 20 were trained to feed upon three patches of different food density, and soon devoted most of their time to the best patch. Twenty minnows in a test shoal were placed in front of the transparent barrier with three equivalently arranged but equally rewarding food patches. Like the goldfish, test minnows chose to feed on the patch adjacent to trained fish, but this result was strongly dependent upon the level of food on their patches. When food was plentiful, or completely absent from their patches, little FAC was observed, even right at the start of a trial. This shows that the success rate experienced by foragers shapes behaviour on a very short time scale. Very low or very high rates of food finding override FAC in these cyprinid shoals.

In this experiment and in the goldfish, FAC reappeared when the test fish were given no patches on which to search. (They foraged on the aquarium gravel.) Controls showed that the fish were not aggregating through timidity for predator advantage in any simple fashion. Since all fish in the experiment had prior experience of the particular design of patch, we can conclude that subtle cognitive cues are used in fish foraging behaviour, as pointed out by Tinbergen (1981) for starling flocks.

Switching of feeding methods

Many fish are able to employ alternative different methods of prey capture. Predators such as pike which usually stalk their prey may sometimes chase or even drive prey from cover by blowing water under rocks. Shoaling fish are often able to switch between filter feeding or biting individual prey (Crowder, 1985). For example, *Oreochromis lidole*, a shoaling cichlid from Lake Malawi, can switch from filter feeding algal food to biting separate prey items.

In a series of elegant experiments covering a range of prey densities, Gibson and Ezzi (1992) showed that switching between filter feeding and biting in shoals of 20 herring was clearly driven by the relative profitability of the two feeding modes. Filter-feeding herring swam faster, using 2–4 times more energy than biters, but at high prey densities this feeding method maximized energy intake. In experiments where the profitabilities of the two feeding methods were similar, both types of feeding were seen in the shoal. The switch was not simultaneous throughout the group; the uniformity of the cruising herring shoal broke up as individuals attempted to maximize their individual gain. Herring that were feeding by biting tended to get left behind by faster-swimming herring that were filtering (Gibson, pers. comm.). These fish stopped biting, caught up with the rest of the group and switched to filtering. So the energy-maximizing rule was constrained by a social decision to join, perhaps to maintain the antipredator advantage of staying with the group.

12.7   TRADE-OFFS BETWEEN PREDATION RISK AND FORAGING

Shoaling minnows were presented with two food patches, one of which was subjected to a sudden simulated 'attack' by a diving kingfisher model (Pitcher *et al.*, 1988). A real predator was not used to avoid the complication of releasing minnow alarm substance (p. 191). A sudden attack was used to avoid the complication of minnows evaluating information using inspection behaviour. When the two patches were equally endowed with food, minnows avoided the risky patch, compared with a pre-attack control trial. But when the risky patch had four times as much food as the safer one, the minnows accepted the risk by continuing to feed normally there (Fig. 12.16).

Fraser and Huntingford (1986) distinguished between alternative strategies of coping with the hazard of a predator attacking while a fish is itself eating. Risk-reckless fish feed regardless of hazard, so food overrides predation threat. Risk avoiders avoid hazard and minimize feeding, so predation threat overrides food. Risk adjusters make a trade-off by reducing feeding in proportion to perceived risk and consequently always feed less under predation threat. Risk balancers adopt the most sophisticated strategy and quantitatively trade off food value against risk, so they would accept more risk for more food.

Since minnows did not diminish their feeding on the well-rewarded patch in face of predation risk, yet stopped feeding there entirely when there was less food, they were probably risk balancers, but with only two food levels it

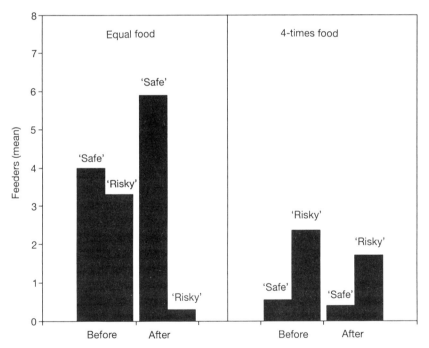

Fig. 12.16 A risk-balancing trade-off between perceived risk of attack by a diving avian predator and foraging in shoaling minnows. Vertical axis shows median number of minnows feeding on either a safe or a risky patch. The risky patch had either the same amount (left) or four times as much food (right) as the safe patch. Scores are shown before (left pairs of bars) and 6 h after (right pairs of bars) a simulated attack by a diving kingfisher model (data from Pitcher *et al.*, 1988).

was difficult to be sure. A further experiment using more food levels with the same protocol as Pitcher *et al.* (1988) (Pitcher and Gummer, unpublished) proved that minnows were genuine risk balancers, almost titrating the amount of food available against the risk of a 'kingfisher' attack.

Risk balancing has been reported from sticklebacks (Milinski, 1985a), coho salmon (Dill, 1983) and American minnows (Cerri and Fraser, 1983; Milinski, 1985b). Lima and Dill (1990) review the evidence that fish make trade-offs between perceived risk of being eaten and likely gains. They conclude that risk of being preyed upon is fundamental to a wide variety of decision-making processes. So we can expect decisions to join, leave and stay with the shoal to be profoundly influenced by predation risk.

Not all trade-offs with predation risk are with foraging in shoals; for example, greater risks are taken by male sticklebacks when egg-guarding (FitzGerald and van Havre, 1985).

Morgan (1988a, b) examined the multiple effects of hunger, shoal size and the presence of a predator on shoal cohesiveness and foraging in bluntnose minnows, *Pimephales notatus*. Cohesiveness, measured by dispersion, stragglers and aggressive interactions among the group, decreased with hunger, but increased with predation risk and shoal size. Foraging generally increased with shoal size and hunger in the absence of a predator. When the predator was present, only fish in large, and presumably safer, shoals fed at high rates. There were limits to the adjustments individuals could make in feeding: for example there is a point at which it is physically impossible to eat more, regardless of opportunity. The results clearly demonstrate that individual shoaling fish include group size and hunger in their trade-offs of security against feeding.

It is possible that there were considerable individual differences in all of these trade-off experiments, but few have employed marked individuals. Godin and Smith (1988) analysed the costs of greater feeding on zooplankton prey for guppies as hunger increased. High feeding rates in hungry fish bore an increased risk of predation. Milinski (1990, pers. comm.) suggests that individual fish may be consistently risk-averse or risk-accepting and so, like the problems with the IFD, a whole range of different outcomes of the trade-off may be possible depending upon the precise initial composition of the group: perhaps a fishy example of chaos.

Moreover, a problem with even the most elegant of experiments using multiple variables of shoal size, hunger and predation risk is that a wide range of outcomes can be determined empirically without increasing our under-standing of the underlying trade-offs to fitness being performed by the individual fish. It is increasingly difficult to do this without sophisticated mathematics. Godin (1990) showed how only a dynamic programming optimization model was able to predict accurately the diet selection of guppies foraging for three sizes of *Daphnia* in the presence and absence of a predator. Guppies attacked or ignored prey according to their state of hunger, the size of the prey and the presence of the predator (a cichlid behind a glass barrier). Simple OFT models, and a model based on visual reactive distance, failed to account for the observed shifts in prey capture. This is the type of carefully analysed experiment which needs to be extended to shoals of fish if our understanding and predictive power in predator influences on trade-offs in shoaling fish are to be deepened.

12.8   SUCCESSFUL PREDATORS ON FISH SHOALS

Despite the effective shoal tactics outlined above, some piscivores appear to be ahead in the arms race and specialize in preying on shoals. Most of them are large relative to their shoaling prey, or have the advantage of long-range detection (visual, olfactory, or acoustic) and fast swimming. Parrish (unpublished data) has documented several hundred attacks by nine species of piscine predator on a school of several hundred thousand flat-iron herring. Although the stalking predators, principally groupers and cornetfish, were the most abundant, the faster, streamlined predators, jacks and tunas, were the most successful at capturing prey (captures/initiated attacks). These predators were also tremendously successful at picking off the occasional straggler, enjoying success rates of over 70%.

Piscine predators such as tunas and jacks may enjoy high success rates when attacking schools because they are able to circumvent several of the school's defences. Cruising predators which accelerate into the interior of the school when striking may not be as susceptible to the confusion effect, either because they do not visually orientate on a single victim (Parrish, unpublished data) or because they break down the local structure of the school entirely, allowing a subsequent chase of individual prey items (Major, 1978). When preying on herring schools, tuna often accelerated into the school, stunning one to several herring by running into them (Parrish, unpublished data). Other species of piscine predator are known to attack prey shoals with accessory structures. Sawfish and teleosts with rostra, such as marlin and swordfish, swim through shoals stunning fish randomly with rapid swings of their saws or swords (Breder, 1967). Thresher sharks also do this with their tails.

Some specialized predators on fish shoals exploit normal school responses to catch their prey. Conspicuously striped colour patterns seem to act to disturb polarization in schools. Wilson *et al.* (1987) noted that 13 out of 14 dolphins and other toothed whales specializing in schooling prey were conspicuously striped, whereas only 3 out of 15 normally countershaded species ate schooling prey fish. One group of penguin species in the genus *Spheniscus* bears similarly striped coloration and specializes in feeding on pelagic schooling prey which polarize when alarmed. Wilson *et al.* tested anchovy schools with two differently coloured penguin models, suspended from an arm rotating at penguin speed. The models simulated a typical back-and-white countershaded penguin, and a striped penguin of the genus *Spheniscus*, which dietary analysis had shown to be a specialist on schooling pelagic prey. The *Spheniscus* model depolarized the anchovy schools in 67% of trials, as opposed to 28% for the normal penguin. The complex striped pattern of such predators appears to overload the prey's optomotor and schooling responses to neighbour fish in the school.

Some predators make use of extra-body means of capturing their shoaling prey. Humpback whales herd shoaling capelin into balls, which they contain by releasing foaming clouds of bubbles in a huge ring from beneath the fish (Norris and Dohl, 1980). Killer whales have been reported to use high-frequency sound to stun shoals of returning salmon (Norris, pers. comm.).

Many successful piscine predators school themselves and may attack in groups. Major (1978) demonstrated that groups of jacks were more successful at capturing individuals from a school of anchovies than were individual jacks. Simultaneous attacks by more than one predator caused the school to break up and the resulting isolated anchovies were struck before they could rejoin their fellows. Potts (1981, 1983) observed groups of jacks hunting several species of shoaling reef fish. The strength of the schools' response to attack was inversely proportional to predator group size, suggesting that larger predator groups are more life-threatening than individuals of the same species. Occasionally non-schooling predators will attack schools of small fish *en masse*. Hobson (1968) observed leopard groupers, usually a solitary stalking predator, simultaneously attacking schools of flat-iron herring.

Gregarious predators hunting as individuals can increase their capture success by making use of local topography to surprise their shoaling prey. Solitary jacks hunting shoals of snappers used rocky outcroppings as cover, allowing the predator to approach the shoal quite closely before it was detected (Potts, 1983). Predators also use relative contrast to their advantage. Many teleost predators attack their shoaling prey at dawn and dusk when light levels are low and the predators have a visual advantage (see Helfman, Chapter 8 this volume; Pitcher and Turner, 1986). Predators may also attack the shoal from below, where the prey are maximally backlit and the predators are minimally so (Munz and McFarland, 1973; Welch and Colgan, 1990). Other predators may provide the prey with cover themselves. Shoals that seek a shady refuge (Helfman, 1981) are exploited by a heron thoughtfully extending a wing to provide such a shady area!

While scant evidence exists, there has been the suggestion that schooling fish predators may actually hunt cooperatively, where hunting-group members take on specific tasks, all of which are needed for a successful hunt. Schmitt and Strand (1982) observed groups of yellowtail herd, and then isolate, part of a school of jack mackerel. After the prey subgroup was surrounded, individual yellowtail made feeding forays while their companions kept the jack mackerel from escapting. A similar degree of coordination was reported for black skipjack feeding on scad (Hiatt and Brock, 1948). Partridge *et al.* (1983) observed groups of bluefin tuna, at the ocean's surface, swimming in concave, parabolic formations. It was suggested that this group structure may allow the tuna to herd and 'seine' their prey. Rigorous tests of how long individual predators associate with each other, as well as determinations of the pay-offs to various hunting-group members, have yet to be performed.

Unlike teleosts, cetacean predators are often reported to forage cooperatively on shoaling prey. Dolphin schools herd their prey into a compact ball before rushing through the shoal to emerge with mouthfuls of up to five fish (Wursig and Wursig, 1980; Bel'kovich *et al.*, 1991). Pelicans acting as a group can herd their shoals of prey, catching alarmed fish at random from the shoal with their scooping beaks.

Finally, many successful predators make use of the confusion created by other simultaneously attacking species. The foraging success of flocking common terns feeding on shoaling prey increases with the number of bluefish also feeding on the same shoal (Safina and Burger, 1989; Safina, 1990). Bluefish drive the prey to the surface where terns can take advantage of the situation. Many species of open ocean seabirds associate with cetaceans as a probable means of finding food. In parts of the eastern tropical Pacific multispecies seabird flocks are found with 60% of the spotted dolpin, and spotted plus spinner dolphin, schools. Note that the cetaceans and the seabirds are taking advantage of shoaling prey driven to the surface by feeding schools of yellowfin tuna (Au and Pitman, 1986).

12.9 POSITION IN THE SHOAL

Once the shoal has been detected by a predator ready to attack, individual group members may still take selfish advantage of the situation by occupying positions of relative safety. Many researchers have documented the fact that individuals straggling from the shoal are preferentially attacked and sustain a much higher risk of death than do their more gregarious counterparts (Morgan and Godin, 1985; Magurran and Pitcher, 1987; Parrish, 1986b; Parrish *et al.*, 1989).

Aberrant individuals within the shoal may also be preferentially attacked, perhaps because they are easier to discern visually, thereby minimizing the confusion effect. Artificially dyed minnows sustained higher rates of attack than their 'natural' conspecifics (Landeau and Terborgh, 1986). Gafftopsail pompano, attacking a mixed school of flat-iron herring and anchovies, preferentially picked off the anchovies, even though the herring constituted the vast majority of the group (Hobson, 1963). Hobson theorizes that the anchovies were much more visible due to the flashes of their gill covers. Even if group members conform morphologically, they may be singled out due to aberrant behaviour. Individuals making 'mistakes' during group avoidance manoeuvres are easily picked off by the attacking predator (Major, 1978; Parrish *et al.*, 1989; Parrish, 1992).

If all prey fish elect to remain in the group there may still be positional effects. Hamilton (1971) explored this concept theoretically in order to demonstrate how predation could select for the evolution of gregarious behaviour. When a hypothetical water snake was constrained to engulf the nearest prey item (a frog), all frogs clumped together in a desperate bid to place as many other individuals as possible between themselves and the predator. Thus peripheral positions carried a high risk value and central locations were relatively safe. In associations between bagrid catfish and cichlid juveniles, McKaye and Oliver (1980) found potential support for the selfish herd. However, the centre of the group was guarded by a parent catfish, which also excluded the cichlid young to the periphery. Parrish (1989b) examined the question of which position in a shoal of Atlantic silversides a black seabass predator preferred to strike, and found that central locations were actually at significantly higher risk than their peripheral couterparts. This result was due, in part, to the ability of the predator to enter the shoal instead of being constrained to only attack edge positions. Field observations of five species of piscine predator attacking a school of flat-iron herring confirmed that the majority of the attacks as well as captures occured in the interior of the school, regardless of the attacking species (Parrish *et al.*, 1989).

Whether individual fish in a shoal actually take advantage of the positional effects by securing positions of highest value, to the detriment of fellow group members, is not known. However, there is some evidence that group members do not occupy all positions equally. Several individuals in a school of ten Atlantic mackerel showed strong preferences for discrete positions within the group, significantly different from a random distribution over all positions (Pitcher *et al.*, 1982b). Individuals in a shoal of herring swimming in a flume showed significant preferences for other group members, regardless of their positions within the group (Parrish, unpublished data).

12.10   SIZE SORTING IN SHOALS

Many studies have remarked upon the size segregation of fish in both mixed-species and monospecific shoals. Typically, smaller fish stay near the surface and the group may gradually grade down to large fish near the bottom of the shoal (Sette, 1950; Breder, 1951). In mixed-species groups there may be actual discontinuities in size at the breaks between monospecific layers of fish (scombrids: Yuen, 1962; clupeoids: Parrish, 1989a). Similarly, monospecific shoals, such as Pacific sardine, *Sardinops sagax*, are often sorted by size, with large fish in the deeper water where they may evade capture by purse seines (Ñiquen, 1986). Size segregation seems to occur because of individual decisions to swim next to neighbours of similar size (mackerel and herring; Pitcher *et al.*, 1985) and size differences of the order of 30% may be measured in different parts of a shoal.

An often-stated explanation of size segregation is that larger fish cruise faster (Muzinic, 1977), but this would soon lead to complete separation of size groups rather than size sorting per se. As fishermen know, this is of course what often happens where fish of greatly differing sizes and ages are concerned.

Detailed laboratory studies have revealed that size sorting can result from differential pay-offs in foraging abilities and risk during predator threat. In the

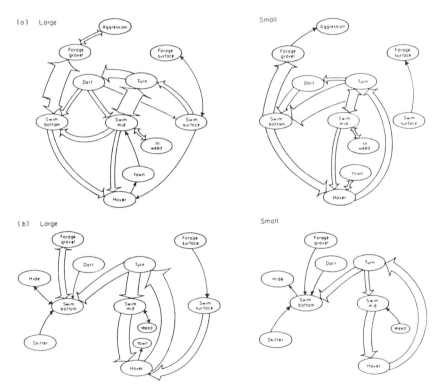

Fig. 12.17 Size-segregating behaviours in response to predation in minnow shoals. Diagrams illustrate organized sequences of behaviour in shoals of 20 mixed-size minnows (10 large and 10 small), (a) before pike exposure for 1 hour (top); (b) after pike exposure (bottom). Arrows indicate transitions between pairs of behaviours significantly greater than random at the 1% level; width of arrows is proportional to the frequency of transitions. Data from 36 pooled sessions recording individually-recognized focal fish. For further details see Pitcher *et al.* (1986b).

absence of predators, minnows will segregate by size into more than one subshoal, with larger fish manipulating access to food sources. However, when the fish are threatened by a predator, the smaller minnows respond by joining the larger-sized shoal (Pitcher *et al.*, 1986b), presumably to minimize their individual risk of attack (Fig. 12.17). There are similar findings in stickleback shoals (Ranta *et al.*, 1992). Muzinic (1977) also reported that the smaller sardines in a size-graded group shoaled independently unless disturbed, and were the first to lose weight and eventually die when food was limiting.

12.11   MIGRATION ADVANTAGES OF SHOALING

Shoaling could increase the accuracy of homing on migration, since the mean direction or route is likely to be a more accurate estimate of the correct destination than any individual's choice (Larkin and Walton, 1969). A similar effect may operate in bird flocks, but there appears to have been no critical test of this idea, which ought to be strongly dependent upon group size. There is now both direct and indirect evidence supporting the migration-enhancement hypothesis.

The first indirect evidence is that adult coho salmon home more accurately to their natal river at higher densities (Quinn and Fresh, 1984). The evidence is based on statistical analysis of tagging returns along with population size estimates, and the figures are subject to considerable variance. The point could be tested directly using the techniques pioneered by Hasler (1983) in proving the home-stream hypothesis.

Secondly, social transmission of information about diurnal migration routes between refuge and feeding sites has been demonstrated in grunt shoals (Helfman and Schultz, 1984). Transplant experiments proved that individuals rapidly acquired knowledge from local residents about routes and shoaling sites. After only a short period with their new conspecifics, isolated transplantees travelled the residents' routes to the correct locations. Such cultural transmission had not been shown in fish before, and is clearly a variation on the information centre idea.

Wijffels *et al.* (1967) showed that groups of eight *Barbus tincto* swam more quickly through a maze than individuals, but careful controls did not confirm Welty's (1934) conclusion that fish in groups learn routes faster than isolates. The fish in Wijffels' study performed better in a group if they had learned the maze in a group, and better alone if they had learned it alone, a general confirmation of the 'context' learning phenomenon analysed by Hinde and Stevenson-Hinde (1973).

Direct support for the migration-enhancement hypothesis comes from painstaking and elegant work by Kils (1986 and pers. comm.), most of which has unfortunately not yet been published. Kils performs field experiments on

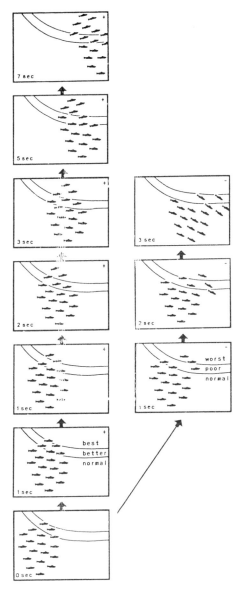

Fig. 12.18 Diagram illustrating Kil's mechanism of synchrokinesis: small movements of individuals copied through the shoal provide an accurate movement towards better conditions. Left column (times labelled from bottom to top) shows fish moving into a favourable area, right column shows them avoiding an unfavourable area. Schematics taken from video frames of herring swimming near the Kiel canal. Reproduced with permission from Kils (1986).

Baltic herring schools from a floating glass-bottomed laboratory (a converted night club) moored near the entrance to the Kiel canal, through which the herring migrate on their way to spawning grounds. The herring use salinity (and probably temperature) cues to orientate their migration to the entrance of the canal. Kils' video measurements on fish in travelling schools, which he can manipulate by introducing water jets beneath his floating laboratory, show how minor course adjustments by neighbouring individuals act to spread the most effective direction of movement through the group.

Kils has extended the idea of more effective course adjustment of fish in groups to include other environmental factors such as patches of planktonic food. He has introduced the new term '**synchrokinesis**' to express this hypothesis (Fig. 12.18; Kils, 1986). The experimental work has not yet shown that course corrections are quantitatively more effective in larger groups, a critical test of the main hypothesis.

12.12   GENES, INHERITANCE AND EVOLUTION IN SHOALING BEHAVIOUR

Minnows from provenances with and without pike exhibited a similar repertoire of antipredator tactics, but those sympatric with the predator performed and integrated the behaviours more effectively (Magurran and Pitcher, 1987). In threespine sticklebacks (Giles and Huntingford, 1984), and in guppies (Seghers, 1974; Magurran and Seghers, 1990a), predators have shaped interpopulation differences in antipredator behaviours. Jakobsson and Järvi (1978) showed that naïve salmon smolt shoals reacted selectively to pike predators, but not to non-predator fish. The basic repertoire of shoaling responses to predators therefore has a genetic basis.

Recent experiments have clarified the genetic and learned basis of shoaling behaviour, and furthermore have shown that there is a genetic basis to improving shoaling responses to a predator. Innate differences in the ability to learn about predators provide an example of what Konrad Lorenz, one of the founders of ethology, termed the 'innate schoolmarm'.

Genetic basis of antipredator behaviour and learning in fish shoals

Minnows were raised from the egg without experience of predators, and their antipredator performance, including inspection behaviour and skittering, was carefully measured at 3 months old and at 2 years old (Magurran, 1989). This was done for minnows from the two provenances mentioned above, one where pike were sympatric (high predator risk: a Dorset river) and one where pike were absent (low predator risk: North Wales). Half of each group experienced

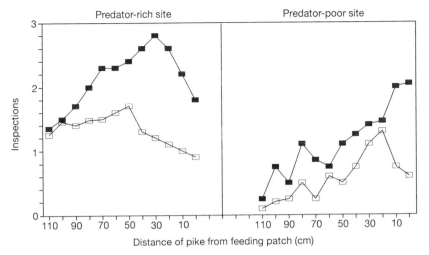

Fig. 12.19 Antipredator shoaling behaviours have a genetic basis, and furthermore, the ability to learn about a predator may also be inherited. Predator inspection behaviour scored in 2-year-old minnows raised from the egg from two provenances in the United Kingdom, one with pike (left graph) and one without pike (right graph). Horizontal axis is distance of a model pike from its garage as its approaches the minnows' feeding patch during a trial (see Fig. 12.9). Vertical axis is mean number of inspections per fish scored during trials. 'Experienced' minnows (filled symbols) had seen a pike model once previously at 3 months old; 'naïve' minnows (open symbols) had had a pikeless control at that time. Predator inspection behaviour was just one of several behaviours showing similar results. For further details see Magurran (1990).

a predator at 3 months old, the other half did not. As with the wild-caught fish, laboratory-raised minnows from the high-risk provenance had more effective antipredator behaviour whatever their history. The most interesting finding from Magurran's elegant experiment was that minnows from the high-risk site proved better able to learn from early experience of a predator than those from the low-risk site (Fig. 12.19). The next step for this type of work will be to disentangle the inheritance of antipredator and shoaling behaviours.

Kinship and altruism

It is possible that, in some fish, altruistic shoal behaviours may have evolved through kin selection. This is one possible explanation for the evolution of fright substance in cyprinids (see Chapter 6 by Hara, this volume). Evidence for kinship in shoals of the freshwater shiner, *Notropis cornutus*, has been produced by Ferguson and Noakes (1981). Quinn and Busack (1985) have demonstrated chemosensory recognition of siblings in shoaling coho salmon juveniles, and Loekle *et al.*, (1982) have similar evidence for poeciliid fish. In three-spined sticklebacks, van Havre and FitzGerald (1988) demonstrated that juveniles prefer to shoal with kin.

Traditionally, it has been argued that kinship could only affect selection in freshwater or reef fish, which do not travel far from a restricted habitat and spawning area, and that such factors could not impinge upon marine species with pelagic larvae. It may be worthwhile to question this assumption, if chemical kin recognition is considered alongside the growing evidence for natal-site reproductive homing in a wide range of teleosts.

Fidelity to particular shoals is a pre-requisite if related fish are to benefit by altruism. Most measurements of school fidelity (e.g. perch: Helfman, 1984) have not produced evidence for high group fidelity. Hilborne (1991), examining data for recaptured individuals in skipjack tuna schools, found very low levels of fidelity to particular groups. But in three-spined sticklebacks, adult females preferred to shoal with familiar companions rather than strange fish (van Havre and FitzGerald, 1988).

Nevertheless, biochemical techniques of investigating close genetic affinity may be expected to provide the critical data. If close kinship, along with the means of kin recognition, can be demonstrated in fish shoals, we may have to revise our ideas of the evolution of the behaviour as exclusively to the advantage of individual fish.

12.13   MIXED-SPECIES SHOALS

Most of the factors favouring larger group size in single-species shoals can, in the simplest analysis, apply equally to mixed-species shoals (Ehrlich and Ehrlich, 1973). These factors are more effective foraging by social observation, better vigilance for predators and greater economy of time budgeting. In mixed shoals, these benefits are likely to be greater the more similar the fish morphologies and diet. In single-species groups, the benefits are countered by increased food competition, but this could be lower for a given group size in mixed shoals. Differences between species allow for exploitation of genuine symbiosis if food items wanted by one species of fish are flushed by the other. Some behaviours, such as predator-evasion tactics, may be more efficiently performed with conspecifics because of similar morphology. Furthermore, small numbers of one species in a mixed shoal may suffer through being conspicuous to predators. Many of these points have been investigated in detail for bird flocks (Krebs and Barnard, 1980), but until recently there has been relatively little detailed research on mixed-species fish shoals.

Mixed shoals of cyprinids have been investigated under semi-natural conditions by Allan (1986), who demonstrated that many of the foraging benefits of larger groups also accrue when species are mixed in one large shoal.

Diet and habitat shifts occur in mixed-species shoals, which appear to minimize competition while retaining the advantage of greater vigilance and search power. Under threat of predation, mixed shoals of cyprinids sort into separate conspecific subgroups (Fig. 12.20) (Allan and Pitcher, 1986). Odd fish with few conspecifics seek refuge, but shoal if enough of their own species are present (Wolf, 1985). Cyprinid species sorting under threat reflect more effective shoal manoeuvres with conspecifics and perhaps an attempt to minimize conspicuousness (Allan and Pitcher, 1986). Piscivores seem to pick out conspicuous individuals for attack, perhaps to overcome the confusion effect. This is exactly the opposite prediction from apostatic selection theory, in which novel conspicuous individuals which stand out are protected and therefore selected for. The theory dealing with circumstances in which conspicuousness does or does not pay therefore bears more careful examination.

In mixed cyprinid shoals comprising feeding minnows, rudd and orfe under threat from a stalking predator, minnows performed about 50% fewer predator inspections per fish, than expected (Pitcher and Graham, unpublished data). Comparison of inspection rates in a series of single-species control experiments with the same total group size as the mixed shoals suggested that the most likely explanation was that minnows are able to take advantage of the inspections of other species. This benefit was not reciprocal for the other

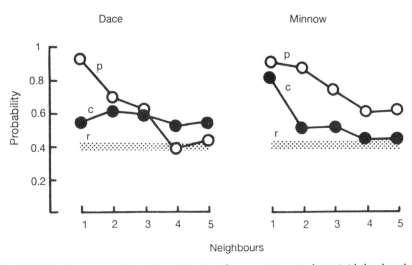

Fig. 12.20 A measure of species sorting in a three-species mixed cyprinid shoal under threat from a model pike. Left: dace, right: minnow. Vertical axis is the probability of the nearest neighbour being the same species as the focal fish. Solid circle show values when pike was absent, open circles when pike was present. Shaded horizontal line indicates random expectation for neighbour identity. Reproduced with permission from Allen and Pitcher (1986).

two cyprinid species. Minnows therefore should be more likely to gain from mixed cyprinid species associations.

On coral reefs, large multispecies shoals composed mainly of herbivorous fishes are frequently found in areas also inhabited by benthic, territorial fishes. In many cases these wandering mixed-species shoals contain a core species, dominating the assemblage, and to several other ancillary species. Often these groups contain members of more than one family, usually scarids (parrotfish), and acanthurids (tangs), and occasionally mullids (goatfish), chaetodontids (butterfly fish), small serranids (typified by hamlets), kyphosids (chubs), and non-benthic pomacentrids (typified by sergeant majors) (Ogden and Buckman, 1973; Itzkowitz, 1977). While the majority of the assemblage feeds on benthic algae, minority members, such as goatfish and hamlets, feed on the inverte-brates and small fishes disturbed by the action of the herbivores (Ogden and Buckman, 1973). Membership in these feeding assemblages is fluid, with individuals joining and leaving as resource levels change (Ogden and Buckman, 1973; Itzkowitz, 1977) or as predators threaten the group (Wolf, 1985).

Krebs (1973) suggested that one of the benefits of mixed-species bird flocking was the greater diversity of food items which became available, not only through effects like flushing, but also by social observation of the foraging specializations of others. Rubenstein *et al.* (1977) have supported this idea by demonstrating a broader diet for individuals in mixed flocks, and equivalent work on fish shoals would be timely. The most rigorous investigation of mixed-species shoaling to date has been performed by Wolf (1983, 1985, 1987). Wolf showed how the costs and benefits of shoaling vary continuously for different species, times and locations in shoals travelling around the reef. The diet and behaviour of three herbivorous fishes (*Sparus, Sparisoma* and *Acanthurus* species) were recorded with group size and composition. Wolf's work raises the possibility of a revealing investigation of conflicts of interest between the species in her mixed-species shoals.

Vine (1974) and Barlow (1974) suggested that one of the benefits to members of feeding assemblages (in their studies composed mainly of acanthurids) was access to the defended algal turfs of territorial pomacentrids. Foster (1985a, b), working on acanthurid-predominated shoals, and Robertson *et al.* (1976), working on scarid-predominated shoals, demonstrated that members of larger feeding groups were able to overwhelm the defences of the damselfish and forage within their territories. This finding was confirmed by Wolf for acanthurids (1987). In an interesting parallel, Marsh and Ribbink (1986) report the same pattern of mixed-species feeding assemblages invading the algal gardens of demersal territorial fish, in the cichlid flock of Lake Malawi. While all authors were able to find a clear benefit to feeding assemblage members, they reported that many members of the larger populations occurred as individuals, and as such were denied access to potential food resources. Reinthal and Lewis (1986), working on acanthurid-predominated shoals, suggested that solitary acanthurids visited wrasse cleaning stations for parasite removal while groups did not. Thus the balance between the costs and benefits of mixed-species group membership may be more subtle than initially suspected, especially as individuals are known to change groups, and perhaps temporarily eschew membership entirely.

In open-water habitats mixed-species shoals may be quite common, but have not received the attention that stream and reef fish assemblages have. This discrepancy may result from the difficulties in collecting observational data and/or performing experiments on large groups of extremely mobile animals. Springer (1957) made observations on several mixed-species shoals composed of varying families of small, silvery, planktivorous fishes (atherinids – silversides, engraulids – anchovies, clupeids – herrings, and carangids – typified by scad) in the Gulf of Mexico. In all species combinations observed, the individuals were similar in size, and the shoal composition was usually heavily weighted towards a single species. Springer reported sightings of shoals with discrete species-specific components, as well as more integrated assemblages. Hobson (1963) noted an integrated school of 90% herring and 10% anchovy in the Sea of Cortez. Parrish (1989a) made detailed observations of hetero-specific assemblages of four species including two clupeids, an engraulid, and an atherinid, over sandflat habitat in Bermuda. Fish assorted by species and size, such that on rare occasions the shoal consisted of up to seven discrete layers. When threatened by predators, the shoal split along species/size layers to perform independent escape manoeuvres (Parrish, unpublished obs.). Unlike previous observations, the shoal was rarely skewed in composition towards any one species. As these fish are all morphologically extremely, and behaviourally superficially, similar, this assemblage may be a further demonstration of the coevolution proposed by Ehrlich and Ehrlich (1973) to explain mixed shoals of haemulids (grunts).

More permanent mixed-species shoals occur in larger, pelagic schooling fishes exemplified by the carangids and scombrids. In the eastern tropical Pacific mixed-species shoals predominantly composed of skipjack and yellowfin tuna are quite common (Matsumoto *et al.*, 1984), with occasional bigeye tuna, bullet mackerel, and frigate mackerel (Hall *et al.*, 1992). Yellowfin and skipjack are found in close association in every oceanic basin in which they are caught, with the smaller skipjack occurring closer to the surface and the larger yellowfin beneath them (Hallier, 1991; Hall and Garcia, 1992). The reasons for this persistent association are not known.

As the species in mixed groups become more different, increasingly subtle relationships are possible. Benefits of larger group size accrue to juvenile French grunts (*Haemulon* sp.) and two species of mysid shrimp on coral reefs (McFarland and Kotchian, 1982). The associations last for only 5 days while the fish grow, but are renewed every 2 weeks or so as fresh cohorts of postlarval grunts recruit from the plankton. Conspicuousness here may be reduced through interspecific mimicry, since the two organisms look remarkably similar. There is a clear differential distribution of benefits, as the young fish prey on their mysid companions as they grow.

In summary, much more work is required on the details of the costs and benefits of mixed-species shoaling. Where conflicts of interest occur, their resolution is likely to lead to profound insight into the behavioural ecology of shoaling, in a similar fashion to recent work on flocking birds.

12.14   OPTIMAL SHOAL SIZE

In most of the preceding discussion, benefits increase with shoal numbers, but this effect occurs most rapidly at small shoal sizes. Some costs, such as intraspecific competition, increase with shoal size as well, and some benefits, such as faster food location, reduce at very large shoal sizes. Cost and benefit curves which intersect at some group size may be envisaged. A simple-minded analysis suggests that there will be an optimal group size where the net benefit is at a maximum (Bertram, 1978).

In the best of possible worlds, we would tend to find groups of this optimal size in the wild. Using different hypothetical models, Sibly (1983), Pulliam and Caraco (1984), and Clark and Mangel (1984) have shown independently that such an optimum is not stable, since individuals tend to join groups because they do better by joining a group than remaining on their own. Group sizes in the wild should therefore be much larger than the 'social' optimum predicted by such averaging. It does not, however, appear that anyone has rigorously tested this prediction, although Clark and Mangel (1984) cite some suggestive evidence.

A simple explanation is based on a direct analogy with the well-known phenomenon of the 'tragedy of the commons' (Hardin, 1968), termed 'common property rent dissipation' by mathematical economists (Clark, 1976). This process leads unregulated 'open-access' resources such as fisheries inexorably to a point where no profit is made, termed the 'zero net revenue' outcome (Pitcher and Hart, 1982). If no restrictions, such as licences, are placed upon access, new boats join a profitable fishery on the expectation of individual gains like those enjoyed by fishermen already working the resource. As successive new boats join, everyone's share of the benefit diminishes, until eventually the zero-profit point is reached. (In some cases the process can overshoot, producing a loss for all.) If there is no alternative resource, the process is automatic and is impossible to halt. Joining of new boats is analogous to fish deciding to join a shoal, and the fishery yields are analogous to the net benefit of being a group member. Huge marine shoals may represent such 'common property' losses, zero-benefit shoaling being favoured over dispersal as a bet hedged against predator encounters. Clark and Mangel (1984) draw the same analogy.

This hypothesis predicts that we should observe groups in which the average net benefit is zero rather than a maximum. The evolution of behaviour which kept a shoal at a certain size by deterring or evicting newcomers would render this analysis invalid in a way analogous to the issue of licences restricting access to a fishery.

Koslow (1981) speculated that the size of large pelagic shoals might be regulated by behaviour optimizing plankton feeding in relation to plankton patch size. This is also unstable for the same reasons as above, unless such altruistic group-size-regulating behaviours occur.

Anderson (1981) provided evidence that a surprising number of shoals off Southern California were of 15 m diameter, and constructed a mathematical model which fitted the observed distribution of shoal sizes. The model was based on two assumptions: (a) that fish joined the group irrespective of shoal numbers, and (b) that fish left the group in proportion to the numbers present. Both of these assumptions seem unrealistic in the light of the evidence we now have about shoal-size-related behaviours.

Do fish exclude newcomers once the social optimum group size is reached? We may speculate that some awareness of the current optimal group size may have evolved, but, clearly, some very carefully controlled experiments need to be devised in order to test this hypothesis.

12.15 SUMMARY

Polarized and synchronized schooling behaviour is regarded as a category of the social grouping termed shoaling. A long search for structure in schools has not contributed much insight of the functions and evolution of shoaling. Homogeneity in fish shoals and schools, previously over-emphasized and mistakenly attributed to altruism, has been shaped by risk of predation. Detailed experimental examination of behaviour during foraging, and under predator threat, reveals tensions which reflect an individual's continual reappraisal of the costs and benefits of joining, staying with, or leaving the social group. Elective group size (EGS) provides a simple and effective measure of shoaling tendency, which may track the environmental factors which shape fish shoaling behaviour.

As a consequence of the efficiency of human predation, fisheries for shoaling species may be inherently unstable. rendering assessment and management of shoaling fish difficult. Therefore the understanding of the dynamics of large-scale changes in shoaling in relation to environmental factors is a major goal for shoaling research.

There is little evidence supporting hydrodynamic advantages in fish shoals, but the accuracy of homing and migration are enhanced through 'synchro-kinesis'.

The chapter reviews critically ideas about how shoaling fish may counter predator attack through avoidance, dilution, abatement, evasion, confusion, detection, mitigation, inspection, inhibition and prediction. Recent experiments have determined an inherited genetic basis for the antipredator components of shoaling behaviour, along with a genetic basis for increasing effectiveness with predator experience.

Predator avoidance and attack dilution are insufficient singly as general advantages to account for shoaling, but the two effects combined provide an ESS favouring shoal behaviour, here termed the 'attack abatement effect'. A cognitive dilution benefit may also accrue in shoals. Shoaling aids early predator detection, and predator confusion is generated both passively by grouping and actively through shoal manoeuvres. Group tactics also aid predator evasion. Skittering in minnows may reveal conflicts between shoal members. Shoal members take risks to perform predator inspection to gain information about the predator. The evolution of inspection behaviour is enigmatic: tit-for-tat and risk dilution may both be involved in mitigating the costs of inspection.

Simple optimal foraging theory (OFT) or the ideal free distribution (IFD) fail to explain foraging in shoals, but recent developments in foraging theory illuminate the constraints and pressures in group foraging. In larger shoals fish find food faster, spend more time feeding despite predator threat, are less timid, more vigilant, sample the habitat more effectively and transfer information about feeding sites more quickly (passive information transfer, PIT). Forage area copying (FAC) is an important component of these shoaling advantages, but the behaviour is contingent upon an individual's feeding rate.

In shoals, information about the type and quality of a food source, number of companions, threat and alternative feeding sites is acquired and decisions are taken surprisingly rapidly. Trade-offs between these factors are commonly made and may differ between individuals. The behavioural plasticity conferred by shoaling may therefore be seen as one of its major general advantages.

Fish in mixed-species shoals attempt to maximize their gains from foraging and vigilance, but, under attack, sort into species and sizes to optimize escape manoeuvres and reduce their conspicuousness among aliens. Mixed-species shoals may show a great complexity of behavioural and ecological interactions, but it seems that the standard costs and benefit rules of living in groups often apply. Many more field measurements on open-water mixed-species shoals are required before any further general principals of this common phenomenon are clear.

Competition for food increases with shoal numbers, but the optimal shoal size that maximizes the net benefit for all is unlikely to be observed since it is unstable for the same reasons as the degradation of common property resources. In the absence of behaviour which actively limits entry, we may expect to observe shoals in which the average net benefit is zero. This means that the plasticity in the face of risk referred to above may have been crucial to the evolution of shoaling. However, recently demonstrated kin recognition, genetic shoal affinity and the evolution of ostensibly altruistic shoal phenomena like 'fright substance' could mean that the search for behaviours which limit shoal size to the 'social optimum' may be worthwhile.

ACKNOWLEDGEMENTS

TJP would like to thank John R. Allan, Paul Hart, Gene Helfman, George Turner, Anne Magurran and the referee for their comments on the first edition and Alida Bundy for help in preparing the second edition. J. M. Cullen kindly allowed discussion of some of his unpublished experiments. We are grateful to Bill and Peggy Hamner for creating the opportunity for the co-authors to meet in Monterey to discuss the second edition of this chapter.

REFERENCES

Abrahams, M. V. (1989) Foraging guppies and the ideal free distribution: influence of information on patch choice. *Ethology*, **82**, 116–26.

Abrahams, M. V. and Colgan, P. (1985) Risk of predation, hydrodynamic efficiency and their influence on school structure. *Env. Biol. Fishes*, **13**, 195–202.

Abrahams, M. V. and Colgan, P. (1987) Fish schools and their hydrodynamic function – a reanalysis. *Env. Biol. Fishes*, **20**, 79–80.

Allan, J. R. (1986) The influence of species composition on behaviour in mixed species cyprinid shoals. *J. Fish Biol.*, **29** (Supp. A), 97–106.

Allan, J. R. and Pitcher, T. J. (1986) Species segregation during predator evasion in cyprinid fish shoals. *Freshwat. Biol.*, **16**, 653–9.

Anderson, J. J. (1981) A stochastic model for the size of fish schools. *Fishery Bull. Fish Wildl. Serv. U.S.*, **79**, 315–23.

Andorfer, K. (1980) The shoal behaviour of *Leucaspius delineatus* (Heckel.) in relation to ambient space and the presence of a pike, *Esox lucius*. *Oecologia*, **47**, 137–40.

Au, D. W. K. and Pitman, R. L. (1986) Seabird interactions with dolphins and tuna in the eastern tropical Pacific. *Condor*, **88**, 304–17.

Barlow, G. W. (1974) Extraspecific imposition of social grouping among surgeonfishes (Pisces, Acanthuridae). *J. Zool., Lond.*, **174**, 333–40.

Barnard, C. J. (1983) *Animal Behaviour: Ecology and Evolution*, Croom Helm, Beckenham, U.K., 339 pp.

Barnard, C. J. (ed.) (1984a) *Producers and Scroungers*, Croom Helm, Beckenham, U.K., 303 pp.

Barnard, C. J. (1984b) The evolution of food scrounging strategies within and between species, in *Producers and Scroungers* (ed. C. J. Barnard), Croom Helm, Beckenham, pp. 95–126.

Barnard, C. J. and Sibly, R. (1981) Producers and scroungers: a general model and its application to captive flocks of house sparrows. *Anim. Behav.*, **29**, 543–50.

Barnard, C. J., Thompson, D. B. A. and Stephens, H. (1982) Time budgets, feeding efficiency and flock dynamics in mixed species flocks of lapwings, golden plovers and gulls. *Behaviour*, **80**, 44–69.

Barrie, B. D., Huntingford, F. A., Lazarus, J. and Webb, S. (1992) A dynamic analysis of cooperative predator inspection in sticklebacks. *Animal Behaviour* (in press).

Bel'kovich, V. M., Ivanova, E. E., Yefremenkova, O. V., Koarovitsky, L. B. and Kharitonov, S. P. (1991) Searching and hunting behaviour in the bottlenose dolphin (*Tursiops truncatus*) in the Black Sea, in *Dolphin Societies: Discoveries and Puzzles* (eds K. Pryor and K. S. Norris), Univ. California Press, Los Angeles, pp. 38–67.

Belyayev, N. and Zuyev, G. V. (1969) Hydrodynamic hypothesis of schooling in fishes. *J. Ichthyol.*, **9**, 578–84.

Bergelson, J. M., Willis, J. H. and Robakiewicz, P. E. (1986) Variance in search time: do groups always reduce risk? *Anim. Behav.*, **34**, 289–91.

Bertram, B. C. R. (1978) Living in groups: predators and prey, in *Behavioural Ecology*, 1st edn (eds. J. R. Krebs and N. B. Davies), Blackwell, Oxford, U.K., pp. 64–96.

Bertram, B. C. R. (1980) Vigilance and group size in ostriches. *Anim. Behav.*, **28**, 278–86.

Blake, R. W. (1983) *Fish Locomotion*. Cambridge Univ. Press, Cambridge.

Blaxter, J. H. S., Gray, J. A. B. and Denton, E. J. (1981) Sound and startle responses in herring shoals. *J. mar. biol. Ass. U.K.* **61**, 851–69.

Breder, C. M., jun. (1951) Studies on the structure of fish shoals. *Bull. Am. Mus. nat. Hist.*, **98**, 1–27.

Breder, C. M., jun. (1954) Equations descriptive of fish schools and other animal aggregations. *Ecology*, **35**, 361–70.

Breder, C. M., jun. (1959) Studies on social grouping in fishes. *Bull. Am. Mus. nat. Hist.*, **117**, 393–482.

Breder, C. M., jun. (1965) Vortices and fish schools. *Zoologica*, **50**, 97–114.

Breder, C. M., jun. (1967) On the survival value of fish schools. *Zoologica*, **52**, 25–40.

Breder, C. M., jun. (1976) Fish schools as operational structures. *Fishery Bull. Fish Wildl. Serv.*, U.S., **74**, 471–502.

Broadbent, D. E. (1965) Information processing in the nervous system. *Science*, **150**, 457–62.

Brock, V. E. and Riffenburgh, R. H. (1960) Fish schooling: a possible factor in reducing predation. *J. Con. perm. int. Explor. Mer*, **25**, 307–17.

Brockmann, H. J. and Barnard, C. J. (1979) Kleptoparasitism in birds. *Anim. Behav.*, **27**, 487–514.

Caraco, T. (1980) Stochastic dynamics of avian foraging flocks. *Am. Nat.*, **115**, 262–75.

Caro, T. (1986) The functions of stotting: a review of the hypotheses. *Anim. Behav.*, **34**, 663–84.

Cerri, R. D. (1983) The effect of light intensity on predator and prey behaviour in cyprinid fish: factors that influence prey risk. *Anim. Behav.*, **31**, 736–42.

Cerri, R. D. and Fraser, D. F. (1983) Predation and risk in foraging minnows: balancing conflicting demands. *Am. Nat.*, **121**, 552–61.

Charnov, E. L. (1976) Optimal foraging: the marginal value theorem. *Theo. Pop. Biol.* **9**, 129–36.

Clark, C. (1974) Possible effects of schooling on the dynamics of exploited fish populations. *J. Cons. perm. int. Explor. Mer*, **36**, 7–14.

Clark, C. (1976) Mathematical bioeconomics: the optimal management of renewable resources. Wiley, New York, USA. 352 pp.

Clark, C. and Mangel, M. (1984) Foraging and flocking strategies: information in an uncertain environment. *Am. Nat.*, **123**, 626–41.

Cowie, R. J. and Krebs, J. R. (1979) Optimal foraging in patchy environments, in *Population Dynamics* (eds R. M. Anderson, B. D. Turner and R. L. Taylor), Blackwell, Oxford, U.K., pp. 183–205.

Crowder, L. B. (1985) Optimal foraging and feeding mode shifts in fishes. *Env. Biol. Fishes*, **12**, 57–62.

Croy, M. (1990) Information across fish shoals about danger. Unpublished report to ASAB, 10 pp.

Csanyi, V. (1985) Ethological analysis of predator avoidance by the paradise fish (*Macropodus opercularis* L.) 1. Recognition and learning of predators. *Behaviour*, **92**, 227–40.

Csirke, J. (1988) Small shoaling pelagic fish stocks, in *Fish Population Dynamics* (ed. J. A. Gulland), Wiley, London, pp. 271–302.

Cui, Y. and Wootton, R. J. (1989) Grouping fails to affect the growth and energy budget of a cyprinid, *Phoxinus phoxinus* (L.). *J. Fish Biol.*, **35**, 795–8.

Cullen, J. M., Shaw, E. and Baldwin, H. (1965) Methods for measuring the three-dimensional structure of fish schools. *Anim. Behav.* **13**, 534–43.

Cushing, D. H. and Harden-Jones, F. R. (1968) Why do fish school? *Nature, Lond.*, **218**, 918–20.

da Silva, J. and Terhune, J. M. (1988) Harbour seal grouping as an anti-predator strategy. *Anim. Behav.*, **36**, 1309–16.

Dawkins, R. and Krebs, J. R. (1979) Arms races between and within species. *Proc. R. Soc.*, **205B**, 489–511.

Dill, L. M. (1983) Adaptive flexibility in the foraging behavior of fishes. *Can. J. Fish. aquat. Sci.*, **40**, 398–408.

Dugatkin, L. A. (1988) Do guppies play tit-for-tat during predator inspection visits? *Behav. Ecol. Sociobiol.*, **23**, 395–9.

Dugatkin, L. A. (1990) N-person games and the evolution of cooperation: a model based on predator inspection in fish. *J. theor. Biol.*, **142**, 123–35.

Dugatkin, L. A. (1991) Predator inspection, tit-for-tat and shoaling: a comment on Masters and Waite. *Anim. Behav.*, **41**, 898–9.

Dugatkin, L. A. and Alfieri, M. (1991a) Tit-for-Tat in guppies (*Poecilia reticulata*): the relative nature of cooperation and defection during predator inspection. *Evol. Ecol.*, **5**, 300–309.

Dugatkin, L. A. and Alfieri, M. (1991b) Guppies and the tit-for-tat strategy: preference based on past interaction. *Behav. Ecol. Sociobiol.*, **28**, 243–6.

Eaton, R. C., Bombardieri, R. A. and Meyer, D. L. (1977) Teleost startle responses. *J. exp. Biol.*, **66**, 65–81.

Eggers, D. M. (1976) Theoretical effect of schooling by planktivorous fish predators on rate of prey consumption. *J. Fish. Res. Bd Can.*, **33**, 1964–71.

Ehrlich, P. R. and Ehrlich, A. H. (1973) Coevolution: heterospecific schooling in Caribbean reef fishes. *Am. Nat.*, **107**, 157–60.

Ferguson, M. M. and Noakes, D. L. G. (1981) Social grouping and genetic variation in the common shiner *Notropis cornutus*. *Env. Biol. Fishes*, **6**, 357–60.

FitzGerald, G. J. and van Havre, N. (1985) Flight, fright and shoaling in sticklebacks (Gasterosteidae). *Biol. Behav.*, **10**, 321–31.

Foster, S. A. (1985a) Group foraging by a coral reef fish: a mechanism for gaining access to defended resources. *Anim. Behav.*, **33**, 782–92.

Foster, S. A. (1985b) Size-dependent territory defense by a damselfish. *Oecologia*, **67**, 499–505.

Foster, W. A. and Treherne, J. E. (1981) Evidence for the dilution effect in the selfish herd from fish predation on a marine insect. *Nature, Lond.*, **293**, 466–7.

Fraser, D. F. and Huntingford, F. A. (1986) Feeding and avoiding predation hazard:

the behavioural response of the prey. *Ethology*, **73**, 56–68.

Freon, P. (1991) Seasonal and interannual variations of mean catch per set in the Senegalese sardine fisheries: fish behaviour or fishing strategy? in *Long-term Variability of Pelagic Fish Populations and their Environment* (eds T. Kawasaki, S. Tanaka, Y. Toba, and A. Taniguchi), Pergamon Press, Oxford, pp. 135–45.

Fretwell, S. D. and Lucas, H. L. (1970) On territorial behaviour and other factors influencing habitat distribution in birds. *Acta Biotheor.*, **19**, 16–36.

George, C. J. W. (1960) Behavioural interaction of the pickerel (*Esox niger* and *Esox americanus*) and the mosquitofish (Gambusia patruclis). PhD thesis, Harvard University.

Gibson, R. N. and Ezzi, I. A. (1992) The relative profitability of particulate and filter feeding in the herring *Clupea harengus* L. *J. Fish Biol.*, **40**, 577–90.

Giles, N. and Huntingford, F. A. (1984) Predation risk and interpopulation variation in anti-predator behaviour in the three spined stickleback. *Anim. Behav.*, **32**, 264–75.

Giraldeau, L.-A. (1984) Group foraging: the skill-pool effect of frequency dependent learning. *Am. Nat.*, **124**, 72–9.

Godin, J.-G. J. (1986a) Antipredator function of shoaling in teleost fishes: a selective review. *Naturaliste can.*, **113**, 241–50.

Godin, J.-G. J. (1986b) Risk of predation and foraging behaviour in shoaling banded killifish (*Fundulus diaphanus*). *Can. J. Zool.*, **64**, 1675–8.

Godin, J.-G. J. (1990) Diet selection under risk of predation, in *Behavioural Mechanisms of Food Selection* (NATO ASI 20) (ed. R. N. Hughes), Springer-Verlag, Berlin, pp. 739–69.

Godin, J.-G. J. and Keenleyside, M. H. A. (1984) Foraging on patchily distributed prey by a cichlid fish: a test of the ideal free distribution theory. *Anim. Bhev.*, **32**, 120–31.

Godin, J.-G. J. and Morgan, M. J. (1985) Predator avoidance and school size in a cyprinodontid fish, the banded killifish (*Fundulus diaphanus* Lesueur). *Behav. Ecol. Sociobiol.*, **16**, 105–10.

Godin, J.-G. J. and Smith, S. A. (1988) A fitness cost of foraging in the guppy. *Nature, Lond.*, **333**, 69–71.

Godin, J.-G. J., Classon, L. J. and Abrahams, M. V. (1988) Group vigilance and shoal size in a small characin fish. *Behaviour*, **104**, 29–40.

Goodey, W. and Liley, N. R. (1985) Grouping fails to influence the escape behaviour of the guppy (*Poecilia reticulata*). *Anim. Behav.*, **33**, 120–31.

Gray, J. A. B. and Denton, E. J. (1991) Fast pressure pulses and communication between fish. *J. mar. biol. Ass. U.K.*, **71**, 83–106.

Groot, P. de (1980) A study of the acquisition of information concerning resources by individuals in small groups of red-billed weaver birds. Unpublished PhD thesis, University of Bristol, U.K.: cited in J. R. Krebs and N. B. Davies (1981) *Introduction to Behavioural Ecology*, Blackwell, Oxford, U.K.

Grover, J. T. and Olla, B. (1983) The role of the rhinoceros auklet *Cerorhinca monocerata* in mixed species feeding assemblages of seabirds in the Strait of Juan de Fuca, Washington. *Auk*, **100**, 979–82.

Guilford, T. and Dawkins, M. (1991) Receiver psychology and the evolution of animal signals. *Anim. Behav.*, **42**, 1–14.

Guthrie, D. M. (1980) *Neuroethology: an Introduction*, Blackwell, Oxford, U.K., 221 pp.

Hager, M. C. and Helfman, G. S. (1991) Safety in numbers: shoal size choice by minnows under predatory threat. *Behav. Ecol. Sociobiol.*, **29**, 271–6.

Hall, M. and Garcia, M. (1992) The association of tunas with floating objects and dolphins in the eastern tropical Pacific: study of repeated sets on the same object. *I.A.T.T.C. Work Doc.* no. 4 (16 pp.)

Hall, M., Lennert, C. and Arenas, P. (1992) The association of tunas with floating objects and dolphins in the eastern tropical Pacific: the purse-seine fishery for tunas in the eastern tropical Pacific. *I.A.T.T.C. Work Doc.* no. 2 (58 pp.).

Hall, S. J., Wardle, C. S. and MacLennan, D. N. (1986) Predator evasion in a fish school: test of a model for the fountain effect. *Mar. Biol.*, **91**, 143–8.

Hallier, J.-P. (1991) Tuna fishing on log-associated schools in the western Indian ocean: an aggregation behavior. *I.P.T.P. Coll. Vol. Work Doc.*, **4**, 325–42.

Hamilton, W. D. (1971) Geometry for the selfish herd. *J. theor. Biol.*, **31**, 295–311.

Hancock, J., Antezana, T. and Hart, P. J. B. (1992) Searching behaviour and catch of horse mackerel *Trachurus murphyi* by industrial purse seiners off south-central Chile. *Fisheries Research* (in press).

Hardin, G. (1968) The tragedy of the commons. *Science*, **162**, 1243–8.

Hart, P. J. B. and Connellan, B. (1984) The cost of prey capture, growth rate, and ration size in pike as functions of prey weight. *J. Fish Biol.*, **25**, 279–91.

Hasler, A. D. (1983) Synthetic chemicals and pheromones in homing salmon, in *Control Processes in Fish Physiology* (eds J. C. Rankin, T. J. Pitcher and R. T. Duggan), Croom Helm, Beckenham, U.K., pp. 103–16.

Healey, M. C. and Prieston, R. (1973) The interrelationships among individuals in a fish school. *Tech. Rep. Fish. Res. Bd Can.*, **389**, 1–15.

Helfman, G. S. (1981) The advantage to fishes of hovering in shade. *Copeia*, **1981**, 392–400.

Helfman, G. S. (1984) School fidelity in fishes: the yellow perch patten. *Anim. Behav.* **32**, 663–72.

Helfman, G. S. and Schultz, E. T. (1984) Social transmission of behavioural traditions in a coral reef fish. *Anim. Behav.*, **32**, 379–84.

Hemmings, C. C. (1966) Olfaction and vision in schooling. *J. exp. Biol.*, **45**, 449–64.

Hiatt, R. W. and Brock, V. E. (1948) On the herding of prey and the schooling of the black skipjack, *Euthynnus yaito* Kishinouye. *Pacif. Sci.*, **2**, 297–8.

Hilborne, R. (1991) Modelling the stability of fish schools: exchange of individual fish between schools of skipjack tuna (*Katsuwonus pelamis*). *Can. J. Fish. aquat. Sci.*, **48**, 1081–91.

Hinde, R. A. and Stevenson-Hinde, J. (1973) *Constraints on Learning*, Academic Press, London, 488 pp.

Hobson, E. S. (1963) Selective Feeding by the gafftopsail pompano *Trachinotus rhodopus* (Gill), in mixed schools of herring and anchovies in the Gulf of California. *Copiea*, **1963**, 595–6.

Hobson, E. S. (1968) Predatory behaviour of some shore fishes in the Gulf of California. *Bur. Sport Fish. Wild. Res. Rep.*, **73**, 1–92.

Howell, D. J. (1979) Flock foraging in nectar-eating bats: advantages to the bats and to the host plants. *Am. Nat.*, **114**, 23–49.

Hunter, J. R. (1966) Procedure for the analysis of schooling behaviour. *J. Fish. Res. Bd Can.*, **23**, 547–62.

Inagaki, T., Sakamoto, W., Aoki, I. and Kuroki, T. (1976) Studies on the schooling behaviour of fish. III. Mutual relationships between speed and form in schooling. *Bull. Jap. Soc. Sci. Fish.* **42**, 629–35.

Inman, A. J. and Krebs, J. R. (1987) Predation and group living. *Trends Ecol. Evol.*, **2**, 31–2.

Itazawa, Y., Matsumoto, T. and Kanda, T. (1978) Group effects on physiological and ecological phenomena in fish I – Group effect on the oxygen consumption of the rainbow trout and the medaka. *Bull. Jap. Soc. scient. Fish.*, **44**, 965–9.

Itzkowitz, M. (1977) Social dynamics of mixed-species groups of Jamaican reef fishes. *Behav. Ecol. Sociobiol.*, **2**, 361–84.

Jakobsson, S. and Järvi, T. (1978) Antipredator behaviour of 2-year hatchery reared

Atlantic salmon *Salmo salar* and a description of the predatory behaviour of burbot *Lota lota.* smolts. *Zool. Revy.*, **38**(3), 57–70.

John, K. R. (1964) Illumination and vision and the schooling behaviour of *Astyanax mexicalis. J. Fish. Res. Bd Can.*, **21**, 1453–73.

Kanda, T. and Itazawa, Y. (1978) Group effect on the growth of medaka. *Bull. Jap. Soc. scient. Fish.*, **44**, 1197–1200.

Keenleyside, M. H. A. (1955) Some aspects of the schooling behaviour of fish. *Behaviour*, **8**, 83–248.

Kennedy, G. J. A. and Pitcher, T. J. (1975) Experiments on homing in shoals of the European minnow, *Phoxinus phoxinus* (L.). *Trans. Am. Fish. Soc.*, **104**, 452–5.

Kenward, R. E. (1978) Hawks and doves: attack success and selection in goshawk flights at wood pigeons. *J. Anim. Ecol.*, **47**, 449–60.

Kils, U. (1986) Verhaltensphysiologie Untersuchungen an pelagischen Schwarmen Schwarmbildung als Strategie zur Orientierung in Umwelt-Gradienten Bedeutung der Schwarmbildung in der Aquakultur. Habilitationsschrift, Institut fuer Meereskunde, Mathematisch–Naturwissenschaftliche Fakultat, Christian-Albrechts-Universitat, Kiel, Germany, 168 pp.

Klimley, A. P. (1985) Schooling in *Sphyrna lewini*, a species with low risk of predation: a non-egalitarian state. *Z. Tierpsychol.*, **70**, 297–319.

Koslow, J. A. (1981) Feeding selectivity in schools of northern anchovy in the southern Californian Bight. *Fishery Bull. Fish Wildl. Serv. U.S.*, **79**, 131–42.

Krebs, J. R. (1973) Social learning and the significance of mixed-species flocks of chickadees (*Parus* spp). *Can J. Zool.*, **51**, 1275–88.

Krebs, J. R. (1974) Colonial nesting and social feeding as strategies for exploiting food resources in the great blue heron, *Ardea herodius. Behaviour*, **51**, 99–134.

Krebs, J. R. (1978) Optimal foraging: decision rules for predators, *Behavioural Ecology*, 1st edn (eds J. R. Krebs and N. B. Davies), Blackwell, Oxford, pp. 23–63.

Krebs, J. R. and Barnard, C. J. (1980) Comments on the functions of flocking in birds, in *Acta XVIII Cong. int. Ornithology, Berlin 1978*, pp. 795–9.

Krebs, J. R. and Davies, N. B. (eds) (1978) *Behavioural Ecology*, 1st edn, Blackwell, Oxford, 494 pp.

Krebs, J. R. and Dawkins, R. (1984) Animal signals: mindreading and manipulation, *Behavioural Ecology*, 2nd edn (eds J. R. Krebs and N. B. Davies), Blackwell, Oxford, pp. 380–402.

Krebs, J. R., MacRoberts, M. and Cullen, J. M. (1972) Flocking and feeding in the great tit *Parus major*: an experimental study. *Ibis*, **114**, 507–30.

Kruuk, H. (1976) The biological function of gulls attraction towards predators. *Anim. Behav.*, **24**, 146–53.

Landeau, L. and Terborgh, J. (1986) Oddity and the 'confusion effect' in predation. *Anim. Behav.*, **34**, 1372–80.

Larkin, P. A. and Walton, A. (1969) Fish school size and migration. *J. Fish. Res. Bd Can.*, **26**, 1372–4.

Lauder, G. V. (1983) Food capture, in *Fish Biomechanics* (eds P. W. Webb and D. Weihs), Praeger, New York, pp. 280–311.

Lazarus, J. (1979) The early warning function of flocking in birds: an experimental study with captive *Quelea. Anim. Behav.*, **27**, 855–65.

Lazarus, J. (1985) Responsiveness: the missing dimension in vigilance studies. Unpublished poster paper: Int. Ethological Conf., Toulouse, 1985.

Lazarus, J. and Metcalfe, N. B. (1990) Tit-for-tat cooperation in sticklebacks: a critique of Milinski. *Anim. Behav.* **39**, 987–8.

Lendrem, D. (1983) Predation risk and vigilance in the blue tit. *Behav. Ecol. Sociobiol.*, **13**, 9–13.

Lester, N. P. (1981) Feeding decisions in goldfish: testing a model of time allocation. Unpublished PhD thesis, Univ. Sussex, UK.

Lester, N. P. (1984) The feed–feed decision: how goldfish solve the patch depletion problem. *Behaviour*, **89**, 175–99.

Lima, S. L. and Dill, L. M. (1990) Behavioural decisions made under risk of predation. *Can. J. Zool.*, **68**, 619–40.

Loekle, D. M., Madison, D. M. and Christian, J. J. (1982) Time dependency and kin recognition of cannibalistic behaviour among poecilid fishes. *Behav. Neurol. Biol.* **35**(3), 315–18.

MacCall, A. D. (1976) Density dependence of the catchability coefficient in the Californian Pacific sardine *Sardinops sagax caerula*, purse seine fishery. *Calif. Coop. Oceanic Fish. Invest. Rep.*, **18**, 136–48.

McFarland, W. N. and Kotchian, M. (1982) Interaction between schools of fish and mysids. *Behav. Ecol. Sociobiol.*, **11**, 71–6.

McKaye, K. R. and Oliver, M. K. (1980) Geometry of a selfish school: defence of cichlid young by a bagrid catfish in Lake Malawi, Africa. *Anim. Behav.*, **28**, 1278–90.

McNicol, R. G. and Noakes, D. L. G. (1984) Environmental influences on territoriality of juvenile brook char *Salvelinus fontinalis* in a stream environment. *Env. Biol. Fishes*, **10**, 29–42.

Magurran, A. E. (1984) Gregarious goldfish. *New Scient.*, 9 Aug. 1984, 32–3.

Magurran, A. E. (1989) The inheritance and development of minnow anti-predator behaviour. *Anim. Behav.*, **39**, 834–42.

Magurran, A. E. (1990) The adaptive significance of schooling as an anti-predator defence in fish. *Annals zool. Fenn.*, **27**, 51–66.

Magurran, A. E. and Bendelow, J. A. (1990) Conflict and cooperation in White Cloud Mountain minnow schools. *J. Fish Biol.*, **37**, 77–83.

Magurran, A. E. and Girling, S. (1986) Predator recognition and response habituation in shoaling minnows. *Anim. Behav.*, **34**, 510–18.

Magurran, A. E. and Higham, A. (1988) Information transfer across fish shoals under predator threat. *Ethology*, **78**, 153–8.

Magurran, A. E. and Pitcher, T. J. (1983) Foraging, timidity and shoal size in minnows and goldfish. *Behav. Ecol. Sociobiol.*, **12**, 142–52.

Magurran, A. E. and Pitcher, T. J. (1987) Provenance, shoal size and the sociobiology of predator evasion behaviour in minnow shoals. *Proc. R. Soc.*, **229B**, 439–65.

Magurran, A. E. and Seghers, B. H. (1990a) Population differences in predator recognition and attack cone avoidance in the guppy. *Anim. Behav.*, **40**, 443–52.

Magurran, A. E. and Seghers, B. H. (1990b) Population differences in the schooling behaviour of newborn guppies, *Poecilia reticulata. Ethology*, **84**, 334–42.

Magurran, A. E., Oulton, W. and Pitcher, T. J. (1985) Vigilant behaviour and shoal size in minnows. *Z. Tierpsychol.*, **67**, 167–78.

Major, P. F. (1978) Predator–prey interactions in two schooling fishes, *Caranx ignobilis* and *Stolephorus purpureus. Anim. Behav.*, **26**, 760–77.

Mangel, M. (1990) Resource divisibility, predation and group formation. *Anim. Behav.*, **39**, 1163–72.

Marsh, A. C. and Ribbink, A. J. (1986) Feeding schools among Lake Malawi cichlid fishes. *Env. Biol. Fishes*, **15**, 75–9.

Masters, W. M. and Waite, T. A. (1990) Tit-for-tat during predator inspection, or shoaling? *Anim. Behav.*, **39**, 603–4.

Matsumoto, W. M., Skillman, R. A. and Dizon, A. E. (1984) Synopsis of Biological Data on Skipjack Tuna, *Katsuwonus pelamis*. NOAA Tech. Rept. NMFS Circ. 451, *FAO Fisheries Synopsis*, **136**, 46–53.

Milinski, M. (1979) An evolutionarily stable feeding strategy in sticklebacks. *Z. Tierpsychol.*, **51**, 36–40.

Milinski, M. (1984) Competitive resource sharing: an experimental test of a learning rule for ESSs. *Anim. Behav.*, **32**, 233–42.

Milinski, M. (1985a) Risk of predation taken by parasitised sticklebacks under competition for food. *Behaviour*, **93**, 203–16.

Milinski, M. (1985b) The patch choice model: no alternative to balancing. *Am. Nat*, **125**, 317–20.

Milinski, M. (1987) Tit-for-tat in sticklebacks and the evolution of cooperation. *Nature, Lond.*, **325**, 433–7.

Milinski, M. (1988) Games fish play: making decisions as a social forager. *Trends Ecol. Evol.* **3**, 325–30.

Milinski, M. (1990a) On cooperation in sticklebacks. *Anim. Behav.*, **40**, 1190–91.

Milinski, M. (1990b) Information overload and food selection, in *Behavioural Mechanisms of Food Selection* (NATO ASI 20) (Ed. R. N. Hughes), Springer-Verlag, Berlin, pp. 721–36.

Milinski, M., Kulling, D. and Kettler, R. (1990a) Tit for tat: sticklebacks trusting a cooperating partner. *Behav. Ecol.*, **1**, 7–11.

Milinski, M., Pfluger, D., Kulling, D. and Kettler, R. (1990b) Do sticklebacks cooperate repeatedly in reciprocal pairs? *Behav. Ecol. Sociobiol.*, **27**, 17–21.

Misund, O. A. (1991) Dynamics of moving masses; variability in packing density, shape and size among pelagic shoals, in Swimming Behaviour of Schools Related to Fish Capture and Acoustic Abundance Estimation, Dr Philos. thesis, Bergen, Norway (132 pp.), pp. 105–32.

Morgan, M. J. (1988a) The effect of hunger, shoal size and the presence of a predator on shoal cohesiveness in bluntnose minnows, *Pimephales notatus* Rafinesque. *J. Fish Biol.*, **32**, 963–71.

Morgan, M. J. (1988b) The effect of hunger, shoal size and predator presence on foraging in bluntnose minnows. *Anim. Behav.*, **36**, 1317–22.

Morgan, M. J. and Colgan, P. W. (1987) The effects of predator presence and shoal size on foraging in bluntnose minnows, *Pimephales notatus*. *Environmental Biology of Fishes*, **20**, 105–11.

Morgan, M. J. and Godin, J.-G. J. (1985) Antipredator benefits of schooling behaviour in a cyprinodontid fish, the banded killifish (*Fundulus diaphanus*). *Z. Tierpsychol.*, **70**, 236–46.

Motta, P. J. (1983) Response by potential prey to coral reef fish predators. *Anim. Behav.*, **31**, 1257–8.

Munz, F. W. and McFarland, W. N. (1973) The significance of spectral position in the rhodopsins of tropical marine fishes. *Vision Res.*, **13**, 1829–74.

Murphy, G. I. (1980) Schooling and the ecology and management of marine fish, in *Fish Behaviour and its Use in Capture and Culture of Fishes* (eds J. E. Bardach, J. J. Magnuson, R. C. May and J. M. Reinhart), ICLARM, Manila, Philippines, pp. 400–414.

Murphy, K. E. and Pitcher, T. J. (1991) Individual behavioural strategies associated with predator inspection in minnow shoals. *Ethology*, **88**, 307–19.

Muzinic, R. (1977) On the shoaling behaviour of sardines (*Sardina pilchardus*) in aquaria. *J. Cons. perm. int. Explor. Mer*, **37**, 147–55.

Neill, S. R. StJ. and Cullen, J. M. (1974) Experiments on whether shooling by their prey affects the hunting behaviour of cephalopod and fish predators. *J. Zool., Lond.*, **172**, 549–69.

Ñiquen, M. (1986) Informe de la compaña de pesca de la M/P 'Rio Damuji' (Enero-15

Febrero 1986). Inf. interno Inst. Mar Perú, Callao: 53 pp. (available from IMARPE, PO Box 22, Callao, Perú).

Norris, K. S. and Dohl, T. P. (1980) The structure and functions of cetacean schools, in *Cetacean Behaviour: Mechanisms and Processes* (ed. L. M. Herman), John Wiley & Sons, NY, pp. 211–68.

Norris, K. S. and Schilt, C. R. (1988) Cooperative societies in three-dimensionsal space: on the origins of aggregations, flocks, and schools with special reference to dolphins and fish. *Ethol. and Sociobiol.*, **9**, 149–79.

Nursall, J. R. (1973) Some behavioural interactions of spottail shiners (*Notropis hudsonius*), yellow perch (*Perca flavescens*) and northern pike (*Esox lucius*). *J. Fish. Res. Bd Can.*, **30**, 1161–78.

Ogden, J. C. and Buckman, N. S. (1973) Movements, foraging groups, and diurnal migrations of the striped parrotfish *Scarus croicensis* Bloch (Scaridae). *Ecology*, **54**, 589–96.

Ohguchi, O. (1981) Prey density and selection against oddity by three-spined sticklebacks. *Adv. Ethol.*, **23**, 1–79.

Olsen, F. C. W. (1964) The survival value of fish schooling. *J. Cons. perm. int. Explor. Mer*, **29**, 115–16.

Parker, F. R. (1973) Reduced metabolic rate in fishes as a result of induced schooling. *Trans. Am. Fish. Soc.*, **102**, 125–31.

Parker, G. A. (1978) Searching for mates, in *Behavioural Ecology*, 1st edn (eds J. R. Krebs and N. B. Davies), Blackwells, Oxford, pp. 214–44.

Parker, G. A. (1984) Evolutionarily stable strategies, in *Behavioural Ecology*, 2nd edn (eds J. R. Krebs and N. B. Davies), Blackwells, Oxford, pp. 30–61.

Parrish, J. K. (1989a) Layering with depth in a heterospecific fish aggregation. *Env. Biol. Fishes*, **26**, 79–86.

Parrish, J. K. (1989b) Re-examining the selfish herd: are central fish safer? *Anim. Behav.*, **38**, 1048–53.

Parrish, J K. (1992) Do predators shape fish schools? interactions between predators and their schooling prey. *Neth J. Zool.* (in press.)

Parrish, J. K. and Kroen, W. K. (1988) Sloughed mucus and drag-reduction in a school of Atlantic silversides, *Menidia menidia. Mar. Biol.*, **97**, 165–9.

Parrish, J. K. Strand, S. W. and Lott, J. L. (1989) Predation on a school of flat-iron herring, *Harengula thrissina. Copiea*, **1989**, 1089–91.

Partridge, B. L. (1982a) Structure and function of fish schools. *Scient. Am.*, **245**, 114–23.

Partridge, B. L. (1982b) Rigid definitions of schooling behaviour are inadequate. *Anim. Behav.*, **30**, 298–9.

Partridge, B. L. and Pitcher, T. J. (1979) Evidence against a hydrodynamic function of fish schools. *Nature, Lond.*, **279**, 418–19.

Partridge, B. L. and Pitcher, T. J. (1980) The sensory basis of fish schools: relative roles of lateral line and vision. *J. comp. Physiol.* **135A**, 315–25.

Partridge, B. L., Johansson, J. and Kalish, J. (1983) The structure of schools of giant bluefin tuna in Cape Cod Bay. *Env. Biol. Fishes*, **9**, 253–62.

Partridge, B. L., Pitcher, T. J., Cullen, J. M. and Wilson, J. P. F. (1980) The three-dimensional structure of fish schools. *Behav. Ecol. Sociobiol.*, **6**, 277–88.

Pitcher, T. J. (1973) The three-dimensional structure of schools in the minnow, *Phoxinus phoxinus* (L.). *Anim. Behav.*, **21**, 673–86.

Pitcher, T. J. (1975) A periscopic method for determining the three dimensional positions of fish in schools. *J. Fish. Res. Bd Can.*, **32**, 1533–8.

Pitcher, T. J. (1979a) The role of schooling in fish capture. *Int. Comm. Explor. Sea CM* 1979/B:5: 1–12.

Pitcher, T. J. (1979b) Sensory information and the organisation of behaviour in a shoaling cyprinid. *Anim. Behav.* **27**, 126–49.

Pitcher, T. J. (1980) Some ecological consequences of fish school volumes. *Freshwat. Biol.*, **10**, 539–44.

Pitcher, T. J. (1983) Heuristic definitions of shoaling behaviour. *Anim. Behav.*, **31**, 611–13.

Pitcher, T. J. (1992) Who dares wins: the function and evolution of predator inspection behaviour in fish shoals. *Neth. J. Zool.* (in press).

Pitcher, T. J. and Hart, P. J. B. (1982) *Fisheries Ecology*, Croom Helm, London, 414 pp.

Pitcher, T. J. and House, A. C. (1987) Foraging rules for group feeders: area copying depends upon food density in shoaling goldfish. *Ethology*, **76**, 161–7.

Pitcher, T. J. and Lawrence, J. E. T. (1985) A simple stereo television system with application to the measurement of coordinates of fish in schools. *Behav. Res. Meth. Instrum. Computers*, **16**, 495–501.

Pitcher, T. J. and Magurran, A. E. (1983) Shoal size, patch profitability and information exchange in foraging goldfish. *Anim. Behav.*, **31**, 546–55.

Pitcher, T. J. and Partridge, B. L. (1979) Fish school density and volume. *Mar. Biol.*, **54**, 383–94.

Pitcher, T. J. and Turner, J. R. (1986) Danger at dawn: experimental support for the twilight hypothesis in shoaling minnows. *J. Fish Biol.*, **29** (Supp. A), 59–70.

Pitcher, T. J. and Wyche, C. J. (1983) Predator avoidance behaviour of sand-eel schools: why schools seldom split? in *Predators and Prey in Fishes* (eds D. L. G. Noakes, B. G. Lindquist, G. S. Helfman and J. A. Ward), Junk, The Hague, pp. 193–204.

Pitcher, T. J., Green, D. and Magurran, A. E. (1986b) Dicing with death: predator inspection behaviour in minnow shoals. *J. Fish Biol.*, **28**, 439–48.

Pitcher, T. J., Lang, S. H. and Turner, J. R. (1988) A risk-balancing tradeoff between foraging rewards and predation hazard in a shoaling fish. *Behav. Ecol. Sociobiol.*, **22**, 225–8.

Pitcher, T. J., Magurran, A. E. and Allan, J. R. (1983) Shifts of behaviour with shoal size in cyprinids. *Proc. Br. Freshwat. Fish. Conf.*, **3**, 220–28.

Pitcher, T. J., Magurran, A. E. and Allan, J. R. (1986a) Size segregative behaviour in minnow shoals. *J. Fish Biol.*, **29** (Supp. A), 83–96.

Pitcher, T. J., Magurran, A. E. and Edwards, J. I. (1985) Schooling mackerel and herring choose neighbours of similar size. *Mar. Biol.*, **86**, 319–22.

Pitcher, T. J., Magurran, A. E. and Winfield, I. (1982a) Fish in larger shoals find food faster. *Behav. Ecol. Sociobiol.*, **10**, 149–51.

Pitcher, T. J., Wyche, C. J. and Magurran, A. E. (1982b) Evidence for position preferences in schooling mackerel. *Anim. Behav.*, **30**, 932–4.

Poole, T. B. and Dunstone, N. (1975) Underwater predatory behaviour of the American mink *Mustela vison. J. Zool., Lond.*, **178**, 395–412.

Potts, G. W. (1970) The schooling ethology of *Lutjianus monostigma* in the shallow reef environment of Aldabra. *J. Zool., Lond.*, **161**, 223–35.

Potts, G. W. (1981) Behavioural interactions between the Carangidae (Pisces) and their prey on the fore-reef slope of Aldabra, with notes on other predators. *J. Zool., Lond.*, **195**, 385–404.

Potts, G. W. (1983) The predatory tactics of *Caranx melapygus* and the response of its prey, in *Predators and Prey in Fishes* (eds D. L. G. Noakes, B. G. Lindquist, G. S. Helfman and J. A. Ward), Junk, The Hague, pp. 181–91.

Potts, W. K. (1984) The chorus line hypothesis of manoeuver coordination in avian flocks. *Nature Lond.*, **309**, 344–5.

Poulin, R. (1991) Group-living and the richness of the parasite fauna in Canadian freshwater fishes. *Oecologia*, **86**, 390–4.

Poulin, R. and FitzGerald, G. J. (1989a) Shoaling as an anti-ectoparasite mechanism in juvenile sticklebacks. *Behav. Ecol. Sociobiol.*, **24**, 251–5.

Poulin, R. and FitzGerald, G. J. (1989b) Risk of parasitism and microhabitat selection in juvenile sticklebacks. *Can. J. Zool.*, **67**, 14–18.

Powell, G. V. N. (1974) Experimental analysis of the social value of flocking by starlings (*Sturnus vulgaris*) in relation to predation and foraging. *Anim. Behav.*, **23**, 504–8.

Pulliam, H. R. and Caraco, T. (1984) Living in groups: is there an optimal group size? *Behavioural Ecology* 2nd edn (eds J. R. Krebs and N. B. Davies), Blackwells, Oxford, pp. 122–47.

Quinn, T. P. and Busack, C. A. (1985) Chemosensory recognition of siblings in juvenile coho salmon *Oncorhynchus kisutch*. *Anim. Behav.*, **33**, 51–6.

Quinn, T. P. and Fresh, K. (1984) Homing and straying in chinook salmon *Oncorhynchus tschawytscha* from Cowlitz River hatchery, Washington. *Can. J. Fish. aquat. Sci.*, **41**, 1078–82.

Radakov, D. V. (1973) *Schooling in the Ecology of Fish*. Israel Programme for Scientific Translations, Wiley, New York, 173 pp.

Ranta, E., Lindstrom, K. and Peuhkuri, N. (1992) Size matters when three-spined sticklebacks go to school. *Anim. Behav.*, **43**, 160–62.

Reinthal, P. N. and Lewis, S. M. (1986) Social behaviour, foraging efficiency and habitat utilization in a group of tropical herbivorous fish. *Anim. Behav.*, **34**, 1687–1693.

Robertson, D. R., Sweatman, H. P. A., Fletcher, G. A. and Cleland, M. G. (1976) Schooling as a means of circumventing the territoriality of competitors. *Ecology*, **57**, 1208–20.

Robinson, C. M. (1991) Schooling behaviour and hunger, unpublished PhD thesis, University of Wales. (175 pp.)

Robinson, C. M. and Pitcher, T. J. (1989a) The influence of hunger and ration level on shoal density, polarisation and swimming speed of herring, *Clupea harengus* L. *J. Fish Biol.*, **34**, 631–3.

Robinson, C. M. and Pitcher, T. J. (1989b) Hunger as a promoter of different behaviours within a shoal of herring: selection for homogeneity in a fish shoal. *J. Fish Biol.*, **35**, 459–60.

Rogers, C., Roden, C., Lohoefener, R., Mullin, K. and Hoggard, W. (1990) Behavior, distribution, and relative abundance of cownose ray schools *Rhinoptera bonasus* in the Northern Gulf of Mexico. *NE Gulf Sci.*, **11**, 69–76.

Rubenstein, D. I., Barnett, R. J. Ridgely, R. S. and Klopfer, P. H. (1977) Adaptive advantages of mixed species feeding flocks among seed-eating finches in Costa Rica. *Ibis*, **119**, 10–21.

Ruppell, G. and Gosswein, E. (1972) Die Schwarme von *Leucaspius delineatus* (Cyprinidae, Teleostei) bei Gefahr im Hellen und im Dunkeln. *Z. vergl. Physiol.*, **76**, 333–40.

Safina, C. (1990) Bluefish mediation of foraging competition between roseate and common terns. *Ecology*, **71**, 1804–9.

Safina, C. and Burger, J. (1989) Population interactions among free-living bluefish and prey fish in an ocean environment. *Oecologia*, **79**, 91–5.

Savino, J. F. and Stein, R. A. (1982) Predator–prey interaction between largemouth bass and bluegills as influenced by simulated submersed vegetation. *Trans. Am. Fish. Soc.*, **11**, 255–66.

Schmitt, R. J. and Strand, S. W. (1982) Cooperative foraging by yellowtail, *Seriola lalandei* (Carangidae), on two species of fish prey. *Copeia*, **1982**, 714–17.

Seghers, B. H. (1974) Schooling behaviour in the guppy (*Poecilia reticulata*): an evolutionary response to predation. *Evolution*, **28**, 486–9.

Seghers, B. H. (1981) Facultative shooling behaviour in the spottail shiner (*Notropis hudsonius*): possible costs and benefits. *Env. Biol. Fishes*, **6**, 21–4.

Sette, O. E. (1950) Biology of the Atlantic mackerel (*Scomber scombrus*) of North America. 2. Migration and habits. *Fish. Bull. Nat. Mar. Fish. Serv. U.S.*, **51**, 251–358.

Shaw, E. (1962) The schooling of fishes. *Scient. Am.*, **206**, 128–38.

Shaw, E. (1965) The optomotor response and the schooling of fishes. *Int. Comm. Northwest Atl. Fish. Spec. Publ.*, **6**, 753–79.

Shaw, E. (1978) Schooling fishes. *Am. Scient.*, **66**, 166–75.

Shaw, E. and Tucker, A. (1965) The optomotor reaction of schooling carangid fishes. *Anim. Behav.*, **13**, 330–66.

Sibly, R. M. (1983) Optimal group size is unstable. *Anim. Behav.*, **31**, 947–8.

Sih, A. (1980) Optimal foraging: can foragers balance two conflicting demands. *Science*, **210**, 1041–3.

Smythe, N. (1977) The function of mammalian alarm advertising: social signals or pursuit invitation? *Am. Nat.*, **111**, 191–4.

Springer, S. (1957) Some observations on the behaviour of schools of fishes in the Gulf of Mexico and adjacent waters. *Ecology*, **38**, 166–71.

Stephens, D. W. and Krebs, J. R. (1986) *Foraging Theory*, Princeton Univ. Press, Princeton, NJ, 247 pp.

Street, N. G. and Hart, P. J. B. (1985) Group size and patch location by the stoneloach, *Noemacheilus barbatulus*, a non-visually foraging predator. *J. Fish Biol.*, **217**, 785–92.

Street, N. G., Magurran, A. E. and Pitcher, T. J. (1984) The effects of increasing shoal size on handling time in goldfish *Crassius auratus*. *J. Fish Biol.*, **25**, 561–6.

Sutherland, W. S. (1983) Aggregation and the 'ideal free distribution'. *J. Anim. Ecol.*, **52**, 821–8.

Taylor, R. J. (1984) *Predation*, Chapman and Hall, London, 166 pp.

Tinbergen, L. (1981) Foraging decisions in starlings. *Ardea*, **69**, 1–67.

Treherne, J. E. and Foster, W. A. (1981) Group transmission of predator avoidance in a marine insect: the Trafalgar effect. *Anim. Behav.*, **29**, 911–17.

Tremblay, D. and FitzGerald, G. J. (1979) Social organisation as an antipredator strategy in fish. *Naturaliste can.*, **105**, 411–13.

Treisman, M. (1975) Predation and the evolution of gregariousness. I. Models of concealment and evasion. *Anim. Behav.*, **23**, 779–800.

Turner, G. F. and Pitcher, T. J. (1986) Attack abatement: a model for group protection by combined avoidance and dilution. *Am. Nat.*, **128**, 228–40.

Uematsu, T. and Takamori, J. (1976) Social facilitation in feeding behaviour of the himedaka *Oryzias latipes*. I. Continuous observation during a short period. *Jap. J. Ecol.*, **26**, 135–40.

Van Havre, N. and FitzGerald, G. J. (1988) Shoaling and kin recognition in the threespine stickleback (*Gasterosteus aculeatus* L.). *Biol. Behav.*, **13**, 190–201.

Videler, J. J. and Wardle, C. S. (1991) Fish swimming stride by stride: speed limits and endurance. *Rev. Fish Biol. Fish.*, **1**, 23–40.

Vine, I. (1971) Risk of visual detection and pursuit by a predator and the selective advantage of flocking behaviour. *J. theor. Biol.*, **30**, 405–22.

Vine, P. J. (1974) Effects of algal grazing and aggressive behavior of the fishes *Pomacentrus lividus* and *Acanthurus sohal* on coral reef ecology. *Mar. Biol.*, **24**, 131–46.

Ward, P. and Zahavi, A. (1973) The importance of certain assemblages of birds as information centres for food finding. *Ibis*, **115**, 517–34.

Webb, P. W. (1976) The effect of size on the fast-start performance of rainbow trout and a consideration of piscivorous predator–prey interactions. *J. exp. Biol.*, **65**, 157–77.

Webb, P. W. (1980) Does schooling reduce fast start response latencies in teleosts? *Comp. Biochem. Physiol.*, **65A**, 231–34.

Weihs, D. (1973) Hydromechanics and fish schooling. *Nature, Lond.*, **241**, 290–91.

Weihs, D. (1975) Some hydrodynamical aspects of fish schooling, in *Symposium on Swimming and Flying in Nature* (eds T. Y. Wu, C. J. Broklaw and C. Brennan), Plenum Press, New York, pp. 703–18.

Weihs, D. and Webb, P. W. (1983) Optimisation of locomotion, in *Fish Biomechanics* (eds P. W. Webb and D. Weihs), Preager Press, New York, pp. 339–71.

Welch, C. and Colgan, P. (1990) The effects of contrast and position on habituation to models of predators in eastern banded killifish (*Fundulus diaphanus*). *Behav. Process*, **22**, 61–71.

Welty, J. L. C. (1934) Experiments on group behaviour of fishes. *Physiol. Zool.*, **B7**, 85–128.

Wijffels, W., Thines, G. Dijkgraaf, S. and Verheijen, F. J. (1967) Apprentissage d'un labyrinthe simple par des Teleosteans isoles ou groupes de l'espece *Barbus tincto*. *Arch. neerl. Zool.*, **3**, 376–402.

Williams, G. C. (1964) Measurements of consociation among fishes and comments on the evolution of schooling. *Publs Mich. St. Univ. Mus. biol. Ser.*, **2**, 351–83.

Wilson, E. O. (1975) *Sociobiology: the New Synthesis*, Harvard Univ. Press, Cambridge, MA, 697 pp.

Wilson, R. P., Ryan, P. G., James, A. and Wilson, M. P.-T. (1987) Conspicuous coloration may enhance prey capture in some piscivores. *Anim. Behav.*, **35**, 1558–60.

Wolf, N. G. (1983) The behavioural ecology of herbivorous fishes in mixed species groups, unpublished PhD thesis, Cornell University, Ithaca, New York. (144 pp.)

Wolf, N. G. (1985) Odd fish abandon mixed-species groups when threatened. *Behav. Ecol. Sociobiol.*, **17**, 47–52.

Wolf, N. G. (1987) Schooling tendency and foraging benefit in the ocean surgeonfish. *Behav. Ecol. Sociobiol.*, **21**, 59–63.

Wursig, B. and Wursig, M. (1980) Behavior and ecology of the dusky dolphin. *Fish. Bull.*, **77**, 871–90.

Yates, G. T. (1983) Hydrodynamics of body and caudal fin propulsion, *Fish Biomechanics* (eds P. W. Webb and D. Weihs), Praeger Press, New York, pp. 177–213.

Ydenberg, R. C. and Dill, L. M. (1986) The economics of fleeing from predators. *Adv. Study Behav.*, **16**, 229–49.

Yuen, H. S. H. (1962) Schooling behaviour within aggregations composed of yellowfin and skipjack Tuna. *F.A.O. Fish. Rep.*, **6**(3), 1419–29.

Zuyev, G. V. and Belyayav, V. V. (1970) An experimental study of the swimming of fish in groups as exemplified by the horse mackerel (*Trachurus mediterraneus ponticus* Aleev). *J. Ichthyol.*, **10**, 545–9.

Chapter thirteen

Individual differences and alternative behaviours

Anne E. Magurran

"No one supposes that all individuals of the same species are cast in the very same mould" (Darwin, 1859)

13.1 INTRODUCTION

Although Darwin (1859) clearly established the importance of individual differences, it is only in recent years that behavioural ecologists have begun to explore the evolutionary significance of intraspecific variation in behaviour (e.g. Lott, 1984). Previously, individual variation was treated as white noise, or viewed as a maladaptive deviation from an optimal strategy (discussion, Ringler, 1983; Arak, 1984). Many classical ethology textbooks proceeded with the implicit assumption that individual differences in behaviour are less important and less interesting than differences amongst species. Fish behaviour, where most students began by looking at the stereotyped response of the breeding male stickleback, *Gasterosteus aculeatus*, to red objects (Tinbergen, 1951) was no exception. Despite first impressions, rigid behaviour patterns are of course not the rule. This chapter will show that there can be considerable variation between (and even within) individual fish in a whole host of behaviours, including the methods they use to find food, avoid predators or secure a mating. Ringler (1983), for example, lists 19 studies documenting individual variation in foraging tactics. There are even individual differences in the behaviour of members of fish schools (Pitcher *et al.*, 1982; Helfman, 1984; and see Chapter 12 by Pitcher and Parrish, this volume), long considered the most egalitarian and uniform of piscine societies (Shaw, 1970), as well as in the way that male sticklebacks attack red dummies (Rowland, 1982).

Behaviour of Teleost Fishes 2nd edn. Edited by Tony J. Pitcher. Published in 1993 by Chapman & Hall. ISBN 0 412 42930 6 (HB) and 0 412 42940 3 (PB).

Why are individual differences so important? Darwin (1859) identified variation amongst individuals as the raw material for evolution. Although Darwin was primarily concerned with morphological variation, his arguments apply equally to physiological and behavioural variation; indeed, behaviour, physiology and morphology are often linked and may have profound influences on one another. Contemporary evolutionary biology, which stresses the importance of the individual (and ultimately the gene) rather than the group or species as the unit of selection (Dawkins, 1989), has reawakened an interest in individual differences. The recent emphasis on behavioural trade-offs (Lima and Dill, 1990), where individuals must reconcile conflicting demands, has promoted the study of decision-making and flexible behaviour. In addition, game theory and associated concepts such as the evolutionarily stable strategy or ESS (Maynard Smith, 1982; Parker, 1984) provide a framework within which the costs and benefits of alternative behaviours can be assessed.

The aim of this chapter is to explore the origins and consequences of individual differences in teleost behaviour and to discuss their evolutionary significance. It will begin with a brief review of the mechanisms proposed to account for individual variation in behaviour. The next section will examine behavioural flexibility within and amongst individuals. Many of the examples are drawn from studies of foraging, mating and antipredator behaviour since these are areas where individual differences have been particularly well documented. After considering the genetic and developmental basis of individual differences, the chapter will conclude by identifying ways in which behavioural variation can influence ecological and evolutionary processes such as population dynamics and speciation.

13.2   WHY DO INDIVIDUAL DIFFERENCES ARISE?

There are many ways in which the behaviour of individual animals can vary and many reasons why such variation occurs. Individual differences in behaviour may, for instance, reflect sex or body size or be a response to food availability, predation pressure, parasitism or competition from conspecifics. These differences can be genetically determined, or result from development, environment or experience. Broadly speaking, however, individual differences in behaviour can be viewed as a product of three mechanisms (Partridge and Green, 1985). These are a variable environment, phenotypic differences and the behaviour of other individuals.

A variable environment. Behavioural flexibility is clearly appropriate in an environment that varies in space or time. For example fish must adjust their behaviour to take account of changes in food availability, predator pressure or mating opportunities.

Phenotypic differences. An animal's phenotype may constrain its behaviour. This can operate in one of two ways. First, different sizes, sexes or morphs can have different requirements in terms of food, cover etc. For instance three morphs of brown trout, *Salmo trutta*, occur sympatrically in Lough Melvin, a small (22 km^2) lake in north-western Ireland. The three types of trout look distinct (each has its own local name) and have different foraging techniques: the sonaghen feeds in open water, the gillaroo feeds on the benthos, while the ferox enjoys a predatory lifestyle and specializes on eating other fish (Ferguson and Mason, 1981; Ferguson, 1986, 1989). Second, if an animal is competitively inferior it may be forced to modify its behaviour in order to 'make the best of a bad job' (Maynard Smith, 1982; Partridge, 1988). This situation, where less successful individuals have reduced access to a resource or lower mating success, contrasts with the last category, where the 'best' behaviour depends on what others are doing.

Behaviour of other individuals. Individuals can employ different strategies to achieve the same end. It appears, for example, that territory holders and sneak males in the bluegill sunfish, *Lepomis macrochirus*, enjoy, on average, approximately equal mating success (Gross, 1982). Despite the fact that short-term benefits may be unequal, different behaviour patterns can be selected if they bring equivalent lifetime success (Dunbar, 1982). Since the pay-off of a given behaviour pattern will be affected by what other individuals are doing (for example, the benefits of being a sneaky male diminish if too many other males adopt the same policy), the proportions of the various strategies in a population will be maintained through frequency dependent selection (Gross, 1984, 1985).

Additionally, the behaviour of other individuals may influence the short-term costs and benefits of particular activities. For instance, an individual's pay-off from schooling in terms of protection against predators will tend to increase with the number of participants. Similarly, Whoriskey and FitzGerald (1985) showed that cannibalistic attacks on male territories by female threespine sticklebacks were only likely to be successful if 20 or more females combined to form a raiding school.

These three mechanisms do not necessarily operate independently since an individual's phenotype will determine its reaction to a variable environment as well as its response to the behaviour of others. Thus, an individual fish's size will influence the outcome of contests with conspecifics, its vulnerability to predators as well as its use of the available habitat.

Austad (1984) suggested that alternative behaviours could be subdivided according to whether they have equal or unequal pay-offs, are reversible or irreversible, and are genetically or environmentally determined. Unfortunately, as Caro and Bateson (1986) point out, it is often impossible to slot behaviours neatly into this classification. Lifetime reproductive success is difficult to measure, it can be problematic to prove that behaviours are irreversible and, as a later section will show, genetic factors often interact with environmental ones.

So far this chapter has not distinguished changes in an individual's behaviour from variation amongst individuals. For each species the overall tapestry of behavioural variation can be unravelled to reveal behavioural flexibility at the level of the individual as well as suites of behaviour that distinguish different individuals. It is now time to look explicitly at these scales of variation.

13.3   DIFFERENCES WITHIN AN INDIVIDUAL

Behavioural flexibility and decision making

It is clear that animals need to perform a variety of behaviours in order to survive and ultimately reproduce. For instance foraging and predator avoidance are activities that consume considerable portions of time during the life of a typical fish. It is also obvious that the benefits of each behaviour cannot be viewed independently but must be weighted against the relative merits and demands of other activities. Consequently, hiding at the slightest hint of danger might be a superb antipredator tactic but it severely limits the time available for finding mates or food. It is essential therefore that individuals are able to accurately assess current conditions and modify their behaviour accordingly. Excellent reviews of behavioural decision making, particularly in the context of predation risk, have recently been produced by Dill (1987) and by Lima and Dill (1990).

Most investigations of behavioural trade-offs to date have concentrated on the conflicting demands of foraging and predator avoidance, and it is now widely accepted that fish are adept at reaching a compromise between the two activities. For instance, Metcalfe *et al.* (1987) showed that juvenile Atlantic salmon, *Salmo salar*, reduce their foraging for up to 2 h after a brief sighting of a predator (brown trout, *S. trutta*) model. Less effort has been devoted to exploring the even more fascinating dilemma faced by animals that must trade off courtship and reproduction against predator avoidance (Magnhagen, 1991). Courting males often develop bright coloration and adopt eye-catching displays that seem as likely to attract a potential predator as the intended mate. One example of a mating/risk trade-off is however provided by male guppies, *Poecilia reticulata*, which switch from conspicuous displays to sneaky mating attempts in the presence of a predator (Endler, 1987; Magurran and Seghers, 1990a; Magurran and Nowak, 1991; p. 453).

Short-term changes in foraging or predator avoidance tactics may be linked to motivational state (see Chapter 2 by Colgan, this volume). Hungry sticklebacks, for instance, are less selective in their choice of prey (Beukema, 1968), and, like hungry guppies (Godin and Smith, 1988), are more vulnerable to capture by a predator (Milinski, 1984a). Nest-guarding male sticklebacks are reluctant to desert if threatened by a pike, *Esox lucius* (Huntingford, 1976). Learning and experience are important in bringing about longer-term changes (Giraldeau, 1984; Dill, 1983) and experienced fish search more and with greater efficiency (Beukema, 1968; Atema *et al.*, 1980; Potts, 1980), attack more speedily (Godin, 1978) and are most successful in capturing prey (Webb and Skadsen, 1980; Vinyard, 1982).

The degree to which individuals are flexible in their behaviour can vary across populations and species. This flexibility (or lack of it) will often be linked to morphology, predation regime or social structure. Reist (1983) found that brook sticklebacks, *Culaea inconstans*, in Canada differ with respect to their pelvic skeleton and associated spines, varying from forms where it is complete through a range of intermediates to forms where it is totally absent. Pike prefer the large, spineless sticklebacks, and Reist observed that these fish are most likely to spend time in cover to avoid initial detection and to retreat if discovered. Reist scored a number of behaviours and classified them either as adaptive, that is reducing risk of detection and increasing chance of evasion, or non-adaptive, defined as those behaviours that do not reduce a fish's chance of being caught. Sticklebacks without spines are more likely to show adaptive behaviour, and individual differences in behaviour are least prominent in these vulnerable fish. Savino and Stein (1989) showed that bluegill sunfish modified their behaviour to a greater extent than fathead minnows, *Pimephales promelas*, when exposed to predators (largemouth bass, *Micropterus salmoides*, and pike). Fathead minnows tend to occur in habitats with no predators. Magurran and Pitcher (1983) noted that individual differences are more pronounced in goldfish, *Carassius auratus*, than in European minnows, *Phoxinus phoxinus*. Minnows have the higher schooling tendency. Since schooling is only an efficient antipredator device when all individuals are of similar appearance and behave in a coordinated fashion, one consequence of selection for schooling is the reduction of variable behaviour, particularly the opportunities of competing for limited resources (Magurran and Seghers, 1991).

Phenotypic changes

It is inevitable that pronounced behavioural changes will accompany morphological and physiological development. Although it is beyond the scope of this chapter to review behavioural ontogeny (or age-related tactics: Caro and Bateson, 1986) in detail (but see Chapter 3 by Huntingford for a fuller discussion), the following examples bear out the behavioural consequences of phenotypic changes.

French grunts, *Haemulon flavolineatum*, change colour, habitat preference, diel foraging pattern and twilight migratory behaviour as they develop (Helfman *et al.*, 1982). When they initially recruit from the plankton as post-larvae, the grunts resemble mysids (*Mysidium* spp.) in appearance. At this stage they school with the mysids as well as exploiting small mysids as food (McFarland and Kotchian, 1982). Migratory behaviour develops in the pre-juvenile stage when the fish become pigmented, acquire a lateral melanistic strip, and reach 10–15 mm in length (Helfman *et al.*, 1982). As the fish grow they also switch from diurnal to nocturnal foraging and form larger schools. This change in foraging appears to be linked to development of the visual system (Helfman *et al.*, 1982). Younger schools are located near sea urchins in the sand halo around a feeding reef. In the larger, nocturnally feeding fish, daytime schools are formed over coral heads.

The idea of ontogenetic niche shifts has been addressed formally by the work of Werner and Gilliam (1984) who explored behavioural changes associated with growth in bluegill sunfish in North America (see also Osenberg *et al.*, 1988). These fish have three distinct phases of development. When the larvae first leave their nests they are small, relatively inconspicuous to fish predators and consequently able to forage on zooplankton in open water. As they grow larger the juvenile bluegills become a target for fish predators such as the largemouth bass, *Micropterus salmoides*, and take refuge in vegetated areas, even when the open water is a more profitable feeding zone (Werner *et al.* 1983a, b). When the bluegills finally reach the size at which predators cease to be a serious threat they can again move freely between habitats in response to food availability. In addition to superior strength, large fish may also enjoy improved powers of food detection. Breck and Gitter (1983), working on bluegill sunfish, were able to show that an increase in fish size is accompanied by an increase in lens diameter and lens acuity, which in turn are associated with an increase in reactive distance, the distance at which fish respond to prey. Since larger bluegills are also able to swim faster while searching for food, they encounter more prey than smaller fish (Mittlebach, 1981). These benefits are enhanced by an exponential decrease in handling time (above a critical ratio of prey length/fish length) with increasing fish size.

At the other end of the spectrum, longevity and life expectancy may influence behaviour. Magnhagen (1990) examined the reproductive behaviour of the black goby, *Gobius niger*, in the presence and absence of a cod, *Gadus morhua*. The maximum life span of these fish is about 5 years. Young (2–3-year-old) male gobies significantly reduced their nest building and spawning activity when exposed to the predator. Only one out of the seven pairs tested was willing to spawn in such circumstances. By contrast 86% of 4–5-year-old males ignored the predator and continued to spawn as normal. Part of this behavioural change may be due to the fact that older fish are larger and consequently less vulnerable to predation. Nevertheless, it seems likely that a diminishing expectation of reproduction is the main reason why older fish disregard danger.

Clark and Ehlinger (1987) distinguish behavioural flexibility from developmental plasticity. Developmental plasticity reflects the potential of an individual to develop into one out of a range of possible phenotypes given the appropriate experience or environmental conditions during ontogeny. Behavioural flexibility, on the other hand, is a measure of the range of tactics available to an individual at any one point during its life.

Response to conspecifics

Schooling

Individual schooling tendency can vary as the costs and benefits of belonging to a school change. The proximity of a predator almost always leads fish (of a schooling species) to congregate (Magurran 1990a). For instance, Magurran and Pitcher (1987) found that European minnows, *Phoxinus phoxinus*, switched from being solitary or part of a small, loosely organized shoal to forming a large, polarized school in the presence of a pike. More individuals participate in inspection behaviour and form larger inspecting groups when the risk of predation increases, for instance when a predator advances (Fig. 13.1). School membership is undoubtedly advantageous in such circumstances. Neill and Cullen (1974) found that attack success per strike diminished for both cephalopod and fish predators as prey school size increased from one to 20. Due to the selfish herd effect the benefits of schooling are not equal for all participants. Hamilton (1971) pointed out that individuals positioning themselves close to other members of a group are less at risk of being captured by a randomly striking predator. Morgan and Godin (1985) confirmed that straggling killifish, *Fundulus diaphanus*, are indeed more likely to be attacked by a predator (in this case the white perch, *Morone americana*). However, in a recent re-examination of the selfish herd concept, Parrish (1989) found, contrary to expectation, that central members of schools of 25 Atlantic silversides, *Menidia menidia*, suffered more attacks from black seabass, *Centropristis striata*, than peripheral fish.

It can equally be argued that the size of school that an individual joins influences its behaviour. Thus Magurran and Pitcher (1987) showed that minnows in groups of 10 are more likely to abandon a school and seek cover if attacked than minnows with 50 or even 20 colleagues in the vicinity. Even when a predator is absent the behaviour of individual fish is contingent on group size. This effect has been demonstrated for two species of cyprinid, European minnows and goldfish. In small shoals, fish of both species are timid, do little foraging and spend time hiding in weed beds; as shoal size and

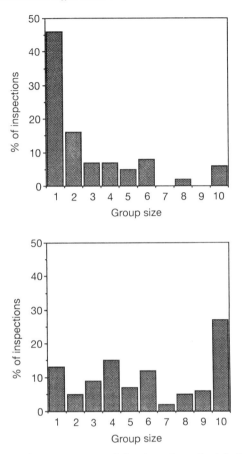

Fig. 13.1 Group size during inspection behaviour. In each trial of this experiment 10 European minnows (from a population sympatric with pike) were allowed to feed on an artificial food patch. Without warning, a 21 cm (realistically painted) model pike would emerge from its hide and 'stalk' towards the feeding patch. Minnows began 'inspecting' the pike as soon as they detected it. During an inspection one or more minnows approached the model before turning and swimming away from it. In the first third of the stalk (top), nearly half of the total 130 inspections observed (over 20 trials) were by single minnows. By contrast, in the final third of the stalk (bottom) almost 30% of the 188 inspections consisted of groups of 10 fish. The figure is based on data collected for the experiment described by Magurran (1986).

corporate vigilance increase, the same individuals become less timid and devote progressively more time to foraging (Magurran and Pitcher, 1983).

There are circumstances where it pays an individual to remain solitary. Savino and Stein (1982) have shown that bluegill sunfish under predator threat are less likely to school if they are in dense vegetation. In this case the advantage of being able to hide alone in thick cover outweighs the benefits of belonging to a more detectable school.

Sex changes

One of the most dramatic behavioural shifts of all arises when an individual changes sex (Krebs and Davies, 1987; see Shapiro, 1979, for a review of sex changes in fish). Sequential hermaphroditism occurs widely (it is found in 13 families) and provides a mechanism for individuals to adjust their reproductive output to the social structure of the group they belong to (Wootton, 1990). One species where the phenomenon has been particularly well documented is the blue-headed wrasse, *Thalassoma bifasciatum*. These wrasse are found on coral reefs where large (terminal phase) males, which are brightly coloured, defend mating sites (Warner, 1988). Mating activity is concentrated on these sites for short periods each afternoon, and terminal-phase males enjoy up to 40 times as many matings per day as smaller initial-phase males. It is often advantageous for individuals to commence life as females, only switching gender when they are big enough to defend a mating site. This switch is socially controlled since removal of large males is sufficient to initiate a sex change in the largest females (Warner *et al.*, 1975). Primary (initial phase) males tend to be found in large populations where there is a greater chance of achieving a sneak mating.

The alternative change, from male to female, is less common and likely to occur only where there is a premium on female size. The anemone fish, *Amphiprion akallopsis*, provides one example. These fish live as pairs on sea anemones. Since their reproductive success depends on the ability of females to produce eggs it is advantageous for the larger fish to be female (Fricke, 1979).

13.4 DIFFERENCES BETWEEN INDIVIDUALS

A variable environment

Population differences in behaviour

When populations are isolated, either geographically or reproductively, and subject to a variety of selection pressures, a mosaic of behavioural variation will emerge. Threespine sticklebacks (Lavin and McPhail, 1985, 1986), European minnows (Magurran, 1986; Magurran and Pitcher, 1987), sailfin mollies, *Poecilia latipinna* (Trexler, 1988, 1989a–c) and brown trout (Ferguson, 1989; L'Abée-Lund and Hindar, 1990) are but four of many species where population differentiation has been documented. A particularly well-explored case of variation is provided by the guppy, *Poecilia reticulata*. Guppies are small poeciliid fishes endemic to NE South America and the adjacent islands. In Trinidad, where the species is abundant and widely distributed, populations vary with respect to male coloration (Endler, 1978, 1983), morphology (Liley and Seghers, 1975), life-history strategy (Reznick, 1982; Reznick and Endler, 1982; Reznick *et al.*, 1990) and behaviour (Seghers, 1974; Luyten and Liley, 1985; Fraser and Gilliam, 1987; Houde, 1988; Houde and Endler, 1990; Licht, 1989; Magurran and Seghers, 1990a–c, 1991). Perhaps the most remarkable aspect of population differentiation in guppies from Trinidad is that it exists within a very limited geographical area. Divergent populations can be separated by as little as a few kilometres or, in some cases, metres.

Many of the population differences can be attributed to contrasting predation regimes. For example, in some rivers guppies coexist with a range of characin and cichlid predators including the pike cichlid, *Crenicichla alta*. Guppies in these rivers form cohesive schools (Seghers, 1974) and have better-developed predator avoidance behaviour, including higher reaction distance, faster transmission of alarm responses and different habitat usage when compared with fish from low-risk populations (Liley and Seghers, 1975). Breeding experiments have confirmed that schooling tendency is heritable (Seghers, 1974; Breden *et al.*, 1987).

Schooling is not advantageous in all situations where predators are present. In some rivers, particularly those draining the northern slopes of the Northern Range Mountains, palaemonid prawns from the genus *Macrobrachium* are abundant. Although there has been some dispute about the feeding behaviour of these prawns (Luyten and Liley, 1985; Rodd and Reznick, 1991), recent observations (Magurran and Seghers, 1990c) support Endler's contention (1978, 1980, 1983) that they are indeed guppy predators. Prawns become active at dusk (Seghers and Magurran, unpublished data) and use tactile and chemical signals in addition to vision when detecting prey. This behaviour makes schooling a liability since clusters of prey are more likely to attract attention than solitary individuals. Field observations in March 1990 (Magurran and Seghers, 1991) indicated that nearly 70% of guppies recorded in a section of the Paria River, where *Macrobrachium* spp. occur in high densities, were not members of schools or shoals. By comparison only 7% of guppies in the Lower Aripo (a river also containing pike cichlid) were solitary. In the context of the Paria River, a low schooling tendency does not equate with poor predator assessment. Guppies from this location show clear attack cone avoidance when inspecting a prawn (Fig. 13.2).

Ovoviviparity results in the production of relatively mature offspring, and guppies are capable of forming polarized and cohesive schools from birth (Magurran and Seghers, 1990b). Although there are pronounced population differences in the schooling tendencies of new-born fish, the observed patterns do not necessarily mirror those observed in adults (Magurran and Seghers, 1990b). For instance, new-born guppies from both the Upper and Lower Aripo

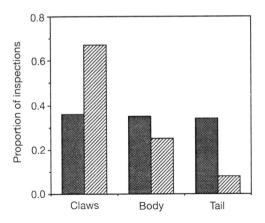

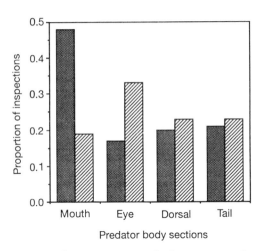

Predator body sections

Fig. 13.2 The proportion of inspections towards three sections of a predatory prawn, *Macrobrachium crenulatum* (top) and a predatory fish, *Rivulus hartii* (bottom) by guppies from two Trinidad populations. Lower Aripo guppies (hatched columns) are subject to predation by fish but not by prawns. The reverse is true for Paria guppies (dark columns). Guppies avoid the most dangerous zone, the attack cone, of the predator type that poses the greatest threat in their particular habitat. Redrawn from Magurran and Seghers (1990a); note differing vertical scales.

populations form cohesive schools whereas only adult Lower Aripo schools are well organized (Seghers, 1974; Breden *et al.*, 1987). *C. alta* does not occur in the Upper Aripo. The discrepancy in the behaviour of new-born and adult fish raises two intriguing possibilities. The first is that the selection pressures that operate on the adult guppies and new-born guppies are discontinuous. Thus immature fish could be vulnerable to cannibalism or to predators (such as the cyprinodont *Rivulus hartii*) that are inept at capturing larger fish (Liley and Seghers, 1975). The second explanation is that the observed differences are independent of risk but do reflect different developmental trajectories in isolated populations. The high levels of schooling behaviour by new-born guppies in both portions of the Aripo could be a consequence of shared ancestry. New-born guppies from the Paria River (part of a separate drainage system) resemble adults in that they, too, have a reduced schooling tendency.

Reznick and Endler (1982) have shown that guppies experiencing high natural levels of predation increase their investment in reproduction relative to other populations. The increased investment is brought about by devoting proportionately more body weight to reproductive effort, commencing reproduction at a smaller size and producing larger numbers of smaller offspring. The pivotal role of predation in inducing these life-history shifts has been elegantly demonstrated in a long-term study by Reznick *et al.* (1990). In 1976, 200 adult guppies were transplanted from the Lower Aripo (a control site with *C. alta*) to a tributary where guppies had not previously been present. The only other fish species in the tributary was *R. hartii*. Guppies in both the introduction site and the control site were then monitored for 11 years (representing between 30 and 60 generations). Significant differences between the two sets of females were already apparent within 2 years of the transplant. As predicted, fish from the introduction site now delayed reproduction until they were larger and decreased their reproductive allotment (embryo weight/ total body weight). The life-history patterns of guppies from the two sites continued to diverge over the course of the 11 year study. Reznick *et al.* (1990) confirmed that the differences were genetically based by raising guppies from both the control and introduction sites in a common environment for two generations.

Guppies have a promiscuous mating system in which female preference for particular male colour patterns plays a major role (Houde, 1987, 1988). Predators also key in on male colour, and the interaction of sexual selection and natural selection results in striking population variation in colour pattern (Endler, 1983). In some rivers, such as the Paria, males have conspicuous orange/red spots (Houde, 1988). In other rivers, particularly those where *R. hartii* is the only potential predator, iridescent blues and greens are common (Endler, 1983). Barrier waterfalls, which prevent the migration of fish predators, often mark the boundary between brightly coloured upstream fish and their drabber downstream counterparts (Haskins *et al.*, 1961). Recent experiments have demonstrated that female choice and male colour covary across populations (Breden and Stoner, 1987; Stoner and Breden, 1988; Houde, 1988; Houde and Endler, 1990). The debate on whether female choice is 'adapative' (correlated with male quality) or 'arbitrary' is one of the central issues in behavioural ecology today (Kodric-Brown, 1990). In guppies, at least, there is evidence that females may be choosing 'good genes'. The intensity of carotenoid colours is influenced by diet, and Kodric-Brown (1989) showed that females preferred the brighter of two males that were otherwise matched for size and colour pattern. The significance of male colour in those populations where red is not a key component remains a mystery and may well be an important focus of research over the coming decade. There is, of course, no reason why female guppies in different populations should use colour as their sole criterion in mate choice. Intriguingly, Reynolds (1990) has recently discovered that female guppies in another river, the Quare, appear to choose males on the basis of body size and tail length. As a daughter's growth rate is significantly correlated with the length of her father's tail, these Quare River females are evidently selecting males with 'good genes'.

Guppy populations vary not only in their characteristic behaviours but also in their behavioural flexibility. Male guppies have two main methods of achieving a mating. They may either opt for a sigmoid display (during which the body is arched into an S-shape and quivered) in the hope of attracting a receptive female, or employ a sneaky mating attempt (Liley, 1966). Not only do males from different populations vary in their use of these tactics (Luyten and Liley, 1985, 1991), but they also differ in the extent to which they modify their courtship behaviour as a consequence of predation risk. Magurran and Seghers (1990a) showed that Upper Aripo males were risk-reckless, and failed to modify their behaviour in the presence of predators. Lower Aripo males by comparison were sensitive to risk and reduced their display rate in favour of sneaky mating attempts if threatened.

Dramatic population variation tends to overshadow individual differences within populations. Yet a single summary statistic such as a population mean can mask considerable individual variation. European minnows provide an apt example. As in guppies, minnow antipredator behaviour can be related to the predation regime. Minnows from a population sympatric with pike form more cohesive schools and have better-integrated predator evasion tactics than fish from a population where pike are absent (Magurran and Pitcher, 1987). One component of antipredator behaviour, inspection, provides a particularly striking comparison. Minnows from the high-risk population commence inspection of an approaching pike earlier and make more repeated inspections than their low-risk counterparts. As a consequence of this inspection, minnows lose foraging opportunities (Magurran, 1986). The population differences in the trade-off of feeding against predator assessment are clearly revealed in Fig. 13.3. Superimposed on the dichotomy are pronounced individual differences within the two populations. As the participating fish were of similar size and (since they had been fed equally prior to the experimental trials) presumably of similar motivation, the explanation for this within-population individual variation is unknown.

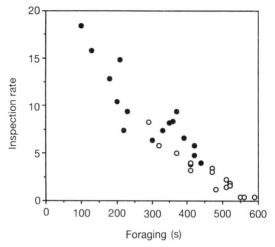

Fig. 13.3 A trade-off between foraging and inspection behaviour in minnows from high-risk and low-risk populations. In each case 16 minnows were individually marked. This graph shows the mean time spent foraging and the mean number of inspections (averaged over $n = 5$ trials) of these 32 minnows during the approach of a model pike (see caption of Fig. 13.1 for details). The low-risk population ($\bigcirc$) came from Llyn Padarn, North Wales, where pike are absent. The high-risk minnows ($\bullet$) originated in the River Frome, Dorset, where pike abundant. Data were collected for the experiment described by Magurran (1986).

Geographic variation in behaviour can be related to abiotic factors as well as biotic ones. For example, life-history patterns are often linked to climatic conditions (Mann *et al.*, 1984). Mills (1988) found that, in higher latitudes, the growth rate of European minnows is lower and sexual maturity is delayed while Vøllestad and L'Abée-Lund (1990) showed that female roach, *Rutilus rutilus*, produced fewer, heavier eggs in harsh climates.

Stochastic factors

So far deterministic processes have been invoked to explain population differences in behaviour. However, in certain circumstances, stochastic factors may be responsible for initiating or exaggerating population variation (Clark and Ehlinger, 1987; Wilson, 1989). Genetic bottlenecks, drift (Hedrick, 1983), founder effects (Knight *et al.*, 1987; Johnson, 1988) and environmental disturbance may well play a significant role in differentiating populations, particularly when demes are small and gene flow is restricted. Surprisingly little effort has been devoted to exploring the relative contribution of stochastic and deterministic processes to the differentiation of fish populations. Trexler (1989a,b) observed considerable levels of regional variation in the genetic structure of populations of the sailfin molly in Georgia and Florida. Since the demes within a region exhibited only a small proportion of the observed variation, Trexler (1988) concluded that the population differentiation had not arisen through Wright's (1970) shifting balance process. On the other hand Stearns (1983) could not exclude the possibility that founder effects and genetic drift had influenced population variation in life-history traits in the mosquitofish, *Gambusia affinis*, and Carvalho *et al.* (1991) suggested that founder effects had contributed to the marked genetic divergence of Trinidad guppy populations.

Phenotypic differences

Sex

The presence of two sexes inevitably leads to variable behaviour. Males and females, for instance, have different methods of achieving a mating and often contribute unequally to brood care. Gender differences can, however, influence behaviour in spheres not immediately associated with reproduction. Being male or female may, for example, have important repercussions for behavioural trade-offs. Abrahams and Dill (1989) looked at the interaction between foraging and risk of predation in both male and female guppies. In the experiment single-sex groups of guppies were given a choice of two feeding patches, one of which entailed a risk of predation. When both patches were equally profitable the guppies chose to feed on the safer of the pair. However, guppies could be induced to feed on the risky patch when extra food was supplied there. This trade-off of extra food in exchange for risk was not adopted equally by the two sexes. Only females seemed willing to accept the greater danger in exchange for higher energetic rewards. This is probably due to the fact that male guppies virtually stop growing at maturity while females continue to get larger. Dussault and Kramer (1981) observed that females fed at six times the rate of male guppies over a 12 h period. Abrahams (1989) also noted sex differences when male and female guppies (again tested in single-sex groups) were presented with two feeding patches with no predator present. Males sampled more frequently whereas females relied to a greater extent on experience.

Sex-specific risk-taking and foraging strategies are also found in the coho salmon, *Oncorhynchus kisutch* (Holtby and Healey, 1990). In this species the degree of sexual dimorphism varies across populations on the West Coast of the US and Canada. Holtby and Healey (1990) relate such variation to physical differences in the breeding environment, spawning density and the relative risk attached to the ocean nursery areas.

Morphology

Many species have morphologically distinct forms. For instance, threespine sticklebacks have variable jaw morphology. Ibrahim and Huntingford (1988) found that fish from Loch Lomond, Scotland had a smaller gape (relative to body size) and more closely spaced gill rakers than sticklebacks from the Balmaha Pond (also near Glasgow). Zooplankton are abundant in Loch Lomond (where they form 95% of the stickleback diet) but rare in Balmaha Pond. These morphological differences are adaptive; laboratory tests showed that Loch Lomond sticklebacks fed more efficiently on zooplankton than did the Balmaha Pond fish. The Scottish sticklebacks provide an example of morphological differences between populations. Morphological and behavioural variation can also be observed within a single population. For example, a Mexican cichlid (the Cuarto Cienegas cichlid, *Cichlasoma minckleyi*) occurs as two distinct morphs with different dentition and different diets (Kornfield *et al.*, 1982; Kornfield and Taylor, 1983). One morph, the 'small tooth' form, has papilliform pharyngeal teeth and feeds on plant material; the other, 'large tooth' morph, with molariform pharyngeal dentition, specializes on snails. These morphological differences led the two forms to be initially misclassified as separate species, an error rectified by electrophoretic analysis.

Recent work (Ehlinger, 1989, 1990; Ehlinger and Wilson, 1988) has revealed fascinating morphological and behavioural variation in the bluegill sunfish. This work clearly demonstrates why it is misleading to frame optimal arguments in terms of the so-called average individual. Bluegills captured from vegetated areas of lakes are deeper bodied and have longer pelvic and pectoral fins than the streamlined fish that occur in open-water habitats. These body plans promote efficient locomotion within the respective habitats (Webb, 1984; Wilson, 1989) and also translate into functional differences in foraging abilities. Deeper-bodied fish move slowly and stealthily and then remain motionless (hover) while they search for prey. This technique is ideal for capturing the cryptic prey associated with vegetated habitats. Open-water forms are more mobile and combine faster swimming with shorter hover durations (Fig. 13.4). Individual bluegills can gradually adjust their searching tactics (Ehlinger, 1989) in laboratory tests although field observations suggest that wild fish are inflexible in their foraging behaviour (Wilson, 1989). It is likely that diet choice is also related to gut morphology (Partridge and Green, 1985) and that digestive efficiency is compromised if an individual switches to a new food item (Windell, 1978).

Dominance and status

In many species of fish, dominant individuals differ in colour or markings from those of inferior status. One of the best-known examples is that of the Midas cichlid, *Cichlasoma citrinellum*, found in lakes in Nicaragua (Barlow, 1983). In

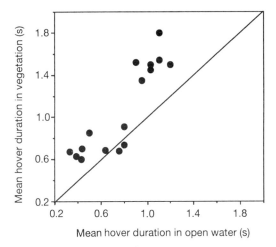

Fig. 13.4 Behavioural variation in the bluegill sunfish. Hover duration of 16 individuals was measured both in the open water (x-axis) and in vegetation (y-axis). Individuals were either long or short hover types. Two discrete clusters of points are therefore evident. Nearly all fish spent longer hovering in the vegetation – this explains why the majority of points lie to the left of the diagonal line. Redrawn from Ehlinger (1990).

this species, which is polychromatic, the majority of the adult fish are a grey, cryptic morph, known as normal and the remainder, some 8%, as conspicuous morph known as gold. The gold morphs, which lack melanophores, range in colour from yellow to red with orange the predominant hue. Although these brightly coloured fish are at a disadvantage as they are more visible to predators, being gold brings benefits since gold morphs are dominant over normal morphs of similar size (Barlow and Ballin, 1976). Golds also suffer less fin damage in contests than normals (Barlow, 1983), and as a result of their dominance over normals, gold juveniles grow at a faster rate (Barlow, 1983; Barlow and McKaye, 1982). Surprisingly, golds are not intrinsically more aggressive. Rather, they become dominant by virtue of the effect of their colour on opponents. This appears to be the result of the relative rarity of gold morphs (Barlow *et al.*, 1975; Barlow, 1983) combined with the intimidating nature of the gold colour itself. Normal Midas cichlids take on a red, orange or yellow coloration during breeding and Barlow suggests that the effectiveness of the gold colour stems from its association with threat displays. The gold effect seems to be enhanced if the gold morph is in a competitive situation in which it is already at an advantage, for example if it is actively defending a territory (Barlow, 1983).

Contests between individuals are energetically costly. During the breeding season, male threespine sticklebacks experience, on average, a 91% reduction in liver glycogen and a 73% drop in liver lipid (Chellappa *et al.*, 1989). The consequences of reduced energy reserves emerged in an experiment by Chellapa and Huntingford (1989) in which males, matched for both size and colour, were allowed to fight. The results showed that levels of liver glycogen (measured during post-mortem examination) were significantly higher in the victors than in their defeated rivals.

Individual differences become especially important when limited resources increase levels of intraspecific competition. Food, like mates, may well be in short supply, and inevitably body size is one factor influencing the success of individuals (Wilson, 1975; Ranta and Lindström, 1990). However, as Keen (1982) showed for juvenile brown bullhead, *Ictalurus nebulosus*, larger size may need to be coupled with higher levels of aggression before a fish can capture a disproportionate share of the resource. Conversely, early success in social interactions may enhance status, allow an individual to become dominant, and as a consequence, secure more food and grow faster. Huntingford *et al.* (1990) concluded that competitive ability, termed 'fierceness', determined the outcome of early interactions in juvenile Atlantic salmon. In the early tests (5–12 weeks after first feeding), the larger member of a pair was dominant in only 54% of cases. As the fish became older a correlation between dominance and size emerged. After 5 months of feeding competitively, the larger fish was the dominant in 77% of trials. When status and size differences were eliminated by allowing individual salmon to compete with their mirror image, it emerged that the more-slowly developing fish (which delay smolting) were most likely to give way to a competitor (Metcalfe, 1989; Metcalfe *et al.*, 1989). Both size and prior residency were found to be important in determining the outcome of contests among minnows for feeding sites (Pitcher *et al.*, 1986) – see Table 13.1 for details.

Rubenstein (1981), in a series of experiments with the Everglades pygmy sunfish, *Elassoma evergladi*, found that individual differences in growth rate were magnified under competitive conditions. In Rubenstein's work the sunfish were raised in densities of 1, 4, 8 and 16 fish per laboratory tank or field cage. Food supply to the whole tank or cage was kept constant to ensure more vigorous competition in the larger group sizes. Rubenstein showed that as fish density increased from 4 to 16 fish, the growth rate of all fish fell. However, the growth of the best competitors increased much more slowly than the growth of the average fish and was not adversely affected by the higher fish density. The coefficient of variation of growth rate increased from 35% for solitary fish and 45% for fish in a group of four to 209% in the largest group. Average fecundity of both males and females was also reduced in larger groups and, as with growth rate, individual differences in fecundity were exaggerated. Reduced growth rates of subordinate individuals have also been found in bluegill sunfish (Drager and Chizar, 1982) and in rainbow trout, *Oncorhynchus mykiss* (Metcalfe, 1986). In rainbow trout, dominant fish obtain a significantly greater food intake for a given expenditure of energy.

Table 13.1 Feeding interactions of large and small minnows. Twenty large (mean length 60 mm) and 20 small (mean length 40 mm) minnows foraged together on an artificial food patch consisting of 90 gravel-filled pots, each pot big enough to accommodate only one fish. Nine pots contained food. Interactions between a minnow in possession of a pot (the owner) challenged for occupancy by another fish (the intruder) were scored. The results (summarizing data from eight 15 minute trials) show that size of fish and ownership of a pot were important in determining the outcome of contests. It is also interesting to note that minnows invested most effort in defending those pots that held food. With the exception of small owners, which were always evicted by large intruders, there was no significant difference in the number of times minnows won or lost possession of a no-food pot (see also Pitcher *et al.*, 1986). ***, $P < 0.001$; NS, not significant

| | Owner | | | |
| | Food | | No food | |
Intruder	*Large*	*Small*	*Large*	*Small*
Large	Owner wins ***	Owner loses ***	Outcome uncertain NS	Owner loses ***
Small	Owner wins ***	Owner wins ***	Outcome uncertain NS	Outcome uncertain NS

Rubenstein provides a clear demonstration of the consequences of inequalities in competitive ability. Milinski's (1982, 1984a–c) work on sticklebacks shows precisely how the foraging behaviour of good and poor competitors differs, and indicates the methods that poor competitors adopt in an attempt to compensate for these inequalities. Milinski (1982) fed daphnia, *Daphnia magna*, to pairs of sticklebacks and used the proportion of daphnia that each fish caught as a measure of its relative competitive ability. Each pair of sticklebacks was then presented with daphnia belonging to two size classes. Since handling time was the same for both prey sizes classes, the larger daphnia were more profitable. In each pair of sticklebacks tested, poorer competitors caught significantly fewer large daphnia. Surprisingly, when allowed to forage alone, these poorer competitors continued to direct over half of their attacks towards small daphnia; good competitors by contrast still preferred to attack large daphnia. Milinski argues that lower success rate in capturing large daphnia makes it advantageous for the less successful sticklebacks to incorporate more small daphnia in their diet. Metcalfe (1986) suggested that the optimum strategy for a subordinate rainbow trout is to minimize energy expenditure rather than to maximize food intake. Behavioural responses that benefit competitively inferior individuals have also been considered by Parker (1982), Sutherland

and Parker (1985), and Parker and Sutherland (1986). In each case the best strategy will depend on the precise mix of competitors present.

Direct competition for the most profitable prey item is not the only problem that poor competitors meet. It is usual for food to be patchily distributed, for example marine plankton (Steele, 1976) and freshwater invertebrates (Allan, 1975; Brooker and Morris, 1980). Individuals must therefore decide how long to stay on one patch before moving on to another (see Chapter 8 by Hart, this volume). Milinski (1984c) investigated the foraging behaviour of sticklebacks in a patchy environment. In this experiment shoals of six sticklebacks were allowed to forage on two patches of cryptic daphnia. Daphnia were pipetted into these patches at a known rate. Since the patches were nearly 0.5 m apart, fish moving from one to the other incurred a cost in travel time. The best patch was twice as profitable and fish allocated themselves between patches in a ratio of 2:1, apparently in accordance with the predictions of the ideal free distribution (Fretwell and Lucas, 1970). As Fig. 13.5 shows, however, foraging returns were not equal and there was a positive and significant relationship between competitive rank and number of daphnia captured. Milinski found that the poorer competitors took longer at the beginning of a trial to ascertain the profitabilities of the patches, and switched patches more frequently (see also Abrahams, 1989).

Subordinate status is not always a handicap. An explicit demonstration of the benefits that can accrue to subordinate fish is provided by Taborsky (1984). Taborsky investigated the behaviour of *Lamprologus brichardi*, a Lake

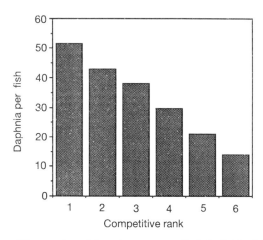

Fig. 13.5   Competitive rank and foraging success. Each individual in a group of six sticklebacks was assigned a competitive rank on the basis of number of daphnia captured during a trial. Nine trials were conducted in all. This figure shows the mean number of daphnia consumed by the nine most successful fish (rank 1), the nine second-most successful fish, and so on. Redrawn from Milinski (1984a).

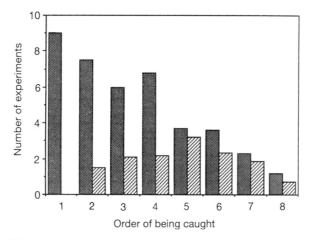

Fig. 13.6 Helpers and protection from predation. This graph shows the order in which helpers (hatched columns) and control fish (dark columns) of the same size were caught in nine experiments. A total of 55 fish were captured. Redrawn from Taborsky (1984).

Tanganyika cichlid. In this species, adult pairs defend territories whereas young of previous broods either remain with their parents as helpers, participating in brood care and territorial defence, or join non-breeding family-independent aggregations. Due to their low rank in the family hierarchy, helpers grow at a slower rate than non-helpers in the aggregations. Compensating for this cost are two benefits. First, helpers contribute to the reproductive success of their parents, ensuring the production of more siblings. Second, membership of the territorial unit gives helpers protection against predators. Taborsky ran an experiment in which groups of helpers, with their parents, and equal-sized groups of controls (non-helper, aggregation members) were exposed to one of their natural predators, *Lamprologus elongatus*. All fish had access to cover. Controls were caught earlier and more frequently than helpers (Fig. 13.6).

Disease

Diseased individuals are without doubt disadvantaged. Parasite infestation, for example, reduces the intensity of the red coloration in breeding male sticklebacks and makes them less attractive to females (Milinski and Bakker, 1990). Parasitized individuals are also likely to have a lower competitive ability. Crowden and Broom (1980) showed that the reactive distance of dace, *Leuciscus leuciscus*, to *Gammarus* was greater for fish heavily infected with eye flukes, *Diplostomom* sp. Milinski (1984b) found that sticklebacks parasitized with a microsporidian sporozoan, *Glugea anomola*, or a cestode, *Schistocephalus solidus*, attacked small daphnia more often than unparasitized fish did. The percentage of attacks directed at small prey was highest in sticklebacks infected with both parasites. By shifting their diet away from the larger, more profitable daphnia, the parasitized sticklebacks avoided direct confrontation and maximized their food intake. Giles (1983) found that parasitized sticklebacks started foraging sooner after exposure to a frightening overhead stimulus than unparasitized fish. Giles suggests that this behaviour is a response to the metabolic demands that the parasite places on the host. Further evidence that parasites manipulate their host's behaviour is supplied by Milinski (1985, 1986) and Godin and Sproul (1988). Milinski discovered that sticklebacks infested with *S. solidus* fed much closer to a threatening cichlid than sticklebacks where *G. anomola* was the sole parasite. This behaviour is adaptive for the parasites. Since the stickleback is the only host for *G. anomola*, both gain from effective antipredator behaviour. On the other hand *S. solidus* benefits if the stickleback is captured since fish-eating birds are its definitive host (Wootton, 1984).

Frequency dependent mating behaviour?

The remarkable diversity of teleost mating systems across species (see Chapter 10 by Turner, this volume) is mirrored by considerable intraspecific variation in reproductive behaviour. Fish are a particularly interesting group since their alternative mating strategies can involve separate developmental pathways, distinct morphologies and different behaviour patterns. Differences in male behaviour are most likely to occur when male reproductive success is dependent on aggression and male–male competition (Dominey, 1981). Given the logistical problems of assessing reproductive success in natural populations, it is extremely difficult to establish whether alternative strategies are equally successful (and maintained through frequency dependent selection: Partridge, 1988) or represent a 'best of a bad job' scenario. The major challenge of providing unequivocal evidence for frequency dependent reproductive behaviour and measuring lifetime reproductive success remains unanswered (Partridge, 1988). Recently developed techniques such as DNA fingerprinting could be a valuable asset in this endeavour.

One fascinating but often neglected case of alternative mating strategies involves the Atlantic salmon, *Salmo salar*. It has been known for at least 150 years that male salmon parr mature precociously, and experiments by Orton *et al.* (1938) and Jones and King (1950) showed conclusively that these fish are reproductively active and able to fertilize eggs. Precocious males are found across the geographic range of the salmon (Jones and Orton, 1940) although the proportion of precociously maturing males varies between stocks (Schaffer and Elson, 1975). In Newfoundland rivers for instance, the incidence of precociousness ranges from 12% to 100% with a mean of 73% (Dalley *et al.*,

1983). Salmon parr must compete for females with adult males that have been to sea and returned to the spawning ground. In some cases females appear reluctant to spawn in the presence of the precocious males (Jones, 1959), though in situations where adult males are absent (such as in the Exploits River, Newfoundland: Myers and Hutchings, 1987) females will spawn successfully with mature parr. The usual technique of the precocious male is to take up position under the bellies of the spawning adult male and female, to remain as inconspicuous as possible, and to fertilize eggs as they are laid in the gravel. Little is known about the factors that control the development of sexual precociousness in these males, though there is evidence that age of first spawning may be heritable (Schaffer and Elson, 1975; Thorpe and Morgan, 1980; Thorpe *et al.*, 1983) and that males that mature precociously are generally the largest fish (Rowe and Thorpe, 1990). Information on the subsequent reproductive success and migration patterns is incomplete but it is becoming clear that precocious maturation conflicts with efficient smolting. Langdon and Thorpe (1985) and Thorpe (1987) found that Atlantic salmon that had matured at 1.5 years showed poorer physological adaptation to seawater, and Eriksson *et al.* (1987) concluded that mature parr had a reduced rate of survival when compared with immature smolts. In the sea trout (*S. trutta*), where precocious males also occur, mature parr remain small and resident for the duration of their lives (Dellefors and Faremo, 1988; Bohlin *et al.*, 1990).

Male salmon parr can be classified as sneaky breeders (Arak, 1984), a widespread strategy in fish. In the typical case the sneaky male (sometimes referred to as the cuckolder (Gross, 1982), submissive male (Dominey, 1981) or scrounger (Barnard, 1984)) reduces the cost of mating, such as attracting females, defending territories or being more visible to predators, by stealing fertilizations from dominant fish. These lower costs must be balanced against reduced attractiveness to females and possible lower reproductive success. An excellent example is provided by Constanz (1975) who worked on the gila topminnow, *Poeciliopsis occidentalis*. These fish belong to a family in which male growth ceases at sexual maturity (Constanz, 1989). Since position in the dominance hierarchy is related to size, small fish are condemned to remain subordinate. Large dominant males form territories and court females; small males remain at a distance from the females, seeking an opportunity to enter the territory and perform sneak copulations. The differences between the males and the costs and benefits of the two strategies are summarized in Table 13.2. If the large males are removed, small males will, within a matter of minutes, take on the dominant colour pattern and start defending territories and courting females. Although it is likely that a male's propensity to become a sub-dominant sneak is at least in part genetically based, these topminnows are not locked into one behaviour pattern. Bluegill sunfish, by contrast, are committed early on in their development to become either a parental male or

Table 13.2 Costs and benefits of territorial and sneaky mating strategies in *Poeciliopsis occidentalis* (after Constanz, 1975)

Characteristic	Large territorial male			Small sneaky male		
	Strategy	Cost	Benefit	Strategy	Cost	Benefit
Body size	Large	Delayed sexual maturity	Higher competitive ability	Small	Lower competitive ability	Quicker maturation
Body colour	Jet black	More visible to predators	Encourages submissive to flee	Light coloured	Greater energy expenditure	Not so visible to predators; trespassing into territories easier
Agonistic behaviour and territoriality	Aggressive, holds territory	Time and energy expended in territory defence	Territory permits relatively undisturbed courtship	Submissive, no territory	Movement inhibited; less knowledge of the geography of territory	Time saved in territory defence potentially available for mating attempts
Courtship and copulation	Courts female, longer copulation		Female easier target; mating less likely to be disturbed	Does not court female; shorter copulation	Female evasive target; mating activities readily disrupted	
Gonopodial length	Short	possibly less accuracy in contacting moving female	Less investment in gonopodium	Long	Greater investment in gonopodium	Greater accuracy for moving target

a sneak (Gross, 1982, 1984). Parental males, which only gain sexual maturity at 7 years of age, build nests, attract females and care for the eggs after fertilization. Sneaks become reproductively active at 2 years and commence breeding attempts by dashing into the nest and releasing sperm as the female spawns. At a later stage of their development they take on adult female size, colour and behaviour and swim between the spawning female and parental male. Dominey (1981) has put forward two reasons why these female mimics are tolerated in nests. Spawning females are attracted to areas where other females are aggregating. By having female mimics around him, the parental male might give real females the impression that his nest is a centre of spawning activity. Alternatively, parental males that allocate too much time to evicting female mimics decrease the success of their encounters with functional females.

Approximately 20% of male bluegill sunfish become sneaks and are responsible for fertilizing 20% of the eggs (Gross, 1982). There is some evidence that an evolutionarily stable equilibrium of sneaky males and parental males is maintained through frequency-dependent pay-offs (Gross, 1984). At low densities sneaky males have many opportunities to fertilize eggs, but as their proportion in the population rises, the reproductive success of each sneak will diminish. Recent works suggest that the sneaky and parental male phenotypes have a genetic basis (M. R. Gross, personal communication) and precociously maturing (i.e. sneaky) bluegill males differ in morphology and feeding behaviour from the males destined to become parentals (Ehlinger, 1990). As Ehlinger (1990) notes, teasing out potential linkages between reproduction and feeding tactics represents an exciting avenue for future research.

13.5 ACQUISITION OF ALTERNATIVE BEHAVIOURS

Many investigations attempt to categorize individual differences as either inherited or learnt. In some situations, for example the Lough Melvin trout (Ferguson, 1989), behavioural differences are undoubtedly of genetic origin. In other cases the consequences of differential experience are evident. Hatchery-reared salmon are notoriously vulnerable to predators when they are released. For instance, Patten (1977) notes that wild coho salmon, *Oncorhynchus kisutch*, are subject to predation by sculpins (*Cottus* sp.) until they reach a length of 45 mm. Hatchery-reared fry, however, remain at risk until they grow to 60 mm. Experience also mediates the foraging behaviour of hatchery-reared Atlantic salmon parr, even after their release into the wild. Sosiak *et al.* (1979) showed that wild parr had fuller stomachs and consumed a wider diversity of prey taxa than hatchery-reared parr that had spent 1–3 months in the same stream. Cultural transmission is another non-genetic source of variation in individual behaviour patterns (Helfman and Schultz, 1984). Warner (1988) found that the mating sites used by bluehead wrasse, *Thalassoma bifasciatum*, remain constant over several generations. In this species tradition overrides resource assessment, and it is only when a new group of fish colonizes a coral reef that the quality of potential mating sites is taken into account (Warner, 1990).

Inheritance and experience often act synergistically and it can be difficult, and sometimes pointless, to attempt to disentangle their separate contributions (Caro and Bateson, 1986; see also Chapter 3 by Huntingford, this volume). A number of recent studies have underlined the interaction of the two processes. Magurran (1990b), for instance, noted that a genetic predisposition to learn, as well as the appropriate experience, was necessary for the complete development of antipredator behaviour in European minnows. Minnows from a high-risk population (where pike were present) showed a greater increase in inspection behaviour following limited early experience of simulated predatory attack, when compared with fish from a low-risk population. Huntingford and Wright (1989) showed that populations of threespine sticklebacks differed in the speed with which they learnt to recognize a dangerous feeding patch. Barlow *et al.* (1990) discovered that assortative mating in the Midas cichlid is the consequence of a number of factors, including an inherent preference for the ancestral colour type (i.e. normal males) and the colours of the parents and companions to which the individual has previously been exposed. As Slater (1983) points out, the degree of variability of a behaviour is no guide to whether it is environmentally or genetically determined. In several species, for example, the presence of distinct stereotyped behaviour patterns has erroneously been used as evidence for the genetic basis to behaviour (Dunbar, 1982).

Caro and Bateson (1986) list five ways in which the development of alternative behaviours is mediated. These factors can be either genetic or environmental in origin. **Canalizing** factors buffer external events and may help reduce individual variability. Shaw (1970), for instance, found that schooling behaviour develops irrespective of social experience. Fish raised in isolation are able to school when first given the opportunity. **Facilitating** effects accelerate development. For example, reproduction occurs earlier in the season in threespine sticklebacks occurring in southerly latitudes where temperatures are higher (Wootton, 1984, 1990). **Maintaining** factors allow the retention of behaviour patterns that would otherwise vanish from the repertoire. Threespine sticklebacks from high-risk sites, where predators are abundant, retain a high level of inspection behaviour over a series of encounters with a large, but non-attacking fish (Huntingford and Coulter, 1989). Such maintaining factors must be absent in sticklebacks from a low-risk population because these fish rapidly habituate to the same stimulus. **Enabling** or **predisposing** factors permit the action of an event that would not otherwise influence the behaviour in question. Tulley and Huntingford (1987), for example, found that 'fathering', where male sticklebacks retrieve offspring that stray from the nest, influences the subsequent development of predator avoidance behaviour, and Goodey and Liley (1986) showed that cannibalistic attacks on new-born guppies increase an individual's ability to avoid predators encountered months later. Finally, **initiating** effects lead to the expression of a behaviour that would not otherwise develop. The sex of larvae of the Atlantic silverside, *Menidia menidia*, is determined by temperature (Conover, 1984). Larvae that develop at the beginning of the season, while the water is still cold, become females. The longer time available for growth allows females to become larger and produce more eggs.

13.6 ECOLOGICAL AND EVOLUTIONARY SIGNIFICANCE OF INDIVIDUAL DIFFERENCES

Competitive differences in conjunction with a variable environment, different individual requirements and a variety of tactics that can be used to achieve the same ends make individual differences in behaviour inevitable. Many of the examples of variable behaviour discussed in this chapter have resulted from the ability of individuals to capture a disproportionate share of limited resources. Rather than compete directly for scarce resources, subordinate individuals may find it advantageous to adjust their behaviour. As a result, it is often incorrect to assume that optimal behaviour is the same for all individuals, even when every attempt has been made to control for potential sources of variability such as size differences or variation in motivation. It is also naïve to consider populations as groups of identical individuals. Ecologists, like ethologists, have at last become aware of the issue and now explicitly include individual differences in their analyses. Many of the current ideas have been brought together by Łomnicki (1988), in his extensive review of the population ecology of individuals. Parker (1982, 1985), for instance, has addressed the idea of phenotype-limited evolutionarily stable strategies while Hassell and May (1985) and Smith and Sibly (1985) have considered ways in which population dynamics are influenced by individual behaviour. One case in point is despotic behaviour, which limits the number of territories and hence the number of breeding individuals, thereby reducing population density. Community structure may also be influenced by individual differences (Ringler, 1983). Werner (1984), for instance, discusses how interspecific competition is affected by ontogenetic niche shifts.

Behavioural differences can be instrumental in bringing about evolutionary changes, including speciation (Wilson, 1989). Endler (1989) has suggested that the observed population differentiation in guppies represents the early stages of speciation, an assertion supported by an electrophoretic analysis which revealed allellic substitution at allozymic loci amongst populations (Carvalho *et al.*, 1991). Population-specific patterns of mate choice are likely to play a key role in the development of reproductive isolation (Houde and Endler, 1990). In guppy populations in Trinidad this reproductive isolation is often allied to geographic isolation. Yet stocks or morphs need not be physically isolated for reproductive isolation to occur. It seems probable that the three sympatric stocks of brown trout in Lough Melvin are the product of events during the last glaciation 13 000 years age, when ice movements partitioned existing populations and facilitated new colonizations (Ferguson, 1989). Whatever the precise origin of the three stocks, it is clear that their genetic integrities are now maintained through reproductive isolation. The gillaroo spawns in the lake and outflowing river, the sonaghen in small inflowing rivers and the ferox in the deepest section of the largest inflowing river.

13.7   SUMMARY

Recent advances in evolutionary biology have highlighted the importance of individual differences in behaviour. By reviewing examples drawn from investigations of foraging behaviour, predator avoidance and mating tactics, this chapter has shown that intraspecific variation in fish behaviour is widespread. Three mechanisms are generally proposed to account for individual differences in behaviour: a patchy environment, phenotypic differences and the behaviour of other individuals. These factors promote not only differences amongst individuals but also behavioural flexibility within a single individual. Since the appropriate behaviour for an individual fish will depend on the nature of its phenotype in conjunction with the variability of the environment and the activities of others, it can be misleading to frame optimal-behaviour arguments in terms of the average individual.

ACKNOWLEDGEMENT

Financial support from The Royal Society (London) is gratefully acknowledged.

REFERENCES

Abrahams, M. V. (1989) Foraging guppies and the ideal free distribution: the influence of information on patch choice. *Ethology*, **82**, 116–26.
Abrahams, M. V. and Dill, L. M. (1989) A determination of the energetic equivalence of the risk of predation. *Ecology*, **70**, 999–1007.

Allan, J. D. (1975) The distributional ecology and diversity of benthic insects in Cement Creek, Colorado. *Ecology*, **56**, 1040–53.

Arak, A. (1984) Sneaky breeders, in *Producers and Scroungers* (ed. C. J. Barnard), Croom Helm, London, pp. 154–94.

Atema, J., Holland, K. and Ikehara, W. (1980) Olfactory responses of yellowfin tuna (*Thunnus albacares*) to prey odors: chemical search image. *J. Chem. Ecol.*, **6**, 457–65.

Austad, S. N. (1984) A classification of alternative reproductive behaviors and methods of field-testing ESS models. *Am. Zool.*, **24**, 309–19.

Barlow, G. W. (1983) The benefits of being gold: behavioural consequences of polychromatism in the midas cichlid *Cichlasoma citrinellum*, in *Predators and Prey in Fishes* (eds D. L. G. Noakes, D. G. Lindquist, G. S. Helfman and J. A. Ward), Junk, The Hague, pp. 73–86.

Barlow, G. W. and Ballin, P. J. (1976) Predicting and assessing dominance from size and coloration in the polychromatic midas cichlid. *Anim. Behav.*, **24**, 793–814.

Barlow, G. W. and McKaye, K. R. (1982) A comparison of feeding, spacing and aggression in color morphs of the midas cichlid. II. After 24 hours without food. *Behaviour*, **80**, 127–42.

Barlow, G. W., Bauer, D. H. and McKaye, K. R. (1975) A comparison of feeding spacing and aggression in color morphs of the midas cichlid. I. Food continuously present. *Behaviour*, **54**, 72–96.

Barlow, G. W., Francis, R. C. and Baumgartner, J. V. (1990) Do the colours of parents, companions and self influence assortative mating in the polychromatic Midas cichlid. *Anim. Behav.*, **40**, 713–22.

Barnard, C. J. (1984) When cheats may prosper, in *Producers and Scroungers* (ed. C. J. Barnard), Croom Helm, London, pp. 6–33.

Beukema, J. J. (1968) Predation by the three-spined stickleback (*Gasterosteus aculeatus* L.): the influence of hunger and experience. *Behaviour*, **31**, 1–126.

Bohlin, T., Dellefors, C. and Faremo, U. (1990) Large or small at maturity – theories on the choice of alternative male strategies in anadromous salmonids. *Annls zool. fenn.*, **27**, 139–48.

Breck, J. E. and Gitter, M. J. (1983) Effect of fish size on the reactive distance of bluegill (*Lepomis macrochirus*) sunfish. *Can. J. Fish. Aquat. Sci.*, **40**, 162–7.

Breden, F. and Stoner, G. (1987) Male predation risk determines female preference in the Trinidad guppy. *Nature, Lond.*, **329**, 831–3.

Breden, F., Scott, M. and Michel, E. (1987) Genetic differentiation for anti-predator behaviour in the Trinidad guppy. *Poecilia reticulata. Anim. Behav.*, **35**, 618–20.

Brooker, M. P. and Morris, D. L. (1980) A survey of the macroinvertebrate riffle fauna of the River Wye. *Freshwat. Biol.*, **10**, 437–58.

Caro, T. M. and Bateson, P. (1986) Organization and ontogeny of alternative tactics. *Anim. Behav.*, **34**, 1483–99.

Carvalho, G. R., Shaw, P. W., Magurran, A. E. and Seghers, B. H. (1991) Marked genetic divergence revealed by allozymes among populations of the guppy, *Poecilia reticulata* (Poeciliidae) in Trinidad. *Biol. J. Linn. Soc.*, **42**, 389–405.

Chellappa, S. and Huntingford, F. A. (1989) Depletion of energy reserves during reproductive aggression in male three-spined stickleback, *Gasterosteus aculeatus*, L. *J. Fish Biol.*, **35**, 315–16.

Chellappa, S., Huntingford, F. A., Strang, R. H. C. and Thomson, R. Y. (1989) Annual variation in energy reserves in male three-spined stickleback, *Gasterosteus aculeatus* L. (Pisces, Gasterosteidae) *J. Fish Biol.*, **35**, 275–86.

Clark, A. B. and Ehlinger, T. J. (1987) Pattern and adaptation in individual behavioral differences, in *Perspectives in Ethology*, Vol. 7, (eds P. P. G. Bateson and P. H. Klopfer), Plenum, New York, pp. 1–47.

Conover, D. A. (1984) Adaptive significance of temperature-dependent sex determination in a fish. *Am. Nat.*, **123**, 297–313.

Constanz, G. D. (1975) Behavioral ecology of the mating behavior of the male gila topminnow *Poeciliopsis occidentalis* (Cyprinodontiformes: Poecilidae). *Ecology*, **56**, 966–73.

Constanz, G. D. (1989) Reproductive biology of poeciliid fishes, in *Ecology and Evolution of Livebearing Fishes* (eds G. K. Meefe and F. F. Snelson), Prentice Hall, Englewood Cliffs, NJ, pp. 33–50.

Crowden, A. E. and Broom, D. M. (1980) Effects of the eyefluke, *Diplostomum spathaceum*, on the behaviour of dace, *Leuciscus leuciscus*. *Anim. Behav.*, **28**, 287–94.

Dalley, E. L., Andrews, C. W. and Green, J. M. (1983) Precocious male Atlantic salmon parr (*Salmo salar*) in insular Newfoundland. *Can. J. Fish. Aquat. Sci.*, **40**, 647–52.

Darwin, C. (1859) *The Origin of Species* (1968 edition), Penguin, London, 476 pp.

Dawkins, C. R. (1989) *The Selfish Gene*, Oxford University Press, Oxford, 352 pp.

Dellefors, C. and Faremo, U. (1988) Early sexual maturation in males of wild sea trout, *Salmo trutta* L., inhibits smoltification. *J. Fish Biol.*, **33**, 741–50.

Dill, L. M. (1983) Adaptive flexibility in the foraging behaviour of fishes. *Can. J. Fish. Aquat. Sci.*, **40**, 398–408.

Dill, L. M. (1987) Animal decision making and its ecological consequences: the future of aquatic ecology and behaviour. *Can. J. Zool.*, **65**, 803–11.

Dominey, W. J. (1981) Maintenance of female mimicry as a reproductive strategy in bluegill sunfish (*Lepomis macrochirus*), in *Ecology and Ethology of Fishes* (eds D. L. G. Noakes and J. A. Ward), Junk, The Hague, pp. 59–64.

Drager, B. and Chizar, D. (1982) Growth rate of bluegill sunfish (*Lepomis macrochirus*) maintained in groups and in isolation. *Bull. Psychon. Soc.*, **20**, 284–6.

Dunbar, R. I. M (1982) Intraspecific variation in mating strategy, in *Perspectives in Ethology*, Vol. 5 (eds P. P. G. Bateson and P. Klopfer), Plenum Press, New York, pp. 385–431.

Dussault, G. V. and Kramer, D. L. (1981) Food and feeding behavior of the guppy *Poecilia reticulata* (Pisces: Poeciliidae). *Can. J. Zool.*, **59**, 684–701.

Ehlinger, T. J. (1989) Learning and individual variation in bluegill foraging: habitat specific techniques. *Anim. Behav.*, **38**, 643–58.

Ehlinger, T. J. (1990) Habitat choice and phenotype-limited feeding efficiency in bluegill: individual differences and trophic polymorphism. *Ecology*, **71**, 886–96.

Ehlinger, T. J. and Wilson, D. S. (1988) A complex foraging polymorphism in bluegill sunfish. *Proc. natn Acad. Sci. (U.S.A.)*, **85**, 1878–82.

Endler, J. A. (1978) A predator's view of animal color patterns. *Evol. Biol.*, **11**, 319–64.

Endler, J. A. (1980) Natural selection on color patterns in *Poecilia reticulata*. *Evolution*, **34**, 76–91.

Endler, J. A. (1983) Natural and sexual selection on color patterns in poeciliid fishes. *Env. Biol. Fishes*, **9**, 173–90.

Endler, J. A. (1987) Predation, light intensity and courtship behaviour in *Poecilia reticulata* (Pisces: Poeciliidae). *Anim. Behav.*, **35**, 1376–85.

Endler, J. A. (1989) Conceptual and other problems in speciation, in *Speciation and its Consequences*, (eds D. Otte and J. A. Endler), Sinauer, Sunderland, Mass, pp. 625–48.

Eriksson, T., Eriksson, L.-O. and Lundqvist, H. (1987) Adaptive flexibility in life history tactics of mature Baltic salmon parr in relation to body size and environment. *Am. Fish. Soc. Symp.*, **1**, 236–43.

Ferguson, A. (1986) Lough Melvin, a unique fish community. *Occ. Paps. Irish Sci. Technol., R. Dublin Soc.*, **1**, 1–17.

Ferguson, A. (1989) Genetic differences among brown trout, *Salmo trutta*, stocks and their importance for conservation and management of the species. *Freshwat. Biol.*, **21**, 35–46.

Ferguson, A. and Mason, F. M. (1981) Allozyme evidence for reproductively isolated sympatric populations of brown trout, *Salmo trutta* L., in Lough Melvin, Ireland. *J. Fish Biol.*, **18**, 629–42.

Fraser, D. F. and Gilliam, J. F. (1987) Feeding under predation hazard: response of the guppy and Hart's rivulus from sites with contrasting predation hazard. *Behav. Ecol. Sociobiol.*, **21**, 203–9.

Fretwell, S. D. and Lucas, H. L. (1970) On territorial behaviour and other factors influencing habitat distribution in birds. *Acta Biotheor.*, **19**, 16–36.

Fricke, H. W. (1979) Mating system, resource defence and sex changes in the anemone fish *Amphiprion akallopsis*. *Z. Tierpsychol.*, **50**, 313–26.

Giles, N. (1983) Behavioural effects of the parasite, *Schistocephalus solidus*, (Cestoda) on an intermediate host, the three-spined stickleback, *Gasterosteus aculeatus* L. *Anim. Behav.*, **31**, 1192–4.

Giraldeau, L.-A. (1984) Group foraging: the skill pool effect and frequency-dependent learning. *Am. Nat.*, **124**, 72–9.

Godin, J.-G. J. (1978) Behavior of juvenile pink salmon (*Oncorhynchus gorbuscha* Walbaum) toward novel prey: influence of ontogeny and experience. *Env. Biol. Fishes*, **3**, 261–6.

Godin, J.-G. J. and Smith, S. A. (1988) A fitness cost of foraging in the guppy. *Nature, Lond.*, **333**, 69–71.

Godin, J.-G. and Sproul, C. D. (1988) Risk taking in parasitized sticklebacks under threat of predation: effects of energetic need and food availability. *Can. J. Zool.*, **66**, 2360–67.

Goodey, W. and Liley, N. R. (1986) The influence of early experience on escape behaviour in the guppy (*Poecilia reticulata*). *Can. J. Zool.*, **64**, 885–8.

Gross, M. R. (1982) Sneakers, satellites and parentals: polymorphic mating strategies in North American sunfish. *Z. Tierpsychol.*, **60**, 1–26.

Gross, M. R. (1984) Sunfish, salmon and the evolution of alternative reproductive strategies and tactics in fish, in *Fish Reproduction: Strategies and Tactics* (eds G. Potts and R. J. Wootton), Academic Press, London, pp. 55–75.

Gross, M. R. (1985) Disruptive selection for alternative life histories in salmon. *Nature, Lond.*, **313**, 47–8.

Hamilton, W. D. (1971) Geometry for the selfish herd. *J. theor. Biol.*, **31**, 295–311.

Haskins, C. P., Haskins, E. F., McLaughlin, J. J. A. and Hewitt, R. E. (1961) Polymorphism and population structure in *Lebistes reticulatus*, a population study, in *Vertebrate Speciation* (ed. W. F. Blair), University of Texas Press, Austin, pp. 320–95.

Hassell, M. P. and May, R. M. (1985) From individual behaviour to population dynamics, in *Behavioural Ecology: Ecological Consequences of Adaptive Behaviour* (eds R. M. Sibly and R. H. Smith), Blackwell, Oxford, pp. 3–32.

Hedrick, P. W. (1983) *Genetics of Populations*, Science Books, Boston, MA.

Helfman, G. S. (1984) School fidelity in fishes: the yellow perch pattern. *Anim. Behav.*, **32**, 663–72.

Helfman, G. S. and Schultz, E. T. (1984) Social transmission of behavioural traditions in a coral reef fish. *Anim. Behav.*, **32**, 379–84.

Helfman, G. S., Meyer, J. L. and McFarland, W. N. (1982) The ontogeny of twilight migration patterns in grunts (Pisces: Haemulidae) *Anim. Behav.*, **30**, 317–26.

Holtby, L. B. and Healey, M. C. (1990) Sex-specific life history tactics and risk-taking in coho salmon. *Ecology*, **71**, 678–90.

Houde, A. E. (1987) Mate choice based upon naturally occurring color pattern variation in a guppy population. *Evolution*, **41**, 1–10.

Houde, A. E. (1988) Genetic difference in female choice between two guppy populations. *Anim. Behav.*, **36**, 510–16.

Houde, A. E. and Endler, J. A. (1990) Correlated evolution of female mating preferences and male color patterns in the guppy *Poecilia reticulata. Science*, **248**, 1405–8.

Huntingford, F. A. (1976) A comparison of the reaction of sticklebacks in different reproductive conditions towards conspecifics and a predator. *Anim. Behav.*, **24**, 694–7.

Huntingford, F. A. and Coulter, R. M. (1989) Habituation of predator inspection in the three-spined stickleback, *Gasterosteus aculeatus* L. *J. Fish Biol.*, **35**, 153–4.

Huntingford, F. A. and Wright, P. J. (1989) How sticklebacks learn to avoid dangerous feeding patches. *Behav. Process.*, **19**, 181–9.

Huntingford, F. A., Metcalfe, N. B., Thorpe, J. E., Graham, W. D. and Adams, C. E. (1990) Social dominance and body size in Atlantic salmon parr, *Salmo salar* L. *J. Fish Biol.*, **36**, 877–82.

Ibrahim, A. A. and Huntingford, F. A. (1988) Foraging efficiency in relation to within-species variation in morphology in three-spined sticklebacks, *Gasterosteus aculeatus. J. Fish Biol.*, **33**, 823–4.

Johnson, M. S. (1988) Founder effects and geographic variation in the land snail, *Theba pisana. Heredity*, **61**, 133–42.

Jones, J. W. (1959) *The Salmon*, Collins, London.

Jones, J. W. and King, G. M. (1950) Further experimental observation of the spawning behaviour of Atlantic salmon. *Proc. zool. Soc. Lond.*, **120**, 317–23.

Jones, J. W. and Orton, J. H. (1940) The paedogenetic male cycle in *Salmo salar* L. *Proc. R. Soc.*, **128B**, 485–99.

Keen, W. H. (1982) Behavioral interactions and body size differences in competition for food among juvenile brown bullhead. *Can. J. Fish. Aqua. Sci.*, **39**, 316–20.

Knight, A. J., Hughes, R. N. and Ward, R. N. (1987) A striking example of the founder effect in the mollusc *Littorina saxatilis. Biol. J. Linn. Soc.*, **32**, 417–26.

Kodric-Brown, A. (1989) Dietary carotenoids and male mating success: an environmental component to female choice. *Behav. Ecol. Sociobiol.*, **25**, 393–401.

Kodric-Brown. A. (1990) Mechanisms of sexual selection: insights from fishes. *Annls zool. fenn.*, **27**, 87–100.

Kornfield, I. and Taylor, J. N. (1983) A new species of polymorphic fish, *Cichlasoma minckleyi* from Cuarto Cienegas, Mexico (Teleostei: Cichlidae). *Proc. Biol. Soc. Wash.*, **96**, 253–69.

Kornfield, I., Smith, D. C., Gagnon, P. S. and Taylor, J. N. (1982) The cichlid fish of Cuarto Cienegas, Mexico: direct evidence of conspecificity among distinct trophic morphs. *Evolution*, **36**, 658–64.

Krebs, J. R. and Davies, N. B. (1987) *An Introduction to Behavioural Ecology*, Blackwell, Oxford, 389 pp.

L'Abée-Lund, J. H. and Hindar, K. (1990) Interpopulation variation in reproductive traits of anadromous brown trout. *J. Fish Biol.*, **37**, 755–64.

Langdon, J. S. and Thorpe, J. E. (1985) The ontogeny of smoltification: developmental patterns of gill Na/K-ATPase, SDH and chloride cells in juvenile Atlantic salmon, *Salmo salar. Aquaculture*, **45**, 43–95.

Lavin, P. A. and McPhail, J. D. (1985) The evolution of freshwater diversity in the threespine stickleback (*Gasterosteus aculeatus*): site specific differentiation of trophic morphology. *Can. J. Zool.*, **63**, 2632–43.

Lavin, P. A. and McPhail, J. D. (1986) Adaptive divergence of trophic phenotype among freshwater populations of the three-spined stickleback (*Gasterosteus aculeatus*). *Can. J. Fish. Aquat. Sci.*, **43**, 2455–63.

Licht, T. (1989) Discriminating between hungry and satiated predators: the response of guppies (*Poecilia reticulata*) from high and low predation sites. *Ethology*, **82**, 238–42.

Liley, N. R. (1966) Ethological isolating mechanisms in four sympatric species of poeciliid fishes. *Behaviour (Supp.)*, **13**, 1–197.

Liley, N. R. and Seghers, B. H. (1975) Factors affecting the morphology and behaviour of guppies in Trinidad, in *Function and Evolution in Behaviour* (eds G. P. Baerends, C. Beer and A. Manning), Clarendon Press, Oxford, pp. 92–118.

Lima, S. L. and Dill, L. M. (1990) Behavioral decisions made under risk of predation: a review and prospectus. *Can. J. Zool.*, **68**, 619–40.

Lomnicki, A. (1988) *Population Ecology of Individuals*, Princeton University Press, Princeton, NJ.

Lott, D. F. (1984) Intraspecific variation in the social systems of wild vertebrates. *Behaviour*, **88**, 266–325.

Luyten, P. H. and Liley, N. R. (1985) Geographic variation in the sexual behaviour of the guppy, *Poecilia reticulata* (Peters). *Behaviour*, **95**, 164–79.

Luyten, P. H. and Liley, N. R. (1991) Sexual selection and competitive mating success of male guppies (*Poecilia reticulata*) from four Trinidad populations. *Behav. Ecol. Sociobiol.*, **28**, 329–36.

McFarland, W. N. and Kotchian, N. M. (1982) Interaction between schools of fish and mysids. *Behav. Ecol. Sociobiol.*, **11**, 71–6.

Magnhagen, C. (1990) Reproduction under predation risk in the sand goby, *Pomatoschistus minutus*, and the black goby, *Gobius niger:* the effect of age and longevity. *Behav. Ecol. Sociobiol.*, **26**, 331–5.

Magnhagen, C. (1991) Predation risk as a cost of reproduction. *Trends Ecol. Evol.*, **6**, 183–6.

Magurran, A. E. (1986) Predator inspection behaviour in minnow shoals: differences between population and individuals. *Behav. Ecol. Sociobiol.*, **19**, 267–73.

Magurran, A. E. (1990a) The adaptive significance of schooling as an anti-predator defence in fish. *Annls zool. fenn.*, **27**, 51–66.

Magurran, A. E. (1990b) The inheritance and development of minnow anti-predator behaviour. *Anim. Behav.*, **39**, 834–42.

Magurran, A. E. and Nowak, M. N. (1991) Another battle of the sexes: the consequences of sexual asymmetry in mating costs and predation risk in the guppy, *Poecilia reticulata. Proc. R. Soc.*, **246B**, 31–8.

Magurran, A. E. and Pitcher, T. J. (1983) Foraging, timidity and shoal size in minnows and goldfish. *Behav. Ecol. Sociobiol.*, **12**, 147–52.

Magurran, A. E. and Pitcher, T. J. (1987) Provenance, shoal size and the sociobiology of predator evasion behaviour in minnow shoals. *Proc. R. Soc.*, **229B**, 439–63.

Magurran, A. E. and Seghers, B. H. (1990a) Risk sensitive courtship in the guppy (*Poecilia reticulata*). *Behaviour*, **112**, 194–201.

Magurran, A. E. and Seghers, B. H. (1990b) Population differences in the schooling behaviour of newborn guppies. *Poecilia reticulata. Ethology*, **84**, 334–42.

Magurran, A. E. and Seghers, B. H. (1990c) Population differences in predator recognition and attack cone avoidance in the guppy *Poecilia reticulata. Anim. Behav.*, **40**, 443–52.

Magurran, A. E. and Seghers, B. H. (1991) Variation in schooling and aggression amongst guppy (*Poecilia reticulata*) populations in Trinidad. *Behaviour*, **118**, 214–234.

Mann, R. H. K., Mills, C. A. and Crisp, D. T. (1984) Geographical variation in the life-history tactics of some species of freshwater fish, in *Fish Reproduction. Strategies and Tactics* (eds G. W. Potts and R. J. Wootton), London, Academic Press, pp. 171–86.

Maynard Smith, J. (1982) *Evolution and the Theory of Games*, Cambridge University Press, Cambridge.

Metcalfe, N. B. (1986) Intraspecific variation in competitive ability and food intake in salmonids: consequences for energy budgets and growth rates. *J. Fish Biol.*, **28**, 525–32.

Metcalfe, N. B. (1989) Differential response to a competitor by Atlantic salmon adopting alternative life-history strategies. *Proc. R. Soc.*, **236B**, 21–7.

Metcalfe, N. B., Huntingford, F. A., Graham, W. D. and Thorpe, J. E. (1989) Early social status and the development of life-history strategies in Atlantic salmon. *Proc. R. Soc.*, **236B**, 7–19.

Metcalfe, N. B., Huntingford, F. A. and Thorpe, J. E. (1987) The influence of predation risk on the feeding motivation and foraging strategy of juvenile Atlantic salmon. *Anim. Behav.*, **35**, 901–11.

Milinski, M. (1982) Optimal foraging: the influence of intraspecific competition on diet selection. *Behav. Ecol. Sociobiol.*, **11**, 109–15.

Milinski, M. (1984a) A predator's costs of overcoming the confusion effect of swarming prey. *Anim. Behav.*, **32**, 1157–62.

Milinski, M. (1984b) Parasites determine a predator's optimal feeding strategy. *Behav. Ecol. Sociobiol.*, **15**, 35–7.

Milinski, M. (1984c) Competitive resource sharing: an experimental test of a learning rule for ESSs. *Anim. Behav.*, **32**, 233–42.

Milinski, M. (1985) Risk of predation of parasitized sticklebacks (*Gasterosteus aculeatus* L.) under competition for food. *Behaviour*, **93**, 203–16.

Milinski, M. (1986) A review of competitive resource sharing under constraints in sticklebacks. *J. Fish Biol.*, **29** (Supp. A), 1–14.

Milinski, M. and Bakker, T. C. M. (1990) Female sticklebacks use male coloration in mate choice and hence avoid parasitized males. *Nature, Lond.*, **344**, 330–33.

Mills, C. A. (1988) The effect of extreme northerly climatic conditions on the life history of the minnow, *Phoxinus phoxinus* (L.). *J. Fish Biol.*, **33**, 545–61.

Mittlebach, G. G. (1981) Foraging efficiency and body size: a study of optimal diet and habitat use by bluegills. *Ecology*, **62**, 1370–86.

Morgan, M. J. and Godin, J.-G. (1985) Antipredator benefits of schooling behaviour in a cyprinidontid fish, the banded killifish (*Fundulus diaphanus*). *Z. Tierpsychol.*, **70**, 236–46.

Myers, R. A. and Hutchings, J. A. (1987) Mating of anadromous Atlantic salmon. *Salmo salar* L., with mature male parr. *J. Fish Biol.*, **31**, 143–6.

Neill, S. R. StJ. and Cullen, M. (1974) Experiments on whether schooling by their prey affects the hunting behaviour of cephalopod and fish predators. *J. Zool., Lond.*, **172**, 549–69.

Orton, J. H., Jones, J. W. and King, G. M. (1938) The male sexual stage in salmon parr (*Salmo salar* L. juv.). *Proc. R. Soc.*, **125B**, 103–14.

Osenberg, C. W., Werner, E. E., Mittlebach, G. G. and Hall, D. J. (1988) Growth patterns in bluegill (*Lepomis macrochirus*) and pumpkinseed (*L. gibbosus*) sunfish: environmental variation and the importance of ontogenetic niche shifts. *Can. J. Fish. Aquat. Sci.*, **45**, 17–26.

Parker, G. A. (1982) Phenotype-limited evolutionarily stable strategies, in *Current Problems in Sociobiology* (ed. King's College Sociobiology Group), Cambridge University Press, Cambridge, pp. 65–89.

Parker, G. A. (1984) Evolutionarily stable strategies, in *Behavioural Ecology*, 2nd edn (eds J. R. Krebs and N. B. Davies), Blackwell, Oxford, pp. 30–61.

Parker, G. A. (1985) Population consequences of evolutionary stable strategies, in *Behavioural Ecology: Ecological Consequences of Adaptive Behaviour* (eds R. M. Sibly and R. H. Smith), Blackwell, Oxford, pp. 33–58.

Parker, G. A. and Sutherland, W. J. (1986) Ideal free distributions when individuals differ in competitive ability: phenotype limited ideal free models. *Anim. Behav.*, **34**, 1222–42.

Parrish, J. (1989) Re-examining the selfish herd: are central fish safer? *Anim. Behav.*, **38**, 1048–53.

Partridge, L. (1988) The rare-male effect: what is its evolutionary significance? *Phil. Trans. R. Soc.*, **319B**, 525–39.

Partridge, L. and Green, P. (1985) Intraspecific feeding specializations and population dynamics, in *Behavioural Ecology: Ecological Consequences of Adaptive Behaviour* (eds R. Sibly and R. H. Smith), Blackwell, Oxford, pp. 207–26.

Patten, B. G. (1977) Body size and learned avoidance as factors affecting predation on coho salmon fry, *Oncorhynchus kisutch*, by torrent sculpin, *Cottus rhotheus*. *Fishery Bull.*, *Fish Wildl. Serv. U.S.*, **75**, 457–9.

Pitcher, T. J., Magurran, A. E and Allan, J. A. (1986) Size segregative behaviour in minnow shoals. *J. Fish Biol.*, **29** (Supp. A), 83–96.

Pitcher, T. J., Wyche, C. and Magurran, A. E. (1982) Evidence for position preferences in mackerel schools. *Anim. Behav.*, **30**, 932–4.

Potts, G. W. (1980) The predatory behaviour of *Caranx melamphygus* (Pisces) in the channel environment of the Aldabra Atoll (Indian Ocean). *J. Zool., (Lond.,)* **192**, 323–50.

Ranta, E. and Lindström, K. (1990) Assortative schooling in three-spined sticklebacks. *Ann. zool. fenn.*, **27**, 67–75.

Reist, J. (1983) Behavioural variation in pelvic phenotypes of brook stickleback, *Culaea inconstans*, in response to predation by Northern pike, *Esox lucius*, in *Predators and Prey in Fishes* (eds D. L. G. Noakes, D. G. Lindquist, G. S. Helfman and J. A. Ward), Junk, The Hague, pp. 93–105.

Reynolds, J. D. (1990) The evolution of female choice and male courtship: theory and tests with Trinidadian guppies, PhD thesis, University of Toronto.

Reznick, D. N. (1982) The impact of predation on life-history evolution in Trinidadian guppies: genetic basis of observed life history patterns. *Evolution*, **36**, 1236–59.

Reznick, D. N. and Endler, J. A. (1982) The impact of predation on life history evolution in Trinidadian guppies (*Poecilia reticulata*). *Evolution*, **36**, 160–77.

Reznick, D. N., Bryga, H. and Endler, J. A. (1990) Experimentally induced life-history evolution in a natural population. *Nature, Lond.*, **346**, 357–9.

Ringler, N. H. (1983) Variation in foraging tactics of fishes, in *Predators and Prey in Fishes* (eds D. L. G. Noakes, D. G. Lindquist, G. S. Helfman and J. A. Ward), Junk, The Hague, pp. 159–72.

Rodd, F. H. and Reznick, D. A. (1991) Life history evolution in guppies: III. The impact of prawn predation on guppy life histories. *Oikos*, **62**, 13–19.

Rowe, D. K. and Thorpe, J. E. (1990) Differences in growth between maturing and non-maturing male Atlantic salmon, *Salmo salar* L. parr. *J. Fish Biol.*, **36**, 643–59.

Rowland, W. J. (1982) The effects of male nuptial coloration on stickleback aggression: a re-examination. *Behaviour*, **80**, 118–26.

Rubenstein, D. I. (1981) Individual variation and competition in the Everglades pygmy sunfish. *J. Anim. Ecol.*, **50**, 337–50.

Savino, J. F. and Stein, R. A. (1982) Predator prey interactions between largemouth bass and bluegills as influenced by simulated submerged vegetation. *Trans. Am. Fish. Soc.*, **111**, 255–66.

Savino, J. F. and Stein, R. A. (1989) Behavioural interactions between fish predators and their prey: effect of plant density. *Anim. Behav.*, **37**, 311–21.

Schaffer, W. M. and Elson, P. F. (1975) The adaptive significance of variation in life history among local populations of Atlantic salmon in North America. *Ecology*, **56**, 577–90.

Seghers, B. H. (1974) Schooling behavior in the guppy (*Poecilia reticulata*): an evolutionary response to predation. *Evolution*, **28**, 486–9.

Shapiro, D. Y. (1979) Social behavior, group structure and the control of sex reversal in a hermaphroditic fish, in *Advances in the Study of Behavior*, Vol. 10 (eds J. S. Rosenblatt, R. A. Hinde, C. Beer and M.-C. Busnel), Academic Press, New York, pp. 43–102.

Shaw, E. (1970) Schooling in fishes: critique and review, in *Development and Evolution of Behaviour* (eds. L. R. Aronson, B. Tobach, D. S. Lehrman and J. S. Rosenblatt), W. H. Freeman, San Francisco, pp. 452–80.

Slater, P. J. B. (1983) *The development of individual behaviour in animal behaviour 3: genes, development and learning* (eds T. R. Halliday and P. J. B. Slater), Blackwell, Oxford, pp. 82–113.

Smith, R. H. and Sibly, R. (1985) Behavioural ecology and population dynamics: towards a synthesis, in *Behavioural Ecology: Ecological Consequences of Adaptive Behaviour* (eds R. M. Sibly and R. H. Smith), Blackwell, Oxford, pp. 577–92.

Sosiak, A. J., Randall, R. G. and McKenzie, J. A. (1979) Feeding by hatchery-reared and wild Atlantic salmon (*Salmo salar*) parr in streams. *J. Fish. Res. Bd Can.*, **36**, 1408–12.

Stearns, S. C. (1983) A natural experiment in life-history evolution: field data on the introduction of mosquitofish (*Gambusia affinis*) to Hawaii. *Evolution*, **37**, 601–17.

Steele, J. H. (1976) Patchiness, in *The Ecology of the Seas* (eds D. H. Cushing and J. J. Walsh) Blackwell, Oxford, pp. 98–115.

Stoner, G. and Breden, F. (1988) Phenotypic differentiation in female preference related to geographic variation in male predation risk in the Trinidad guppy (*Poecilia reticulata*). *Behav. Ecol. Sociobiol.*, **22**, 285–91.

Sutherland, W. J. and Parker, G. A. (1985) Distribution of unequal competitors, in *Behavioural Ecology: Ecological Consequences of Adaptive Behaviour* (eds R. M. Sibly and R. H. Smith), Blackwell, Oxford, pp. 255–74.

Taborsky, M. (1984) Broodcare helpers in the cichlid fish *Lamprologus brichardi*: their costs and benefits. *Anim. Behav.*, **32**, 1236–52.

Thorpe, J. E. (1987) Smolting versus residency: developmental conflict in salmonids. *Am. Fish. Soc. Symp.*, **1**, 244–52.

Thorpe, J. E. and Morgan, R. I. G. (1980) Growth-rate and smolting-rate of progeny of male Atlantic salmon parr, *Salmo salar*. L. *J. Fish Biol.*, **17**, 451–60.

Thorpe, J. E., Morgan, R. I. G., Talbot, C. and Miles, M. S. (1983) Inheritance of development rates in Atlantic salmon. *Aquaculture*, **33**, 123–32.

Tinbergen, N. (1951) *The Study of Instinct*, Oxford University Press, Oxford.

Trexler, J. C. (1988) Hierarchical organization of genetic variation in the sailfin molly, *Poecilia latipinna* (Pisces: Poeciliidae). *Evolution*, **42**, 1006–17.

Trexler, J. C. (1989a) Phenotypic plasticity in the sailfin molly, *Poecilia latipinna* (Pisces: Poeciliidae). I Field experiments. *Evolution*, **44**, 143–56.

Trexler, J. C. (1989b) Phenotypic plasticity in the sailfin molly, *Poecilia latipinna* (Pisces: Poeciliidae). II Laboratory experiments. *Evolution*, **44**, 157–67.

Trexler, J. C. (1989c) Phenotypic plasticity in poeciliid life histories, in *Ecology and Evolution of Livebearing Fishes (Poeciliidae)* (eds G. K. Meefe and F. F. Snelson), Prentice Hall, Englewood cliffs, NJ, pp. 201–14.

Tulley, J. J. and Huntingford, F. A. (1987) Parental care and the development of adaptive variation in anti-predator responses in sticklebacks. *Anim. Behav.*, **35**, 1570–72.

Vinyard, G. L. (1982) Variable kinematics of Sacramento perch (*Archoplites interruptus*) capturing evasive and non-evasive prey. *Can. J. Fish. Aquat. Sci.*, **39**, 208–11.

Vøllestad, L. A. and L'Abée-Lund, J. H. (1990) Geographic variation in life-history strategy of female roach. *Rutilus rutilus* L. *J. Fish Biol.*, **37**, 853–64.

Warner, R. R. (1988) Traditionality of mating-site preferences in a coral reef fish. *Nature Lond.*, **335**, 719–21.

Warner, R. R. (1990) Resource assessment versus tradition in mating-site determination. *Am. Nat.*, **135**, 205–17.

Warner, R. R., Robertson, D. R. and Leigh, E. G. (1975) Sex change and sexual selection. *Science*, **190**, 934–44.

Webb, P. W. (1984) Body form, locomotion and foraging in aquatic vertebrates. *Am. Zool.*, **24**, 107–20.

Webb, P. W. and Skadsen, J. M. (1980) Strike tactics of *Esox. Can. J. Zool.*, **58**, 1462–9.

Werner, E. E. (1984) The mechanisms of species interactions and community organization in fish, in *Ecological Communities: Conceptual Issues and the Evidence* (eds D. R. Strong, D. Simberloff, L. G. Abele and A. B. Thistle), Princeton University Press, Princeton, NJ, pp. 360–82.

Werner, E. E. and Gilliam, J. F. (1984) The ontogenetic niche and species interactions in size-structured populations. *A. Rev. Ecol. Syst.*, **15**, 393–425.

Werner, E. E., Gilliam, J. F., Hall, D. J. and Mittlebach, G. G. (1983a) An experimental test of the effects of predation risk on habitat use in fish. *Ecology*, **64**, 1540–48.

Werner, E. E., Mittlebach, G. G., Hall, D. J. and Gilliam, J. F. (1983b) Experimental tests of optimal habitat use in fish: the role of relative habitat profitability *Ecology*, **64**, 1525–39.

Whoriskey, F. G. and FitzGerald, G. J. (1985) Sex, cannibalism and sticklebacks. *Behav. Ecol. Sociobiol.*, **18**, 15–18.

Wilson, D. S. (1975) The adequacy of body size as a niche difference. *Am. Nat.*, **109**, 769–84.

Wilson, D. S. (1989) The diversification of single gene pools by density- and frequency-dependent selection, in *Speciation and Its Consequences* (eds D. Otte and J. A. Endler), Sinauer, Sunderland, Mass, pp. 366–85.

Windell, J. T. (1978) Digestion and the daily ration of fishes, in *Ecology of Freshwater Fish Production* (ed S. D. Gerking), Blackwell, Oxford, pp. 159–83.

Wootton, R. J. (1984) *A Functional Biology of Sticklebacks*, Croom Helm, London, 265 pp.

Wootton, R. J. (1990) *Ecology of Teleost Fishes*, Chapman and Hall, London, 404 pp.

Wright, S. (1970) Random drift and the shifting balance theory of evolution, in *Mathematical Topics in Population Genetics* (ed K. Kojima), Springer-Verlag, New York, pp. 1–31.

Chapter fourteen

Fish behaviour by day, night and twilight

Gene S. Helfman

14.1 INTRODUCTION

The simple diel cycle of rising and setting of the sun imposes on the behaviour and activity of fishes a dramatic, overriding set of predictable constraints. As a direct result, many kinds of behaviour and the species that engage in them follow characteristic convergent patterns that transcend geographic and taxonomic boundaries. These patterns can be recognized in such fundamental activities as the times when fishes feed, breed, aggregate and rest, in the transitions between activities, in food types, and in the ways in which fishes feed and avoid being eaten. The objectives of this chapter are to review the available information concerning the influences of day, night and twilight on various classes of fish behaviour; to delimit general diel activity patterns that characterize fishes in different habitat types; and to explore the environmental, ecological, physiological and developmental factors that interact with the cycle of daylight and darkness in determining diel patterns of fish behaviour. Throughout the chapter, day and daytime refer to daylight hours; night and night-time refer to periods of darkness; crepuscular refers to twilight periods of dusk and dawn (sunset and sunrise); and diel refers to the 24-hour cycle.

Scope of the chapter

The emphasis in this chapter is on direct field observations of diel patterns in the wild, particularly as influenced by changing ambient illumination. Circadian rhythms of fish activity (24-hour patterns driven by endogenous factors with external cues often involving daylength and daylight) will not be considered, nor will endogenous influences of brain secretions (e.g. Kavaliers, 1989). However, their importance should not be discounted (Thorpe, 1978). Other exogenous factors besides photoperiod also affect activity cycles in fishes.

Behaviour of Teleost Fishes 2nd edn. Edited by Tony J. Pitcher. Published in 1993 by Chapman & Hall. ISBN 0 412 42930 6 (HB) and 0 412 42940 3 (PB).

Table 14.1 Diel activity patterns, defined as major feeding period, of better-known groups and families of teleost fishes. In many cases, information is available for only a fraction of the species in a group. Some large families appear in more than one column. Families are listed alphabetically within columns: common names are in parenthesis. See Hobson (1965), Helfman (1978), Lowe-McConnell (1987), and Potts (1990) for references and details on most groups

All or most species studied diurnal	All or most species studied nocturnal	Both diurnal and nocturnal species	Several crepuscular species*	Several species without well-defined activity periods
Acanthuridae (surgeonfishes)	Amiidae (bowfin)	Carangidae (jacks)	Carangidae (jacks)	Aulostomidae (trumpetfishes)
Ammodytidae (sandeels)	Anguillidae (eels)	Catostomidae (suckers)	Elopidae (tarpon)	Muraenidae (moray eels)
Anthiinae	Anomalopidae (flashlightfishes)	Centrarchidae (sunfishes)	Fistulariidae (cornetfishes)	Scombridae (mackerels)
Characoidei (characins)	Apogonidae (cardinalfishes)	Congridae (conger eels)	Gadoidei (cod)	Scorpaenidae (scorpionfishes)
Chaetodontidae (butterflyfishes)	Batrachoididae (toadfishes)	Cyprinidae (minnows)	Lutjanidae (snappers)	Serranidae (groupers)
Cichlidae (cichlids)	Clupeidae (herring)	Gadoidei (cod)	Serranidae (groupers)	
Cirrhitidae (hawkfishes)	Diodontidae (porcupinefishes)	Leiognathidae (ponyfishes)		
Cyprinodontidae (killifishes)		Mullidae (goatfishes)		

Embiotocidae (surfperch)	Grammistidae (soapfishes)	Pleuronectiformes (flatfishes)
Esocidae (pike)	Gymnotoidei (knifefishes)	Salmonidae (salmonids)
Gobiidae (gobies)	Haemulidae (grunts)	Serranidae (groupers)
Kyphosidae (sea chubs)	Holocentridae (squirrelfishes)	Sphyraenidae (barracuda)
Labridae (wrasses)	Kuhliidae (aholeholes)	
Mugilidae (mullets)	Lutjanidae (snappers)	
Mullidae (goatfishes)	Mormyridae (mormyrs)	
Percidae (perches)	Ophichthidae (snake eels)	
Pomacanthidae (angelfishes)	Ophidiidae (cusk eels)	
Pomacentridae (damselfishes)	Pempheridae (sweepers)	
Scaridae (parrotfishes)	Sciaenidae (drums)	
Siganidae (rabbitfishes)	Siluriformes (catfishes)	
Synodontidae (lizardfishes)		

*Also active at other times.

Seasonal progressions involve changing temperature, with 'normal' diel patterns disrupted during the cold-water period of later autumn, winter and early spring in temperate and polar regions (Emery, 1978; Muller, 1978). Tide also influences and sometimes overrides diel activity cycles (see Chapter 15 by Gibson, this volume). Certain environments lack daily light cues and contain fishes that are continually active or arrhythmic. The deep sea is one example, but little is known of the diel activities of bathypelagic fishes. The behaviour of cave-dwelling fishes is covered by Parzefall in Chapter 17, this volume. Diel vertical migrations, particularly characteristic of temperate marine species in the Atlantic, have been thoroughly reviewed by Woodhead (1966) and Blaxter (1976).

14.2   ACTIVITY PERIODS IN FISHES

Feeding

Feeding involves several categories of behaviour, but as used here generally refers to the major period of foraging, including movement between foraging areas and other sites. In the wild, most of a fish's day appears to be spent either pursuing food or avoiding predators; many fishes appear to separate the day into an active, food-gathering phase and a relatively inactive, resting phase that is intimately linked with predator avoidance. Territoriality, comfort, and other maintenance activities are generally restricted to the foraging period.

The majority of fishes at tropical and temperate latitudes feed primarily during either the day or the night, a smaller number foraging during the crepuscular or twilight periods of dusk and dawn (Gushima *et al.*, 1977; Emery, 1978). In general, the timing of activity is a familial characteristic (Table 14.1), but exceptions to this generalization are certainly common. For example, many species of nominally nocturnal fishes will feed during the day if food is available, as any fisherman knows. Such an exception reflects the well-known opportunism that characterizes many fishes, particularly predatory forms (Larkin, 1979); it also suggests that activity patterns in fishes may be strongly determined by the activity patterns of their prey, a topic sorely in need of quantitative investigation (e.g. Naud and Magnan, 1988; Glozier and Culp, 1989).

The relative numbers of fishes active at different periods of the diel cycle in different assemblages are remarkably similar. A survey of such 'temporal ratios', defined as the percentage of species active by day, night or twilight in different fish assemblages, shows that about one-half to two-thirds of the species in most assemblages are diurnal, one-quarter to one-third are nocturnal, and the remaining 10% or so are primarily crepuscular (Helfman, 1978; Ebeling and Hixon, 1991). These temporal ratios are probably a result of the relationship between phylogenetic position, temporal habits and trophic patterns among teleost fishes, as noted by Hobson (1974) and Hobson and Chess (1976). More primitive or generalized fishes are typically large-mouthed, nocturnal or crepuscular predators, whereas more advanced species have often specialized towards diurnality and feeding on smaller animals or on plants. Based on the food web or pyramid that typifies most animal communities, one would expect herbivores to be most numerous, and specialized carnivores least numerous as individuals in an assemblage.

Day and night are not necessarily synonymous with light and dark in aquatic environments. In clear habitats where light penetrates well, diurnally active fishes may maintain activity on bright, moonlit nights (e.g. threespine stickleback, Allen and Wooton, 1984). Alternatively, in many lakes and nearshore marine habitats, suspended sediments and plankton productivity can lead to lightless conditions at midday in relatively shallow depths. Diel periodicity in the feeding behaviour of fishes in such habitats must have a strong endogenous basis and prey capture will depend on non-visual sensory capabilities (e.g. Hoekstra and Janssen, 1985). For visual feeders, success in environments that vary in turbidity and light penetration may depend on relative abilities to forage under low light conditions (e.g. Grecay and Targett, in press). Successional changes from percid-dominated to cyprinid-dominated assemblages in temperate lakes that have undergone eutrophication have been linked directly to the relative importance of vision in the feeding of different species (e.g. Diehl, 1988).

Breeding

A notable exception to characteristic diel activity patterns involves the relative breakdown of normal patterns during spawning periods (e.g. Nash, 1982). Many species that usually feed during the day and then seek shelter at or shortly after sunset may breed actively at sunset (Neudecker and Lobel, 1982; Thresher, 1984). Some strongly diurnal species will even spawn at night (e.g. yellow perch, *Perca flavescens*, Helfman, 1981). Such night spawning may be even more common than we suspect, given our inability to make detailed nocturnal observations. Normal rhythms of activity are also disturbed during the reproductive season. 'Nocturnal' or 'diurnal' fish may be continuously active during the day and night when engaged in parental care (e.g. diurnal damselfishes, Emery, 1973b; the nocturnal ictalurid, *Ictalurus nebulosus*, Helfman, 1981; but see Culp, 1989, on longnose dace, *Rhinichthys cataractae*). A breakdown in normal activity patterns also applies to strongly nocturnal species. Although most cardinalfishes are nocturnal, spawning often occurs by day (Thresher, 1984). Even in the nocturnally spawning *Cheilodipterus quinquelineatus*, pair formation and courtship begin about 1 h before sunset, long before the usual onset of foraging activity (Kuwamura, 1987).

Reebs and Colgan (1991) found that four species of normally diurnal cichlids fan their eggs actively at night, an activity change that has also been reported for male threespine and black-spotted sticklebacks, *Gasterosteus aculeatus* and *G. wheatlandi* (Reebs *et al.*, 1984) and sergeant major damselfish, *Abudefduf saxatilis* (Albrecht, 1969). Focusing on this behaviour in female convict cichlids, *Cichlasoma nigrofasciatum*, Reebs and Colgan determined that increased nocturnal fanning was in part a direct response to darkness, as fish engaged in more fanning during 1 h pulses of darkness during the day. However, some of the increased fanning at night can be attributed to a circadian rhythm in fanning behaviour. Fish kept in constant dark or constant light still showed elevated levels of fanning during what would have been the normal darkness period. Reebs and Colgan concluded that nocturnal fanning by diurnal fish made adaptive sense, since egg respiration is not going to stop after sunset. It would be interesting to compare daytime v. night-time respiration and metabolism of eggs and young in related fishes that do and do not fan at night (e.g. *Abudefduf saxatilis* (Albrecht, 1969) v. *Amphiprion* spp., (Moyer and Bell, 1976; Ross, 1978)).

A simple explanation for the apparent breakdown in diel patterns during breeding is not immediately evident. Obviously, parental care cannot be limited to just one period of the day, particularly if the young are relatively mobile or if eggs occur in relatively exposed locales. But what of diurnal species that spawn in the evening? (See Chapter 10 by Turner, this volume.) A possible explanation is that adults release their gametes into the water column late in the day because both light levels and numbers of active planktivores are reduced at that time. Floating eggs, because of their relatively small size, may also be less subject to predation by zooplanktivores at night (Hobson and Chess, 1978). Parents are thus maximizing the likelihood of initial survival of their offspring, before the fertilized eggs are dispersed by water movements (Johannes, 1978).

Reduction of predation on floating eggs may explain why 'broadcast' spawners breed at dusk. However, a large class of benthic spawners have a breeding peak at dawn (Thresher, 1984; Kohda, 1988). Included in this group are many diurnal, permanently territorial, herbivorous damselfishes, in which females leave their territories to deposit eggs in the territories of males. Kohda (1988) investigated the timing of feeding activity of other herbivorous reef fishes. He concluded that female damselfishes spawned at dawn to minimize loss of food within their territories to the many diurnal competitors which had not yet initiated activity.

Breeding at twilight would appear to entail some additional hazards for the adults engaged in spawning, given the abundance and activity of predatory fishes at that time (p. 492). However, many fishes show a surprising disregard for their own safety during spawning periods, a phenomenon termed 'spawning stupor' (Johannes, 1978). Twilight spawning could represent an instance where adult survival is jeopardized in favour of maximizing survival of offspring (but see Kuwamura, 1987).

14.3   DIEL PATTERNS IN DIFFERENT HABITAT TYPES

Our knowledge of activity patterns in fishes in the wild is strongly influenced by factors that affect our ability to conduct research, particularly water clarity and water temperature, not to mention proximity of study locales to centres of research. Not surprisingly, the best-known species occur on coral reefs, followed by temperate lakes and temperate marine assemblages. The following accounts summarize information on diel activity patterns in these different habitat types, focusing on the families of fishes active at different times, the activities in which they engage, and the behavioural, ecological and evolutionary factors that may have influenced the observed patterns. Emphasis will be placed on coral-reef habitats, and information from other systems will be summarized and compared. Geographically, attention is focused on tropical Pacific and Caribbean faunas, temperate lakes of North America, and California kelp-bed fishes. Diel patterns have received less study than other ecological traits in tropical fresh waters; see Lowe-McConnell (1975, 1987) for a review of general day/night differences in tropical lakes and rivers.

14.4   CORAL-REEF FISHES

Daytime

Coral reefs during daytime are characterized by a great diversity of forms and activity among the fishes. Diurnal fishes on coral reefs include many boldly 'poster-coloured' forms such as butterflyfishes, angelfishes, damselfishes, wrasses, parrotfishes, surgeonfishes and triggerfishes (Fig. 14.1). The functions of these colour patterns have been variously attributed to inter- and intra-specific territoriality, species recognition, aposematic properties, crypsis, or combinations of the above (e.g. Hamilton and Peterman, 1971; Kelly and Hourigan, 1983; Thresher, 1984). Nevertheless, these striking colour patterns probably relate to the bright light/clear water conditions that prevail on coral reefs during the day and the apparent importance of vision and visual displays to these fishes (Levine *et al.*, 1980; see Chapter 4 by Guthrie and Muntz, this volume).

Foraging by diurnal reef fishes is carried out by both solitary and grouped individuals. Many of the group feeders are herbivores. In fact, practically all herbivorous fishes on coral reefs are diurnal. Solitary herbivores are generally stationary territory holders, whereas shoaling species are usually roving and non-territorial. In some species, the distinction between shoaling and soli-tariness is an ontogenetic one, with younger fishes forming shoals and older fishes holding territories that serve as both feeding and breeding sites (Ogden

Midday

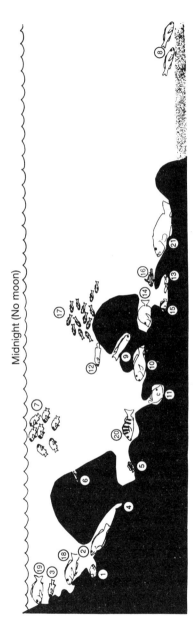

Midnight (No moon)

Fig. 14.1 Relative day v. night (moonless) distribution of common tropical marine fishes, southern Gulf of California. Emphasis is on positions relative to the rocky reef bottom and solitary v. social nature. Note that many nocturnal species form daytime resting schools near cover and then disperse at night, whereas diurnal fishes that school in the water column by day rest individually in cover at night. Diurnal families shown are (1, 5, 13) Pomacentridae, (4, 9) Labridae, (6) Blenniidae, (10) Scaridae, (11) Balistidae, (14) Acanthuridae, and (15) Chaetodontidae; nocturnal families are (2, 21) Serranidae, (3, 7) Holocentridae, (8, 19, 20) Haemulidae, (12) Grammistidae, (16) Sciaenidae, (17) Apogonidae, and (18) Lutjanidae. Modified slightly from Hobson (1965); used with permission.

and Buckman, 1973). Shoals among herbivorous fishes probably serve as predator protection, but may also help overcome the territorial defence of algal patches by herbivorous damselfishes and parrotfishes (e.g. Ehrlich, 1975; Foster, 1985; see Chapter 12 by Pitcher and Parrish, this volume).

Some of the most abundant diurnal fishes on reefs are zooplanktivorous, the smaller individuals and species of which generally feed as parts of large, relatively stationary aggregations. This group includes many damselfishes and streamlined groupers (e.g. *Anthias* spp.), snappers and wrasses. The most abundant diurnal zooplanktivores occur over drop-off regions such as outer reef faces or in lagoon areas where at least moderate tidal currents occur (Hobson and Chess, 1978). The food of these fishes is generally quite small (< 1 mm), consisting largely of calanoid copepods, cladocerans and larval invertebrates transported by currents and originating upstream from the feeding locale. Group formation in these fishes probably serves both for predator protection and in maintaining social and sexual hierarchies (Shapiro, 1986; Ross, 1990). Both functions appear to be intimately related to vision and high levels of ambient illumination.

These distantly related zooplanktivores have remarkably convergent morphologies and behaviours, evidently influenced by predator avoidance interacting with the distance above the refuge sites over which they feed, the strength of water currents and the common need to take tiny prey (Davis and Birdsong, 1973; Hobson and Chess, 1978). The most streamlined forms typically forage higher above the bottom and in stronger currents. Zooplanktivores, as well as other diurnal reef fishes, have also evolved visual capabilities that permit both high resolution of small prey items (e.g. small zooplankters) as well as sensitive detection of motion of prey or predators (Munz and McFarland, 1973). These visual tasks are again dependent on the high light levels characteristics of daytime on coral reefs. Because of the 'many watchful eyes' nature of such aggregations, detection of a predator and flight by one prey individual generally results in the entire group diving rapidly to the protective refuge of the reef below (see Chapter 12 by Pitcher and Parrish, this volume).

Other feeding guilds of coral-reef fishes active by day include relatively solitary species that feed on either mobile or sessile invertebrates, such as corals, hydrozoans, sponges and sea urchins (Hobson, 1974, 1975). Some of these fishes have striking colorations or shapes, including butterflyfishes, angelfishes, triggerfishes and pufferfishes. These fishes may rely on anatomical characteristics such as deep bodies, stout spines, tough skins, toxins and relatively large size, which allow them to feed in exposed locations in broad daylight while overcoming the defences evolved by their prey. Bright coloration in these fishes may serve as optical reinforcement of relative inedibility to visually hunting predators.

Fishes in sand and sea-grass areas during daylight hours are relatively scarce and difficult to distinguish from the background of white sand, green grass or algae. These small fishes (pipefishes, wrasses, parrotfishes, pike blennies, dragonets, gobies) generally feed on small invertebrates. Larger fishes in the grass beds either have exceptional escape abilities, such as the sand-diving razorfish *Hemipteronotus*, or have specialized abilities for probing in the sand in pursuit of buried prey (e.g. goatfishes), or are able to blow sand away from prey with expelled jets of water (e.g. boxfish).

The activities of cleanerfishes, species that feed on parasites and necrotic tissue of other fishes, appear to be limited to daylight. These cleaners are usually wrasses and gobies or juvenile angelfishes and butterflyfishes. They often defend relatively permanent 'cleaning stations' to which host fishes are attracted (Gorlick *et al.*, 1978). Visual cues, including distinctive striped 'guild' coloration and a zigzag/jerky swimming pattern, are probably used by some cleaners to attract and reduce aggression in host fishes; hosts in turn often use colour changes and unusual body postures when soliciting cleaning (Kuwamura, 1976). Since the taxonomic groups and life-history stages of the cleaners are predominantly diurnal, and since visual displays play an important role in stimulating cleaning interactions, it appears likely that cleaning is a diurnal phenomenon. Cleaning could occur at night (possibly by invertebrate cleaners), but investigations of nocturnal behaviour in fishes and invertebrates are relatively scarce.

During the day, nocturnal fishes rest on or in reef structures. Some medium- and large-sized nocturnal fishes occur singly in holes or under ledges (e.g. drums, bigeyes, porcupine fishes). Most nocturnal fishes form daytime resting aggregations (Starck and Davis, 1966; Hobson, 1975). Many small species (squirrelfishes, cardinalfishes, sweepers) enter holes and small caverns. Others form resting aggregations under ledges, at the mouths of caves, or over coral heads. This latter category includes some of the most abundant fishes on reefs, such as anchovies, herrings, silversides, Indo-Pacific apogonids, and grunts.

Predators that feed during the day tend to be cryptic, benthic species (e.g. lizardfishes, scorpionfishes, flatheads, flatfishes) or relatively stationary, water-column hoverers (e.g. groupers, barracuda). Some active predators, such as the jacks and mackerels, may also form travelling, synchronized schools (Potts, 1980) and make periodic, medium-speed passes at prey, perhaps testing their responsiveness (Smith, 1978). Only a few predators, such as trumpetfishes, can be frequently observed stalking prey in broad daylight (Helfman, 1989). They employ mimetic coloration and behaviour (Kaufman, 1976; Aronson, 1983), often matching the bold hues of the prey themselves. They may exploit colour vision in the prey, a strategy limited to daytime conditions.

Given the acute vision of diurnal prey fishes, the likelihood of detection of a predator involved in a bold, frontal attack is high. As a consequence, most diurnal predators rely on infrequent errors in judgment on the part of the prey for success (Hobson, 1979): a prey fish that moves too close to a hovering or concealed predator, or that leaves its aggregation, or that habituates to the presence of a predator and is distracted by other events, is a candidate for an opportunistic attack. In sum, because of differential visual capabilities, the protection afforded by shoaling, and in some cases defensive anatomical specializations, prey fishes have a behavioural and physiological advantage during the bright light of daytime.

Night-time

Information on fish activities at night is often inferred from static observation of distribution or analyses of stomach contents (e.g. Starck and Davis, 1966; Vivien, 1973; Hobson, 1974; Gladfelter and Johnson, 1983). Nonetheless, a fairly clear picture of the locales and activities of the more abundant species on coral reefs has emerged (Fig. 14.1).

Many diurnal species have analogues among the night-active fishes. Such 'replacement sets' include all functional groups except the herbivores and cleanerfishes. Probably the most conspicuous of such replacement sets consists of the many silvery-red/brown, large-mouthed, large-eyed zooplanktivores that hover above the reef (squirrelfishes, bigeyes, cardinalfishes, sweepers). Their prey is usually larger than that encountered by diurnal planktivores and often originates more locally, consisting of invertebrates (polychaete worms, amphipods, isopods) that spend daytime secluded in the reef or buried in the sand and emerge at dusk (Hobson and Chess, 1978). Typically, weak current areas are more densely populated with planktivores at night than during the day. The fishes involved are relatively deep bodied and robust, in contrast to their more streamlined diurnal counterparts. This lack of streamlining may reflect reduced predation at night, but is probably also linked to the weak currents that typify the foraging areas of these fishes.

Fish abundance often increases over sandy areas at night. For example, Caribbean sea-grass beds at night abound with squirrelfishes, cardinalfishes and grunts that have migrated away from the reef during twilight and which feed on invertebrates that are now relatively exposed, moving over the sand and vegetation (Robblee and Zieman, 1984). Preliminary distributional information indicates the possibility of nocturnal feeding territories, at least among grunts (McFarland and Hillis, 1982).

Piscivores and invertebrate feeders also frequent the reef at night. These nocturnal predators (squirrelfishes, scorpionfishes, groupers, snappers) rely on good night vision to find moving, relatively large crustaceans or resting fishes. Olfaction and tactile sensation are important in food detection for the moray and snake eels that move about the reef and grass beds at night. Some parrotfishes sleep in mucous cocoons, which presumably seal in both olfactory and tactile cues that might be used by foraging eels (Winn and Bardach, 1959) or predatory molluscs.

One striking difference between day and night is the dearth of group-associated activities in the dark. Diurnal fishes that have sought refuge are seldom found in groups at night, and nocturnal fishes that had formed distinctive resting aggregations by day occur either singly or in much looser aggregations while foraging at night. Shoaling depends partly on visual signals, but olfaction and detected water turbulence are also important (e.g. Pitcher *et al.*, 1976). Therefore a reduction in shoaling at night probably results not from an inability of shoalmates to see one another, but from a reduced incidence of predation and hence less need for the predator-deterrent benefits of aggregating. Any competitive interactions that occurred between foraging nocturnal fishes would also lead to increased spacing, particularly if social aggregations by day resulted from a balance that favoured predator avoidance over competition (e.g. diurnal shoaling in grunts, McFarland *et al.*, 1979; p. 493). The major exception to the generalization of group break-up at night is the behavior of flashlight fishes (e.g. *Photoblepharon palpebratus*, Morin *et al.*, 1975). These small reef fishes bioluminesce using a light produced by symbiotic bacteria contained in a subocular organ. Field observations indicate that individuals maintain close proximity by flashing their lights on and off, and that a group may use its summed light emissions to attract prey.

An additional difference concerns the coloration of nocturnally active fishes as well as resting diurnal fishes. The bold colours of daytime in general give way to solid washes of brown, red and silver, often in combination. Nocturnal fishes that were relatively colourful by day take on a paler appearance, losing their bold stripes, bars and other marks. Diurnal fishes tend to assume a muted coloration, either uniformly pale or somewhat blotchy (Emery, 1973a; Hobson, 1975). The absence of colourful patterns is probably related to poor colour vision via retinal cones at night (see Chapter 4 by Guthrie and Muntz, this volume). The absence of bold contrasting marks in many species suggests that contrast visual signals are no longer useful in shoal maintenance, or are discouraged for reasons of predator avoidance. (The detection of outline silhouettes against a light background is probably more important for these large-eyed nocturnal fishes. Ed.) They may also be abandoned in favour of other sensory modes for social communication, such as acoustic cues (i.e. nocturnal foragers such as squirrelfishes, drums and grunts). (See Chapter 5 by Hawkins, this volume.)

Twilight

Two twilight, or crepuscular, periods occur during each diel cycle: at dusk and dawn. Evening twilight lasts from sunset until the sun is at a certain angle below the horizon (18° below for astronomical twilight). At tropical latitudes, twilight generally lasts between 70 and 85 min, depending on the time of year. Taken together, dusk and dawn amount to only 5% of the 24 hour diel cycle, but ecologically, twilight may play a role in the lives of fishes out of proportion to the actual time involved. Two keys to understanding this role are first, that twilight is a period of environmental, behavioural and ecological transition, and, second, that predators exploit the transitional nature of these periods.

Environmental transitions primarily involve changes in light. Illumination levels at the surface fall from approximately 100 lx at sunset to about 0.01 lx 30 min later. During this period, diurnal fishes 'change over' from their daytime activity mode to their night-time inactivity mode, whereas nocturnal fishes initiate activity. During twilight, predators seem to be maximally active and successful. Twilight activities of coral-reef fishes have been studied in the Gulf of California, Hawaii, Australia and the Caribbean (Helfman, 1981). Despite substantial faunal differences, events at the different locales are surprisingly similar, suggesting similar selective pressures and behavioural convergence in twilight activities. The most complete studies are those by Hobson (1968, 1972), who found a sequence of five overlapping events at evening twilight that were essentially repeated in reverse at dawn. The events are (1) vertical and horizontal migrations of diurnal fishes; (2) cover-seeking of diurnal fishes; (3) evacuation of the water column; (4) emergence of nocturnal fishes; and (5) vertical and horizontal migrations of nocturnal fishes.

Migrations of diurnal fishes

During the hour preceding sunset, zooplanktivorous fishes descend to locations immediately above the reef. Smaller individuals and species move earlier. Many fishes, such as surgeonfishes and parrotfishes, also move horizontally from feeding areas to resting areas along predictable paths (Hobson, 1973; Dubin and Baker, 1982).

Cover-seeking of diurnal fishes

Beginning shortly before sunset and continuing over the next 15–20 min (i.e. overlapping with the next period), diurnal fishes seek shelter for the night. This event defines 'changeover' in most studies (Helfman, 1981). Smaller fishes in particular enter holes or cracks in the reef and are lost from sight. Larger species, such as butterflyfishes, parrotfishes, surgeonfishes and rabbitfishes, often rest in slight depressions or at the bases of overhanging coral heads. Some resting-site fidelity occurs, an individual occupying the same site on successive evenings (Hobson, 1972; Ehrlich *et al.*, 1977; Robertson and Sheldon, 1979). Agonistic interactions observed at this time suggest that sites constitute resting territories.

A distinctive sequence in cover-seeking appears to exist, with non-overlapping times of changeover characteristic of each group of species. For example, wrasses usually change over first, followed approximately by zooplanktivorous damselfishes, butterflyfishes, larger damselfishes, surgeon-fishes and parrotfishes. The seeming predictability of this sequence led Domm and Domm (1973) to postulate that it existed to minimize confusion resulting from fights over limited resting sites. Such confusion could be exploited by predators, but this hypothesis has not been adequately tested. One problem is that the reported 'species' sequence may be an artefact of having watched the same individual on different evenings and then extrapolated to species. Individual fish may have relatively precise, repeated times at which they seek shelter, but the range of changeover times may in fact overlap considerably among species.

Evidence that the sequence exists and relates to the vulnerability of prey fishes to reef predators comes from the work of Potts (1980, 1981, 1983). Potts found that prey fishes threatened by predators during the daytime retreated to the safety of the reef in the same order that they sought shelter at twilight, and that this order reflected relative vulnerabilities among species, with more vulnerable species retreating first.

Evacuation of the water column: the quiet period

At about 10–15 min after sunset, the degree of activity above the reef rapidly decreases. The number of small and medium-sized fishes that have been hovering over the reef quickly diminishes. For approximately 15–20 min, the water column is essentially devoid of small fishes, and movement above the reef has come to a standstill. It is an eerie period for a human observer, which Hobson (1972) has aptly named the quiet period, the time when neither diurnal nor nocturnal fishes are truly active.

The quiet period is, however, a time of major activity for larger reef predators, such as groupers, jacks and snappers (Hobson, 1968; Major, 1977; see also Protasov, 1966). These fishes typically lurk just over the bottom, striking at prey above that remain in the water column. Such actions give predators a considerable strategic advantage. With the sun below the horizon, light levels are relatively dim. Dark-coloured predators moving close to the darkened reef face are particularly difficult to see from above. However, prey in the water column are silhouetted against the backlighting of the evening sky. These predators typically strike upwards at their prey, which may have difficulty detecting the predator's initial movement off the bottom. Hobson (1968, 1972) gained direct evidence of use of this tactic by recording the times at which predators broke the water surface during strikes. He found that the number of such attacks increased dramatically during the quiet period (see also Major, 1977). It seems likely that the quiet period of inactivity by small diurnal and nocturnal fishes is an avoidance response to the potential of being eaten by such predators that are maximally active at this time.

Emergence of the nocturnal fishes

Although some nocturnal fishes are visible, most remain in hiding through the quiet period. The quiet period ends as most nocturnal forms initiate activity about half an hour after sunset, near the limits of comfortable vision for humans. Some species move up into the water column, whereas others undertake extensive horizontal migrations between resting and feeding areas.

Migrations of nocturnal fishes

Species most often involved in intra-reef movements are squirrelfishes, copper sweepers and grunts (Ogden and Ehrlich, 1977; Gladfelter, 1979; McFarland *et al.*, 1979). The twilight migrations of juvenile grunts in the Caribbean are the most intensively studied of any twilight behaviours and exemplify the behavioural and evolutionary forces that have influenced such migrations.

Most grunts spend daylight hours in resting aggregations over reefs. At night they feed on invertebrates in sand or sea-grass areas. At dusk, they migrate from day time to night-time locales and the schools disband in the grass beds. At dawn, shoals of the same individuals migrate back to the same coral head they occupied the previous day, using the same route as the previous evening. Resting site locations, migration times and migration routes are all remarkably precise and predictable. Differences between shoals can often be explained on the basis of age differences of group members.

The dusk migration itself has several components that occur between 5 and 20 min after sunset (i.e. overlapping the quiet period). After a period of noticeable restlessness, the travelling school converges on a staging area, usually at the base of the coral head nearest the grass bed. Next, individuals make short excursions (< 1 m) along the migration route, but quickly return to the group ('ambivalence'). Finally, at about 20 min after sunset, the school streams off the reef and into the grass bed, where individuals progressively leave the shoal and commence foraging. Times of migration can differ by less than a minute on successive evenings if cloud cover is similar. The route used on different nights also differs by only a few centimetres. It is not unusual for fish to travel several hundred metres during migration. McFarland *et al.* (1979) concluded that the timing of migrations was driven by changing light.

Helfman *et al.* (1982) addressed the question of whether precise knowledge of migration times and routes can change during development, and if such knowledge is open to modification by experience. They found that larger juveniles were more precise, that migrations occurred later, and that variability in times decreased with increasing age. Possible influences on these changes included (1) development of the eye, which allowed larger fish to migrate at lower light levels (i.e. later times); (2) social influences, whereby young fish

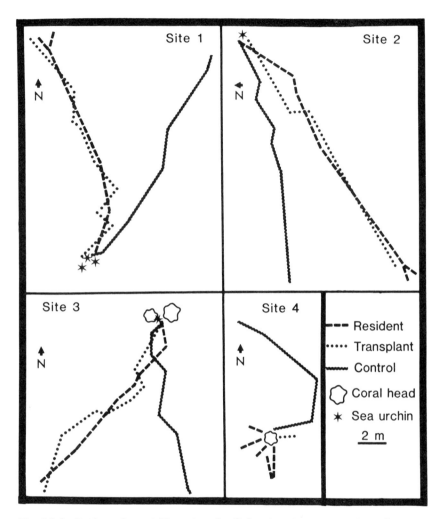

Fig. 14.2 Evidence for social learning of twilight migration routes in juvenile grunts. Migration routes are shown for four experimental sites in St. Croix, U.S. Virgin Islands. 'Resident' fish were those at a locale with an established migration route (dashed lines). 'Transplant' fish were moved to the resident sites and had an opportunity to follow residents during evening and morning migrations; residents were then removed from the site and the route taken by transplants that evening is shown (dotted lines). 'Control' fish were also moved to resident sites, but after all other fish were removed; hence controls had no opportunity to learn migration routes. Control migrations (wavy lines) were significantly different from those of residents and transplants, whereas resident and transplant migrations were not different, indicating that transplants had learned the migration routes of the residents. At Site 4, residents showed no definitive migration; transplants also wandered a short distance and stopped. Controls undertook a migration similar to the one established at the site from which they were captured. Modified from Helfman and Schultz (1984); used with permission.

joining a shoal might learn migration routes by following larger fish; and (3) crepuscular predators, which selected out young juveniles that migrated at the wrong time or over the wrong terrain.

The above descriptions characterize the migrations of juvenile grunts 40–120 mm long and 2 months to 2 years old. The remarkable precision and constancy between nights in these juveniles also has a longer-term basis: resting-site locales and migration routes are social traditions, remaining relatively constant over periods of more than 3 years, even though shoal members seldom exceed 2 years in age. The question of possible learning of routes was further examined by Helfman and Schultz (1984), who transplanted small juveniles between sites and found that transplanted fish could in fact learn a new migration route by following resident fish at the new site (Fig. 14.2). These findings also indicated that fishes can acquire knowledge about social traditions via learning, i.e. can engage in cultural behaviour.

Determinants of twilight behaviour in coral-reef fishes

The changing nature of light during twilight, and the increased activity of predators at that time, suggest a direct link between predation and vision during crepuscular periods. Munz and McFarland (1973), and McFarland and Munz (1976) found strong correlates between the structure of reef-fish eyes, the timing of twilight activities, and the behaviour of predators (Fig. 14.3; see also Lythgoe, 1979; Pankhurst, 1989). Diurnal fishes typically have eyes with many small cone cells in their retinae that function optimally at high light levels; this maximizes resolution and motion detection at the expense of night-time vision (see Chapter 4 by Guthrie and Muntz, this volume). Nocturnal fishes typically contain rod-dominated eyes with fewer, but larger, cones. These eyes maximize light capture, but sacrifice resolution and motion detection and are overloaded by the intense light of daytime. Surprisingly, both diurnal and nocturnal fishes have rod visual pigments that are most sensitive to light at around 490 nm, which is a better match to prevailing wavelengths during twilight than to night-time conditions. Both diurnal and nocturnal forms may sacrifice some nocturnal abilities in favour of better vision during twilight. The most likely selective force driving this twilight match is crepuscular predation. Twilight-active piscivores possess intermediate eyes with an intermediate number of large cones. The predators' eyes function poorly relative to potential prey capabilities during daylight and darkness, but may function better than either a diurnal or a nocturnal eye during twilight. The quiet period corresponds closely to the time when predators have both a relative visual and a behavioural advantage over prey. This behavioural advantage at twilight has been shown experimentally to include a decrease in the distance at which prey take evasive action (Pitcher and Turner, 1986; see also Howick and O'Brien, 1983). The result is that few small fishes can

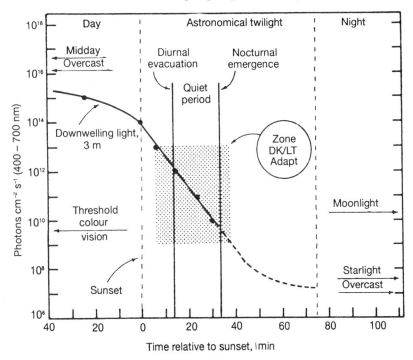

Fig. 14.3 The relationship between changing light and fish behaviour at dusk on a coral reef. The rate at which light is reduced is maximal during the 15–30 min period after sunset, as the fish eye adapts to the dark. Neither cones nor rods are maximally functional during this period (stippled area). This corresponds to the period of major activity of many reef predators, as well as predators in other aquatic systems. The 'Quiet Period', when neither diurnal nor nocturnal species are active, occurs at this time. Inactivity is evidently a response to the increased vulnerability that many reef fishes experience as light levels fall from photopic to scotopic levels. Approximate lux values: 10^{14} photons = 10,000 lx, 10^{12} = 200 lx, 10^{10} = 0.3 lx, 10^{8} = 0.0001 lx. Reproduced with permission from Munz and McFarland (1973).

afford to be in the water column during the quiet period, when the rate of change of light is most rapid and illumination is at levels at which both diurnal and nocturnal eyes function poorly. This may also explain why attacks during daylight are limited to behavioural errors by prey, and why there are relatively few nocturnal piscivores.

A note of caution should be interjected in this discussion. Although many workers agree that predation in fish assemblages reaches a peak during crepuscular periods, predatory events are certainly not confined to, nor may they be maximal during, these preiods (e.g. Major, 1978; Potts, 1980, 1981; Endler, 1987; Helfman, 1989). Predation is an infrequent event; it is usually difficult to obtain enough observations to allow statistical comparisons among time periods. Interestingly, one of the few investigations with sufficient data to allow quantitative comparison found no evident crepuscular peak. Parrish (1992) looked at the timing of predation on flat-iron herring, *Harengula thrissina*, by four species of piscivores in the Gulf of California (see also Parrish *et al.*, 1989). All four species (a serranid, a carangid, a scombrid, and a fistularid) are generally considered as crepuscular predators (Table 14.1). The herring occurred in large (100 000 + individuals), daytime resting schools and undertook an evening, offshore migration. Parrish found no evidence of a crepuscular peak in predation; all four species attacked prey throughout daylight hours, with only a slight increase in the number of successful attacks during late afternoon. Comparative studies are needed to pursue the ideas proposed by Parrish, who unfortunately did not make observations during the quiet periods. Perhaps where prey are superabundant, as in large clupeid, atherinid, and engraulid schools, predator success rates are high enough during the day to eliminate the need for feeding during twilight (J. L. Parrish, personal communication).

14.5 DIEL PATTERNS IN OTHER HABITAT TYPES

Extensive observational studies of day/night patterns of fish behaviour have not been numerous in habitats other than coral reefs. A brief summary of the literature on fish activities in temperate lakes and kelp beds is given below, emphasizing assemblage-level characteristics, and comparing diel patterns among the habitat types.

Temperate lakes

Knowledge of day/night patterns among temperate lake fish assemblages comes largely from observational studies in North America (Emery, 1973a; Helfman, 1981). As with coral reefs, temperate lake assemblages consist mostly of diurnal and nocturnal fishes, although the distinction between the two blurs in comparison with coral-reef fishes (review, Helfman, 1981; see also Diehl, 1988, and Jamet *et al.*, 1990, on European species). Phylogenetic patterns of diel behaviour are also less distinct: families of temperate lake fishes are more likely to contain both diurnal and nocturnal species.

Diurnal guilds in lakes include zooplanktivores (minnows, sunfishes, perches), which often form large feeding aggregations. Herbivorous fishes are relatively lacking in North American lakes, although they are more common in European lentic environments (Presj, 1984). Feeders on benthic or plant-associated invertebrates are abundant, including members of the minnow, sunfish and perch families as well as mudminnows, suckers, and topminnows. Diurnal piscivores include pike and pickerel and larger black basses; the latter are also active at twilight and during the night. Cleaning behaviour exists but is not as highly evolved as in tropical marine fishes. The dominant cleaner in North American lakes is the bluegill sunfish, *Lepomis macrochirus*; much of its cleaning activity is performed during twilight (Helfman, 1981). Nocturnal fishes generally rest during the day, either singly on the bottom in vegetation or other structure (e.g. ictalurid catfishes and the percid *Stizostedion*), or in daytime resting shoals associated with various kinds of structure.

Nocturnal fishes in temperate lakes are abundant, replacing most of the diurnal trophic categories. Zooplanktivores swim more in open-water limnetic regions than above littoral regions (Hall *et al.*, 1979; Bohl, 1980). Abundant zooplanktivores occur in the whitefish family Coregonidae and also among the minnows. Fishes that feed on invertebrates forage mostly near the bottom and also take small, resting fishes at night. This group includes eels, catfishes, trout, sculpins, sunfishes, and drums. The catfish, *Ictalurus nebulosus*, also feeds on filamentous algae, which makes it a nocturnal herbivore, another contrast with coral-reef patterns. American eels, *Anguilla rostrata*, show distinctive day v. night patterns of movement and activity in a variety of habitats that indicate the influence of light, tide, and endogenous rhythms (Helfman *et al.*, 1983; Helfman, 1986; Fig. 14.4). Nocturnal piscivores include bottom-feeding forms such as bowfin and burbot, as well as those that also swim in the water column (e.g. salmonids, temperate basses, sunfishes, percids). In contrast to coral-reef fishes which seek shelter at night, diurnal lake fishes tend to rest in relatively exposed locales on barren or sparsely vegetated bottoms, or in clearings amidst vegetation.

As with coral-reef fishes, twilight in temperate lakes is a period of transition when diurnal fishes cease activity and nocturnal fishes initiate feeding. In fact, twilight events in temperate lakes bear a number of marked similarities to coral reefs. Helfman (1981) found a series of six general activity patterns among the fishes that were repeated in reverse order at dawn: increased movement of diurnal fishes, group break-up, cessation of feeding, first stop or slowdown, initiation of activity by nocturnal fishes, and cessation of activity of diurnal fishes. Increased movement involved some migrations from feeding to resting areas as well as descent from the water column of smaller planktivores. Shoals disbanded in both habitat types and diurnal fishes then ceased feeding. Cessation of activity probably corresponds to cover-seeking in coral-reef species, the main difference being that few lake fishes sought shelter for the evening.

Size and age differences in timing were also found in both systems. Larger individuals were often active later in the evening than smaller individuals. In some species, diurnal juveniles changed to nocturnally foraging adults (see also Helfman, 1978; see Sazima and Machado, 1990, for similar size differences in activity in tropical freshwater fishes). Young lake fishes had a greater affinity for structure while resting, a greater likelihood of defending

NOON

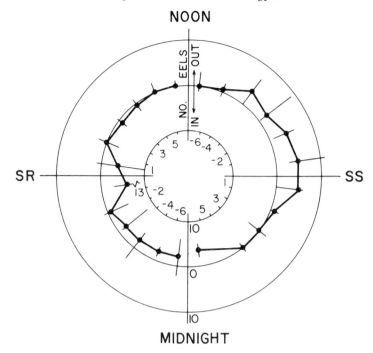

MIDNIGHT

Fig. 14.4 Movement patterns of American eels in a cave-spring in northern Florida. The heavy black line indicates the hourly medians and ranges of number of animals passing a video camera positioned 17 m deep at the mouth of a tunnel leading into deeper portions of the cave. Values external to the '0' circle represent eels leaving the tunnel, those internal to the '0' circle are re-entering the tunnel. Eels in this system occupied deep, sunless portions of the cave during the day, emerged from the cave prior to dusk to feed in the more productive basin region, and returned to the tunnel during early morning hours. SR, sunrise; SS, sunset. Light levels in excess of 10 lx appeared to inhibit eel activity (see also Culp, 1989, for a similar inhibitory illumination level). From Helfman (1986); used with permission.

resting sites, and faster shoal formation at dawn. Such ontogenetic differences may relate to greater predation pressure on smaller, younger fish, as well as a foraging shift to the larger prey active at night which are only available to larger fish (Hobson, 1972; Magnan and FitzGerald, 1984).

Other similarities between the two systems include: (1) dawn activities occurring at lower light levels than corresponding dusk activities; (2) upsurge in non-feeding activities (e.g. agonistic behaviour) at dusk; (3) ecological replacement sets, shifting to older individuals and larger-mouthed species at dusk; (4) a shift from group to individual activities in the evening; (5) a breakdown in characteristic twilight activities during the breeding season; and (6) an upsurge in predation at twilight.

These similarities imply that similar selective pressures have operated with respect to behaviour at twilight. However, several differences between the assemblages indicate potentially important differences in the determinants of diel patterns. Differences in the lake fishes included: (1) less resting-site constancy and structural affinity in diurnal lake fishes; (2) fewer twilight-migrating groups; (3) increased cleaning activity at dusk; (4) increased intraspecific and ontogenetic variability in timing; (5) later times and less light for most twilight activities of diurnal species at dusk; (6) more prolonged changeover; (7) greater between-species overlap in changeover time; and (8) lack of a quiet period.

The last four categories indicate less precision and predictability of twilight events in the temperate lake fauna. Helfman (1981) postulated that differences in the lake fishes might result from the combined influence of lower and variable species diversity, reduced crepuscular predation, variable and often poor underwater visibility, variable and longer twilight length, and long-term and short-term climatic instability. Variability in all these parameters would select for fishes with an ability to change their behaviour as a function of immediate conditions, i.e. that were behaviourally plastic. Helfman predicted that the twilight activities of temperate lake fishes would converge towards those of coral-reef fishes in locales where species diversity, water clarity, twilight length and crepuscular predation levels more closely resembled those found in the coral-reef environment.

Some evidence supports at least the contention that diel activity patterns in temperate freshwater assemblages will vary in response to predation pressure. Hanych *et al.* (1983) found twilight migratory patterns of one minnow species in a Minnesota lake that differed from patterns of the same species reported by Helfman (1981). The authors suggested that the differences could be explained as a plastic response by the species to disparate levels of predation in the two lakes. In a related study, Naud and Magnan (1988) concluded that evening offshore migrations by northern redbelly dace, *Phoxinus eos*, resulted in part from a need to form daytime shoals near structure to avoid predation by brook charr, *Salvelinus fontinalis*, despite greater abundance of zooplankton prey in unstructured limnetic habitats. Such diurnal predation constrained dace from spending daylight hours in refugeless limnetic regions where their zooplankton prey were more abundant. Finally, Tonn and Paszkowski (1987) also found greater activity at twilight by yellow perch in a predator-free lake when compared with perch activity in lakes containing crepuscular predators. However, central mudminnow, *Umbra limi*, activity showed no differences in response to predators. In the lake where perch, a mudminnow predator, were active at twilight, mudminnows did not shift away from twilight activity, as would have been predicted by Helfman (1981).

Temperate marine fishes

Studies by Ebeling and Bray (1976) in south-central California and by Hobson and Chess (1976) and Hobson *et al.* (1981) in Southern California have explored diel patterns in kelp-bed-associated assemblages and compared them with tropical regions (review, Ebeling and Hixon, 1991). The faunas studied contained 25 to 27 species, 14 in common, with affinities to tropical or temperate groups. The southern California assemblage contained more 'tropical derivative' species and the south-central California assemblage contained more temperate derivatives. This zoogeographic/phylogenetic component may reveal possible interactions between ecological and historical factors in determining activity patterns.

Both faunas contained diurnal, nocturnal and crepuscular species, although some species were difficult to characterize with respect to temporal patterns. As with coral-reef assemblages, diurnal species included abundant shoaling zooplanktivores (Atherinidae, Serranidae, Embiotocidae, Pomacentridae, Labridae, Scorpaenidae); foragers on small, hidden invertebrates (Serannidae, Embiotocidae, Clinidae, Gobiidae); and cleanerfishes (Labridae, Embiotocidae). As in temperate lakes, herbivores were relatively rare (two kyphosid species). Many nocturnal species in southern California formed daytime resting aggregations, but these were absent in the south-central California assemblage. When active, the nocturnal species (Scorpaenidae, Haemulidae, Sciaenidae, Embiotocidae) fed primarily on zooplankton or other invertebrates that were larger than the food of their diurnal counterparts. Cleanerfish, and probably herbivores, were inactive at night; one herbivore, *Girella nigricans*, may have shifted to feeding on invertebrates after sunset. At night, some diurnal fishes rested in holes or sought other shelter, whereas many species rested in comparatively exposed locales. Piscivores (Serranidae, Scorpaenidae, Hexagrammidae) were primarily nocturnal or crepuscular.

As elsewhere, twilight was a period of transition. In southern California, diurnal fishes migrated from foraging areas to resting areas, some along established routes but at less predictable times than coral-reef fishes. Ebeling and Bray (1976) felt that such migrations were uncommon at their site. As in both coral reefs and temperate lakes, twilight movements were more common in nocturnal fishes; some migrations were extensive (e.g. the embiotocid, *Hyperprosopon argenteum*, moved 1.6 km: Ebeling and Bray, 1976; the haemulid, *Xenistius californiensis*, moved > 400 m: Hobson *et al.*, 1981). Two ontogenetic patterns also found elsewhere were shifts from diurnal to nocturnal foraging with growth in some species, and a tendency for small zooplanktivores to remain relatively close to cover. Resting-site fidelity for diurnal fishes appeared slight, perhaps more like the temperate lake than the tropical reef situation. Interestingly, two primarily diurnal species (*Brachyistius frenatus* and *Chromis punctipinnis*) aggregated at night, a pattern not observed in either coal reefs or temperate lakes.

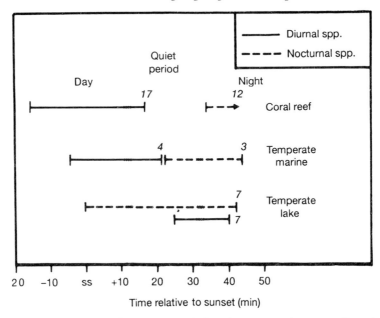

Fig. 14.5 Comparison of evening twilight changeover between diurnal and nocturnal fishes in different habitat types. Data are the ranges of mean times at which diurnal fishes cease moving or seek shelter and at which nocturnal fishes initiate movement. The quiet period, when neither diurnal nor nocturnal fishes on coral reefs are active, occurs when light levels are changing most rapidly and when dark adaptation of the fish eye is taking place (Fig. 14.3). Note particularly the greater overlap of diurnal and nocturnal species in temperate regions. Number of species is shown to the right of each line. Times for temperate lake fishes are standardized to a 100-min astronomical twilight length to account for seasonal changes in twilight lengths; absolute times relative to sunset would generally be 5 min later. SS, sunset. Data from Hobson (1972), Munz and McFarland (1973), Helfman (1981) and Hobson *et al.* (1981).

Diel patterns at these two warm temperate locales share obvious similarities with other habitat types, suggesting that certain patterns can be expected in almost any fish assemblage (Fig. 14.5). However, some obvious differences occurred between the two sites, as did disagreement on interpretation of results. Hobson *et al.* (1981) emphasized the similarities between kelp-bed and coral-reef fishes with respect to diel replacement sets, daytime resting aggregations and twilight migrations. Ebeling and Bray (1976) felt that these phenomena were noticeably lacking at their site, particularly when nocturnal zooplanktivores were concerned. They stated that "after the dusk period of intensified activity, the notably lackluster night life gives the kelp forest an aura of desolation" (page 714), and emphasized that turbid water conditions might discourage nocturnal zooplanktivory (Fig. 14.6).

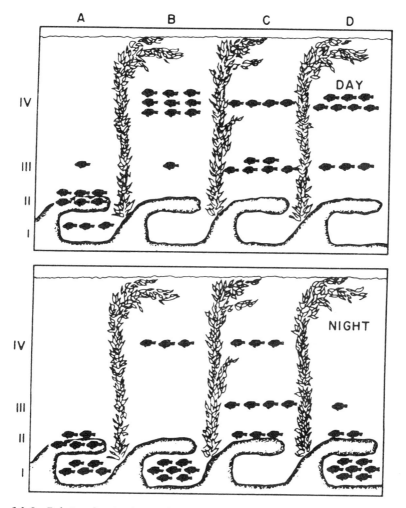

Fig. 14.6 Relative day (top) v. night (bottom) distributions of functional groups of fishes inhabiting kelp beds near St. Barbara, California (34° N). Vertical zones are depicted for fishes that are (ɪ) sheltered in holes, crevices, etc.; (ɪɪ) exposed but on the bottom; (ɪɪɪ) within 1 m of the bottom; and (ɪᴠ) > 1 m above the bottom. The four ecological groups are (A) demersal species; (B) large-mouthed generalized predators with temperate origins; (C) surfperches, also with temperate origins; and (D) small-mouthed grazing and picking tropical derivatives. Note that temperate groups B and C, the large-mouthed predators and particularly the microcarnivorous surfperches, continue to occupy the water column at night or rest exposed on the bottom. In contrast, the tropically-derived group D descends to the bottom and takes shelter at night. Each fish symbol represents approx. 10% of the total individuals in the group. Reproduced with permission from Ebeling and Bray (1976).

Some of the disagreement may relate to zoogeographic and phylogenetic differences: six tropical derivative species occurred at the southern California site only, four of which migrated at twilight or fed in the water column at night (see Hobson *et al.*, 1981, Table 4). Ebeling and Bray (1976) interpreted differences in cover-seeking on the basis of phylogenetic affinities: tropical derivative species sought shelter at night, whereas temperate derivatives rested in relatively exposed locations, perhaps because "tropical derivatives compete more successfully against the primarily temperate species for shelter" (page 714). Hobson *et al.* (1981) countered that the same temperate species rest in exposed locales at higher latitudes where tropical derivative species are lacking. They also pointed out that on coral reefs larger species tended to rest in relatively exposed locales, suggesting that the exposed resting fishes in south-central California were relatively large.

The various authors generally agree on the distinctiveness of diel patterns. The overall assemblage patterns of activity in warm temperate marine fishes appear to be less clearly defined than on coral reefs, more like the temperate lake pattern. Ebeling and Bray (1976: 715) concluded that "the general program of diel activity in the kelp forest appears to be comparatively unstructured". Hobson *et al.* (1981: 22) found "little evidence of the detailed community transition patterns that typically characterise these phenomena on tropical reefs". This lack of distinctiveness results from several factors, including: (1) relatively high variance in times of changeover; (2) species that feed both by day and by night; (3) diurnal species that cease activity relatively late, and nocturnal species that initiate activity relatively early, producing an indistinct or nonexistent quiet period (Fig. 14.5); and (4) possible incompleteness of replacement sets. A trend toward decreasing temporal organization as one proceeds from low to high latitudes is further supported by observations in kelp forests of Washington State (47° N). Twilight changeover is least structured for kelp-bed fishes in cold temperate Puget Sound, Washington, where neither warm temperate nor subtropical species exist (Moulton, 1977, cited in Ebeling and Hixon, 1991).

The causes underlying attributes of indistinctness during changeover remain unresolved. In addition to the zoogeographic and phylogenetic considerations summarized above is a comparative lack of crepuscular predation in warm temperate faunas. Predators appear to be a primary selective force promoting precision among coral-reef fishes (p. 495); reduced crepuscular predation in kelp-bed assemblages could lead to increased variability in twilight activities.

A third explanation for temperate indistinctness is that the longer twilight of temperate regions results in a relaxation of visual constraints on twilight behaviours. Crisply distinct behaviours in coral-reef fishes could be a response to short twilight, when light levels decrease at a rate equal to or faster than the rate at which the fish eye dark-adapts. The longer twilight lengths at temperate latitudes may exceed the dark-adaptation rate, and thereby remove the necessity for completing twilight activities in a minimum of time with a minimum of variability (Helfman, 1981). In a related explanation based on eye morphology, Pankhurst (1989) proposed that anomalous visual character-istics of temperate marine fishes in New Zealand might induce a lack of clearcut temporal patterns.

Support for the idea that duration of twilight behaviours is directly related to length of twilight comes from observations of predator–prey interactions in Alaska at latitude 56° N (Hobson, 1986). Hobson found that sand lances, *Ammodytes hexapterus*, experienced intensified predation by four species of benthic piscivores when the prey sought night-time shelter in sediments through much of a 2.5 h twilight changeover period. The period of danger during twilight apparently increases in relation to the length of twilight, at least where predators are abundant. Evidence of an influence of predation at high latitudes during twilight also exists in European waters, where pollack, *Pollachius pollachius*, feed on two-spot goby, *Gobiusculus flavescens* (Edwards and Potts, unpublished, cited in Potts, 1990). Both species are diurnal, but predatory activity increases during evening twilight in the pollack, while the goby seeks shelter among cracks in kelp holdfasts. Gobies re-emerge at dawn and are subjected to increased predatory attacks by pollack.

Two additional factors may influence diel patterns in temperate as well as tropical assemblages. Hobson *et al.* (1981) studied the retinal pigments of the fishes at their locale and compared their findings with expectations based on Munz and McFarland's 'twilight hypothesis' (Munz and McFarland, 1973; McFarland and Munz, 1976; p. 495). This hypothesis predicts that, because crepuscular periods are particularly dangerous, visual adaptations will arise to counter crepuscular predation at the cost of nocturnal function. As with coral-reef fishes, the spectral sensitivity of rod pigments in kelp-bed fishes matched twilight visual conditions better than nocturnal conditions. However, Hobson *et al.* found that pigment characteristics were an even better match to the light generated by bioluminescent organisms, which typically luminesce when disturbed by movement in the water column. They felt that natural selection would strongly favour sensitivity to bioluminescence because such flashes of light in the dark would often signify the movement of both predators and prey. It would be difficult to separate the selective influence of light quality from that related to bioluminescence, since a pigment with an absorption maximum around 500 nm, as found in kelp-bed fishes, would be maximally sensitive to both.

A final confounding factor concerns the invertebrates active at night in temperate kelp beds. Stepien and Brusca (1985) found that cirolanid isopods, species listed as prey of some nocturnal kelp-bed fishes by Hobson and Chess (1976), may in fact also be predators on diurnal fishes that rest in the kelp beds. Caged fishes placed in the kelp beds at night are attacked and often killed by these invertebrates, particularly if the fish have sustained some mechanical injury. The disputed topic of different resting locales of different species may be influenced by the density of these killer isopods and the manoeuvers a fish must employ to avoid them. The entire subject of how parasites might affect diel activity cycles of fishes in general (e.g. Cochran, 1986) remains unexplored.

14.6   FUTURE RESEARCH

Comparative studies of diel cycles are largely lacking from several major habitats. Our knowledge of daily activity patterns of tropical freshwater fishes is rudimentary (e.g. Lowe-McConnell, 1987; Lewis *et al.*, 1986), limited primarily to studies of a few species (e.g. Endler, 1987; Gilliam and Fraser, 1988). Both tropical lake and stream habitats are amenable to observational studies (e.g. Sazima, 1986), and our understanding of the processes that structure tropical assemblages would undoubtedly be advanced by such studies. For example, both direct observational and seining data indicate that the activities of fin-eating, scale-eating and piscivorous piranhas (*Serrasalmus* spp., *Pygocentrus* spp., Characidae) severely limit the number of fishes that frequent open-water habitats by day in Venezuela and Brazil (Winemiller, 1989; Sazima and Machado, 1990). Similar predator-induced shifts in distribution interact with twilight activities in guppies (Gilliam and Fraser, 1988). Day-night coloration changes in Amazonian fishes may relate to conspicuousness by day and predator avoidance by night, as suggested above for coral reef fishes (Lythgoe and Shand, 1983). Access to tropical areas is complicated by political and logistical difficulties, but growing concern over reduction of tropical diversity should facilitate more research in these locales.

More surprising is the lack of community-level studies of activity cycles in temperate streams and rivers. Intriguing patterns have emerged from the few investigations of individual species that have been published. For example, juvenile coho salmon, *Oncorhynchus kisutch*, aggregated during winter in woody debris, presumably because it offers protection from predators and prevents downstream displacement at high water discharges (McMahon and Hartman, 1989). Movements out of these areas were largely restricted to twilight periods, perhaps in avoidance of visual predators.

This final example illustrates a potentially hazardous outgrowth of the research emphasis in this field to date. Many studies of diel activities have focused on the influence that predation may have on activity cycles, often where little evidence for predation exists (Helfman, 1986, and above). However, activity cycles are also influenced by other biotic factors, including prey availability (Naud and Magnan, 1988; Glozier and Culp, 1989), parasite activity (Stepien and Brusca, 1985; Cochran, 1986; Poulin and FitzGerald, 1988), and competition (Tonn and Paszkowski, 1987). Additionally, strong phylogenetic patterns among evolutionarily related but ecologically dissimilar fishes (as emphasized in Table 14.1), suggest that major activity patterns in many fishes may be historical accidents with minimal relation to current biotic and environmental stimuli. All these possibilities should be kept in mind when drawing conclusions about causative factors of observed cycles of activity.

14.7 SUMMARY

Most fish species forage primarily during the day or night, with a few species primarily active during crepuscular periods of dawn and dusk. Several similarities in these diel patterns of activity occur in very different habitat types. In coral reef, temperate lake, and temperate marine kelp-bed assemblages, diurnal fishes include herbivores, zooplanktivores, invertebrate feeders, piscivores and cleaners. 'Replacement sets' of these various functional groups are active at night, except for herbivores and cleaners. Twilight is a time of transition between daytime and night-time modes, when diurnal fishes cease activity and nocturnal fishes initiate feeding. It is also a period of major activity and success of piscivores. Predator/prey interactions at twilight have had an apparently strong influence on the behaviour and visual physiology of both diurnal and nocturnal fishes. Predators are relatively rare in some temperate assemblages, which may explain a lower degree of precision in the twilight activities of temperate species. Differences in water clarity, species diversity, phylogenetic constraints and twilight length may also have influenced behavioural differences in temperate versus tropical habitats.

ACKNOWLEDGEMENT

Dr E. S. Hobson kindly commented on a draft manuscript of this chapter.

REFERENCES

Albrecht, H. (1969) Behaviour of four species of Atlantic damselfishes from Colombia, South America (*Abudefduf saxatilis, A. taurus, Chomis multilineata, C. cyanea*; Pisces, Pomacentridae). *Z. Tierpsychol.*, **26**, 662–76.

Allen, J. R. M. and Wootton, R. J. (1984) Temporal patterns in diet and rate of food consumption of the three-spined stickleback (*Gasterosteus aculeatus*) in Llyn Frongoch, an upland Welsh lake. *Freshwat. Biol.*, **14**, 335–46.

Aronson, R. B. (1983) Foraging behavior of the West Atlantic trumpetfish. *Aulostomus maculatus:* use of large, herbivorous reef fishes as camouflage. *Bull. Mar. Sci.*, **33**, 166–71.

Blaxter, J. H. S. (1976) The role of light in the vertical migration of fish – a review, in

Light as an Ecological Factor: II (eds G. C. Evans, R. Bainbridge and O. Rackham), Blackwell, London, pp. 189–210.

Bohl, E. (1980) Diel pattern of pelagic distribution and feeding in planktivorous fish. *Oecologia*, **44**, 368–75.

Cochran, P. A. (1986) The daily timing of lamprey attacks. *Env. Biol Fishes*, **16**, 325–9.

Culp, J. M. (1989) Nocturnally constrained foraging of a lotic minnow (*Rhinichthys cataractae*). *Can. J. Zool.*, **67**, 2008–12.

Davis, W. P. and Birdsong, R. (1973) Coral reef fishes which forage in the water column. *Helgolander wiss. Meeresunters.*, **24**, 292–306.

Diehl, S. (1988) Foraging efficiency of three freshwater fishes: effects of structural complexity and light. *Oikos*, **53**, 207–14.

Domm, S. B. and Domm, A. J. (1973) The sequence of appearance at dawn and disappearance at dusk of some coral reef fishes. *Pacif. Sci.*, **27**, 128–35.

Dubin, R. E. and Baker, J. E. (1982) Two types of cover-seeking at sunset by the princess parrotfish, *Scarus taeniopterus*, at Barbados, West Indies. *Bull. Mar. Sci.*, **32**, 572–83.

Ebeling, A. W. and Bray, R. N. (1976) Day versus night activity of reef fishes in a kelp forest off Santa Barbara, California, *U.S. Fish. Bull.*, **74**, 703–17.

Ebeling, A. W. and Hixon, M. A (1991) Tropical and temperate reef fishes: comparison of community structures, in *The Ecology of Coral Reef Fishes* (ed. P. F. Sale), Academic Press, pp. 509–563.

Ehrlich, P. R. (1975) The population biology of coral reef fishes. *A. Rev. Ecol. Syst.*, **6**, 211–47.

Ehrlich, P. R., Talbot, F. H., Russell, B. C. and Anderson, G. R. V. (1977) The behaviour of chaetodontid fishes with special reference to Lorenz' 'Poster Colouration' Hypothesis. *J. Zool., Lond.*, **138**, 213–28.

Emery, A. R. (1973a) Preliminary comparisons of day and night habits of freshwater fish in Ontario lakes. *J. Fish. Res. Bd Can.*, **30**, 761–74.

Emery, A. R. (1973b) Comparative ecology and functional osteology of fourteen species of damselfish (Pisces: Pomacentridae) at Alligator Reef, Florida Keys. *Bull. Mar. Sci.*, **23**, 649–770.

Emery, A. R. (1978) The basis of fish community structure: marine and freshwater comparisons. *Env. Biol. Fishes*, **3**, 33–47.

Endler, J. A. (1987) Predation, light intensity and courtship behaviour in *Poecilia reticulata* (Pisces: Poeciliidae). *Anim. Behav.*, **35**, 1376–85.

Foster, S. A (1985) Group foraging by a coral reef fish: a mechanism for gaining access to defended resources. *Anim. Behav.*, **33**, 782–92.

Gilliam, J. F. and Fraser, D. F. (1988) Resource depletion and habitat segration by competitors under predation hazard, in *Size-structured Populations* (eds B. Ebenman and L. Persson), Springer-Verlag, Berlin, pp. 173–84.

Gladfelter, W. B. (1979) Twilight migrations and foraging activities of the copper sweeper, *Pempheris schomburgki* (Teleostei, Pempheridae). *Mar. Biol.*, **51**, 109–19.

Gladfelter, W. B. and Johnson, W. S. (1983) Feeding niche separation in a guild of tropical reef fishes (Holocentridae). *Ecology*, **64**, 552–63.

Glozier, N. E. and Culp, J. M. (1989) Experimental investigations of diel vertical movements by lotic mayflies over substrate surfaces. *Freshwat. Biol.*, **21**, 253–60.

Gorlick, D. L., Atkins, P. D. and Losey, G. S. (1978) Cleaning stations as water holes, garbage dumps, and sites for the evolution of reciprocal altruism? *Am. Nat.*, **112**, 341–53.

Grecay, P. A. and Targett, T. E. (in press) The effect of turbidity, light level and prey concentration on feeding success of juvenile weakfish (*Cynoscion regalis*). *Mar. Ecol. Prog. Ser.*

Gushima, K., Kondou, K. and Murakami, Y. (1977) Diel change in family composition of reef fishes. *J. Fac. Fish. Anim. Husb., Hiroshima Univ.*, **16**, 151–6.

Hall, D. J., Werner, E. E., Gilliam, J. F., Mittelbach, G. G., Howard, D., Doner, C. G., Dickerman, J. A. and Steward, A. J. (1979) Diel foraging behaviour and prey selection in the golden shiner (*Notemigonus chrysoleucas*). *J. Fish. Res. Bd Can.*, **36**, 1029–39.

Hamilton, W. J. and Peterman, R. M. (1971) Countershading in the colourful reef fish *Chaetodon lunula*: concealment, communication or both. *Anim. Behav.*, **19**, 357–64.

Hanych, D. A., Ross, M. R., Magnien, R. E. and Suggars, A. L. (1983) Nocturnal inshore movement of the mimic shiner (*Notropis volucellus*): a possible predator avoidance behavior. *Can. J. Fish. Aquat. Sci.*, **40**, 888–94.

Helfman, G. S. (1978) Patterns of community structure in fishes: summary and overview. *Env. Biol. Fishes*, **3**, 129–48.

Helfman, G. S. (1981) Twilight activities and temporal structure in a freshwater fish community. *Can. J. Fish. Aquat. Sci.*, **38**, 1405–20.

Helfman, G. S. (1986) Diel distribution and activity of American eels (*Anguilla rostrata*) in a cave-spring. *Can. J. Fish. Aquat. Sci.*, **43**, 1595–1605.

Helfman, G. S. (1989) Threat-sensitive predator avoidance in damselfish–trumpetfish interactions. *Behav. Ecol. Sociobiol.*, **24**, 47–58.

Helfman, G. S. and Schultz, E. T. (1984) Social transmission of behavioural traditions in a coral reef fish. *Anim. Behav.*, **32**, 379–84.

Helfman, G. S., Meyer, J. L. and McFarland, W. N. (1982) The ontogeny of twilight migration patterns in grunts (Pisces: Haemulidae). *Anim. Behav.*, **30**, 317–26.

Helfman, G. S., Stoneburner, D. L., Bozeman, E. L., Whalen, R. and Christian, P. A. (1983) Ultrasonic telemetry of American eel movements in a tidal creek. *Trans. Am. Fish. Soc.*, **112**, 105–10.

Hobson, E. S. (1965) Diurnal–nocturnal activity of some inshore fishes in the Gulf of California. *Copeia*, **1965**, 291–302.

Hobson, E. S. (1968) Predatory behavior of some shore fishes in the Gulf of California. *U.S. Bur. Sport Fish. Wildl., Res. Rep.*, **73**, 1–92.

Hobson, E. S. (1972) Activity of Hawaiian reef fishes during evening and morning transitions between daylight and darkness. *U.S. Fish. Bull.*, **70**, 715–40.

Hobson, E. S. (1973) Diel feeding migrations in tropical reef fishes. *Helgolander wiss. Meeresunters.*, **24**, 361–70.

Hobson, E. S. (1974) Feeding relationships of teleostean fishes on coral reefs in Kona, Hawaii. *U.S. Fish. Bull.*, **72**, 915–1031.

Hobson, E. S. (1975) Feeding patterns among tropical reef fishes. *Am. Sci.*, **63**, 382–92.

Hobson, E. S. (1979) Interactions between piscivorous fishes and their prey, in *Predator–Prey Systems in Fisheries Management* (ed. H. Clepper), Sport Fishing Institute, Washington, DC, pp. 231–42.

Hobson, E. S. (1986) Predation on the Pacific sand lance, *Ammodytes hexapterus* (Pisces: Ammodytidae), during the transition between day and night in southeastern Alaska. *Copeia*, **1986**, 223–6.

Hobson, E. S. and Chess, J. R. (1976) Trophic interactions among fishes and zooplankters near shore at Santa Catalina Island, California. *U.S. Fish. Bull.*, **74**, 567–98.

Hobson, E. S. and Chess, J. R. (1978) Trophic relationships among fishes and plankton in the lagoon at Enewetak Atoll, Marshall Islands. *U.S. Fish. Bull.*, **76**, 133–53.

Hobson, E. S., Chess, J. R. and McFarland, W. N. (1981) Crepuscular and nocturnal activities of California nearshore fishes, with consideration of their scotopic visual pigments and the photic environment. *U.S. Fish. Bull.*, **79**, 1–30.

Hoekstra, D. and Janssen, J. (1985) Non-visual feeding behavior of the mottled sculpin, *Cottus bairdi*, in Lake Michigan. *Env. Biol. Fishes*, **12**, 111–17.

Howick, G. L. and O'Brien, W. J. (1983) Piscivorous feeding behavior of largemouth bass: an experimental analysis. *Trans. Am. Fish. Soc.*, **112**, 508–16.

Jamet, J. L., Gres, P., Lair, N. and Lasserre, G. (1990) Diel feeding cycle of roach (*Rutilus rutilus*, L.) in eutrophic Lake Aydat (Massif Central, France). *Arch. Hydrobiol.*, **118**, 371–82.

Johannes, R. E. (1978) Reproductive strategies of coastal marine fishes in the tropics. *Env. Biol. Fishes*, **3**, 65–84.

Kaufman, L. (1976) Feeding behavior and functional coloration of the Atlantic trumpetfish, *Aulostomus maculatus*. *Copeia*, **1976**, 377–8.

Kavaliers, M. (1989) Day–night rhythms of shoaling behavior in goldfish: opioid and pineal involvement. *Physiol. Behav.*, **46**, 167–72.

Kelly, C. D. and Hourigan, T. F. (1983) The function of conspicuous coloration in chaetodontid fishes: a new hypothesis. *Anim. Behav.*, **31**, 615–17.

Kohda, M. (1988) Diurnal periodicity of spawning activity of permanently territorial damselfishes (*Teleostei: Pomacentridae*). *Env. Biol. Fishes*, **21**, 91–100.

Kuwamura, T. (1976) Different responses of inshore fishes to the cleaning wrasse, *Labroides dimidiatus*, as observed in Sirahama. *Publs. Seto. mar. biol. Lab.*, **23**, 119–44.

Kuwamura, T. (1987) Night spawning and paternal mouthbrooding of the cardinalfish *Cheilodipterus quinquelineatus*. *Jap. J. Ichthyol.*, **33**, 431–4.

Larkin, P. A. (1979) Predator–prey relations in fishes: an overview of the theory, in *Predator–Prey Systems in Fisheries Management* (ed. H. Clepper), Sport Fishing Institute, Washington, DC, pp. 13–22.

Levine, J. S., Lobel, P. S. and MacNichol, E. F. (1980) Visual communication in fishes, in *Environmental Physiology of Fishes* (ed. M. A. Ali), Plenum, New York, pp. 447–75.

Lewis, D., Reinthal, P. and Trendall, J. (1986) *A Guide to the Fishes of Lake Malawi National Park*. World Wildlife Fund, Gland, Switzerland, 71 pp.

Lowe-McConnell, R. H. (1975) *Fish Communities in Tropical Freshwaters: their Distribution, Ecology and Evolution*, Longman, London, 304 pp.

Lowe-McConnell, R. H. (1987) *Ecological Studies in Tropical Fish Communities*, Cambridge Univ. Press, Cambridge, 369 pp.

Lythgoe, J. N. (1979) *The Ecology of Vision*, Clarendon Press, Oxford, 226 pp.

Lythgoe, J. N. and Shand, J. (1983) Diel colour changes in the neon tetra *Paracheirodon innesi*. *Env. Biol. Fishes*, **8**, 249–54.

McFarland, W. N. and Hillis, Z.-M. (1982) Observations on agonistic behaviour between members of juvenile French and white grunts – family Haemulidae. *Bull. Mar. Sci.*, **32**, 255–68.

McFarland, W. N. and Munz, F. W. (1976) The visible spectrum during twilight and its implications to vision, in *Light as an Ecological Factor: II* (eds G. C. Evans, R. Bainbridge and O. Rackham), Blackwell, Oxford, pp. 249–70.

McFarland, W. N., Ogden, J. C. and Lythgoe, J. N. (1979) The influence of light on the twilight migrations of grunts. *Env. Biol. Fishes*, **4**, 9–22.

McMahon, T. W. and Hartman, G. F. (1989) Influence of cover complexity and current velocity on winter habitat use by juvenile coho salmon (*Oncorhynchus kisutch*). *Can. J. Fish. Aquat. Sci.*, **46**, 1551–7.

Magnan, P. and FitzGerald, G. J. (1984) Ontogenetic changes in diel activity, food habits and spatial distribution of juvenile and adult creek chub, *Semotilus atromaculatus*. *Env. Biol. Fishes*, **11**, 301–7.

Major, P. F. (1977) Predator–prey interactions in schooling fishes during periods of twilight: a study of the silverside *Pranesus insularum* in Hawaii. *U.S. Fish. Bull.*, **75**, 415–26.

Major, P. F. (1978) Predator–prey interactions in two schooling fishes, *Caranx ignobilis* and *Stolephorus purpureus*. *Anim. Behav.*, **26**, 760–77.

Morin, J. G., Harrington, A., Nealson, K., Krieger, N., Baldwin, T. O. and Hastings, J. W. (1975) Light for all reasons: versatility in the behavioral repertoire of the flashlight fish. *Science*, **190**, 74–6.

Moulton, L. L. (1977) An ecological analysis of fishes inhabiting the rocky nearshore regions of Northern Puget Sound, Washington. PhD thesis, Univ. Washington, 194 pp. (Not seen, cited in Ebeling and Hixon, 1991.)

Moyer, J. T. and Bell, L. J. (1976) Reproductive behavior of the anemonefish *Amphiprion clarkii* at Miyake-Jima, Japan. *Jap. J. Ichthyol.*, **23**, 23–32.

Muller, K. (1978) The flexibility of circadian rhythms shown by comparative studies at different latitudes, in *Rhythmic Activity of Fishes* (ed. J. E. Thorpe), Academic Press, London, pp. 91–104.

Munz, F. W. and McFarland, W. N. (1973) The significance of spectral position in the rhodopsins of tropical marine fishes. *Vision Res.*, **13**, 1829–74.

Nash, R. D. M. (1982) The diel behaviour of small demersal fish on soft sediments on the west coast of Scotland using a variety of techniques: with special reference to *Lesuerigobious friesii* (Pisces: Gobiidae). *Mar. Ecol.*, **3**, 161–78.

Naud, M. and Magnan, P. (1988) Diel onshore–offshore migrations in northern redbelly dace, *Phoxinus eos* (Cope), in relation to prey distribution in a small oligotrophic lake. *Can. J. Zool.*, **66**, 1249–53.

Neudecker, S. and Lobel, P. S. (1982) Mating systems of chaetodontid and pomacanthid fishes at St. Croix. *Z. Tierpsychol.*, **59**, 299–318.

Ogden, J. C. and Buckman, N. S. (1973) Movements, foraging groups, and diurnal migrations of the striped parrotfish *Scarus croicensis* Block (Scaridae). *Ecology*, **54**, 589–96.

Ogden, J. C. and Ehrlich, P. R. (1977) The behavior of heterotypic resting schools of juvenile grunts (Pomadasyidae). *Mar. Biol.*, **42**, 273–80.

Pankhurst, N. W. (1989) The relationship of ocular morphology to feeding mode and activity periods in shallow marine teleosts from New Zealand. *Env. Biol. Fishes*, **26**, 201–11.

Parrish, J. K. (1992) Levels of diurnal predation on a school of flat-iron herring, *Harengula thrissina*. *Env. Biol. Fishes*, **34**, 257–63.

Parrish, J. K., Strand, S. W. and Lott, J. L. (1989) Predation on a school of flat-iron herring, *Harengula thrissina*. *Copeia*, **1989**, 1089–91.

Pitcher, T. J. and Turner, J. R. (1986) Danger at dawn: experimental support for the twilight hypothesis in shoaling minnows. *J. Fish Biol.*, **29**, (Supp. A), 59–70.

Pitcher, T. J., Partridge, B. L. and Wardle, C. S. (1976) A blind fish can school. *Science*, **194**, 963–5.

Potts, G. W. (1980) The predatory behaviour of *Caranx melampygus* (Pisces) in the channel environment of Aldabra Atoll (Indian Ocean). *J. Zool. Lond.*, **192**, 323–50.

Potts, G. W. (1981) Behavioural interactions between the Carangidae (Pisces) and their prey on the fore-reef slope of Aldabra, with notes on other predators. *J. Zool., Lond.*, **195**, 385–404.

Potts, G. W. (1983) The predatory tactics of *Caranx melampygus* and the response of its prey, in *Predators and Prey in Fishes* (eds D. L. G. Noakes, D. G. Lindquist, G. S. Helfman and J. A. Ward), Dr W. Junk. The Hague, pp. 181–91.

Potts, G. W. (1990) Crepuscular behaviour of marine fishes, in *Light and Life in the Sea* (eds P. J. Herring, A. K. Campbell, M. Whitfield and L. Maddock), Cambridge Univ. Press, Cambridge, pp. 221–7.

Poulin, R. and FitzGerald, G. J. (1988) Water temperature, vertical distribution, and risk of ectoparasitism in juvenile sticklebacks. *Can. J. Zool.*, **66**, 2002–5.

Presj, A. (1984) Herbivory by temperate freshwater fishes and its consequences. *Env. Biol. Fishes*, **10**, 281–96.

Protasov, V. R. (1966) *Vision and Near Orientation of Fish* (transl. M. Raveh), Israel Program for Scientific Translation, IPST No. 5738, U.S. Dept Interior, Washington, DC.

Reebs, S. G. and Colgan, P. W. (1991) Nocturnal care of eggs and circadian rhythms of fanning activity in two normally diurnal cichlid fishes, *Cichlasoma nigrofasciatum* and *Herotilapia multispinosa*. *Anim Behav.*, **41**, 303–311.

Reebs, S. G., Whoriskey, F. G. and FitzGerald, G. J. (1984) Diel patterns of fanning activity, egg respiration, and the nocturnal behavior of male three-spined stickle-backs, *Gasterosteus aculeatus* L. (f. *trachurus*). *Can. J. Zool.*, **62**, 329–34.

Robblee, M. B. and Zieman, J. C. (1984) Diel variation in the fish fauna of a tropical seagrass feeding ground. *Bull. Mar. Sci.*, **34**, 335–45.

Robertson, D. R. and Sheldon, J. M. (1979) Competitive interactions and the availability of sleeping sites for a diurnal coral reef fish. *J. exp. mar. Biol. Ecol.*, **40**, 285–98.

Ross, R. M. (1978) Reproductive behavior of the anemonefish *Amphiprion melanopus* on Guam. *Copeia*, **1978**, 103–7.

Ross, R. M. (1990) Evolution of sex-change mechanisms in fishes. *Env. Biol. Fishes*, **29**, 81–93.

Sazima, I. (1986) Similarities in feeding behaviour between some marine and freshwater fishes in two tropical communities. *J. Fish Biol.*, **29**, 53–65.

Sazima, I. F. and Machado, F. A. (1990) Underwater observations of piranhas in western Brazil. *Env. Biol. Fishes*, **28**, 17–31.

Shapiro, D. Y. (1986) Intra-group home ranges in a female-biased group of sex-changing fish. *Anim. Behav.*, **34**, 865–70.

Smith, C. L. (1978) Coral reef fish communities: a compromise view. *Env. Biol. Fishes*, **3**, 109–28.

Starck, W. A. II and Davis, W. P. (1966) Night habits of fishes of Alligator Reef, Florida. *Ichthyologia*, **38**, 313–56.

Stepien, C. A. and Brusca, R. C. (1985) Nocturnal attacks on nearshore fishes in southern California by crustacean zooplankton. *Mar. Ecol. Prog. Ser.*, **25**, 91–105.

Thorpe, J. E. (1978) *Rhythmic Activity of Fishes*, Academic Press, London, 288 pp.

Thresher, R. E. (1984) *Reproduction in Reef Fishes*, TFH Publications, Neptune City, NJ, 399 pp.

Tonn, W. M. and Paszkowski, C. A. (1987) Habitat use of the central mudminnow (*Umbra limi*) and yellow perch (*Perca flavescens*) in *Umbra–Perca* assemblages: the roles of competition, predation, and the abiotic environment. *Can. J. Zool.*, **65**, 862–70.

Vivien, M. L. (1973) Contribution a la connaissance de l'ethologie alimentaire de l'ichtyofaune de platier interne des recifs coralliens de tulear (Madagascar). *Tethys, Supp.*, **5**, 221–308.

Winemiller, K. O. (1989) Ontogenetic diet shifts and resource partitioning among piscivorous fishes in the Venezuelan ilanos. *Env. Biol. Fishes*, **26**, 177–99.

Winn, H. E. and Bardach, J. E. (1959) Differential food selection by moray eels and a possible role of the mucous envelope of parrotfishes in reduction of predation. *Ecology*, **40**, 296–8.

Woodhead, P. M. J. (1966) The behaviour of fish in relation to light the sea. *Oceanogr. Mar. Biol. Ann. Rev.*, **4**, 337–403.

Chapter fifteen

Intertidal teleosts: life in a fluctuating environment

R. N. Gibson

15.1   INTRODUCTION

Teleost fishes are numerous and diverse and can be found in a wide range of habitats. Some of these habitats, like caves and the deepest parts of the sea, change relatively slowly in their physico-chemical characteristics. Others are subject to marked diel and seasonal variation. In the narrow area between the tidemarks, the situation is further complicated by the ebb and flow of the tides. For fishes unprotected by the thick exoskeleton of many intertidal invertebrates, the sea-shore is a difficult region to occupy, although the rewards of colonization may be high because there are likely to be few competitors. Naturally, there are also disadvantages, such as the risk of predation by both terrestrial and aquatic predators, regular and possibly rapid changes in environmental conditions and, perhaps most drastic of all for fishes, the almost complete removal of their normal locomotory medium for hours at a time.

Fishes that utilize the intertidal zone have overcome these problems in two ways. The first is to avoid the zone at low tide and make use of its resources, usually food, only when it is submerged. The strategy of migration in and out with the tide is employed by large numbers of fishes on shores throughout the world. It is particularly favoured by juveniles, which may find, in addition to food, refuge from larger predators in the shallow water. Consequently, intertidal and shallow subtidal zones frequently constitute important 'nursery grounds'. Movement into and out of the intertidal zone over a seasonal time scale can be seen in some species whose distribution is mainly subtidal. In these forms the juveniles are found intertidally for the first few months of their life but gradually move into deeper water as they grow. This shift in

Behaviour of Teleost Fishes 2nd edn. Edited by Tony J. Pitcher. Published in 1993 by Chapman & Hall. ISBN   0 412 42930 6 (HB) and 0 412 42940 3 (PB).

distribution with size may also be a response to the relatively greater risk of predation for small fishes in deeper water. Such a relationship has been demonstrated for killifishes, *Fundulus* spp., in salt marshes (Kneib, 1984, 1987). It results in many species only being present on the shore at certain times of the year. On sandy shores these seasonal visitors, of which juvenile flatfishes are good examples, migrate up and down the beach with each tide (Tyler, 1971; Kuipers, 1973; Ansell and Gibson, 1990). In some areas flatfish larvae settle in intertidal pools and migration only begins when physical conditions in the pools become unfavourable (Van der Veer and Bergman, 1986).

The second strategy is to remain in the intertidal zone at low tide. This strategy requires the presence of some form of shelter to alleviate the dangers of exposure, of which predation and desiccation are the most severe. Most fishes that remain between the tidemarks do so mainly on rocky shores where shelter is provided by space beneath boulders or algae, crevices in rocks and the temporary refuges of rock pools. Important exceptions to this generalization are the mudskippers on tropical shores. They may be present in very large numbers on mudflats where they construct their own shelter in the form of burrows in the mud.

It is possible, therefore, to make a basic distinction between the two types of fishes that can be found intertidally: there are the 'visitors' or 'transients', which are present only for part of the year or the tidal cycle, and the 'residents', which spend most, if not all, of their life there. This chapter describes the behavioural adaptations of resident fishes to life on the sea-shore and begins by describing the nature of the environmental changes the fishes are likely to experience and the structure of the fishes themselves. Detailed reviews of the biology of intertidal fishes are given by Gibson (1969, 1982).

15.2   MORPHOLOGICAL AND PHYSIOLOGICAL ADAPTATIONS TO INTERTIDAL LIFE

At high tide, conditions in the intertidal zone will reflect those of the sea and will be relatively uniform as regards such factors as temperature, salinity, pH and oxygen content of the water. The ebbing tide exposes the shore to air, however, and the environment changes suddenly from an aquatic to a virtually terrestrial one. The location of a fish after the tide has receded may therefore be extremely critical. If it is in a rock pool, then ambient conditions may change little over the few hours the pool is isolated from the sea. On the other hand, depending on the air temperature, time of day and prevailing weather conditions, the water in the pool may be heated or cooled, its salinity reduced or increased. At night, its pH and oxygen content may be drastically lowered by respiring algae (Truchot and Duhamel-Jouve, 1980; Morris and Taylor,

1983). Under such conditions of reduced oxygen, some species possess the ability to move out of pools and respire aerially (Davenport and Woolmington, 1981). When out of the water, the fish can take advantage of the greater concentration of oxygen in air, but at the same time it becomes exposed to the dangers of desiccation. The duration of exposure to such hazards will depend on the vertical position of an individual on the shore, with those at higher levels being subjected to more severe conditions for longer periods than those at lower levels. Consequently resident species are physiologically very versatile. They are eurythermal and euryhaline and often possess the ability to respire equally well in moist air or water (see references in Gibson, 1969, 1982; Bridges, 1988). Many species can survive out of the water for several hours and tolerate a loss of water equivalent to 20% or more of their body weight (Horn and Riegle, 1981). In this respect they are more tolerant of desiccation than many amphibians (Horn and Gibson, 1988).

Turbulence is a further factor of major significance in the lives of intertidal fishes. In contrast to most other factors, which are likely to be stressful at low tide, the turbulence caused by wave action will affect fishes only when the intertidal zone is submerged. Only on the calmest day in very sheltered areas is turbulence likely to be absent. On most shores some degree of water movement is always present, and it may vary from a gentle surge to the full force of breaking waves. Life in such conditions imposes certain constraints on morphology and behaviour if the fish are not to be swept away from their habitat or subject to damage by being dashed against rocks. Consequently it is possible to recognize considerable similarity in the body forms of those families that are especially frequent as intertidal residents.

The most important families that have colonized the intertidal zone are the blennies and their relatives the gunnels (Pholidae), pricklebacks (Stichaeidae), triplefins (Tripterygiidae) and Clinidae, the gobies (Gobiidae), the clingfishes (Gobiesocidae) and sculpins (Cottidae). The main features that these groups have in common (Fig. 15.1) are small size (rarely more than 15–20 cm), negative buoyancy resulting from the absence or reduction of the swim bladder, compressed or depressed body form, and some modifications of the fins which act as attachment devices. In the gobies and clingfishes, the pelvic fins are united to form suction pads, whereas in the blennies the rays of the pectoral, pelvic and anal fins are often curved at their distal ends, allowing them to cling to rough surfaces. All these features combined enable the fish to keep close to the bottom where water velocity is lowest, to resist sudden surges and to take advantage of small apertures for shelter if conditions become too harsh. Thigmotaxy, the distinctive behavioural tendency to keep in contact with solid objects, also reduces the amount of time the fish spend away from shelter and ensures that when not active they come to rest in holes, crevices or beneath stones. Coupled with this 'cryptobenthic' mode of life is a style of locomotion in which the fish is propelled as much by its large pectoral

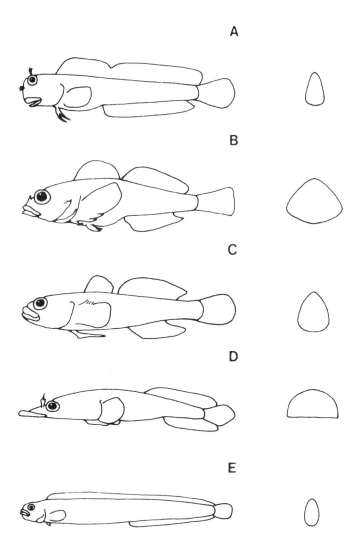

Fig. 15.1 Drawings of selected intertidal fishes with diagrammatic cross-sections of the body at its deepest point. Lengths are maximum total lengths. (A) *Istiblennius lineatus* (Blenniidae, Indian Ocean, 16 cm). (B) *Oligocottus maculosus* (Cottidae, north-eastern Pacific, 9 cm). (C) *Gobius paganellus* (Gobiidae, western Europe and Mediterranean, 12 cm). (D) *Lepadogaster lepadogaster* (Gobiesocidae, south-western Europe and Mediterranean, 6.5 cm). (E) *Pholis gunnellus* (Pholidae, northern Atlantic and Arctic, 25 cm).

fins as by its tail. Movement is therefore in short hops and darts along the bottom, and the fish rarely undertake sustained bouts of swimming in open water.

15.3 HABITAT SELECTION

One of the keys to survival for mobile foraging species in the intertidal zone is the ability to find and occupy a favourable position at low tide. Many behaviour patterns of intertidal fishes can be interpreted as adaptations for protecting either the individual or its progeny from adverse conditions when the tide is out.

Perhaps the most general way of ensuring that a favourable position is obtained is to select a level on the shore where conditions are likely to correspond to the physiological capabilities which a fish possesses. Species capable of resisting desiccation or high temperatures, for example, will be able to survive at higher levels than others which lack this ability (Barton, 1988). Differences in physiological performance coupled with differences in behaviour are thus responsible for the differential distribution or zonation of species which have been described in many parts of the world. On the other hand, it is possible that predation pressure significantly influences the distribution patterns of some species (Yoshiyama, 1981).

The two examples of zonation patterns illustrated in Fig. 15.2 are taken from regions with a marked tidal range, but even in the Mediterranean, where fluctuations in water level due to tides are small, field observations have shown that blennies occupy distinct zones and microhabitats within 1m or 2m depth (Abel, 1962; Zander, 1972, 1980).

For the most part, the behavioural mechanisms underlying these zonation patterns are unknown. Field observations indicate that larvae may resist displacement into deeper water after hatching and prior to settlement, and so maintain their proximity to the adult habitat (Marliave, 1986). There is also some experimental evidence to suggest that the larvae of certain species are able to select particular substrata on which to settle, and so might be responsible for the initial choice of habitat (Marliave, 1977). Such settling behaviour of the larvae could be responsible for the initial and fairly coarse choice of habitat. It is not yet clear whether the final zonation patterns of adults are the result of (1) precise initial settlement, or (2) random settlement followed by active choice of the exact zone to be occupied, or (3) random settlement and subsequent passive elimination of individuals from unfavourable zones, or all three mechanisms.

Much more information is available on the behaviour involved in maintaining position in a habitat once selected. Numerous studies (Gibson, 1982; Bennett and Griffiths, 1984) have shown that cover of some form is of

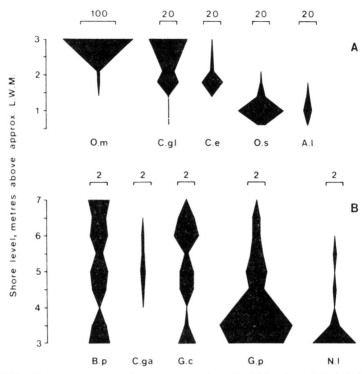

Fig. 15.2 Zonation patterns of some common intertidal fishes from British Columbia (A, from data in Green, 1971a) and the Atlantic coast of France (B, redrawn from Gibson, 1972). In (A) the total numbers of fish from a series of tide pools are plotted. In (B) the scales represent the mean number of fish per pool over the range examined. In both cases species which occupy the upper shore levels can be distinguished from those mainly inhabiting the lower levels. Species abbreviations: A.l, *Artedius lateralis*; B.p, *Lipophrys* (*Blennius*) *pholis*; C.e, *Clinottus embryum*; C.ga, *Coryphoblennius galerita*; C.gl, *Clinocottus globiceps*; G.c, *Gobius cobitis*; G.p, *Gobius paganellus*; N.l, *Nerophis lumbriciformis*; O.m, *Oligocottus maculosus*; O.s, *Oligocottus snyderi*.

overwhelming importance, and in a general sense this serves to keep the fish in areas where shelter is likely to be available over the low-tide period. Requirements for a specific type of cover have also been demonstrated, and by choosing to shelter in an alga, for example, which itself grows at a particular level on the shore, the fish ensures that an appropriate zone is selected.

A demonstration of how different behavioural reponses to the same stimulus can lead to a separation of species on the shore is provided by Nakamura's experiments with two sculpins (Nakamura, 1976). The common tidepool sculpin *Oligocottus maculosus* and its relative *O. snyderi* can both be found in tide pools on the Pacific coast of North America. The former inhabits the upper part of the shore whereas *O. snyderi* tends to be limited to vegetated pools at lower levels and is also found subtidally (Fig. 15.2). When given a choice of three habitat types in aquaria (eelgrass, rocks or open sand), *O. maculosus* showed a strong preference for rocks, and *O. snyderi* greatly preferred the eelgrass (Fig. 15.3(A)). The results were the same whether the species were tested separately or together, showing that the differences in distribution were not caused by aggressive interactions or competition for a particular habitat type. Furthermore, when tested in an experimental depth gradient with a simulated tidal cycle and artificial 'pools', *O. maculosus* was consistently found in the shallower depths whereas *O. snyderi* showed a random distribution with depth (Fig. 15.3(B)). The effect of temperature on distribution was also tested with *O. snyderi*, the more stenothermal of the two species. Raising the temperature in the 'pools' at 'low tide' altered the 'tidepool' distribution to lower levels on subsequent 'tides'. These results, together with experiments on the temperature tolerances of the two species, help to explain their observed distribution. The more eurythermal *O. maculosus*, with its preference for shallow water and generalized rock habitats, occupies the upper shore. The stenothermal *O. snyderi*, in contrast, avoids the upper pools because of their unstable temperature regime, even though suitable vegetation may be present, and is consequently restricted to lower levels. Sculpins may be unusual in that they seem to show no intra- or interspecific aggressive behaviour. In other studies of habitat selection, such interactions have been observed. The gobies *Gobiosoma robustum* and *G. bosci*, for example, which live in eelgrass and among oysters, alter their habitat preference in the presence of the other species (Hoese, 1966).

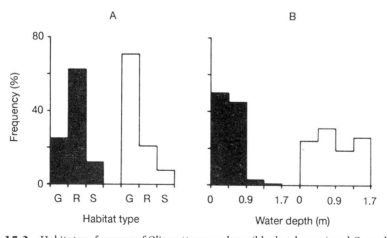

Fig. 15.3 Habitat preferences of *Oligocottus maculosus* (black columns) and *O. snyderi* (open columns) in aquaria. (A) Distribution of the two species when presented with a choice of eelgrass (G), rocks (R), and open sand (S). (B) Distribution with respect to water depth. From data in Nakamura (1976).

Comparable results were found with the Californian blennies *Hypsoblennius jenkinsi* and *H. gilberti*. The latter lives intertidally and subtidally on rocky shores and wanders over a large home range. In contrast, *H. jenkinsi* is found only subtidally and inhabits the holes of boring molluscs, gastropod tubes or mussel beds, where it is sedentary and highly territorial. When tested alone in aquaria, *H. jenkinsi* was found frequently among the calcareous tubes of the gastropod *Serpulorbis*. *H. gilberti* usually occupied spaces at the bases of the tubes or open sand. When the two species were mixed, these preferences were greatly reinforced (Fig. 15.4). The change in distribution when the two species were together is partly explicable by the greater territorial tendencies and aggressive superiority of *H. jenkinsi*. It rarely moved from the immediate area of its 'home' tube, could readily displace *H. gilberti* from its home position, and won over 90% of aggressive encounters between the two species (Stephens *et al.*, 1970). A similar situation has been described for the syntopic Mediterranean blennies *Blennius canevae* and *B. incognitus* (Koppel, 1988). Both live in very shallow water and use the empty holes of boring mussels as refuges and breeding sites. *B. canevae* selects smaller, horizontal holes that correlate with its body size, whereas *B. incognitus* will utilize larger holes of any inclination whose diameter has no correlation with body size. These differences,

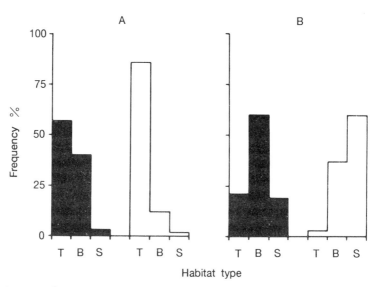

Fig. 15.4 Habitat preferences of *Hypsoblennius jenkinsi* (A) and *H. gilberti* (B) when tested in aquaria alone (black columns) and together (open columns). The histograms plot the percentage of observations in which the fish were seen among the tubes of *Serpulorbis* (T), at the base of the tubes (B), or on the open sand (S). Redrawn from Stephens *et al.* (1970).

together with differences in territorial behaviour, may reduce competition for refuges between the specialized *B. canevae* and the opportunistic *B. incognitus.*

15.4   HABITAT MAINTENANCE

Once established in a favourable zone it is clearly an advantage to maintain position there, and the simplest way of doing that is not to move very far. If a fish has to move over relatively large distances from its refuge, to obtain sufficient food, for example (Bennett *et al.*, 1983), then it is advantageous to be able to return to the place formerly occupied. Restricted movement is a common feature of intertidal fishes (Gibson, 1988) and many studies have demonstrated that some individuals can be found in the same pool at low tide on numerous consecutive occasions. Figure 15.5 illustrates the results of two such studies from widely separated areas. *Clinocottus analis* clearly tended to stay longer in pools than *Taurulus (Acanthocottus) bubalis*. The difference in the patterns of residence of the two species illustrated does not necessarily represent a real interspecific difference because several other factors can affect how long a fish inhabits a particular pool. One important factor regulating residence time is the amount of disturbance to which the pool is subjected. Disturbance is usually caused by wave action which, when severe, can totally

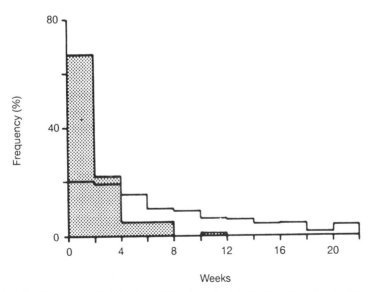

Fig. 15.5 Frequency distribution of the time that individual fish stay in tide pools. Open columns, *Clinocottus analis* in California (from data in Richkus, 1978); stippled columns, *Taurulus (Acanthocottus) bubalis* in Britain. Redrawn from Gibson (1967).

alter the suitability of a pool, mostly by removing or altering the configuration of cover provided by rocks and boulders.

There is, however, a biological factor which affects residence time or fidelity to a particular pool and that is a fish's age. Numerous observations suggest that young fishes are more mobile and do not stay in the same pool as long as older ones. The reason for the greater mobility of young fish seems to be that they are at the colonizing phase of their life history. When pools are depopulated, either naturally or experimentally, it is usually the young stages that are the first to repopulate the vacant areas (Gibson, 1988). It also seems likely, for one species at least (*Oligocottus maculosus*: Craik, 1981), that the young stages during their initial period of extensive movement are acquiring

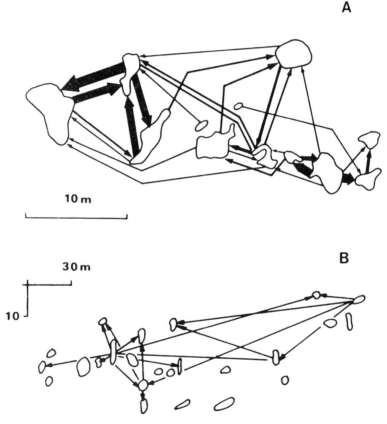

Fig. 15.6. Extent of movement of fish between tide pools. (A) *Clinocottus analis* over 20 weeks; the thicker the arrows, the greater the number of fish that moved between pools. The thinnest line represents two fish. Simplified from Richkus (1978). (B) *Lipophrys (Blennius) pholis* over 15 months. Redrawn from Gibson (1967). Note the different scales in the two diagrams.

a knowledge of their immediate locality, and that this knowledge is retained for future use in homing as discussed in the next section. The general pattern that emerges from such studies is that individuals tend to move over a rather restricted range which is not necessarily centred upon one particular pool or low-tide location (Fig. 15.6).

Homing

It seems that these restricted patterns of movement cannot be accounted for solely by poor locomotory abilities because many species from different families in widely separated geographical areas return to their original location when experimentally displaced. Examples of species that will home to their pool of origin are known in the Cottidae, Blenniidae, Gobiidae and Kyphosidae, from places as far apart as Europe, the Caribbean and the Pacific coast of North America (Gibson, 1988). Such experimental displacement probably mimics the fishes' high-water excursions and provides evidence that they are able to recognize and learn topographical details of their immediate environment and use them in navigating back to their home site. The possession of a topographic memory seems to be a common feature of blennies and gobies, and a detailed knowledge of the environment can be rapidly acquired.

Convincing evidence for both the rapid learning and the long-term retention of the knowledge of an area comes from experiments on the small tropical goby *Bathygobius soporator* (Aronson, 1951, 1971). This fish can be found either on sandy beaches or in rock pools. When stimulated, fish in pools at low tide will jump out of their pool into another close by with remarkable accuracy and with very few failures (Fig. 15.7). The function of the jumping behaviour remains unknown but it provides a convenient means of assessing the learning and orientational capabilities of the fish. Those accustomed to their surroundings will jump readily from one pool to the next, but others transferred to the pool from elsewhere will not unless allowed to familiarize themselves with the area first. Experiments in artificial pools showed that familiarization is obtained by swimming over the pool surroundings at high tide. Only one tide is necessary for this familiarization process, and the topographic knowledge so acquired can be retained for at least 40 days.

A knowledge of local geography, presumably acquired visually, does not completely account for homing behaviour, because some fish can still home even when displaced over distances greater than they would normally be expected to travel. Experiments with *Oligocottus maculosus* have shown that vision is not essential for this fish to home because blinded individuals can return to their home pool when displaced, although not as successfully as normal ones. In this species olfaction also apparently plays a large role in homing because destruction of the olfactory organs greatly impairs the ability to return to the place formerly occupied (Khoo, 1974). These results provide

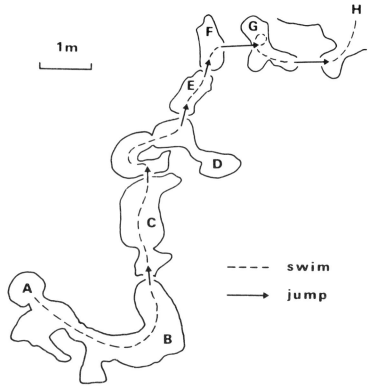

Fig. 15.7 Results of an experiment demonstrating the topographic knowledge of *Bathygobius soporator*. A fish was disturbed in pool A and after continual stimulation swam through pools B to G jumping over intervening ledges until it reached the open sea at H. Modified from Aronson (1951).

a comparison over short distances with the long-range olfactory homing mechanisms of salmonids (Brannon, 1981), and with short-range homing by cyprinids in the laboratory (Kennedy and Pitcher, 1975).

The function of homing and fidelity to a particular area seems to be to enable a fish to utilize and return to a location that provides a suitable refuge over the critical low tide period. Tide pools are the most obvious refuges, and all homing studies have been conducted with tide-pool species, mainly, perhaps, because the pools act as natural reservoirs or collecting basins for the fish as well as convenient sampling points for the investigator. The fish are not passively collected in pools by the outgoing tide, however, as the homing studies show, and homing behaviour ensures that the fish return only to those pools that do provide a favourable environment at low tide. To some extent this interpretation of the function of homing by tide-pool fishes is borne out by the few observations that have been made on the species that also occupy habitats outside pools. Such fish, the gunnels and pricklebacks for example, do not seem to show a fidelity to one place for any length of time, and have a much wider range of habitat preference. Hence the need to return to one particular location is not as great.

Territoriality

One other form of restricted movement which ensures that a favourable location is maintained is territoriality, in which a specific area, usually much smaller than a home range, is actively defended against others of the same and sometimes different species (Keenleyside, 1979). The possession of a territory guarantees to the holder exclusive rights to feeding, shelter and a spawning site in the defended area. Although food and reproduction are of obvious importance, shelter as a protection against turbulence and predation is equally vital for these small fishes. Territoriality, dominance hierarchies and the aggressive behaviour that accompanies them are particularly common among gobies, blennies, clingfishes and mudskippers, although the degree to which it is developed varies between species and sometimes even between the sexes of one species. Some species chase others away from their immediate vicinity wherever they happen to be; others are rigidly territorial and defend a well-defined area. Several studies have demonstrated experimentally that dominant fish, or those possessing territories, have better access to food and shelter than subordinates. In the striped blenny, *Chasmodes bosquianus*, this priority of access to shelter greatly decreases the risk of predation (Phillips and Swears, 1979).

Remarkable examples of territoriality are known among the mudskippers of tropical and subtropical shores. In Madagascar, *Periophthalmus sobrinus* can be found on the banks of tidal channels among mangrove swamps. At high tide the fish occupy a weakly defended territory at the top of the bank where their main burrow is situated. As the tide recedes, the fish move down the banks of the channel, some defending their areas of descent, until they reach the floor of the channel which is mostly left empty by the outgoing tide. Here they have other strongly defended territories in which they feed until the tide returns and they move back up to their burrows (Brillet, 1975). In Kuwait, another species, *Boleophthalmus boddarti*, lives on shallow tidal flats in large numbers and demarcates its territory by building mud walls along the boundaries (Fig. 15.8). The walls are formed from mouthfuls of mud which the fish carry from their burrows and deposit at the edge of their territory. In some densely populated areas the boundaries about one another on all sides and form a striking polygonal mosaic (Clayton, 1987). The occurrence of wall building is density dependent, and removal and replacement experiments have demonstrated that the walls act as visual barriers and reduce aggression between neighbours (Clayton, 1987; Clayton and Vaughan, 1986).

Fig. 15.8 Pentagonal territories of the mudskipper *Boleophthalmus boddarti* on a mudflat in Kuwait. The territorial boundaries and the openings to the burrows are clearly visible. The mud walls are 3–4 cm high. Reproduced with permission from Clayton (1987).

15.5 REPRODUCTION

Although many intertidal species defend territories throughout the year, territoriality is much more common during the reproductive season. In virtually all resident species that have been studied, the male selects and defends a territory, where spawning takes place. Spawning females of blennies, gobies, clingfishes and sculpins usually lay eggs in batches, which are deposited individually in a single layer and adhere to the substratum by means of adhesive filaments or special attachment areas. The gunnels and pricklebacks also lay eggs in batches, but these adhere to one another and are not attached to the substratum itself. This habit of laying demersal eggs is common to many shallow-water fishes and may serve as a mechanism for retaining the offspring in or close to their optimum environment. It does mean, however, that the static egg masses of species that spawn in this way are vulnerable to both benthic and pelagic predators, to damage by turbulence and to all the other physical risks that intertidal life imposes on the adults. Consequently the eggs are usually laid in sheltered places where these risks are minimized. Suitable sites are in holes, crevices and on the undersides of stones. It has also been suggested (Wirtz, 1978) that such sites offer the added advantage in turbulent water of increasing the probability of sperm reaching and fertilizing the ova.

Such cryptic spawning sites, although reducing stress from physical factors, pose several other problems not met by free-spawning pelagic fishes. The males, which generally seem to be responsible for selecting the spawning site, must attract females to the chosen location before spawning can begin. Successful rapid attraction of a mate may be particularly important for intertidal fishes because their reproductive activities are limited by the tidal cycle as well as the daily fluctuations in light intensity. This limitation arises partly because the time available for mating may be curtailed over the low-tide period for those individuals which are left out of water, but also because the majority of species are diurnal. Consequently, courtship can only take place during daylight because it relies so heavily on visual communication between the partners. In most groups this initial attractive phase of reproductive behaviour consists of some form of movement off the bottom, which advertises the male's position and is in marked contrast to the normal bottom-orientated methods of locomotion. The mudskippers, for example, leap into the air off the mud (Brillet, 1969), and some blennies swim vertically upwards and may 'loop the loop' (Phillips, 1977; Wirtz, 1978). In other blennies these vertical movements are reduced to lifting the anterior portion of the body or rearing up. The head in particular is an important signalling organ during the later phases of courtship in hole-dwelling blennies, and frequently bears dramatic colour patterns of contrasting light and dark areas (Fig. 15.9). The striking appearance of the head is often further enhanced by the presence of crests and tentacles. In addition to these visual signals, there is some evidence to suggest that certain blennies may also use pheromones secreted from special glands on the anal fin rays to attract females (Laumen *et al.*, 1974; Zander, 1975).

Once one or more females have been attracted and induced to spawn in the nest site, the male guards the eggs until they hatch. Guarding provides the obvious benefit of deterring predation, but the general parental activities of the male also include active care of the eggs themselves. This care involves removing dead eggs, keeping developing eggs clean and free from detritus, and

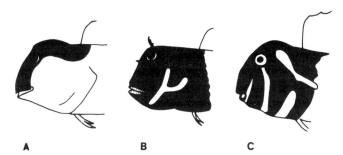

Fig. 15.9 Diagrammatic representation of head markings of male blennies in the breeding season. (A) *Blennius canevae.* (B) *Entomacrodus vermiculatus.* (C) *Lipophrys velifer.* Redrawn from Zander (1975), Abel (1973), and Wirtz (1980), respectively.

providing an adequate supply of oxygen by fanning water over the egg mass. Fanning is particularly important when the eggs are laid in confined spaces where water circulation may be restricted.

In general, paternal care of attached egg masses is common among intertidal fishes and the wider significance of the phenomenon has been discussed by Ridley (1978) and Blumer (1979). Both authors consider that it is frequently associated with external fertilization and male site attachment. External fertilization ensures that the male has a high probability of genetic relatedness to the offspring he is to guard, and site attachment enables the male to attract a succession of females and fertilize their eggs while guarding those from previous spawnings.

15.6 SYNCHRONIZATION OF BEHAVIOUR WITH THE TIDES

The fluctuating nature of the environment in the intertidal zone imposes on its inhabitants a continual rhythm of change. Certain phases of this rhythm may be more favourable than others for the performance of particular activities. Feeding, for example, is probably best undertaken at high water when movement is less restricted and a greater variety of food items may be accessible. Although the small size and turbulent habitat of most intertidal fishes makes them difficult to observe for extended periods in the sea, a few studies have been made of their natural activity patterns. The tidal migrations of many 'transient' species were mentioned in an earlier section (p.514). In areas of large tidal amplitude, 'resident' species also synchronize their activity with particular phases of the tidal cycle. On rockyshores, activity is probably greatest at or before high tide, as shown by a study of the large stichaeid *Cebidichthys violaceus* (Fig. 15.10). This species is large enough for a miniaturized ultrasonic transmitter to be attached to it and its detailed movements

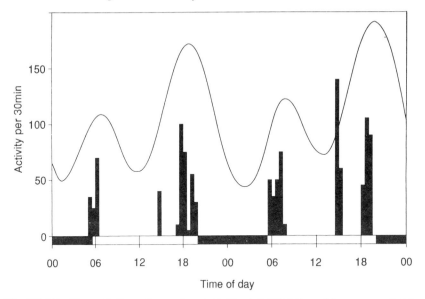

Fig. 15.10   The activity pattern of an individual *Cebidichthys violaceus* recorded by ultrasonic telemetry on the shore. The form of the tidal cycle is shown by the continuous line and the times of darkness by the black bars on the horizontal axis. The fish was most active on the flooding tide. Modified from Ralston and Horn (1986).

in space and time monitored by receivers stationed on the shore (Ralston and Horn, 1986). In areas where the tidal range is small, in the Mediterranean for example, activity patterns are synchronized with the light/dark cycle rather than with the tides (Taborsky and Limberger, 1980; Koppel, 1988). Some form of synchronization with the changes in the environment (which to a large extent are predictable) is therefore likely to make more efficient use of the limited periods when conditions are favourable.

Synchronization can be achieved by a direct response to change; flooding of a pool or sudden immersion by the rising tide, for example, could be used to signal the start of that tide's activities, but many species possess an inherent timing mechanism, a 'biological clock', which is phased with, and can operate independently of, current environmental conditions. In the laboratory the presence of this clock is manifest as a locomotory rhythm in which periods of rest alternate with periods of high activity (Fig. 15.11). Such rhythms, which seem to be a common feature of mobile intertidal animals (Palmer, 1974), can be further modulated on a seasonal, lunar or diel basis or by interaction with an inherent circadian rhythm. In intertidal fishes, whose rhythmic behaviour is much less well known than that of intertidal invertebrates, examples of each type of modulation have been described (Gibson, 1978, 1982, 1992; Northcott *et al.*, 1990). In at least one species, this endogenous rhythm seems to be

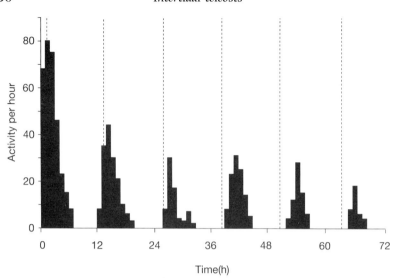

Fig. 15.11 An example of an endogenous tidal rhythm. The activity pattern of a single *Lipophrys pholis* recorded in continuous light in the laboratory. The vertical dotted lines indicate the predicted times of the high tide on the shore where the fish was captured. The timing of the activity peaks relative to high tide gradually changes because the endogenous rhythm 'free runs' at a period greater than that of the tidal cycle (12.4 h). Modified from Northcott *et al.* (1990).

entrained by cycles of hydrostatic pressure associated with the rise and fall of the tides (Northcott *et al.*, 1991).

The role of endogenous rhythmicity in a fish's day-to-day life is unknown, but the pattern of activity it shows in the laboratory, where most extraneous stimuli are deliberately excluded, is unlikely to be identical to the one exhibited in nature. Several studies (Green, 1971b; Gibson, 1975; Taborsky and Limberger, 1980) have shown discrepancies between field and laboratory activity patterns. The differences may be caused by the effect of artificial laboratory conditions on the expression of the activity pattern. On the other hand, exact correspondence should not be expected, because the pattern shown in the field is likely to be a combination of direct responses to concurrent stimuli underlain by the basic rhythm which regulates the overall level of behaviour, whatever the behaviour may be. If conditions are exceptionally severe so that no activity is possible, during storms for example, the rhythm may serve as a true clock so that when conditions return to normal the animal can perform the activity appropriate to the state of the tide. In general, however, the main function of the clock is probably for tidal prediction, allowing individuals to be prepared for changes in environmental conditions that take place with tidal frequency.

15.7 AMPHIBIOUS BEHAVIOUR

Many of the characteristics of intertidal fishes, particularly their ability to survive emersion and their morphological adaptations to benthic life, have enabled them to leave the water and continue their activities on land. This amphibious behaviour varies in duration from a few seconds to several hours at a time and may be active or passive. In its extreme form, the behaviour results in a distinct preference for land over water, as in the Chinese mudskipper *Periophthalmus cantonensis* (Gordon *et al.*, 1985) and, if the substratum is wet, the rockskipper *Alticus kirki* (Brown *et al.*, 1991).

The principal adaptations for a terrestrial mode of life relate to respiration and the maintenance of water and temperature balance. Some mudskippers, for example, regulate their body temperature by altering the colour of their skin or by actively selecting sites in their habitat that provide optimum thermal conditions (Tytler and Vaughan, 1983). The number of gill lamellae in amphibious species is sometimes reduced to prevent both gill collapse and evaporation. The skin and bucco-pharyngeal epithelium are often heavily vascularized to promote aerial respiration and the loss of carbon dioxide, the fish never allowing these surfaces to dry. Mudskippers and some blennies frequently roll onto their sides on moist substrata, or return to the water to wet their skin and renew the water held in their mouth and gill cavity. Other less amphibious species simply remain in moist conditions or just above the water-line, where they are continually wetted by waves and spray. Keeping the skin wet also promotes evaporative cooling, keeping the body temperature below lethal levels.

It is interesting that aerial respiration and amphibious behaviour seem to have evolved in marine and freshwater fishes for different reasons (Graham, 1976). The great majority of marine air breathers live in an environment where the water is normoxic for most of the time. Their requirement for an air-breathing capability is mainly to survive exposure at low tide and perhaps only secondarily to avoid hypoxic conditions that may arise occasionally in pools. Actively amphibious marine forms have been able to exploit a habitat where there are few competitors. Air breathing in freshwater fishes, on the other hand, assists survival in habitats where the water frequently becomes hypoxic and may even dry up for long periods. Few freshwater species are amphibious and this may be because the environment around fresh waters has already been fully exploited by the Amphibia themselves. Further discussion of air breathing in marine and freshwater fishes can be found in Graham (1976) and Johansen (1970), respectively.

Those species which are active out of water have to contend with the problem of the much lower density of air compared with that of water. Air provides no buoyancy, and greater muscular strength is required to move equivalent distances on land. Mudskippers overcome this problem by using their strong pectoral fins to propel themselves forward, and their method of progression is analogous to that of a man on crutches; the pectoral fins are the 'crutches' and the pelvic fins the 'legs'. This is a slow means of locomotion in which the tail plays little part. If danger threatens, however, the fish escape by skipping rapidly over the ground, propelled forward and upwards by sudden vigorous thrusts of their tail and pelvic fins (Harris, 1960). Amphibious blennies such as the rockskipper *Alticops (Alticus)* (Zander, 1967), the pearl blenny *Entomacrodus* (Graham *et al.*, 1985) and the clinid *Mnierpes* (Graham, 1970) also move on land using alternate strokes of their long tails, and are capable of jumping many times their own body length when disturbed. They are aided in this form of locomotion by the curved tips of the anal and pectoral fin rays.

Species that make extended, active excursions onto land are found principally in the tropics and subtropics, although such behaviour has also been observed in a rudimentary form in two Mediterranean blennies, *Coryphoblennius galerita* and *Blennius trigloides*, that leave the water at night to 'sleep' (Zander, 1983; Loisy, 1987), and in the cool temperate Chilean clingfish *Sicyases sanguineus* (Cancino and Castilla, 1987). The selection pressures that have led to the evolution of such behaviour in marine fishes have not been investigated in detail, but various suggestions have been put forward, including avoidance of adverse aquatic conditions, competition for food and space in the intertidal zone, and the avoidance of aquatic predators (Sayer and Davenport, 1991). There is little evidence to support any of these suggestions, but the trait has arisen independently in both rocky-shore species (blennies, clinids, clingfishes) and those inhabiting muddy shores (mudskippers, some gobies). This taxonomic diversity of amphibious species is matched by the variety of their ecophysiological and behavioural adaptations to life on land, enabling them to occupy and exploit resources unavailable to the great majority of fishes. Some of these adaptations may be similar to those used by ancestral forms during the original colonization of the land in the Devonian period (Brown *et al.*, 1991).

15.8 SUMMARY

Many teleost fishes utilize the intertidal zone either as a permanent habitat or as a feeding ground when it is submerged. Those that live there permanently show many morphological, physiological and behavioural adaptations which enable them to survive and reproduce in a habitat subject to regular change. They are small, negatively buoyant and frequently possess fin modifications which allow them to cling to the bottom. To survive the critical low-tide period they are eurythermal and euryhaline; they can also respire efficiently out of water and tolerate considerable water loss. Some have these abilities developed to such a high degree that they lead a semi-terrestrial life. The great majority lay demersal eggs which are attached to the substratum and guarded by the male until they hatch. Behavioural patterns that ensure they acquire and maintain suitable low-tide refuges include thigmotaxy and the ability to select zones and microhabitats on the shore which are compatible with their physiological capabilities. They can learn and remember details of their environment, limit their movements to a restricted area within the selected zone, and return (home) to such areas if displaced. Many species possess an inherent 'biological clock' which is phased with the tides and probably enables them to synchronize their activities with the tidal cycle.

ACKNOWLEDGEMENTS

I am grateful to David Clayton, Malcolm Gordon, Sally Northcott and Martin Sayer for their various contributions to this revised chapter and to Linda Robb for her help with the figures.

REFERENCES

Abel, E. F. (1962) Freiwasserbeobachtungen an Fischen in Golf von Neapel als Beitrag zur Kenntnis ihrer Okologie und ihres Verhaltens. *Int. Revue ges. Hydrobiol. Hydrog.*, **47**, 219–90.

Abel, E. F. (1973) Zur Oko-Ethologie des amphibisch lebenden Fisches *Alticus saliens* (Forster) und von *Entomacrodus vermiculatus* (Val.) (Blennioidea, Salariidae) unter besondere Berucksichtigung des Fortpflanzungsverhaltens. *Sber. öst. Akad. Wiss. Abteil I*, **181**, 137–53.

Ansell, A. D. and Gibson, R. N. (1990) Patterns of feeding and movement of juvenile flatfishes on an open sandy beach, in *Trophic Relationships in the Marine Environment* (eds M. Barnes and R. N. Gibson), Aberdeen University Press, Aberdeen, pp. 191–207.

Aronson, L. R. (1951) Orientation and jumping behaviour in the gobiid fish *Bathygobius soporator*. *Am. Mus. Novit.*, No. 1486, 22 pp.

Aronson, L. R. (1971) Further studies on orientation and jumping behaviour in the gobiid fish *Bathygobius soporator*. *Ann. N.Y. Acad. Sci.*, **188**, 378–92.

Barton, M. (1988) Response of two amphibious stichaeoid fishes to temperature fluctuations in an intertidal habitat, *Hydrobiologia*, **120**, 151–7.

Bennett, B. A. and Griffiths, C. L. (1984) Factors affecting the distribution, abundance and diversity of rock-pool fishes on the Cape Peninsula, South Africa. *S. Afr. J. Zool.*, **19**, 97–104.

Bennett, B., Griffiths, C. L. and Penrith, M.-L. (1983) The diets of littoral fish from the Cape Peninsula. *S. Afr. J. Zool.*, **18**, 343–52.

Blumer, L. S. (1979) Male parental care in the bony fishes. *Q. Rev. Biol.*, **54**, 149–61.

Brannon, E. L. (1981) Orientation mechanisms of homing salmonids, in *Proc. Salmon Trout Migratory Behav. Symp.* (eds E. L. Brannon and E. L. Sale), University of Washington, Seattle, pp. 219–27.

Bridges. C. R. (1988) Respiratory adaptations in intertidal fish. *Am. Zool.*, **28**, 79–96.

Brillet, C. (1969) Première description de la parade nuptiale des poissons amphibies periophthalmidae. *C. r. hebd. Séanc. Acad. Sci.*, Paris, **269D**, 1889–92.

Brillet, C. (1975) Relations entre territoire et comportement aggressif chez *Periophthalmus sobrinus* Eggert (Pisces, Periophthalmidae) au laboratoire et en milieu natural. *Z. Tierpsychol.*, **39**, 283–331.

Brown, C. R., Gordon, M. S. and Chin, H. G. (1991) Field and laboratory observations on microhabitat selection in the amphibious Red Sea rockskipper fish, *Alticus kirki* (Family Blenniidae). *Mar. Behav. Physiol.*, **19**, 1–13.

Cancino, J. M. and Castilla, J. C. (1987) Emersion behaviour and foraging ecology of the common Chilean clingfish *Sicyases sanguineus* (Pisces: Gobiesocidae). *J. nat. Hist.*, **22**, 249–61.

Clayton, D. A. (1987) Why mudskippers build walls. *Behaviour*, **102**, 185–95.

Clayton, D. A. and Vaughan, T. C. (1986) Territorial acquisition in the mudskipper *Boleophthalmus boddarti* (Teleostei, Gobiidae) on the mudflats of Kuwait. *J. Zool., Lond.*, **209A**, 501–19.

Craik, G. J. S. (1981) The effects of age and length on homing performance in the intertidal cottid, *Oligocottus maculosus* Girard. *Can. J. Zool.*, **59**, 598–604.

Davenport, J. and Woolmington, A. D. (1981) Behavioural responses of some rocky shore fish exposed to adverse environmental conditions. *Mar. Behav. Physiol.*, **8**, 1–12.

Gibson, R. N. (1967) Studies on the movements of littoral fish. *J. Anim. Ecol.*, **36**, 215–34.

Gibson, R. N. (1969) The biology and behaviour of littoral fish. *Oceanogr. Mar. Biol. Ann. Rev.*, **7**, 367–410.

Gibson, R. N. (1972) The vertical distribution and feeding relationships of intertidal fish on the Atlantic coast of France. *J. Anim. Ecol.*, **41**, 189–207.

Gibson, R. N. (1975) A comparison of the field and laboratory activity patterns of juvenile plaice, in *Proc. 9th Europ. Mar. Biol. Symp.* (ed. H. Barnes), Aberdeen University Press, Aberdeen, pp. 13–28.

Gibson, R. N. (1978) Lunar and tidal rhythms in fish, in *Rhythmic Activity in Fishes* (ed. J. E. Thorpe), Academic Press, London, pp. 201–13.

Gibson, R. N. (1982) Recent studies on the biology of intertidal fishes. *Oceanogr. Mar. Biol. Ann. Rev.*, **20**, 363–414.

Gibson, R. N. (1988) Patterns of movement in intertidal fishes, in *Behavioural Adaptation to Intertidal Life* (eds G. Chelazzi and M. Vanini), Plenum Press, London, pp. 55–63.

Gibson, R. N. (1992) Tidally-synchronised behaviour in marine fishes, in *Rhythms in Fishes* (ed. M. A. Ali), NATO-ASI Series A, Plenum Press, New York, pp. 67–86.

Gordon, M. S., Gabaldon, D. J. and Yip, A. Y. (1985) Exploratory observations on microhabitat selection within the intertidal zone by the Chinese mudskipper fish *Periophthalmus cantonensis*. *Mar. Biol.*, **85**, 209–15.

Graham, J. B. (1970) Preliminary studies on the biology of the amphibious clinid *Mnierpes macrocephalus*. *Mar. Biol.*, **5**, 136–40.

Graham, J. B. (1976) Respiratory adaptations of marine air-breathing fishes, in *Respiration of Amphibious Vertebrates* (ed. G. M. Hughes), Academic Press, London, pp. 165–87.

Graham, J. B., Jones, C. B. and Rubinoff, I. (1985) Behavioural, physiological and ecological aspects of the amphibious life of the pearl blenny *Entomacrodus nigricans* Gill. *J. exp. mar. Biol. Ecol.*, **89**, 255–68.

Green, J. M. (1971a) Local distribution of *Oligocottus maculosus* Girard and other tidepool cottids on the west coast of Vancouver Island, British Columbia. *Can. J. Zool.*, **49**, 1111–28.

Green, J. M. (1971b) Field and laboratory activity patterns of the tidepool cottid *Oligocottus maculosus* Girard. *Can. J. Zool.*, **49**, 255–64.

Harris, V. A. (1960) On the locomotion of the mudskipper *Periophthalmus koelreuteri* (Pallas) Gobiidae. *Proc. zool. Soc. Lond.*, **134**, 107–35.

Hoese, H. D. (1966) Habitat segregation in aquaria between two sympatric species of *Gobiosoma. Publs Inst. mar. Sci. Univ. Tex.*, **11**, 7–11.

Horn, M. H. and Gibson, R. N. (1988) Intertidal fishes. *Scient. Am.*, **258**, 64–70.

Horn, M. H. and Riegle, K. C. (1981) Evaporative water loss and intertidal vertical distribution in relation to body size and morphology of stichaeoid fishes from California. *J. exp. mar. Biol. Ecol.*, **50**, 273–88.

Johansen, K (1970) Air breathing in fishes, in *Fish Physiology*, Vol. 4, (eds W. S. Hoar and D. J. Randall), Academic Press, London, pp. 361–411.

Keenleyside, M. H. A. (1979) *Diversity and Adaptation in Fish Behaviour*, Springer-Verlag, Berlin, 208 pp.

Kennedy, G. J. A. and Pitcher, T. J. (1975) Experiments on homing in shoals of the European minnow, *Phoxinus phoxinus (L.)*. *Trans. Am. Fish. Soc.*, **104**, 454–7.

Khoo, H. W. (1974) Sensory basis of homing in the intertidal fish *Oligocottus maculosus* Girard. *Can. J. Zool.*, **52**, 1023–9.

Kneib, R. T. (1984) Patterns in the utilization of the intertidal salt marsh by larvae and juveniles of *Fundulus heteroclitus* (Linnaeus) and *Fundulus luciae* (Baird). *J. exp. mar. Biol. Ecol.*, **83**, 41–51.

Kneib, R. T. (1987) Predation risk and the use of intertidal habitats by young fishes and shrimp. *Ecology*, **68**, 379–86.

Koppel, V. H. (1988) Habitat selection and space partitioning among two Mediterranean blenniid species. *Mar. Ecol. Pubbl. Staz. Zool. Napoli.*, **9**, 329–46.

Kuipers, B. (1973) On the tidal migration of young plaice (*Pleuronectes platessa* L.) in the Wadden Sea. *Neth, J. Sea Res.*, **6**, 376–88.

Laumen, J., Pern, U. and Blum, V. (1974) Investigations on the function and hormonal regulation of the anal appendices in *Blennius pavo* (Risso). *J. exp. Zool.*, **190**, 47–56.

Loisy, P. (1987) Observations sur l'emersion nocturne de deux blennies mediter-ranéennes: *Coryphoblennius galerita* et *Blennius trigloides* (Pisces, Perciformes). *Cybium*, **11**, 55–73.

Marliave, J. B. (1977) Substratum preferences of settling larvae of marine fishes reared in the laboratory. *J. exp. mar. Biol. Ecol.*, **27**, 47–60.

Marliave, J. B. (1986) Lack of planktonic dispersal of rocky intertidal fish larvae. *Trans. Am. Fish. Soc.*, **115**, 149–54.

Morris, S. and Taylor, A. C. (1983) Diurnal and seasonal variations in physico-chemical conditions within intertidal rock pools. *Est. Coast. Shelf Sci.*, **17**, 339–55.

Nakamura, R. (1976) Experimental assessment of factors influencing microhabitat selection by the two tidepool fishes *Oligocottus maculosus* and *O. snyderi. Mar. Biol.*, **37**, 97–104.

Northcott, S. J., Gibson, R. N. and Morgan, E. (1990) The persistence and modulation of endogenous circatidal rhythmicity in *Lipophrys pholis* (Teleostei). *J. mar. biol. Ass. U.K.*, **70**, 815–27.

Northcott, S. J., Gibson. R. N. and Morgan, E. (1991) The effect of tidal cycles of hydrostatic pressure on the activity of *Lipophrys pholis* L. (Teleostei). *J. exp. mar. Biol. Ecol.*, **148**, 45–57.

Palmer, J. D. (1974) *Biological Clocks in Marine Organisms*, Wiley, London, 173 pp.

Phillips, R. R. (1977) Behavioural field study of the Hawaiian rockskipper *Istiblennius zebra* (Teleostei, Blenniidae) I. Ethogram. *Z. Tierpsychol.*, **32**, 1–22.

Phillips. R. R. and Swears, S. B. (1979) Social hierarchy, shelter use, and avoidance of predatory toadfish (*Opsanus tau*) by the striped blenny (*Chasmodes bosquianus*). *Anim. Behav.*, **27**, 1113–21.

Ralston, S. L. and Horn, M. H. (1986) High tide movements of the temperate-zone herbivorous fish *Cebidichthys violaceus* (Girard) as determined by ultrasonic telemetry. *J. exp. mar. Biol. Ecol.*, **98**, 35–50.

Richkus, W. A. (1978) A quantitative study of intertidepool movement of the wooly sculpin *Clinocottus analis. Mar. Biol.*, **49**, 277–84.

Ridley, M. (1978) Paternal care. *Anim. Behav.*, **26**, 904–32.

Sayer. M. D. J. and Davenport, J. (1991) Amphibious fish: why do they leave the water? *Rev. Fish Biol. Fish.*, **1**, 159–81.

Stephens, J. S., jun, Johnson, R. K., Key, G. S. and McCosker, J. E. (1970) The comparative ecology of three sympatric species of the genus *Hypsoblennius* Gill (Teleostomi, Blenniidae). *Ecol. Monogr.*, **40**, 213–33.

Taborsky, M. and Limberger, D. (1980) The activity rhythm of *Blennius sanguinolentus* Pallas, an adaptation to its food source? *Mar. Ecol. Pubbl. Staz. Zool. Napoli*, **1**, 143–53.

Truchot, J. P. and Duhamel-Jouve, A. (1980) Oxygen and carbon dioxide in the marine environment: diurnal and tidal changes in rock pools, *Respir. Physiol.*, **39**, 241–54.

Tyler, A. V. (1971) Surges of winter flounder, *Pseudopleuronectes americanus*, into the intertidal zone. *J. Fish. Res. Bd Can.*, **28**, 1727–32.

Tytler, P. and Vaughan, T. (1983) Thermal ecology of the mudskipper *Periophthalmus koelreuteri* (Pallas) and *Boleophthalmus boddarti* (Pallas) of Kuwait Bay. *J. Fish Biol.*, **23**, 327–37.

Van der Veer, H. W. and Bergman, M. J. N. (1986) Development of tidally related behaviour of a newly settled 0-group plaice (*Pleuronectes platessa*) population in the western Wadden Sea. *Mar. Ecol. Progr. Ser.*, **32**, 121–9.

Wirtz, P. (1978) The behaviour of the Mediterranean *Tripterygion* species (Pisces, Blennioidei). *Z. Tierpsychol.*, **48**, 142–74.

Wirtz, P. (1980) A revision of the eastern-Atlantic Tripterygiidae (Pisces, Blennioidei) and notes on some west African blennioid fish. *Cybium*, 3rd series, 1980, 83–101.

Yoshiyama, R. M. (1981) Distribution and abundance patterns of rocky intertidal fishes in central California. *Env. Biol. Fishes*, **6**, 315–32.

Zander, C. D. (1967) Beitrage zur Okologie und Biologie littoralbewohnender Salariidae und Gobiidae (Pisces) aus dem Roten Meer. *'Meteor' Forschungsergebnisse, D(2)*, 69–84.

Zander, C. D. (1972) Beitrage zur Okologie und Biologie von Blenniidae (Pisces) des Mittelmeeres. *Helgoländer wiss. Meeresunters.*, **23**, 193–231.

Zander, C. D. (1975) Secondary sex characteristics of Blennioid fishes (Perciformes). *Pubbl. Staz. zool. Napoli*, **39**, (*Supplement*), 717–27.

Zander, C. D. (1980) Morphological and ecological investigations on sympatric *Lipophrys* species (Blenniidae, Pisces). *Helgoländer wiss. Meeresunters.*, **34**, 91–110.

Zander, C. D. (1983) Terrestrial sojourns of two Mediterranean blennioid fish (Pisces, Blennioidei, Blenniidae). *Senckenberg. Marit.*, **15**, 19–26.

Chapter sixteen

The behavioural ecology of sticklebacks

G. J. FitzGerald and R. J. Wootton

16.1   INTRODUCTION

Four problems currently dominate the study of animal ecology. First, electrophoretic studies of proteins and mitochondrial DNA (e.g. Withler and McPhail, 1985; Gach and Reimchen, 1989) expose high levels of genetic variation both within and between populations. But there is controversy about the processes which maintain this variation (Lewontin, 1974). Secondly, natural populations fluctuate in abundance. Some fluctuations are caused by abiotic factors while others reflect the effects of biotic interactions such as predation and parasitism (Krebs, 1978). The relative importance of abiotic and biotic factors and the effect of a population's own density on its changes in abundance have to be estimated. Age-specific birth and death rates categorize the life-history pattern of a species. This pattern is assumed to have adaptive significance. The third problem is to develop a theory of life histories which will identify that significance and yield predictions of how a life-history pattern is likely to alter with changes in the environment (Charlesworth, 1980). The fourth problem is to determine the factors that control the number of species that can coexist in a given area, particularly to assess the relative roles of deterministic and stochastic processes in the pattern of colonizations and extinctions in a community (Schoener, 1987). Because the behaviour of an animal mediates its interaction with the environment, solutions to these problems will demand a knowledge of the behavioural responses to the abiotic and biotic environment (see also Pitcher *et al.*, 1979). The sticklebacks (Gasterosteidae) provide unusually favourable material for revealing the importance of the study of behaviour to the analysis of ecological problems. The general biology of the sticklebacks has been reviewed by Wootton (1976, 1984). These small fish have a wide geographical distribution. They frequently adapt well to laboratory conditions and they have a sufficiently short life span that all stages in their ontogeny

Behaviour of Teleost Fishes 2nd edn. Edited by Tony J. Pitcher. Published in 1993 by Chapman & Hall. ISBN   0 412 42930 6 (HB) and 0 412 42940 3 (PB).

Table 16.1   The sticklebacks (Gasterosteidae)

Species	Common names	Distribution
Apeltes quadracus	Fourspine	North America
Culaea inconstans	Fivespine, Brook	North America
Gasterosteus aculeatus	Threespine	North America and Eurasia
G. wheatlandi	Blackspotted	North America
Pungitius pungitius	Ninespine, Tenspine	North America and Eurasia
P. platygaster	Ukrainian	Black and Aral Sea
Spinachia spinachia	Sea, Fifteenspine	Europe

can be studied experimentally. Their behaviour not only provides a behavioural ecologist with the raw material for research; it also yields a continuing aesthetic pleasure.

The Gasterosteidae form a natural (monophyletic) family containing five genera (Table 16.1) (Figs. 16.1–16.6), whose evolutionary relationship to other teleosts is obscure. The geographical distribution of the family is entirely in the northern Hemisphere between about 35° and 74° N. In many areas two species are sympatric, and in north-eastern America four or five species may be sympatric. All have similar life histories characterized by a short life span and elaborate reproductive behaviour, although there are differences in their life-history traits (e.g. Table 16.2). During the breeding season in spring and summer, the reproductively active male defends a territory within which it builds a nest. Once eggs are laid in the nest, they are guarded and ventilated by the male. In most species, the female can spawn several times during a breeding season. Some species have populations that are anadromous, moving from coastal waters into fresh or brackish water in spring or early summer to breed, with the young and any surviving adults migrating seaward in autumn.

Sticklebacks show considerable intra- and inter-population variation in morphological characteristics. Breeding experiments show that for many of the characteristics, the phenotypic variability reflects an underlying genetic variation. A well-studied example is the lateral-plate polymorphism of the threespine stickleback, *Gasterosteus aculeatus* (Fig. 16.7). Sticklebacks lack scales, but the flanks of the body may be protected by bony lateral plates. In *G. aculeatus* three distinct plate morphs are recognized. The completely plated morph (*trachurus*) has a full row of plates that runs from just behind the head on to a keeled caudal peduncle; the partially plated morph (*semiarmatus*) has an anterior row and a caudal row separated by a gap: the low-plated morph (*leiurus*) has only an anterior row and an unkeeled caudal peduncle. Within each morph, there is variation in the number of plates, both within and between populations. Plate morph and life history are correlated. Anadromous populations of *G. aculeatus* are usually monomorphic for, or contain a high proportion of, the completely plated morph. Resident populations in fresh

Table 16.2 Selected life-history traits of four sympatric sticklebacks from the Isle Verte area of the St Lawrence Estuary, Quebec (modified from Craig and FitzGerald, 1982). Means ± SD given; details of sample sizes, ranges etc. are given in Craig and FitzGerald (1982)

Species	Standard length (mm)	Wet wt (g)	Egg number (clutch size)	Egg diameter (mm)	Longevity (years)	Breeding season	Post-hatching care of fry
Gasterosteus aculeatus	64 ± 4.0	3.958 ± 0.917	366 ± 175	1.39 ± 0.06	2+	May and June	Yes
G. wheatlandi	33 ± 2.9	0.596 ± 0.189	80.1 ± 20.3	1.25 ± 0.10	1+	May and June	Yes
Pungitius pungitius	44 ± 3.7	0.933 ± 0.250	76.1 ± 22.4	1.14 ± 0.08	1+	May and June	Yes
Apeltes quadracus	41 ± 6.9	1.042 ± 0.466	36.1 ± 11.9	1.40 ± 0.06	2+	May to end of July	No

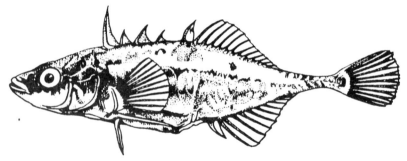

Fig. 16.1 Fourspined stickleback. *Apeltes quadracus.* 57 mm.

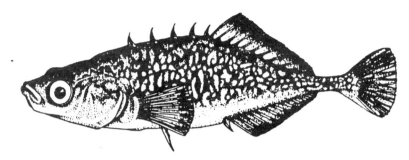

Fig. 16.2 Brook stickleback, *Culaea inconstans.* 50 mm

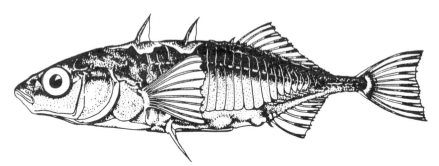

Fig. 16.3 Threespined stickleback, *Gasterosteus aculeatus* (fully-plated form) 60 mm.

Fig. 16.4 Blackspotted stickleback, *Gasterosteus wheatlandi.* 38 mm.

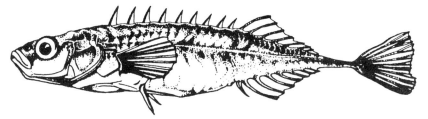

Fig. 16.5 Ninespined stickleback, *Pungitius pungitius*. 57 mm.

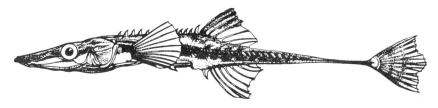

Fig. 16.6 Fifteenspined stickleback, *Spinachia spinachia*. 95 mm.

water are commonly monomorphic for the low-plated morph, though in parts of the geographical range, resident freshwater populations monomorphic for the completely plated morph are common (Bell, 1976, 1984). Some river systems contain anadromous and resident populations which coexist during the summer period with little or no gene flow between them (Bell, 1984).

The behavioural processes that are relevant to the four problems outlined above include the habitat selection, foraging, defence against predators and parasites and reproductive behaviour.

16.2 HABITAT SELECTION

The habitat of an animal is the complex of physical and biotic factors that determines or describes the place where an animal lives (Partridge, 1978). The chosen habitat depends on the response of the animal not only to abiotic factors but also to the biotic factors of food supply, intra- and interspecific competition, predation, parasitism and disease. The ability to select preferred habitats can lead to selective segregation between sympatric species that are potential competitors (Nilsson, 1967). Such segregation minimizes interference competition between the species. The use of a restricted habitat may also allow further adaptive specialization to, and efficient exploitation of, that habitat, reinforcing the distinctiveness between sympatric taxa (review, Wootton, 1990).

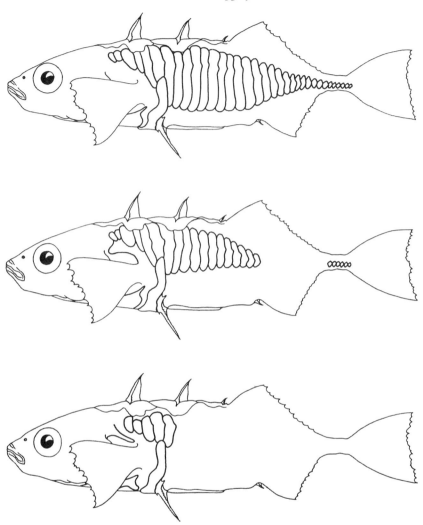

Fig. 16.7 Common plate morphs of *G. aculeatus*. Top, completely plated; middle, partially plated; bottom, low plated (from Wootton, 1984).

Both *G. aculeatus* and *Pungitius pungitius*, the most widely distributed species, occur in streams, ponds, small and large lakes, estuaries, tidal marshes, littoral pools and coastal waters. They are split into many genetic units which are reproductively isolated from each other (Bell, 1984). Within a genetic unit, the habitat preference may be more restricted than is suggested by the range of habitats in which the species is found. *Apeltes quadracus* and *G. wheatlandi* are confined to coastal waters, bays, lagoons and a few freshwater sites on the north-eastern seaboard of North America. *Culaea inconstans* is restricted to inland waters of North America east of the Rocky Mountains and north from the Great Lakes. *Spinachia spinachia* is found only in coastal waters of western Europe. Although *S. spinachia* lives in low salinities in the Baltic Sea, it is the only stickleback with no freshwater populations (Wootton, 1976).

Habitat cues

Salinity

Sticklebacks are euryhaline. The least tolerant of high salinities, *C. inconstans*, can live in alkaline prairie lakes in salinities as high as 20 ppt (seawater is about 35 ppt). The other species will tolerate salinities up to and greater than seawater. Within a species, tolerance varies between populations, and within populations it can vary seasonally (Baggerman, 1957).

Rivière des Vases is a tidal river on the south shore of the St Lawrence Estuary near Isle Verte, Québec. During the breeding season, *G. aculeatus* and *P. pungitius* occur throughout a 2.5 km stretch of the river including upstream freshwater stretches. *A. quadracus* and especially *G. wheatlandi* are found only in the downstream brackish water. When tested in a salinity gradient which ranged from 0 to 35 ppt, sexually mature males and females of the four species preferred salinities that corresponded to the distribution of species in the river (Audet *et al.*, 1985).

The ability to select a position along a salinity gradient may have consequences for reproductive success. Kedney *et al.*, (1987) found that *G. aculeatus* males breeding in fresh water in Rivière des Vases had higher egg mortalities and took longer to hatch their eggs than males breeding in nearby salt marsh pools. However, the fry from eggs incubated in fresh water had significantly lower mortality in fresh water than the fry from eggs incubated at 20 ppt (Bélanger *et al.*, 1987).

Temperature

For ectothermic animals, temperature is a major factor limiting their distribution. Although the upper or lower lethal temperatures may be rarely experienced, the effect of temperature on metabolism will have consequences for feeding, growth and reproduction (Elliott, 1981).

Conditioning experiments showed that *S. spinachia* could detect temperature increments as small as 0.05 °C (Bull, 1957). Such sensitivity would allow a stickleback to choose where it lives in a temperature gradient.

In tidal salt marsh pools at Isle Verte, *G. aculeatus*, *G. wheatlandi* and *P. pungitius* ceased breeding when the temperature rose above 30 °C (FitzGerald, 1983). Some *P. pungitius* remained in the pools throughout July and August, whereas the adult *Gasterosteus* moved back to the estuary. Sexually mature sticklebacks collected at Isle Verte in May and tested in a horizontal temperature gradient which ranged from 5 to 30 °C showed interspecific differences in their temperature preferences (Lachance *et al.*, 1987). *G. aculeatus* and *G. wheatlandi* preferred similar ranges of 9–12 and 11–14 °C respectively. But *P. pungitius* showed a bimodal preference of 9–10 and 15–16 °C. Whoriskey *et al.* (1986) had noted that in an unusually cold May, *P. pungitius* bred while the *Gasterosteus* species stopped most of their activities. Usually *P. pungitius* started breeding later in May than the *Gasterosteus* species.

In the Camargue, a coastal wetland in southern France, *G. aculeatus* breed at 10 °C in March in freshwater marshes. Juveniles move out of the marshes via drainage canals to saline lagoons as the temperatures rise later in spring. Juveniles that fail to move are trapped in marshes that dry out during the summer (Crivelli and Britton, 1987). Mori (1985) describes populations of low-plated *G. aculeatus* in Japan in which some breeding took place all the year round, although with a peak in April and May. The fish were living in water derived from springs at an almost constant temperature of 15 °C.

C. inconstans reproduces at temperatures between 15 and 19 °C, with reproduction inhibited at temperatures above 19 °C. In a temperature gradient, sexually mature *C. inconstans* selected a range of 14.9–20.2 °C, but a range of 8.9–25.6 °C after spawning (MacLean and Gee, 1971).

Dissolved oxygen

Sticklebacks may encounter low oxygen conditions in summer in productive lakes and in winter when the lake surface becomes covered by a blanket of ice and snow. In deoxygenated water, they become agitated and the resulting movements may take them into more oxygenated water (Jones, 1964). In ice-covered lakes, *C. inconstans* and *P. pungitius* use localized regions of high oxygen concentration at cracks in the ice or bubbles at the ice/water interface (Klinger *et al.*, 1982). In warm, hypoxic waters, sticklebacks will show aquatic surface respiration exploiting the thin oxygen-rich layer at the air/water interface (Gee *et al.*, 1978; Whoriskey *et al.*, 1985). Such behaviour may increase the risk of bird predation.

Vegetation and substrate

In virtually all *Gasterosteus* and in a few populations of *Pungitius*, males build their nests on the substratum. In the other species and the exceptional *Gasterosteus* and typical *Pungitius* populations, the males nest in vegetation. Some selection of vegetation type by *A. quadracus* has been demonstrated experimentally. A freshwater population showed a preference for *Elodea* over *Potamogeton* (Baker, 1971). Tidal pool fish avoided nesting in *Fucus* that was covered with filamentous algae (Courtenay and Keenleyside, 1983).

Such preferences for types of vegetation or other habitat features may be important in allowing the coexistence of closely related but genetically distinct populations or species in small areas.

Two phenotypes of *G. aculeatus* are found in Enos Lake on Vancouver Island (Bentzen and McPhail, 1984; Bentzen *et al.*, 1984; Ridgway and McPhail, 1984). The two forms, 'benthics' and 'limnetics', coexist in this 17.6 ha lake, but they are reproductively isolated. Although both forms use the littoral of the lake for breeding, the limnetic males tend to nest in open areas and the benthic males in dense vegetation.

Blouw and Hagen (1990) describe two forms of *G. aculeatus* coexisting in coastal habitats in Nova Scotia. In one form the breeding male has the red throat and bluish back typical of *G. aculeatus*. The form nests on the substrate. The breeding male of the second form (the 'white stickleback') has an iridescent white-green back. It nests off the substrate in thick filamentous algae. Against the background of the green algae, the white males stand out with startling clarity. The proportion of the white form in the total population declines sharply with a decrease in algal cover.

In the Isle Verte salt marsh pools, *P. pungitius* nests in the algae, whereas in the same pools, *G. aculeatus* and *G. wheatlandi* often nest in open areas (FitzGerald, 1983). A similar preference of male *P. pungitius* for nesting in vegetation and male *G. aculeatus* in more open areas was described for fish taken from a Belgian stream in which the two species lived sympatrically (Ketele and Verheyen, 1985).

At Isle Verte both species of *Gasterosteus* and *Pungitius* tended to avoid entering pools that would dry out in the 2 week period between flood tides (Whoriskey and FitzGerald, 1989). The cues the fish were using are not known, but there was no effect of the density or species composition of the fish that were already in the pools. Indeed, the three species tended to select the same pools.

Behavioural interactions

Little is known of the use of space by non-breeding sticklebacks, particularly in the winter (Blouw and Hagen, 1981). Anadromous *G. aculeatus* probably shoal (*sensu* Pitcher, 1983), but MacLean (1980) described some fish in a resident population holding feeding territories outside the breeding season. Territoriality in juvenile *G. aculeatus* was described by Bakker and Feuth-De Bruijn (1988). This territoriality was more strongly expressed in juveniles from a resident population than in juveniles from an anadromous population.

Indirect evidence suggests that habitat selection may play a role in reducing intraspecific competition between age classes during the breeding season. Picard *et al.* (1990) found that at the Isle Verte site both 1 + and 2 + *G. aculeatus* bred. But the proportion of 1 + fish was much higher in a nearby river than in the salt marsh pools. The proportion of 1 + fish was also low in the St Lawrence estuary waters adjacent to breeding sites.

Interspecific aggression may play a key role in the habitat segregation of sympatric species. *G. aculeatus* could displace both *G. wheatlandi* and *A. quadracus* from their nest sites in laboratory aquaria (Rowland, 1983a, b). A field experiment at the Isle Verte site showed that *G. aculeatus* lowered the reproductive success of *G. wheatlandi* breeding in the same pools (FitzGerald and Whoriskey, 1985).

Interspecific interactions between sticklebacks and other species may also be a factor in habitat selection. In the Matamek River, Québec, territorial male *P. pungitius* attacked juvenile brook charr, *Salvelinus fontinalis*, usually when the charr approached the nest of fry (Gaudreault *et al.*, 1986). In a stream tank, female *P. pungitius* were also observed attacking the charr.

In the Chehalis River in Washington State (USA), males of an endemic, freshwater esocid fish, *Novumbra hubbsi*, take up breeding territories in the summer. There is probably interspecific competition for space between male *N. hubbsi* and male *G. aculeatus* at this time. This competition may have been the major causal factor in the evolution of the unusual breeding colours of male *G. aculeatus* that are sympatric with *N. hubbsi*. Such males become black during the breeding season rather than developing the typical breeding colours of red throat and blue-green back (McPhail, 1969). Territorial male *N. hubbsi* also adopt a black or dark brown breeding coloration. Experimental evidence showed that 'black' sticklebacks suffered fewer territorial intrusions by *N. hubbsi* than typically coloured males (Hagen and Moodie, 1979; Hagen *et al.*, 1980).

16.3   FORAGING BEHAVIOUR

Despite the wide geographical distribution and habitat diversity of stickle-backs, the same prey types form the major part of the diet. The commonest prey include copepods, cladocerans, ostracods, isopods, amphipods, chirono-mid larvae and pupae, and other insect larvae and nymphs (Wootton, 1976).

Detection and selection of prey

Visual cues

Sticklebacks are primarily visual predators. The cues most important for a foraging stickleback are prey colour relative to the background, movement, size and shape. Experimental studies on *S. spinachia* (Kislalioglu and Gibson, 1976a, b) and *G. aculeatus* (Ibrahim and Huntingford, 1989a,b) have assessed the relative importance of these cues. For *S. spinachia* attacking mysids, the relative importance of the cues were ordered in the sequence: movement >

length > colour > shape. For *G. aculeatus* attacking chironomid larvae, the order of importance of the stimuli was: colour > movement > shape > size when the larvae were red, but movement > colour > shape > size when the chronomids were pale. *G. aculeatus* tends to prefer prey that are red, long and thin and moving at about $4-7 \text{ cm s}^{-1}$.

For *G. aculeatus* using these cues in laboratory tests, the prey chosen was not invariably the most profitable. (Profitability can be measured as the net energy return per unit time – see Hart, Chapter 8.) But a field experiment did suggest that under more natural conditions, the cues of colour and movement did enable the *G. aculeatus* to gain a high rate of energy intake (Ibrahim and Huntingford, 1989a).

Palatability

Once the prey has been detected, its palatability will determine whether it is eaten or rejected. *G. aculeatus* feeding on *Tubifex* tended to restrict their search to the area where they had eaten, but if a worm was rejected, the fish moved out of the area (Thomas, 1977). When *G. aculeatus* that had been feeding on *Tubifex* were given highly palatable enchytraeid worms as well as *Tubifex*, the risk of the *Tubifex* being eaten declined sharply once the sticklebacks learned to take the enchytraeids (Beukema, 1968).

Apostatic selection

Although in some circumstances sticklebacks prey preferentially on odd or conspicuous prey (Ohguchi, 1981), in other situations they preferentially attack the commonest type of prey (Visser, 1981). Such apostatic selection gives relative protection to the rarer morph of the prey and can result in frequency-dependent selection. *G aculeatus* feeding on pale and dark *Asellus* preferentially attacked the most common form (Maskell *et al.*, 1977).

Effect of prey density

Experiments in which *G. aculeatus* were allowed to attack swarms of *Daphnia* showed that a hungry fish attacked the densest part of the swarm, but as the fish became less hungry it attacked less dense parts (Milinski, 1977b). High densities of prey allow a higher attack rate by the stickleback, but also seem to exert a 'confusion cost' which could make the fish less vigilant to other cues such as the approach of a predator (see below).

Prey density also affects the profitability of prey of different sizes because of the relationship between density and rate of encounter. Experiments in which two sizes of *Daphnia*, 'large' and 'small', were presented to *G. aculeatus* at different densities showed that at low densities the fish behaved as though they selected the prey with the largest 'apparent size'. 'Apparent size' depends on absolute prey size and the distance of the prey from the fish. A small prey close to the fish can have a larger 'apparent size' than a bigger prey further from the fish. But at high densities the sticklebacks behaved as though they were selecting the most profitable prey, as would be predicted by the optimal foraging hypothesis (see Hart, Chapter 8; Gibson, 1980).

Effect of stickleback density

Sticklebacks frequently forage in loose shoals, which tend to become looser as the fish become hungrier. When food is discovered by one fish in the shoal, the others will rush to the feeding fish and start searching in the same area (Keenleyside, 1955). This social facilitation of foraging can allow the stickle-backs to exploit locally abundant supplies of food. In some populations, female and non-reproductive male *G. aculeatus* form shoals and raid the nests of reproductively active males. The shoaling fish destroy the nest and eat the eggs contained therein (Whoriskey and FitzGerald, 1985a, b). Although a territorial male stickleback can evict a solitary raiding fish, the male's defence is overwhelmed by weight of numbers when a shoal attacks. However, at least in some populations, the male will show a display which distracts the shoal from the nest to another site (Whoriskey and FitzGerald, 1985b; Foster, 1988; Ridgway and McPhail, 1988).

The decision by an individual fish of where to forage is related to its rate of feeding which in turn will be related to the number of fish trying to feed in the same area. Experimental studies in which groups of *G. aculeatus* were presented with *Daphnia* at different rates at the two ends of an aquarium showed that the fish distributed themselves between the two patches in a ratio that approximately reflected the profitability of the two patches (Milinski, 1984a). Despite this distribution, the fish did not all capture *Daphnia* at the same rate. Some fish were better competitors than others. The fish seemed to be using some 'rule-of-thumb' learning rule so they moved between the two patches in a way that enabled them to do as well as they could, given their differing competitive abilities.

The effect of competition for food can extend beyond the period when the competing fish are interacting (Milinski, 1982). When *G. aculeatus* were fed on *Daphnia*, the better competitors took a higher proportion of the large *Daphnia* than the poorer competitors. Subsequently, even when tested on their own, the poorer competitors failed to increase the proportion of large *Daphnia* in the diet.

Effect of predators and parasites

Foraging behaviour is also modified by the presence of predators and parasites. In the presence of a potential predatory fish, *G. aculeatus* fed more slowly on the *Tubifex* provided and avoided those worms that were close to the predator, although the predator was behind a glass partition (Milinski, 1985). Stickle-backs, frightened by exposure to a model of a piscivorous bird, attacked a low-density swarm of *Daphnia* whereas unfrightened fish attacked the densest swarm (Milinski and Heller, 1978; Heller and Milinski, 1979) (see also Chapter 9). In a field experiment, the presence of a piscivorous trout caused *G. aculeatus* to spend less time in the water column than fish not exposed to the trout. The exposed fish also consumed less food and shifted from feeding on profitable but more difficult-to-handle prey (chironomid larvae) to less profitable but easier-to-handle cladocerans (Ibrahim and Huntingford, 1989c). A laboratory experiment suggested that in the presence of different numbers of trout, sticklebacks adjusted their feeding rate in relation to the level of predation risk, and the reduction was the same irrespective of the food availability (risk adjustment). Although there was some individual variation, they did not accept a greater risk at higher food levels (risk balancing) (Fraser and Huntingford, 1986; see also Werner and Gilliam, 1984).

These effects of predators on foraging behaviour may be modified if the sticklebacks are infested by some parasites. Sticklebacks infested with the cestode *Schistocephalus solidus* approached the predator more closely and fed at a faster rate than uninfested fish. Infested fish recovered and started to forage again more rapidly than uninfested fish after being frightened by a sudden exposure to a model of a heron (Giles, 1987a). *S. solidus* makes heavy demands for nutrients on its host; infested fish seem to have a higher feeding motivation than uninfested fish, and this leads to them feeding in 'riskier' areas than the latter. In the absence of the predator, the infested fish are poor competitors in the presence of uninfested fish (Milinski, 1984b).

Phenotypic differences in foraging success

G. aculeatus provides evidence of differences in foraging behaviour correlated with morphological differences. These differences are found for populations both within individual lakes (Bentzen and McPhail, 1984) and within individual drainage systems (Lavin and McPhail, 1986).

The 'benthic' and 'limnetic' forms of *G. aculeatus* in Enos Lake are named because of their usual foraging habitats (Bentzen and McPhail, 1984). The limnetics have relatively smaller mouths than the benthics but more gill rakers. In experimental tests, the benthics took relatively larger prey than the limnetics and were more successful at foraging on invertebrates in a detritus substratum than limnetic males. Limnetic females did not forage on the substratum. Limnetics were more successfull at feeding on zooplankton than the benthics (Bentzen and McPhail, 1984). The limnetic males have an intermediate trophic status. They must use the substratum of the littoral zone when they nest, and at least during the reproductive season will be dependent on the substratum for some of their food.

In lakes in the Cowichan River drainage on Vancouver Island, three phenotypes, 'open water', 'intermediate' and 'littoral' have been recognized (Lavin and McPhail, 1985, 1986, 1987). Diet is correlated with phenotype. The littoral form has relatively fewer and shorter gill rakers than the open water morph, but the former has a longer upper jaw. The littorals can feed on significantly larger prey and are more effective at handling benthic prey than the two other morphs. However, the open water and intermediate forms forage more effectively on prey in the water column.

Intertaxon competition for food

Although there are similarities between the diets of the sticklebacks, there is little field evidence of competition for food between the taxa when they are living sympatrically. Delbeek and Williams (1987) compared the diets of juvenile and adult *A. quadracus*, *G. aculeatus*, *G. wheatlandi* and *P. pungitius* living sympatrically at coastal sites in New Brunswick, Canada. They suggested that a degree of habitat partitioning between the species together with the abundance of potential prey meant that competition for food was not a factor in allowing coexistence. Field experiments in which pairs of species have been kept together in enclosures have also failed to provide clear evidence of such competition. In one such experiment, juvenile *G. aculeatus* and *G. wheatlandi* were kept together in enclosures in pools at the Isle Verte site (Poulin and FitzGerald, 1989a). Although *G. wheatlandi* did show a reduction in its condition in some enclosures, there was no unequivocal evidence for interspecific competition. Earlier enclosure studies with adult sticklebacks at the same site had failed to find evidence that *G. aculeatus*, *G. wheatlandi* or *P. pungitius* changed their diet when kept enclosed with one of the other species (Walsh and FitzGerald, 1984). When high densities of sticklebacks were kept in enclosures for 2 months, there was no evidence that they had a significant impact on the abundance of the macrobenthos (Ward and FitzGerald, 1983a).

One field study has provided some evidence of interspecific competition for food between *G. aculeatus* and juvenile sockeye salmon, *Oncorhynchus nerka*, in an oligotrophic, coastal lake in British Columbia (O'Neill and Hyatt, 1987). The presence of the two species in enclosures in the lake caused reductions in the abundance of large zooplankters and produced a zooplankton community dominated by nauplii and rotifers, which were seldom eaten by the fish. O'Neill and Hyatt (1987) suggest that in such oligotrophic lakes, the sticklebacks and sockeye compete for suitable zooplankton prey when food is in short supply.

Experiments with *G. aculeatus* have shown that growth rate and the total fecundity of the females are functions of the rate of food consumption (Wootton, 1984). If food was in short supply, the presence of other species competing for food could affect both growth and fecundity – two important components of fitness.

16.4   PREDATION AND PARASITISM

Defence against predators

Adult sticklebacks are eaten by piscivorous birds, mammals, reptiles and large carnivorous fishes. Eggs, larvae and juveniles are also prey of carnivorous invertebrates such as dragonfly nymphs. Defence against this range of predators has both structural and behavioural components.

The structural defense has two elements – cryptic coloration and skeletal traits. All the sticklebacks are cryptically coloured except for males during the breeding season which adopt distinctive breeding colorations (male *S. spinachia* are an exception). In *G. aculeatus*, there is both inter- and intrapopulation variation in the development of the red coloration on the anterior ventral surface which is characteristic of breeding males. In some populations the reduction or absence of the red colour in breeding males has been attributed to the presence of visually hunting fish predators such as trout (Semler, 1971). But Reimchen (1989) has suggested that in waters that are discoloured by humic substances, a black rather than a red throat may provide more visual contrast because of the colour-filtering properties of such waters. Reimchen (1989) argues that behavioural studies that examine the effect of the red colour on rates of predation or on mate choice should be carried out in water that has the same light-transmission properties as the water in which the sticklebacks live.

The skeletal defences are the spines and their associated skeletal elements together with the lateral plates. The anterior plates probably stabilize the erect spines if the latter are displaced by a predator (Reimchen, 1983). The spines can be locked in an erect position, so a predator cannot depress but may break them if it can exert sufficient force. There is good evidence that the spines are effective in deterring some vertebrate predators (reviews, Bell, 1984; Wootton, 1984).

Behavioural defence has two components: behaviour that reduces the chances of the stickleback being detected, and behaviour that reduces the chance of capture after detection. *A. quadracus, P. pungitius* and *C. inconstans* are often associated with vegetation in which they can hide. At Isle Verte, *P. pungitius* hid in algae in response to movement above the pool (Whoriskey and FitzGerald, 1985c). *P. pungitius* and *C. inconstans* are less well protected because their spines are shorter than those of *Gasterosteus* spp. Small piscivorous fish will take *P. pungitius* in preference to *G. aculeatus* (Hoogland *et al.*, 1957).

The balance between behavioural and morphological defences can also vary intraspecifically. Some populations of *G. aculeatus, P. pungitius* and *C. inconstans* show a reduction or complete loss of spines and associated protective skeletal structures. Such populations are not restricted to habitats that are free of predators. Experiments with *C. inconstans* suggest that the behaviour of the spineless forms makes them less likely to be captured. The reduction or loss of skeletal armour may make the fish more streamlined and manoeuvrable (Reimchen, 1980; Reist, 1980a,b, 1983).

When sticklebacks in a shoal are disturbed the fish clump more closely together and often swim away from the source of the disturbance in a polarized shoal (Keenleyside, 1955; Whoriskey and FitzGerald, 1985c). At the Isle Verte site, FitzGerald and van Havre (1985) simulated the flight of a predatory bird by skimming a 'frisbee' over the pools. *P. pungitius* hid in algal thickets. Female and non-territorial male *Gasterosteus* formed shoals. The average shoal size increased with an increase in fish density up to an asymptotic size of about 40 fish (Fig. 16.8) Sticklebacks in large shoals resumed normal activities sooner after the flight of the 'frisbee' than those in smaller shoals (Fig. 16.9). In a shoal the individual stickleback may be at less risk because of the dilution effect, the confusion effect and the possibility of engaging in complex antipredator tactics (see Pitcher and Parrish, Chapter 12). The maximum shoal size noted by FitzGerald and van Havre (1985) may represent the size at which the advantages of increased protection from predation are balanced by an increased interindividual competition for food.

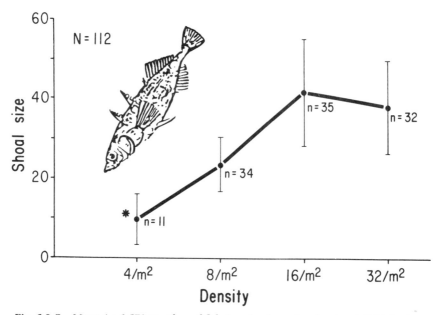

Fig. 16.8   Mean ($\pm$ 1 SD) number of fish in mixed species shoals of sticklebacks in tide pools containing fish of different densities. There were two pools at each density except at a density of 4 fish/m^2.* 1 pool only. (Redrawn from FitzGerald and van Havre 1985).

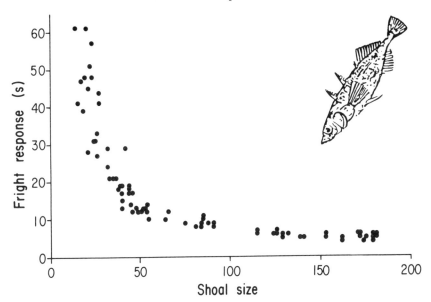

Fig. 16.9 Fright response of mixed species shoaling fish. Time to resume normal activity after a disturbance as a function of shoal size. (Redrawn from FitzGerald and van Havre, 1985, see original paper for details).

On detecting the presence of a potential predatory fish, some sticklebacks will approach it (Huntingford, 1976). Such 'investigatory' behaviour may allow the stickleback to obtain information about the motivational state and attack distance of the predator. An ingenious experiment using mirrors showed that a stickleback was more likely to approach closely to the predator if apparently accompanied by another fish than if the accompanying stickleback appeared to desert the investigator (Milinski, 1987). Such observations have been used to support the tit for tat model of the evolution of cooperative behaviour (Axelrod and Hamilton, 1981; but cf. Lazarus and Metcalfe, 1990).

In the presence of a piscivorous fish, a stickleback displays a distinct behavioural repertoire. Typically, the stickleback stops and fixes the predator with its binocular vision. It may then approach the predator in a series of jerky movements (investigation). Two extreme types of escape behaviour can be shown. The stickleback may flee, jumping away in an unpredictable direction (protean behaviour) or it may freeze, becoming motionless except for slight opercular movements. Spines are usually erected during the encounter, reinforcing the behavioural tactics with a morphological defence. Analysis of the behaviour of *G. aculeatus* being stalked by a satiated pike suggested that there are individual differences in the antipredator behaviour displayed (Huntingford and Giles, 1987).

Antipredator behaviour is influenced by the prior experience of the stickleback. Fish that have survived an earlier encounter with a predator react at a greater distance than naïve fish. Nor does the previous stimulus have to be predator-specific. Juvenile *G. aculeatus* that had experienced normal paternal care had a greater response to the presence of a potential predator (model of a stalking pike) than fish reared after hatching without a father (Tulley and Huntingford, 1987a). This effect was observed in a population from a site where predators were abundant, but not in a population from a site where predators were absent.

The interaction of experience and heredity is an important factor in determining the response of sticklebacks to a predator. For several populations of *G. aculeatus* that differed in their exposure to predation, the responses to presentations of potential fish or bird predators have been compared (Giles and Huntingford, 1984; Giles, 1984a,b; Tulley and Huntingford, 1987a,b,1988). The response to the predator tended to be stronger in fish from populations that were exposed to predation. And in one such population, the response to the model of a bird predator developed independently of specific experience of the predator. As a consequence, for fish 30 mm in length, the response of wild-caught and laboratory-reared fish was similar. However, for fish from a population not exposed to predation, the laboratory-reared fish failed to develop a protective response. In both populations the young juveniles (10 mm long) showed only a jumping-away response to the predator, and even then only about 50% of the juveniles responded. These experiments and other observations suggest that there are genetically based interpopulation differences in the response to predators in *G. aculeatus* populations, but that these responses are modified by experiential factors.

Huntingford (1976, 1977) also found evidence of a positive correlation between the level of aggression shown by male sticklebacks in the breeding season and the 'boldness' the males showed in the presence of predators outside the breeding season. Subsequent analysis has suggested that within populations, this correlation relates only to the boldness shown after exposure to a predator. The positive correlation may be caused by a common factor in both situations – a fear-induced suppression of ongoing behaviour (Tulley and Huntingford, 1988).

Boldness towards potential predators also increases when the males are parental. Male *G. aculeatus* with eggs or fry were less likely to desert when a predatory rainbow trout, *Salmo gairdneri*, was manoeuvred near them than males with empty nests (Kynard, 1979a). The chance of a male *G. aculeatus* deserting his nest when presented with a model of a predatory fish decreased with an increase in the number of eggs and with an increase in the age of the eggs (Pressley, 1981). Male *G. aculeatus* with eggs in their nests maintained a higher rate of attacks on a male intruder when in sight of piscivorous fish (trout) than males with empty nests (Ukegbu and Huntingford, 1988). This effect of eggs in the nest declined for tests carried out late in the breeding season. The degree of parental male boldness may relate to the value of the brood being defended.

How effective are the morphological and behavioural defences against predation? About 30% of the sticklebacks entering the Isle Verte salt marsh pools were eaten by bird predators (Whoriskey and FitzGerald, 1985c). Significantly more male *G. aculeatus* and *G. wheatlandi* were taken than would be expected from their frequency in the community. Over 13% of fish sampled from a population of *G. aculeatus* from a lake on the Queen Charlotte Islands carried marks suggesting they had escaped after predatory attacks by birds or fish (Reimchen, 1988). Broken spines, the imprints of the bills of fish-eating birds and skin lacerations were the commonest injuries.

Defence against parasites

Some ectoparasites, most notably the fish louse, *Argulus canadensis*, attack sticklebacks like predators. Poulin and FitzGerald (1988, 1989b) suggest that the choice of microhabitat can influence the chances of such attacks. In the salt marsh pools at Isle Verte, the juvenile *Gasterosteus* tend to swim in the surface waters in the early morning. But in the afternoon, as the water temperatures get higher, the sticklebacks move towards the bottom. The *Argulus* rest on the bottom between attacks, and so the sticklebacks are more exposed to attack in the afternoons and indeed show heavier levels of infestation at this time. Shoaling also reduces the risk of attack for a fish (Poulin and FitzGerald, 1989c).

Interactions between parasitism and predation

Parasitic infestation, especially by *S. solidus*, may also make sticklebacks more susceptible to predation. The effect of *S. solidus* seems to be mediated by three factors. The parasite increases the metabolic rate of the stickleback and so increases the fish's oxygen requirements. At low oxygen concentrations, sticklebacks move to the surface and show aquatic surface respiration. Infested sticklebacks show this behaviour at higher oxygen concentrations than uninfested fish (Giles; 1987b; Smith and Kramer, 1987). Moving to the surface is likely to put the infested stickleback at greater risk of being eaten by a bird. The effect of infestation on foraging behaviour is again likely to put the infested fish at greater risk. Finally, the distension of the body caused by a heavy infestation of *S. solidus* may make the fish less effective at swimming and manoeuvring.

16.5 BEHAVIOURAL ECOLOGY OF REPRODUCTION

There is a division of behavioural roles between male and female sticklebacks. The male constructs and defends a nest in which the eggs are concealed from predators. The male's ventilatory behaviour keeps the eggs provided with a flow of oxygenated water and clear of detritus. Courtship takes place within a male's territory and provides an opportunity for mate choice. Apart from courtship, the female is free to forage for food and so increase her fecundity during the breeding season (Wootton, 1984). Females may also be more cautious than males in the presence of potential predators because they do not have to defend a nest and its contents from conspecifics (Giles, 1984b; FitzGerald and van Havre, 1985).

Mate choice

Between morphs

Courtship provides a mechanism by which reproductive isolation between closely related taxa can be maintained without the wastage of gametes. Mate choice tests have shown that the resident and anadromous forms of *G. aculeatus* breeding in the same river show a preference for mating with a partner of the same morph (Hay and McPhail, 1975; McPhail and Hay, 1983). The 'limnetic' and 'benthic' forms in Enos Lake also show a preference for a mate of the same form (Ridgway and McPhail, 1984). In both these examples, the most pronounced behavioural differences between the forms occurred early in the courtship interaction. In Lake Azabach'ye, Kamachatka, an anadromous (migratory) completely plated morph usually breeds earlier than the low-plated, resident form. By artificially accelerating the reproductive development of the resident form with hormone injections, Zyuganov and Bugayev (1988) allowed the two forms to breed at the same time. Various combinations of the sexes of the two forms were then introduced into quarries and the type of offspring was recorded. Under conditions of free choice of their own kind, reproductive isolation was complete. But in quarries where only the other form was available, hybridization occurred. Positive assortative mating was also observed between typical and the 'white' form of *G. aculeatus* where they occur sympatrically in Nova Scotia (Blouw and Hagen, 1990). Choice tests in the laboratory and field observations showed that the two are reproductively isolated. Crosses between the forms are fertile (D. M. Blouw, personal communication), indicating that the two forms are part of a species complex.

Between species

Courtship also contributes to the reproductive isolation between the taxonomically recognized species. Differences in male breeding colour, form of the male's courtship dance to the female and the way in which the male leads the female to the nest are all cues that may be used to prevent crosses between the species. Attempts at interspecific crosses usually break down early in courtship, frequently with the female failing to respond to the approach of a heterospecific male (Wootton, 1984).

Female choice

Female *G. aculeatus*, in addition to mating preferentially with their own morph will also prefer males with eggs already in the nest (Ridley and Rechten, 1981; Jamieson and Colgan, 1989; Belles-Isles *et al.*, 1990), males with the brightest nuptial coloration (e.g. McPhail, 1969; Semler, 1971; Milinski and Bakker, 1990), and males whose nests are located in shelter (Sargent and Gebler, 1980).

Ridley and Rechten (1981) found that females were more likely to enter a nest and spawn, and less likely to refuse a mating, if the nest contained eggs. Jamieson and Colgan (1989) confirmed that males that had recently spawned and had eggs in their nests were significantly more likely to obtain a second spawning than males that had not spawned and had no eggs in their nests. But, there was no evidence of females visiting nests before spawning and rejecting males with no eggs. Consequently, Jamieson and Colgan (1989) disagreed with Ridley and Rechten's original conclusion that females actively choose males with eggs, though a field study tended to support the latter's interpretation (Goldschmidt and Bakker, 1990). In both the studies, the females were given a choice between spawning in an empty nest or in nests containing one clutch. In the field, males may have up to eleven clutches in a nest (e.g. Kedney *et al.*, 1987). When females were presented with a choice between empty nests and those with one, two, three, four or five clutches, they were more likely to spawn in a nest containing one clutch (Belles-Isles *et al.*, 1990) (Fig. 16.10). The chance that a female would spawn with a male with eggs decreased as the number of eggs in the nest increased. This may occur because of egg crowding in nests with more than one clutch, or because males become less receptive to females after receiving one clutch (van Iersel, 1953).

Some aspects of courtship behaviour can be interpreted in terms of sexual selection. In a detailed study of nuptial coloration in male *G. aculeatus*, McLennan and McPhail (1989a, b) describe the coloration as a complex mosaic of three colours: blue eyes, black dorsal–lateral body colour and red ventral–lateral body colour. Hue intensity and the pattern of coloration is highly variable among males. Gravid *G. aculeatus* females choose males on the basis of subtle differences in the intensity of the males' red coloration (Milinski and Bakker, 1990). Males infested with parasites are less brightly coloured and in poorer physical condition than uninfested males. Milinski and Bakker (1990) argue that females use the red coloration, a secondary sexual characteristic,

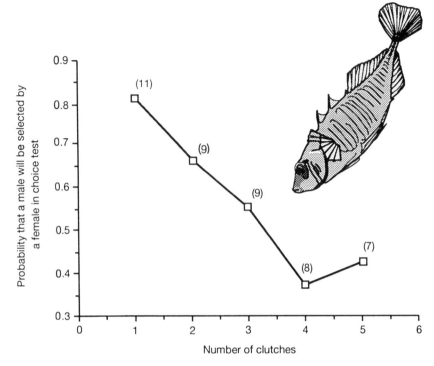

Fig. 16.10 Probability that males with different numbers of eggs are chosen by female three-spined sticklebacks in a simultaneous choice test. Sample sizes are given in parentheses. (Redrawn from Belles-Isles *et al.*, 1990).

as an index of a male's physical condition. In those populations of G. *aculeatus* in which males may not develop a full expression of the ventral red coloration, the relative roles of female mate choice, the presence of visually hunting predators, the light-transmitting properties of the water and the supply of carotenoids (red pigment) in the diet on the expression of nuptial colours have still to be clarified (Reimchen, 1989).

Even in populations where males are brightly coloured, females may choose using other criteria. At Isle Verte, all sexually mature male G. *aculeatus* are red to some extent and most males harbour few parasites. Nevertheless, there is great variation in male reproductive success measured by the number of eggs a male obtains. At this site, successful and unsuccessful males do not differ obviously in the quality of their nest sites. FitzGerald (1983) found a negative correlation between the number of eggs per nest and male aggressiveness. In laboratory choice experiments, females initially responded equally to males of all levels of aggression. But highly aggressive males obtained fewer matings than less aggressive males. The former would end a courtship bout in order to attack neighbouring males sooner than the latter (Ward and FitzGerald, 1987). Females may also choose males on the basis of body size (Rowland, 1989a). A female sexual response was elicited more effectively by the larger of two models of males. The female did not court the larger model exclusively but allocated courtship in proportion to the size of the two models. There are several possible advantages of mating with larger males. If larger fish are less vulnerable to predation, the offspring of larger males are less likely to experience loss of parental care before they become independent. Larger males might also defend eggs more effectively. A laboratory experiment showed that when the size differences were sufficient, larger males were more successful in the competition for breeding territories (Rowland, 1989a). But in a field study at Isle Verte, the mating success (number of eggs obtained) of small $1+$ and large $2+$ *G. aculeatus* did not differ, although more $2+$ than $1+$ fish nested (Dufresne *et al.*, 1990).

A female's preference for a male may be thwarted if, after she spawns with the male of her choice, her eggs are fertilized by another, 'sneaking' male (van den Assem, 1967; Rico *et al.*, 1991).

Male choice

The courtship male *G. aculeatus* allocated to models of gravid females was related to their size (Rowland, 1982, 1989b; Sargent *et al.*, 1986). In some populations, females pose a risk to a male's offspring. A guardian male can generally defend his nest against raids by one or a few females by biting and chasing them off his territory (Ridgway and McPhail, 1988). But males must court to obtain eggs and most raids occur on nests when a male is courting rather than performing nest-directed activities such as fanning eggs (F. G. Whoriskey, personal communication). Males tending eggs preferentially court single females rather than females in larger groups (Belles-Isles *et al.*, 1990). Males without eggs also tend to avoid courting females in groups, perhaps because the other females might raid the nest containing the newly spawned eggs.

Territorial, nesting and parental behaviour

Territoriality and nest sites

A territory is essential for the reproductive success of a male. Although territorial males may sneak fertilizations in the nests of other males, as yet there are no reports of non-territorial males achieving this (Wootton, 1976).

The choice of a nest site may also be crucial for a male's reproductive success. Sargent and Gebler (1980) allowed male *G. aculeatus* to compete for a limited number of clay flower pots as nest sites in wading pools. Males nesting inside the pots spawned earlier and more often, had a higher mean and lower variance for hatching success and suffered fewer stolen fertilizations, nest raids, and territorial encounters than males nesting outside the pots. In natural populations of *G. aculeatus*, concealed nests are more likely to contain fry than nests in open areas, probably because the concealed nests suffer less intra-specific egg predation (Moodie, 1972a; Kynard, 1979a, b). In tide pools at Isle Verte, females and non-territorial males are the main egg predators, and the most vulnerable nests are those in deep water, far from the pool banks and without cover (FitzGerald *et al.*, 1992).

Experimental and field studies on *G. aculeatus* suggest that phenotypes of the low-plated morph with different numbers of lateral plates nest in different microhabitats because of variation in their competitive abilities. Some plate phenotypes are more likely to nest in the preferred sites which are in or near cover and in relatively deep water (Moodie, 1972a; Kynard, 1978, 1979b). Plate phenotypes may also differ in aggressiveness (Huntingford, 1981). Body size also plays a role in competition for territories (Rowland, 1989b; Dufresne *et al.*, 1990). Rowland (1989b) found that a 15% weight difference was sufficient to provide a competitive advantage to the heavier male when two males were allowed to compete for limited space in an aquarium. But note that FitzGerald and Kedney (1987) found that a male's fighting ability in dyadic combats conducted in an aquarium did not correlate with its ability to obtain a territory in a group of males in a wading pool. Here, a much greater weight difference (50%) was needed by a male to outcompete a smaller rival for space (see also Dufresne *et al.*, 1990).

Suitable nest sites may be in short supply in some habitats. In tidal salt marsh pools along the eastern seaboard of North America, three and sometimes four species of stickleback may be trying to build nests simul-taneously. FitzGerald (1983) found that in May, sexually mature *P. pungitius* did not usually reproduce in pools, presumably because of a lack of algae, whereas *G. aculeatus* and *G. wheatlandi* nested in open areas. When algae appeared in the pools in early June, four or five *P. pungitius* m^{-2} were able to establish territories and spawn. In experimental tanks, *P. pungitius* were at a competitive disadvantage to *G. aculeatus* in sparsely vegetated areas (Ketele and Verhayen, 1985).

Sometimes sticklebacks tend more than one nest at a time (Courtenay, 1985; Mori, 1987). Half of the male *A. quadracus* observed nesting at Courtenay's field site were simultaneously tending two or more nests scattered throughout their territories. Multinesting males with several clutches in their care had the clutches distributed amongst their several nests. Courtenay hypothesized that multinesting is an adaptive response to the pressure of egg predation in the complex and variable intertidal zone in which *A. quadracus* lives. *A. quadracus* also builds multitiered nests where all the clutches acquired by a male are housed in separate sections of the same nest (Courtenay, 1985). Under what circumstances do males construct the two types of nest?

In the Tsuya River, Japan, most *G. aculeatus* built nests on the stream bottom, but some nests were built on a vertical wall (Mori, 1985). These nests were probably built in response to high male densities when some males were unable to obtain a ground territory or when the water was too deep to build normal nests. Two of the 12 vertical nests contained eggs, but they failed to hatch.

The density of males and food supply can also affect the proportion of males that can establish territories and build nests. Experiments with *G. aculeatus* suggests that if the area available for nests is restricted, the proportion of males that become territorial and successfully build nests is related to their density (van den Assem, 1967; Stanley and Wootton, 1986). At low food levels, the proportion of males with territories is also reduced. At high densities and food levels, some males that are physiologically capable of building nests fail to do so, although if isolated, they will then build. At low food levels, some males become physiologically unable to build nests even in isolation (Stanley and Wootton, 1986). However, once males have a territory, even if they are starved, they are able to complete at least one parental cycle (FitzGerald *et al.*, 1989).

In a homogeneous environment, breeding success can be correlated with territory size (van den Assem, 1967; Goldschmidt and Bakker, 1990). Laboratory studies showed that male *G. aculeatus* with large territories were more likely to be successful in courtship, suffered fewer egg losses and had a better chance of hatching their clutches than males with small territories (van den Assem, 1967; Black, 1971). Thus both nest location and territory size contribute to male reproductive success (Sargent and Gebler, 1980; Sargent, 1982).

The importance of nest location was also illustrated by Kynard's (1979a) study of a population of *G. aculeatus* in Lake Wapato (Washington State). Males nesting in rocky areas had fewer eggs in their nests than males in vegetated areas. The reproductively active males tended to clump in the preferred habitats, so some areas which superficially looked suitable for nesting were underused. A similar clumping of *G. aculeatus* and *G. wheatlandi* nests was observed in tidal pools, which again suggests that nest location is more important than territory size (FitzGerald, 1983). In the pools there was a negative correlation between internest distance and the number of eggs in the nest, whereas in Lake Wapato this correlation was positive (Kynard, 1979a). In the pools at Isle Verte, some males that were not observed in aggressive interactions with other males nevertheless had high numbers of eggs in their nests (Whoriskey, 1984).

Whoriskey and FitzGerald (1985a) analysed the patterns of nest destruction following removal of the nest-guarding male *G. aculeatus* at Isle Verte. Nest-site variables per se offered little protection against egg predators in the absence of paternal defence. Site quality and male quality may interact to determine the number of eggs a male can hatch.

There is some evidence for both intra- and interspecific competition in tidal pools, but the relative intensity seems to differ between *G. wheatlandi* and *G. aculeatus*. When the densities were manipulated in pools that contained only one of the two species, there was no consistent relationship between the number of eggs per square meter and fish density over a range of densities from 4 to 32 fish m^{-2} (sex ratio 1:1) for both species (Whoriskey and FitzGerald, 1987). The proportion of egg biomass that was consumed by sticklebacks increased with fish density in pools that contained only *G. aculeatus* but not in pools containing *G. wheatlandi* (Whoriskey and FitzGerald, 1985b). A similar analysis of interspecific competition showed that *G. aculeatus* attacked and destroyed *G. wheatlandi* nests, but there was no evidence of the smaller *G. wheatlandi* destroying *G. aculeatus* nests (FitzGerald and Whoriskey, 1985; Gaudreault and FitzGerald, 1985).

Parental care

The principal activity during the parental phase is egg ventilation – by fanning in all species except *A. quadracus*. Fanning behaviour (total time fanning, number of fanning bouts and fanning beat rate) is responsive to environmental factors in some species including *G. aculeatus* (Reebs *et al.*, 1984), but less so in *S. spinachia* (Potts *et al.*, 1988). For *G. aculeatus* in salt marsh pools, there was a positive correlation between the respiratory rate of the eggs, egg age, water temperature and dissolved oxygen levels. The abiotic conditions in the pools are highly variable, whereas *S. spinachia* nests in turbulent, shallow coastal waters in which it is likely that neither the oxygen concentration of the water nor its temperature show rapid changes (Potts *et al.*, 1988).

The parental success of a male will depend on his ventilatory behaviour. Experimental studies of *G. aculeatus* suggest that the average length of the fanning bouts and an even temporal distribution of bouts and interfanning intervals may contribute to hatching success (van Iersel, 1953; Sargent and Gebler, 1980). Thus the optimal allocation of available time between nest ventilation, nest defence and foraging during the parental phase of a male poses a significant problem for analysis.

Cannibalism and reproduction

The reproductive biology of sticklebacks includes the phenomenon of nest-raiding and cannibalism (review, FitzGerald, 1991). Both breeding and non-breeding males and females will raid nests and eat eggs.

Male cannibalism

Sometimes a reproductively active male will carry stolen eggs back to his nest, but usually they are eaten (e.g. Wootton, 1976; Salfert and Moodie, 1985; Whoriskey and FitzGerald, 1985a; Hyatt and Ringler, 1989a, b). Parental males eat their own eggs, but the relative frequency of filial and hetero-cannibalism may vary among populations of the same species and according to circumstances such as the availability of alternative food. Because the eggs in the stomachs of parental males were usually at a different stage of development from those in the nest, both Kynard (1979a) and Hyatt and Ringler (1989a, b) concluded that there was no evidence to support Rohwer's (1978) speculation that males consume some eggs from their own nests to keep themselves in good condition. Hyatt and Ringler (1989b) considered that male heterocannibalism was a key factor regulating the population abundance of *G. aculeatus* in a British Columbia lake. Belles-Isles and FitzGerald (1991) concluded that filial cannibalism was relatively common in both laboratory and field populations of *G. aculeatus* and *G. wheatlandi*. They found that parental males were more cannibalistic than non-parental males. The filial cannibalism by male *G. aculeatus* was not related to food ration (*G. wheatlandi* was not studied). For more information on filial cannibalism in fishes, see FitzGerald (1992).

Female cannibalism

Eggs certainly provide female raiders with food, but it is unlikely that this is the only function of cannibalism. In the pools at Isle Verte, there are abundant supplies of alternative prey including zooplankton and macroinvertebrates (Ward and FitzGerald, 1983a, b; Castonguay and FitzGerald, 1990). Vickery *et al.*, (1988) speculated that nest raiding by females at Isle Verte evolved primarily because of competition for suitable mates and not because of nutri-tional or energetic advantages. It was assumed that high-quality males or nest sites are limiting resources for which females compete. There is some evidence in support of this model (FitzGerald and Whoriskey, 1992). But nutritional advantages may also be important (Belles-Isles and FitzGerald, in press). The total seasonal egg production of female *G. aculeatus* on a diet including eggs was compared with that of females on a non-egg diet. The cannibals produced 1.5 times as many eggs as the non-cannibals over the 2 month breeding season.

Anti-cannibalism behaviour

Foster *et al.* (1988) argued that cannibalism by adults was the driving force leading to the evolution of ontogenetic changes in habitat use in *G. aculeatus*. Small juveniles (< 15 mm) are concentrated in the vegetation whereas larger juveniles are pelagic. Juveniles of all sizes feed primarily on zooplankton, even when in vegetation. The juveniles may be capable of rapidly assessing predation risk and of altering their behaviour accordingly.

The distraction displays that males use to lure groups of females away from the nest have already been mentioned. The preference that females show for spawning in a nest that already contains a clutch may be related to the increased 'boldness' that parental males show. It may also reduce the risk of cannibalism by a dilution effect. Under artificially high densities for a single species (32 fish m^{-2}), nest raiding by females can lead to the destruction of all the nests in *G. aculeatus* pools (Whoriskey and FitzGerald, 1985b). The ecological significance of nest raiding and cannibalism has still to be clarified. More studies such as those of Hyatt and Ringler (1989b) are needed to determine the demographic consequences of these phenomena.

Despite these depredations by cannibals, Kynard (1979a) drew attention to the high production of young by *G. aculeatus* in Lake Wapato during the early part of the breeding season. Males were rearing a high percentage (76.8%) of their eggs through to fry. On average each male was producing about 400 free-swimming fry. The potential for numerical increase by *G. aculeatus* was illustrated when 4000 were put in Marion Lake, a small forest lake in British Columbia. The following year, the population was estimated at 120 000 (McPhail, cited in Krebs, 1978).

16.6   CONCLUSION

Many of the major themes of behavioural ecology are illustrated by research on sticklebacks – which is contributing important results to the resolution of the four problems outlined at the start of this chapter. Progress has been made in understanding habitat selection, foraging behaviour, defence against predators and parasites, and reproductive behaviour. The suitability of the sticklebacks for both laboratory and field studies helps to ensure a close relationship between theory, experiment and observation. Early studies on the behaviour of sticklebacks concentrated on the problem of causation. More recently, the ecological and evolutionary significance of their behaviour has been appreciated. They provide such suitable material that their study will yield as much significant information for behavioural ecology and sociobiology as the original studies did for the then-young discipline of ethology.

ACKNOWLEDGEMENTS

We thank H. Guderley and F. G. Whoriskey for their valuable comments on the original version of this chapter.

REFERENCES

Assem, J. van den (1967) Territory in the three-spined stickleback, *Gasterosteus aculeatus* L. An experimental study in intra-specific competition. *Behaviour* (Supp.), **16**, 1–164.

Audet, C., FitzGerald, G. J. and Guderley, H. (1985) Salinity preferences of four sympatric sticklebacks (Pisces: Gasterosteidae) during their reproductive season. *Copeia*, **1985**, 209–13.

Axelrod, R. and Hamilton, W. D. (1981) The evolution of cooperation. *Science*, **211**, 1390–96.

Baggerman, B. (1957) An experimental study on the timing of breeding and migration in the three-spined stickleback. *Archs néerl. Zool.*, **12**, 105–317.

Baker, M. C. (1971) Habitat selection in fourspine sticklebacks (*Apeltes quadracus*). *Am. Midl. Nat.*, **85**, 239–42.

Bakker, Th. C. M. and Feuth-de Bruijn, E. (1988) Juvenile territoriality in stickleback *Gasterosteus aculeatus* L. *Anim. Behav.*, **36**, 1556–8.

Bélanger, G., Guderley, H. and FitzGerald, G. J. (1987) Salinity during embryonic development influences the response to salinity of *Gasterosteus aculeatus* L. (*trachurus*). *Can. J. Zool.*, **65**, 451–4.

Bell, M. A. (1976) Evolution of phenotypic diversity in *Gasterosteus aculeatus* super-species on the Pacific coast of North America. *Syst. Zool.*, **25**, 211–27.

Bell, M. A. (1984) Evolutionary phenetics and genetics: threespine stickleback, *Gasterosteus aculeatus*, and related species, in *Evolutionary Genetics of Fishes* (ed. B. J. Turner), Plenum, New York, pp. 431–528.

Belles-Isles, J.-C. and FitzGerald, G. J. (1991) Filial cannibalism in sticklebacks: a reproductive management strategy? *Ethol. Ecol. Evol.*, **3**, 49–62.

Belles-Isles, J.-C. and FitzGerald, G. J. (in press) A fitness advantage of cannibalism in female sticklebacks (*Gasterosteus aculeatus* L.) *Ethol. Ecol. Evol.*, in press.

Belles-Isles, J.-C., Cloutier, D. and FitzGerald, G. J. (1990) Female cannibalism and male courtship tactics in threespine sticklebacks. *Behav. Ecol. Sociobiol.*, **26**, 363–8.

Bentzen, P. and McPhail, J. D. (1984) Ecology and evolution of sympatric sticklebacks (*Gasterosteus*): specialization for alternative trophic niches in the Enos Lake species pair. *Can. J. Zool.*, **62**, 2280–86.

Bentzen, P., Ridgway, M. S. and McPhail, J. D. (1984) Ecology and evolution of sympatric sticklebacks (*Gasterosteus*): spatial segregation and seasonal habitat shifts in the Enos Lake species pair. *Can. J. Zool.*, **62**, 2436–9.

Beukema, J. J. (1968) Predation by the three-spined stickleback (*Gasterosteus aculeatus*): the influence of hunger and experience. *Behaviour*, **31**, 1–126.

Black, R. (1971) Hatching success in the three-spined stickleback (*Gasterosteus aculeatus*) during the parental phase. *Anim. Behav.*, **19**, 532–41.

Blouw, D. M. and Hagen, D. W. (1981) Ecology of the fourspine stickleback, *Apeltes quadracus*, with respect to a polymorphism for dorsal spine number. *Can. J. Zool.*, **59**, 1777.

Blouw, D. M. and Hagen, D. W. (1990) Breeding ecology and evidence of reproductive isolation of a widespread stickleback fish (Gasterosteidae) in Nova Scotia, Canada. *Biol. J. Linn. Soc.*, **39**, 195–217.

Bull, H. O. (1957) Behaviour: conditioned responses, in *The Physiology of Fishes* (ed. M. E. Brown), Academic Press, London, pp. 211–28.

Castonguay, M. and FitzGerald, G. J. (1990) The ecology of the calanoid copepod *Eurytemora affinis* in an inland saltmarsh. *Hydrobiologia*, **202**, 124–32.

Charlesworth, B. (1980) *Evolution in Age-structured Populations*, Cambridge University Press, Cambridge, 300 pp.

Courtenay, S. (1985) Simultaneous multinesting by the fourspine stickleback, *Apeltes quadracus*. *Can. Fld Nat.*, **99**, 360–63.

Courtenay, S. and Keenleyside, M. H. A. (1983) Nest site selection by the fourspine stickleback, *Apeltes quadracus*, *Can. J. Zool.*, **61**, 1443–7.

Craig, D. and FitzGerald, G. J. (1982) Reproductive tactics of four sympatric sticklebacks (Gasterosteidae). *Env. Biol. Fishes*, **7**, 369–75.

Crivelli, A. J. and Britton, R. H. (1987) Life history adaptations of *Gasterosteus aculeatus* in a Mediterranean wetland. *Env. Biol. Fishes*, **18**, 109–25.

Delbeek, J. C. and Williams, D. D. (1987) Food resource partitioning between sympatric populations of brackish water sticklebacks. *J. Anim. Ecol.*, **56**, 949–68.

Dufresne, F., FitzGerald, G. J. and Lachance, S. (1990) Age and size related differences in reproductive success and reproductive costs in threespine sticklebacks. *Behav. Ecol.*, **1**, 140–7.

Elliott, J. M. (1981) Some aspects of thermal stress on freshwater teleosts, in *Stress and Fish* (ed. A. D. Pickering), Academic Press, London, pp. 209–45.

FitzGerald, G. J. (1983) The reproductive ecology and behaviour of three sympatric sticklebacks (Gasterosteidae) in a saltmarsh. *Biol. Behav.*, **8**, 67–79.

FitzGerald, G. J. (1991) The role of cannibalism in the reproductive ecology of the threespine stickleback. *Ethology*, **89**, 177–94.

FitzGerald, G. J. (1992) Filial cannibalism in fishes: why do parents eat their offspring? *TREE*, **7**, 7–10.

FitzGerald, G. J. and van Havre, N. (1985) Flight, fright and shoaling in sticklebacks (Gasterosteidae). *Biol. Behav.*, **10**, 321–31.

FitzGerald, G. J. and Kedney, G. I. (1987) Aggression, fighting and territoriality in sticklebacks: three different phenomena. *Biol. Behav.*, **12**, 186–95.

FitzGerald, G. J. and Whoriskey, F. G. (1985) The effects of interspecific interactions upon male reproductive success in two sympatric sticklebacks, *Gasterosteus aculeatus* and *G. wheatlandi*. *Behaviour*, **93**, 112–25.

FitzGerald, G. J. and Whoriskey, F. G. (1992) Cannibalism in fish: empirical studies, in *Cannibalism Ecology and Evolution Among Diverse Taxa* (eds M. A. Elgar and B. J. Crespi), Oxford University Press, Oxford, pp. 238–55.

FitzGerald, G. J., Guderley, H. and Picard, P. (1989). Hidden reproductive costs in the three-spined stickleback (*Gasterosteus aculeatus*). *Exp. Biol.*, **48**, 295–300.

FitzGerald, G. J., Whoriskey, F. G., Morrissette, J. and Harding, M. (1992). Habitat scale, female cannibalism, and male reproductive success in threespine sticklebacks (*Gasterosteus aculeatus*). *Behav. Ecol.*, **3**, 141–7.

Foster, S. A. (1988) Diversionary displays of paternal sticklebacks: defenses against cannibalistic groups. *Behav. Ecol. Sociobiol.*, **22**, 335–40.

Foster, S. A., Garcia, B. and Town, M. Y. (1988) Cannibalism as the cause of ontogenetic niche shift in habitat use by fry of the threespine stickleback. *Oecologia*, **74**, 577–85.

Fraser, D. F. and Huntingford, F. A. (1986) Feeding and avoiding predation hazard: the behavioural response of prey. *Ethology*, **73**, 56–68.

Gach, M. H. and Reimchen, T. E. (1989) Mitochondrial DNA patterns among endemic stickleback from the Queen Charlotte Islands: a preliminary study. *Can. J. Zool.*, **67**, 1324–8.

Gaudreault, A. and FitzGerald, G. J. (1985) Field observations of intraspecific and interspecific aggression among sticklebacks (Gasterosteidae). *Behaviour*, **94**, 203–11.

Gaudreault, A., Miller, T., Montgomery, W. L. and FitzGerald, G. J. (1986) Interspecific interactions and diet of sympatric juvenile brook charr, *Salvelinus fontinalis*, and adult ninespine sticklebacks, *Pungitius pungitius*. *J. Fish. Biol.*, **28**, 133–40.

Gee, J. H., Tallman, R. F. and Smart, H. J. (1978) Reactions of some Great Plains fishes to progressive hypoxia. *Can. J. Zool.*, **56**, 1962–6.

Gibson, R. N. (1980) Optimum prey-size by three-spined sticklebacks (*Gasterosteus aculeatus*): a test of the apparent-size hypothesis. *Z. Tierpsychol.*, **52**, 291–307.

Giles, N. (1984a) Development of the overhead fright response in wild and predator-naive three-spined sticklebacks, *Gasterosteus aculeatus* L. *Anim. Behav.*, **32**, 276–9.

Giles, N. (1984b) Implications of parental care for the anti-predator behaviour of adult male and female three-spined sticklebacks, *Gasterosteus aculeatus*, in *Fish Reproduction: Strategies and Tactics* (eds G. W. Potts and R. J. Wootton), Academic Press, London, pp. 275–89.

Giles, N. (1987a) Predation risk and reduced foraging activity in fish: experiments with parasitized and non-parasitized three-spined sticklebacks, *Gasterosteus aculeatus* L. *J. Fish Biol.*, **31** 37–44.

Giles, N. (1987b) A comparison of the behavioural responses of parasitized and non-parasitized three-spined sticklebacks, *Gasterosteus aculeatus* L., to progressive hypoxia. *J. Fish Biol.*, **30**, 631–8.

Giles, N. and Huntingford, F. A. (1984) Predation risk and interpopulation variation in anti-predator behaviour in the three-spined stickleback, *Gasterosteus aculeatus* L. *Anim. Behav.*, **32**, 264–75.

Goldschmidt, T. and Bakker, T. C. M. (1990) Determinants of reproductive success of male sticklebacks in the field and in the laboratory. *Neth. J. Zoo.*, **40**, 664–87.

Hagen, D. W. and Moodie, G. E. E. (1979) Polymorphism for breeding colours in *Gasterosteus aculeatus*. I. Their genetics and geographical distribution. *Evolution*, **33**, 641–8.

Hagen, D. W., Moodie, G. E. E. and Moodie, P. F. (1980) Polymorphism for breeding colours in *Gasterosteus aculeatus*. II. Reproductive success as a result of convergence for threat display. *Evolution*, **34**, 1050–59.

Hay, D. E. and McPhail, J. D. (1975) Mate selection in threespine sticklebacks (*Gasterosteus*). *Can. J. Zool.*, **53**, 441–50.

Heller, R. and Milinski, M. (1979) Optimal foraging of sticklebacks on swarming prey. *Anim. Behav.*, **27**, 1127–41.

Hoogland, R. D., Mooris, D. and Tinbergen, N. (1957) The spines of sticklebacks (*Gasterosteus* and *Pygosteus*) as a means of defence against predators (*Perca* and *Esox*). *Behaviour*, **10**, 205–37.

Huntingford, F. A. (1976) The relationship between anti-predator behaviour and aggression among conspecifics in the three-spined stickleback, *Gasterosteus aculeatus*. *Anim. Behav.*, **24**, 245–60.

Huntingford, F. A. (1977) Inter- and Intraspecific aggression in male sticklebacks. *Copeia*, **1977**, 245–60.

Huntingford, F. A. (1981) Further evidence for an association between lateral scute number and aggressiveness in the threespine stickleback, *Gasterosteus aculeatus*. *Copeia*, **1981**, 717–19.

Huntingford, F. A. and Giles, N. (1987) Individual variation in anti-predator responses in the three-spined stickleback (*Gasterosteus aculeatus* L.). *Ethology*, **74**, 205–10.

Hyatt, K. and Ringler, N. H. (1989a) The role of nest raiding and egg predation in regulating population density of threespine sticklebacks (*Gasterosteus aculeatus*) in a coastal British Columbian lake. *Can. J. Fish. Aquat. Sci.*, **46**, 372–83.

Hyatt, K. and Ringler, N. H. (1989b) Egg cannibalism and the reproductive strategies of threespine sticklebacks (*Gasterosteus aculeatus*) in a coastal British Columbia lake. *Can. J. Zool.*, **67**, 2036–46.

Ibrahim, A. A. and Huntingford, F. A. (1989a) The role of visual cues in prey selection in three-spined sticklebacks (*Gasterosteus aculeatus*). *Ethology*, **81**, 265–72.

Ibrahim, A. A. and Huntingford, F. A. (1989b) Laboratory and field studies on diet choice in three-spined sticklebacks, *Gasterosteus aculeatus* L., in relation to profitability and visual features of prey. *J. Fish Biol.*, **34**, 245–57.

Ibrahim, A. A. and Huntingford, F. A. (1989c) Laboratory and field studies of the effect of predation risk on foraging in three-spined sticklebacks (*Gasterosteus aculeatus* L.). *Behaviour*, **109**, 46–57.

Iersel, J. J. A. van (1953) An analysis of the parental behaviour of the male three-spined stickleback (*Gasterosteus aculeatus* L.). *Behaviour* (Supp.), **3**, 1–159.

Jamieson, I. and Colgan, P. (1989) Eggs in the nests of males and their effect on mate choice in three-spined sticklebacks. *Anim. Behav.*, **38**, 859–65.

Jones, J. R. E. (1964) *Fish and River Pollution*, Butterworths, London, 203 pp.

Kedney, G. I., Boulé, V. and FitzGerald, G. J. (1987) The reproductive ecology of threespine sticklebacks breeding in fresh and brackish water. *Am. Fish. Soc. Symp.*, **1**, 151–61.

Keenleyside, M. H. A. (1955) Some aspects of schooling behaviour of fish. *Behaviour*, **8**, 183–248.

Ketele, A. G. L. and Verheyen, R. F. (1985) Competition for space between the three-spined stickleback, *Gasterosteus aculeatus* L. f. *leiura* and the nine-spined stickleback, *Pungitius pungitius* (L.). *Behaviour*, **93**, 127–38.

Kislalioglu, M. and Gibson, R. N. (1976a) Prey "handling time" and its importance in food selection by the 15-spined stickleback, *Spinachia spinachia* (L.). *J. exp. mar. Biol. Ecol.*, **25**, 151–8.

Kislalioglu, M. and Gibson, R. N. (1976b) Some factors governing prey selection by the 15-spined stickleback, *Spinachia spinachia* (L.). *J. exp. mar. Biol. Ecol.*, **25**, 159–70.

Klinger, S. A., Magnuson, J. J. and Gallepp, G. W. (1982) Survival mechanisms of the central mudminnow (*Umbra limi*), fathead minnow (*Pimephales promelas*) and brook stickleback (*Culaea inconstans*) for low oxygen in winter. *Env. Biol. Fishes*, **7**, 113–20.

Krebs, C. J. (1978) *Ecology: Experimental Analysis of Distribution and Abundance*, 2nd edn, Harper and Row, New York, 678 pp.

Kynard, B. E. (1979a) Breeding behaviour of a lacustrine population of threespine sticklebacks (*Gasterosteus aculeatus*). *Behaviour*, **67**, 178–207.

Kynard, B. E. (1979b) Nest habitat preference of low plate number morphs in threespine sticklebacks (*Gasterosteus aculeatus*). *Copeia*, **1979**, 525–8.

Lachance, S., Magnan, P. and FitzGerald, G. J. (1987) Temperature preferences of three sympatric sticklebacks (*Gasterosteidae*). *Can. J. Zool.*, **65**, 1573–6.

Lavin, P. A. and McPhail, J. D. (1985) The evolution of freshwater diversity in the threespine stickleback (*Gasterosteus aculeatus*): site-specific differentiation of trophic morphology. *Can. J. Zool.*, **63**, 2632–8.

Lavin, P. A. and McPhail, J. D. (1986) Adaptive divergence of trophic phenotype among freshwater populations of the threespine stickleback (*Gasterosteus aculeatus*). *Can. J. Fish. Aquat. Sci.*, **43**, 2455–63.

Lavin, P. A. and McPhail, J. D. (1987) Morphological divergence and the organization of trophic characters among lacustrine populations of the threespine stickleback (*Gasterosteus aculeatus*). *Can. J. Fish. Aquat. Sci.*, **44**, 1820–29.

Lazarus, J. and Metcalfe, N. B. (1990) Tit-for-tat cooperation in sticklebacks: a critique of Milinski. *Anim. Behav.*, **39**, 987–8.

Lewontin, R. C. (1974) *The Genetic Basis of Evolutionary Change*, Columbia University Press, New York, 346 pp.

MacLean, J. (1980) Ecological genetics of threespine sticklebacks in Heisholt Lake. *Can. J. Zool.*, **58**, 2026–39.

MacLean, J. A. and Gee, J. H. (1971) Effects of temperature on movements of prespawning brook sticklebacks, *Culaea inconstants*, in the Roseau River, Manitoba. *J. Fish. Res. Bd Can.*, **28**, 919–23.

McLennan, D. A. and McPhail, J. D. (1989a) Experimental investigations of the evolutionary significance of sexually dimorphic nuptial colouration in *Gasterosteus aculeatus* L.: the relationship between male colour and male behaviour. *Can. J. Zool.*, **67**, 1778–82.

McLennan, D. A. and McPhail, J. D. (1989b) Experimental investigations of the evolutionary significance of sexually dimorphic nuptial colouration in *Gasterosteus aculeatus* L.: temporal changes in the structure of the male mosaic signal. *Can. J. Zool.*, **67**, 1767–77.

McPhail, J. D. (1969) Predation and the evolution of a stickleback (*Gasterosteus*). *J. Fish. Res. Bd Can.*, **26**, 3183–208.

McPhail, J. D. and Hay, D. E. (1983) Differences in male courtship in freshwater and marine sticklebacks (*Gasterosteus aculeatus*). *Can. J. Zool.*, **61**, 292–7.

Maskell, M., Parkin, D. T. and Verspoor, E. (1977) Apostatic selection by sticklebacks upon a dimorphic prey. *Heredity, Lond.*, **39**, 83–9.

Milinski, M. (1977a) Experiments on the selection by predators on the spatial oddity of their prey. *Z. Tierpsychol.*, **43**, 311–25.

Milinski, M. (1977b) Do all members of a swarm suffer the same predation? *Z. Tierpsychol.*, **45**, 373–88.

Milinski, M. (1982) Optimal foraging: the influence of intraspecific competition on diet selection. *Behav. Ecol. Sociobiol.*, **11**, 109–15.

Milinski, M. (1984a) Competitive resource sharing: an experimental test of a learning rule for ESSs'. *Anim. Behav.*, **32**, 233–42.

Milinski, M. (1984b) Parasites determine a predator's optimal feeding strategy. *Behav. Ecol. Sociobiol.*, **15**, 35–7.

Milinski, M. (1985) Risk of predation of parasitized sticklebacks (*Gasterosteus aculeatus* L.) under competition for food. *Behaviour*, **93**, 203–16.

Milinski, M. (1987) TIT FOR TAT in sticklebacks and the evolution of cooperation. *Nature, Lond.*, **325**, 433–5.

Milinski, M. and Bakker, T. C. H. (1990) Female sticklebacks use male colouration in mate choice and hence avoid parasitized males. *Nature, Lond.*, **344**, 330–33.

Milinski, M. and Heller, R. (1978) Influence of a predator on the optimal foraging of sticklebacks (*Gasterosteus aculeatus* L.) *Nature, Lond.*, **275**, 642–4.

Moodie, G. E. E. (1972a) Morphology, life history and ecology of an unusual stickleback (*Gasterosteus aculeatus*) in the Queen Charlotte Islands, Canada. *Can. J. Zool.*, **50**, 721–32.

Moodie, G. E. E (1972b) Predation, natural selection and adaptation in an unusual stickleback. *Heredity, Lond.*, **28**, 155–67.

Mori, S. (1985) Reproductive behaviour of the landlocked three-spined stickleback, *Gasterosteus aculeatus microcephalus. Behaviour*, **93**, 21–35.

Mori, S. (1987) Multinesting behaviour by the freshwater three-spined stickleback, *Gasterosteus aculeatus. J. Ethol.*, **5**, 199–202.

Nilsson, N.-A. (1967) Interactive segregation between fish species, in *Ecology of Freshwater Fish Production* (ed. S. D. Gerking), Blackwell, Oxford, pp. 295–313.

Ohguchi, O. (1981) Prey density and selection against oddity by three-spined sticklebacks. *Adv. Ethol.*, **23**, 1–79.

O'Neill, S. M. and Hyatt, K. D. (1987) An experimental study of competition for food between sockeye salmon (*Oncorhynchus nerka*) and threespine sticklebacks (*Gasterosteus aculeatus*) in a British Columbia coastal lake. *Can. Spec. Publ. Fish. Aquat. Sci.*, **96**, 143–60.

Partridge, L. (1978) Habitat selection, in *Behavioural Ecology* (eds J. R. Krebs and N. B. Davies), Blackwell, Oxford, pp. 351–76.

Picard, P., jun., Dodson, J. J. and FitzGerald, G. J. (1990) Habitat segregation among the age groups of *Gasterosteus aculeatus* (Pisces: Gasterosteidae) in the middle St. Lawrence estuary, Canada. *Can. J. Zool.*, **68**, 1202–8.

Pitcher, T. J. (1983) Heuristic definitions of fish shoaling behaviour. *Anim. Behav.*, **31**, 611–3.

Pitcher, T. J., Kennedy, G. J. A. and Wirjoatmodjo, S. (1979) Links between the behaviour and ecology of fishes. *Proc. 1st Br. Freshwat. Fish. Conf.*, pp. 162–75.

Potts, G. W., Keenleyside, M. H. A. and Edwards, J. M. (1988) The effect of silt on the parental behaviour of the sea stickleback, *Spinachia spinachia. J. mar. biol. Ass. U.K.*, **62**, 329–34.

Poulin, R. and FitzGerald, G. J. (1988) Water temperature, vertical distribution, and risk of ectoparasitism in juvenile sticklebacks. *Can. J. Zool.*, **66**, 2002–5.

Poulin, R. and FitzGerald, G. J. (1989a) Early life histories of three sympatric sticklebacks in a salt marsh. *J. Fish. Biol.*, **34**, 207–21.

Poulin, R. and FitzGerald, G. J. (1989b) Risk of parasitism and microhabitat selection in juvenile sticklebacks. *Can. J. Zool.*, **67**, 14–18.

Poulin, R. and FitzGerald, G. J. (1989c) Shoaling as an anti-ectoparasite mechanism in juvenile sticklebacks (*Gasterosteus* spp.). *Behav. Ecol. Sociobiol.*, **24**, 251–5.

Pressley, P. H. (1981) Parental effort and the evolution of nest guarding tactics in the threespine stickleback, *Gasterosteus aculeatus. Evolution*, **35**, 282–95.

Reebs, S. G., Whoriskey, F. G. and FitzGerald, G. J. (1984) Diel patterns of fanning activity, egg respiration and the nocturnal behaviour of male threespine sticklebacks, *Gasterosteus aculeatus* L. (from *trachurus*). *Can. J. Zool.*, **62**, 320–41.

Reimchen, T. E. (1980) Spine deficiency and polymorphism in a population of *Gasterosteus aculeatus*, an adaptation to predators? *Can. J. Zool.*, **58**, 1232–44.

Reimchen, T. E. (1983) Structural relationships between spines and lateral plates in threespine stickleback (*Gasterosteus aculeatus*). *Evolution*, **37**, 931–46.

Reimchen, T. E. (1988) Inefficient predators and prey injuries in a population of giant stickleback. *Can. J. Zool.*, **66**, 2036–44.

Reimchen, T. E. (1989) Loss of nuptial colouration in threespine sticklebacks (*Gasterosteus aculeatus*). *Evolution*, **43**, 450–60.

Reist, J. D. (1980a) Selective predation upon pelvic phenotypes of brook stickleback, *Culaea inconstans*, by northern pike, *Esox lucius. Can. J. Zool.*, **58**, 1245–52.

Reist, J. D. (1980b) Predation upon pelvic phenotypes of brook stickleback, *Culaea inconstans*, by selected invertebrates. *Can. J. Zool.*, **58**, 1253–8.

Reist, J. D. (1983) Behavioural variation in pelvic phenotypes of brook stickleback, *Culaea inconstans*, in response to predation by northern pike, *Esox lucius. Env. Biol. Fishes*, **8**, 255–67.

Rico, C., Kuhnlein, U. and FitzGerald, G. J. (1991) Spawning patterns in the three-spined stickleback (*Gasterosteus aculeatus*): an evaluation by DNA fingerprinting. *J. Fish Biol.*, **39**, Suppl. A, 151–58.

Ridgway, M. S. and McPhail, J. D. (1984) Ecology and evolution of sympatric sticklebacks (*Gasterosteus*): mate choice and reproductive isolation in the Enos Lake species pair. *Can. J. Zool.*, **62**, 1813–18.

Ridgway, M. and McPhail, J. D. (1988) Raiding shoal size and a distraction display in male sticklebacks (*Gasterosteus*). *Can. J. Zool.*, **66**, 201–5.

Ridley, M. and Rechten. C. (1981) Female sticklebacks prefer to spawn with males whose nest contains eggs. *Behaviour*, **76**, 152–61.

Rohwer, S. (1978) Parental cannibalism of offspring and egg raiding as a courtship strategy. *Am. Nat.*, **112**, 429–40.

Rowland, W. J. (1982) Mate choice by male sticklebacks, *Gasterosteus aculeatus*. *Anim. Behav.*, **30**, 1093–8.

Rowland, W. J. (1983a) Interspecific aggression and dominance in *Gasterosteus*. *Env. Biol. Fishes*, **8**, 269–77.

Rowland, W. J. (1983b) Interspecific aggression in sticklebacks – *Gasterosteus aculeatus* displaces *Apeltes quadracus*. *Copeia*, **1983**, 541–4.

Rowland, W. J. (1989a) The effect of body size, aggression, and nuptial colouration on competition for territories in male three-spined sticklebacks, *Gasterosteus aculeatus*. *Anim. Behav.*, **37**, 282–9.

Rowland, W. J. (1989b) Mate choice and supernormality effect in female sticklebacks (*Gasterosteus aculeatus*). *Behav. Ecol. Sobiol.*, **24**, 433–8.

Salfert, I. and Moodie, G. E. E. (1985) Filial egg-cannibalism in the brook stickleback, *Culaea inconstans* (Kirtland). *Behaviour*, **93**, 82–100.

Sargent, R. C. (1982) Territory quality, male quality, courtship intrusions, and female choice in the threespine stickleback, *Gasterosteus aculeatus*. *Anim. Behav.*, **30**, 346–74.

Sargent, R. C. and Gebler, J. B. (1980) Effects of nest site concealment on hatching success, reproductive success, and paternal behaviour of the threespine stickleback, *Gasterosteus aculeatus*. *Behav. Ecol. Sociobiol.*, **7**, 137–42.

Sargent, R. C., van den Berghe, E. and Gross, M. (1986) Male mate choice in fishes. *Anim. Behav.*, **34**, 545–50.

Schoener, T. W. (1987) Axes of controversy in community ecology, in *Community and Evolutionary Ecology of North American Stream Fishes* (eds W. J. Matthews and D. C. Heins), University of Oklahoma Press, Normal, pp. 8–16.

Semler, D. E. (1971) Some aspects of adaptation in a polymorphism for breeding colours in the threespine stickleback (*Gasterosteus aculeatus*). *J. Zool., Lond.*, **165**, 291–302.

Smith, R. S. and Kramer, D. L. (1987) Effects of a cestode (*Schistocephalus* sp.) on the response of ninespine sticklebacks (*Pungitius pungitius*) to aquatic hypoxia. *Can. J. Zool.*, **65**, 1862–5.

Stanley, B. V. and Wootton, R. J. (1986) Effects of ration and male density on the territoriality and nest-building of male three-spined sticklebacks. (*Gasterosteus aculeatus*). *Anim. Behav.*, **34**, 527–35.

Thomas, G. (1977) The influence of eating and rejecting prey items upon feeding and food searching behaviour in *Gasterosteus aculeatus* L. *Anim. Behav.*, **25**, 52–66.

Tulley, J. J. and Huntingford, F. A. (1987a) Age, experience and the development of adaptive variation in anti-predator responses in three-spined sticklebacks (*Gasterosteus aculeatus*). *Ethology*, **75**, 285–90.

Tulley, J. J. and Huntingford, F. A. (1987b) Paternal care and the development of adaptive variation in antipredator responses in sticklebacks. *Anim. Behav.*, **35**, 1570–72.

Tulley, J. J. and Huntingford, F. A. (1988) Additional information on the relationship between intra-specific aggression and anti-predator behaviour in the three-spined stickleback, *Gasterosteus aculeatus, Ethology*, **78**, 219–22.

Ukegbu, A. A. and Huntingford, F. A. (1988) Brood value and life expectancy as determinants of parental investment in male three-spined sticklebacks, *Gasterosteus aculeatus. Ethology*, **78**, 72–82.

Vickery, W., Whoriskey, F. G. and FitzGerald, G. J. (1988) On the evolution of nest raiding and male defensive behaviour in sticklebacks (Pisces: Gasterosteidae). *Behav. Ecol. Sociobiol.*, **22**, 185–93.

Visser, M. (1981) Prediction of switching and counter switching based on optimal foraging. *Z. Tierpsychol.*, **55**, 129–38.

Walsh, G. and FitzGerald, G. J. (1984) Resource utilization and coexistence of three species of sticklebacks (Gasterosteidae) in tidal salt marsh pools. *J. Fish. Biol.*, **25**, 405–20.

Ward, G. and FitzGerald, G. J. (1983a) Fish predation on the macrobenthos of tidal salt marsh pools. *Can. J. Zool.*, **61**, 1358–61.

Ward, G. and FitzGerald, G. J. (1983b) Macrobenthic abundance and distribution in tidal pools of a Quebec saltmarsh. *Can. J. Zool.*, **61**, 1071–85.

Ward, G. and FitzGerald, G. J. (1987) Male aggression and female mate choice in the threespine stickleback, *Gasterosteus aculeatus. J. Fish Biol.*, **30**, 679–90.

Werner, E. E. and Gilliam, J. F. (1984) The ontogenetic niche and species interactions in size-structured populations. *A. Rev. Ecol. Syst.*, **15**, 393–425.

Whoriskey, F. G. (1984) Le rôle de facteurs choisis, biotique et abiotiques dans la structuration d'une communauté d'épinoche (Pisces: Gasterosteidae). PhD Thesis, Université Laval, Quebec, 183 pp.

Whoriskey, F. G. and FitzGerald, G. J. (1985a) Nest sites of the threespine stickleback: can site characteristics alone protect the nest against egg predators and are nest sites a limiting resource? *Can. J. Zool.*, **63**, 1991–4.

Whoriskey, F. G. and FitzGerald, G. J. (1985b) Sex, cannibalism, and sticklebacks. *Behav. Ecol. Sociobiol.*, **18**, 15–18.

Whoriskey, F. G. and FitzGerald, G. J. (1985c) The effect of bird predation on an estuarine stickleback (Pisces: Gasterosteidae) community. *Can. J. Zool.*, **63**, 301–7.

Whoriskey, F. G. and FitzGerald, G. J. (1987) Intraspecific competition in sticklebacks (Gasterosteidae: Pisces): does Mother Nature concur? *J. Anim. Ecol.*, **56**, 939–47.

Whoriskey, F. G. and FitzGerald, G. J. (1989) Breeding-season habitat use by sticklebacks (Pisces: Gasterosteidae) at Isle Verte, Quebec. *Can. J. Zool.*, **67**, 2126–30.

Whoriskey, F. G., FitzGerald, G. J. and Reebs, S. G. (1986) The breeding season population structure of three sympatric territorial sticklebacks (Pisces: Gasterosteidae). *J. Fish. Biol.*, **29**, 635–48.

Whoriskey, F. G., Gaudreault, A., Martel, N., Campeau, S. and FitzGerald, G. J. (1985) The activity budget and behaviour patterns of female threespine sticklebacks, *Gasterosteus aculeatus* L. in a Quebec tidal salt marsh. *Naturaliste can.*, **112**, 112–18.

Withler, R. E. and McPhail, J. D. (1985) Genetic variability in freshwater and anadromous sticklebacks (*Gasterosteus aculeatus*) of southern British Columbia. *Can. J. Zool.*, **63**, 528–33.

Wootton, R. J. (1976) *The Biology of the Sticklebacks*, Academic Press, London, 387 pp.

Wootton, R. J. (1984) *The Functional Biology of Sticklebacks*, Croom Helm, London, 265 pp.

Wootton, R. J. (1990) *Ecology of Teleost Fishes*, Chapman and Hall, London, 404 pp.

Zyuganov, V. V. and Bugaev, V. F. (1988) Isolating mechanisms between spawning populations of the threespine stickleback, *Gasterosteus aculeatus*, of Lake Azabach'ye, Kamchatka. *Vop. Ikhtiol.*, **28**, 322–5 (English translation).

Chapter seventeen

Behavioural ecology of cave-dwelling fishes

Jakob Parzefall

17.1 INTRODUCTION

The ecological conditions in caves are characterized by two main factors: nearly all caves have complete darkness and more or less constant temperature. The animals found in this habitat form a heterogeneous assembly. Some animals use caves only occasionally to avoid unfavourable conditions outside. Others, such as bats, enter caves regularly to rest during the day and in winter. But there are also many species which live permanently in caves. Omitting all the different classifications (Vandel, 1965) of cave-living animals, we can call these true cave-dwellers 'troglobionts'. Their striking morphological differences in comparison with their epigean relatives concern the reduction of the eye and dark pigmentation. These reduction phenomena can be observed in many groups of animals. The degree of reduction in different species studied seems to be connected with the phylogenetic age of cave colonization (Wilkens, 1982).

In the teleostean fish, members of about 14 families have colonized caves successfully (Table 17.1). In nine of these families, epigean members are known to be nocturnal and so it is not surprising that those species with a preference for activity in darkness enter caves. But among the ancestors of cave fish we also note Characidae, Poeciliidae and Cyprinidae, which include many species with clear diurnal activity.

What enables certain families of fish to survive and to reproduce in caves? We must look to see whether all these families show behavioural preadaptations to a life in darkness. Comparing the cave-living fishes with their epigean relatives it is interesting to study the possible behavioural adaptations to the cave habitat, but this is difficult because the behaviour of cave-dwelling fishes has not been very well examined. This chapter therefore concentrates mainly

Behaviour of Teleost Fishes 2nd edn. Edited by Tony J. Pitcher. Published in 1993 by Chapman & Hall. ISBN 0 412 42930 6 (HB) and 0 412 42940 3 (PB).

Table 17.1   Families with obligate cave-dwelling fishes. F, freshwater; M, marine. Data from Thines (1955, 1969), Thines and Proudlove (1989)*

Family	Populations F	Populations M	Species F	Species M	Epigean relatives F	Epigean relatives M	Diurnal	Nocturnal
				Number of			*General activity*	
Amblyopsidae	–	–	4ᵃ	–	+	–	–	+ᵈ
Astroblepidae (Loricariidae)	–	–	1	–	+	–	–	+
Brotulidae (Ophidiidae)	–	–	6	–	–	+ᵇ	–	+ᵇ
Characidae	29ᶜ	–	2	–	+	–	+	–
Clariidae	–	–	5	–	+	–	–	+
Cobitidae	–	–	4	–	+	–	+	–
Cyprinidae	–	–	9	–	+	–	+	–
Gobiidae	–	–	6	2	–	+	+	–
Ictaluridae	–	–	4	–	+	–	–	+
Mastacembalidae	–	–	1	–	+	–	–	+
Pimelodidae	–	–	3	–	+	–	–	+
Poeciliidae	1ᵉ	–	–	–	+	–	+	–
Trichomycteridae	–	–	5	–	+	–	–	+
Synbranchidae	–	–	6	–	+	–	–	+

Additional data: ᵃCooper and Kuehne (1974); ᵇRiedl (1966); ᶜMitchell, Russel and Elliot (1977); ᵈPoulson (1963); ᵉGordon and Rosen (1962).

on the three best-studied families: the Amblyopsidae, the Characidae and the Poeciliidae.

17.2   COMPARISON OF BEHAVIOUR PATTERNS IN CAVE-DWELLING FISHES AND THEIR EPIGEAN RELATIVES

Potential cave-dwellers must have the sense organs and behaviour necessary to find food and to reproduce in caves. Such animals may be said to be preadapted to cave life. In this chapter we will first examine feeding and reproductive behaviour, and then look at other behaviour patterns that have been studied.

Feeding behaviour

Suitable food sources and quantity vary from cave to cave. Since there are no green plant producers in the dark, cave fish depend upon food brought in from the outside by animals or by floods. Compared with above-ground habitats, most caves do not have an abundance of food.

An exception is the sulphur cave where livebearing toothcarp, *Poecilia mexicana* (formerly described as *P. sphenops* (Schultz and Miller, 1971)), exist at high population density (Parzefall, 1974, 1979) and which has an exceptional abundance of food (Table 17.2). These fishes mainly feed on white material of different algae and bacteria (J. Parzefall, unpublished) which develops in connection with hydrogen-sulphide springs and covers all the rocks in this cave. They obtain this in the same way as their epigean conspecifics, which scrape off green algae from rocks. The rich food production in the sulphur cave leads to a higher population density in the cave form of *Poecilia mexicana* than in epigean populations. In another part of the cave with a bat colony, the fish also feed on guano or invertebrates living on the guano (Gordon and Rosen, 1962; Peters *et al.*, 1973). In this cave system, the abundance of food means that no changes in food detection and feeding are required.

The feeding behaviour of various Mexican cave populations of *Astyanax mexicanus* (= *fasciatus*) differs from that of the omnivorous fish of epigean habitats, which generally occur at higher population density (Table 17.2). In the various cave habitats studied, all animals swim without contact in slow

Table 17.2 Population density and food sources of cave-dwelling fishes and their epigean relatives

Family and species*	Density (fish m^{-2})	Food sources
Poeciliidae:		
Poecilia mexicana[a]		
Epigean fish	2–50	Algae
Cave fish	100–200	Bat guano, sulphur bacteria and algae
Characidae:		
Astyanax fasciatus[b]		
Epigean fish	15–200	Omnivorous
Cave fish	5–15	Bat guano
Amblyopsidae:[c]		
Chologaster cornuta, C. agassizi	0.001	Invertebrates
Cave habitat	0.005	Invertebrates
Epigean habitat	0.01	Invertebrates
Typhlichthys subterraneus	0.03	Invertebrates
Amblyopsis spelaea	0.05	Invertebrates
Amblyopsis rosae	0.15	Invertebrates, young conspecifics

*Data from: [a]Parzefall (1979), [b]Parzefall (1983), [c]Poulson (1963, 1969).

zigzag movements dispersed over the pool. Reactions to clay balls and food balls of 5 mm diameter have been tested in Pachon, Micos and Chica Caves, in test areas of 1 m². In Pachon and Micos Cave fish a falling clay ball induced higher swimming activity within a distance of 0.80 m and animals started searching for food near the bottom or on the water surface. However, there was only a slight increase in fish in the test area here (Fig. 17.1(a) and (b)).

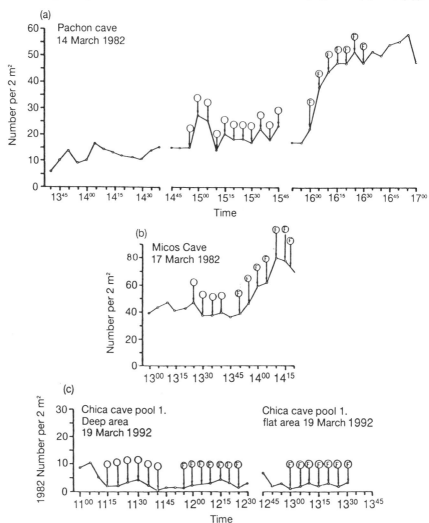

Fig. 17.1 Density of cave-dwelling *Astyanax fasciatus* counted every 5 min in a test area. Each arrow marks the number after the offer of a clay ball (empty circle), or a food ball (circled F) of 5 mm diameter. Reproduced with permission from Parzefall (1983).

On the other hand, in Chica Cave no reaction was observed (Fig. 17.1(c)), and this population showed no reaction to food balls either. In the two other populations food balls produced a significant increase of individuals in the test area. The fish bit off pieces with a head-shaking movement or tried to carry the food ball away, followed by others. After the food offer in Pachon Cave was stopped, most fish continued searching in the test area for some hours. In epigean habitats, the eyed fish react to a similar protocol with a rapid increase in numbers during the feeding period and a rapid decrease without food in the test area (Fig. 17.2).

The epigean forms normally school near the river bank down to a depth of 2 m. In a small pool of the Rio Naranjos, which was separated by flat passages of the river, territoriality has been observed. Schools with smaller fish stay near the water surface; the larger ones school in deeper water. The fish follow with rapid swimming movements the smaller particles that arrive, and test them by direct contact. They also try to feed on larger immobile prey. All smaller animals such as insects which fall into the water or which swim unprotected (e.g. young fishes) are eaten within seconds. The smaller fish near the water surface catch the food first. When it sinks into deeper water, the larger *Astyanax* schooling there catch it. In test areas of 1 m water depth, sinking food seldom reached the bottom. In general, the epigean fish get their food visually in open water, whereas cave fish are guided to food by water movements and chemical traces from the food (Parzefall, 1983).

Laboratory studies in darkness with the aid of infra-red video recording showed that there was a clear difference in the method of picking up food from the bottom (Schemmel, 1980). The cave fish fed at an angle of about $45°$ to the substratum whereas the epigean form stood vertically on its head. This

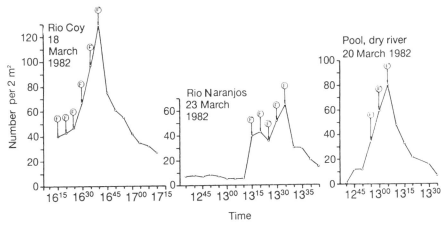

Fig. 17.2 Density of epigean *Astyanax fasciatus* counted in different habitats. For further explanation see Fig. 17.1. Reproduced with permission from Parzefall (1983).

difference in behaviour depends on genetic factors (Fig. 17.3). Environmental factors cannot explain the difference because the possibility of visual orientation is excluded in darkness and the bottom is flat. Genetic analysis shows that a trifactorial but essentially Mendelian inheritance explains the data from backcrosses (Wilkens, 1988).

This behavioural difference is inherited independently of eye size and number of melanophores. In addition the taste buds, which are restricted to

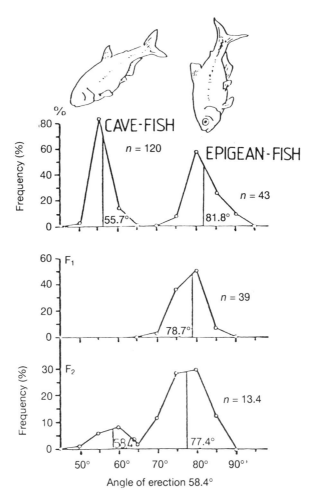

Fig. 17.3 Feeding behaviour in *Astyanax fasciatus*. Frequency distribution of the angle of feeding in darkness in a cave population, an epigean one, and in the hybrids. Note differing vertical scales in hybrids. Redrawn from Schemmel (1980).

the mouth region in epigean fish, are found over the lower jaw and cover ventral areas of the head in the cave fish. Schemmel (1980) states that such an evolutionary improvement of the gustatory equipment needs a lower angle of the body position during feeding, and both these traits could be achieved by small genetic steps during a process of integrative interaction.

In competition experiments with small pieces of meat on the bottom in total darkness, 80% of the food was found by hypogean specimens of the Pachon population, but only 20% by epigean fish (Hüppop, 1987). Cave and epigean populations reared with dry food or living prey did not show a preference for the food type that had been offered during a 10 month period. However, both populations preferred living prey. The fish reared with living prey grew significantly faster than the fish fed with dry food. So fish of both populations must have the ability to evaluate the better food (Klimpel and Parzefall, 1990). In addition, the cave fish are able to build up enormous fat reserves. One-year-old cave fish fed *ad libitum* had a mean fat content of 37% of fresh body mass compared with 9% in epigean fish under the same conditions (Hüppop, 1989).

The Amblyopsidae comprise six species in four genera. *Chologaster cornuta* is nocturnal and lives in swamps in the Atlantic coastal plain of the United States; *Chologaster agassizi* is found in springs and caves, more commonly in springs, and the remaining four species (Table 17.2) are limited to caves. The fish population density in caves is extremely low (Poulson, 1969). Their food, consisting of aquatic invertebrates such as isopods, amphipods and copepods, is also very scarce. As a consequence cave fish species must swim a great deal to get enough food. Comparing this behaviour in the partial troglodyte *C. agassizi* and in the hypogean *Typhlichthys subterraneus*, it was shown that swimming behaviour is more efficient in *Typhlichthys*. In addition the maximal prey detection distance is greater in the cave species: *Daphnia* was detected by *C. agassizi* within 10 mm and in *T. subterraneus* within 30–40 mm (Poulson, 1963). Food-finding ability at low prey densities in the dark in *Amblyopsis spelaea* is much better than in *C. agassizi*. When one *Daphnia* was introduced into a 100 l aquarium in the dark, *A. spelaea* found a prey hours before *C. agassizi* did. In contrast *C. agassizi* ate all ten *Daphnia* introduced in a 5 l aquarium before *A. spelaea* had eaten half of them (Poulson and White, 1969). These results at high prey densities probably result from a low maximum food intake in the cave fish. The behavioural changes are combined with adaptive changes in the system of free neuromasts and brain anatomy. As a result of these changes, obstacle avoidance and spatial memory are also enhanced in the cave-limited species (Poulson, 1963). The *A. spelaea* studied in Upper Twin Cave, Indiana (Mohr and Poulson, 1966) are distributed irregularly over the cave area. The biggest fish occur in deeper water, especially along underwater ledges. Each swims and feeds regularly in a limited area, moving upstream along a ledge for about 20 m and floating downstream to its starting point.

Upstream-moving fish search under the rocks for prey. The time spent under a rock was directly related to the number of isopods to be found there. Defence of food territories has not been observed, although the food is so scarce that adult fish get barely enough to stay alive.

The behaviour of *Speoplatyrhinus poulsoni* (Cooper and Kuehne, 1974), an endemic species of a cave in north-western Alabama, has not been studied.

Sexual behaviour

Having found enough food to reach maturity, the next problem to be solved by cave fish is finding a sexual partner in darkness. Subsequently they need behaviour patterns that provide effective fertilization in the absence of any visual orientation.

Male *Poecilia mexicana*, which actively seek mates, can readily find females because of the very high population density in the cave. The male checks conspecifics by nipping at the genital region. Females are recognized by a species-specific chemical substance (Zeiske, 1971) which is produced continuously by mature females. In addition to this species-specific substance, the female produces a chemical substance in the genital region and becomes attractive to males, but only for several day after the birth of young. If a male comes into contact with such a female, his nipping behaviour becomes faster and he tries to copulate. The attractive female can only be recognized through direct nipping contact because the chemical signals cannot be transferred through water (Parzefall, 1973).

Comparative laboratory and field studies have shown that the epigean populations of this species have similar sexual behaviour: in a school of adult *P. mexicana*, the females are checked continuously with nipping by dominant males. On contact with an attractive female, the colour of the male fish darkens. The male then tries to separate the attractive female from the school to defend her against other males, and to copulate. This behaviour gives the bigger dominant males a reproductive advantage. The smallest males counteract this one-sided selective advantage by ambushing the attractive females and attempting to copulate without any aggressive interaction against the defending dominant male. For this reason a small male waits with an inconspicuous body coloration in typical head-down position near an attractive female for the possibility of quick copulation. The best moment to do so is during an aggressive interaction between the dominant male and another male (Parzefall, 1969, 1979).

In contrast to epigean fish, cave females are not defended by males, and small males do not perform ambushing behaviour. The consequences of these behavioural differences for the population structure in cave males will be discussed below.

Comparative studies with the closely related species *P. velifera* and *P.*

latipinna show that they possess a visual display in addition to the chemical signals. All the Poeciliidae studied, with the exception of *P. mexicana*, *P. sphenops* and *P. vivipara*, perform similar species-specific swimming movements with erect fins in front of or around the female (Parzefall 1969). The absence of such a visual display in *P. mexicana* is clearly associated with successful reproduction in a lightless environment. The prior existence of the chemical communication system does not require any further special adaptations of reproductive behaviour in order to operate in darkness.

In the characid fish *Astyanax fasciatus*, with its several blind cave populations which have been studied in the laboratory, we have a similar situation (Wilkens, 1972). A female ready for reproduction produces a chemical substance in the genital region and remains swimming in a small area. After the first contact between a male and an attractive female, the male maintains contact by touching the genital region of the female. Whenever the male loses contact with the female, he searches for her actively. The male seems only to be able to identify the female through nipping contact in the genital region. After some contacts the male tries to swim into a position parallel to the female. In this position the animals turn their ventral side very rapidly to the water surface and then release the genital products. The female seems to be able to produce the chemical substance without being stimulated by males. This substance stimulates only males, and is perceived by them with the olfactory sense. There is also weak evidence that females sometimes mark the substrate by short contacts with the genital region. Sexual activities have been released experimentally by putting the fish in aquaria with fresh water. Sometimes, after the first spawning of one pair, some other fish started sexual activities. Comparing an epigean population with hypogean ones, Wilkens (1972) could neither show differences concerning sexual behaviour, not find any visual display in the epigean population.

In natural habitats sexual behaviour has not yet been observed (Parzefall, 1983), probably because most field studies have been conducted during the dry season. In all populations examined, young animals were absent. It is therefore likely that reproductive activity in *A. fasciatus* takes place during the rainy season, perhaps in connection with an increase of food supply and rising water levels.

Such an annual reproductive cycle is known in the Amblyopsidae (Poulson, 1963). The swampfish, *Chologaster cornuta*, leaves the cypress swamp habitats in early April when it spawns. It comes into open streams for the rich food along edges of submerged weed banks. *C. Agassizi*, the spring cave fish, seems to spawn in caves in February when water levels are at the year's maximum. A yearly cycle exists also in the three cave-dwelling species. Ova reach mature size and breeding occurs during high water from February through April. *Amblyopsis spelaea* shows an especially well-defined cycle. In this species the females carry the eggs in their gill cavities until hatching and carry the young

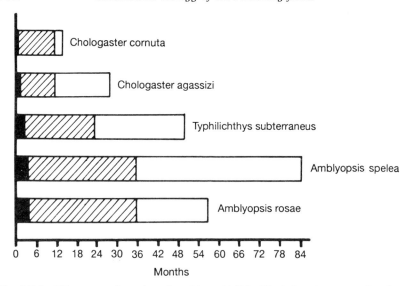

Fig. 17.4 Life spans of species of amblyopsid fish. Black bar is time to hatching, striped bar is time from hatching to first reproduction, open bar is reproductive life span (data from Poulson, 1963; Culver, 1982).

until they lose their yolk sacs, a total period of 4–5 months. Thus the young appear in late summer and early autumn (Poulson, 1963). In Upper Twin Cave, Indiana, the larger *A. spelaea* in the early winter appear to be spread out along the stream passage in pairs. Male and female seem to establish a territory. But neither Poulson (1963) nor Bechler (1983) have observed agonistic behaviour in the field. During a 7 year study (Poulson, 1963), the reproduction rate in this cave was low. In the population, which never had fewer than 81 fish nor more than 130, on average about five females a year produced 40 to 60 large, heavily yolked eggs each. Other females did not breed at all. After a year of better food supply and higher reproductive rate by more egg-producing females, the young fish were reduced through cannibalism by adults.

Comparing the life spans of the different amblyopsid species, we can observe that, with increasing evolutionary time in caves, the times to hatching and the attainment of sexual maturity are increased (Fig. 17.4), the mean number of ova declines, and the volume of ova becomes larger. This corresponds to the classic pattern of K-selected species (Culver, 1982).

Agonistic behaviour

Agonistic behaviour includes aggressive and submissive patterns. In some cave-dwelling fish, agonistic behaviour remains unchanged, but in others a striking reduction of this complex behaviour has been observed.

Most of the cave-dwelling fish studied seem to have agonistic behaviour. In the Clariidae, one of the five cave-dwelling species, *Uegitglanis zammaranoi*, an anophthalmic phreatic fish from Somalia, shows an aggressive behaviour which is clearly based on a dominance relationship. When two specimens are placed together in an aquarium, they start fighting with gasping movements of the mouth, yawning, chasing, butting with the head or mouth, biting and mouth-locking. The dominant fish tends to occupy the lower part of the aquarium. The subordinate fish exhibits rubbing of the bottom, a vertical position or rapid flight reactions. The intensity of fights increases when a dominant fish is introduced into the aquarium of a subordinate one (Ercolini *et al.*, 1981). Some of these aggressive patterns have been observed in the natural habitat where the specimens studied have been collected (Berti and Ercolini, 1979). A comparable aggressive behaviour has been described for two pimelodid catfish: *Pimelodella kronei* (Trajano, 1991) and an undescribed cave form of *Rhamdia guatemalensis* (Bormann, 1989). Though these blind species have no possibility for visual communication, complicated intraspecific aggressive behaviour still exists.

For some other cave-living fish we have only a few data, and detailed studies are lacking. Studied in aquaria, the blind cyprinid *Caecobarbus geertsi*, from caves in the lower Congo, attacks conspecifics by blows in the middle of the body. When two fish alternate these attacks, they perform circling movements (Thines, 1969). In another blind cyprinid, *Garra barreimiae*, from Oman, adult fish defend territories (Fausel, 1990). The blind intertidal *Typhlogobius californiensis*, which lives in pairs in burrows built by the ghost shrimp, *Callianassa affinis*, attacks intruding conspecifics. The aggressive behaviour is released by chemical signals (MacGinitie, 1939). The synbranchid eel *Furmastix infernalis* studied in Hoctun Cave (Yucatan, Mexico) and in aquaria seems to have individual territories; biting and tail-beating against intruders have been observed in aquaria (J. Parzefall, unpublished). During a short observation period of the blind species *Typhliasina pearsei* (Brotulidae) in the Cueva del Pochote and in aquaria, head-shaking movements between animals at a distance of 10–20 cm have been registered which release flight reactions, but no other aggressive patterns have been found (J. Parzefall, unpublished; Schemmel 1977).

In more detailed, comparative studies, the reduction of agonistic behaviour in caves has been demonstrated in three examples. Bechler (1983) has compared this behaviour in five amblyopsid species. Six aggressive acts and two submissive acts have been exhibited (Table 17.3). Males and females of the four species that behaved agonistically did not appear different. The fifth species, the swampfish *Chologaster cornuta*, did not show any agonistic behaviour. In the highly adapted cave species *Amblyopsis rosae*, only tail-beating and the submissive acts are still existent. Detailed analysis of the behavioural diversity demonstrates that *Chologaster agassizi* and *Amblyopsis*

Table 17.3 Aggressive repertoire for each amblyopsid species. +, Act observed in a species; −, act not observed (from Bechler, 1983)

Behaviour	Chologaster cornuta	Chologaster agassizi	Typhlichthys subterraneus	Amblyopsis spelaea	Amblypsis rosae
Aggressive acts					
Tail beat	−	+	+	+	+
Head butt	−	+	+	+	−
Attack	−	+	+	+	−
Bite	−	+	+	+	−
Chase	−	+	+	+	−
Jaw lock	−	+	−	+	−
Submissive acts					
Freeze	−	+	+	+	+
Escape	−	+	+	+	+

spelaea, the least cave-adapted subterranean species, in each subfamilial lineage, engage in relatively intense, complex agonistic bouts. In contrast, the more highly cave-adapted species *Typhlichthys subterraneus* and *A. rosae* engage in simpler, less intense bouts, which are considerably shorter in length. The most frequent act by any of the species studied was tail-beating. Reduced selective pressure with increasing adaptation to cave life is the most probable explanation for the observed reduction in agonistic behaviour. Concluding the field data, Bechler (1983) states that food is a primary selective force in amblyopsids. In adaptation to this factor, metabolic rate and fecundity decrease in the cave, longevity is increased and swimming efficiency is improved. These adaptations confer the advantage of energy conservation on the more highly evolved subterranean species. It is suggested that this conservation of energy serves to reduce selection pressures produced by a scarce food supply, and allows for a reduction in overt agonistic behaviour.

Another case of reduction of aggressive behaviour is found in cave-living populations of the characid *Astyanax fasciatus*. Epigean fish usually live in schools, and territoriality only exists as a special case in small pools. Except for weak ramming attacks in the epigean fish, no aggression has been observed in field studies conducted during the dry season in various natural habitats in Mexico (Parzefall, 1983). Laboratory studies have demonstrated that, with increasing size of the aquaria, the frequency of ramming attacks decreases and the territorial fish tend to school (Fig. 17.5). Ramming and biting in some epigean populations can lead to the death of smaller animals after several attacks, especially in small aquaria where they cannot escape. There is no difference in aggression between male and female. Hungry fish become much more aggressive (Burchards *et al.*, 1985). In the blind cave populations,

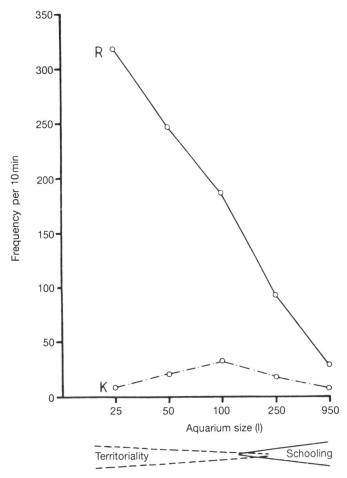

Fig. 17.5 Frequency of two aggressive patterns (R, ramming movements; K, circling) and the tendency for territoriality and schooling of epigean *Astyanax fasciatus* in aquaria of different sizes. Reproduced with permission from Burchards *et al.* (1985).

however, fights are very rare and no animal has ever been seen to die because of attack.

To obtain quantitative data, three different populations have been tested in visible light and in the infra-red (Fig. 17.6). The epigean fish originates from Teapao River. The Pachon Cave population is completely blind, and the Micos Cave population reared in light is variable in eye size, indicating that this population is a phylogenetically young one (Wilkens, 1976). The Pachon Cave population, which is believed to be a phylogenetically old cave form, shows a loss of aggressive behaviour. The cave form from Micos Cave performs

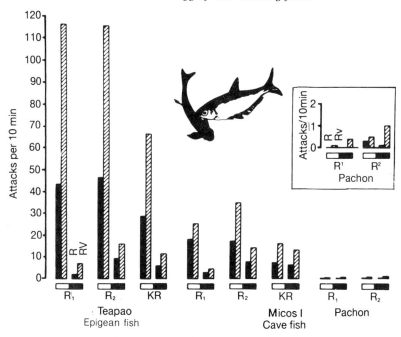

Fig. 17.6 Frequency of ramming attacks during a light–dark (LD) cycle in two cave populations and an epigean one of *Astyanax fasciatus*. R. Ramming; RV, ramming attempt; R_1, R_2, test groups with LD 3:3: KR, test group with LD12:12 (data from Burchards *et al.*, 1985). Inset graph: Pachon Cave fish, with expanded vertical scale.

aggressive patterns in light and darkness. However, there is less aggression in darkness. In addition the Micos Cave form selected for functional eyes still has a significantly lower aggression level when tested in light (Hofmann, 1990). These results demonstrate first, that absence of visual communication only reduces the aggression to a certain degree, and secondly that this behaviour seems to undergo a reduction process in the cave biotope. In a subsequent study, ramming and attempted ramming were compared in two epigean populations, one blind cave population and the hybrid generations (Fig. 17.7). In darkness, all fish tested exhibit the same level of aggression. In visible light, there is no difference between the two epigean populations either. But there is a significant difference from the hybrid generations. The average in F_1 generations in a test group with a good optomotor response (minimum separable 15' of arc) cannot be distinguished from a group with a lower response. The significantly lower level of aggression in the F_1 and F_2 generations, despite a well-developed eye, suggests that the low aggression in the cave fish is based on a genetically controlled reduction of this behaviour. We do not know

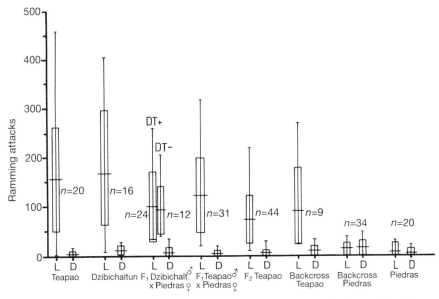

Fig. 17.7 Ramming and ramming attempts in light (L) and darkness (D) of two epigean populations (Rio Teapao, Cenote Dzibichaltun), the Piedras Cave population and their hybrids. Each fish was tested against an opponent of the Teapao River in light (L) and darkness (D). DT, Optomotor response positive (+) or negative (−). Reproduced with permission from Burchards *et al.* (1985).

enough about the function of the aggressive behaviour in epigean *Astyanax*. For the cave-dwelling populations we do not know enough about density and food sources to explain this reduction phenomenon.

In *Poecilia mexicana*, the third example of reduction in aggressive behaviour, the cave population still has eyes. Aggressive behaviour gradually reduces and the eye diameter decreases from the entrance to the end of the cave (Parzefall, 1974). But there is no topographic, physical, chemical or biotic factor (including light) to be found that can explain this gradient (Peters *et al.*, 1973). Comparative studies in aggressive behaviour have been concentrated on the cave form collected near the end of the cave (chamber XIII after Gordon and Rosen, 1962). In the epigean fish, aggressive behaviour has the function of establishing a size-dependent rank order in the males within the schools of adults. As already described (p. 580), the males separate especially attractive females from the school and defend them vehemently against rivals. In the cave, where the fish live at much higher density (Table 17.2), no aggressive behaviour was to be observed. After cave specimens that showed a good visual orientation with males of their own or of the epigean population were tested in visible light, the reduction of aggression remained unchanged (Fig. 17.8). In

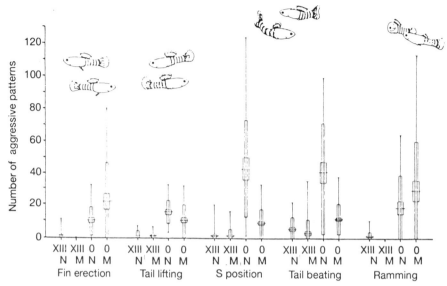

Fig. 17.8 Aggressive patterns in males of *Poecilia mexicana*. XIII, Cave population; O, epigean population; N, standard test of a male against an opponent of the same population; M, male tested against an opponent of the other population. The total variability, mean, standard error and standard deviation are shown. Reproduced with permission from Parzefall (1979).

all parts of the test, the mean number of aggressive acts was about zero. In the mixed tests the majority of the cave males answered the initial attacks of the epigean males with sexual behaviour. Only a few reacted defensively for a short time. No cave male showed an attack reaction. Tests in darkness with epigean males, carried out with the aid of infra-red, showed one aggressive pattern, namely tail beating. Aggressive behaviour in *P. mexicana* therefore seems to be based mainly on visual communication. For this reason, the reduction of aggressiveness in the cave could be the result of reduced visual orientation. If this is the case, tail beating should not be reduced as it can be perceived non-visually. But there is no significant difference between the number of tail beatings and other aggressive movements in the cave fish. This reduction must therefore be considered genetic for all aggressive patterns.

To confirm the hypothesis, cave and epigean populations were crossbred and all generations were tested for aggressiveness. In Figs 17.9 and 17.10 the frequency distribution for fin erection and S position are presented. The parental generations are clearly separated. The F_1 and F_2 crosses show a more or less intermediate value for the average (shown by the arrow). The expected separation in the backcrosses is only present in fin erection. Without detailed examination it can be stated that these data confirm the hypothesis.

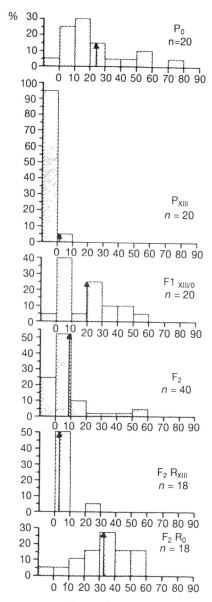

Fig. 17.9 Aggressive fin erection of *Poecilia mexicana*. Frequency distribution in the epigean population (P_0), the cave population (P_{XIII}) and hybrid generations. Arrows denote mean values. Reproduced with permission from Parzefall (1979).

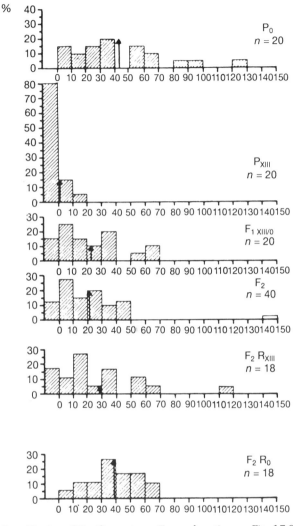

Fig. 17.10 S-positioning of *Poecilia mexicana*. For explanation see Fig. 17.9. Reproduced with permission from Parzefall (1979).

Follow-up tests of the F_2 generation failed to find a correlation between eye diameter and the aggressiveness of the males. However, despite free recombination, a positive correlation exists in the F_2 generation for several aggressive patterns found in the epigean form. The reduction of all aggressive patterns as compared with surface dwellers in the cave population is about equal (Parzefall, 1979).

From the findings it was concluded that a closely linked polygenic system exhibiting additive gene interaction controls the inheritance of aggressive behaviour in *P. mexicana*.

As a result of the reduction in aggression, we should expect a change in the average size of males, because there is no longer any reproductive advantage in being large. Plotting the total length of different samples from outside to the very back of the cave, we can indeed see a diminution in the size of males in the cave (Fig. 17.11).

There are two hypotheses explaining the reduced aggression in *P. mexicana* in this cave. Aggressive behaviour patterns cannot be seen in darkness. For this reason, stabilizing selection acting on aggressive behaviour is no longer effective. Over a long period, this leads to a degeneration of the aggression. But one can argue that tail beating is still performed and could be detected (by the lateral line) in darkness, and so could serve as a basis for necessary aggression in the cave. The fact that tail beating shows the same degree of reduction as the other aggressive patterns, and the much higher degree of reduction in aggressive behaviour in comparison with the slightly reduced eye, suggest that these changes could be better explained by a selection pressure acting directly against aggression. Selection in general can cause

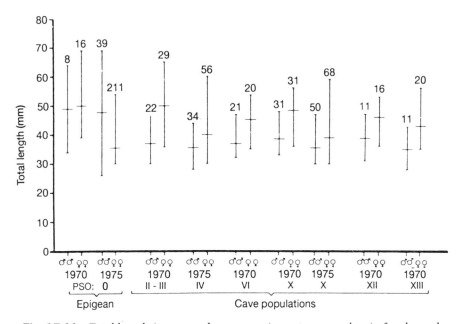

Fig. 17.11 Total length (means and ranges; *n* given atop range bars) of males and females in different samples of *Poecilia mexicana* collected in the natural habitat. Reproduced with permission from Parzefall (1979).

changes more rapidly than can the absence of stabilizing selection. The latter generally causes the eye reduction in caves.

Schooling behaviour

For two of the species already presented, *Astyanax fasciatus* and *Poecilia mexicana*, schooling has been reported as common organization in the epigean habitat. There is some evidence from field and laboratory observations, that the epigean *A. fasciatus* in the absence of perturbation caused e.g. by predators or observers tends to defend small individual territories. This has to be confirmed by further studies. However, in the cave populations this behaviour has not been observed (Parzefall, 1979, 1983). In the following cave-fish species studied by Berti and Thines (1980) and Jankowska and Thines (1982), schooling behaviour was also absent: *Caecobarbus geertsi* (Cyprinidae), *Barbobsis devecchii* (Cyprinidae) and *Uegitglanis zammaranoi* (Clariidae). All these cave-dwelling fish show a random spatial distribution in their habitats and in aquaria. In *P. mexicana* and in *A. fasciatus* the fish swim without contact, dispersed throughout the cave pool. This is more striking in *Astyanax* with its low population density. An absence of schooling has also been found in one epigean population of *A. fasciatus* which lives near a cave entrance and shows an affinity to the cave (Romero, 1983).

From different studies on schooling behaviour (Partridge and Pitcher, 1980) the important role of visual orientation is well known. Individual saithe, *Pollachius virens*, that had been blinded with opaque eye covers were able to join schools of 25 normal saithe, probably by using their lateral line organs. But this result does not mean that blind saithe would school in the wild (Pitcher *et al.*, 1976), and cave fish do not seem to operate in this manner.

Epigean *A. fasciatus* observed in darkness with infra-red cease schooling. Therefore the loss of visual orientation could be the direct reason for the absence of schooling in the cave. The other possibility is that the genetic basis for this behaviour has changed after the long period of cave life without any stabilizing selection for schooling.

To answer these questions, the tendency of *A. fasciatus* to school has been tested in hybrids between the blind Piedras Cave population and the epigean one of the Teapao River. For these experiments, only hybrids that had demonstrated a good optomotor response (minimum separable of 15′ of arc) were used. Despite good visual orientation, the F_2 hybrids and the eyed Micos Cave fish showed a weaker tendency to school (Fig. 17.12) than epigean fish, and an increased variability. Similar experiments (Senkel, 1983) have been conducted with animals of the phylogenetically young cave fish from Micos Cave after selective breeding for functional eyes. In these tests the tendency for schooling was weaker than in the F_2 cross. From the data presented one can conclude that in cave-dwelling *A. fasciatus* there seems to exist a genetically based reduction of schooling behaviour.

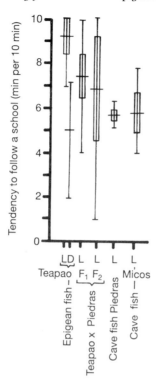

Fig. 17.12 Tendency to follow a school of *Astyanax fasciatus* in min per 10 min. Tests of single specimens of the epigean Teapao population, the hypogean Micos and Piedras populations, the F_1 and F_2 hybrids. In the hybrids, only specimens with a good optomotor response were used. L, light; D, dark; mean ($n = 10$ for each), standard deviation (boxes) and total variability (vertical lines) are shown. Reproduced with permission from Parzefall (in press).

In *Poecilia mexicana* only subadult fish were tested, in order to avoid sexual attraction as a factor. During these tests the epigean fish of the Rio Teapao (Fig. 17.13, PMT) orientated itself to the school in visible light. Compared with this population, the tendency of conspecifics to follow the school was already significantly reduced in an epigean population (PMO) caught in the milky sulphur-containing river coming out of the cave. There is no difference in comparison with the cave population (Fig. 17.13 PSxIII). In both these populations there was strikingly more variability than in the epigean fish. However, in darkness, no preference for the school was registered. In addition hybrids between epigean fish and cave fish have been tested. The F_1 hybrids were at the same level as the epigean fish. However, the F_2 generation differed from the epigean fish in the mean and the higher variability. In the backcross there is a significant difference in B_2 from the epigean fish. So we also can state

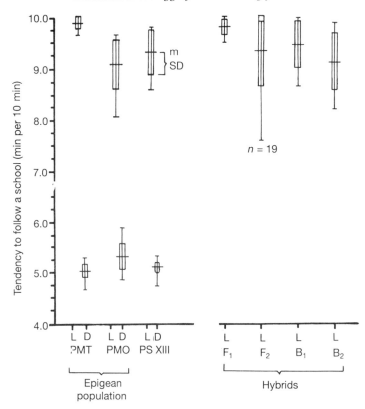

Fig. 17.13 Tendency to follow a school of subadult *Poecilia mexicanus* in min per 10 min. Results are displayed as in Fig. 17.12. (*n* = 10 except as marked). Tests of single speciments in light (L) and darkness (D) of two epigean, one cave population and hybrids. Reproduced with permission from Parzefall (in press).

that a genetically based reduction of the tendency to school in *P. mexicana* already took place in the epigean population near the cave entrance. The reason could be that the milky water causes difficulties in visual orientation (Parzefall, in press).

Alarm reaction

The alarm or fright reaction appears to be confined to the Ostariophysi (Pfeiffer, 1977). 'Schreckstoff' is an alarm pheromone released from club cells in the epidermis when it is damaged. The alarmed conspecifics may react by seeking cover, closer crowding, rapid swimming or immobility. Up to now in cave-dwelling fishes, only *Astyanax fasciatus* and *Caecobarbus geertsi* have been studied. In the *A. fasciatus* population of Chica Cave tested by Pfeiffer (1963) and Thines and Legrain (1973), no alarm reaction was found against own-skin extract or alarm substance, of the epigean *A. fasciatus*. However, it was demonstrated that the cave population does have the alarm substance, because the epigean fish reacted well to extracts from the skin of the Chica fish. It is very probable that the cave population can perceive alarm substance with its well-developed olfactory system, since it immediately searches for food when skin extract is poured into the aquarium. Thines and Legrain (1973) describe a similar reaction in *C. geertsi*: alarm substance causes foraging orientated towards the bottom.

The inheritance of the alarm reaction has been studied in hybrids between an epigean population and the Pachon Cave population of *A. fasciatus* (Pfeiffer, 1966). All F_1 hybrids tested responded to alarm substance. The segregation rate in the F_2 and F_2R generations leads to the interpretation that two dominant factors are responsible for the fright reaction. But the results with the backcross to the cave fish do not fit completely with this hypothesis.

Pfeiffer (1963) tried to explain the loss of the fright reaction by the absence of predators in the cave habitat. Thines and Legrain (1973) follow this explanation for *Caecobarbus geertsi*.

In a detailed study (Fricke, 1988; Parzefall and Fricke, 1989) it could be demonstrated that only some patterns of the alarm behaviour are reduced in the cave-dwelling *A. fasciatus*: cover-seeking at the bottom, rapid swimming (= zigzags) and immobility. However, all the cave populations that were tested avoided the site where the alarm substance was released (Fig. 17.14). In addition, in the cave fish a higher swimming activity is released. It seems that the alarm behaviour has been changed during adaptation to life in darkness: visual alarm signals are reduced. However, the avoidance reaction, being effective in darkness, still exists. The higher swimming activity probably leads the animals away from the predation area.

Unfortunately, in all cave habitats of *Astyanax* studied, no predator has been detected (Mitchell *et al.*, 1977; Parzefall, 1983). It is quite probable that in caves with many fish, bats (*Noctilio leporinus*) live at least in part on cave fish (Wilkens, 1988). However, cannibalism occurs. It has been observed in Micos Cave that the bigger fish show predatory behaviour against smaller ones (Parzefall, 1983). This is in accordance with studies by Wilkens and Burns (1972), who noted the presence of small *Astyanax* in the stomach of the Micos fish. The alarm behaviour is age-dependent: it can first be observed in fish at age 46 days (Pfeiffer, 1966), so young *A. fasciatus* are not protected by the alarm reaction.

Phototactic behaviour

The light reaction of both epigean and cave forms in *Astyanax fasciatus* has been the subject of different studies (Langecker, 1989, 1990). Recently

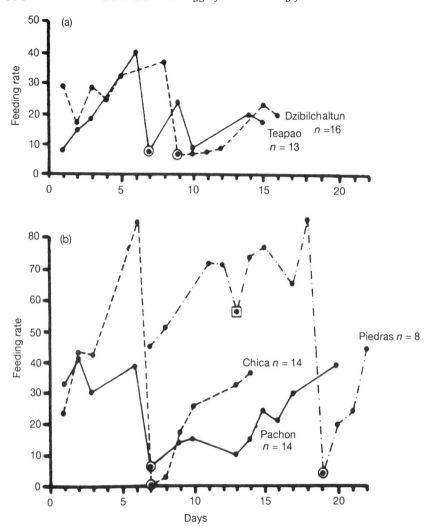

Fig. 17.14 Feeding rate at the water surface in (a) two epigean and (b) three cave populations of *Astyanax fasciatus*. ● Denotes addition of food; ◉ encircled, addition of alarm substance; ▣ boxed, addition of *Petrophyllum* skin extract (without alarm substance). Reproduced with permission from Fricke (1988).

Langecker (1989, 1990) demonstrated in choice-preference experiments that the phototactic response in the intact epigean fish is highly variable. These fish prefer twilight to lower or higher light intensity. Juvenile fish show a photopositive response under all tested light conditions while the adult fish are slightly photonegative (Fig. 17.15). The photonegative behaviour is

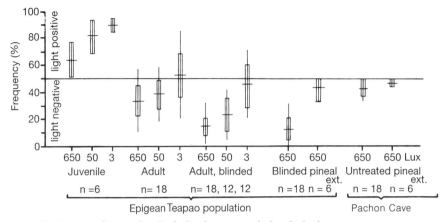

Fig. 17.15 Preference for the lighted area in a light–dark choice experiment using *Astyanax fasciatus*. Mean and variability of eyed, blinded and pinealectomized epigean fish and blind Pachon Cave fish under different light conditions (data from Langecker, 1988, 1990).

greatly increased in blinded fish. Blinded epigean fish that have been pine-alectomized reveal the importance of the pineal organ for light perception and phototactic behaviour. The phototactic index is at the same level as for the cave populations tested (Fig. 17.15). In contrast to the epigean fish, removal of the pineal organ has no significant effect on the cave fish. So, the differences in phototactic behaviour in the cave fish could be caused by a reduction in parts of the pineal organ. But ultrastructurally there is surprisingly little change in this organ. In addition the degree of differentiation of the photo-receptor cells indicates that they are still sensitive (Langecker, 1990). Therefore the different phototactic response of the cave fish must be a true behavioural regression. The persistence of the photoreceptor cells in the pineal organ of even the phylogenetically old Pachon fish suggests that there is a biological significance which is so far unknown.

Dorsal light reaction

With some exceptions nearly all the fish swim with their back towards the light. This vertical orientation of the dorso–ventral axis is triggered by the direction of light and gravity. If the light direction changes, the fish show a compensatory reaction. This reaction results in a deviation from the normal vertical position and can be measured experimentally as an angle of inclination. The angle of inclination is species specific and genetically controlled (Bogenschütz, 1961). The dorsal light reaction depends on the functioning of the eyes.

In a behavioural study of *Astyanax fasciatus* using cross breeding in fish with

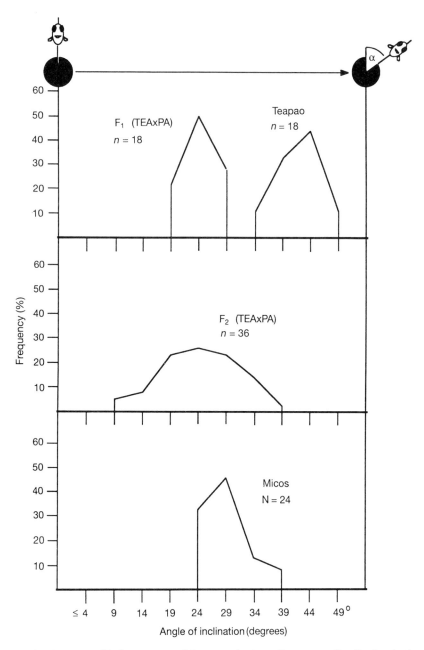

Fig. 17.16 Dorsal light reaction of *Astyanax fasciatus*. Frequency distribution in the epigean population, F_1 and F_2 hybrids with the blind Pachon Cave fish, and in the Micos Cave fish. All specimens tested had functional eyes (data from Langecker, 1990).

functional eyes, Langecker (1990) found that the angle of inclination must have changed in the cave fish (Fig. 17.16). Using the phylogenetically young cave population of Micos Cave after selection for functional eyes he obtained comparable results (Fig. 17.16). This system, independently from the eyes, has almost completely regressed in the cave fish. The regression is genetically based on a polygenetic system of at least three genetic factors (Langecker, 1990).

Circadian clock

The circadian clock in animals has to be synchronized by external stimuli called zeitgeber (forcing signals), principally light and temperature (Bünning, 1973). These stimuli are normally absent in caves. Epigean animals tested in darkness show an endogenous free-running daily rhythm in their locomotory activity. Therefore the question arises as to whether, in cave-dwellers that have already been living in darkness and more or less constant temperature for a long time, such an endogenous circadian rhythm still exists.

In a group of bind *Astyanax fasciatus* of unknown origin, Erckens and Weber (1976) frequently found one to four postoscillations after the animals had been transferred from light–dark cycles (LD) into the darkness. After the postoscillations had damped out, the animals seemed to become aperiodically active. In subsequent studies with the Pachon Cave fish, weak circadian periodicities have been detected in constant darkness after a light–dark cycle of 12h:12h only. Using cycles of other lengths, this phenomenon was lacking. These periodicities were detected by the mathematical procedure of complex demodulation. In the eyed epigean *A. fasciatus*, circadian activity cycles were quite obvious in darkness (Erckens, 1981; Erckens and Martin, 1982a, b). Thines and Weyers (1978) have described an aperiodic locomotory behaviour in darkness in the blind Pachon fish.

In the Amblyopsidae three species have been tested (Poulson and Jegla, 1969). The springfish, *Chologaster agassizi*, which is functionally blind and cannot focus on an object, possesses a circadian clock. This species is nocturnal in epigean habitats and goes underground in winter and breeds there. In contrast, the obligate cave-dwellers *Typhlichthys subterraneus* and *Amblyopsis rosae* show no circadian components in their behaviour in darkness.

These two examples of cave-living fishes reflect the general situation in all obligate cave-dwelling animals. More extremely specialized cave-dwelling animals exhibit no, or doubtful, locomotor periodicities, or periodicities of very poor precision. As the change between activity and rest reflects the most drastic changes in metabolism, Lamprecht and Weber (1982) conclude that a periodic organization of resting and activity metabolism is not necessary, perhaps even unfavourable, inside caves.

17.3   REGRESSIVE EVOLUTION AND BEHAVIOUR

The discussion about regressive evolution mainly centres on the reduction of eyes and pigmentation. But regressive evolution is by no means unique to caves. There are also examples in parasites or animals that become sessile (Dzwillo, 1984).

What makes regressive evolution in cave animals especially interesting is the possibility that selection may play a minor role compared with the accumulation of neutral mutations and genetic drift. The question of whether selection or neutral mutation plays the dominant role cannot be answered on the basis of our current knowledge (Culver, 1982). Kosswig (1948, 1963) has demonstrated that in animals entering caves, various patterns (e.g. eyes and pigmentation) first show a high variability. His hypothesis is that the absence of stabilizing selection allows an accumulation of selectively neutral mutations, and that variability decreases during cave life within a population because of mutation pressure. In phylogenetically old cave animals this leads to genetic homozygosity and diminished variability for the reduced patterns. Other authors favour selective explanations involving selection for increased metabolic economy (Poulson, 1963) or indirect effects of pleiotropy (Barr, 1968). There is only a general consensus that differentiation of cave populations, especially with regard to regressive evolution, cannot occur when there is gene exchange with surface populations.

The regressive evolution in the different behavioural traits presented cannot be explained by only one of the above-mentioned hypotheses. The neutral-mutation hypothesis seems to be the most plausible explanation for the reduction of the circadian clock, dorsal light reaction, phototactic behaviour and the lack of schooling behaviour. To test this hypothesis, more comparative data on cave-dwellers of different phylogenetic age are necessary. If it is valid, it should be possible to demonstrate higher variability for these behaviour patterns in phylogenetically younger cave populations in parallel with the degree of eye reduction. Circadian activity, schooling and dorsal light reaction are relatively independent of motivational changes and are therefore good examples for testing. One cannot, however, neglect the opposing arguments that these characters of little or no selective value in caves change as a pleiotropic by-product of selection for adaptively advantageous characters (Barr, 1968), or, on the other hand, that the reduction of these characters is an economic advantage: increased metabolic economy conserves energy for cave-living organisms (Poulson, 1963). So the compensation for behavioural traits that are unnecessary in caves can be adaptive.

On the other hand, the reduction of agonistic behaviour seems more likely to be caused by selection. The great reduction of agonistic behaviour in comparison to the eye, which is only slightly reduced, favours the explanation of selection against aggression in *Poecilia mexicana*. The aggressive males seem to have no more reproductive advantage because of the risk of losing contact with the female in darkness during an aggressive encounter against an opponent. This risk is increased by the fact that only tail beating, which requires a short distance to the opponent, is effective in darkness. The most successful strategy in darkness should be a quick copulation after a short nipping contact. Such a behaviour already exists among small males in epigean habitats. The absence of bigger males in the cave population supports this hypothesis (Parzefall, 1979).

The abundance of food in the *Poecilia* cave allows us to neglect energy conservation as another reason for the reduction of aggression. However, the study of Bechler (1983) on the Amlyopsidae has pointed to food as a primary selective force. But several facts argue against the hypothesis that the reduction of agonistic behaviour serves specifically to conserve energy. So this reduction can be explained as a consequence of the decreased metabolic rate and fecundity, which allows a reduction in overt agonistic behaviour with continuing adaptation to cave life.

For *Astyanax fasciatus* more data from different cave populations are necessary in order to understand the reduction of agonistic behaviour. In epigean fish, ramming an opponent occurs even in darkness, and increases in hungry animals studied in light (Brust-Burchards, 1980; Dölle, 1981). So *A. mexicanus* should be able to defend food sources in the cave. But in Pachon Cave the fish fed together without any aggression, and in Chica Cave there was no reaction to food offered (Parzefall, 1983). Up to now we can only speculate that the low density of *A. mexicanus* in the caves, in connection with enough food in the form of bat guano, does not promote agonistic behaviour in this habitat.

The changes of head-standing behaviour and of the alarm reaction in *A. fasciatus* are surely adaptive. The evolutionary improvement of the gustatory equipment presupposes changes in the head-standing movement (Schemmel, 1980). If the food is mainly distributed on the bottom of the habitat, it seems to be a useful tactic to search at an angle of about 45° in slow zigzag movements (Parzefall, 1983). In the alarm behaviour, only those patterns persist which are effective in darkness.

17.4 CONCLUSION

In the examples of different behaviour traits presented, chemical communication is the most important mode of communication in caves. Acoustic and tactile stimuli also exist, but here again few exact data are available.

It is striking that, despite regressive evolution for agonistic behaviour, schooling, fright reaction and circadian activity, no example exists that shows the gradual reduction of a visual display among cave-dwelling animals. For this reason the absence of visual displays is probably a necessary assumption for the successful colonization of a lightless habitat. In every case studied, the type of communication that is realized in the cave form exists already in its epigean relatives.

In some of the caves, food is extremely scarce. Cave-dwelling fish respond to severe food limitation by a decreased metabolic rate and fecundity. Many individuals of the cave-limited species in the Amblyopsidae fail to reproduce.

17.5   SUMMARY

In the teleost fish, members of about 14 families have colonized caves successfully. Mainly three of these families have been studied for possible behavioural adaptations to the cave habitat. Potential cave-dwellers seem to need a preadaptation for cave life in their sexual behaviour which is mainly based on chemical communication. The most striking phenomenon is reduction of different behavioural traits ranging from circadian locomotory activity, fright reaction, schooling and aggressive behaviour to parts of the feeding behaviour.

These behavioural differences between cave-dwellers and their epigean relatives allow the use of these animals for studies on evolutionary genetics in behaviour because of fertile cross-breeding. On the basis of the present data it is very likely in cave animals that, with some important exceptions, such as aggressive behaviour, selection may play a minor role compared with the accumulation of neutral mutations. For these reasons field and laboratory studies on cave-dwelling fish seem to hold great promise.

ACKNOWLEDGEMENTS

The author would like to thank the DFG for financial support. He is also indebted to M. Irentschiuck and A. Schlupp for technical assistance, and M. Hanel for drawing some figures.

REFERENCES

Barr, T. C. (1968) Cave ecology and the evolution of troglobites. *Evol. Biol.*, **2**, 35–102.
Bechler, D. L. (1983) The evolution of agonistic behaviour in amblyopsid fishes. *Behav. Ecol. Sociobiol.*, **12**, 35–42.
Berti, R. and Ercolini, A. (1979) Aggressive behaviour in the anophthalmic phreatic fish *Uegitglanis zammaranoi* Gianferrari (Clariidae, Siluriformes). *Monitore zool. ital.*, **13**, 197.
Berti, R. and Thines, G. (1980) Influences of chemical signals on the topographic

orientation of the cave fish *Caecobarbus geertsi* Boulenger (Pisces, Cyprinidae). *Experientia*, **36**, 1384–5.

Bogenschütz, H. (1961) Vergleichende Untersuchung über die optische Komponente der Gleichgewichtshaltung bei Fischen. *Z. vergl. Physiol.*, **44**, 626–55.

Bormann, J. (1989) Das Sozialverhalten von *Rhamdia guatemalensis* (Pimelodidae, Pisces), Vergleich der Höhlenpopulation mit der oberirdisch lebenden Population. Diploma thesis, University of Hamburg, 126 pp.

Brust-Burchards, H. (1980) Das Aggressionsverhalten von Fischen. Eine vergleichende Betrachtung unter besonderer Berücksichtigung von *Astyanax mexicanus*. Diploma thesis, University of Hamburg, 74 pp.

Bünning, E. (1973) *The Physiological Clock*, Vol. 3, Auflage Springer, Berlin, 71–112.

Burchards, H., Dölle, A. and Parzefall, J. (1985) The aggressive behaviour of an epigean population of *Astyanax mexicanus* (Characidae, Pisces) and some observations of three subterranean populations. *Behav. Process.*, **11**, 225–35.

Cooper, J. E. and Kuehne, R. A. (1974) *Speoplatyrhinus poulsoni*, a new genus and species of subterranean fish from Alabama. *Copeia*, **1974**, 486–93.

Culver, D. C. (1982) *Cave Life, Evolution and Ecology*, Harvard University Press, Cambridge, MA, 46–47.

Dölle, A. (1981) Über Ablauf und Funktion des Aggressionsverhaltens von *Astyanax mexicanus* (Characidae, Pisces) unter Berücksichtigung zweier Höhlenpopulationen. Diploma thesis, University of Hamburg, 67 pp.

Dzwillo, M. (1984) Regressive Evolution in der Phylogenese des Tierreiches. *Fortschr. Zool. Syst. Evolutionsforsch.*, **3**, 115–26.

Erckens, W. (1981) The activity controlling time-system in epigean and hypogean populations of *Astyanax mexicanus* (Characidae, Pisces). *Proc. Eighth Int. Congr. Speleology*, **2**, 796–7.

Erckens, W. and Martin, W. (1982a) Exogenous and endogenous control of swimming activity in *Astyanax mexicanus* (Characidae, Pisces) by direct light response and by a circadian oscillator. I Analyses of the time-control systems of an epigean river population. *Z. Naturforsch.*, **37C**, 1253–65.

Erckens, W. and Martin, W. (1982b) Exogenous and endogenous control of swimming activity in *Astyanax mexicanus* (Characidae, Pisces) by direct light response and by a circadian oscillator. II Features of time-controlled behaviour of a cave population and their comparison to an epigean ancestral form *Z. Naturforsch.*, **37C**, 1266–73.

Erckens, W. and Weber, F. (1976) Rudiments of an ability for time measurements in the cavernicolous fish *Anoptichthys jordani* Hubbs and Innes (Pisces, Characidae). *Experientia*, **32**, 1297–9.

Ercolini, A., Berti, R. and Cianfanelli A. (1981) Aggressive behaviour in *Uegitglanis zammaranoi* Gianferrari (Clariidae, Siluriformes) an anophthalmic phreatic fish from Somalia. *Monitore zool. ital.*, **5**, 39–56.

Fausel, P. (1990) Das Sozialverhalten der Höhlenbarbe *Garra barreimiae* (Cyrinidae, Pisces). Diploma thesis, University of Hamburg, 120 pp.

Fricke, D. (1988) Reaction to alarm substance in cave populations of *Astyanax mexicanus* (Characidae, Pisces). *Ethology*, **76**, 305–8.

Gordon, M. S. and Rosen, D. E. (1962) A cavernicolous form of the Poeciliid Fish *Poecilia sphenops* from Tabasco, Mexico, *Copeia*, **1962**, 360–68.

Hofmann, S. (1990) Vergleichende Untersuchung zum Aggressionsverhalten einer oberirdischen und einer unterirdischen Population von *Astyanax fasciatus* (Characidae, Pisces). Diploma thesis, University of Hamburg, 72 pp.

Hüppop, K. (1987) Food finding ability in cave fish (*Astyanax fasciatus*) *Int. J. Speleol.*, **16**, 59–66.

Hüppop, K. (1989) Genetic analysis of oxygen consumption rate in cave and surface fish of *Astyanax fasciatus* (Characidae, Pisces). Further support for the neutral mutation theory. *Mém. Biospéléol.*, **16**, 163–9.

Jankowska, M. and Thines, G. (1982) Etude comparative de la densité de groupes de poissons cavernicoles et epiges (Characidae, Cyprinidae, Clariidae). *Behav. Process.*, **7**, 281–94.

Klimpel, B. and Parzefall, J. (1990) Comparative study of predatory behaviour in cave and river populations of *Astyanax fasciatus* (Characidae, Pisces). *Mém. Biospéléol.*, **17**, 27–30.

Kosswig, C. (1948) Genetische Beiträge zur Präadaptationstheorie. *Rev. Fac. Sci. (Istanbul)*, Ser. **B**, **5**, 176–209.

Kosswig, C. (1963) Genetische Analyse konstruktiver und degenerativer Evolutionsprozesse. *Z. Zool. Syst. Evolutionsforsch.*, **1**, 205–39.

Lamprecht, G. and Weber, F. (1982) A test for the biological significance of circadian clocks: evolutionary regression of the time measuring ability in cavernicolous animals, in *Environmental Adaptation and Evolution* (eds D. Mossakowski and G. Roth), Fischer Verlag, Stuttgart, pp. 151–78.

Langecker, T. G. (1989) Studies on the light reaction of epigean and cave populations of *Astyanax fasciatus* (Characidae, Pisces). *Mém. Biospéléol.*, **16**, 169–76.

Langecker, T. G. (1990) Das Licht als Evolutionsfaktor bei Höhlentieren – untersucht an ober- und unterirdisch lebenden Populationen des *Astyanax fasciatus* (Characidae, Pisces). PhD thesis, University of Hamburg, 88 pp.

MacGinitie, G. E. (1939) The natural history of the blind goby *Typhlogobius californiensis* Steinbacher. *Am. Midl. Nat.*, **21**, 489–505.

Mitchell, R. W., Russel, W. H. and Elliot, W. R. (1977) Mexican eyeless characin fishes, genus *Astyanax*, environment, distribution and evolution. *Spec Publ. Mus. Texas Tech. Univ.*, **12**, 1–89.

Mohr, Ch. M. and Poulson, T. L. (1966) *The Life of the Cave*, McGraw-Hill, New York, 92–97 pp.

Partridge, B. L. and Pitcher, T. J. (1980) The sensory basis of fish schools: relative roles of lateral line and vision. *J. Comp. Physiol.*, **135**, 315–25.

Parzefall, J. (1969) Zur vergleichenden Ethologie verschiedener *Mollienesia*-Arten einschließlich einer Höhlenform von *M. sphenops*. *Behaviour*, **33**, 1–36.

Parzefall, J. (1973) Attraction and sexual cycle of Poeciliidae, in *Genetics and Mutagenesis of Fish* (ed. J. H. Schröder), Springer Verlag, Berlin, pp. 177–83.

Parzefall, J. (1974) Rückbildung aggressiver Verhaltensweisen bei einer Höhlenform von *Poecilia sphenops* (Pisces, Poeciliidae). *Z. Tierpsychol.*, **35**, 66–82.

Parzefall, J. (1979) Zur Genetik und biologischen Bedeutung des Aggressionsverhaltens von *Poecilia sphenops* (Pisces, Poeciliidae). *Z. Tierpsychol.*, **50**, 399–422.

Parzefall, J. (1983) Field observation in epigean and cave populations of Mexican characid *Astyanax mexicanus* (Pisces, Characidae). *Mém. Biospéléol.*, **10**, 171–6.

Parzefall, J. (in press) Schooling behaviour in population-hybrids of *Astyanax fasciatus* and *Poecilia mexicana* (Pisces, Characidae), in *New Trends in Ichthyology*, (ed. J. H. Schröder), Parey Verlag.

Parzefall, J. and Fricke, D. (1989) Alarm reaction and schooling in population hybrids of *Astyanax fasciatus* (Pisces, Characidae). *Mém. Biospéléol.*, **16**, 177–82.

Peters, N., Peters, G., Parzefall, J. and Wilkens, H. (1973) Über degenerative und konstruktive Merkmale bei einer phylogenetisch jungen Höhlenform von *Poecilia sphenops* (Pisces, Poeciliidae). *Int. Rev. ges. Hydrobiol.*, **58**, 417–36.

Pfeiffer, W. (1963) Vergleichende Untersuchung über die Schreckreaktion und den Schreckstoff der Ostariophysen. *Z. vergl. Physiol.*, **47**, 111–47.

Pfeiffer, W. (1966) Über die Vererbung der Schreckreaktion bei *Astyanax* (Characidae, Pisces). *Z. Vererbungsl.*, **98**, 97–105.

Pfeiffer, W. (1977) The distribution of fright reaction and alarm substance cells in fishes. *Copeia*, **1977**, 653–65.

Pitcher, T. J., Partridge, B. L. and Wardle, C. S. (1976) A blind fish can school. *Science*, **194**, 963–5.

Poulson, T. L. (1963) Cave adaptation in amblyopsid fishes. *Am. Midl. Nat.*, **70**, 257–90.

Poulson, T. L. (1969) Population size, density and regulation in cave fishes. *Acts Fourth Int. Congr. Speleology, Ljubljana, Yugoslavia*, **4–5**, 189–92.

Poulson, T. L. and Jegla, T. C. (1969) Circadian rhythms in cave animals. *Actes Fourth Int. Congr. Speleology, Ljubljana Yugoslavia*, **4–5**, , 193–5.

Poulson, T. L. and White, W. B. (1969) The cave environment. *Science*, **165**, 971–81.

Riedl, R. (1966) *Biologie der Meereshöhlen*, Parey Verlag, Hamburg, 195–240 pp.

Romero, A. (1983) Behaviour in an "intermediate" population of the subterranean-dwelling characid *Astyanax fasciatus*. *Env. Biol. Fishes*, **10**, 203–8.

Schemmel, Ch. (1977) Zur Morphologie und Funktion der Sinnesorgane von *Typhliasina pearsei* (Hubbs) (Ophidioidea, Teleostei). *Zoomorphol.*, **87**, 191–202.

Schemmel, Ch. (1980) Studies on the genetics of feeding behaviour in the cave fish *Astyanax mexicanus* f. anoptichthys. *Z. Tierpsychol.*, **53**, 9–22.

Schultz, R. L. and Miller, R. R. (1971) Species of the *Poecilia sphenops* complex in Mexico. *Copeia*, **1971**, 282–90.

Senkel, S. (1983) Zum Schwarmverhalten von Bastarden zwischen Fluss- und Höhlenpopulationen bei *Astyanax mexicanus* (Pisces, Characidae). Diploma thesis, University of Hamburg, 69 pp.

Thines, G. (1955) Les poissons aveugles (I). Origine, taxonomie, répartition géographique, comportment. *Annls. Soc. r. zool. Belg.*, **86**, 1–128.

Thines, G. (1969) *L'Evolution Regressive des Poissons Cavernicoles et Abyssaux*, Masson, Paris, 323–35 pp.

Thines, G. and Legrain, J. M. (1973) Effects de la substance d'alarme sur le compartement des poissons cavernicoles *Anoptichthys jordani* (Characidae) et *Caecobarbus geertsi* (Cyprinidae). *Ann. Spéléol.*, **28**, 291–7.

Thines, G. and Proudlove, G. (1986) Pisces, in *Stygofauna Mundi* (ed. L. Botosaneanu). Brill, Leiden, pp. 709–34.

Thines, G. and Weyers, M. (1978) Responses locomotrices du poisson cavernicole *Astyanax mexicanus* (Pisces, Characidae) à des signaux périodiques et apériodiques de lumière et de température. *Int. J. Speleol.*, **10**, 35–55.

Trajano, E.(1991) Agonistic behaviour of *Pimelodella kronei*. *Behav. Process.*, **23**, 113–240.

Vandel, A. (1965) *Biospeleology*, Pergamon Press, London, 23–25 pp.

Wilkens, H. (1972) Über Präadaptationen für das Höhlenleben, untersucht am Laichverhalten ober- und unterirdischer Populationen des *Astyanax mexicanus* (Pisces). *Zool. Anz.*, **188**, 1–11.

Wilkens, H. (1976) Genotypic and phenotypic variability in cave animals. Studies on a phylogenetically young cave population of *Astyanax mexicanus* (Filippi) (Characidae, Pisces). *Ann. Spéléol.*, **31**, 137–48.

Wilkens, H. (1982) Regressive evolution and phylogenetic age: the history of colonization of freshwaters of Yucatan by fish and crustacea. *Texas Mem. Mus. Bull.*, **28**, 237–43.

Wilkens, H. (1988) Evolution and genetics of epigean and cave *Astyanax fasciatus* (Characidae, Pisces). *Evol. Biol.*, **23**, 271–367.

Wilkens, H. and Burns, R. J. (1972) A new *Anoptichthys* cave population (Characidae, Pisces). *Ann Spéléol.*, **27**, 263–70.

Zeiske, E. (1971) Ethologische Mechanismen als Voraussetzung für einen Übergang zum Höhlenleben. Untersuchungen an Kaspar-Hauser-Männchen von *Poecilia sphenops* (Pisces, Poeciliidae). *Forma Functio*, **4**, 387–93.

Part four

Applied Fish Behaviour

INTRODUCTION

The final section of this book contains two chapters on applied aspects of fish behaviour: the capture of fishes by fishing gear, and the management of freshwater fisheries.

In Chapter 18, Clem Wardle gives a detailed description of how fishing gear exploits the behaviour of fishes, with the aid of some superb underwater photographs of fish interacting with commercial fishing gear from his work in Scotland. After a brief review of FADs (Fish Aggregation Devices), the main types of fishing gear and their general relation to fish behaviour, Wardle concentrates on how fish respond to the various elements of a towed trawl, the most widely-used commercial fishing gear in the world. The chapter includes an explanation of the often-reported fountain manoeuvre in fish schools and a detailed appraisal of just what is visible to a fish entering fishing gear and the effect of different light levels in the mass movement of fish schools, revisiting elements of fish vision from Chapter 3 and shoaling from Chapter 12. Wardle concludes by describing modifications to trawls that use his research findings, for example, trawls which separate different commercial species in the catch.

Ken O'Hara, in the last chapter in the book, considers how fish behaviour impinges upon the management of freshwater fisheries for recreation and food. By and large, the author thinks that the major impact of behavioural studies is yet to come, partly because freshwater fishery management is often a highly empirical art. O'Hara reviews techniques of tracking and telemetry. He describes how management may take advantage of information about behaviour in feeding, migration, fish passes, vegetation, water quality, stock manipulations, habitat improvement, habitat management and mitigation. O'Hara ends with plea for the genetic conservation of wild fish stocks. Practical fishery managers tend only to believe field-based research, but they can only gain from the current upsurge in knowledge, 'even if some myths about behaviour have to be jettisoned en route.'

One topic not covered by this section is fish behaviour in aquaculture (see Muir and Roberts 1985).

REFERENCE

Muir, J. F. and Roberts, R. J. (1985) (eds) *Recent Advances in Aquaculture, Vol 2*, Croom Helm, London, 300 pp.

Chapter eighteen

Fish behaviour and fishing gear

C. S. Wardle

18.1 INTRODUCTION

Throughout history human hunters for fish have made use of their knowledge of the behaviour of fish in order to make catches of them. There are more than 20 000 different species of teleost fishes, each with its own characteristic world of reaction and behaviour, so that numerous appropriate fish-capture systems have been invented. Our knowledge of the sensory abilities and behaviour of fishes has been summarized in the earlier chapters of this book. Fish behaviour is involved in catching fish, both on the oceanic scale, where the annual cycles of maturity cause migrations so that fish are found in different locations that become known to the fisherman by observation of their availability, and on a smaller scale, where the reactions of a fish to each part of an approaching trawl can cause the fish to swim into the cod end. In order to be successful, the fisherman must have local knowledge of the day-to-day movements of the fish and of their likely distribution. In all fisheries, one of the most important of the fisherman's skills is to use the appropriate gear at the right time in the right place.

18.2 REVIEW OF FISHING METHODS

Although the main discussion of this chapter will be concerned with the reaction and behaviour of fish in towed trawls, it is worth considering other fishing methods briefly to realize that each system has its own complex inter-action of technology and fish behaviour. The purse seine, because of its huge size, is one of the most complicated nets to operate at sea, yet it is extremely simple in its concept of, first, rapidly sinking an inpenetrable wall of netting

Behaviour of Teleost Fishes 2nd edn. Edited by Tony J. Pitcher. Published in 1993 by Chapman & Hall. ISBN 0 412 42930 6 (HB) and 0 412 42940 3 (PB).

around a group of fish, then closing the bottom edge with purse strings, and finally pumping or scooping the concentrated fish aboard the vessel. Some examples of the way in which the physiology and behaviour of tuna affect their availability to this fishing method at the oceanic scale in the Pacific are described by Sharp (1978). Tuna, herring, mackerel and other pelagic fishes of the ocean are caught in huge numbers by purse seine nets when they aggregate at certain times of the year, usually near coasts, for spawning or feeding.

Fish aggregation devices (FADs) make easier the catching of fish in warmer waters where natural aggregations do not occur. FADs are made by hanging tree branches or tent-shaped sheets of plastic beneath rafts, and they are very effective in certain areas of the sea in aggregating fish. The fish are then gathered by purse seine, baited hooks or gill nets (Hunter and Mitchell, 1968). FADs have been found, by their stimulation of aggregation behaviour, to increase the productivity of local fisheries for very little expenditure of energy (Preston, 1982). Natural reefs and artificial reefs such as shipwrecks, oil rigs and pipelines all influence the behaviour of those fish that are in or moving through an area by aggregating certain species and making local fishing easier.

A length of line ending with a hook embedded in a tasty morsel of bait forms another simple fishing gear that has lead to numerous variations. Fish hooks have been discovered that date from the Stone Age, and they were probably used to catch fish long before then. Many more types of hook are available than there are species of fish, and each type of hook has been evolved by humans matching the mechanics of the device to the behaviour, shape and size of the fish (Hurum, 1977). As in all other forms of fishing, using a line and baited hook involves a degree of luck, but the odds are shortened by careful study of the behaviour and likely distribution of the species to be caught. The intricacies of angling are the source for many well-known volumes of anecdotes and wise sayings on ways to hook the elusive fish. There is a world-wide research effort into finding simple, reproducible artificial baits for commercial long-line fishing. The research relies heavily on experiments investigating the feeding behaviour of fish (Mackie *et al.*, 1980; Johnstone and Hawkins, 1981).

Fish traps are constructed by humans in great variety, often unbaited, and they operate on some seasonal facet of the animal's behaviour that causes fish to move through a local area of coast or river. Salmon are caught around the coast of the UK by so-called beach engines or stake-nets in which walls of netting supported on stakes driven into the beach lead the fish through a series of non-return funnels into a holding chamber that can be lifted and emptied at the surface. Studies of the behaviour of fish carrying ultrasonic transmitters have shown that many migrating salmon can find their way along a beach through extensive lines of these commercial stake-net traps without being caught (Hawkins *et al.*, 1979). Larger and more elaborate traps of this type are used to catch blue-fin tuna around the coasts of the Mediterranean Sea and its islands, and off the Atlantic coast of Spain (Rodriguez-Roda, 1964). Migratory fish species are caught by various ingenious traps in all parts of the world. Spears, hand nets, snares and invisible nets that gill or entangle the fish are skilfully used in many parts of the world. These types of fishing gear, as well as poisons and explosives, work only when they stimulate no behavioural response of the fish. Gill nets, drift nets and trammel nets rely on the fish blundering into the mesh and being unaware of its presence until too late (Dickson, 1989a, 1989b).

Trawls and Danish seine nets have developed from the historical techniques of beach seining and beam trawling. Beach seines and beam trawls rely on a minimum of reaction from the fish. In the beach seine, like the purse seine which was probably developed from it, fish are herded and surrounded with a wall of netting guiding them into a short funnel and cod-end bag. In a beam trawl the fish are overrun by the rigid mouth of the net and trapped in a much longer funnel and cod-end. During the evolution of modern trawls and the Danish seine net, it is apparent, as we shall see in the rest of this chapter, that more and more of the fishes' repertoire of behaviour has been involved as these gears have become larger, more versatile and more effective.

Evolution of gear in the North Sea

The evolution of each of these gears is outlined in Fig. 18.1. This evolution demonstrates a close link between technological change and human knowledge of fish and their reactions in fishing gear. For example, in 1848 Jan Vaevers found that by adding longer ropes and a large anchor to his beach-seine net and fishing it on the sea-bed in deeper water, he could catch a previously unexploited abundant stock of flatfish. Danish anchor seining was born, and developed into the modern, man- and energy-efficient, Danish seine-netting technique. It was natural that beach seines were also towed to sea, the wings held open by two sailing boats, and so a beach seine was developed to create a light pair trawl that could be fished on the surface or the sea-bed. Single-boat sea-bed fishing in sailing vessels, with their unreliable towing power, used beam trawls, their size limited by the length of beam that could be stowed along the side of the vessel. Once the steady, reliable pulling power of the steam engine was built into fishing vessels, otter boards were invented to replace the beam and this allowed the nets to be made much wider, and to be folded and stowed.

Humans have a natural curiosity to know how their devices work and an urge to make them better if they do not seem quite right. However, once they work, great care is taken to avoid any change. One can imagine that the beach seine and beam trawls were often used in shallow clear water where their

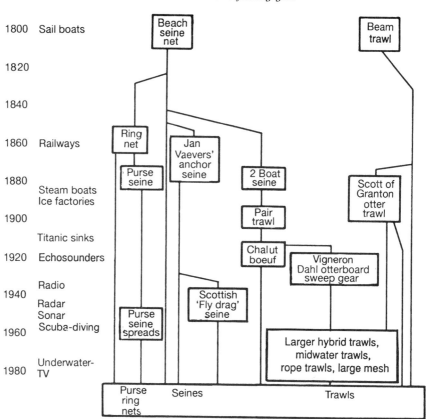

Fig. 18.1 Chronological chart showing an outline of the evolution of the beach seine and the beam trawl. The chart reflects a diversification of fishing methods increasing the exploitation of fish behaviour as technology changes. Some of the technology that has influenced the gears is shown in the left-hand margin. For more details of the gears named, see Le Gall (1931), Davis (1958), Thomson (1969) and Wardle (1983).

effect in catching fish could be watched. As soon as gears are fished in deeper water, they are out of sight and the operator must judge his net by indirect evidence such as the presence of polished areas on chains or the size of the catch for the effort made. The Danish seine net invented by Jan Vaevers demonstrated a quite dramatic advance on his countrymen's efforts with the beach seine because he caught many more fish. Similarly the long-wing haddock seine used between 1930 and the 1950s was replaced 'overnight' in 1953 by the commercial introduction of the so-called Vinge trawl as a seine net. When this net was fished, it had twice the headline height and it caught twice as many haddock for the same effort (Thomson, 1969).

As can be seen from Fig. 18.1, gears continued to evolve as technology made new materials, new ship designs, new navigation aids, new charts, echosounders and radar available to the fisherman. It is inevitable that humans as hunters continue to apply their knowledge of fish behaviour to the design and development of fishing techniques, while at the same time they

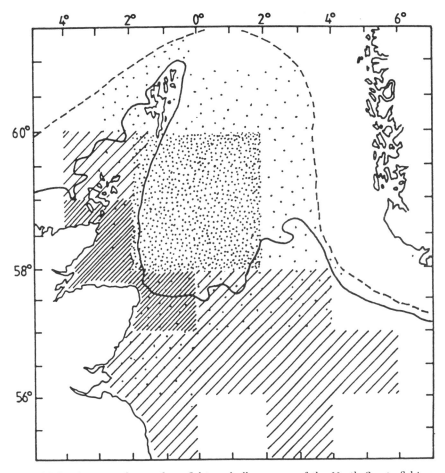

Fig. 18.2 A major change from fishing shallow areas of the North Sea to fishing areas deeper than 100 m was made possible by the introduction of the otter trawl patented by Scott of Granton in 1894. The chart shows those areas from which fish were landed by steam trawlers at Aberdeen market for the first three months of the years 1891 (hatched) and 1901 (stippled); principal areas are darker. In 1891 using only beam trawls there were 760 landings averaging 3328 kg, and in 1901 using only otter trawls 644 landings of 6344 kg. Note the depth contours of 100 m (continuous line) and 200 m (dashed line). Redrawn from Plates II and III and p. 140 of Wemys Fulton (1902).

make use of new technology. The growth of railways and the introduction of steam power to fishing caused a general mobilization of the fishing effort away from local traditional grounds to intrude on fishing grounds of others. It also meant fishing quite new areas. For example, with the introduction of the otter trawl and the steam trawler, areas of the North Sea deeper than 100 m were made fishable; areas where, before that time, no fish had ever been taken by trawler (Fig. 18.2). The increased range and power of the fishing fleet had generated local disputes all around the coast of the UK by 1880, when a Royal Commission was set up to investigate complaints of damage to gear and grounds by these new trawlers. The Royal Commission on Trawling (Dalhousie Commission, 1885) recommended, among other things, that money should be set aside for the first time by government to support scientific investigation into how these new gears worked and to find out how they affected stocks and the sea-bed. One hundred years later we are still asking similar questions for similar reasons.

Modern trawls

Modern trawls are made in many sizes to allow different types of fishing boat to tow them at a maximum speed: between 3 and 4 knots when using full

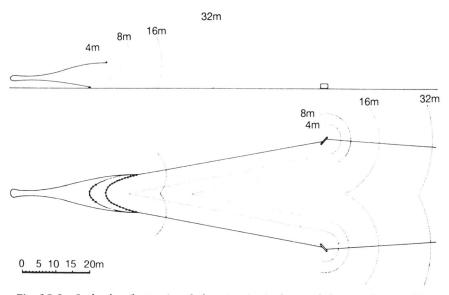

Fig. 18.3 Scale plan (bottom) and elevation (top) of a trawl that can be towed by a 600 hp vessel. The parts of the gear can be identified from Fig. 18.4. Dotted lines indicate the various ranges of visibility discussed in the text in relation to the reaction behaviour of fish.

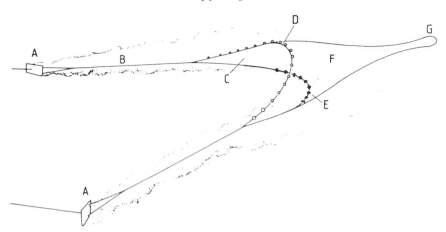

Fig. 18.4 Sketch of a trawl in action. A, otter board; B, sweep wire; C, wing; D, headline; E, ground gear and fishing line; F, funnel of the net; G, cod end.

power. A typical otter trawl, shown as a scale diagram in Fig. 18.3 and as a sketch in Fig. 18.4, can be towed by a 600 hp vessel at $2\,\mathrm{m\,s^{-1}}$ (4 knots) and maintains relatively constant geometry as indicated in Fig. 18.3. The $2.5\,\mathrm{m} \times 1.5\,\mathrm{m}$ otter boards are made of wood or steel and are towed on the trawl warps (narrow steel cables) with an angle of attack that causes them to spread away from each other, opening the front of the net. Otter boards remain about 30 m apart when the trawl is towed. Each wing of the net is linked to an otter board by sweeps (steel cables) 55 m long. The tips of the wings of the net are 10 m apart, so the sweeps run at an angle to the forward motion of about $10°$. Sand clouds, thrown up by the turbulent swirls of water sucked in behind the two otter boards, are left as visually opaque trails spreading inwards and forming walls along the line of the sweeps (Fig. 18.5). In this way a walled passage is formed on the sea-bed, 30 m wide at the otter boards and narrowing to 10 m wide at the net mouth. The upper lip of the trawl mouth, known as the headline, is held up by fifty 200 mm floats. It arches 6 m above the sea-bed and the net tapers to the cod end some 40 m behind. When a trawl is fished on smooth sand, the lower lip of the mouth of the net, called the fishing line, is weighted with chain and holds the edge of the net close to the sea-bed. When fishing stony bottoms, various heavy ground gears are attached below this fishing line and can add considerably to both the visual contrast and the intensity of the noise stimulus of this zone of the net. The heaviest ground gears are made up of rubber discs and wheels threaded onto chains and wires, all chosen to be extremely tough and ease the relatively fragile advancing net over any stones and boulders that may be in its track.

Fig. 18.5 A polyvalent otter board is towed by the warp (bottom left) towards the camera. The trawl mouth (not visible) is to the left of the sand cloud some 50 m behind the otter board. Note how the sand cloud spreads in along the sweep, forming a wall of turbulent, opaque water all the way to the wing end of the net. Reproduced with permission from Main and Sangster (1981a). (Crown copyright).

18.3   FISH BEHAVIOUR IN TRAWLS

Observations of the reactions of fish to trawls have been made by unique diver-operated and remotely controlled vehicle techniques developed and used by scientists at the Marine Laboratory at Aberdeen since 1975. These observations are recorded using a TV camera and video tape and over this period a general pattern of reactions has been built up. A series of nine reports describe the techniques and many of the observations discussed in this chapter (Main and Sangster, 1978a to 1983b). The development of the techniques, and the main conclusions of the work, are outlined by Wardle (1983, 1987, 1989). The fish behaviour patterns observed in trawls and recorded during this period confirmed earlier observations of the Danish seine net made between 1965 and 1975 by scuba diving (Hemmings, 1973). The various parts of the trawl that will be discussed as stimulators of fish reaction are identified in Fig. 18.4.

The present account attempts to interpret the reactions of fish observed during the process of capture by a trawl by applying knowledge of fish behaviour. To do this, fish will be followed through a trawl and their behaviour analysed at each stage. The stages of capture are outlined in Fig. 18.6.

At the approach of a fishing boat towing a trawl, the first indication to the fish of an intrusion will inevitably be the sound of the engines and propeller of the boat. The hearing ability of fish has been discussed by Hawkins in Chapter 5 of this volume. Noise from the engines of motorized vessels towing at near full power will be heard by fish well beyond the range at which the vessel could be seen. In clear water when fish are no deeper than 40 m, a ship may be seen as a silhouette on the surface by fish beneath it. At greater distances the ship will certainly be heard but not seen. It is well known that vessel noise can cause fish to change their depth. Fish have been observed to move deeper, and even down to the sea-bed, as a vessel passes. Unnecessary noise is carefully avoided on purse-seining vessels for this reason, and the skipper knows that sudden noises made on or by the vessel will frighten those fish he is pursing and make them dive away from the ship and net.

In general, vertebrates use sounds for communication and listen for unfamiliar sounds that warn of dangers. Sound sources arouse curiosity or enhance sensitivity such that a visual explanation is expected. An animal is alerted by the sound, and the source is watched for and reacted to more effectively if it arrives. Although the fishing ship and gear are rarely seen by the fish, sounds from ships and fishing gears will often be heard due to the differences in range of these two senses underwater (Hawkins, 1973).

As the sounds of the towing ship begin to fade, a new sound from the otter boards will start to grow. This sound is generated by scraping and knocking on the sea-bed, and will vary according to the nature of the sea-bed. Observations of fish in the region of otter boards suggest that the visible range determines the distance at which a fish first reacts to the otter board. For example, in conditions of poor visibility, fish can be observed close to the board, only just avoiding collision, whereas in clear water, fish are seen reacting well away from the board, skirting the area at about the distance one would estimate they first see the board. Unfortunately it is not yet possible to observe at high enough resolution, by any sensing device independent of vision, the reactions of fish outside the visible range.

When the otter board is viewed from the sea-bed directly ahead in its track, the sand cloud which is illuminated by downwelling light from above is seen as a bright halo-like margin around the darker board (Fig. 18.7). The otter board is a good example of a high-contrast image, with added sound attracting attention to its approach. Four different reactions to the board, depending on the position of the fish in relation to the track of the board, are shown as A1

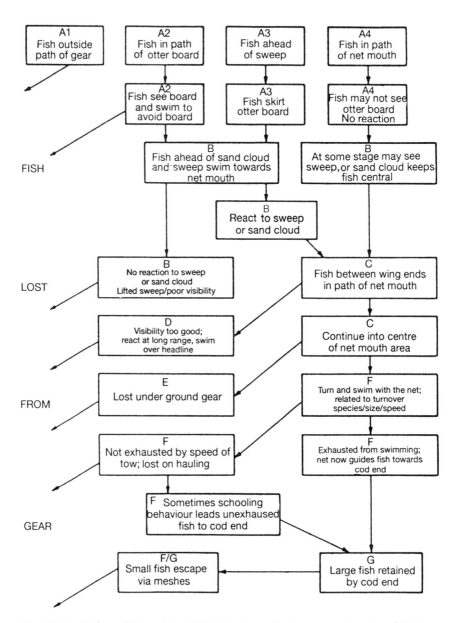

Fig. 18.6 Outline of the points of fish behaviour during capture in a trawl. Letters denote positions in Fig. 18.4. Arrows indicate alternative reactions of fish at each of the positions in the gear; they do not show the relative importance of these pathways.

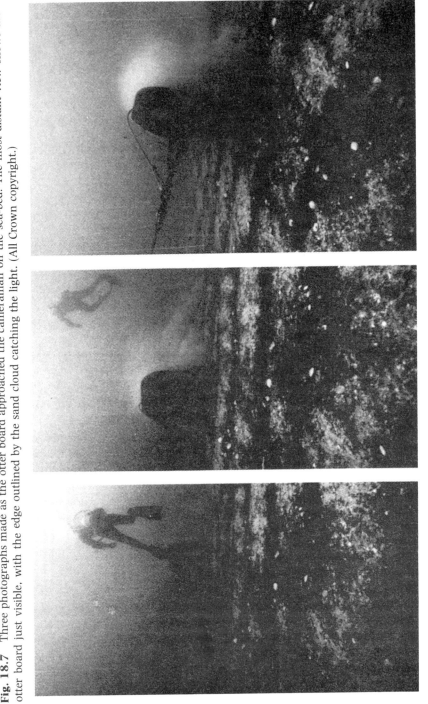

Fig. 18.7 Three photographs made as the otter board approached the cameraman on the sea-bed. The most distant view shows the otter board just visible, with the edge outlined by the sand cloud catching the light. (All Crown copyright.)

to A4 in Fig. 18.6. It is assumed from the observations that the reactions of fish, although affected by sound, are primarily in response to the presence of the visual stimulus. The reactions to the otter board will now be analysed on this basis.

Reactions to otter board

In the majority of roundfish species, each eye is able to see a field of vision of 170° to 180°. The two fields overlap in front of the fish giving binocular vision, and to the rear of the fish there is a blind zone of about 20–30° on either side of the fish's tail (Walls, 1942) (Fig. 18.8). A fish positioned on the sea-bed in the path of the otter board reacts to the board as soon as its presence is sensed visually. Observations have shown us that fish swim around the otter board at a distance just within the range of visibility. With a first limit set by the

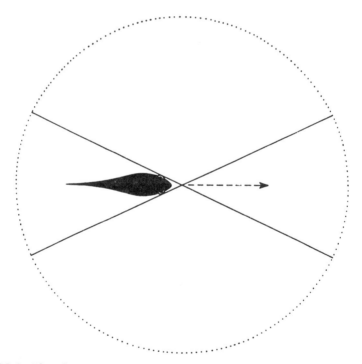

Fig. 18.8 Plan showing the two hemispherical fields of vision of a roundfish like a cod. The line forming a tangent to each eye delimits the outer edge of the field of view of each eye as the fish passes through the centre of the circle. The sector immediately behind the fish is the blind zone; the sector ahead of the fish is a zone of binocular vision. The dotted circle represents a limit to range of visibility. See text for discussion.

visual field of the fish, a second limit set by the visual range of the approaching
otter board and a third limit setting a swimming speed, the most direct escape
or avoidance route can be predicted as in Fig. 18.9(a). It is assumed that the
blind zone extends to 25° on either side of the fish's swimming track, and the
most direct escape route, while keeping the otter board just in the rear edge
of the field of view, is a course whereby the otter board (Fig. 18.9(a), horizontal

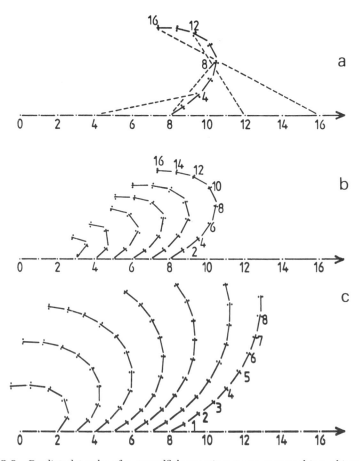

Fig. 18.9 Predicted tracks of a roundfish reacting to an approaching object such
as an otter board. The horizontal line is the track of the otter board, where numbers
indicate time in seconds. The curve of the fish track is set by keeping the otter board
at each position just at the edge of the visual field (dashed lines in (a)). The figure
shows (a) how the track is determined; (b) the swimming speed of the fish set at half
the approach speed of the otter board (fish tracks shown start from points 3 to 8
ahead of the otter board); (c) a similar set of tracks when swimming speed is equal
to the otter board approach speed. Numbers on the curves identify coincident positions
of fish and otter board.

line) is kept at 155° to the swimming direction. In the scale plan of the reaction of a fish to an otter board (Fig. 18.9(a)), the figures indicate time in seconds for coincident positions of the otter board and the fish. The fish in this example chooses a swimming speed which matches that of the otter board. Underwater visibility can vary greatly, and in Fig. 18.9(b) are shown the escape routes assuming first reaction of the fish at maximum visible ranges of between 2 and 8 m from the board. Figures 18.9(b) and (c) compare the resulting track of a fish swimming at half the speed of, and at a speed equal to that of, the approaching otter board, respectively.

This model can be extended to describe the reaction of a group of fish to the otter board. In the diagram (Fig. 18.10(a)), the otter board moves in a straight line from left to right and each fish shown in the figure is keeping one eye on the board while maintaining a swimming track close to an angle of 155° to the current position of the board. In order to maintain visual contact with the board, each fish decreases this angle as it approaches the maximum visible range (the circle drawn around the otter board in Fig. 18.10(a)). In this way the group of fish gathers together again automatically behind the board. The same behaviour, the 'fountain manoeuvre', is seen as a reaction to any similar moving object.

In an experiment designed to test this observation a group of 200 whiting, *Merlangius merlangus* L. (length 8–12 cm), reacted by swimming away from a black ball drawn slowly through the group. Each reacting fish maintained an angle, body to ball of 135° and a distance just within the maximum visibility range (Hall *et al.*, 1986). The result suggested that each fish in the group maintained the predator at a point near to the rear edge of the field of view of one eye during the evasion manoeuvre. It was concluded that the fountain manoeuvre is delimited by the visual field and visual range of the fish.

A human observer moving through a group of fish can observe this zigzag orientation of those fish seen directly ahead, and as they split off to the left and the right, one eye of each fish is always just visible (Fig. 4.7 in Wardle, 1983). The fish pass on either side and rejoin as a group behind. However, the sweep and the sandcloud following the otter board comes between the resulting two groups of fish (Fig. 18.10(b)). The spreading sand cloud (Fig. 18.5) takes over as the next stimulus, herding the inside fish towards the track of the mouth of the trawl and the outside fish to freedom.

The observed swimming speed used by the fish when skirting the otter board is slow and close to that of the approaching board: why is this? Studies of fish swimming have shown that fish are able and willing to cruise for long periods at slow speeds. Their endurance is virtually unlimited as long as the muscular contractions involve aerobic respiration, releasing energy using continuously replaceable metabolites and oxygen (Wardle, 1977; Bone, 1978). On the other hand, fish can reach high swimming speeds for limited periods by contracting the large white lateral muscles, which are fuelled by the rapid conversion of

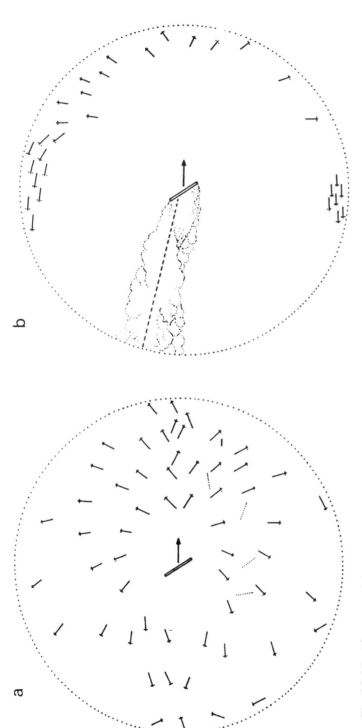

Fig. 18.10 Plan view of the predicted reaction of fish due to visual stimulus from (a) an isolated moving object like an otter board or human swimmer; and (b) an otter board with its trailing wall of sand cloud and sweep wire. The dotted circle indicates a maximum visible range of 8 m.

a limited muscle-glycogen store to lactic acid, for which no oxygen is required (Black *et al.*, 1961; Wardle, 1975). However, if the fish uses its high-speed swimming, it is obliged after only minutes to seek shelter and to rest for as long as 24 h while its glycogen store is rebuilt. During much of this period the fish is without adequate escape ability and vulnerable to attack (Batty and Wardle, 1979). Fish in general are observed to use the minimum swimming speed in order to maintain a safe distance between themselves and an identified threatening object. In this way they maximize their endurance for swimming and maintain their anaerobic reserves for real emergencies. This limit to the fishes' swimming response, together with the limit imposed by their visual field, seems to result in a characteristic but effective avoidance strategy when an otter board threatens to overrun them.

The dimensions of a typical trawl can be related to the visible range of objects under water, which can vary from zero to perhaps a maximum of 40 m (see Chapter 4 by Guthrie and Muntz, this volume). In Fig. 18.3 visibility ranges of 2, 4, 8 and 16 m are indicated by the dotted contour lines. Consider then a group of fish spread across the track of the gear, in a line some distance ahead of the otter board (Fig. 18.11). The first fish to react visually to the otter board will be in a position directly ahead of the board at a point where, on Fig. 18.11, the circumference of visible range (16 m) meets the band of fish. The model (Fig. 18.9) describing the escape reaction route is now applied and these fish are drawn in Fig. 18.11 moving away at $+155°$ or $-155°$. If each fish uses the minimum speed needed to maintain the otter board just within visible range, it should be expected to follow a course similar to that plotted in Fig. 18.9(a), but starting at a first visual reaction distance of 16 m. For the purpose of illustrating this point on Fig. 18.11, very clear water with a visible range of 16 m is chosen. However, by referring to Fig. 18.3 it is clear that with only 4 m or 8 m visible range, those fish situated further than 4 m or 8 m from the board as it passes will not have identified the visual stimulus of the board. They might to some extent react to the movements of neighbours that move in their direction, aided by the sound, but might otherwise show no reaction until later, when some other part of the gear approaches near enough for them to see it (Fig. 18.6).

Behaviour near trawl mouth

During the 20 s after the otter board has passed, those fish swimming gently towards the as yet invisible mouth of the net are guided in by the sweeps and sand cloud, but they are then quite suddenly surrounded by the visible array of netting, ropes, floats and bobbins of the trawl gear (C, D and E in Figs 18.4 and 18.6). In the 600 hp net (Fig. 18.3), with 8 m visibility both of the wings will be visible from the centre of the track of the net. The buoyed headline will be seen as a high-contrast silhouette as it passes overhead but not much before

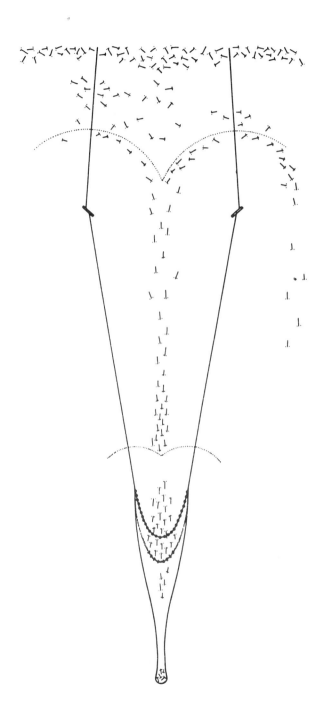

Fig. 18.11 Plan showing the reaction of fish predicted by visual reaction with a visible range of 16 cm (dotted lines). For discussion see text.

Fig. 18.12 Fish typically hold station with the trawl, swimming for long periods in the mouth area. The mackerel (top) were photographed at 100 m deep with a remote-controlled vehicle. Flash was used for the photograph; the TV camera could just see the fish by natural light at this depth. The saithe (bottom) were photographed by natural light by a diver. Photo reproduced with permission from Main and Sangster (1981b). (Both Crown copyright).

it does so. At about the same time the ground-gear bobbins, black against the white sand and outlined by the wisps of sand thrown up between them, form a further strong visual stimulus coming directly towards the fish across the sea-bed. The fish usually turn abruptly at this point in the mouth of the trawl and swim forwards, just matching the towing speed of the gear (Figs 18.11, 18.12). The visual field of those fish swimming forwards at this point contains monocular images of netting on either side, and chains and bobbins intruding into the rear view from both sides (Fig. 18.13). It is not surprising that the fish now try to hold, for as long as possible, a stable unchanging position relative to the fast-moving visual stimulus of the gear. This behaviour appears to be an optomotor response, and Hemmings (1973) demonstrated that haddock swim holding station with the mouth of a net even when the netting behind the mouth is completely removed. For further discussion of the optomotor reflex and its relation to rheotaxis and the behaviour of fish in trawls, see Harden Jones (1963), Arnold (1974) and Wardle (1983). (See also Chapter 12 by Pitcher and Parrish, this volume).

Direct observations have repeatedly shown that larger fish such as saithe, cod and haddock swim for very long periods in the mouth of trawls and Danish seine nets, whereas the smaller species such as sprats and sandeels try hard

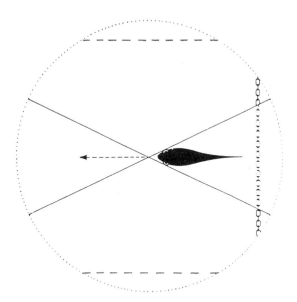

Fig. 18.13 Plan suggesting the contents of the visual fields of the left and right eyes of a roundfish swimming forwards at the same speed as the net inside the mouth of a trawl. Note that the chain in the blind zone (dotted segment of chain) is not seen by the fish. The net panels (dashed lines) on the opposite sides of the gear disappear outside the visible range (dotted circle).

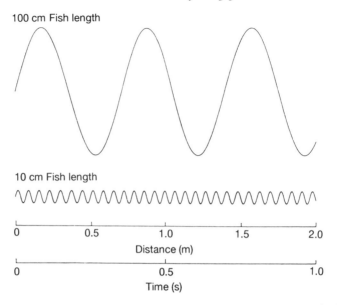

Fig. 18.14 Tracks of the tail tip of a 100 cm and a 10 cm fish. These tracks illustrate why a different frequency of tailbeat is required of 10 cm and 100 cm fish swimming at $2\,\mathrm{m\,s^{-1}}$ (4 knots). For discussion see text.

to maintain their position but after only a few minutes give up and allow the net to pass them. These observations illustrate very directly the scale effect in fish swimming performance (Wardle, 1977). At this position in the trawl mouth, all sizes of fish are being stimulated visually to swim at exactly the towing speed, say of $2\,\mathrm{m\,s^{-1}}$. It is well established that a typical teleost fish moves forwards 0.7 of its body length for each completed tailbeat cycle (Wardle and Videler, 1980). A cod 1 m long therefore needs 2.9 tailbeats per second to maintain its position in this situation, whereas a 10 cm cod or sprat is struggling to swim at the same speed by beating its tail at a frequency of 28.6 Hz (Fig. 18.14). The large fish is easily cruising aerobically and could continue for hours, whereas the small fish is at or near its maximum speed using its anaerobic muscle power output, and is soon exhausted. The effect of fish size on the tailbeat frequency needed to swim at 1 and 4 knots is shown in Fig. 18.15. The maximum tailbeat frequency of each length of fish at three temperatures is also shown in this figure. These maximum performance figures are calculated from measurements of the contraction time of the swimming muscle (after Wardle, 1975).

   Whatever the size of the fish, there is an enforced change in the behavioural response to the net if the fish becomes exhausted. This change in behaviour may be stimulated by the accumulation of lactic acid in the muscle tissue, or

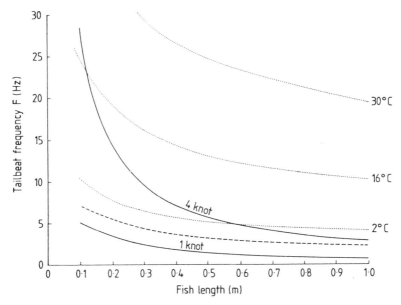

Fig. 18.15 Tailbeat frequency of fish of different sizes swimming at 0.5 and 2 m s^{-1} (1 and 4 knots, solid lines). Dotted lines show the maximum tailbeat frequencies of fish at the indicated temperatures. Dashed line shows the tailbeat frequencies at maximum aerobic cruising speed. See text for discussion.

the corresponding depletion of glycogen, or it may simply be the result of loss of muscle power. The effect is an enforced change of tactics where the behaviour of maintaining station with the gear gives way to turning or dropping back. The fish now takes an increased risk of entering an unknown area and dodging through the threatening array of visual patterns.

Looking again at the scale plan of the net (Fig. 18.3), it is seen that the distance from the mouth area to the cod end is some 30 m. With a visible range of 8 m, the fish in the mouth area will see, as it turns, what looks like a clear passage surrounded by a circle of netting, and the vanishing lines of the thicker seams of the netting where the panels are joined together (Fig. 18.16, top). In practice the larger fish as they become exhausted turn and swim, keeping clear of the net walls, and are guided down the funnel of the net. They are sometimes seen to delay their passage towards the cod end by turning and attempting to swim forwards, particularly where the net narrows or is obstructed, but these fish eventually end up exhausted and trapped in the cod end. Smaller fish swim only briefly, keeping station in the mouth area, and then turn back into the funnel. They are often seen to swim directly towards the meshes at various points along the net walls and particularly the top of the net funnel. Many of the smaller fish do reach the

Fig. 18.16 (Top) Saithe are seen swimming towards the cod end, which is not visible but is the darker zone (middle right) where the lines of the funnel of the net vanish. The impression is of a clear passage through the net owing to the underwater visibility. Some of the saithe appear to be feeding on the sea-bed. Photograph reproduced with permission from Main and Sangster (1983a). (Bottom) Mesh selection at the cod end. Small fish are escaping from among the larger trapped fish. (Both Crown copyright.)

cod end, where they also pass out through the meshes just ahead of the catch (Fig. 18.16, bottom). The scale effect described here results in a large turnover of small fish passing into, through and out of the trawl, a very small proportion of these fish being represented in the catch. At the other extreme, groups of large fish have been observed swimming in the net mouth for long periods but do not become exhausted and are not caught; they swim away when the net is hauled from the sea-bed.

Gear components as visual stimuli

Fish vision has been discussed by Guthrie and Muntz in Chapter 4 in this volume, and we can now consider parts of the fishing gear as visual stimuli. In general, fish are specialists at seeing low-contrast images, and it was suggested earlier that the role of sound might be to attract the fishes' attention to these images. Images formed by the approach of fishing gear will nearly always be of low contrast when they are first presented to the fish at maximum visible range, but some, like the otter boards and ground-gear bobbins, rapidly become of high contrast. The distance at which the image is first seen varies, and can lead to quite different reaction behaviour in different combinations of visible range and swimming performance of the fish.

Water depth has a large effect on the colour and contrast of materials and structures used to make up fishing gears, and so modifies the gear as a visible stimulus. For example, the bright orange twines often used to construct the panels of trawls look to the human eye a shade of grey-green when seen below 20 m depth. When viewed horizontally against the grey-green water background, even when close up in good visibility, such a panel has a very low contrast. Images of low contrast are detected at much shorter range than images of high contrast. Green netting becomes nearly as bright in this light as white, whereas red netting looks quite black when viewed at the same depths. These appearances of the various colours of materials can be used in appropriate ways to make the various parts of a trawl have well-defined functions as a stimulus to the fish. From these observations the net builder should aim to make the visible stimulus of the net less vulnerable to the variations that occur in the optical properties of the water background. The maximum visible range is achieved by putting highly reflective white material next to black. Observations of fish reactions indicate that there are areas of the fishing gear where application of high-contrast patterns will make the stimulus clearer to the fish. For example, the otter boards should be black with white reflective borders so that they are always seen at maximum range.

In good visibility, fish on the ground ahead of a trawl mouth have been observed to rise over the headline. To avoid fish reacting in this way, the range at which the headline is first seen can be dramatically reduced by the use of grey netting and countershaded floats. However, when this headline passes overhead, the floats and netting become silhouetted against the downwelling light and so change from least to most visible and help to keep the fish near the sea-bed. The net viewed from just inside the mouth region should have maximum contrast in order to define an area where the fish will swim forwards. Strong vertical black-and-white-striped patterns would create the maximum stimulus in all visibility conditions. Behind this region of collection and exhaustion, the netting should have the least visibility when the fish turns and looks back towards the cod end. Grey or orange netting used here would encourage fish to swim down this tunnel even in clear water conditions. A net having some of the proposed properties mentioned in this section is sketched in Fig. 18.17.

The colour of the sea-bed can vary and will alter the contrast, and so the appearance of the lower panels of a bottom trawl. A midwater or pelagic trawl is usually towed well clear of the sea-bed and out of visible range of the sea-bed. From inside the mouth of a pelagic trawl, the top netting panels are seen by the fish against the light from above, and the lower panels are seen against a black background. In order to create a surrounding pattern of maximum contrast when seen from inside the mouth, the top of the pelagic net must be made in black, the base in white, and the sides in black-and-white stripes. To make the rear parts invisible, so that exhausted fish are stimulated to turn towards the cod end, the top should be white, the base black and the sides grey.

Direct observations of nets show us that the angular size of images at different ranges has important effects. For example, when viewed with only one eye it is difficult to distinguish a mesh 7 m across made from 23 mm twine seen at a range of 40 m from a 1.7 m mesh of 6 mm rope seen at 10 m. Both mesh sizes give a $10°$ image at the eye. In many trawls the small mesh used in the main net panels contributes nothing to the distant view, and the image seen is the pattern of strengthening ropes that form the seams and framework of the net (for examples see Fig. 18.12, bottom, and Fig. 18.16, top). Large meshes made from thick ropes have been found extremely effective in

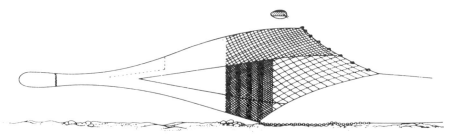

Fig. 18.17 Artist's impression of a trawl design suggested by the fish behaviour observations discussed in the text. An enlarged view of the countershaded float is shown above the headline.

midwater trawls of large size. These large meshes have important properties of creating panels of large area with low hydrodynamic drag, allowing the same ship to tow a bigger net. The image of the large mesh made from thick rope will only stimulate from maximum visible range when attention is paid to using the appropriate colour or shade in relation to the angle of viewing. To preserve the appropriate stimulus when new designs are considered, attention must be given to contrast and dimensions in relation to the variable water conditions. In this way the fish reactions may be controlled more precisely.

The stimulus to behaviour generated by the sense of vision is limited by the light level, and it is important to consider how the relation between the stimulus and the reaction may change as the light level falls.

Blaxter (1970) reviews the abilities of fish to see. He suggests that many fish, particularly those from the deep sea, might have potential abilities 10 to 100 times better than humans, when the large pupil sizes and high retinal pigment densities are considered. There are, however, very few studies demonstrating threshold light levels for fish behavioural responses, and our main frame of reference for considering the behaviour of fish is set by knowledge of owls, cats, humans and pigeons (Martin, 1983). Martin points out that the behavioural threshold is at extremely low light levels of 10^{-6} to 10^{-7} lux for cats, owls and men. The low-light TV camera, used to observe fish reactions to fishing gear, ceases to form images at levels below 10^{-3} lux and bioluminescence may then take over and be seen as flashes. The milky-green light of disturbed dinoflagellates can reveal moving net panels (Wardle, 1983). As light level falls, white objects become more and more like black and they are eventually not seen against a black background. All objects become less visible due to this reduction in contrast. Reaction distance is reduced as the contrast is reduced (Anthony, 1981). The loss of behavioural response to visual stimulus is related to the inability of the eye to pick up these differences in contrast. Of the few animals carefully studied, pigeons show a behaviour threshold at 10^{-4} lux, humans and owls at 10^{-6} lux, and cats at 10^{-7} lux (Martin, 1983, Figure 7).

A number of studies have shown light level thresholds for schooling in fish to be between 10^{-6} and 10^{-7} lux and the obligate schooling of mackerel ceases at light levels below 10^{-6} lux (Glass *et al.*, 1986). It is argued here that the main reactions of fish in the trawl are due to the stimulus of behaviour by visible images. Convincing proof of this argument would be the observation of no reaction when light was below the threshold for stimulating the behaviour of the fish. There are major problems in obtaining such observations; first, how to observe the fish, secondly how to know the light level when the observations are made, and thirdly to find conditions where there is no bioluminescence and low-enough ambient light levels. Flash photographs taken from a remotely controlled vehicle positioned inside the mouth of a trawl at midnight in February off Orkney, UK, have consistently shown examples of

Fig. 18.18 Are these photographs evidence of no reaction when fish are in darkness? (Three flash photos taken inside the mouth of a trawl towing at $1.5\,\mathrm{m\,s^{-1}}$ in February 1983.) The bobbin rig of the trawl ground gear is just on the right-hand edge of each photo, moving to the left. In the top photo, taken at 13.50 GMT, the sandeels are orientated and swimming ahead of the ground gear. In the middle photo, taken at 22.43 GMT, sandeels and other fish are pointing in all directions. In the lowest photo, at 00.05 GMT, three roundfish are pointing in three directions, as are the less visible sandeels. (All Crown copyright.)

fish showing no reaction to the large ground gear bobbin wheels (Fig. 18.18). Further illustrations and conclusions confirming these observations can be found in Glass and Wardle (1989). This sort of evidence tells us that none of the other senses attributed to the fish can be used to orientate or react to fishing gears within the short time period available at these towing speeds.

A shortening of the fish's reaction distance can be observed by filming them in fishing gears as light level drops or by careful control of light level in tank experiments (Cui *et al.*, 1991). In brightest daylight, in clear water with 40 m visible range, reaction distance may be 40 m, whereas at night, in the same water conditions, fish are seen to approach within 1 or 2 m of the gear before showing any signs that they are aware of its presence. Fishermen know that in daylight they do not catch fish in certain areas whereas at night at the same position they are successful. Filming at 100 m depth in daylight in such an area off Fair Isle, UK, showed large numbers of haddock swimming over the headline of a 600 hp trawl similar in scale to that shown in Fig. 18.3: none was caught during a 6 h tow. At night, however, large numbers were seen in the net mouth close to the bobbin region and a large catch was accumulated. Referring to Fig. 18.3, the effect of darkness was to reduce the visible range from 40 m to less than 2 m. The effect on a fish of the reaction distance changing can be considered by estimating the time from first seeing the object to collision with it (Fig. 18.19) and then noting the distance that must be swum in this time, i.e. the speed needed to pass around the object (Fig. 18.20). For example the headline of a net first seen at 40 m range and approaching at $2 \, \text{m s}^{-1}$ takes 20 s to collision. To avoid the 5 m-high headline the fish needs to swim up from the sea-bed at only $0.25 \, \text{m s}^{-1}$, but if the headline is first seen at 5 m, a swimming speed of $2 \, \text{m s}^{-1}$ will be needed: a speed close to the maximum speed of many smaller fish.

Observations of fish both in the wild and in large tanks indicate that many species develop home grounds, which are areas where the fish spends most of its time. Fish are extremely cautious when new strange objects enter their home ground. Cod, haddock and saithe can be trained to race between feeding lights through an area with which they are familiar. The same fish will not race into an area that has not been explored. Tank experiments in which these species were trained to race 8 m between feeding lights demonstrated a timidity of fish to pass a new object, such as a rope lain across the tank floor while the fish were feeding at one of the lights. When the other light was flashed, the fish started rapidly in its direction but swerved aside when they came up to the rope. Several minutes were spent patrolling before they cautiously crossed the rope and raced to the calling light and food. Replacing the rope with a large-mesh net through which these fish could easily swim caused longer delays. It is equally interesting that if the large-mesh net was left in position, the fish would race through it after a day or so without hesitation. The acceptance by the fish of the intruding object (that is, when timidity is

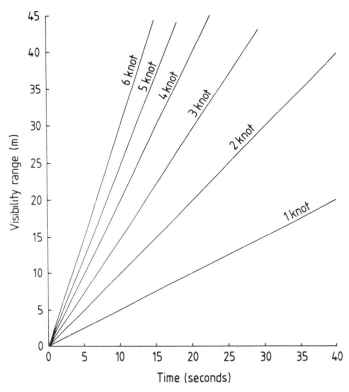

Fig. 18.19 Time to collision with an object first seen at visible range approaching at the speeds indicated on the lines. (One knot is approximately $0.52\,\mathrm{m\,s^{-1}}$.)

lost) might be considered a process of habituation. These experiments indicate a relatively long period of timidity stimulated by quite simple objects intruding into the fish's home ground, and have some significance in the context of a fast-moving gear where required reaction times are very short. Moving away from, or holding station with, the approaching fishing gear indicates a timidity to the approaching object and reluctance to allow the device to pass, and this leads to the device herding the fish. Observations have repeatedly shown that a rope towed across the sea-bed will herd fish ahead of it. If the rope is angled to the forward motion, the fish slide along the rope and will be concentrated at one end. This action leads fish along the trawl sweep or the Danish seine net rope into the mouth of the net from a wide-swept area of sea-bed.

Those fish that school tend to behave *en masse* when responding to such stimuli (see Chapter 12 by Pitcher and Parrish, this volume). The erratic decision-making of individuals responding to stimuli slightly differently from their neighbours is smoothed out by the general reaction of the school. The

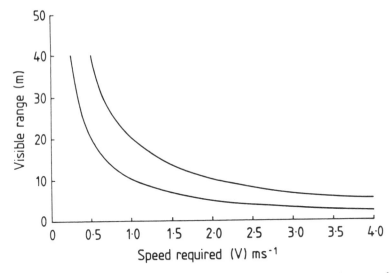

Fig. 18.20 Swimming speed required to clear the headline of a trawl approaching at $2\,\mathrm{m\,s}^{-1}$ (4 knots). The headline heights drawn are $5\,\mathrm{m}$ (lower curve) and $10\,\mathrm{m}$ (upper curve). The fish starts to swim upwards from the sea-bed at the different visible ranges indicated.

erratic individual returns to the school after a brief lonely excursion. Schooling sandeels have been observed to be herded by rope arrays, and as they exhaust trying to swim ahead of the ropes, the whole group flows back, keeping clear of the ropes. Sandeels, sprats, mackerel and saithe have all shown this mass decision when in the mouth of trawls. A large group of saithe swimming quite easily in the mouth of a trawl and predictably able to outswim the net owing to their larger size all turned and swam from the mouth to the cod end, apparently owing to this *en masse* decision generated by their schooling behaviour.

Mackerel (length 33 cm) can swim continuously at speeds between 0.4 and 3.5 body lengths s^{-1} (i.e. up to $1.15\,\mathrm{m\,s}^{-1}$ or 2.2 knots). If they are forced to swim at 5 body lengths s^{-1} (3.2 knots) they can swim for only 10 minutes and must drop back (Fig. 2 in He and Wardle, 1988). Such basic findings of the performance abilities help explain why for example when observed in trawls they are seen to be caught because they become exhausted swimming in the mouth area of the net (Fig. 18.12). At speeds slower than 2.2 knots, mackerel explore all areas of the net in small schools. They have been seen swimming forwards steadily from well back in the funnel of the net, overtaking the net and leaving through the mouth. They have also been observed feeding on sandeels while swimming in the trawl mouth.

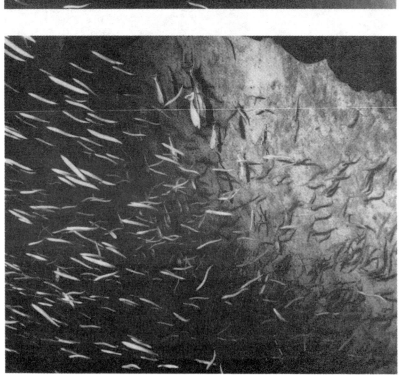

Fig. 18.21 The sandeels separate completely from the other species still swimming ahead of the ground gear in the mouth of a trawl. (Both Crown copyright.)

18.4 CONCLUSIONS

The more information we can gather of this sort, the more we can think in terms of making use of these patterns of behaviour to make a gear more precise in its action. This does not necessarily mean more efficient in a crude sense of simply catching more fish. Although a sequence of general reaction patterns has been described here, we are beginning to notice that each species and each size group of fish has some distinct and specific reactions when passing through a trawl. Sandeels are often seen in trawls, and because of their short endurance, small size and schooling behaviour are often seen to separate from other species as they respond to the fishing gear (Fig. 18.21). Such separation of species can be exploited. It has, for example, become clear that haddock and cod with their different behaviour can be separated (Main and Sangster, 1982b). When haddock become exhausted in the mouth of the net, they drop back into the net, rising high over the ground line into the top part of the net mouth (Fig. 18.22), whereas cod remain low near the sea-bed. A horizontal panel of netting, with its leading edge dividing the net mouth, can guide fish to different cod ends, and such a design has been found quite practical in separating cod and haddock. Each of the cod ends can have different mesh sizes to select a different size range. *Nephrops norwegicus*, the scampi or Norwegian prawn, stays low in a trawl and can be separated from most fish, except cod and flatfish, by a similar approach (Main and Sangster, 1982a). At present *Nephrops* trawls in Scotland are allowed to have meshes as small as 70 mm and these nets sometimes catch a great number of young roundfish. The *Nephrops* never rise more than 70 cm from the sea-bed, whereas the small roundfish will tend to rise as they tire and pass into the upper zone of a separator net. The upper cod end can have a large mesh, only retaining the marketable sizes of these fish.

The possibilities for increasing the precision of fishing grow as we begin to see these patterns of behaviour, as we develop explanations and as we apply appropriate stimuli in new designs of fishing gear.

18.5 SUMMARY

The fishing methods devised by humans to hunt the world's stocks of fish involve many aspects of the behaviour of the different species caught. This chapter considers modern trawls and the sequence of stimuli they present to fish. At the start, long-range sound is followed by visual stimuli at a range dependent upon water conditions. Direct observation of fish in the trawl has shown a sequence of typical responses as fish progress towards the cod end. First, an avoidance reaction observed at the otter board may be modelled using visual distance, swimming performance and limits of the visual field,

Fig. 18.22 (Top) The haddock (black spots, middle, top, left) are just starting to rise as they tire and stop swimming in the mouth of a trawl. Cod, flatfish and saithe are seen nearer the sea-bed. (Bottom) Haddock and sandeels rise up above the level of a separating panel away from cod swimming close to the sea-bed. When the cod tire, they fall back beneath the separating panel. Reproduced with permission from Main and Sangster (1983a) (top) and Main and Sangster (1982a) (bottom). (Both Crown copyright.)

explaining how fish are funnelled towards the mouth of the trawl. Next, the optomotor reflex, timidness, swimming performance and size explain how fish that reach the mouth of the trawl turn forwards within the net. Finally, following a period of swimming at the mouth, exhaustion leads to a change in behaviour, and the fish allow the net to overtake them, turn and seek a clear visual path back through the centre of the long funnel that leads to the cod end. Although sound is important in drawing attention to the approaching gear, observations suggest that vision is the predominant sense driving precise, high-speed reactions at close range within the trawl. Selecting net construction materials to form appropriate visual stimuli can make the performance of gear more consistent in variable water conditions. Furthermore, trawls that can separate their catch into species have been designed to exploit detailed behavioural differences in the responses of some species to components of the fishing gear.

REFERENCES

Anthony, P. D. (1981) Visual contrast thresholds in the cod *Gadus morhua* L. *J. Fish Biol.*, **19**, 87–103.

Arnold, G. P. (1974) Rheotropism in fishes. *Biol. Rev.*, **49**, 515–76.

Batty, R. S. and Wardle, C. S. (1979) Restoration of glycogen from lactic acid in the anaerobic swimming muscle of plaice *Pleuronectes platessa* L. *J. Fish Biol.*, **15**, 509–19.

Black, E. C., Robertson, A. C. and Parker, R. R. (1961) Some aspects of carbohydrate metabolism in fish, in *Comparative Physiology of Carbohydrate Metabolism in Heterothermic Animals* (ed. A. W. Martin), University of Washington Press, Seattle, pp. 89–122.

Blaxter, J. H. S. (1970) Light, animals, fishes. *Mar. Ecol.*, **1**(1), 213–321.

Bone, Q. (1978) Locomotor muscle, in *Fish Physiology*, Vol. VII, *Locomotion*, (eds W. S. Hoar and D. J. Randall), Academic Press, New York, pp. 361–424.

Cui, G., Wardle, C. S., Glass, C. W., Johnstone, A. D. F. and Mojsiewicz, W. R. (1991) Light level thresholds for visual reaction of mackerel, *Scomber scombrus* L., to coloured monofilament nylon gillnet materials. *Fish. Res.*, **10**, 255–63.

Dalhousie Commission (1885) Trawl net and beam trawl fishing. *Report of the commissioners appointed to inquire and report on the complaints that have been made by line and drift net fishermen of injuries sustained by them in their calling owing to the use of the trawl net and beam trawl, in territorial waters of the United Kingdom.* Eyre and Spottiswoode, London, C. 4324, 517 pp.

Davis, F. M. (1958) An account of the fishing gear of England and Wales. *MAFF Fish. Invest.*, Ser. II, **21** (8), 1–165.

Dickson, W. (1989a) Cod gillnet effectiveness related to local abundance, availability and fish movement. *Fish. Res.*, **7**, 127–48.

Dickson, W. (1989b) Cod gillnet simulation model. *Fish. Res.*, **7**, 149–74.

Glass, C. W. and Wardle, C. S. (1989) Comparison of the reactions of fish to a trawl gear, at high and low light intensities. *Fish. Res.*, **7**, 249–66.

Glass, C. W., Wardle, C. S. and Mojsiewicz, W. R. (1986) A light intensity threshold for schooling in the Atlantic mackerel, *Scomber scombrus* L. *J. Fish Biol.*, **29** (Supp. A), 71–81.

Hall, S. J., Wardle, C. S. and MacLennan, D. N. (1986) Predator evasion in a fish school: test of a model for the fountain effect, *Mar. Biol.*, **91**, 143–8.

Harden Jones, F. R. (1963) The reaction of fish to moving backgrounds. *J. exp. Biol.*, **40**, 437–46.

Hawkins, A. D. (1973) The sensitivity of fish to sounds. *Oceanogr. Mar. Biol. Ann. Rev.*, **11**, 291–340.

Hawkins, A. D., Urquhart, G. G. and Shearer, W. M. (1979) The coastal movements of returning Atlantic salmon, *Salmo salar* L. *Scott. Fish. Res. Rep.*, **15**, 1–14.

He, P. and Wardle, C. S. (1988) Endurance at intermediate swimming speeds of Atlantic mackerel, *Scomber scombrus* L., herring, *Clupea harengus* L., and saithe, *Pollachius virens* L. *J. Fish Biol.*, **33**, 255–66.

Hemmings, C. C. (1973) Direct observation of the behaviour of fish in relation to fishing gear. *Helgolander wiss. Meeresunters.*, **24**, 348–60.

Hunter, J. R. and Mitchell, C. T. (1968) Field experiments on the attraction of pelagic fish to floating objects. *J. Cons. perm. int. Explor. Mer.*, **31**, 427–34.

Hurum, H. J. (1977) *A History of the Fish Hook*, Adam and Charles Black, London, 148 pp.

Johnstone, A. D. F. and Hawkins, A. D. (1981) A method for testing the effectiveness of different fishing baits in the sea. *Scott. Fish. Info. Pamphlet*, **3**, 1–7.

Le Gall, J. (1931) Les principales pêches maritimes de la France: les filets et engins qui y sont employés. *Pêche Maritime*, **666**, 181–200.

Mackie, A. M., Adron, J. W. and Grant, P. T. (1980) Chemical nature of feeding stimulants for the juvenile Dover sole, *Solea solea* (L.). *J. Fish Biol.*, **16**, 701–8.

Main, J. and Sangster, G. I. (1978a) The value of direct observation techniques by divers in fishing gear research. *Scott. Fish. Res. Rep.*, **12**, 1–15.

Main, J. and Sangster, G. I. (1978b) A new method of observing fishing gear using a towed wet submersible. *Progr. Underwat. Sci.*, **3**, 259–67.

Main, J. and Sangster, G. I. (1979) A study of bottom trawling gear on both sand and hard ground. *Scott. Fish. Res. Rep.*, **14**, 1–15.

Main, J. and Sangster, G. I. (1981a) A study on the sand clouds produced by trawl boards and their possible effect on fish capture. *Scott. Fish. Res. Rep.*, **20**, 1–20.

Main, J. and Sangster, G. I. (1981b) A study on the fish capture process in a bottom trawl by direct observations from a towed underwater vehicle. *Scott. Fish. Res. Rep.*, **23**, 1–24.

Main, J. and Sangster, G. I. (1982a) A study of separating fish from *Nephrops norwegicus* L. in a bottom trawl. *Scott. Fish. Res. Rep.*, **24**, 1–9.

Main, J. and Sangster, G. I. (1982b) A study of a multi-level bottom trawl for species separation using direct observation techniques. *Scott. Fish. Res. Rep.*, **26**, 1–17.

Main, J. and Sangster, G. I. (1983a) Fish reactions to trawl gear – a study comparing light and heavy ground gear. *Scott. Fish. Res. Rep.*, **27**, 1–17.

Main, J. and Sangster, G. I. (1983b) TUV II – a towed wet submersible for use in fishing gear research. *Scott. Fish. Res. Rep.*, **29**, 1–19.

Martin, G. R. (1983) Schematic eye models in vertebrates. *Prog. sens. Physiol.*, **4**, 43–81.

Preston, G. (1982) The Fijian experience in the utilisation of fish aggregation devices, in *14th reg. tech. Meeting Fish. S. Pacific Comm.*, pp 1–61.

Rodriguez-Roda, J. (1964) Biologia del atun, *Thunnus thynnus* (L.), de la costa sudatlantica de Espana. *Invest. Pesquera*, **25**, 33–146.

Sharp, G. D. (1978) Behavioural and physiological properties of tunas and their effects on vulnerability to fishing gear, in *The Physiological Ecology of Tunas* (eds G. D. Sharp and A. E. Dizon), Academic Press, New York, pp. 397–449.

Thomson, D. (1969) *The Seine Net: its Origin, Evolution and Use*, Fishing News (Books) Ltd, London, 192 pp.

Walls, G. L. (1942) *The Vertebrate Eye and its Adaptive Radiation*, Bull. Cranbrook Inst. Sci., **19**, 375–376.

Wardle, C. S. (1975) Limit of fish swimming speed. *Nature, Lond.*, **225**, 725–7.

Wardle, C. S. (1977) Effects of size on swimming speeds of fish, in *Scale Effects in Animal Locomotion* (ed. T. J. Pedley), Academic Press, New York, pp. 299–313.

Wardle, C. S. (1983) Fish reactions to towed fishing gears, in *Experimental Biology at Sea* (eds A. Macdonald and I. G. Priede), Academic Press, New York, pp. 167–95.

Wardle, C. S. (1987) Investigating the behaviour of fish during capture, in *Development in Fisheries Research in Scotland* (eds R. S. Bailey and B. B. Parrish), Fishing News Books Ltd, Farnham, UK, pp. 139–55.

Wardle, C. S. (1989) Understanding fish behaviour can lead to more selective fishing gears, in *Proc. World Symp. Fishing Gear Fishing Vessel Design* (eds S. G. Fox and J. Huntington), Newfoundland and Labrador Institute of Fisheries and Marine Technology, St. John's, Newfoundland, Canada, pp. 12–18.

Wardle, C. S. and Videler, J. J. (1980) Fish Swimming, in *Aspects of Animal Movement* (eds H. Y. Elder and E. R. Trueman), Cambridge University Press, Cambridge, pp. 125–50.

Wemys Fulton, T. (1902) North Sea investigations, in *20th Ann. Rep. Fishery Bd Scotland*, Part III, HMSO, London, pp. 73–227.

Chapter nineteen

Fish behaviour and the management of freshwater fisheries

K. O'Hara

19.1 INTRODUCTION

The management of freshwater fish for human benefit is undertaken on differing levels throughout the world. In the industrialized countries fresh-water fish are often principally exploited for recreational purposes whilst in developing countries food production is paramount. In this latter case the detail of what constitutes a fishery, and what are the requirements of the user (Larkin, 1980), may be secondary to the more formal management objectives of assessing the fish yield of a water body. Nevertheless, all managers must be aware of the behaviour of their target species, whether this relates to, for example, how vulnerable the fish are to angling, or the success of an artificial reef in acting as a fish attractor.

The investigation of animal behaviour, particularly from a theoretical viewpoint, has recently been one of the principal areas of biological research interest and the subject has a wider connotation than the traditional ethological approach. A thorough review of current aspects of the behavioural ecology of teleost fishes is provided by Wootton (1990). Such developments may not have been wholeheartedly embraced by all fisheries managers, but whether such information is a necessity to the manager could be a similar debating point to that of the relevance of the more theoretical aspects of ecology to fisheries ecologists (Kerr, 1980; Werner, 1980). It would certainly be true to say that the application of some behavioural theories, such as optimal foraging, has not been warmly received by all fisheries biologists (see Regier *et al.*, 1979, and discussions in Stroud and Clepper, 1979).

Behaviour of Teleost Fishes 2nd edn. Edited by Tony J. Pitcher. Published in 1993 by Chapman & Hall. ISBN   0 412 42930 6 (HB) and 0 412 42940 3 (PB).

Although fisheries management must rest ultimately on the application of ecological principles, social and financial factors are major considerations (Royce, 1987). The marrying of these often conflicting aspects may prove to be difficult, particularly where economically important species are concerned. Harris (1978), for example, notes the lack of any rigorous attempt to assess the numerous stockings of Atlantic salmon, (*Salmo salar*), eggs and juveniles in the rivers of England and Wales. By comparison, it is relatively straightforward to evaluate the success or failure of supplemental stocking using the 'put and take' approach, where fish of a desired catch size are added to a water body and are caught and removed quickly by anglers so that the natural constraints of population regulation are bypassed. Although management objectives that set and meet a target of satisfying the demands of the recreational fishery consumer in terms of catch rate may totally disregard scientifically derived information on fish behaviour, such approaches may nevertheless be successful. However, in those areas such as the improvement of degraded habitats, and supplemental stocking of a species where natural recruitment is occurring, there is a necessity for careful consideration of population biology, and fish behaviour therefore becomes important.

There are no universal criteria for management practices, and cultural differences mean that some of the widely disseminated fisheries literature from the North American sport fishery field, which centres on piscivorous fishes, may not be directly applicable to some European recreational fisheries. Taking just two examples to illustrate this point, the pike (*Esox lucius*) and the carp (*Cyprinus carpio*) are often managed completely differently. In Britain pike may be subject to vigorous, if often unsuccessful, attempts to reduce their density in both salmonid and non-salmonid waters because they are considered by some to decrease the abundance of more preferred species. On the other hand pike are actively and widely stocked in North America. The reverse situation pertains to carp, an introduced species which is much vilified in North America and subject to active removal, whilst in much of Europe the fish is widely eaten, and is also an important recreational species stocked in many countries. However, if such differences are borne in mind, the transposition of knowledge can usefully be made in attempts to enhance or reduce fish numbers in the manner required by management.

This account will centre mainly on recreational fisheries and no major attempt will be made to discuss aquaculture in relation to food production, although there is equally a need for an appreciation of behaviour (Bardach *et al.*, 1980). Freshwater life history stages of diadromous fishes will be considered since they are of major importance, particularly anadromous species, and because of the considerable management effort expended on stock enhancement. Given the broad scope of the subject of fisheries management, some selection of the information available has been a necessity. Inevitably there are differences in depth of coverage, but four subject areas – general aspects, movements and migrations, habitat and direct stock manipulation – have been recognized in which fish behaviour can be directly related to management.

19.2   TECHNIQUES

Studies of fish behaviour can be accomplished both in the field and in the laboratory. Since the approach taken in this chapter is essentially related to applied aspects and attempts to relate behaviour to the natural environment, a field-based emphasis is provided. The assessment of behaviour in the field rested until comparatively recently on the extrapolation of results using direct capture methods and the aquatic biologist did not have the facility to observe the animal with the ease of his terrestrial counterpart. Field techniques for investigating fish behaviour have been given an effective coverage by Helfman (1983) and Winter (1983). Developments in scuba diving, telemetry and improved sonar have opened whole new vistas for the manager in assessing habitat usage by fish. This of course does not mean that there will not be a continued need for more traditional approaches, and the careful study of diet from stomach samples can still give information on interactions that may be undetected by other methods, particularly where laboratory studies would prove difficult (see Nilsson, 1978, for examples). Two papers, both concerning brown trout, (*Salmo trutta*), provide pleasing contrasts of approach. Tytler and Holliday (1984) used a sophisticated sonic tracking method to study trout behaviour in a lake, whereas Bachman (1984) directly observed stream-dwelling trout from a tower, identifying individual fish from their markings. The new techniques have removed some of the constraints that the very nature of the aquatic environment imposes on progress, but they have not substituted the need for careful observation.

19.3   BEHAVIOUR AND GENERAL ASPECTS OF FISHERIES MANAGEMENT

Estimation of various stock parameters including age, growth and population size are prerequisites to successful management. Allied fisheries legislation may or may not be designed specifically to protect stocks (Everhart and Youngs, 1981).

Age and growth determination

The study of growth in fishes has exerted a dominant role in investigations. This is because of the great variation in size found between different populations of the same species and the associated flexibility of growth patterns (Weatherley and Gill, 1987).

Age determination mostly relies on interpreting changes in the hard structures of fish and depends in temperate climates on abiotic factors particularly temperature. There are many ageing studies that require some understanding of fish behaviour for a correct interpretation of growth patterns. To take the most notable example, the interpretation of Atlantic salmon scales assumes a knowledge of the times of migration and period of sea life (Mills, 1989). Similarly the descent of brown trout into lakes from nursery streams is an event that leaves a record on the scales because of increased growth. The importance of accurately ageing fish is a prelude to many management manipulations on stocks, particularly exploitation rates. In tropical climates where seasonal temperature variations are too limited to be reflected in growth patterns, various behavioural events can leave a regular pattern which is used for interpretative purposes (Bagenal and Tesch, 1978).

Population estimation

Accurate assessment of animal population size presents aquatic biologists with difficulties and because of their mobility, problems in making quantitative estimates are often magnified for fishes. All capture of fish by Man assumes some knowledge of their behaviour (see Chapter 18 by Wardle, this volume). An excellent example is the longstanding investigation of perch, (*Perca fluviatilis*) population numbers in Lake Windermere, England, which takes place annually at spawning time when the fish congregate in the lake shallows and are vulnerable to fish traps (Le Cren *et al.*, 1977).

Absolute population estimates, both capture–recapture and catch–effort methods, incorporate assumptions about the behaviour of the animal (Seber, 1973). Managers are aware that all fish are not equally available for capture, particularly with respect to size. This may be due to gear selectivity or it may be caused by differential habitat selection, for example larger fish are known often to inhabit deeper areas of rivers and may prove difficult to capture by electrofishing. To overcome the problem of such behavioural differences, fish may be segregated into size groups. The division of fish populations into mobile and static components (Hunt and Jones, 1974b) has obvious implications for population estimation since the two groups will have different probabilities of recapture (Roff, 1973). A thorough and critical account of a population estimate that was considered inaccurate due to a number of failures to fulfil the assumptions of the models, partially as a result of the behaviour of the fish, is provided by Hunt and Jones (1974a).

Catch–effort methods can be equally prone to error because of failure to meet the underlying assumptions. Stott and Russell (1979) describe an estimate of a grass carp, *Ctenopharyngodon idella*, population that proved to be wrong since the fish were less vulnerable to capture on the second fishing because of modified behaviour after their first exposure to electrofishing and netting. Peterson and Cederholm (1984) compared a removals technique with mark recapture for juvenile coho salmon, *Oncorhynchus kisutch*. They found that the removals technique sometimes provided unreliable estimates and this was partly due to behavioural conditioning to electrofishing. Similarly, behavioural factors were cited by Zalewski (1983) as being responsible for innaccuracy in catch–effort estimates on small riverine fish.

It is important to know if fish are shoaling, solitary or territorial in their behaviour, since if small sampling sections are used it is quite possible for the complete home range of a shoaling species not to be encompassed (Hart and Pitcher 1973). Problems of sampling are compounded by the very small volume occupied by shoals in relation to the total available space (Pitcher, 1980).

Legislation

Fishing regulations are usually designed to protect desired species and are not necessarily based on fish behaviour, although the timing of particular events in fish life history stages will influence activities (such as harvesting anadromous fishes) which are directly dependent on a detailed knowledge of migration times (Mundy, 1983). Interestingly, many non-game fish such as carp are afforded no protection by legislation in North America. In fact the opposite situation pertains and exploitation is encouraged. Game fishes, on the other hand, have close seasons and other restrictive practices imposed for their conservation (Borgeson, 1979).

In England and Wales, game and non-game fisheries have close seasons resulting from statute in the *Salmon and Freshwater Fisheries Act* of 1975. These close seasons are designed to give protection to adults whilst spawning, but implicit in the legislation is the fact that fish may be more vulnerable to capture as they aggregate for spawning. Behavioural alterations as a result of handling can cause stress; problems stemming from such treatment are well documented (Schreck, 1981). Even when fish are returned after capture, handling stress (in addition to that imposed by reproduction) is likely to be deleterious, and protective measures are therefore probably sensible in situations where a large proportion of the stock are regularly caught.

19.4 MOVEMENTS AND MIGRATIONS

Precise definitions of movement and migration do not concern us here and for a full assessment of this subject reference should be made to Baker (1978). Aspects of migration are reviewed by Gauthereaux (1980), McKeown (1984),

McLeave *et al.* (1984) and McDowell (1988). A thorough treatment of migration, including the various patterns shown by freshwater fishes with specific relation to production is provided by Northcote (1978). The relevance of an understanding of migratory behaviour to successful management cannot be overemphasized particularly for the important anadromous salmonids (Mundy, 1983).

The migratory homing of salmon has been described in detail by Hasler and Scholz (1983), upstream migration was extensively reviewed by Banks (1969) and salmonid migration by Brannon and Salo (1982). The fact that salmon can be imprinted to return to a home river is being utilized in salmon ranching (Thorpe, 1980).

Migratory behaviour of sea trout, (*Salmo trutta*), is inadequately understood in comparison with the Atlantic salmon (Harris, 1978). It has always been assumed that sea trout do not show the same degree of accuracy in homing as Atlantic salmon (Mills, 1971), and the results of Pratten and Shearer (1983a) would tend to support this. However, Sambrook (1983) reported extremely precise returns of sea trout, and for management purposes the question remains open. The difficulties of enhancing sea trout stocks are further compounded by the lack of understanding of the relationship of the migratory form (the sea trout) with the non-migratory form (the brown trout). In particular the question as to whether the migratory habit is under environmental or genetic control (Pratten and Shearer, 1983b) remains unresolved.

Upstream movement itself appears to be initiated by a complex of factors (Banks, 1969), of which river flow appears to be of particular importance. River regulation schemes have often therefore been designed to include artificial freshets at intervals to induce salmon movement.

Catadromous fishes have not attracted the research attention devoted to anadromous species, but in Europe and Japan eels are an important food resource. Both the juvenile (elver) and adult stages are subjected to capture by Man, the former principally for stocking, and much of the elver catch of the River Severn, England, is exported to mainland Europe. An understanding of both upstream and downstream migrations is an important prerequisite to the effective management of stocks. Tesch (1977) provides an excellent account of eel biology, including good descriptions of migration. Increased interest in the eel as a resoure is stimulating study, notably species previously unfished such as *Anguilla australis australis* in Tasmania (Sloane, 1984).

Migrations are also undertaken by many primary freshwater fishes, both in rivers (Hynes, 1970) and in lakes (Northcote, 1978). However, these freshwater behaviour patterns are often ignored during river modification schemes such as dam construction, particularly for non-sport fishes. Where the species are of apparently lesser importance such as detritivorous fishes, then little consideration may be given, although the ecological impact of destroying traditional migratory routes may be equally serious (Bowen, 1983).

The movement and dispersal of fishes within lakes and rivers affect population features such as abundance and have aroused considerable interest with respect to the behaviour of new or supplemental stocked fish. Hunt and Jones (1974b) attributed the spread of barbel, (*Barbus barbus*), within the River Severn, England, to the mobile component of the population. Stevens and Miller (1983) suggested that the dispersal of four species of riverine fishes was influenced by the water flow in a river system, and good recruitment years were correlated with high flows because increased dispersal reduced density-dependent mortality.

Movement of stocked fish is often a problem for management since it represents a wastage of resources. Introduced catchable-sized brown trout apparently move less than rainbow trout, *Oncorhynchus mykiss*, or brook trout, *Salvelinus fontinalis* (Cresswell, 1981). Moring (1982) suggests that hatchery-reared rainbow trout that had an anadromous component in the stock history were more likely to move than two other strains and gave poorer angler returns. An interesting finding from long-term studies on Atlantic salmon in Ireland was that hatchery fish introduced as smolts, in addition to having lower survival to adult than wild fish also gave poorer angling returns. One reason suggested is that the fish could not migrate to their homing point and consequently were 'unsettled', spending considerable time in searching behaviour (Mills and Piggins 1983).

Some aspects of behavioural response, including movement, of prey to predators and the management implications have been considered by Stein (1979).

The development of telemetric tracking techniques has proved to be of great benefit in examining the migration patterns of anadromous salmonids, both adult and juvenile. Movement studies on non-diadromous fish are also attracting considerable interest through the use of radio and sonic tagging studies, some of which are of great value to managers. Hockin *et al.* (1988) demonstrated that radio-tracked grass carp, *Ctenopharyngodon idella*, (Fig. 19.1) in a linear water system showed limited movements and fed within small areas (Fig. 19.2); such results have obvious implications for the use of this fish as a weed-control agent.

These tags have been developed in order to provide measurement of abiotic variables in addition to locating the position of the fish (Haynes and Gerber, 1989).

19.5 MANAGEMENT OF FISH HABITAT

Any animal requires a place in which to live, obtain food and reproduce; considerable efforts are expended by fisheries managers in meeting these needs (Milner *et al.*, 1985). The provision of living places in newly created aquatic

Fig. 19.1   Radio tracking grass carp on a canal.

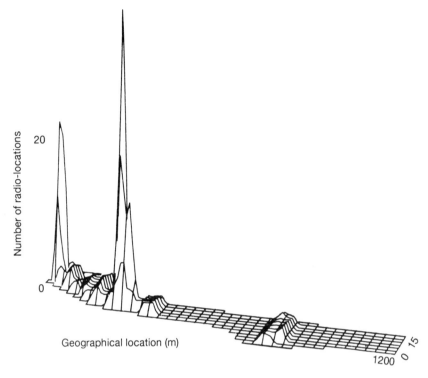

Number of radio-locations

20

0

Geographical location (m)

1200

15

0

Fig. 19.2 Movements of five grass carp recorded by radio tracking in a 1.2 km section of canal. Reproduced with permission from Hockin *et al.* (1988).

habitats, attempts to improve existing habitats and the restoration of degraded environments are subjects that have produced a large literature. Habitat improvement also encompasses practices such as the provision of fish passes, creation of spawning areas and the amelioration of pollution and water quality problems. Physical improvement structures placed in lotic (Swales and O'Hara 1980) and lentic habitats (Prince and Maughan 1978) are popular measures, possibly not only because they are a directly observable attempt at management but also because beneficial effects have been shown to accrue on populations (Hunt, 1969).

Habitat improvement – rivers and streams

The fact that river engineering projects associated with land drainage, flood alleviation and navigation are often detrimental to fishes is without dispute (Swales, 1982a; Brookes *et al.*, 1983). Simplification and removal of habitat heterogeneity effectively reduces the places in which fish can live, the result often being a fall in species diversity, species shifts and altered abundance (Swales 1982a).

Habitat improvements are used in some developing countries as a strategy to increase fish stocks (Cooper, 1980), but it is the efforts used to increase numbers in sport fisheries that have been most extensively documented. In particular there is a long history in management of the use of habitat improvement devices to enhance trout populations (Swales and O'Hara, 1980). Cover within a stream can be provided by instream structures such as rocks, vegetation and tree roots, but is also afforded indirectly by the effects of overhanging vegetation, deep water and surface turbulence (Binns and Eiserman, 1979; Swales and O'Hara, 1980). Since it has been found that trout spend a considerable amount of time under cover, the increases in biomass of trout found after the provision of additional cover may be mediated through this behavioural response (Hartzler, 1983).

Implicit in these conclusions is the supposition that trout are territorial, and the provision of enhanced suitable habitat increases the number of potential territories. The assumption that there is a density-dependent limit to the number of resident salmonids a stream can hold mediated through territorial behaviour (Solomon, 1985) is the basis of many management decisions. Bachman (1984) has thoroughly reviewed the subject of territoriality in trout and concluded that in his study stream brown trout did not maintain a territory. Social structure was best described as "a cost minimising, size dependent, linear dominance hierarchy of individuals having overlapping home ranges". Whatever the precise definition of the behavioural interactions displayed by trout, there is nevertheless an upper limit to the number of trout a given section of river can hold on a sustained basis. Hartzler (1983) demonstrated that the inclusion of half-log covers in a stream where habitat cover was already good did not increase the number of brown trout, although biomass increased. He correctly concludes that supplemental cover is most likely to prove effective in fertile systems where existing cover is low. Similar benefits are likely to accrue where extensive drainage schemes have been performed. Kennedy (1984) has suggested that one of the factors involved in deterioration in numbers of Atlantic salmon and brown trout juveniles following drainage schemes may be the removal of suitable-size stones from the stream bed. Again it matters little in management terms whether these changes are due to the removal of the visual basis of territory borders or whether rocks afford more energy-saving sites: the end result is the same.

If, as Shirvell and Dungey (1983) suggest, there are distinct microhabitat preferences for trout according to differing activities and the most limiting of these could regulate population size, then more enhancement programmes should take cognizance of microhabitat requirements. Practical strategies for enhancement of juvenile Atlantic salmon using stones to provide various microhabitat needs are described by Rimmer *et al.* (1983).

Instream flow requirements have attracted considerable attention because the altered conditions resulting from abstraction and other water-supply schemes can influence fish (Fraser, 1975). In addition to the anticipated direct effect of water volume on the amount of stream-bed habitat available to fish, behavioural changes in territorial interactions have been documented. Brook trout, *Salvelinus fontinalis*, for example, have increased territory sizes at lower flow velocities (McNicol and Noakes, 1984).

Whereas such interpretations might hold for species with territorial or restricted home range behaviour, it is difficult to apply them directly to shoaling fishes such as many of the cyprinids. These have been shown to possess a home range, but they are much wider ranging in habit than salmonids (Hunt and Jones, 1974b). Edwards *et al.* (1984) demonstrated that channelization of a stream increased the abundance of cyprinid fish in a warm-water community to the detriment of more desired sport fish. Similarly, Karr (1981) has reported that degradation of streams will cause species shifts to give dominance to omnivorous fish such as cyprinids. However, in Britain, where the ichthyofauna is relatively species depauperate, cyprinids are the dominant fish family and the roach, *Rutilus rutilus*, is the most popular angled species. Habitat restoration and conservation schemes must therefore of necessity include this family. Swales and O'Hara (1983) described the effects of a habitat improvement scheme on a stream community in which the most recreationally important species were dace, *Leuciscus leuciscus*, and chub, *L. cephalus*. They demonstrated that weirs (Fig. 19.3) and groynes attracted both dace and chub, but overhead cover was particularly important for chub (Fig. 19.4). Although these two species are closely related, chub grow much larger. Recapture rates were much higher for marked chub than dace, and larger chub were apparently more solitary in behaviour than dace or young chub that shoal. The effectiveness of cover in attracting and holding chub may be due to this solitary behaviour.

Habitat improvement – lakes and ponds

The addition of habitat improvement structures to lentic habitats is well documented, but the techniques are slightly different from stream improvements in that submerged structures have found most favour. These are thoroughly described by Nelson *et al.* (1978). Tyre reefs and related structures are primarily intended as fish attractors providing cover, usually in relatively bare environments such as newly constructed reservoirs (Prince and Maughan, 1978); (see also Chapter 18 by Wardle, this volume). Their primary function is to concentrate fish for angling capture, but the structures quickly become colonized by periphyton and other organisms which act as a food source for fish. Small reefs that provide cover for only a few fish are utilized effectively because captured fish are quickly replaced by new inhabitants (Noble, 1980).

Fig. 19.3   A weir installed in a small lowland stream as part of a habitat improvement programme.

A sensible procedure that is often adopted is to leave as much natural cover material as possible on the bed of new impoundments during construction.

Oxygen levels in lakes can deteriorate to lethal conditions associated with stratification and the formation of ice cover in winter. Artificial destratification by pumps in summer can render an anoxic hypolimnion habitable, but it also affects the thermal stratification of the lake and thus the fish distribution, since fish can regulate their spatial position to remain within their area of thermal preference (Magnuson *et al.*, 1979). Care must be exercised with stenothermal species such as trout which have a low upper lethal level, and 'two-tier' fisheries have been effected in this situation. Cold, low-oxygen hypolimnial water is pumped to the surface to aerate it, and this is returned to the hypolimnion without disturbing the stratification, allowing trout to survive in the hypolimnion and maintaining thermally tolerant species in the

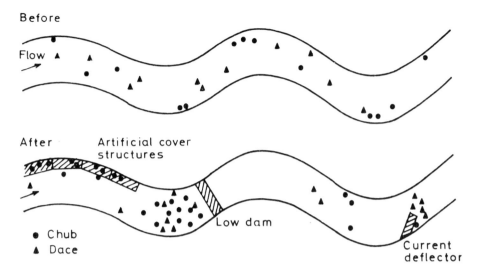

Fig. 19.4 The influence of habitat improvement structures on the distribution of dace and chub in a small lowland river. Modified from Swales and O'Hara (1983).

epilimnion. The distribution of fishes within a thermally stratified lake was considered a contributory factor in angler yield, and artificial destratification increased the catch of channel catfish, *Ictalurus punctatus*, probably as a result of more benthic habitat becoming available (Mosher, 1983). The seasonal migration of fish to wintering areas (Northcote, 1978), often deeper regions, is catered for and used as a measure to protect fish from ice by including a deeper hole when constructing small impoundments (Bennett, 1971).

Vegetation and management

Weed cover and bankside vegetation have very important roles to play in the lives of many fish (Killgore *et al.*, 1989). The floodplains of the potamon zone of rivers are exploited by many fishes which show a behavioural seasonality in moving onto these areas to breed and feed (Welcomme, 1979). Failure to protect these areas from the encroachment of man has meant that river channels have been constrained, with the consequent loss of floodplain area and thus fish production.

Similar types of movement are recorded in lakes, and are used by managers to enhance or reduce population sizes according to preference. Noble (1980) has documented the approach of raising water levels in reservoirs to enable northern pike to move into marshy spawning areas, and lowering levels to expose carp eggs. Providing cover in shoreline areas of lakes in the form of weed or other vegetation has been found to enhance abundance of juvenile largemouth bass, *Micropterus salmoides*, in these areas (McCammon and Geldern, 1979). Aquatic plants can provide cover for prey fishes and may be extremely important to predator–prey interactions. Durocher *et al.* (1984) found that a reduction of submerged vegetation to below 20% cover was detrimental to largemouth bass stocks. A similar result was found for the same species by Wiley *et al.* (1984), except that high weed levels were also detrimental. Grimm (1981a, b) has demonstrated that macrophyte cover is an important regulatory factor in the population dynamics of pike, with one of the determining factors associated with increased vegetation cover being a reduction in intraspecific predation. A telemetric study on pike (Chapman and Mackay, 1984) confirmed Grimm's findings that pike smaller than 41 cm in length remain in vegetated areas whereas large (> 54 cm) pike use both open and vegetated areas. Encouraging the growth of vegetation in barren lakes can therefore be used as a management technique to enhance sport fishes, whether these be piscivorous or non-piscivorous.

Fewer studies appear to have been conducted on the importance of instream weed to running-water fishes. Weed has been included as a physical cover factor in a number of salmonid abundance studies, and fish abundance is determined by such factors (Hermanson and Krog, 1984). However, Swales (1982b) was unable to demonstrate any measurable detrimental effect of weed cutting on non-salmonid species, although Mortensen (1977) has found a marked influence on brown trout. This may be a real difference or may possibly be caused by the problems of accurately measuring population densities in shoaling species. A recent investigation on a large lowland cyprinid-dominated river demonstrated the importance of macrophytes as a habitat for the early developmental stages of roach (Copp, 1990).

Fish passes

Rheotropic responses to water current are practically utilized in a number of situations, particularly fish-guiding devices and fish passes. Mills (1989) describes the practical aspects of these structures for migrating salmonids, and Arnold (1974) has thoroughly reviewed the subject of rheotropism in fish. When new fish passes are installed for salmonids, the provision of rheotactically attractive water flows at fish passes is one of the most difficult design criteria to achieve. These can be used to enable counts of upstream migrants to be made as the fish migrate through.

Water quality

A full description of behavioural responses to water quality parameters or particular pollutants is beyond the scope of the present chapter, and reference should be made to Alabaster and Lloyd (1983). However, the fisheries manager should be aware of behavioural changes that may be induced in fish by toxic agents. Marcucella and Abramson (1978) have used the term 'behavioural toxicology' to describe these effects, and emphasize their importance in reducing the adaptive response of fish to the environment.

One subject area that has attracted considerable attention is that of thermal influences, because of the widespread occurrence of heated power-station effluents. The effects of electricity generation on the biota, including thermal effluents are comprehensively reviewed by Langford (1983). The thermal niche of fishes was described conceptually by Magnuson *et al.* (1979), and the behavioural response is of clear importance when assessing the impact of heated waters on different species (Coutant, 1987).

Feeding

An awareness of feeding behaviour is often a prerequisite to a knowledge of managing stocks. Predator–prey interactions are an obvious application of such information (Stroud and Clepper, 1979). Because of the importance of fish size in many sport fisheries, growth maximization is desirable, as is production in commercially important species. Diet studies provide some elucidation of likely constraints, and recently the application of optimal foraging theory has been used to aid interpretation (Wootton, 1990).

The feeding behaviour of Atlantic salmon juveniles was found to be broadly consistent with their selecting a diet which maximizes growth (Wankowsky, 1981). It has been suggested that one of the reasons why stocked trout in streams show a low yield may be that they do not feed efficiently. Ersbak and Haase (1983) showed that hatchery-reared brook trout were conditioned to respond to a fish pellet and could not therefore forage effectively. Kennedy *et al.* (1984) were unable to demonstrate any differences in feeding between hatchery-reared Atlantic salmon smolts and wild fish, and attributed this to prior exposure to invertebrate food items because the hatchery supply was unfiltered river water.

Stein (1979) has reviewed the influence of piscivores on the behaviour of prey, and the importance of a predator on prey behaviour was shown by Werner *et al.* (1983b). The impact of predation on prey behaviour must be understood before effective manipulation can be effected. One area that has attracted considerable attention is the provision of forage fishes, the desired attributes of which were listed by Ney (1980) and include the behavioural vulnerability of the prey to the predator. An example of an ideal forage fish is the threadfin shad, *Dorosoma petuense*. This abundant species resides in open water and is vulnerable to bass predation (Heidinger, 1977).

Some aspects of the feeding biology of the roach and perch (examined from both intra- and interspecific competitive viewpoints) have been reported by Persson (1983a, b). He suggests that increased eutrophication is likely to encourage roach populations over perch because of their ability to switch feeding to algae/detritus when animal food is in short supply.

The feeding behaviour of the common carp has been implicated, because of its habit of rooting in the bottom mud, in encouraging water turbidity and increasing eutrophication. Selective predation by fish of large zoo-plankters has been suggested as a cause of increased phytoplankton levels in some lakes (Leah *et al.*, 1980). There has recently been an upsurge in consideration of biomanipulation, including selective removal of fish as an aid to the control of eutrophication.

19.6   MANAGEMENT AND DIRECT STOCK MANIPULATIONS

Considerable efforts are expended by fish managers in attempts to alter the composition of fish populations. Usually this involves the enhancement of desired species, and often the reduction of perceived predators or competitors. Many such programmes are doomed to failure from the outset because of a failure to recognize the constraints that exist, even though these are often the result of behavioural traits that are well appreciated. A thorough review of European stock enhancement practices which makes an interesting comparison with North American procedures is provided by Welcomme *et al.* (1983).

Stocking

There is considerable management justification for the creation of fisheries where none previously existed, for example in reservoir tailwaters. Other recreational practices can involve a put-and-take approach which ignores the normal constraints of carrying capacity: the fish are often removed so quickly that it matters little how they behave other than that they are vulnerable to capture by rod and line.

Among the commonest attempts at stock enhancement is the introduction of young stages of salmonids, particularly the salmons, both Pacific and Atlantic. The experience of stocking Atlantic salmon eggs, fry or parr in the British Isles is not one that suggests any great impact on the abundance of adult stocks (Harris, 1978). Many failures are attributed to the addition of new individuals to areas where the stream bed is already fully occupied by naturally recruited fish. There is a considerable body of evidence to suggest that hatchery-reared salmonids are at a considerable disadvantage to indigenous fish in streams, and often this disadvantage can be attributed to differences in behaviour (Ersbak and Haase, 1983). Analysis of stocking policies by Solomon (1985) suggests that Atlantic salmon fry do not disperse as widely as previous studies had suggested (Kennedy, 1987). Therefore it is unproductive to make plantings of eggs or juveniles in large numbers at a few sites, but spot stockings at short intervals may be more effective.

Attempts at establishing Pacific salmon in Europe have essentially failed (Welcomme *et al.*,1983). One possible reason for this is the lack of behavioural traits programmed to the new migratory cues. More probably, however, the introduced juveniles have been poorly suited to their new environment and few attempts have been made to breed from the small numbers of fish which did return (D. J. Solomon, personal communication).

The influence of olfactory imprinting and homing in salmon has been reviewed by Hasler and Scholz (1983), and distinct management potential exists for enhancing stocks by directing the site of return and re-establishing spawning streams.

Stocking non-diadromous salmonids in lentic waters has often proved to be more successful than in running water (Ersbak and Haase, 1983). There is evidence that brown trout are solitary in lakes, for example Tytler and Holliday (1984) showed that they exhibited restricted movements, but the ranges of individual fish did overlap. In lake-dwelling salmonids such as juvenile sockeye salmon, *Oncorhynchus nerka*, shoaling behaviour is typical rather than the territoriality exhibited by many stream-living juvenile salmonids (Keenleyside, 1979). Donald and Anderson (1982) have noted that rainbow trout can be stocked up to a certain density in lakes without mortality occurring, but growth deteriorates and above this level, density-dependent mortality occurs.

The behaviour of hatchery-reared fish can be deliberately selected to be different (Babey and Berry, 1989). In this category is the production of sunfish hybrids that have a greater vulnerability to angling capture because of their aggressiveness (Kurzawski and Heidinger 1982). The susceptibility of fish to angling capture may change as a result of learning (Beukema, 1970), and this could influence the performance of a fishery where fish are returned after capture, the general practice with coarse fish in Britain. Using controlled angling in ponds, Raat (1985) was able to demonstrate that carp learned to avoid capture after one experience of hooking, but that there was variability within the population in their susceptibility to angling capture. Similarly Pawson and Purdom (1987) statistically demonstrated a difference between catchability of three strains of rainbow trout.

For non-territorial species, stocked fish may not be as subject to density-dependent constraints that result in their mortality, as territorial species. However, compensatory changes must be exhibited at some level whether this is, for example, mediated through reduced fecundity via reduced growth or survival of progeny (Goodyear, 1980). The stocking of cyprinids in the Netherlands has been so successful in numbers that reduced growth rates are apparent and other management techniques are being sought (Welcomme *et al.*, 1983).

Competition and other interactions

A common management practice is to remove undesirable fish from waters. This has often been applied to salmonid streams, where non-game fish are often regarded as the cause of reduced game fish catches. The subject of competition is contentious, none more so than in fisheries management where many species are actively removed on the flimsiest of evidence. Countless examples of removals could doubtless be located: in opposition the scientific evaluations are few but often telling where they do exist. Baltz and Moyle (1984) review this area and subscribe to the view that many declines in game fish are often due to changed environmental conditions (see also Karr, 1981). The irony of the situation is that taxonomically closely related game species have been found in some studies to be in competition, rather than other fish (Marrin and Erman, 1982). Brown trout are more aggressive in behaviour than Atlantic salmon, and sympatrically occupy deeper pool areas whilst salmon are restricted to riffles. Whether this is detrimental to salmon is open to question since they appear to be anatomically better adapted to these zones: larger pectoral fins allow the salmon to remain close to the bottom in riffles, thus avoiding energetically expensive swimming into the current (Jones, 1975). This division of the habitat is an example of interactive segregation (Nilsson, 1978), a term which avoids the implied deleterious connotations of competition.

Nevertheless, some managers maintain that brown trout and salmon juveniles interact to the detriment of salmon, and efforts are often made to reduce densities of brown trout. This is normally a fruitless exercise as these are quickly replaced by new individuals. This is not a surprising result, given the density-dependent nature of the population-regulatory mechanisms operating in stream salmonid communities (Solomon, 1985). The complexity of intra- and interspecific interactions was well demonstrated by Kennedy and Strange (1980), and extrapolation from this exemplary study illustrates how and why the application of unresearched management techniques can lead to impractical approaches.

In many non-salmonids, stunting is a feature of overstocked situations, and culling will increase growth rates (Welcomme *et al.*, 1983). Werner *et al.* (1983a, b) have persuasively argued from optimal foraging theory, optimal habitat use and the influence of a predatory fish that the growth of bluegills, *Lepomis macrochirus*, can be restricted through the creation of 'competitive bottlenecks'. One of the usual management practices when bass/bluegill combinations 'go out of balance' (Bennett, 1971) is to reduce the density of bluegills which have become stunted. This clearly would have the effect of reducing such bottlenecks. An alternative has been to produce virtual monosex populations of sunfish hybrids to prevent or reduce reproduction (Kurzawski and Heidinger, 1982).

Intraspecific competitive effects have often proved easier to demonstrate than interspecific changes, because changes in growth in fish are often detected. Several examples of such changes, and a perceptive account of competition, are provided by Weatherley and Gill (1987). There are in fact many documented shifts in feeding behaviour when new species have been added to a water body (Nilsson, 1978). This is one of the criteria taken as indicative of competition and interestingly, as Werner *et al.* (1980) pointed out, a source of information that has generally been missed by fisheries ecologists.

Stockings of salmonids of the *Oncorhynchus* genus are often spectacularly successful, particularly for those species like pink, *O. gorbuscha*, and chum, *O. keta*, in which the fry migrate directly to sea after a very short freshwater residence period, and therefore avoid the restrictions of the limited freshwater habitat in terms of space or food.

Stock preservation

Some fisheries biologists are becoming increasingly concerned that insufficient attention has been given to the genetical aspects of fish populations by fisheries managers (Allendorf *et al.*, 1986). Horn and Rubenstein (1984) have discussed behavioural adaptations and life histories including fisheries examples. As well as the likely influence of exploitation on the adjustment of sea ages of salmonid stocks (Schaffer and Elson, 1975), it is possible that stocking with home river progeny fish of a specific sea age will give fish with the same migratory pattern if this trait is genetically determined. It appears that the time of return, at least in Atlantic salmon, is both genetically and environmentally determined (Gardner, 1976). However, Gardner suggests that there is sufficient genetic influence to justify using parents of specific sea ages in attempts to influence the age of return of the progeny. A thorough overview of these and related topics in respect of the management of Pacific salmon is provided by Waples *et al.* (1990). Atlantic salmon show iteroparity in that a number of fish survive spawning to reproduce again. Whereas it is useless to attempt to preserve spent Pacific salmon, since they all display a semelparous life history, the conservation of Atlantic salmon kelts (spent fish returning to the sea) is one option in stock enhancement programmes (Mills, 1989).

19.7 SUMMARY

Fisheries management seeks to maximize user appreciation of the resource, whether this be in terms of catch rate, aesthetics of the experience or food production. As such it must take cognizance of all aspects of fish biology, including behaviour (and some of human behaviour!) (Pawson, 1986; Chapman and Helfrich, 1988). Many fisheries management practices cannot be placed within an ethological or theoretical framework because they are often empirically derived from practical experience. However, the extensive and developing literature on fish behaviour which includes a management objective suggests that the two areas are converging. It is evident that fishes differ from many other animals in that they show indeterminate growth, and population constraints (such as competition), many mediated through behaviour, yield outcomes other than those which may be predicted from theory (Weatherley and Gill, 1987).

If we are to understand how factors such as food and space can influence growth and the population dynamics of fish, then ethological studies allied to ecological principles will be of great benefit. Care must be exercised to ensure that such information is not used to further increase pressures on already overfished stocks, but it should be applied constructively as with the clear benefits that are beginning to occur in our understanding of stream-dwelling juvenile salmonids. Fisheries managers can only gain from the current upsurge in knowledge of the behaviour of their charges, even if some myths about behaviour have to be jettisoned *en route*.

ACKNOWLEDGEMENTS

Thanks for photographic material are due to D. C. Hockin and S. Swales, and for preparation of the plates to B. Lewis. The patience and help of Ms. A. Callaghan with typing are gratefully recorded.

REFERENCES

Alabaster, J. S. and Lloyd, R. (1983) *Water Quality Criteria for Freshwater Fish*, Butterworths, London.

Allendorf, F. W., Ryman, N. and Utter, F. W. (1986) Genetics and fishery management, in *Population Genetics and Fishery Management* (eds N. Ryman and F. Utter), University of Washington Press, Seattle, pp. 1–19.

Arnold, G. P. (1974) Rheotropism in fishes. *Biol. Rev.*, **49**, 515–76.

Babey, G. J. and Berry, C. R. (1989) Post-stacking performance of three strains of rainbow trout in a reservoir. *N. Am. J. Fish. Manage.*, **9**, 309–15.

Bachman, R. A. (1984) Foraging behaviour of free-ranging wild and hatchery brown trout in a stream, *Trans. Am. Fish. Soc.*, **113**, 1–32.

Bagenal, T. B. and Tesch, F. W. (1978) Age and growth, in *Methods for the Assessment of Fish Production in Fresh Waters* (ed. T. B. Bagenal), Blackwell, Oxford, pp. 101–36.

Baker, R. R. (1978) *The Evolutionary Ecology of Animal Migration*, Hodder and Stoughton, London.

Baltz, D. M. and Moyle, P. B. (1984) Segregation by species and size classes of rainbow trout *Salmo gairdneri*, and Sacramento sucker, *Catostomus occidentalis*, in three California streams. *Env. Biol. Fishes.*, **10**, 101–10.

Banks, J. W. (1969) A review of the literature on the upstream migrations of adult salmonids. *J. Fish Biol.*, **1**, 85–136.

Bardach, J. E., Magnusson, J. J., May, R. B. and Reinhart, J. N. (eds) (1980) *Fish Behaviour and its Use in the Capture and Culture of Fishes*, International Center for Living Aquatic Resources Management, Manila, Philippines.

Bennett, G. W. (1971) *Management of Lakes and Ponds*, Van Nostrand Reinhold, New York.

Beukema, J. J. (1970) Angling experiments with carp, 2. Decreasing catchability through one-trial learning. *Neth. J. Zool.*, **20**, 81–92.

Binns, N. A. and Eiserman, F. M. (1979) Quantification of fluvial trout habitat in Wyoming. *Trans. Am. Fish. Soc.*, **108**, 215–28.

Borgeson, D. P. (1979) Controlling predator–prey relationships in streams, in *Predator–Prey Systems in Fisheries Management* (eds R. H. Stroud and H. Clepper), Sport Fishing Institute, Washington, DC, pp. 425–30.

Bowen, S. H. (1983) Detritivory in Neotropical fish communities, *Env. Biol. Fishes*, **9**, 137–44.

Brannon, E. L. and Salo, E. O. (eds) (1982) *Proceedings of the Salmon and Trout Migratory Behavior Symposium*, University of Washington, Seattle.

Brookes, S., Gregory, K. J. and Dawson, F. H. (1983) As assessment of river channelization in England and Wales. *Sci. Total Environ.*, **27**, 97–111.

Chapman, B. D. and Helfrich, L. A. (1988) Recreational specializations and motivations of Virginia river anglers. *N. Am. J. Fish. Manage.*, **8**, 390–98.

Chapman, C. A. and Mackay, W. C. (1984) Versatility in habitat use by a top aquatic predator, *Esox lucius* L. *J. Fish Biol.*, **25**, 109–15.

Cooper, E. L. (1980) Fisheries management in streams, in *Fisheries Management* (eds R. T. Lackey and L. A. Neilsen), Blackwell, Oxford, pp. 297–322.

Copp, G. H. (1990) Shifts in the microhabitat of larval and juvenile roach, *Rutilus rutilus* (L.), in a floodplain channel. *J. Fish Biol.*, **36**, 683–92.

Coutant, C. C. (1987) Thermal preference: when does an asset become a liability? *Env. Biol. Fishes*, **18**, 161–72.

Cresswell, R. C. (1981) Post-stocking movements and recapture of hatchery-reared trout released into flowing waters – a review. *J. Fish Biol.*, **18**, 429–42.

Donald, D. B. and Anderson, R. S. (1982) Importance of environment and stocking density for growth of rainbow trout in mountain lakes. *Trans. Am. Fish. Soc.*, **111**, 675–80.

Durocher, P. P., Provine, W. C. and Kraai, J. E. (1984) Relationship between Abundance of largemouth bass and submerged vegetation in Texas reservoirs. *N. Am. J. Fish. Manage.*, **4**, 84–8.

Edwards, C. J., Griswold, B. L., Tubb, R. A., Weber, E. C. and Woods, L. C. (1984) Mitigating effects of artifical riffles and pools on the fauna of a channelized warmwater stream. *N. Am. J. Fish. Manage.*, **4**, 194–203.

Ersbak, K. and Haase, B. L. (1983) Nutritional deprivation after stocking as a possible mechanism leading to mortality in stream-stocked brook trout. *N. Am. J. Fish. Manage.*, **3**, 142–51.

Everhart, W. H. and Youngs, W. D. (1981) *Principles of Fishery Science*, Cornell University Press, Ithaca, NY.

Fraser, J. C. (1975) Determining discharges for fluvial resources, *FAO Fish. Tech. Pap.*, no. 143.

Gardner, M. L. G. (1976) A review of factors which may influence the sea-age and maturation of atlantic salmon *Salmo salar* L. *J. Fish Biol.*, **9**, 289–327.

Gauthereaux, S. A. (1980) *Animal Migration, Orientation and Navigation*, Academic Press, New York.

Goodyear, C. P. (1980) Compensation in fish populations, in *Biological Monitoring of Fish* (eds C. H. Hocutt and J. R. Stauffer), Lexington Books, Toronto, pp. 253–80.

Grimm, M. P. (1981a) The composition of northern pike (*Esox lucius* L.) populations in four shallow waters in The Netherlands, with special reference to factors influencing 0 + pike biomass. *Fish. Manage.*, **12**, 61–76.

Grimm, M. P. (1981b) Intraspecific predation as a principal factor controlling the biomass of northern pike (*Esox lucius* L.). *Fish. Manage.*, **12**, 77–9.

Harris, G. S. (ed.) (1978) *Salmon Propagation in England and Wales*, National Water Council, 1 Queen Anne's Gate, London.

Hart, P. J. B. and Pitcher, T. J. (1973) Population densities and growth of five species of fish in the River Nene, Northamptonshire. *Fish. Manage.*, **4**, 69–86.

Hartzler, J. R. (1983) The effects of half-log covers on angler harvest and standing crop of brown trout in McMichaels Creek, Pennsylvania. *N. Am. J. Fish. Manage.*, **31**, 228–38.

Hasler, A. D. and Scholz, A. T. (1983) *Olfactory Imprinting and Homing in Salmon*, Springer-Verlag, Berlin.

Haynes, J. M. and Gerber, G. P. (1989) Movements and temperatures of radiotagged salmonines in Lake Ontario and comparisons with other large aquatic systems. *J. Freshwat. Ecol.*, **5**, 197–204.

Heidinger, R. C. (1977) Potential of the threadfin shad as a forage fish in Midwestern power cooling reservoirs. *Trans. Illinois State Acad. Sci.*, **70**, 15–25.

Helfman, G. S. (1983) Underwater methods, in *Fisheries Techniques* (eds L. A. Nielsen and D. L. Johnson), American Fisheries Society, Bethesda, MD, pp. 349–69.

Hermansen, H. and Krog, C. (1984) Influence of physical factors on density of stocked brown trout (*Salmo trutta fario* L.) in a Danish lowland stream. *Fish. Manage.*, **15**, 107–15.

Hockin, D. C., O'Hara, K. and Eaton, J. W. (1988) A radiotelemetric study of the movements of grass carp in a British Canal. *Fish. Res.*, **7**, 73–84.

Horn, H. S. and Rubenstein, D. I. (1984) Behavioural adaptations and life history, in *Behavioural Ecology: an Evolutionary Approach*, 2nd edn. (eds J. R. Krebs and N. B. Davies), Blackwell, Oxford, pp. 279–98.

Hunt, P. C. and Jones, J. W. (1974a) A population study of *Barbus barbus* (L.) in the River Severn, England. I. Densities, *J. Fish Biol.*, **6**, 255–67.

Hunt, P. C. and Jones, J. W. (1974b) A population study of *Barbus barbus* (L.) in the River Severn, England. II. Movements. *J. Fish Biol.*, **6**, 269–78.

Hunt, R. L. (1969) Effects of habitat alteration on production standing crops and yield of brook trout in Lawrence Creek, Wisconsin, in *Symposium on Salmon and Trout in Streams* (H. R. MacMillan Lectures in Fisheries, 1968), University of British Columbia, Vancouver, Canada, pp. 281–312.

Hynes, H. B. N. (1970) *The Ecology of Running Waters*, Liverpool University Press, Liverpool.

Jones, A. N. (1975) A preliminary study of fish segregation in salmon spawning streams. *J. Fish Biol.*, **7**, 95–104.

Karr, J. R. (1981) Assessment of biotic integrity using fish communities. *Fisheries, Bethesda, MD*, **6**, (6) 21–7.

Keenleyside, M. H. A. (1979) Diversity and Adaptation in Fish Behaviour, Springer-Verlag, Berlin.

Kennedy, G. J. A. (1984) The ecology of salmonid habitat re-instatement following river drainage schemes, in *Proc. Inst. Fishery Manage. Study Course. N. Ireland Branch*, New University of Ulster, Coleraine, pp. 1–25.

Kennedy, G. J. A. (1987) Factors affecting the survival and distribution of salmon (*Salmo salar* L.) stocked in upland trout (*Salmo trutta* L.), streams in Northern Ireland, *EIFAC Tech. Pab.*, **42(1)**, 227–42.

Kennedy, G. J. A. and Strange, C. D. (1980) Population changes after two years of salmon (*Salmo salar* L.) stocking in an upland trout (*Salmo trutta* L.) stream. *J. Fish Biol.*, **17**, 577–86.

Kennedy, G. J. A., Strange, C. D., Anderson, R. J. D. and Johnston, P. M. (1984) Experiments on the descent and feeding of hatchery-reared salmon smolts (*Salmo salar* L.) in the river Bush. *Fish. Manage.*, **15**, 15–25.

Kerr, S. R. and Werner, E. E. (1980) Niche theory in fisheries ecology. *Trans. Am. Fish. Soc.*, **109**, 254–60.

Killgore, K. J., Morgan, R. P. and Rybicki, N. B. (1989) Distribution and abundance of fishes associated with submerged aquatic plants and the Potomac River. *N. Am. J. Fish. Manage.*, **9**, 101–11.

Kurzawski, K. F. and Heidinger, R. C. (1982) The cyclic stocking of parentals in a farm pond to produce a population of male bluegill × female green sunfish F_1 hybrids and male redear sunfish × females green sunfish F_1 hybrids. *N. Am. J. Fish. Manage.*, **2**, 188–92.

Langford, T. E. (1983) *Electricity Generation and the Ecology of Natural Waters*, Liverpool University Press, Liverpool.

Larkin, P. A. (1980) Objectives of management, in *Fisheries Management* (eds R. T. Lackey and L. A. Nielsen), Blackwell, Oxford, pp. 245–62.

Leah, R. T., Moss, B. and Forrest, D. E. (1980) The role of predation in causing major changes in the limnology of a hyper-eutrophic lake. *Int. Rev. Ges. Hydrobiol. Hydrog.*, **65**, 223–47.

Le Cren, E. D., Kipling, C. and McCormack, J. C. (1977) A study of the number, biomass and year-classes strengths of perch (*Perca fluviatilis* L.) in Windermere from 1941 to 1966. *J. Anim. Ecol.*, **46**, 281–307.

McCammon, G. W. and Geldern, C. R. von, jun. (1979) Predator–prey Systems in Large reservoirs, in *Predator–Prey Systems in Fisheries Management* (eds R. H. Stroud and H. Cleper), Sport Fishing Institute, Washington, DC, pp. 431–42.

McDowell, R. M. (1988) *Diadromy in Fishes*, Croom Helm, London, 308 pp.

McKeown, B. A. (1984) *Fish Migration*, Croom Helm, London.

McLeave, J. D., Arnold, G. P., Dodson, J. J. and Neill, W. H. (eds) (1984) *Mechanisms of Migration in Fishes*, Plenum Press, New York.

McNicol, R. E. and Noakes, D. L. G. (1984) Environmental influences on territoriality of juvenile brook charr, *Salvelinus fontinalis*, in a stream environment. *Env. Biol. Fishes*, **10**, 29–42.

Magnuson, J. J., Crowder, L. B. and Medvick, P. A. (1979) Temperature as an ecological resource. *Am. Zool.*, **19**, 331–43.

Marcucella, H. and Abramson, C. I. (1978) Behavioral toxicology and teleost fish, in *The Behavior of Fish and other Aquatic Animals* (ed. D. I. Mostopsky), Academic Press, London, pp 33–7.

Marrin, D. L. and Erman, E. C. (1982) Evidence against competition between trout and nongame fishes in Stampede Reservoir, California. *N. Am. J. Fish. Manage.*, **2**, 262–9.

Mills, C. P. R. and Piggins, D. J. (1983) The release of reared salmon smolts (*Salmo salar*) into the Burrishoole River System (Western Ireland) and their contribution to the rod and line fishery, *Fish. Manage.*, **14**, 165–75.

Mills, D. (1971) *Salmon and Trout*, Oliver and Boyd, Edinburgh.

Mills, D. H. (1989) *Ecology and Management of Atlantic Salmon*, Chapman and Hall, London.

Milner, M. J., Hemsworth, R. J. and Jones, B. E. (1985) Habitat evaluation as a fisheries management tool. *J. Fish. Biol.*, **27**(Supp. A), 85–108.

Moring, J. R. (1982) An efficient hatchery strain of rainbow trout for stocking Oregon streams. *N. Am. J. Fish. Manage.*, **2**, 209–15.

Mortensen, E. (1977) Density dependent mortality of trout fry (*Salmo trutta* L.) and its relationship to the management of small streams. *J. Fish Biol.*, **11**, 613–17.

Mosher, T. D. (1983) Effects of artificial circulation on fish distribution and angling success for channel catfish in a small prairie lake. *N. Am. J. Fish. Manage.*, **3**, 403–9.

Mundy, P. R. (1982) Computation of migratory timing statistics for adult chinook salmon in the Yukon River, Alaska, and their relevance to fisheries management. *N. Am. J. Fish. Manage.*, **2**, 359–70.

Nelson, R. W., Horak, G. C. and Olson, J. E. (1978) *Western Reservoir and Stream Habitat Improvements Handbook*, Fish and Wildlife Service, US Dep. Interior, Washington, DC.

Ney, J. J. (1980) Evolution of forage fish management in lakes and reservoirs. *Trans. Am. Fish. Soc.*, **110**, 725–8.

Nilsson, N.-A. (1978) The role of size-biased predation in competition and interactive segregation in fish, in *Ecology of Freshwater Fish Production* (ed. S. D. Gerking), Blackwell, Oxford, pp. 303–25.

Noble, R. L. (1980) Management of lakes, reservoirs and ponds, in *Fisheries Management* (eds R. T. Lackey and L. A. Neilsen), Blackwell, Oxford, pp. 265–95.

Northcote, T. G. (1978) Migratory strategies and production in freshwater fishes, in *Ecology of Freshwater Fish Production* (ed. S. D. Gerking), Blackwell, Oxford, pp. 326–59.

Pawson, M. G. (1986) Performance of rainbow trout, *Salmo gairdneri* Richardson, in a put-and-take fishery, and the influence of anglers' behaviour on catchability. *Aquacult. Fish. Manage.*, **17**, 59–73.

Pawson, M. G. and Purdom, C. E. (1987) Relative catchability and performance of three studies of rainbow trout, *Salmo gairdneri* Richardson, in a small fishery. *Aquacult. Fish. Manage.* **18**, 173–86.

Persson, L. (1983a) Food consumption and the significance of detritus and algae to intraspecific competition in roach *Rutilus rutilus* in a shallow eutrophic lake. *Oikos*, **41**, 118–25.

Persson, L. (1983b) Effects of intra- and interspecific competition on dynamics and size structure of a perch *Perca fluviatilis* and roach *Rutilus rutilus* population. *Oikos*, **41**, 126–32.

Peterson, N. C. and Cederholm, C. J. (1984) A comparison of the removal and mark–recapture methods of population estimation for juvenile coho salmon in small streams. *N. Am. J. Fish. Manage.*, **4**, 99–102.

Pitcher, T. J. (1980) Some Ecological Consequences of Fish School Volumes. *Freshwater Biol.*, **10**, 539–44.

Pratten, D. J. and Shearer, W. M. (1983a) Sea trout of the River North Esk. *Fish. Manage.*, **14**, 49–65.

Pratten, D. J. and Shearer, W. M. (1983b) The migration of North Esk sea trout. *Fish. Manage.*, **14**, 99–113.

Prince, E. D. and Maughan, O. E. (1978) Freshwater Artificial Reefs: Biology and Economics. *Fisheries*, (*Bethesda, MD*), **3**(1), 5–9.

Raat, A. J. P. (1985) Analysis of angling vulnerability of common carp, *Cyprinus carpio* L., in catch-and-release angling in ponds. *Aquacult. Fish. Manage.*, **16**, 171–87.

Regier, H. A., Paloheimo, J. E. and Gallucci, Y. F. (1979) Factors that influence the abundance of large piscivorous fish, in *Predator–Prey Systems in Fisheries Management* (eds R. H. Stroud and H. Clepper), Sport Fishing Institute, Washington, DC, pp. 333–41.

Rimmer, D. M., Paim, V. and Saunders, R. L. (1984) Changes in the selection of micro habitat by juvenile Atlantic salmon (*Salmo salar*) at the summer–autumn transition in a small river. *Can. J. Fish. Aquat. Sci.*, **41**, 469–75.

Roff, D. A. (1973) An examination of some statistical tests used in the analysis of mark–recapture data. *Oecologia*, **12**, 35–54.

Royce, W. F. (1987) *Fishery Development*, Academic Press, London.

Sambrook, H. (1983) Homing of sea trout in the River Fowey catchment, Cornwall, in *Proc. Third Br. Freshwat. Fish. Conf.*, Liverpool University, England, pp. 30–40.

Schaffer, W. M. and Elson, P. F. (1975) The adaptive significance of variations in life history among local populations of Atlantic salmon in North America. *Ecology*, **56**, 577–90.

Schreck, C. B. (1981) Stress and compensation in teleostean fishes: response to social and physical factors, in *Stress and Fish* (ed. A. D. Pickering), Academic Press, London, pp. 295–321.

Seber, G. A. F. (1973) *The Estimation of Animal Abundance and Related Parameters*, Griffin, London.

Shirvell, C. S. and Dungey, R. G. (1983) Microhabitats chosen by brown trout for feeding and spawning in rivers. *Trans. Am. Fish. Soc.*, **112**, 355–67.

Sloane, R. D. (1984) Preliminary observations of the migrating adult freshwater eels (*Anguilla australis australis* Richardson) in Tasmania. *Aust. J. Mar. Freshwat. Res.*, **35**, 471–6.

Solomon, D. J. (1985) Salmon stock and recruitment, and stock enhancement. *J. Fish. Biol.*, **27**, (Supp. A), 45–57.

Stein, R. A. (1979) Behavioral response of prey to fish predators. in *Predator–Prey Systems in Fisheries Management*, (eds R. H. Stroud and H. Clepper), Sport Fishing Institute, Washington, DC, pp. 343–53.

Stevens, D. E. and Miller, L. W. (1983) Effects of river flow on abundance of young chinook salmon, American shad, longfin, smelt and delta smelt in the Sacramento–San Joaquin, river system. *N. Am. J. Fish. Manage.*, **3**, 425–37.

Stott, B. and Russell, I. C. (1979) An estimate of a fish population that proved to be wrong. *Fish. Manage.*, **10**, 169–71.

Stroud, R. H. and Clepper, H. (1979) *Predator–Prey Systems in Fisheries Management*, Sport Fishing Institute, Washington, DC.

Swales, S. (1982a) Environmental effects of river channel works used in land drainage improvement. *J. Env. Manage.*, **14**, 103–26.

Swales, S. (1982b) Impacts of weed-cutting on fisheries: an experimental study in a small lowland river. *Fish. Manage.*, **13**, 125–37.

Swales, S. and O'Hara, K. (1980) Instream habitat improvement devices and their use in freshwater fisheries management. *J. Env. Manage.*, **10**, 167–79.

Swales, S. and O'Hara, K. (1983) A short-term study of the effects of a habitat improvement programme on the distribution and abundance of fish stocks in a small lowland river in Shropshire. *Fish. Manage.*, **14**, 135–44.

Tesch, F.-W. (1977) *The Eel: Biology and Management of Anguillid Eels*, Chapman and Hall, London.

Thorpe, J. E. (ed.) (1980) *Salmon Ranching*, Academic Press, London.

Tytler, P. and Holliday, F. G. T. (1984) Temporal and spatial relationships in the movements of loch dwelling brown trout, *Salmo trutta* L., Recorded by ultrasonic tracking for 24 hours. *J. Fish Biol.*, **24**, 691–702.

Wankowsky, J. W. J. (1981) Behavioural aspects of predation by juvenile Atlantic salmon (*Salmo salar* L.) on particular drifting prey, *Anim. Behav.*, **29**, 557–71.

Waples, R. S., Winans, G. A., Utter, F. W. and Mahuken, C. (1990) Genetic approaches to the management of Pacific salmon fisheries. (Bethesda MD) **15**(5), 19–25.

Weatherley, A. H. and Gill, H. S. (1987) *The Biology of Fish Growth*, Academic Press, London.

Welcomme, R. L. (1979) *Fisheries Ecology of Floodplain Rivers*, Longman, London.

Welcomme, R. L., Kohler, C. C. and Courtenay, W. R. (1983) Stock enhancement in the management of freshwater fisheries: a European perspective. *N. Am. J. Fish. Manage.*, **31**, 265–75.

Werner, E. E., Gilliam, J. F., Hall, D. J. and Mittelbach, G. G. (1983b) An experimental test of the effects of predation risk on habitat use in fish. *Ecology*, **64**, 1540–48.

Werner, E. E., Mittelbach, G. G., Hall, D. J. and Gilliam, J. F. (1983a) Experimental test of optimal habitat use in fish: the role of relative habitat profitability. *Ecology*, **64**, 1525–39.

Wiley, M. J., Gorden, R. W., Waite, S. W. and Powless, T. (1984) The relationship between aquatic macrophytes and sport fish production in Illinois ponds: a simple model. *N. Am. J. Fish. Manage.*, **4**, 111–19.

Winter, J. D. (1983) Underwater biotelemetry, in *Fisheries Techniques* (eds L. A. Nielsen and D. L. Johnson), American Fisheries Society, Bethesda, MD, pp. 371–95.

Wootton, R. J. (1990) *Ecology of Teleost Fishes*, Chapman and Hall, London, 404 pp.

Zalewski, M. (1983) The influence of fish community structure on the efficiency of electrofishing. *Fish. Manage.*, **14**, 177–86.

Author index

Species index

Subject index